PRODUCT WARRANTY HANDBOOK

edited by

Wallace R. Blischke
University of Southern California
Los Angeles, California

D. N. Prabhakar Murthy
The University of Queensland
Brisbane, Queensland, Australia

Marcel Dekker, Inc.   New York • Basel • Hong Kong

Library of Congress Cataloging-in-Publication Data

Product Warranty Handbook / edited by Wallace R. Blischke, D. N. Prabhakar Murthy.
    p. cm.
    Includes index.
    ISBN 0-8247-8955-5 (alk. paper)
    1. Quality of products—Handbooks, manuals, etc.   2. Warranty—Handbooks, manuals, etc. I. Blischke, W. R. II. Murthy, D. N. P.
HF5415.157.P764   1995
658.5'6—dc20   95-40479
                                                          CIP

The publisher offers discounts on this book when ordered in bulk quantities. For more information, write to Special Sales/Professional Marketing at the address below.

This book is printed on acid-free paper.

Marcel Dekker, Inc.
270 Madison Avenue, New York, New York 10016

Current printing (last digit):
10 9 8 7 6 5 4 3 2 1

PRINTED IN THE UNITED STATES OF AMERICA

Preface

Warranty has a very long history, going back several thousand years. Over the last few decades, the study of warranty has received a good deal of attention from researchers and has been pursued from several disciplines. Some of the studies have their origin in practical application; others have been mainly theoretical. A two-way interaction between theory and application is critical for the nurturing and growth of warranty research.

Unfortunately for both researcher and practitioner, a survey of the literature on warranty reveals two glaring deficiencies:

1. The vast literature is disjoint, and investigators from a particular discipline are often unaware of the research carried out by those from other disciplines. A large gap exists between researchers from different disciplines.
2. The gap between theory and practice is even larger, with very few papers from practitioners appearing in the open literature.

The editors became aware of this during their research and preparation for their first book on warranties (*Warranty Cost Analysis,* Marcel Dekker, Inc.). This led to a three-part survey paper on warranty management that appeared in the *European Journal of Operations Research* in late 1992. These papers presented a taxonomy for classifying the different types of warranties and proposed a unified approach to the study of warranty.

The genesis of this book was an idea on the part of the editors to attempt to close the two gaps mentioned. The original plan was to have over 45 chapters written by researchers from different disciplines and by practitioners from different industry and government sectors. In due course, the editors realized that they had undertaken a nearly impossible task. Bridging the first gap would require a multivolume handbook with separate volumes for each of the three different types of products—consumer durables, commercial and industrial products, and government acquisition. The warranties and the issues involved are different for each of these categories. Reducing the second gap proved to be even more difficult, for two important reasons: (1) many practitioners are unaware of the theoretical literature and lack the skills to fully understand it and (2) the great majority of practitioners (or their employers) are very reluctant to release relevant warranty information because of commercial sensitivity.

In light of this we revised our original plan and decided to focus on bridging the first gap and, as far as possible, the second, in the context of consumer durable products. The book deals with consumer product warranties viewed from different perspectives, with each chapter written by an expert from the relevant discipline. Included are chapters on basic warranty concepts and techniques for analysis; history of warranty; warranty legislation and legal actions; statistical, mathematical, and engineering analysis; cost models; and the role of warranty in marketing, management, and society.

In focusing on warranties on consumer products, we are dealing with situations in which there are typically a large number of buyers and few or a modest number of manufacturers. In such cases, individual consumers have no power in setting warranty terms; they are dictated by the manufacturer and market forces. In addition, the buyer has little direct information about product quality and related issues and no opportunity to obtain such information. On the other hand, the manufacturer often has data and tools for analysis of warranty costs (but this information is seldom fully utilized). Because of this inequity with regard to information and analysis, consumer warranties have become a public policy issue. Many of the chapters of the *Handbook* are somewhat theoretical; however, all deal with application issues as well. As such, this handbook can be viewed as an attempt by theory-oriented researchers to address practical issues. In this sense, this compilation differs from many handbooks, which are simply collections of techniques, formulas, and tables meant for the practitioner, that provide little or no depth of understanding. Our intended audience includes both academics and practitioners, including specialists in the many fields in which warranty issues are addressed—engineering, production, marketing, management, accounting, reliability, statistics, consumerism, law, economics, and public policy. It is our hope that those involved in theoretical studies will benefit from the broader perspective gained from the book. We hope that practitioners will also find this book of significant relevance and will, in fact, be motivated to contribute to a future volume that would not only deal with the application of theory to real problems, but also trigger further new theoretical research into warranties for consumer durable products. It is hoped that this book will form the basis for other articles and/or volumes dealing with commercial and industrial products and government acquisition.

The aim of the *Handbook,* then, is to give a broad overview of the study of consumer product warranty carried out by researchers from different disciplines, linking the different research areas. The focus is on concepts to bridge the first gap and guidance to practitioners toward bridging the second. Each chapter concludes with advice and suggestions for practitioners. In addition, areas for future research are indicated.

In presenting the results, particularly those of a more theoretical nature, many technical details are omitted. For details, interested readers may consult the references cited by the authors of the chapters. Also, a bibliography on warranties including many additional references, as well as those given in the chapters, is provided in Chapter 33.

The text of the book is structured into eight parts (Part A through Part H). Each part deals with a particular aspect of warranty and consists of two or more related chapters and an introduction tying them together. The eight parts are shown in the figure on the following page. The book is aimed at a very wide audience, so paths that may be followed by readers from a few particular perspectives are indicated in the figure. Other suggested paths are indicated in the introductions to each section. Specific parts that would be of particular interest to different professional groups are as follows:

Lawyers: Parts B and G
Statisticians: Parts C and D
Operations Analysts: Parts C and F
Engineers: Part F
Accountants: Parts G and H
Marketing Managers: Parts E and H
Public Policy Analysts: Parts B and G
Warranty Managers: Parts D, E, and H

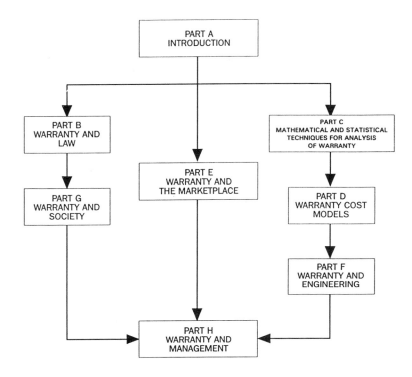

The diagram shows sequential linkings between different parts. Starting with Part A, one can follow the appropriate path depending on the aspect of interest. Thus, for example, for the Public Policy Analyst interested in "Warranty and Society" (Part G), the recommended sequence is Part A → Part B → Part G. Similarly, a reader interested in "Warranty Management" (Part H) may wish to follow the path Part A → Part D → Part E → Part H (with Part C included as an option) in order to fully understand and appreciate warranty and to manage it effectively.

The introduction to each part gives a diagram similar to this one showing the linking between the chapters in the part. These are meant primarily to assist the reader in learning aspects of warranties of particular relevance to a specific information objective.

Topics not covered or mentioned only briefly in the *Handbook* include warranties on commercial/industrial products; warranty in government acquisition; warranties on service; warranties on software; group, fleet, or cumulative warranties; reliability improvement warranties; and multifaceted warranties on complex equipment. In many applications of some of these types of warranties, the rules of the game are quite different;

the buyer often has more power than the seller (e.g., the federal government) and may have as much or more information as well.

The editors thank the contributors to the *Handbook* for their high quality technical contributions and for responding positively to the various comments made on their earlier versions and patiently carrying out revisions. The editors have made a significant attempt at ensuring consistency and smoothness in the flow, but in any volume with numerous authors this is a difficult task. It is hoped that despite the differing styles of presentation and levels of theory, the readers, both academics and practitioners, will find the book interesting and useful.

Wallace R. Blischke
D. N. Prabhakar Murthy

Contents

PART C   MATHEMATICAL AND STATISTICAL TECHNIQUES FOR ANALYSIS OF WARRANTY

PART D   WARRANTY COST MODELS

PART F   WARRANTY AND ENGINEERING

24. Warranty Servicing   621

D. N. Prabhakar Murthy

PART G   WARRANTY AND SOCIETY

25. The Economic Theory of Warranties   659

Nancy A. Lutz

26. Financial Accounting and Reporting for Warranties   675

Richard A. Maschmeyer and
Kashi Ramamurthi Balachandran

PART I   WARRANTY BIBLIOGRAPHY

Contributors

Kashi Ramamurthi Balachandran, B.E.(Hons.), M.S., Ph.D. Department of Accounting, Auditing and Business Law, New York University, New York, New York

Wallace R. Blischke, B.S., M.S., Ph.D. Department of Information and Operations Management, University of Southern California, Los Angeles, California

John R. Burton, B.S., M.A., Ph.D. Department of Consumer Studies and Family Economics, University of Utah, Salt Lake City, Utah

Jingxian Chen, B.E., M.S., D.Sc. Institute for Reliability and Risk Analysis, The George Washington University, Washington, D.C.

Stefanka Chukova, Ph.D. Department of Science and Mathematics, GMI Engineering and Management Institute, Flint, Michigan

Young Hak Chun, Ph.D. Department of Information Systems and Decision Science, College of Business Administration, Louisiana State University, Baton Rouge, Louisiana

Boyan Dimitrov, Ph.D. Department of Science and Mathematics, GMI Engineering and Management Institute, Flint, Michigan

Istiana Djamaludin, M.Eng.Sci., Ph.D. Technology Management Centre, The University of Queensland, Brisbane, Queensland, Australia

Edward W. Frees, Ph.D. School of Business and Department of Statistics, University of Wisconsin, Madison, Wisconsin

Jeffrey J. Hunter, M.Sc.(Hons.), Ph.D. Department of Statistics, Massey University, Palmerston North, New Zealand

J. D. Kalbfleisch, B.Sc., M.Math., Ph.D. Department of Statistics and Actuarial Science, and Dean, Faculty of Mathematics, University of Waterloo, Waterloo, Ontario, Canada

Craig A. Kelley, Ph.D. Department of Management, California State University, Sacramento, Sacramento, California

Rachel S. Kowal, J.D. Department of Accounting, Taxation and Business Law, Leonard N. Stern School of Business, New York University, New York, New York

K. Ravi Kumar, Ph.D. Department of Information and Operations Management, University of Southern California, Los Angeles, California

J. F. Lawless, Ph.D. Department of Statistics and Actuarial Science, University of Waterloo, Waterloo, Ontario, Canada

Arvinder P. S. Loomba, B.Engg., M.B.A., Ph.D. Department of Management, University of Northern Iowa, Cedar Falls, Iowa

Nancy A. Lutz, Ph.D. Department of Economics, Virginia Polytechnic Institute and State University, Blacksburg, Virginia

Nicholas J. Lynn, B.S., M.S. Institute for Reliability and Risk Analysis, The George Washington University, Washington, D.C.

Richard Marcellus, Ph.D. Industrial Engineering Department, Northern Illinois University, Dekalb, Illinois

Richard A. Maschmeyer, B.S., M.Ac., D.B.A. Department of Accounting, School of Business, University of Alaska Anchorage, Anchorage, Alaska

Amitava Mitra, Ph.D. College of Business and Department of Management, Auburn University, Auburn, Alabama

Gregory C. Mosier, J.D., Ed.D. Department of Economics and Legal Studies in Business, College of Business Administration, Oklahoma State University, Stillwater, Oklahoma

Herbert Moskowitz, Ph.D., M.B.A., B.Sc.(Mech.Eng.) Kronnert School of Management, Purdue University, West Lafayette, Indiana

D. N. Prabhakar Murthy, B.E., M.E., M.S., Ph.D. Technology Management Centre, The University of Queensland, Brisbane, Queensland, Australia

V. Padmanabhan, B.Tech., M.S., Ph.D. Graduate School of Business, Stanford University, Stanford, California

Jayprakash G. Patankar, Ph.D. Department of Management, The University of Akron, Akron, Ohio

Ba Pirojboot, M.S. Industrial Engineering Department, Northern Illinois University, DeKalb, Illinois

Nozer D. Singpurwalla, Ph.D. Department of Operations Research, The George Washington University, Washington, D.C.

Joshua Lyle Wiener, Ph.D. Department of Marketing, Oklahoma State University, Stillwater, Oklahoma

Richard J. Wilson, B.Sc. Ph.D.(UNSW) Department of Mathematics, The University of Queensland, Brisbane, Queensland, Australia

Part A

Introduction

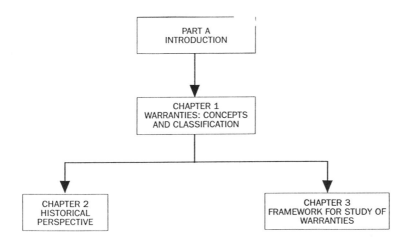

Introduction

Part A serves as an overall introduction to this *Product Warranty Handbook*. It provides background information that is useful in understanding the concept of warranty, its history, and some of the methods of analysis used in the study of the warranty process from the many perspectives represented in the *Handbook*.

Chapter 1, written by the editors, is concerned with the basic warranty concept. Warranty is defined and its roles in various types of transactions are discussed. There are a large number of possible warranty policies; for illustration, 35 different policies are defined in this chapter. To provide structure for this wealth of possibilities, a method of classification is proposed. This results in a taxonomy of warranties useful as both an overview of the warranty process and a starting point in the analysis of warranties from any of the many perspectives.

In Chapter 2, A. P. S. Loomba presents a most interesting study of the history of warranty, beginning in ancient times. Warranty of a sort has been involved as an aspect of trade for a very long time in certain civilizations. Loomba cites evidence dating back over 4000 years. His survey includes early civilizations, the Roman era and subsequent European periods, the Middle Ages, and the pre- and post-industrial era. According to local experts on the history of commerce, this may be the first such study undertaken.

The third chapter of the *Handbook*, also written by the editors, is concerned primarily with the systems approach as a framework for the study of warranty, from the points of view of the stakeholders in the process and of the many disciplines involved in warranty analysis. Stakeholders include manufacturers, consumers, and society; and the models used for analysis from these different perspectives may vary both in structure and content. A number of such models are given. In addition, the models are presented in the context of the many disciplines involved—engineering, production, accounting, marketing, consumerism, and others.

1

Warranties: Concepts and Classification

Wallace R. Blischke

University of Southern California
Los Angeles, California

D. N. Prabhakar Murthy

The University of Queensland
Brisbane, Queensland, Australia

1.1 INTRODUCTION

In today's market, product warranty plays an increasingly important role in both consumer and commercial transactions. The use of warranties is widespread and they serve many purposes, including protection for manufacturer, seller, and buyer; signals of quality; elements of marketing strategy; factors in dispute resolution; and considerations in public policy and legislation.

Because of this diversity of purpose, warranties have been studied by researchers in many different disciplines, including the following: engineering, management, operations analysis, marketing, economics, law, public administration, consumer affairs, statistics, accounting, and quality control. As a result, a vast and disjointed literature on warranty has developed. The major objectives of this book are to structure and unify the many important results which are scattered throughout this literature and to thereby provide an integrated framework for both the study of warranty policies from these different perspectives and the application of the results in real-life situations.

Structure for the many important results on warranty is provided in the *Handbook* by organization of the chapters according to both discipline and area of application. The unifying theme of the book is the systems approach to analysis of warranties. In the analyses, three points of view are recognized: that of the manufacturer (including distributor, retailer, and so forth), that of the consumer (individual, corporation, or government agency), and that of society (including legislators, consumer affairs groups, the courts, and public policy decision-makers).

In this chapter, the subject is introduced by means of a discussion of the basic warranty concept and presentation of a taxonomy for classification of warranties. Some definitions of warranty and its relation to other concepts are discussed in Section 1.2. Section 1.3 deals with the role of warranty in consumer, commercial, and government transactions. The discussion here is introductory in nature and is intended to set the scene for the more detailed, in-depth presentations given in later chapters of the *Handbook*.

Section 1.4 deals with the structure of warranty, that is, the various warranty terms that might be offered, and some of the terminology used in describing warranties. Finally, Section 1.5 is devoted to a warranty taxonomy. There, the most commonly used warranties, as well as many warranties having special or unique features, will be explicitly defined and the many types of warranties will be categorized. This will provide an important foundation for further discussion, particularly for the chapters on cost models and warranty marketing and management.

1.2 DEFINITION OF WARRANTY

A warranty is a seller's assurance to a buyer that a product or service is or shall be as represented. It may be considered to be a contractual agreement between buyer and seller (or manufacturer) which is entered into upon sale of the product or service. A warranty may be implicit or it may be explicitly stated.

In this context, by "seller" is meant the party responsible for assuring that the warranty terms are met. This is ordinarily the manufacturer of the product or a dealer or other retail or discount outlet or the provider of the service. The "buyer" is ordinarily the ultimate paying consumer. In some cases, other parties—for example, an insurer or an independent repair facility—may be involved as well.

In broad terms, the purpose of a warranty is to establish liability among these parties in the event that an item fails or is unable to perform its intended function when properly used, or when a service is improperly performed. The contract specifies both the performance that is to be expected and the redress available to the buyer if a failure occurs or the performance is unsatisfactory.

In this handbook, we shall be concerned almost exclusively with *product* warranties, although warranties on services (for example, auto repairs, home improvements or repairs, delivery services, etc.) will occasionally be mentioned. Furthermore, the emphasis will be on manufactured products (e.g., consumer durables, large purchases such as automobiles, commercial purchases such as aircraft, and so forth), for which failures are reasonably well defined and the time at which a failure occurred can be established with a reasonable degree of accuracy. In these cases, the warranty is intended to assure the buyer that the product will perform its intended function under normal conditions of use for a specified period of time.

Many of the results to be presented in the *Handbook* are also applicable in situations where failures are not as well defined or easily detected, for example, warranties on computer software. In such cases, the main difficulty would be in the application of analytical results, such as the cost models. In fact, special care has to be taken even in defining the concept of a failure and "time" between failures. To date, little analysis of warranties on software has been done, and these are not covered explicitly in the *Handbook*.

The terms warranty and guarantee are often used synonymously. The distinction is that a guarantee is defined to be a pledge or assurance of something; a warranty is a particular type of guarantee, namely, a guarantee concerning goods or services provided by a seller to a buyer.

Another related concept is that of a service contract or "extended service" contract. The difference between a warranty and a service contract is that the latter is entered into voluntarily and is purchased separately—the buyer may even have a choice of terms—whereas a warranty is a part of the product purchase and is an integral part of the sale. Service contracts are covered in the *Handbook* but are not listed separately in the warranty classification scheme given later in this chapter.

At present, nearly everything purchased or leased, whether by an individual, a corporation, or a government agency, is covered by warranty, either express or implied. An *express* warranty is one whose terms are explicitly stated in writing, whereas an *implied* warranty is a contract that is automatically in force upon purchase of an item from a purveyor of such goods. Sales of this type are taken to imply "merchantability" and "fitness for a particular purpose." The distinction between express and implied warranties is specified in the United States in Article 2 of the

Uniform Commercial Code (UCC). For additional details and discussion, see Chapters 4 and 5 of the *Handbook* and Chapter 1 of Ref. 1.

In most of the preceding, the emphasis has been on *consumer* warranties, that is, warranties offered (or implied) in purchases by individuals of one or a few of a particular item, for example, a toaster, television, set of dishes, or pair of replacement automobile tires. Other transactions involving warranties are *commercial* and *government* purchases. A commercial transaction is one involving two (or more) private organizations (usually corporations), for example, an auto manufacturer purchasing machine tools, an aircraft manufacturer purchasing jet engines, or an airline purchasing jet aircraft. Government purchases are purchases made by a government agency from a corporation. Many of the principles and methods of the *Handbook* apply to all three types of transactions. Others are unique to the particular type of application, however, and this will be noted in the remainder of this chapter and in the chapters to follow.

1.3 THE ROLE OF PRODUCT WARRANTY

Warranties are an integral part of nearly all consumer and commercial and many government transactions that involve product purchases. In such transactions, warranties serve a somewhat different purpose for buyer and seller.

1.3.1 Buyer's Point of View

From the buyer's point of view, the main role of a warranty in these transactions is protectional; it provides a means of redress if the item, when properly used, fails to perform as intended or as specified by the seller. Specifically, the warranty assures the buyer that a faulty item will either be repaired or replaced at no cost or at reduced cost.

A second role is informational. Many buyers infer that a product with a relatively long warranty period is a more reliable and long-lasting product than one with a shorter warranty period.

1.3.2 Seller's Point of View

One of the main roles of warranty from the seller's point of view is also protectional. Warranty terms may and often do specify the use and conditions of use for which the product is intended and provide for limited coverage or no coverage at all in the event of misuse of the product. The seller may be provided further protection by specification of requirements for care and maintenance of the product.

A second important purpose of warranty for the seller is promotional. As buyers often infer a more reliable product when a long warranty is offered, this has been used as an effective advertising tool. (See Chapters 22 and 23.) This is often particularly important when marketing new and innovative products, which may be viewed with a degree of uncertainly by many potential consumers. In addition, warranty has become an instrument, similar to product performance and price, used in competition with other manufacturers in the marketplace.

1.3.3 The Role of Warranty in Government Contracting

In simple transactions involving consumer or commercial goods, a government agency may be dealt with in basically the same way as any other customer, obtaining the standard product warranty for the purchased item. Often, however, the government, as a large entity wielding substantial power as well as a very large consumer, will be dealt with considerably differently, with warranty terms negotiated at the time of purchase rather than specified unilaterally by the seller. The role of warranty in these transactions is usually primarily protectional on the part of both parties.

In some instances, particularly in the procurement of complex military equipment, warranties of a certain type play a very different and important role, that of incentivizing the seller to increase the reliability of the items after they are put into service. This is accomplished by requiring that the contractor service the items in the field and make design changes as failures are observed and analyzed. The incentive is an increased fee paid the contractor if it can be demonstrated that the reliability of the item has, in fact, been increased. Warranties of this type are called reliability improvement warranties (RIW) and these are not discussed in this handbook.

1.4 WARRANTY STRUCTURE AND TERMINOLOGY

In order to classify warranties, which will provide structure to their analysis and to the discussions throughout the *Handbook*, it is necessary to identify specific warranty features. This will be further useful to the practitioner in providing a list of the various options that might be available in defining product warranty terms.

1.4.1 Basic Consumer Warranty Terms

The first important characteristic of a warranty is the form of payment to the customer on failure of an item. The most common forms are (1) a lump-sum rebate (e.g., "money-back guarantee"), (2) a free replacement of an item identical to the failed item, (3) a replacement provided at reduced cost to the buyer, and (4) some combination of the preceding terms.

Warranties of type 2 are called free-replacement warranties (FRW). See Chapter 10 for cost models and additional discussion.

For warranties of type 3, the amount of reduction is usually a function of the amount of service received by the buyer up to the time of failure, with decreasing discount as time of service increases. This discount may be a straight percentage of the purchase price that changes one or more times during the warranty period (e.g., 50% discount during the first half and 20% discount during the remainder of the warranty period) or it may be a continuous function of the time remaining in the warranty period. The latter is called a pro-rata warranty (PRW) and is discussed in more detail in Chapter 11.

The most common combination warranty is one that provides for a free replacement up to a specified time and a replacement at prorated cost during the balance of the warranty period. This is called a combination FRW/PRW and is analyzed in Chapter 12.

Note that there are many variations of these basic warranties and many possible combinations. First, there are rebate forms of all of these, under which the buyer is given a partial or full refund. This may or may not be used toward the purchase of a new item, at the buyer's discretion. Under the nonrebate forms, the buyer does not have a choice; the value of the refund may only be applied as a discount on the purchase of a replacement item. Second, in the case of repairable items, the item may be repaired rather than replaced or only some parts replaced or repaired. Again, this may be done at no cost to the buyer or at a reduced cost.

Warranty coverage may also be limited in many ways. For example, certain types of failures or certain parts may be specifically excluded from coverage. Coverage may include parts and labor or parts only, or parts and labor for a portion of the warranty period and parts only thereafter. The variations are nearly endless. **It is very important that the exact terms of a warranty be carefully delineated, particularly when attempting to predict future warranty costs.**

1.4.2 Renewing Warranties

Under a *renewing* warranty, all replaced or repaired items are covered under a warranty that is identical to that of the original purchased item. Under a *nonrenewing* warranty, the coverage extends only over the time remaining in the original warranty period. For example, if an item is sold with an FRW of length 1 year and it fails after 9 months, the replacement item would carry a warranty of 1 year if the FRW were renewing, but it would be warrantied for only 3 months under nonrenewing FRW.

Under a renewing warranty, items are replaced or repaired until the time between two successive failures is at least the length of the warranty period, say *W*, that is, items are replaced until one is found that has a lifetime of at least *W*. Under a nonrenewing warranty, items are replaced or repaired until the *total* service time of the original and all replacement items is at least *W*.

Note that the rebate form of a warranty is automatically nonrenewing because no replacement item is supplied. It would, in effect, be renewing, however, if the buyer immediately used the rebate toward the purchase of an identical new item.

1.4.3 Warranty Dimension

Another important characteristic, particularly for purposes of analysis, is the dimensionality of a warranty. By this is meant the number of variables specified in the warranty terms. The most common warranty is a one-dimensional warranty, and the most commonly used variable is time from purchase. For example, a warranty may guarantee that a product be "free from defects in material and workmanship for a period of one year from date of purchase." Other examples of one-dimensional warranties are those based on usage (e.g., miles driven or milimeters of tire wear) or on "cycles" (e.g., number of take-offs and landings of an aircraft).

Higher-dimensional warranties involve more than one characteristic measuring product service, with guaranteed service amounts specified for each. The most common two-dimensional warranty is one based on both calendar time and usage. This is particularly common in the automobile industry, where cars are typically sold with warranties limited by time from purchase as well as mileage (e.g., 2 years or 24,000 miles, whichever occurs first). Many other products are sold with warranties of this type. Three- and higher-dimensional warranties would involve additional service measures. An example would be calendar time, flight hours, and landings of an aircraft. Because of their increasing complexity, these are used only in a few specialized applications.

1.4.4 Multicharacteristic Warranties

Rather than using multivariate measures of service, many of the more complex warranties guarantee two or more characteristics. Each characteristic may be quantified by means of one or more of the variables discussed previously, and these variables may or may not be related. For example, an engine may be guaranteed to operate without failure for a specified period of time and to provide a specified level of fuel efficiency.

On certain types of products, separate warranties may be provided on individual parts or subsystems. This is true, for example, of many appliances, for example the compressor of a refrigerator or air conditioner carrying different warranty terms from those of rest of the unit. Television picture tubes usually carry a separate warranty.

On complex equipment, this is nearly always the case. For some of these, where parts or subsystems are obtained from sources other than the equipment manufacturer, the suppliers, themselves, often warranty the items they supply. An example of a consumer warranty of this type is an automobile warranty. Typically, items such as tires and batteries are warrantied by the manufacturers of these items rather than by the automobile manufacturer.

1.4.5   Other Types of Warranty

A number of different warranties have been suggested in the context of commercial and government purchasing. One of these is the RIW, mentioned previously. Reliability improvement warranties are typically multivariate and multiattribute warranties, guaranteeing a number of features in addition to service. For example, RIWs usually guarantee not just replacements on failure of a warrantied item but the mean time to failure of a batch of items.

Another type of warranty that has been proposed for use in situations of this type is the *cumulative* warranty. Under this warranty, an entire batch of items is guaranteed to provide a specified total amount of service, without specifying a guarantee on any individual item. For example, rather than guaranteeing that each item in a batch of 100 will operate without failure for 2000 hours, the batch as a whole is guaranteed to provide at least 200,000 hours of service. If after the last item in the batch has failed, the total service time is less than 200,000 hours, items are provided as specified in the warranty (e.g., free of charge or at pro-rata cost) until such time as the total of 200,000 hours is achieved. Warranties of this type are not discussed in the *Handbook*.

1.5   A TAXONOMY FOR CLASSIFICATION OF WARRANTIES

1.5.1   Basis of Classification

The first criterion for classification of a warranty to be used in the taxonomy to be presented in this section is whether or not the warranty requires development after the sale of the product. Policies which do not involve product development can be further divided into two groups—Group A, consisting of policies applicable for single item sales, and Group B, policies used in the sale of groups of items (called *lot* or *batch* sales). This division and the remainder of the taxonomy are shown in Figure 1.1.

Policies in Group A can be subdivided into two subgroups, based on whether the policy is renewing or nonrenewing. A further subdivision comes about in that warranties may be classified as "simple" or "combination." The free-replacement (FRW) and pro-rata (PRW) policies discussed previously are simple policies. A combination policy is a simple policy combined with some additional features, or a policy which combines the terms of two or more simple policies. The resulting four types of policies under category A are labeled A1–A4 in Figure 1.1.

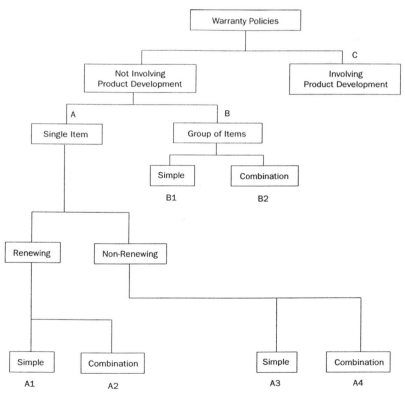

Figure 1.1   Taxonomy for warranty policies.

Each of these four groupings can be further subdivided into two subgroups based on whether the policy is one dimensional or two (or more) dimensional (not shown in Figure 1.1). Examples of policies of both types will be given in Section 1.5.2.

Policies in Group B can be subdivided into two categories based on whether the policy is "simple" or "combination." These are labeled B1 and B2 in Figure 1.1. As in Group A, B1 and B2 can be further subdivided based on whether the policy is one dimensional or two dimensional.

Finally, policies involving product development are labeled Group C. Warranties of this type are typically part of a maintenance contract and are used principally in commercial applications and government acquisition of large, complex items—for example, aircraft or military equipment. Nearly all such warranties involve time and/or some function of time as well as a number of characteristics that may not involve time, for example, fuel efficiency.

1.5.2   Examples of Warranties

A variety of warranty policies will now be described. Typical products sold with some of these warranties will be indicated. The following notation is used:

W = length of warranty period
C = selling price of the item (cost to buyer)
X = time to failure (lifetime) of an item

One-Dimensional Group A Policies

Subgroup A1 (Nonrenewing Simple Policies)

Policies in this group are variations of the basic FRW and PRW. Three versions of the FRW are as follows:

Policy 1

FREE-REPLACEMENT POLICY (FRW): The seller agrees to repair or provide replacements for failed items free of charge up to a time W from the time of the initial purchase. The warranty expires at time W after purchase.

Policy 2

REBATE FRW: The seller agrees to refund an amount αC, where $0 < \alpha < 1$, if the item fails prior to time W from the time of purchase.

Policy 3

> *REBATE FRW ("MONEY-BACK GUARANTEE")*: The seller agrees to refund the full purchase price C if the item fails prior to time W from the time of purchase.

The warranty of Policy 1 is nonrenewing. Thus, in the case of nonrepairable items, should a failure occur at age X (with $X < W$), then the replaced item has a warranty for a period $W - X$—the remaining duration of the original warranty. Should additional failures occur, this process is repeated until the total service time of the original item and its replacements is at least W. In the case of repairable items, repairs are made free of charge until the total service time of the item is at least W.

Policy 2 is technically not an FRW because the refund is not sufficient for purchase of a replacement. It is included here because the structure and analysis of the policy are basically the same as those of the nonrenewing FRW. Policy 3 can be considered an FRW because the buyer has the option of using the rebate for purchase of a replacement item.

Typical applications of these warranties are consumer products, ranging from inexpensive items such as photographic film to relatively expensive repairable items such as automobiles, refrigerators, large-screen color TVs, and so forth, and on expensive nonrepairable items such as microchips and other electronic components as well.

The basic nonrenewing PRW is as follows:

Policy 4

> *PRO-RATA REBATE POLICY*: The seller agrees to refund a fraction of the purchase price should the item fail before time W from the time of the initial purchase. The buyer is not constrained to buy a replacement item.

The refund depends on the age of the item at failure (X) and it can be either a linear or nonlinear function of $W - X$, the remaining time in the warranty period. This defines a family of pro-rata policies which is characterized by the form of the refund function. Three members of this family are as follows:

Policy 5

> *LINEAR PRW*: The seller agrees to refund an amount $[(W - X)/W]C$ should the item fail prior to W after the time of purchase.

Policy 6

> *PROPORTIONAL LINEAR PRW*: The seller agrees to refund an amount $[\alpha(W - X)/W]C$, where $0 < \alpha < 1$, should the item fail prior to W after the time of purchase.

Policy 7

NONLINEAR PRW: The seller agrees to refund an amount $[(W - X)/W]^2 C$ should the item fail prior to W after the time of purchase.

Policies such as 5 and 6 are sometimes offered on relatively inexpensive nonrepairable products such as batteries, tires, ceramics, and so on. Policy 7 features a quadratic rebate function and is an alternative for which the rebate decreases more rapidly.

Subgroup A2 (Nonrenewing Combined Policies)

Warranties of this type typically feature terms which change one or more times during the warranty period (e.g., full to limited warranty, or FRW to PRW). The most common policy of this type is

Policy 8

COMBINATION FRW/PRW: The seller agrees to provide a replacement or repair free of charge up to time W_1 from the initial purchase; any failure in the interval W_1 to W (where $W_1 < W$) results in a prorated refund. The warranty does not renew.

The proration can be either linear or nonlinear. Again, depending on the form of the proration cost function, we have a family of combined free-replacement and pro-rata policies similar to that for the PRW. Warranties of this type are sometimes used to cover replacement parts or components where the original warranty covers an entire system. They are also widely used in sales of consumer products.

Multistage warranties such as the following are also included in this family:

Policy 9

THREE-STAGE WARRANTY: The seller agrees to provide replacements or repairs free of charge up to time W_1 after initial purchase, at cost C_1 if the failure time X is in the interval $(W_1, W_2]$, and at cost C_2 if X is in the interval $(W_2, W]$, where $0 < W_1 < W_2 < W$ and $C_1 < C_2$. The warranty expires at time W after purchase.

Examples are television sets or appliances for which full coverage is provided for an initial period and only partial coverage (e.g., some parts and/or labor) in later periods.

The general form of this type of policy is as follows:

Policy 10

> *MULTISTAGE REBATE WARRANTY*: The seller agrees to provide replacements or repairs free of charge up to time W_1 after initial purchase, at cost $\alpha_1 C(X)$ if the failure time X is in the interval $(W_1, W_2]$, at cost $\alpha_2 C(X)$ if X is in the interval $(W_2, W_3]$, and so forth, up to cost $\alpha_k C(X)$ if X is in the interval $(W_k, W]$, where $\alpha_1 < \alpha_2 < \cdots < \alpha_k$ are the proportions of the current price $C(X)$ at time of failure for failures in each of k time intervals.

An example—a stand-alone fireplace warrantied for 25 years—is discussed in Ref. 1, Chapters 1 and 2.

Another version of this is a combination lump-sum rebate warranty, under which a refund which is a declining proportion of the original purchase price is given rather than a replacement item or a repair. This policy is given by the following:

Policy 11

> *COMBINATION LUMP-SUM REBATE WARRANTY*: A rebate in the amount of $\alpha_1 C$ is given for any item that fails prior to time W_1 from the time of purchase; the rebate is $\alpha_2 C$ for items that fail in the interval $(W_1, W_2]$, $\alpha_3 C$ for items that fail in the interval $(W_2, W_3]$, and so forth, up to a final interval $(W_{k-1}, W]$, in which the rebate is $\alpha_k C$, with $1 \geq \alpha_1 > \alpha_2 > \cdots > \alpha_k > 0$.

Still another policy of this type is a combination lump-sum and pro-rata refund, given in the following:

Policy 12

> *REBATE COMBINATION FRW/PRW*: The seller agrees to provide a full refund of the original purchase price up to time W_1 from the time of initial purchase; any failure in the interval from W_1 to $W (> W_1)$ results in a pro-rata refund. The warranty does not renew.

The following combination policy is a modification of the FRW that is particularly appropriate for items which are sold as spares and hence not used immediately after purchase.

Policy 13

> *WARRANTY WITH STORAGE LIMITATION*: The policy is characterized by two parameters w and W, with $w < W$. Let S denote the time, subsequent to its purchase, that the item is kept unused in storage before being put in operation and X the time to failure of the item after being put into service. The item, upon failure, is covered under warranty only if $X < w$ *and* $X + S <$

W. If this condition is met, the coverage may be under FRW, PRW, or a combination warranty.

Under this policy, a failed item may not be covered by warranty, even though X is *smaller* than W, if S is sufficiently large.

Subgroup A3 (Renewing Simple Policies)

Two types of policies, analogous to those of Subgroup A1, are included in this subgroup. An example of the first of these is as follows:

Policy 14
  RENEWING FRW: Under this policy, the manufacturer agrees to either repair or provide a replacement free of charge up to time W from the initial purchase. Whenever there is a replacement, the failed item is replaced by a new one with a new warranty whose terms are identical to those of the original warranty.

Under this policy, the buyer is assured of one item that operates for a period W without a failure.

  This type of policy is usually offered with inexpensive electrical, electronic, and mechanical products such as coffee grinders, alarm clocks, tools, and so forth, for which the warranty is contained inside the item package. Upon failure, the item is returned to the seller, who merely replaces it with an identical package. If the buyer simply returns the new warranty card, new warranty coverage results with each replacement.

  The second simple renewing policy is as follows:

Policy 15
  RENEWING PRW: Under this policy, the manufacturer agrees to provide a replacement item, at prorated cost, for any item (including the item originally purchased and any replacements made under warranty) which fails to achieve a lifetime of at least W.

Proration can be either a linear or a nonlinear function of $W - X$, with $X < W$. Depending on the proration function, this again defines a family of pro-rata policies.

  Note: The difference between this and the pro-rata warranty of Policy 4 is that under Policy 4 the refund is unconditional, whereas here it is only provided as a discount on the purchase of a replacement item.

  Many nonrepairable items are sold with this type of policy or a combination having this, following an initial FRW period. Most auto tires and batteries are sold under renewing PRW, the buyer being offered a replacement for a failed item at a reduced price, without a cash rebate option.

Subgroup A4 (Renewing Combination Policies)

These are policies based on combinations of the terms of two or more simple policies, at least one of which is renewing. There are many possibilities. For example, a combination FRW/PRW may renew only in the FRW segment, only in the PRW segment, or in both. Policies that renew in both segments are called *fully renewing*; those that do not are called *partially renewing*. The possible policies are as follows:

Policy 16

FULLY RENEWING FRW/PRW: The seller agrees to provide a replacement free of charge up to time W_1 from the initial purchase; any failure in the interval W_1 to $W (> W_1)$ is replaced at a prorated cost. The proration can be either linear or nonlinear. In either case, the replacement item is offered with a new warranty identical to the original one.

Policy 17

PARTIALLY RENEWING COMBINATION FRW/PRW: The seller agrees to provide a replacement free of charge up to time W_1 from the time of the initial purchase. Replacement items in this time period assume the remaining warranty coverage of the original item. Failures in the interval W_1 to $W (> W_1)$ are replaced at pro-rata cost. Replacement items in this interval are provided warranty coverage identical to that of the original item.

Policy 18

PARTIALLY RENEWING COMBINATION FRW/PRW: The seller agrees to provide a replacement free of charge up to time W_1 from the time of the initial purchase. Replacement items in this time period are provided warranty coverage identical to that of the original item. Failures in the interval W_1 to $W (> W_1)$ are replaced at prorata cost. Replacement items in this interval are covered by pro-rata warranty up to time W from the time of the last replacement.

Policy 16 is the most commonly used version of the combination warranty. In multiperiod policies, many additional versions are possible.

Under Policy 17, the warranty renews only in the PRW period. This means that upon failure of an item at, say, $x \leq W_1$, a free replacement is provided and this replacement item is covered under FRW until time $W_1 - x$, and then under PRW until time W from the initial time of purchase. On the other hand, if the failure occurs at time x', with $W_1 < x' \leq W$, the replacement is provided at prorated cost to the buyer and the warranty begins anew. Policy 18 features renewal only in the FRW period and is rarely used.

Two-Dimensional Group A Policies

In the one-dimensional case, discussed in the previous subsection, a policy is characterized by an interval, called the warranty period, which is defined in terms of a single variable—for example, time, age, usage. In the case of two-dimensional warranties, a warranty is characterized by a region in a two-dimensional plane with one axis representing time or age and the other representing item usage. As a result, many different types of warranties, based on the shape of the warranty coverage region, may be defined.

Subgroup A1 (Nonrenewing Simple Two-Dimensional Policies)

The following are five two-dimensional policies of type A1. The warranty coverage regions in the two-dimensional plane for these five policies are indicated as shaded regions in Figure 1.2. The policies are as follows:

Policy 19

TWO-DIMENSIONAL FRW: The seller agrees to repair or provide a replacement for failed items free of charge up to a time W or up to a usage U, whichever occurs first, from the time of the initial purchase. W is called the warranty period and U the usage limit. The warranty region is the rectangle shown in Figure 1.2(a).

Note that under this policy, the buyer is provided warranty coverage for a maximum time period W and a maximum usage U. If the usage is heavy, the warranty can expire well before W, and if the usage is very light, then the warranty can expire well before the limit U is reached. Should a failure occur at age X with usage Y, it is covered by warranty only if X is less than W *and* Y is less than U. If the item is replaced by a new one, the replacement item is warrantied for a time period $W - X$ and for usage $U - Y$. This type of policy is offered by nearly all auto manufacturers, with usage corresponding to distance driven.

The second two-dimensional Group A1 policy is the following:

Policy 20

TWO-DIMENSIONAL FRW: The seller agrees to repair or provide a replacement for failed items free of charge up to a minimum time W from the time of the initial purchase and up to a minimum total usage U. The warranty region is given by two strips, as shown in Figure 1.2(b).

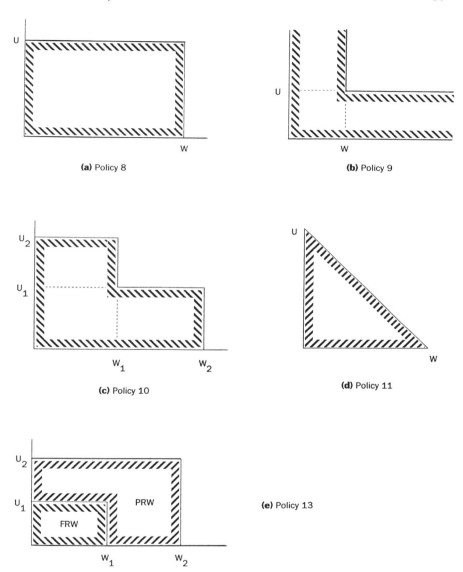

Figure 1.2 Warranty regions for two-dimensional policies (horizontal axis: time; vertical axis: usage).

Note that, under this policy, the buyer is provided warranty coverage for a minimum time period W and for a minimum usage U. If the usage is heavy, the warranty will expire at time W, and if the usage is very light, the warranty will expire only when the total usage reaches the limit U.

Policy 19 tends to favor the manufacturer because it limits the maximum time and usage coverage for the buyer. For a buyer who is a heavy user, the warranty expires before time W due to usage reaching U. Similarly, for a buyer who is a light user, the warranty expires at time W with the total usage below U. In contrast, Policy 20 favors the buyer. Here, a heavy user is covered for a time period W by which time the usage would have well exceeded the limit U and a light user is covered well beyond W, for the policy ceases only when the total usage reaches U. This implies that in the latter case, the manufacturer has to carry spares or replacement units for a time period well beyond W.

The following two policies achieve a compromise between these two extremes:

Policy 21

TWO-DIMENSIONAL FRW: The seller agrees to repair or provide a replacement for failed items free of charge up to a time W_1 from the time of the initial purchase, provided the total usage at failure is below U_2, and up to a time W_2, provided the total usage at failure does not exceed U_1. The warranty region is given by the region shown in Figure 1.2(c).

Note that under this policy, the buyer is provided warranty coverage for a minimum time period W_1 and for a minimum usage U_1. At the same time, the manufacturer is obliged to cover the item for a maximum time period W_2 and for a maximum usage U_2.

The second policy which is a compromise between Policies 19 and 20 is the following:

Policy 22

TWO-DIMENSIONAL FRW: The seller agrees to repair or provide replacements for failed items free of charge up to a maximum time W from the time of the initial purchase and for a maximum total usage of U. Let X be the time since purchase and Y the total usage at failure. The item is covered under warranty if $[Y + (U/W)X] < U$. If $[Y + (U/W)X] \geq U$, then the item is not covered by the warranty. The warranty region is given by the triangle shown in Figure 1.2(d).

Note that under this policy, warranty coverage extends for a maximum time period W and a maximum usage U. The time instant at which warranty ceases depends on the usage rate.

The following are two-dimensional Group A1 policies based on the PRW:

Policy 23

> *TWO-DIMENSIONAL PRW*: The seller agrees to refund to the buyer a fraction of the original purchase price should the item fail before time W from the time of the initial purchase and the total usage at failure is below U. The fraction refunded depends on $W - X$ and $U - Y$.

As in the case of Policy 4, this leads to a family of policies, based on the form of the function which determines the refund. This function can be either linear or nonlinear. Three examples are as follows:

Policy 24

> *TWO-DIMENSIONAL PRW*: The seller agrees to provide a refund in the amount of $(1 - X/W)(1 - Y/U)C$ if the item fails prior to time W from the time of the initial purchase and the total usage is less than U, where X is the lifetime of the item and Y is the total usage.

Policy 25

> *TWO-DIMENSIONAL PRW*: The seller agrees to provide a refund in the amount of $\min\{1 - X/W, 1 - Y/U\}C$ if the item fails prior to time W from the time of the initial purchase and the total usage is less than U, where X is the lifetime of the item and Y is the total usage.

Policy 26

> *TWO-DIMENSIONAL PRW*: The seller agrees to provide a refund in the amount of $[1 - XY/UV]C$ if the item fails prior to time W from the time of the initial purchase and the total usage is less than U, where X is the lifetime of the item and Y is the total usage.

Under Policy 24, the proportion of the purchase price refunded is the product of the proportion of time since the sale to the guaranteed time and the proportion of usage Y to guaranteed usage U. Under Policy 25, it is the minimum of these two proportions. Under Policy 26, it is 1 minus the product of these two proportions. The rebate increases substantially over these three policies.

The warranty coverage region for these three policies is given by the rectangular area shown in Figure 1.2(a). Pro-rata policies for warranty regions shown in Figures 1.2(b)–1.2(d) are defined similarly. In addition, all of these policies may be fully or partially renewing. Because there are many choices for each of these factors, a very large number of policies of this type are possible. Although few of these have been used in practice, this family provides a rich source of possibilities for the practitioner. Caution should be used, however, as a careful cost analysis is necessary before such policies are offered.

Subgroup A2 (Nonrenewing Combination Two-Dimensional Policies)

Many two-dimensional combined versions of the simple warranty policies are possible. The following is one example of this type:

Policy 27
   TWO-DIMENSIONAL COMBINATION FRW/PRW: The seller agrees to provide replacements for failed items free of charge up to a time W_1 from the time of the initial purchase provided the total usage at failure is below U_1. Any failure, with time at failure greater than W_1 but less than W_2 and/or usage at failure greater than U_1 but less than U_2, is replaced at a prorated cost.

The proration can be either a linear or nonlinear function of the warranty parameters and the age and usage at failure. The coverage area is identified in Figure 1.2(e). Many other such policies are possible.

   Two-dimensional versions of Group A3 and A4 policies may be defined similarly.

One-Dimensional Group B Policies

The policies in this group are called cumulative warranties and are applicable only when items are sold as a single lot of n items and the warranty refers to the lot as a whole. The policies are conceptually straightforward extensions of the nonrenewing free-replacement and pro-rata warranties discussed previously. Under a cumulative warranty, the lot of n items is warranted for a total time of nW, with no specific service time guarantee for any individual item.

   Cumulative warranties would quite clearly be appropriate only for commercial and governmental transactions, as individual consumers rarely purchase items by lot. In fact, warranties of this type have been proposed in the United States for use in acquisition of military equipment.

   The rationale for such a policy is as follows. The advantage to the buyer is that multiple-item purchases can be dealt with as a unit rather than having to deal with each item individually under a separate warranty contract. The advantage to the seller is that fewer warranty claims may be expected because longer-lived items can offset early failures.

There are some conceptual difficulties in implementing cumulative warranties and they have not been widely used. (See Ref. 1.) The policies given below are based on the work of Guin [2].

The following notation will be useful for expressing the terms of these policies:

$$X_i = \text{service life of item } i, i = 1, 2, \ldots$$

$$S_i = \sum_{j=1}^{i} X_j$$

Subgroup B1 [Cumulative Simple Policies]

The following two policies are cumulative versions of the FRW.

Policy 28

CUMULATIVE FRW: A lot of n items is warranted for a total (aggregate) period of nW. The n items in the lot are used one at a time. If $S_n < nW$, free-replacement items are supplied, also one at a time, until the first instant when the total lifetimes of all failed items plus the service time of the item then in use is at least nW.

This type of policy is applicable to components of industrial and commercial equipment bought in lots as spares and used one at a time as items fail. Examples of possible applications are mechanical components such as bearings and drill bits. The policy would also be appropriate for military or commercial airline equipment such as mechanical or electronic modules in airborne units.

In the following, it is assumed that more than one item is in use at a given time:

Policy 29

CUMULATIVE FRW: A lot of n items is warranted under cumulative warranty for a total period of nW. Of these, k ($< n$) are put into use simultaneously, with the remaining $n - k$ items being retained as spares. Spares are used one at a time as failures occur. Upon failure of the nth item, free replacements are supplied as necessary until a total service time of nW is achieved.

The following are two cumulative pro-rata warranties:

Policy 30

CUMULATIVE PRW: A lot of n items is purchased at cost nC and warranted for a total period nW. The n items may be used either individually or in batches. S_n, the total service time, is calculated after failure of the last item in the lot. If $S_n < nW$, the buyer is given a refund in the amount of $C(n - S_n/W)$, where C is the unit purchase price of the item.

Rather than waiting for the last failure in the lot before a settlement is made (which could occasionally take a very long time), in the following policy a settlement is made after every Kth item failure. (K is an integer agreed upon by the buyer and the manufacturer and preferably is an integral divisor of n.)

Policy 31

   CUMULATIVE PRW: A lot of n items is warranted for a total time nW. The lot is divided into subsets of size K. Under a cumulative pro-rata warranty, the refund to the buyer at the instant of the Kth failure in each subset is given by $\max\{0, C(K - S_K/W)\}$, where S_K is the sum of the service times of the K failed items in the subset and C is the unit purchase price of the item.

Subgroup B2 (Cumulative Combination Policies)

Again, many combinations could be devised. The following illustrates the many possibilities:

Policy 32

   CUMULATIVE COMBINATION FRW/PRW: A lot of n items is warranted for a total time of nW. Upon failure of the final item in the group, the total service time S_n is calculated. If $S_n < nW_1$, where $W_1 < W$ is a prespecified age, free replacements are provided until a total service time of nW_1 is achieved, say with item $n + J$. Upon failure of this item, the buyer receives a rebate in the amount of $C[\max\{0, (n - S_{n+J}/W)\}]$. If $nW_1 < S_n < nW$, the buyer receives a rebate of $C(n - S_n/W)$.

Group C Policies

Reliability Improvement Warranties

The basic idea of a reliability improvement warranty (RIW) is to extend the notion of a basic consumer warranty (usually the FRW) to include guarantees on the reliability of the item and not just on its immediate or short-term performance. This is particularly appropriate in the purchase of complex, repairable equipment that is intended for relatively long use. The intent of reliability improvement warranties is to negotiate warranty terms that will motivate a manufacturer to continue improvements in reliability after a product is delivered.

Under RIW, the contractor's fee is based on his ability to meet the warranty reliability requirements. These often include a guaranteed MTBF (mean time between failures) as a part of the warranty contract. (See Chapter 6 of Ref. 1.) The following will illustrate the concept:

Policy 33

RELIABILITY IMPROVEMENT WARRANTY: Under this policy, the manufacturer agrees to repair or provide replacements free of charge for any failed parts or units until time W after purchase. In addition, the manufacturer guarantees the MTBF of the purchased equipment to be at least M. If the computed MTBF is less than M, the manufacturer will provide, at no cost to the buyer, (1) engineering analysis to determine the cause of failure to meet the guaranteed MTBF requirement, (2) engineering change proposals, (3) modification of all existing units in accordance with approved engineering changes, and (4) consignment spares for buyer use until such time as it is shown that the MTBF is at least M [3].

The following RIW [4] provides for an initial period during which no MTBF guarantee is in effect, followed by successive periods in which specific improvements in MTBF are required.

Policy 34

RELIABILITY IMPROVEMENT WARRANTY: Under this policy, the manufacturer agrees to repair or provide replacements for any failed parts or units until time W after purchase. In addition, the manufacturer guarantees the MTBF of the purchased equipment to be as follows: No MTBF is guaranteed until time W_1 after the date of first production delivery; during the period from W_1 to W_2 after first delivery, the MTBF is guaranteed to be at least M_1; from W_2 to W_3, the MTBF is guaranteed to be at least M_2; and from W_3 to W, the MTBF is guaranteed to be at least M_3 (with $0 < W_1 < W_2 < W_3 < W$ and $0 < M_1 < M_2 < M_3$). If during any period the MTBF guarantee is not met, the manufacturer will provide, at no cost to the buyer, engineering changes and product modifications as necessary to achieve the MTBF requirements.

A variation of this [5] allows the manufacturer some "free" failures at the outset:

Policy 35

RELIABILITY IMPROVEMENT WARRANTY: A lot of n items is purchased with individual warranty periods W. Items which fail prior to W are repaired or replaced at buyer's expense until k such failures occur, after which the manufacturer will repair or replace failed items until each of the n items in the lot and its replacements achieves a total service time of W.

Cost models for many of the type A (consumer durable) warranties listed in this chapter are given in Part D of the *Handbook*. Not included are RIW and cumulative warranties. For cost models for these, see Ref. 1.

REFERENCES

1. Blischke, W. R., and Murthy, D. N. P. (1993). *Warranty Cost Analysis*, Marcel Dekker, Inc., New York.
2. Guin, L. (1984). *Cumulative Warranties: Conceptualization and Analysis*, Doctoral Dissertation, University of Southern California, Los Angeles, CA.
3. Gandara, A., and Rich, M. D. (1977). *Reliability Improvement Warranties for Military Procurement*, Report No. R-2264-AF, RAND Corp., Santa Monica, CA.
4. Balaban, H. S. (1975). "Guaranteed MTBF for Military Procurement," *Proc. 10th Int. Logistics Symp.*, SOLE.
5. Kruvand, D. H. (1987). Army aviation warranty concepts, *1987 Proc. Annual Reliab. and Maint. Symp.*, 392–394.

2

Historical Perspective on Warranty

Arvinder P. S. Loomba

University of Northern Iowa
Cedar Falls, Iowa

2.1 INTRODUCTION

In order to look at the history of warranty, one has to start by looking at the history of trade itself. Trade, one of the oldest and most predominant of human activities, can be defined as "a business, especially mechanical or mercantile employment, as opposed to profession, carried on as a means of livelihood or profit."

The concept of trade certainly originated as a vehicle to satisfy human needs for numerous goods. Primitive man did try the easy way of need gratification: with the use of brute force. He took what he wanted from those who had it. Over time, the human population increased considerably and brute force techniques gave way to more conservative trade practices that included various modes of product and service exchange, such as the barter system, product sale and purchase using acceptable monetary units, and so on. This evolution in trade practices created a need for some standards, which were developed in some part inadvertently and the rest in a systematic and preconceived manner. Among these, one well-established trade practice that has been evolved over time is known as product warranty.

"Product warranty" can be considered as a statement (written and/or oral) about the nature or quality of the product and its condition or what the warrantor (generally the manufacturer) will do if anything is or goes wrong with the product. A more precise legal definition of product warranty under Magnuson–Moss Warranty–Federal Trade Commission Improvement Act of 1975 is offered by Kelley in Chapter 4. In recent times, most manufacturers offer some form of warranty on their products, usually at no cost to the consumer. However, there are substantial variations in provisions for protection, duration, and service support management of such warranty offerings as are discussed in greater detail by Blischke and Murthy in Chapter 1.

2.2   ORIGIN OF THE WORD "WARRANTY"

The concept of warranty is ancient in origin. At the very outset of mankind's social development, one tribe traded with another, and the quality disputes that arose were similar to the ones still being debated in present-day courtrooms, which is discussed in detail by Kowal in Chapter 5.

Before looking at the historical perspective of warranty, one needs to understand the meaning and origins of the term. The words *warranty* and *guarantee*, known to linguists as "doublets," are derived from same original source but traveling to today's English language by different routes. The origins of the word *warranty* can be traced back to the Old North French word *warant* and *warantie*, to the Old High German word *werēnto* meaning "protector" [1, 2]. During the Middle Ages, the original expressions used included *hoc ex condicione*, *warrantizavit*, *promisit*, and *sub tali plevina* [3, p. 1161].

Warranty has many definitions, depending on the context in which it is used. According to one source:

. . . (i) *Law* in a contract, a promise or binding statement which is non-essential to the main purpose of the contract, so that a failure to honor it does not cause the contract to be ended but may give the other party good reason to claim damages for breach of warranty,

(ii) *Insurance*, a statement made by the insured declaring that facts given by him are true and that insurance contract may be void if any of these facts prove to be untrue,

(iii) *Commercial*, a promise or statement by the seller or the buyer concerning the quality of goods or their fitness for a particular purpose. Without warranty, the goods are being sold on the condition that the seller has no responsibility for any faults or imperfections in the goods, and the buyer has no right to return them or claim damages or any other remedy . . . . [4]

From a legal standpoint, as discussed by Kelley in Chapter 4 and by Kowal in Chapter 5, the issues of negligence, fault, and/or due care are significant in tort liability cases. In contrast, in the case of breach of warranty, these matters are irrelevant. If the goods have some latent defect, the seller is held accountable even though he did not know of and had no means of ascertaining the existence of the defect.

Product warranty has demonstrated its strategic significance to product manufacturers competing in today's industrial marketplace. To acquire a better appreciation for framework to study product warranties, as presented by Murthy and Blischke in Chapter 3, it is imperative that we understand the chronological evolution of this trade practice. In the following sections, we review the development of warranty policies and procedures over time from a historical perspective.

2.3 THE EARLY CIVILIZATIONS

2.3.1 The Babylonian and Assyrian Era

Over the course of human civilization, the issue of warranty has been raised on a consistent basis over a wide variety of products and services ranging from cattle and slaves in the ancient times to automobiles, office equipment, and complex weapon systems in modern times. Some of the earliest documented accounts about trade and customer complaint have been found on the clay tablets from the Ur III dynasty in Babylonia (2128–2004 B.C.). The tablet reads as follows:

. . . Tell Ea-nāsim: Nanni sends the following message:

When you came, you said to me as follows: "I will give Gimil-Sin (when he comes) fine quality copper ingots." You left then but you did not do what you promised. . . . What do you take me for, that you treat somebody like me with such contempt? . . . How have you treated me for that copper? You have withheld my money bag from me in enemy territory; it is now up to you to restore (my money) to me in full. . . . Take cognizance that (from now on) I will not accept here any copper from you that is not of fine quality. I shall (from now on) select and take the ingots individually in my own yard, and I shall exercise against you my

right of rejection because you have treated me with contempt. . . .
[5]

One of the first cases of warranty of fitness has been found recorded
in the Hammurabic Code (circa twentieth century B.C.), dealing with products and services. It provides for an *eye-for-an-eye* type of compensation.
For instance, a house builder, "who has not made strong his work" causing the house to collapse thereby killing the owner, is put to death for his
negligence. If the defective workmanship results in the death of the son
of the owner, then the son of the builder was put to death. However, if
any slaves got killed due to same reasons, the builder has to replace them
and rebuild the house at his expense [6, pp. 83–95]. The code also provided
compensation in the form of "money-back guarantees" for defects discovered in a slave (in this case, the product) after its sale:

> . . . If a man has brought a male or female slave and the slave
> has not fulfilled his month, but the *bennu* disease has fallen upon
> him, he (the buyer) shall return the slave to the seller and the
> buyer shall take back the money he paid. . . . [7, Section 278, p.
> 67]

And:

> . . . If a man has bought a male or female slave and claims been
> raised, the seller shall answer the claim. . . . [7, Section 279, p.
> 67]

Subsequent to the Hammurabi rule, another code appeared around
the second year of the reign of Ashurbânipal, king of Babylon. One frag-
ment of the tablet containing the code provided the following law:

> . . . The man, who has sold a female slave and has had an objec-
> tion made concerning her, shall take her back. The seller shall
> give to the buyer the price named in the deed of sale to its exact
> amount, and shall pay half a shekel of silver for each of the chil-
> dren born to her. . . . [7, Law B, Col. II. 15–23, p. 70]

For these codes, the time period within which a warranty claim (on
a *defective* slave) could be made varied considerably. During the later
Babylonian and Assyrian periods, clauses were inserted in the sale contracts of slaves which provided up to a month to 100 days for fever or
seizures to develop. But the slaves could be returned at any time on
grounds of a *sartu* (vice) plea [7, p. 234].

2.3.2 The Egyptian Era

In spite of the highly developed conceptions of ownership and possession, there does not seem to have existed a correspondingly developed theory of sale, such as the one that existed in Rome. One plausible explanation for this comparative indifference for sale contracts under Egyptian law is the very small amount of commerce which was transacted.

Contracts of sale, as prevalent in Egyptian era (5702–3180 B.C.), were essentially unilateral. In contrast to Roman law, where the purchaser took possession, in Egypt the vendor transferred possession (however, in both cases, the emphasis was placed on the act of one party). Under Egyptian laws, the contracting party, whether in a sale or any other form of contract, detaches from himself a right and conveyed it to the other party. The vendor is obliged to warrant and defend the title which he conveyed. An expression in the warranty requiring the seller to defend the transaction against third-party challenges by "clearing/cleaning" is claimed to origins in Asia and was first introduced into Egyptian formulary in the middle of first millennium B.C. [8, p. 90].

The earliest documents attesting the Arabic sales formularies can be traced back to about A.D. 717–823. The Coptic arbitration documents affirm that in the A.D. 730–740, Arab officials were arbitrating claims between native Egyptians, which included dispute over title to shares in inherited residential property [9].

2.3.3 The Ancient Hindu Period

The law of India is founded in the first instance on the Vedas, which are essentially religious books of the ancient Hindu period. Closely connected to the Vedas are the *Dharma-sutra*, which form the first order of law books, which, in turn, are transformed in *Dharma-sàstras* or legal handbooks used in practice. Among these, one, *Mànava-dharma-sàstra* or *The Ordinance of Manu*, seems to have attained special prominence.

Mànava-dharma-sàstra were originally transcribed in ancient India around the year A.D. 500. In regard to sales, it provided that goods and wares that are damaged, deficient, mixed with another, far away, or concealed should never be sold [10, Section 203, p. 211]. In contrast, the warranty provisions which were provided to a dissatisfied buyer were rather limited in nature:

> . . . Whoever feels regret in this world after buying or selling anything within ten days give (back) or take (back) the goods. But after the period of ten days is passed, he may neither give them

(back) or take them (back); and if he take them (back) or give
them (back), he should be fined six hundred (*panas*) by the
king. . . . [10, Section 222–223, pp. 214–215]

Similar limitations on duration of ''money-back'' warranty existed
for other products and wares. Some examples included fuel wood, grain,
spices, cloth, and metals (same day), iron (1 day), milking cows (3 days),
beasts of burden (5 days), precious stones (7 days), planting seeds (10
days), a male slave (15 days), and a female slave (1 month) [11, Chapter
XI, pp. 174–180].

2.3.4   The Early Islamic Period

The concepts of Islamic laws came into being in the period of the Caliphs
of Medina (circa A.D. 632–661) and were enacted against a varied political
and administrative background primarily during the rule of the Umayyads,
the first dynasty of Islam (circa A.D. 661–750). Islamic laws are a result
of a scrutiny of legal subject matter from a religious angle, comprising
various components of the laws of Arabia and other conquered territories
including eastern Africa and parts of the Indian subcontinent.

The Islamic system of jurisprudence can be traced all the way back
to four ancient schools of law: the Koran (*Qur'ān*), the *sunna* of the
Prophet Muhammad,* the consensus (*Ijmā'*) of the orthodox community,
and the method of analogy (*qiyās*) [12]. The essentials of this theory were
created by Shāfi'ī, whose treatise on jurisprudence is known as *Kitāb al-
Risāla Fī Uṣūl al-Fiqh*, better known as *Kitāb al-Risāla* (the book of
epistles) or simply *al-Risāla* (the epistles) [13].

According to Shāfi'ī doctrines, sale transactions, although allowed,
is not something that is looked upon favorably:

147. (Shāfi'ī said:) And He said:
That is for their having said: Sale is just the same as usury, though
God hath permitted sale and forbidden usury. [13, p. 157]

The fundamental concept of Islamic law, be it concerned with wor-
ship or with legal matters, is the intent (*niyya*). One of the doctrines stated:

If a man buys a slave girl and she has a defect which the seller

---

* Originally, *sunna* was the common law of Arabia before Islam, and it comprised legal and
moral principles. The sunna grew out of the customs of the forefathers and its enforcement
by practice established its legal validity. After the rise of Islam, the term *sunna* came to
mean the sayings, acts, and decisions of the Prophet Muhammad; later on, the substance
of the *sunna*, in addition to the authorities who transmitted it, was called *hadīth* or tradition.

has concealed from him, the case is the same in law, whether the seller did it wittingly or unwittingly, and the seller commits a sin if he does it wittingly . . . the buyer has an equal obligation towards the seller and is not entitled to return her (the slave girl) to the seller after the defect which was developed whilst she was in his (the buyer) possession, just as the seller was not entitled to hold him bound by a sale of an object which had a defect whilst in the seller's possession. [12, *Tr. I*, 6, p. 325]

In addition, both buyer and seller had the option rescinding the sale of slaves or any merchandise [12, *Tr. I*, 12, p. 326], somewhat similar to the concept of "money-back" warranty.

2.4 THE EUROPEAN PERIOD

2.4.1 The Roman Era

During the classical Roman era (circa fifth century B.C.), the fundamental laws of the land were initially formulated under XII (12) tables. These 12 tables were a comprehensive collection or code of rules framed by the officers, called *Decemviri*, specially appointed for the purpose.

This draft of legislation covered laws addressing specific obligations of both vendor and vendee involved in a trade. One of the vendor's obligations was to provide warranty against secret defects in the product. This civil law provided little, if any, protection against tort, because it would be difficult to prove prior knowledge of defects. However, the *Edict of the Aediles* furnished some protection in the context of slave trade (this edict was extended to similar sales of live stock). It provided that in order to sell slaves in the open market, the vendor must declare all mental, moral, and physical defects that the slave had. Further, it must be promised that no such defects (*morbus, vitium*) existed other than those declared [14, Law CLXXII, pp. 488–489].

2.4.2 The Germanic Period

The Germanic people invaded and subsequently settled on territories occupied by the Roman empire in the start of the Christian era (58 B.C.–A.D. 496). These settlers brought along their own laws, judicial system for enforcement of the laws, and punishments for transgressions. The system of legal codes was primarily structured by the Alaman and Bavarian settlers (now Switzerland and southern Germany).

One notable distinction between the Roman codes of sale and its Germanic counterpart was that the Bavarian law of sale had a warranty provision that was *limited* in nature. As noted in the Bavarian laws translated from the Ingolstadt Manuscript:

> ... 9. *Concerning the form of sale.*
> ... But after the sale is transacted, it is not to be changed, unless it happens that one finds a defect, which the seller concealed ... If, however, the seller states the defect, let the sale stand; one cannot change it ... If, however, he does not state it one can change it ... (within three days after the sale) ... And if one has his property for more than three nights, after this he cannot change it, unless perhaps he cannot find the defect in three days. ... [15, p. 162]

In above cases, the buyer can still try to return the product after the "limited" warranty duration expires; however, the seller may refuse to take it back if he testifies that no defect existed on the day of the sale.

2.4.3 The Jewish Period

The Hebrew laws (circa second century A.D.) that followed the codification of the Babylonian and subsequent periods showed a much greater development in ethical content. As declared by Mishnah (*Bava Metzía* IV, 12):

> ... One may not furbish up man or beast or vessels for purposes of sale. One may not, for example, dye a slave's hair black to make him look younger and more vigorous; nor drug an animal to the same end; nor paint old, second-hand implements in order to pass them off as new. ... [16, p. 367]

Deceit as to quality was considered a serious offense. Under the Hebraic laws, nondisclosure was regarded as a fraud and the vendor was obligated to disclose any defects in the merchandise before sale. If the delivered goods or wares fail to meet either the prespecified quantity (*Bava Metzía* 63b) or the description (*Bava Metzía* V, 6) promised at the time of sale, the sale could be rescinded by either the vendor or vendee on the grounds of mistake. This also depended on whether the quality of delivered goods was better or worse than promised at the time of sale.

Even though there were no provisions for express warranty, all land and property transactions carried "implied" warranties that included

1. A guarantee of good title against all the world
2. A warranty that the seller had not encumbered the property
3. A guarantee against any personal claim

These implied warranties were strictly personal and covered only such defects and claims as were established in a court of law at that time.

2.4.4 The Early English Period

One of the earliest series of Kentish laws were established by King Æthelberht I (circa A.D. 597–617) in the lifetime of Augustine. One of his decrees, which alluded to a form of warranty in sale of maidens (by the vendor who, in most cases, would be the father of the maiden) provided the following:

> 77. If a man buys a maiden, the bargain shall stand, if there is no dishonesty.
> §1. If however there is dishonesty, she shall be taken back to her home, and the money shall be returned to him. [17, p. 15]

The earliest laws of Wessex were those enacted by King Ine (circa A.D. 688–725). One of his decrees recognized the concept of implied warranty:

> 56. If anyone buys any sort of beast, and then finds any manner of blemish in it within thirty days, he shall send it back to (its former) owner . . . or (the former owner) shall swear that he knew of no blemish when he sold it him. [17, p. 55]

Some references of warranty (as a guarantee against betrayal) are found in the terms of peace treaty between the English king Alfred, who defeated the Danish king Guthrum in A.D. 878 soon after the great Danish invasion of A.D. 866. One of the conditions that detailed the terms of the treaty stated,

> 5. And we all declared, on the day when the oaths were sworn, that neither slaves nor freemen should be allowed to pass over to the Danish host without permission, any more than that any of them [should come over] to us. If, however, it happens that any of them, in order to satisfy their wants, wish to trade with us, or we [for the same reason wish to trade] with them, in cattle and in goods, it shall be allowed on condition that hostages are given as security for peaceful behavior and as evidence by which it may be known that no treachery is intended. [17, p. 101]

Soon afterward, when peace and friendly relations were established between the English and the Danes, English King Edward and Danish

King Guthrum enacted a series of decrees sometime after A.D. 921. One of their decrees concerning protection of goods and people alike stated,

> 12. If any attempt is made to deprive in any wise a man in orders, or a stranger, of either his goods or his life, the king—or the earl of the province (in which such a deed is done)—and the bishop of the diocese shall act as his kinsmen and protectors, unless he (that person) has some other. And such compensation as is due shall be promptly paid to Christ and the king according to the nature of the offense; or the king within whose dominions the deed is done shall avenge it to the uttermost. [17, p. 109]

2.4.5 The Early Russian Era

A general view on law and justice was developed by the Slavs long before the appearance of the written codes among them. The earliest written documents related to commercial laws can be traced back to the first half of the tenth century (A.D. 911 and A.D. 945) when the first two Russian–Byzantine treaties were drafted. However, with reference to warranty issues and redress, the Charter of the City of Pskov (A.D. 1397–1467) provided:

> . . . And if anyone buys a cow at a price agreed upon, and after purchase (the cow gives birth to a calf), the seller may not sue the buyer for that calf; (on the other hand), if the cow discharges bloody urine, it may be returned and the money is to be refunded. . . . [18, p. 82]

The article did not indicate any specific time limit for this type of money-back warranty provision on the product (in this case, the cow). However, it is logical to assume that any such defects in the "product" will surface, if at all, very soon after its purchase. In addition, the law limited the extent of remedy under this warranty; it did not offer a remedy if the cow successfully delivers a calf.

2.5 THE MIDDLE AGES

2.5.1 The Medieval Times

In the early Medieval Europe during the Middle Ages (A.D. 617–1291), commercial trade and the resulting need for a warranty provision faced a strong setback. During that period, trade was generally denounced because it was not considered "the way of the Christian." Usury was considered a symbol of "covetousness at its lowest." In addition, the quality of the ware and the fullness of its measure held a exalted place. Other prevalent beliefs of the time included the following: Trade was worldly not heavenly; it was sinfully carried on for gain; the merchant could not be pleasing to God [19, pp. 126–132].

Over the course of time, the Church became quite powerful; it established a distinction between a wrongful trade, which was carried on for personal profit, and a rightful trade, which was intended to serve public necessity. Nevertheless, trade remained an instrument of social purpose, although the dealing of the trade had to conform to the standards of the Christian conduct.

The church manuals, codified by Thomas Aquinas, laid down standards of mercantile conduct for Christians during the fourteenth and fifteenth centuries. For example, it is indicated that a sale can be rendered unlawful by a defect in goods or wares that were sold. Irrespective of whether the defect was in kind, in quantity, or in quality, if it was known to the vendor and unrevealed to the buyer, it was considered a sin and fraud, and, accordingly, the sale was considered void [20, Question LXXVII, art. 4, pp. 93–94].

Over the years, in case of the sale of defective products, the prevailing standards suggested resolution which essentially favored buyers more in the spirit of the informal concept of *caveat venditor* or "let the seller beware." Here, if the defect is revealed to the buyer by the vendor at the time of the sale, but the buyer still chooses to purchase the product, the sale becomes final. If, however, the vendor sold a product with a less-obvious defect not known to either the buyer or the seller at the time of the sale, it was mandated that the seller replace the product or refund the buyer's money. On the other hand, if the vendor sold a product with a hidden defect known to him but not known to the buyer, the seller must make good the purchase and expect punishment for "fraud."

2.5.2 English Borough Customs and Guild Practices

In the tradition of guilds of Medieval Ages, the customs of the English Borough favored the buyer. In the early fourteenth century A.D., as the crafts increased in number and became more popular, the scrutiny of the community was progressively extended to ensure an open market for products with fair price, honest measure, and good quality. The conduct of several handicrafts were regulated by their own statutes, which embraced general town ordinances. Goods were required to be publicly displayed so that the buyer can be assured of a warranty of title, which implies that his purchase will not be snatched away by its rightful owner. Marketplaces were appointed for various commodities in order for exchange to be carried out in the open. In addition, the sale of goods in private places and in secret was prohibited. According to English Borough Customs:

> ... the craftsmen are required to keep away from hotels and private houses, save when some great lord should send for them, and required to vend their wares only in their own shops. [21, pp. 354–5 and 360–1]
> ... there were to be no sales by candle light or after the bell had rung for sunset. [21, pp. 141–2, 339, 532–3; 22]
> ... the inspectors of wares and keepers of the market are to discharge their fiduciary offices with diligence and honesty so that the people of the commonality might avoid disorderly and deceitful bargains. [21, p. 73]

The records attest the dominance of the idea of solidarity. Greater emphases were placed on publicity and prevention of fraud than on legal remedy for the buyer, as indicated in the following excerpt:

> ... The deceitful marker and the dishonest vendor were paraded through the streets with their fraudulent wares, exposed in the stocks with their false product burned beneath their feet, and denied the community of their trades and of the liberty. [23, p. 219]

In these times, the buyers relied on the words of sellers, no special collocation of words was necessary to constitute warranty,* an oral understanding was a binding contract, and the merchandise must conform to sample; a confiscation of goods and damages to the wronged were penalties for unfair dealings.

2.6 INDUSTRIAL REVOLUTION ERA AND BEYOND

2.6.1 The Emergence of "Caveat Emptor"

By the dawn of the Industrial Revolution in sixteenth century A.D., the organized salesmanship outmoded the individual seller, and the business practices witnessed a reversal of trade policy. The refusal of public authority, through public legislature and a formal judiciary system, to accord effective protection to the purchaser was reflected by the growing acceptance of *caveat emptor* or "let the buyer beware." The expression *caveat emptor* appeared in print for the first time in the sixteenth century when Sir Anthony Fitzherbert, a well-regarded English judge wrote,

---

* The word "warranty" as used in this context is not an exclusive ceremonial term; it is used as the equivalent of representation, assurance, or pledge.

. . . if he be tamed and have been rydden upon, then *caveat emptor.* . . . [24]

Under the code of *caveat emptor*, buyers were not entitled to receive compensation for any problem associated with the product short of outright fraud on the part of the vendor, unless the vendor had explicitly guaranteed the item in question. In A.D. 1534, Sir Anthony Fitzherbert stated,

. . . If a man sells an unsound horse or unsound wine it behoveth that he warrant the wine to be good and the horse to be sound, otherwise the action will not lie. For if he sells the wine or horse without such warranty, it is at the others' peril and his eyes and his taste ought to be his judges in that case. . . . [25, p. 94C]

Court records show that similar attitudes prevailed in the next century. In an obscure case of *Chandelor* versus *Lopus* in A.D. 1603, the defendant was a goldsmith who knew about precious stones. He sold the plaintiff with affirmation, but without warranty, a stone of a particular kind, which it was not. Although the plaintiff won, the report indicated the unease felt about the final judgment:

. . . for everyone in selling his wares will affirms that his wares are good . . . yet if he does not warrant them to be so, it is no cause of action. . . . [26]

2.6.2 Post-Industrial Revolution Era

Up through the first half of the nineteenth century, *caveat emptor* ruled both in U.S. Courts and communities as well. Sellers rarely offered any sort of formal warranty on their goods. But to most Americans, the need for an express warranty would not have been pressing in any case because products were typically produced and sold by people known personally to the buyer—often by people from the local community. Ordinarily, buyers would have expressed their dissatisfaction on a personal basis, and word of mouth would have be been sufficient to alert potential buyers about the quality of a particular producer's goods or about the trustworthiness of a seller.

During the advent of standardized product warranties late in the nineteenth century, warranties were treated as standardized contracts. In an early study on warranties, it was discovered that most products offered to consumers included a warranty with extremely limited scope. More specifically, a typical product warranty coverage usually excluded remedy for failed component parts, transportation charges, ensuing damages, and so forth. The manufacturer's standpoint in regard to warranty content at the turn of this century exemplified the theory of consumer exploitation [27].

Exploitation theory postulates that relative market power of the manufacturer (or the consumer) dictates what warranty terms on a product are offered. Based on this theory, powerful manufacturers potentially imposed one-sided standardized warranty terms as mechanisms to unilaterally limit their legal obligations to consumers [28]. This theory was widely accepted in the context of standardized warranties and seemed consistent with the warranty case histories late through the nineteenth century. At that time, deceit associated with the sale of goods, such as misbranding, adulteration, and misrepresentation, became widespread. This, in part, led to a trend of dishonest manufacturers offering warranties on products without any intention of honoring them. Because of this, consumers started perceiving a warranty of any sort on products to be an indication of poor quality.

In the next few decades, independent product testing organizations emerged throughout America partly to curb such deceitful practices. Examples of these organizations include Underwriters Laboratory, an independent testing agency sponsored by various insurance companies and underwriters, which tested electrical appliances; the Good Housekeeping Institute, run by the *Good Housekeeping* magazine, which tested household goods; and Consumers' Research, a consumer-sponsored organization, which led to the formation of Consumers Union, publisher of *Consumer Reports*. Seals of approval from organizations such as these served as a symbol of acceptable quality in terms of product reliability and the credibility of the manufacturer's own warranty.

In an effort to gain control over the big businesses, the Federal Trade Commission (FTC) was established in 1914. In addition, the federal government set forth certain codes governing the sale of goods and encouraged all states to adopt them in order to achieve consistency. Several versions of such codes were enacted by congress in the 1930s, particularly the Uniform Sales Act. Under the Uniform Sales Act, an express warranty is defined as

> . . . any affirmation of fact or any promise by the seller relating to the goods . . . if the natural tendency of such affirmation or promise is to induce the buyer to purchase the goods, and if the buyer purchases the goods relying thereon. [29, p. 427]

This definition illustrates the dual nature of the obligation of express warranty. The statute describes two kinds of express warranty: one, which is promissory or contractual in nature,* and the other, which is the non-promissory affirmations of fact.† It should be noted, however, that the implied warranties of quality and of title under the Uniform Sales Act were imposed by law and clearly were nonconsensual [30, p. 420]. The involved parties could potentially use their contractual power by means of a disclaimer to destroy a nonconsensual warranty, but its creation in no way depends on their intentions [30, pp. 230 and 230a].

With growing concerns for buyers' protection, the concept of express warranty was joined by a subversive term *implied warranty*. By 1952, every state in the United States except Louisiana adopted what is termed the Uniform Commercial Code (UCC). This code specifies the obligations of manufacturers, distributors, and any other vendors, with regard to both express and implied warranties. Several forms of legislation have been enacted during the last few decades to regulate warranties on various products, most notably the Magnuson–Moss Warranty–Federal Trade Commission Improvement Act of 1975, which are discussed in greater detail by Kelley in Chapter 4. Consequently, an era of consumerism and warranty protection was initiated in the second half of the twentieth century. This topic is discussed in greater detail by Burton in Chapter 28.

2.7 CONCLUSION

As illustrated in this chapter, most societies have gone through an inevitable development in *product warranty* or "the business practice by which reputable sellers stood behind their goods" and a changing social viewpoint toward the seller's responsibility. As we proceed from Babylonian and Assyrian tablets of the twenty-first century B.C. to the Roman laws of the fifth century B.C., Bavarian laws at the start of the Christian era, Jewish commercial laws of the second century A.D., Hindu religious laws of the fifth, Islamic laws of the eighth, Egyptian formularies of a slightly later period, scattered Russian codes of the early tenth, customs the church rule of the medieval times and customs of the English borough to ultimately the postindustrial era of consumerism and warranty legislation, it is clear that concept of product warranty has maintained a significant position in trade practices of all societies through the ages and will continue to do so for times to come.

---

* Analytically speaking, only the promissory express warranty and those affirmations of fact which constitute an implied-in-fact promise are consensual.

† Here, an actual agreement to contract is not essential and the obligation is imposed by law analogously to the implied-in-law promise.

REFERENCES

1. Barnhart, R. K. (1988). *The Barnhart Dictionary of Etymology*, S. Steinmetz (ed.), H. W. Wilson Co., Bronx, NY.
2. Partridge, E. (1966). *Origins: A Short Etymological Dictionary of Modern English*, 4th ed. Rutledge and K. Paul, London.
3. Hamilton, W. H. (1931). The ancient maxim caveat emptor, *Yale Law Journal*, **XL**(8), 1133–1187.
4. Adams, J. H. (1982). *Longman Dictionary of Business English*, Longman, Harlow, Essex York Press, Beirut.
5. Oppenheim, A. L. (1967). *Letters from Mesopotamia: Official Business and Private Letters on Clay Tablets from Two Millennia*, The University of Chicago Press, Chicago, IL.
6. Driver G. R., and J. C. Miles, (1955). *The Babylonian Laws*, edited with translation and commentaries, Clarendon Press, Oxford.
7. Johns, C. H. W. (1904). *Babylonian and Assyrian Laws, Contracts and Letters*, Charles Scribner's Sons, New York.
8. Yaron, R. (1961). *The Laws of the Aramaic Papyri*, Oxford Press, Oxford.
9. Franz-Murphy, G. (1985). A comparison of the Arabic and Earlier Egyptian Contract formularies, Part II: Terminology in the Arabic warranty and the idiom of clearing/cleaning, *Journal of Near Eastern Studies*, **44**(2), 99–114.
10. Hopkins, E. W. (1891). *The Ordinances of Manu*, translated from Sanskrit, with an introduction by late Arthur Coke Burnett, Kegan Paul, London.
11. Haughton, G. C. (1825). *A Code of Gentoo Laws or Ordination of The Pundits*, London. [s.c.].
12. Schacht, J. (1950). *The Origins of Muhammadan Jurisprudence*, Oxford University Press, Oxford.
13. Khadduri, M. (1961). *Islamic Jurisprudence: Shāfi'ī's al-Risāla*, The Johns Hopkins Press, Baltimore, MD.
14. Buckland, W. W. (1921). *A Text-Book of Roman Law from Augustus to Justinian*, Cambridge University Press, Cambridge.
15. Rivers, T. J. (1977). *Laws of The Alamans and Bavarians*, University of Pennsylvania Press, Philadelphia.
16. Horowitz, G. (1973). *The Spirit of Jewish Law*, Central Book Company, New York.
17. Attenborough F. L. (1922). *The Laws of The Earliest English Kings*, Cambridge University Press, Cambridge.
18. Vernadsky, G. (1979). *Medieval Russian Laws*, Octagon Books, New York.
19. Ashley, W. J. (1910). *An Introduction to English Economic History and Theory*, Rivingtons and Cochran, London.
20. Aquinas, T. St. (1896). *Summa Theologica, Ethicus, II, II*, Rickaby Translation, 2nd ed. Burns and Oates London Benziger Bros., New York.
21. Arthur F. Leach (ed.) (1900). *Beverley Town Documents*, edited for the Selden Society, B. Quaritch, London.
22. Mary Bateson (ed.) (1906). *Borough Customs*, Vol. 2, B. Quaritch, London.
23. Francis B. Bickley (ed.) (1900). *The Little Red Book of Bristol*, published

under the authority of the council of the city and the county of Bristol, W. C. Hemmons et al., Bristol, England.

24. Fitzherbert, A. (1882). *The Book of Husbandry*, Rev. W. W. Skeat (ed.), Trubner and Company, London. reprinted from the edition of 1534.

25. Fitzherbert, A. (1755). *The New Natura Brevium of the most Reverend Judge, Mr. Anthony Fitzherbert,* . . . to which is added a commentary, supposed to be written by the late Lord Chief Justice Hale. Together with the authorities in law, and cases in the books of reports cited in the margin, 8th ed., H. Lintot, London.

26. Chandelor vs. Lopus (1603). *3 Cro. Jac. 4*, 79 English Report 3, court case (Ex. Ch. 1603), London.

27. Bogert, G. G., and Fink, E. E. (1930). Businesses practices regarding warranties in the sale of goods, *Illinois Law Review*, **25**, 400–417.

28. Kessler, F. (1943). Contracts of adhesion—Some thoughts about freedom of contract, *Columbia Law Review*, **43**, 629–642.

29. Vold, L. (1959). *Handbook of the Law of Sales*, 2nd ed. West Publishing Company, St. Paul, MN.

30. Williston, S. (1948). *The Law Governing Sale of Goods at Common Law and Under the Uniform Sales Act*, rev. ed., Baker and Voorhies, New York.

3

A Framework for the Study of Warranty

D. N. Prabhakar Murthy

The University of Queensland
Brisbane, Queensland, Australia

Wallace R. Blischke

University of Southern California
Los Angeles, California

3.1 INTRODUCTION

There are many aspects to product warranty. These have been studied by researchers from different disciplines with limited interaction between disciplines. As a result, the literature on product warranty is not only vast but disjointed. Using the systems approach, Murthy and Blischke [1] developed a framework to study the different aspects of warranty and to integrate this literature. The first step in the systems approach is system characterization. How this is done depends on the perspective for the study. One can define three different perspectives, namely,

1. Consumer perspective
2. Manufacturer perspective
3. Public policy perspective

These three perspectives are discussed in Section 3.2. Following this, salient features of the systems approach are presented in Section

3.3. Sections 3.4 and 3.5 deal with alternate system characterizations to study the different aspects from the three perspectives.

3.2 STAKEHOLDERS IN WARRANTY

3.2.1 Manufacturer's Perspective

A strong motivating factor for any manufacturer is the desire to maximize profits. Offering a warranty results in additional cost due to servicing of the warranty, but, at the same time, if used properly as a marketing tool, it increases sales and, hence, revenue generation. Warranty servicing costs depend on product characteristics and the usage patterns of consumers. If the extra revenue generated exceeds the warranty servicing costs, then it is more sensible to sell the product with warranty. As a result, manufacturers are interested in the study of warranty in order to seek answers to some or all of the following questions:

1. What is the cost of offering a specific warranty policy?
2. How does this compare with other warranty policies?
3. How does the warranty cost change with the parameters (e.g., duration, form of warranty payoff) of the policy?
4. How does one optimize the choice of warranty when multiple business objectives are involved?
5. What is the optimal strategy for servicing warranty? (This would involve, e.g., establishing a policy with regard to repair/replace decisions in the case of repairable items.)
6. What kinds of data (laboratory, field, etc.) are needed and how should the data be analyzed?
7. What are the optimal decisions with regard to product design and manufacture given that the product must be sold with a specific type of warranty policy dictated by the marketplace?

3.2.2 Consumer's Perspective

For a consumer, the difficult problem is the choice between products with different characteristics and different warranty policies. In addition, if the warranty is optional, a buyer would usually like to know if it is worth the additional cost. This is important because there is a growing trend among manufacturers, retailers, and even credit-card issuers to offer extended-term warranties. These always involve additional costs and the terms offered can vary considerably. For example, an extended warranty may cover both labor and parts initially and only cover parts later on, or it may cover only certain parts, and so on. The consumer has to decide, often at the time of purchase and based on very limited information, whether to opt for an extended warranty or not and to determine the best extended terms for his situation when there are multiple options.

3.2.3 Public Policy Perspective

In Western societies, intervention by public policy decision-makers is justified only when the product market is found to be behaving noncompetitively, as noncompetitive prices and terms are generally considered to be substantively unfair. Warranties affect product markets for a variety of reasons, the following two being the most important: First, warranties serve as signals to communicate information about product reliability and performance when consumers cannot evaluate these characteristics easily otherwise, and, second, they act as insurance to protect consumers against being sold defective or poor quality products. It follows that investigation of the way in which the product market behaves under different warranty rules, and the resulting social welfare implications, is essential to the formulation of sensible and effective warranty legislation.

The system characterizations for the study of warranty from the consumer and manufacturer perspectives are similar but differ from the public policy perspective.

3.3 SYSTEMS APPROACH TO THE STUDY OF WARRANTY

The first step in the systems approach is system characterization. This is a process of simplification and involves identifying the important factors of interest and their interrelationships. Each factor, in turn, might involve one or more variables and interactions among the variables. In this chapter, we first discuss a simplified system characterization and, later, a more comprehensive system characterization.

A system characterization can be viewed as a descriptive model of the real world relevant to the study of a particular aspect of warranty and from a particular perspective. A study based solely on the descriptive model at best yields qualitative insights. To get a deeper and more quantitative insight, one needs to translate the descriptive model into a mathematical model. Mathematical modeling for the study of warranty is discussed in Chapter 7. One then carries out an analysis of the model to obtain solutions which, when interpreted properly, yield the desired answers.

3.4 SIMPLIFIED SYSTEM CHARACTERIZATION

A simplified system characterization for the study of warranty from the consumer and manufacturer perspectives is developed in this section.

Later, an alternate characterization for study from the public policy perspective will be presented.

3.4.1   Study from Consumer and Manufacturer Perspectives

The important factors are the following:

1. Manufacturer
2. Consumer
3. Product
4. Product performance
5. Warranty policy

The manufacturer produces products and sells them to consumers with a warranty policy included. Product performance is determined by the interaction between product characteristics (determined primarily by the manufacturer) and product usage (determined by the consumer). When the consumer is not satisfied with product performance, a claim under warranty usually results. The cost of the warranty is the cost incurred by the manufacturer to service a claim under warranty. The magnitude of this cost depends on the terms of the warranty policy. Figure 3.1 shows a schematic representation of the different factors involved and their interrelationships in its most simple form. We discuss each of the factors in some detail.

Manufacturers

Many variables characterize a manufacturer. Some of these follow:

• Size
• Organizational structure

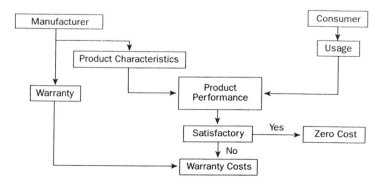

Figure 3.1   A simplified system characterization of the warranty process.

- Management objectives
- Marketing orientation
- Reputation
- Range of products produced
- Quality control philosophy

For the study of warranty all of these are important. Their importance depends on the type of market. Two situations will be considered. The first is a monopolistic, or near-monopolistic, situation characterized by a single manufacturer or a relatively small number of manufacturers. Manufacturers of very specialized equipment with limited sales would fall into this category. Typical examples are manufacturers of (1) heavy earth-moving equipment used in the road construction industry, (2) large electrical machinery such as generators, and (3) manufacturers of weapons systems or components or other defense contractors. The second is a more competitive situation characterized by a large number of manufacturers with very little scope for collusion. Typical examples are manufacturers of consumer durables such as television sets, dishwashers, small appliances, tools, and so on.

In the context of warranty, the latter situation would force manufacturers to offer warranties which are fairer to the consumer due to competition, whereas in the former case, there are no such pressures. A good example is the automobile industry, where in the late sixties and early seventies, warranties were becoming less favorable to the consumer. This trend changed with the penetration of the market by Japanese cars, for which terms more favorable to the consumer were offered. As a result, terms such as 2 or 3 years and unlimited mileage are fairly common.

Consumers

For our purposes, consumers may be grouped into three categories. The first corresponds to the case of a single consumer. A typical example of such a consumer is the federal government as a consumer of certain types of defense-related products such as jeeps, rockets, ships, and so on. Here, the consumer is more powerful than the manufacturer and has a strong input into product development and may dictate product support subsequent to product sale. In fact, until fairly recently, these items were procured without warranty. Over the last 20 years, the U.S. Congress has become progressively more insistent that warranties be part of the procurement process.

The second case corresponds to a relatively small number of consumers for the product. Typically, the consumers are industrial or commercial organizations or government agencies buying very specific industrial or commercial products. Typical examples are (i) airline companies buying aircraft, (ii) car-rental companies purchasing automobiles, and (iii) cities and towns buying fire fighting equipment. The Uniform Commercial Code (UCC) also covers transactions between manufacturer and consumers belonging to this category, including government agencies.

Finally, the last category corresponds to the case where there are a large number of consumers who, in general, are not well informed about the technical aspects of the product. An average citizen buying consumer durables would be a typical example of this category. The Magnuson–Moss Warranty Act was aimed primarily at educating these consumers and looking after their interests. (See Chapters 4, 5, and 27.)

Products

Products can be categorized into four groups as follows:

1. Specialized defense-related products
2. Industrial and commercial products
3. Consumer durables
4. Consumer nondurables

Specialized defense products (e.g., aircraft, ships, tanks, rockets) are characterized by a single consumer (the federal government) and a relatively small number of manufacturers. The products are usually complex and expensive and involve "state-of-art" technology with considerable research and development effort from the manufacturers. As a result, the warranties for such items are typically very complex and involved and are bid and negotiated as a part of the government procurement process.

Industrial and commercial products are characterized by a relatively small number of consumers and manufacturers for such products. The technical complexity of such products and the mode of usage by the user can vary considerably. In general, the consumers are well informed about the technical aspects of the product and, hence, in a better position to negotiate fairer terms of warranty. The products can be either complete units such as cars, trucks, pumps, and so forth, or parts such as batteries, tires, and bulbs needed by a car manufacturer. Quantities purchased can vary from small (hundreds) to very large (orders of millions). Similarly, the price of such products can vary several orders of magnitude.

Consumer durables (typical examples are television sets, automobiles, stereo equipment) and nondurables (typical examples are food items, cosmetics) are consumed by society at large and, hence, characterized by a large number of consumers for the product. The complexity of the product can vary considerably and, in general, the typical consumer is often not well informed about the product.

Consumer durables vary considerably in their functional use and their prices. Small domestic appliances, such as coffee grinders or can openers, cost a few tens of dollars, whereas items such as cars, furniture, and expensive stereo systems can cost many thousands of dollars. It is this category for which the Magnuson–Moss Act ensures that consumer interests are adequately protected.

Many nondurables are relatively cheap (costing less than $5.00) and, as such, are not sold with warranty. However, under law, manufacturers can be held liable under negligence, strict liability, or misrepresentation. Typical examples are actions resulting from (1) food poisoning due to negligence in preparation, (2) failure to warn about possible reactions to a suntan lotion, and so forth.

Warranty Policies

Warranty policies can be either simple or complex, depending on the type of product being covered by the warranty and the coverage offered. Many different types of policies exist. These are discussed in Chapter 1 and in many chapters in the remainder of this Handbook. In the context of consumer durables, two types of warranty policies have been used extensively. These are the free-replacement policy and the pro-rata policy. Under the free-replacement policy, the manufacturer agrees to either repair or provide a replacement at no cost to the consumer up to a specified time from the time of initial purchase. Such a policy is usually offered with products which are repairable—for example, TV sets, automobiles, and appliances. In the pro-rata warranty policy, replacements are provided at pro-rata cost. Such a policy is usually offered with nonrepairable items such as tires, batteries, tools, and so on.

Product Performance

The performance of a product depends on product characteristics and the mode of usage by the consumer. Product characteristics, in turn, depend on the design and manufacturing decisions made by the manufacturer. The mode of usage, in the simplest characterization, can be either normal or abnormal. The latter represents the case where the user uses the product in a manner for which it has not been designed—for example, operating a machine at speeds above the safe limit; using a flashlight to drive a nail; using a microwave to heat food in a metal container. The degree of misuse (or abuse) of a product can vary. Failures occurring due to abnormal use are not usually covered by warranty. However, in certain instances, the manufacturer can be held liable for either not having foreseen a usage which caused the damage or failure to warn of the dangers of certain types of usage.

Product performance can be assessed objectively in some cases, whereas in many others, a fair degree of subjective evaluation of the consumer plays an important role. A typical example involving a high degree of subjective evaluation is the sound quality of a musical instrument. In certain instances, the environment can have a significant impact. An example is poor picture quality in a television set due to geographical location where a hill blocks the signal reaching the set. Note that in this case, the consumer can take action against a vendor under the Implied Warranty of "Fitness for a Particular Use" if the vendor was informed of the problem of reception before the set was bought and the buyer depended on the vendor's judgment.

When a consumer feels dissatisfied with the performance of the item purchased, warranty offers an avenue for redress. When a claim is made under warranty, the manufacturer is legally obliged to respond to it. Failure to do so would constitute a breach of warranty and the consumer then can initiate legal action against the manufacturer.

Warranty Costs

Whenever there is a claim under warranty, the manufacturer incurs a cost. If the claim is not valid, then the only cost is the administrative cost of handling the complaint. A claim is not valid if it not covered by the warranty, if the warranty has expired, if the claim is bogus (i.e., the item has, in fact, *not* failed as claimed), or if the warranty ceases to apply due to consumer misuse of the product. If the claim is valid, there are additional costs. These include the cost of labor and parts for repairable items, replacement by a new item for nonrepairable items, incidental costs such as shipping costs and, in some instances, additional costs to compensate the consumer for having been deprived the use of item while it is undergoing repair. A typical example of such an additional cost is the cost of providing a replacement for the duration of repair.

When a dispute arises between a consumer and a manufacturer, additional costs are incurred by both parties—for example, time and effort spent in resolving the dispute and, in extreme cases, legal fees and court costs.

In general, the total warranty cost for each unit sale is unpredictable because product defects occur randomly and consumer usage varies significantly. McGuire [2, Tables 5 and 7] gives warranty servicing costs, as a percentage of net sales, in various product sectors based on a survey of 369 U.S. manufacturing companies of industrial products. They vary from less than 1% to 10% or more, with the majority of firms having these costs less than 5% for warranty durations varying from 3 months to 5 years.

3.4.2 Study from the Public Policy Perspective

In this section, we discuss a system characterization appropriate for the study of the warranty process from the public policy perspective. The reasons for studying warranties from this perspective are (1) to understand the effect of warranty terms and warranty legislation on product market structure, (2) to evaluate the resulting social welfare implications for society as a whole, and (3) to help in the formulation of legislation relating to warranty.

The product market is determined by the interactions between consumers and manufacturers. Consumers choose products, based on the information they have about prices, quality of products, and warranty terms, to maximize their utility. This generates demand functions for products. Manufacturers decide on product quality levels, prices, and warranty terms to maximize profits based on the demand functions generated by consumers. The outcome of this interaction between consumers and manufacturers defines the actions for each group and characterizes the market outcome. Note that the market consists of two types of economic agents—consumers and manufacturers—and each agent attempts to maximize his/her own interests.

Characterizing each consumer and manufacturer separately leads to a characterization which is too complicated and of limited use. Instead, one characterizes consumers and manufacturers through aggregated group characteristics, resulting in a simplified and more manageable characterization.

If all consumers are similar in their attitude to risk and possess nearly the same information, they may be treated as a single homogeneous group of consumers. If members of the population differ in terms of their attitude to risk (i.e., some are risk-neutral and others are risk-averse), then two or more groups exist and each must be characterized separately. Similarly, if the information (about product quality, price, warranty terms) that consumers possess differs, it is necessary to divide consumers into two or more different groups. Obviously, as the number of groups increases, the characterization becomes more complex and detailed.

Similarly, manufacturers can be characterized through one set of group characteristics if all are similar in terms of their attitude to risk and all produce similar products and have similar cost elements. When this is not the case, we have a heterogeneous population of manufacturers as well. In this case, the number of different groupings needed would depend on the degree of differentiation, with each group being characterized by a different set of characteristics.

Two other factors which play very important roles in determining the market behavior and the actions of consumers and manufacturers are (1) the legislative process and (2) the judicial process. A system characterization along these lines, including all main factors and their interactions, is shown schematically in Figure 3.2.

The study of market behavior involves the concept of market equilibria and this is discussed in Chapter 25.

Social Welfare Implications and Public Policy

Competitive prices are the lowest prices and competitive warranty terms are the fairest terms that the market can sustain. In a noncompetitive market, either the prices are higher, the warranty terms unfair, and/or the product quality inferior (or below competitive level). In a sense, when the market is noncompetitive, the prices and/or warranty terms favor one set of agents at the expense of all others. In Western societies, markets

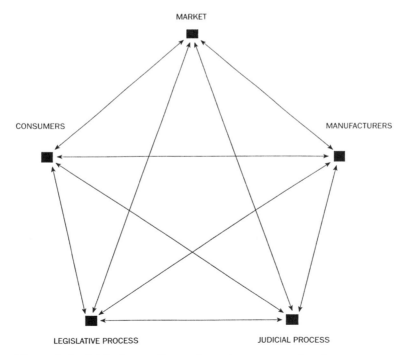

Figure 3.2 Main factors for study at the aggregate level.

behaving in a noncompetitive manner are considered to be substantively unfair and provide justification for action by lawmakers or other public policy decision-makers.

Public policy decision-makers have viewed warranty as a tool for moving the market from a noncompetitive equilibrium to a more competitive equilibrium. Some of the factors which can result in the market being noncompetitive were mentioned earlier. Another important factor which has a effect on market outcome in the context of warranty study is "moral hazard," which is discussed briefly in the next subsection.

Moral Hazard

In many instances, product failure depends on the quality effort of the manufacturer as well as the maintenance effort (or care) on the part of consumer. Unless a manufacturer reveals information about product quality, it is nearly impossible for consumers to obtain this information. As a result, there is asymmetry with regard to this information between the manufacturer and the consumer. Similarly, it is nearly impossible for the manufacturer to observe the maintenance effort of the consumer. Thus, neither manufacturer nor consumer has complete information and they differ in the amount and nature of information that each possesses. This can result in a "moral hazard" for both consumer and manufacturer.

When consumers cannot discern the quality of a product, that is, when they cannot differentiate between high quality products (involving high quality effort on the part of the manufacturer) and low quality products, then there is no incentive for manufacturers to produce high quality products. In this case, manufacturers of low quality products can claim their products to be of high quality and make more profit than manufacturers of high quality products, and, in the worst case, drive out high quality products from the market because "lemons" can be produced at lower cost (see Ref. 3). This creates a manufacturer moral hazard problem. Warranties, however, induce manufacturers to produce high quality products.

On the other hand, if product failure is dependent on consumer maintenance efforts, providing warranties may result in this effort being reduced and, hence, increase the probability of product failure. As such, warranties create a consumer moral hazard problem.

3.5 DETAILED SYSTEM CHARACTERIZATION (CONSUMER/MANUFACTURER PERSPECTIVES)

Figure 3.3 provides a detailed system characterization of the warranty process, containing many more factors than the simple characterization

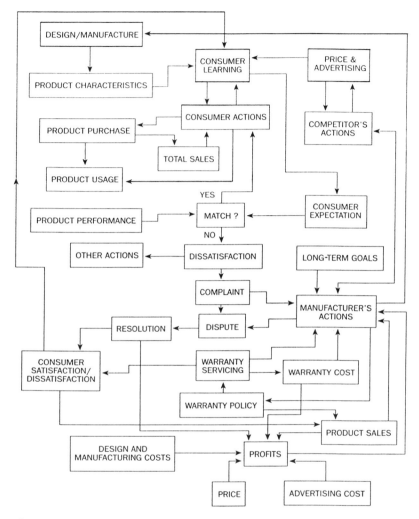

Figure 3.3 Detailed system characterization of the warranty process.

of Figure 3.1. The characterizations of some of the factors (consumer, manufacturer, product, product characteristics, product performance, warranty policy, warranty cost) have been described earlier. In this section, the remaining factors are discussed.

For the purpose of characterization, we can view the warranty process as consisting of three segments:

1. Consumer factors
2. Manufacturer factors
3. Interactions between the two

3.5.1 Consumer Factors

The following three factors regarding consumer actions and reactions are of importance in the analysis of warranties:

1. Product purchase—dealing with the processes leading to a purchase by a consumer
2. Evaluation of product performance—leading to satisfaction or dissatisfaction
3. Reaction to satisfaction or dissatisfaction

Each stage involves many variables and the outcome is dependent on the interaction between these variables. The role which warranty plays in each of the three stages depends on the attitude and behavior of the consumer to warranty.

Attitude and Reaction to Warranty

Based on an empirical study, Darden and Rao [4] report that the attitudes of different consumers to warranty are not the same. They divide the consumer population into seven groups based on attitude to warranty and report that roughly 16.5% are "warranty skeptics," 40.8% are "warranty supporters," and the rest are, in a sense, neutral. "Warranty skeptics" are those who feel that warranties are useless and perceive dealers and retailers as shirking their performance responsibilities by offering a warranty rather than providing a better product.

In a related study, Shimp and Bearden [5] report that some consumers perceive no relationship between warranty and product quality.

Most consumers, however, view warranty in a more positive sense. The role of warranty in their behavior will be discussed in the following subsections.

Purchase Decision

For consumers who view warranty in a negative sense, warranty plays no role in product purchase decisions. Hence, our discussion focuses on consumers who view warranty in a positive sense.

Consumer theory postulates that, for a rational consumer, product purchase decisions involve the following four stages:

Stage 1: Perception of the need for a product

Stage 2: Search for information concerning available product choices
Stage 3: Evaluation of alternative products
Stage 4: Product selection

The first stage deals with issues related to the need for the product. We assume that the consumer recognizes the need and, hence, confine our discussion to the remaining three stages and the role of warranty in each stage.

Search for Information

A consumer search for information can be defined as the mental and physical activities undertaken by the consumer to obtain information about different product attributes such as price, warranty terms, product performance, and so forth.

Research [5,6] has shown that a significant number of consumers view warranty as providing useful information and feel that it is possible to draw inferences about a product from its warranty terms. Consumers believe that a product with a superior warranty will be associated with higher quality and less risk. Empirical studies of Wiener [7] show that this belief is justified and that warranties do act as signals of product reliability.

The amount of effort that a consumer spends on acquiring information depends on

1. The value of the product
2. Availability of the information
3. Satisfaction derived from the search
4. Perceived consequences of the search
5. Cost of the search
6. Perceived usefulness of the information

Most consumers do not act rationally with regard to warranty decisions. Typically, they note the terms of a warranty in a rather cursory manner prior to purchase and read the warranty carefully only when a problem arises during the warranty period, if then.

Evaluation of Alternatives

For consumer durables, the evaluation of a product involves many factors. The two most important of these are cost and performance. Each of these involves many variables. A partial listing would include the following:

Cost: Price, repairs and maintenance, operating costs, opportunity cost

Performance: Capability, durability, efficiency, dependability, economy

Other factors that play a role are as follows:

Manufacturer's reputation
Advertising
Product evaluation by a friend or a third party (e.g., a consumer report)

As indicated earlier, a significant number of consumers view warranty as a good indicator of product quality and durability. To these, a warranty of longer duration tends to convey that a product is more durable. Nordstrom and Metzer [8] report that consumers tend to prefer a product with expressed warranty over one that does not have a warranty other than that implied by law or by consumer regulations.

On the other hand, Shuptrine and Moore [9] state that ". . . despite the fact that consumers perceive expressed warranties as an additional benefit of the product, many consumers have discovered that warranties provide very little recourse when they experience trouble with a product." One reason for this is that consumers often do not read the warranty carefully before they purchase a product. Shuptrine and Moore argue that some warranties create misleading impressions due to the wording and legal terminology involved, which is beyond the comprehension of an average consumer. Their study reveals that for the majority of warranties examined by them, a proper understanding requires some college education. The implication of this is that the warranty may provide little protection and benefit for a large number of individuals because they do not understand the specific terms of the warranty.

Product Choice

This is the last stage in the purchase decision process and involves choosing the most desirable alternative from among those available. The problem is a multiattribute decision problem under uncertainty with warranty being one of the attributes. We focus our attention on a rational consumer's behavior when faced with the decision of choosing between (i) a product sold at a higher price with warranty and (ii) the same product sold without warranty at a lower price. The choice between a product with a warranty and one without depends on several factors. These include the following:

1. The price of the product without warranty
2. The price difference between the product with warranty and one without warranty

3. The perceived likelihood of failure
4. The anticipated cost of repair, in the case of repairable product
5. The refund, in the case of nonrepairable product
6. The consumer's attitude to risk
7. The perceived relationship between warranty and reliability

Consumer Satisfaction/Dissatisfaction

Consumer satisfaction or dissatisfaction depends on how the actual product performance matches the expected performance.

As indicated earlier, product performance can be either simple (e.g., ample, working or not working) or complex, involving many variables. For example, for an automobile engine, performance may include fuel efficiency, quietness of running, mean time to failure, mean time to repair, and so forth). The consumer's expectation about product performance can be either below or above that promised by the manufacturer. In any case, we have a dissatisfied consumer when the actual performance of a product does not meet his or her *own* expectation with regard to that product.

In analyzing warranties, it is the mismatch between actual performance and realistic consumer expectations about what is actually covered by the warranty that is of interest. The actual performance is dependent on product characteristics (and is a function of the design and manufacturing decisions made by the manufacturer) as well as on the usage mode or pattern, which depends on the consumer. Good product design takes into account the differences in usage due to consumer variability. Note that all warranties disallow claims resulting from misuse or abuse of the product.

Typically, a warranty not only covers product performance but also defects in a product. Many serious product deficiencies are virtually impossible for consumers to discern because of a lack of knowledge or perception on the part of the consumer. A few examples are as follows: (1) a defective brake in a car which might not operate properly under certain conditions, (2) flaws in a machined component which can only be detected using specialized equipment such as an X-ray, and (3) improper installation or defective internal parts of an automobile, appliance, or any other complex piece of equipment. Best and Andreasen [10] discuss this topic (namely, perceiving defects) and identify factors which hinder problem perception. These include factors such as buyer's lack of interest in consumer issues, complexity of the problem, socioeconomic status.

An interesting attitude is the feeling on the part of some consumers that it is wrong to suffer from consumer problems. These consumers show a strong disinclination to be identified as being victims of product failure.

Reaction to Dissatisfaction

Figure 3.4 is a schematic representation of the different reactions of dissatisfied consumers. Essentially, a dissatisfied consumer either takes some action or does not take action. Those belonging to the latter category are (1) people who have a strong disinclination to be identified as being victims of consumer-type problems and (2) people who feel that it is not worth taking action for a variety of reasons, for example, feeling that it is not worth the "hassle," or having a negative attitude toward warranties. Dissatisfied consumers belonging to the former category are those who either voice their complaint to seek redress or take some other form of action, such as boycotting a product or company or warning others by word of mouth.

Day and Landon [11] characterize actions taken as either public or private actions. Public action comprises (1) seeking redress from the manufacturer directly, (2) taking legal action, and (3) complaining to some third party (e.g., a consumer affairs agency). Private action comprises (1) boycott of product and/or manufacturer and (2) warning others through word of mouth.

Best and Andreasen [10] report that no action was initiated for nearly two-thirds of dissatisfied consumers. Day and Ash [12] report the percentage of cases where no action of any kind was taken to be 49.6, 29.4, and 23.2 for nondurables, durables, and services, respectively.

Factors Influencing Complaint Behavior

Individual decisions to seek redress are a function of both personal characteristics (e.g., economic status, propensity to complain, awareness) and

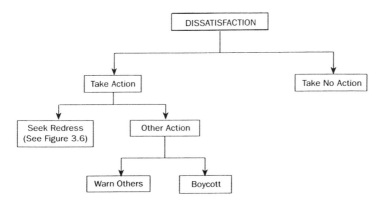

Figure 3.4 Consumer reaction to dissatisfaction.

product-related characteristics (e.g., cost, product importance). Bearden and Oliver [13] characterize the influences of some of the personal and product characteristics on the complaint process through a set of three exogenous variables (income status, propensity to complain, and cost or monetary loss associated with the problem) and three endogenous variables (private complaint, public complaint, and satisfaction in complaint resolution), with a causal relationship between the variables as shown in Figure 3.5. Their empirical study revealed that "(i) greater problem cost appears to stimulate both forms of complaint behavior, (ii) the extent of private complaint is inversely related to satisfaction with firm response, and (iii) public complaint is positively related to resolution satisfaction."

The empirical studies of Best and Andreasen [10] show that consumers who complain tend to be younger, have higher income, and tend to engage in consumerism activities.

Chapters 16 and 17 discuss some of these issues in greater depth.

3.5.2 Manufacturer Factors

As shown in Figure 3.3, the important factors from a manufacturer's point of view are product sales, warranty costs, manufacturing costs, and the resulting profits. Product sales are influenced by the marketing variables (e.g., price, advertising) of the manufacturer and of competitors and by many other external factors (e.g., state of the general economy). For many products, warranty assumes the role of an advertising variable when manufacturers use warranty terms as a promotional tool. Warranty costs depend on product performance, which, in turn, depends on product characteristics and consumer usage and expectations, as discussed in the previous section. Consumer expectations, in turn, are dependent on the

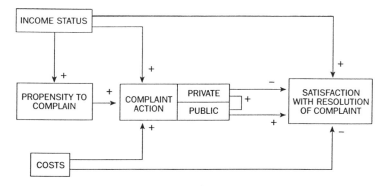

Figure 3.5 Factors influencing complaint behavior.

manufacturer's claims made in using the warranty for promotional purposes. Manufacturing costs are influenced by many other factors, some of which will be indicated later. Thus, there are many factors to be considered and analyzed from the manufacturer's point of view.

A manufacturer must make management decisions with regard to warranty at two levels. The first level is the design (or prelaunch) stage. Here, design and production issues and their costs are the important factors. The second level corresponds to the post-product-launch stage. Here, the important factors are warranty as a marketing variable, the accounting of warranty costs, and managing consumers dissatisfaction. In a sense, decisions at the first level can be viewed as being strategic in nature, whereas those at the second level are more tactical in nature, in that they are responses to market behavior. We next discuss methods of characterization of these factors.

Design and Manufacturing

Product design deals with activities such as conceptual design, materials selection, product development and testing, producibility assessment, and final design specifications. Manufacturing or process design deals with the materials and processes which ensure that the items produced conform to design specifications. Design, manufacturing, and quality control decisions determine product characteristics such as reliability, durability, maintainability, repairability, and so forth. Better product characteristics can be achieved through greater design and manufacturing effort. This reduces the expected cost of warranty, but this reduction is achieved at the expense of increased development and manufacturing costs. As a result, design and manufacturing decisions must be made based on cost considerations which not only include design and manufacturing costs but also take into account the subsequent warranty costs. These costs include the cost of servicing "express warranties" offered by the manufacturer as well as any costs resulting from "implied warranties" of merchantability.

Under "negligence theory," the manufacturer is also responsible for failure to use reasonable care in the design and manufacture of a product when a consumer experiences substantial injury through use of the product. The concept of foreseeability is an important element in negligence theory and refers to circumstances which can be anticipated by a reasonable manufacturer; that is, a consumer may use a product in an unintended way, but such use could reasonably have been foreseen. Thus, the design and the choice of design variables must be done in a manner which takes into consideration possible liabilities resulting from warranties in a wider sense.

Further discussion on warranty and design can be found in Chapter 21 and on warranty and manufacturing in Chapter 23.

Advertising (Promotional) Role of Warranty

The promotional role of warranty is to encourage sales by reducing risks for the consumer. In many warranties, there are disclaimers to guard the manufacturer from unreasonable claims from consumers. In this context, warranties serve a protectional role for the manufacturer. Udell and Anderson [14], based on a study of 365 warranties, report that 39% were primarily protectional, 26% primarily promotional, and the rest about equally protectional and promotional. Their study reveals that the promotional use of warranty is most effective when

1. The retail price of product is high
2. The product is purchased infrequently
3. The consumer views the product as being complex
4. The firm's share of the market is low

Anderson [15] deals with the protectional role of warranty and reports the following:

1. The firm's concern with protecting itself and/or its channels of distribution from consumer claims which it defines as unreasonable varies directly with the complexity of the product.
2. The importance, to the manufacturer, of the protective dimensions of warranty declines as the length of time during which a firm has been producing and marketing a given product increases.
3. The importance of disclaimers and protective warranty will be, in general, greatest for products having a high retail price per unit and least for products with relatively low retail prices.

Further discussion on warranties in the context of marketing can be found in Chapter 15.

Warranty Cost Accounting

Warranty cost accounting is a relatively tricky question. When to recognize warranty cost as an operating expense in a firm's accounting system is dependent on several interrelated accounting concepts, namely, the concept of matching principle, the concept of critical events, and the determination of a firm's liabilities in an accounting sense (see Chapter 26 and Ref. 16).

In the matching principle, costs are recognized in the period in which they contributed to operations and, hence, are matched with revenue realized from operations in that particular period. When one applies the matching principle to warranty cost, one encounters the problem that warranty cost is an "after cost." In this case, there are two possible "critical events" for determining the time at which the cost of providing warranty service should be recognized as an expense. They are (1) the time at which the product is sold and (2) the time at which the actual service is provided.

If it can be argued that the sale would not have been made without the warranty agreement, or the revenue of the sale would have been less without the warranty, then, in a sense, the expected cost of servicing warranty has been incurred and the warranty cost should be charged against the revenue of the period when the product is sold rather than when it is serviced under warranty. In other words, offering a warranty creates an accountable liability.

In contrast, if a warranty is used primarily for promotional purposes, knowing that warranty servicing costs are incurred only in isolated instances, then the cost of providing warranty services is not related to the production of specific revenue but related to a cost of maintaining the image of the firm. In this case, the critical event is the time that the service is provided rather than when the product is sold.

In reality, most warranties have both a promotional role as well as a protective role. The "Statement of Financial Accounting Standards No. 5" [17] outlines the procedure for warranty cost accounting and is contingent on certain conditions (regarding cost estimation and probability of incurring the cost) being met.

This topic is discussed further in Chapter 26.

Managing Consumer Dissatisfaction

From a manufacturer's point of view, managing consumer dissatisfaction is important for two reasons: (1) dissatisfied consumers can stop buying a manufacturer's product and switch to competitors and/or (2) they can influence others not to buy the product. The way in which the factors involved influence one another is shown in Figure 3.6.

Manufacturers can use warranties as an effective tool to resolve the dissatisfaction of consumers and, hence, minimize the loss in sales for the reasons just mentioned. In this context, manufacturers must create avenues which facilitate the voicing of complaints. If this is not done, the complaint process becomes ineffective and sales are affected in a negative

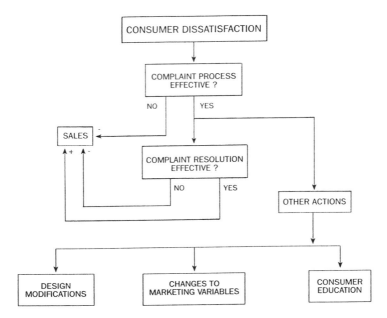

Figure 3.6 Managing consumer dissatisfaction.

manner. Once a complaint is received, the manufacturer must have a mechanism for resolution of the dissatisfaction in an efficient and effective manner. If not, a negative impact on sales will again result. Figure 3.6 shows this schematically.

The Magnuson–Moss Act requires manufacturers to state clearly the mechanism for resolution of consumer dissatisfaction. How effectively a manufacturer uses this feature to market and to ensure proper servicing will influence the company's final profits. Fisk [18] deals with the importance of servicing warranty in a manner to ensure "warranty integrity" and customer satisfaction. Keeping customers satisfied has been recognized in the marketing literature as an effective strategic device and there is a growing emphasis on customer complaint management; see, for example, Refs. 19–21.

Also included in Figure 3.6 are three other actions—modifications to design and manufacturing, making changes to warranty as a marketing variable, and consumer education. These actions will be effective if a majority of complaints are similar and can be rectified by one of these three actions.

Modification to Design and Manufacturing

Modification to design and/or manufacturing is an appropriate action when a specific component of the product can be identified as the cause of a majority of consumer complaints. The decision to initiate a development program to fix the problem costs money and is justified only if it is less than the expected savings in warranty costs. In some instances, to fix a defective component requires a recall of all units sold. This introduces additional cost variables; see, for example, Ref. 22 for a discussion of the economics of the Pinto recall.

Iizuka [23] discusses the linking of complaints with changes to design and manufacturing to achieve an integrated management system.

Modification to Warranty as a Marketing Variable

The cost of warranty depends on the warranty policy. If generous terms of the policy are necessary for the warranty to be an effective marketing tool, the costs can be high unless the product is of very high quality. Thus, if the warranty costs become large and design modifications are expensive, one way of reducing warranty costs is to alter the warranty, offering less generous terms.

Consumer Education

For expensive technical products, such as personal computers, nondefective items are often returned as being defective because the consumer is not competent to use the product properly. One way of reducing the likelihood of this is to ensure proper education through user manuals and introductory courses. Thus, should a majority of complaints be traced to lack of knowledge on the part of consumer, the appropriate action is to revise either the manuals or the courses offered.

3.5.3 Consumer–Manufacturer Interaction

In this subsection, the focus is on the consumer–manufacturer interaction, particularly in the area of disputed warranty claims. The main factors involved are disputes which arise when a consumer and a manufacturer disagree on the validity of a warranty claim by the consumer, and the means of resolving the dispute.

Dispute Resolution

The starting point for our discussion is a complaint to the manufacturer by a dissatisfied consumer seeking redress under warranty. If the manufacturer agrees with the consumer regarding the complaint and rectifies the complaint in a manner that is satisfactory to the consumer, then the problem is solved through a two-party interaction and there is no further dispute. However, if the manufacturer (1) refuses to accept a consumer's claim for redress as being valid or (2) does not carry out the rectification to the satisfaction of the consumer, the problem (at least from the consumer's viewpoint) continues to be unresolved. In some instances, even if the manufacturer discharges his warranty obligations in a correct manner, the consumer might still feel dissatisfied because of false or unreasonable expectations.

When a consumer is not happy with the two-party resolution, he has three options, as indicated in Figure 3.7. The first option is to take no further action. The remaining two options involve a third party in the dispute resolution. In the second option, the consumer complains to a public or private institution (e.g., a consumer affairs group, newspaper,

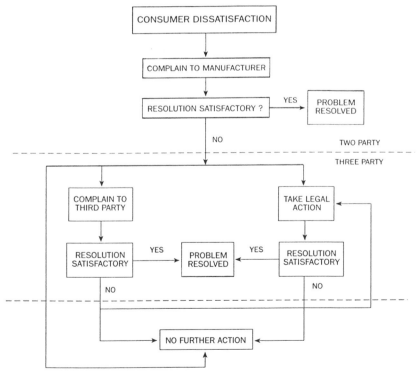

Figure 3.7 Complaint resolution.

or local politician) to seek assistance in resolving the problem; in the third option, he or she initiates legal action. In many cases, consumers seek Option 2 before proceeding to Option 3, whereas in others, they proceed to Option 3 directly when the two-party interaction fails to resolve the problem. The outcome of Options 2 and 3 can result in either the problem being solved to the consumers satisfaction or its not being resolved satisfactorily. Note that in following Option 3, a consumer has a range of institutions upon which to call.

The System of Dispute Management Institutions

When a dispute arises, various institutions may be called forth to resolve it. These range from "formal" to "informal." Steele [24] arranges the different dispute management institutions on a continuum. At the top end—the "formal" pole—is the institution which creates and announces rules of law to channel and regulate behavior so as to prevent dispute. Examples of such laws are the creation of the Federal Trade Commission (FTC) and the passing of the Magnuson–Moss Act by the U.S. Congress. At the bottom end—the "informal" pole—the mechanism of resolving dispute is through informal problem solving: "doing something about specific disputes." In between the "formal" and "informal" poles are a range of institutions, with the FTC initiating action in its capacity as a law-enforcing agency toward the top end, different courts through which individual consumers can initiate action against a manufacturer in the middle, and various public and private complaint bureaus toward the bottom end.

The "informal" pole corresponds to the two-party resolution of disputes, and as one moves from this pole toward the "formal" pole, the presence of a third party increases. Informal dispute resolution involves only the immediate dispute, whereas the more formal approach impacts on other disputes as well. As a result, the time and effort needed to resolve a dispute increases as one moves from the "informal" pole to the "formal" pole.

How well do these mechanisms work? Nowicki ([25], Section 4, Chapter 3) reports that "empirical studies of various alternate dispute resolution programs reveal a mixed bag of success and limitation."

Consumer Decision to Seek Legal Redress

Seeking legal redress involves the use of one or more of the alternate dispute management institutions discussed earlier. Depending on the institution involved, the cost and effort needed on the part of the consumer can vary considerably. Ursic [26] describes the consumer's decision to seek legal redress through four types of variables:

1. Antecedent variables: comprising the nature of prior court experience, amount of prior court experience, availability of the court, social class, and amount of the claim
2. Process variables: comprising anxiety about going to court, perceived effectiveness, utility for money, and the search for evidence
3. Expectation variables: comprising perceived costs, perceived benefits, and perceived probability of occurrence of each of these
4. The dependent variable: "Court Action"

The literature on warranties from a legal perspective is vast. For further details, see Chapters 4 and 5.

Two Additional Factors

For most consumer durables, the consumer purchases goods from a retailer rather than directly from the manufacturer. This is an additional important factor in the study of warranty. Yet another factor is government legislation as it affects warranty.

Fisk [27] has developed a reasonably comprehensive framework for characterization of the different cause and effect relationships in the warranty process, involving consumer, manufacturer, retailer, and the government. Fisk defines 20 variables and 76 cause–effect (or stimulus–response) relationships between the variables. Using this characterization, one can describe complex interactions and study many diverse aspects of warranty.

For additional references relevant to the detailed system characterization, see Murthy and Blischke [1] and Blischke and Murthy [28].

REFERENCES

1. Murthy, D. N. P., and Blischke, W. R. (1992). Product warranty management—II: An integrated framework for study, *European Journal of Operations Research*, **62**, 261–281.
2. McGuire, E. P. (1980). *Industrial Product Warranties: Policies and Practices*, The Conference Board Inc., New York.
3. Akerlof, G. (1970). The market for "Lemons": Quality uncertainty and the market mechanism, *Journal of Economics*, **84**, 488–500.
4. Darden, W. R., and Rao, C. P. (1979). A linear covariate model of warranty attitudes and behaviors, *Journal of Marketing Research*, **16**, 466–477.
5. Shimp, T. A., and Bearden, W. O. (1982). Warranties and other extrinsic

cues effects on consumer's risk perception, *Journal of Consumer Research*, 9(June), 38–46.

6. Perry, M., and Perry, A. (1976). Service contract compared to warranty as a means to reduce consumer's risk, *Journal of Retailing*, **52**, 33–40.

7. Wiener, J. L. (1985). Are warranties accurate signals of product reliability? *Journal of Consumer Research*, **12**, 245–250.

8. Nordstrom, R. D., and Metzer, H. (1976). Warranties: How important as a marketing tool? *Proc. Southern Marketing Assoc.*, 26–28.

9. Shuptrine, F. K., and Moore, E. M. (1980). Even after the Magnuson–Moss Act of 1975, warranties are not easy to understand, *Journal of Consumer Affairs*, **14**, 394–404.

10. Best, A., and Andreasen, A. R. (1977). Consumer response to unsatisfactory purchases: A survey of perceiving defects, voicing of complaints, and obtaining redress, *Law and Society*, **11**, 700–742.

11. Day, R. L., and Landon, E. L. (1977). Towards a theory of consumer complaining behavior, *Consumer and Industrial Buying Behavior*, A. G. Woodside, J. N. Sheth, and P. D. Bennet (eds.), North-Holland, New York.

12. Day, R. L., and Ash, S. B. (1979). Comparison of patterns of satisfaction/dissatisfaction for durables, nondurables, and services, *New Dimensions of Consumer Satisfaction and Complaining Behavior*, R. L. Day and K. H. Hunt (eds.), Division of Business Research, Indiana University, 190–195.

13. Bearden, W. O., and Oliver, R. L. (1985). The role of public and private complaining in satisfaction with problem resolution, *Journal of Consumer Affairs*, **19**, 223–240.

14. Udell, J. G., and Anderson, E. E. (1968). The product warranty as an element of competitive strategy, *Journal of Marketing*, **32**, 1–8.

15. Anderson, E. E. (1973). The protective dimension of product warranty policies and practices, *Journal of Consumer Affairs*, **7**, 111–120.

16. Moellenberndt, R. A. (1977). A critical look at warranty cost recognition, *National Public Accountant*, June, 18–22.

17. FASB (1976). *Financial Accounting Standard Board, Statement No. 5: Accounting for Contingencies*, Financial Accounting Standard Board, Stamford, CT.

18. Fisk, G. (1970). Guidelines for warranty service after sale, *Journal of Marketing*, **34**, 63–67.

19. Kendall, C. L., and Russ, F. A. (1975). Warranty and complaint policies: An opportunity for marketing management, *Journal of Marketing*, **39**, 36–43.

20. Fornell, C., and Wernerfelt, B. (1987). Defensive marketing strategy by customer complaint management: A theoretical analysis, *Journal of Marketing Research*, **24**, 337–346.

21. Fornell, C., and Wernerfelt, B. (1988). A model for customer complaint management, *Marketing Science*, **7**, 287–298.

22. Dardis, R., and Zent, C. (1982). The economics of the Pinto recall, *Journal of Consumer Affairs*, **16**, 261–277.

23. Iizuka, Y. (1988). Adequate claim processing improves quality system, *1988–ASQC Quality Congress Trans.*

24. Steele, E. H. (1975). Fraud, dispute, and the consumer: Responding to consumer complaints, *University of Pennsylvania Law Review*, **123**, 1107–1186.
25. Nowicki, P. (1987). *Regulating and Resolving New Car Warranty Problems and Disputes*, Doctoral Dissertation, Syracuse University.
26. Ursic, M. E. (1985). A model of the consumer decision to seek legal redress, *Journal of Consumer Affairs*, **19**, 21–36.
27. Fisk, G. (1973). Systems perspective on automobile and appliance warranty problems, *Journal of Consumer Affairs*, **7**, 37–54.
28. Blischke, W. R., and Murthy, D. N. P. (1993). *Warranty Cost Analysis*, Marcel Dekker, Inc., New York.

Part B

Warranty and Law

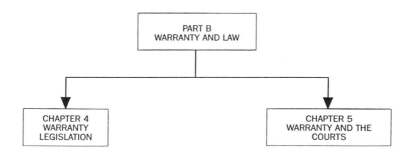

Introduction

Part B of the *Handbook* consists of two chapters that deal with legal aspects of warranty, including both warranty legislation and its enforcement by the courts. Some of the more recent history of warranty is included. The current status of the law and the judicial process is discussed in detail.

In Chapter 4, C. A. Kelley begins with a discussion of the two major pieces of legislation, the Uniform Commercial Code and Magnuson–Moss Warranty and Federal Trade Commission Improvement Act. The concepts of express versus implied and full versus limited warranties are carefully contrasted. The terms of the Magnuson–Moss Act and the degree of their implementation in practice as well as the implications of warranty legislation for business and consumers are discussed.

In Chapter 5, R. S. Kowal discusses legal issues that have arisen when warranty disputes have been litigated in federal and state courts in the United States. Litigation under both the Uniform Commercial Code and the Magnuson–Moss Warranty Act is analyzed from manufacturer, warrantor, and consumer points of view. The significance of warranty legislation and current judicial trends in the interpretation of warranty law are also examined. Future implications of warranty litigation for consumers and business are considered.

4

Warranty Legislation

Craig A. Kelley

California State University, Sacramento
Sacramento, California

4.1 INTRODUCTION

Warranties are governed by various federal and state statutes. An important milestone in the regulation of consumer product warranties occurred in 1939 in the court case of *Baxter* versus *Ford Motor Company* [1]. The case arose after Mr. Baxter had purchased a Ford Town Sedan. This purchase was based in part on information contained in the advertising for the car. The advertising referred to the "Triple Shatter-Proof Glass Windshield," stating that it would not shatter upon impact and, therefore, was an important safety feature because it eliminated the dangers of flying glass. Soon after Mr. Baxter purchased the car, a stone struck the windshield and shattered it, injuring Mr. Baxter. He brought a lawsuit against Ford based on the information in the advertisement. Ford argued that no warranty could exist without privity* and express warranties do not attach

---

* Privity means a direct contractual relationship exists between buyer and seller. Because most products are not purchased directly from the manufacturer, cases were routinely dismissed for the lack of privity.

themselves to the product sold. The Washington State Supreme Court disagreed with Ford and the decision formed the foundation for the establishment of consumer product warranties in the United States.

Several forms of legislation have been enacted to regulate warranties since the *Baxter* versus *Ford Motor Company* decision. The two major pieces of legislation are the Uniform Commercial Code, and the Magnuson–Moss Warranty and Federal Trade Commission Improvement Act of 1975. This chapter discusses each of these pieces of legislation in turn, along with legislation affecting commercial products, international trade, and government procurement. In addition, state legislation regarding warranties is reviewed.

4.2 WARRANTY LEGISLATION

4.2.1 Uniform Commercial Code

The Uniform Commercial Code covers express and implied warranties.* Table 4.1 summarizes the sections of the Uniform Commercial Code that pertain to warranties. Section 2-313 covers the various actions that a seller may take to create an express warranty. Specifically, Section 2-313 states [2]:

    (1)   Express warranties by the seller are created as follows:
          (a)   Any affirmation of fact or promise made by the seller to the buyer which relates to the goods and becomes part of the basis of the bargain creates an express warranty that the goods shall conform to the affirmation or promise.
          (b)   Any description of the goods which is made part of the basis of the bargain creates an express warranty that the goods shall conform to the description.
          (c)   Any sample or model which is made part of the basis of the bargain creates an express warranty that the whole of the goods shall conform to the sample or model.
    (2)   It is not necessary to the creation of an express warranty that the seller use formal words such as "warrant" or

---

* The Uniform Commercial Code is put forth by the National Conference of Commissioners on Uniform State Laws. However, the states have enacted variations of the code. As a result, uniformity has not been achieved.

Table 4.1 Sections of the Uniform Commercial Code

Section	Content
2-312	Warranty of Title and Against Infringement; Buyer's Obligation Against Infringement
2-313	Express Warranties by Affirmation, Promise, Description, and Sample
2-314	Implied Warranty of Merchantability
2-315	Implied Warranty of Fitness
2-316	Exclusion or Modification of Warranties
2-317	Conflict of Express or Implied Warranties
2-318	Third Party Beneficiaries of Express or Implied Warranties

> "guarantee" or that he have a specific intention to make a warranty, but an affirmation merely of the value of the seller's opinion or commendation of the goods does not create a warranty.

An express warranty may be created through a statement of the seller and it does not need to contain the words "warrant" or "guarantee" [3]. In addition, the statement does not have to be in writing, although it does have to make a "promise" rather than be a statement of "opinion."

Another way that an express warranty may be created under the Uniform Commercial Code is when a statement contains a "description or sample" that is "part of the bargain" [3, p. 276]. For a statement to become part of the bargain, a buyer must be able to rely on the statement when making a purchase decision.

According to Section 2-314, an implied warranty of merchantability is inferred in the sale of all goods. Merchantability means the goods are "fit for the ordinary purposes for which such goods are used" [3, p. 276]. An implied warranty also may be created when a buyer relies on the judgment of the seller to suggest goods are fit for a particular purpose. This is known as an implied warranty of fitness. Specifically, Section 2-314 states [2]:

(1) Unless excluded or modified (Section 2-316), a warranty that the goods shall be merchantable is implied in a contact for their sale if the seller is a merchant with respect to goods of that kind. Under this section the serving for value of food or drink to be consumed either on the premises or elsewhere is a sale.

(2) Goods to be merchantable must be at least such as
    (a) pass without objection in the trade under the contract description; and
    (b) in the case of fungible goods, are of fair average quality within the description; and
    (c) are fit for the ordinary purposes for which such goods are used; and
    (d) run, within the variations permitted by the agreement of even kind, quality and quantity within each unit and among all units involved; and
    (e) are adequately contained, packaged, and labeled as the agreement may require; and
    (f) conform to the promises or affirmations of fact made on the container or label if any.
(3) Unless excluded or modified (Section 2-316) other implied warranties may arise from course of dealing or usage of trade.

Section 2-315 defines when an implied warranty of fitness for a particular purpose is established. This section reads as follows [2]:

Where the seller at the time of contracting has reason to know any particular purpose for which the goods are required and that the buyer is relying on the seller's skill or judgment to select or furnish suitable goods, there is unless excluded or modified under the next section an implied warranty that the goods shall be fit for such purpose.

Implied warranties may be excluded under Section 2-316. This section enables a seller to exclude an implied warranty by allowing the buyer to examine the good or a model as thoroughly as he or she desires prior to the purchase. The seller may exclude an implied warranty in situations where the buyer has refused to examine the good. Implied warranties also may be disclaimed by the purchase contract if the individual state laws allow it [3, p. 279]. In the case of implied warranties of merchantability, the disclaimer must conspicuously mention merchantability in writing. To disclaim an implied warranty of fitness under Section 2-316, the disclaimer must be in writing and conspicuously presented but need not mention fitness.

Section 2-316 specifically states [2]:

(1) Words or conduct relevant to the creation of an express warranty tending to negate or limit warranty shall be construed wherever reasonable as consistent with each other;

but subject to the provisions of this Article on parol or extrinsic evidence (Section 2-202) negation or limitation is inoperative to the extent that such construction is unreasonable.

(2) Subject to subsection (3) to exclude or modify the implied warranty of merchantability or any part of it the language must mention merchantability and in case of a writing must be conspicuous, and to exclude or modify any implied warranty of fitness the exclusion must be by a writing and conspicuous. Language to exclude all implied warranties of fitness is sufficient if it states, for example, that "There are no warranties which extend beyond the description on the face hereof."

(3) Notwithstanding subsection (2)

    (a) unless the circumstances indicate otherwise, all warranties are excluded by expressions like "as is," "with all faults" or other language which in common understanding calls the buyer's attention to the exclusion of warranties and makes plain that there is no implied warranty; and

    (b) when the buyer before entering into the contract has examined the goods or the sample or model as fully as he desired or has refused to examine the goods, there is no implied warranty with regard to defects which an examination ought in the circumstances to have revealed to him; and

    (c) an implied warranty can also be excluded or modified by course of dealing or course of performance or usage.

(4) Remedies for breach of warranty can be limited in accordance with the provisions of this Article on liquidation or limitation of damages and on contractual modification of remedy.

A seller may reduce implied warranty liability without disclaiming the implied warranty. Under Section 2-719, contractual limitations may be put on remedies that are available to the buyer [3, p. 280]. Section 2-719 reads as follows [2]:

(1) Subject to the provisions of subsections (2) and (3) of this section and of the preceding section on liquidation and limitation of damages,

    (a) the agreement may provide for remedies in addition to or in substitution for those provided in this Article and may limit or alter the measure of damages recoverable under this Article, as by limiting the buyer's remedies to return of the goods, and repayment of the price or to repair and replacement of non-conforming goods or parts; and

(b) resort to a remedy as provided is optional unless the remedy is expressly agreed to be exclusive, in which case it is the sole remedy.

(2) Where circumstances cause an exclusive or limited remedy to fail of its essential purpose, remedy may be had as provided in this Act.

(3) Consequential damages may be limited or excluded unless the limitation or exclusion is unconscionable. Limitation of consequential damages for injury to the person in the case of consumer goods is prima facie unconscionable, but limitation of damages were the loss is commercial is not.

4.2.2 Magnuson–Moss Warranty and Federal Trade Commission Improvement Act of 1975

Although the Uniform Commercial Code was meant to provide uniform application from state to state, it failed to "standardize" terminology used in warranties. This lead to a great deal of consumer confusion about consumer rights with regard to consumer product warranties. To clarify the usage of warranties, Congress enacted the Magnuson–Moss Warranty and Federal Trade Commission Improvement Act of 1975 [4]. Passage of the act culminated more than 10 years of studies, proposals, and hearings aimed at improving consumer product warranties [5]. Through studies and hearings prior to enactment, a number of problems concerning consumer products came to light, including excessive use of disclaimers, inadequate warranty coverage, consumer difficulty in obtaining warranty service, and complex warranty language [6].

The Magnuson–Moss Warranty Act does not require a seller to provide a warranty. It only applies to sellers that offer a written warranty on consumer products. In addition, the act applies only to the sale of goods. The sale of a service is not covered by the provisions of the act.

The Magnuson–Moss Warranty Act defines a written warranty as:

A. A written affirmation of fact or written promise made in connection with the sale of a consumer product by supplier to a buyer which relates to the nature of the material or workmanship and affirms or promises that such material or workmanship is defect free or will meet a specified level of performance over a specified period of time; or

B. Any undertaking in writing in connection with the sale by a supplier of a consumer product to refund, repair, replace or take other remedial action with respect to such product in the event that such product fails to meet the specifications set forth in the undertaking. [7]

Every written warranty must contain the following sentence: "This warranty gives you specific legal rights, and you may also have other rights which vary from state to state" [8]. In addition, the rules that govern consumer product warranties promulgated by the Federal Trade Commission after the act was passed require the warranty to include the following information:

1. Who can enforce the warranty
2. A clear description and identification of the products, parts, and components covered by or excluded from the warranty
3. What the seller will and will not do
4. The starting date and duration of the warranty
5. An explanation of the process the consumer needs to follow to obtain service or other warranty performance
6. Information concerning any informal dispute settlement procedure
7. Any limitations of the duration of implied warranties
8. Any exclusions or limitations on relief [9]

The sections of the Magnuson–Moss Warranty Act are summarized in Table 4.2. A key section of the Act requires the terms and conditions of a warranty to be disclosed in "simple and readily understood language" and be made available to the consumer prior to the sale of any product with a retail price of $15 or more [10]. A seller may select one or more of the following methods to make their warranty available to a consumer prior to a purchase. The seller may elect to

1. Display the text of the warranty with the product
2. Display the warranty on the package of the product
3. Place signs disclosing the warranty terms "in close proximity" to the product
4. Maintain a binder or series of binders with copies of the warranties of the products sold in each department where the product is sold [11]

Table 4.2 Sections of the Magnuson–Moss Warranty Act of 1975

Section	Content
2301	Definitions of consumer product, consumer, supplier, warrantor, written warranty, implied warranty, and other terms.
2302	Rules governing contents of warranties. (a) Full and conspicuous disclosure of warranty terms and conditions. (b) Availability of warranty terms to the consumer; methods for presenting and displaying information and duration of a written warranty. (c) Prohibition on conditions for written or implied warranties. (d) Detailed substantive warranty provisions. (e) Applicability of the act to consumer products costing more than $15.00.
2303	Designation of written warranties. (a) Full or limited warranty. (b) Applicability of requirements, standards to representations or statements of customer satisfaction. (c) Exemptions by the Federal Trade Commission. (d) Applicability of the act to consumer products costing more than $15.00 and not designated as full warranties.
2304	Federal minimum standards for warranties. (a) Remedies under written warranty; duration of implied warranty; exclusion or limitation on consequential damages for breach of written or implied warranty; election of refund or replacement. (b) Duties and conditions imposed on the consumer by the warrantor. (c) Waiver of standards. (d) Remedy with charge. (e) Incorporation of standards to products designated with full warranty for purposes of judicial actions.
2305	Full and limited warranties.
2306	Service contracts; rules for full, clear, and conscious disclosure of terms and conditions.
2307	Designation of representatives by warrantor to perform duties under a written or implied warranty.
2308	Implied warranties. (a) Restrictions on disclaimers or modifications. (b) Limitation on duration. (c) Effectiveness of disclaimers, modifications, or limitations.
2309	Procedures applicable to promulgation of rules by the Federal Trade Commission.

Table 4.2 Continued

Section	Content
2310	Remedies in consumer disputes.
	(a) Informal dispute settlement procedures.
	(b) Prohibited acts.
	(c) Injunction proceedings by Attorney General or Federal Trade Commission.
	(d) Civil action by consumer for damages; jurisdiction; recovery of costs and expenses.
	(e) Warrantors subject to enforcement of remedies.
2311	Applicability to other laws.
	(a) Federal Trade Commission Act.
	(b) Rights, remedies, and liabilities.
	(c) State warranty Laws.
	(d) Other Federal warranty laws.
2312	Effective Dates

Source: 15 U.S.C.A. sections 2301–2312 (1992).

The disclosure and availability requirements are applicable to many forms of written warranty information, including information contained in advertisements, labels, and point of purchase presentations.

Special rules apply to catalog and mail-order sales and door-to-door sales. In the case of a catalog or mail-order sale, the full text of any express written warranty must be presented on the page that describes the product or on the facing page, or in an information section referenced by page number. The express warranty also may be disclosed by having the prospective buyer write for a copy of the warranty free of charge [12]. Both oral and written warranty statements must be given to the consumer prior to the consummation of a door-to-door sale [13].

Another key provision of the Magnuson–Moss Warranty Act addresses the disclosure of what is covered in the written warranty. The act requires the written warranties on consumer products with a retail price of $15 or more to be conspicuously labeled "Full" or "Limited." Table 4.3 summarizes the differences between a full and limited warranty. Generally, to be considered a full warranty, the warranty must meet a set of standards, including free replacement or full refund if the product continues to malfunction after a reasonable number of attempts to have it fixed [14]. In addition, full warranties may not disclaim or limit implied warranties and limited warranties.

Table 4.3 Full and Limited Warranties

Full warranty	Limited warranty

• Generally, defective products will be repaired or replaced at no cost to the consumer.
• Repair or replacement of a product will be done within a reasonable time after the consumer reports the defect.
• The consumer will force no unreasonable burdens on the consumer in order to obtain warranty service.
• Returning a registration card is not required to establish warranty coverage.
• Warranty coverage extends to subsequent owners of the product.
• If defective products cannot be repaired after a reasonable number of attempts, they will be replaced or refunded.
• Implied warranties many not be disclaimed or limited to the duration of the full warranty.
	• The seller places limitations on warranty coverage. Both limited and full warranties for different parts may be offered.
	• Return of a registration card may be required with a limited warranty but only if the seller does not provide service on products for which the card has not been returned.
	• Implied warranties may be limited to the duration of the limited warranty, provided the time is not too short or in violation of state law.
	• Acceptable limitations on warranties may include the following:
	  • Parts only.
	  • Labor only.
	  • Customer must return product to a store, factory, or service center.
	  • Warranty is nontransferrable.
	  • Customer must pay handling or transportation charges.

Source: Ref. 23.

In cases where warranties do not contain the designation "Full" or "Limited," courts have interpreted the act to mean the designation (usually Full) that is most favorable to the consumer [15].

Under Section 2308 of the Magnuson–Moss Warranty Act, implied warranties may not be disclaimed or modified unless a written limited warranty is offered. In this case, implied warranties may be limited to the duration of the limited warranty, if the duration is reasonable. Limitations on the duration of an implied warranty must be prominently disclosed on the face of the express warranty in simple, understandable language [16].

Two provisions of the Magnuson–Moss Warranty Act focus on consumer remedies if a seller does not honor its warranty. The act encourages a seller and consumer to resolve disputes without litigation through the use of "Informal Dispute Settlement Mechanisms" sanctioned by the Federal Trade Commission [17]. If this informal mechanism does not resolve the problem, the act allows a consumer to bring about an action in state court and recover damages plus attorney's fees and reasonable costs incurred in bringing the action [18]. This issue of warranty and legal action is discussed in Chapter 5.

Often consumer complaints result from a failure to satisfactorily honor one's warranty. In these cases, an organization may devise a "formal" and "informal" method of resolving disputes [19]. Formal methods involve more than one dispute and the aim is to develop general rules of conduct. These methods may range from private bargaining by the grieving party to court action. Informal procedures address problems of a single dispute at the microlevel.

4.2.3 Commercial Warranty Legislation

Commercial product warranties are regulated by Sections 2-312 through 2-315 of the Uniform Commercial Code [20]. These sections cover warranties of title, express warranties, implied warranties of merchantability, and implied warranties of fitness. Sections 2-313 through 2-315 of the code were discussed previously in Section 4.2.1. Section 2-312 provides for a warranty of title and against infringement. In a warranty of title the seller specifies that the title conveyed is clear and the goods are free of liens unknown to the buyer.

The Magnuson–Moss Warranty Act only covers warranties on products that are normally used for consumer purposes, regardless of whether the product is used for commercial purposes also [21]. The act has had an influence on the warranties attached to commercial products where sellers develop a single warranty based on the requirements of the act when products are sold to both consumer and industrial markets. On occasions where component parts are sold to assemblers of consumer products, suppliers may view their components as a consumer product and make certain that their warranties conform to the provisions of the act.

Retailers who buy consumer products for resale from a manufacturer are not considered "consumers" under the act [22].

4.2.4 State Warranty Legislation

Some form of the Uniform Commercial Code is the basis for regulating warranties at the state level in all 50 states. Many states have enacted legislation similar to the Federal Trade Commission Act to supplement the Uniform Commercial Code. These "little FTC acts" and the Fair

Business Practice Statutes in the various states apply to sellers who provide warranties on consumer and commercial products [23].

Generally, the Magnuson–Moss Warranty Act does not supersede state laws that regulate warranties [24]; rather the act was intended to govern transactions involving formal written warranties.

4.2.5   International Warranty Legislation

Warranties may be part of any product purchase and, therefore, are part of the purchase contract. The governance of contracts varies widely from country to country. Contracts are enforced in countries either by common law or civil law. Under common law, a promise (e.g., a warranty) must be part of the contract, whereas under civil law, contracts carry implicit promises that may be enforced [25]. Common law contracts are more detailed and usually specify definitions, relationships, and liabilities. A breach of contract claim under a warranty most often occurs when the terms of the contract are misunderstood or misinterpreted [25]. When dealing in foreign markets, sellers should always consult with legal consul that is familiar with the laws and regulations of a specific country.

4.2.6   Warranties and Government Procurement

Generally, warranties in the consumer, commercial, and government sectors are very similar. They must define what is covered by the warranty, the duration of the coverage, the coverage that is provided, and the seller's liability under the warranty. However, there are some significant differences that should be noted. In government procurement, the customers have a very specific set of requirements that usually do not have applications to other buyers. This is not true for buyers of consumer products. In addition, service of a product under warranty is often performed by the government entity. Therefore, government suppliers must often meet a set of specific requirements covered by a specific law.

4.3   LITERATURE REVIEW OF THE EFFECTS OF WARRANTY LEGISLATION

Most of the studies conducted prior to the passage of the Magnuson–Moss Warranty Act focused on the role that warranties play in influencing consumers' product choices (see Chapter 16). No studies were identified that specifically examined the effects of the sections of the Uniform Commercial Code regulation of warranties. Research on the effectiveness of the regulation of consumer product warranties increased dramatically after the Magnuson–Moss Warranty Act was passed. These studies may be broadly categorized as warranty studies related to readability, disclosure, comprehension, and the overall effectiveness of the Magnuson–Moss Warranty Act. The results of these studies are presented in the following subsections.

4.3.1 Readability of Warranties

One of the major provisions of the Magnuson–Moss Warranty Act was to have sellers write their warranties in simple and easily understandable language. Several studies, utilizing various measures of readability, have been conducted to determine if consumer product warranties were more or less readable after the act went into effect. Generally, these studies concluded that there was a slight improvement in the readability of post-Magnuson–Moss Warranty Act warranties, but that warranties were still difficult to understand [26–30]. Chapters 16, 27, and 28 discuss these studies in greater detail.

4.3.2 Disclosure of Warranty Terms

A few studies have examined potential consumer reaction to the disclosure of warranty terms. Specifically, the interpretation of the words "Full" and "Limited" have been investigated. One study found consumers were aware that a full warranty was "better" than a limited warranty but could not explain how the two differed from a legal point of view [31]. Another study concluded that one possible reason why consumers do not know the difference between a full and limited warranty is that consumers process the duration of the warranty terms (e.g., 3 months or 1 year), not the word full or limited [32]. Yet a third study reported consumers identified warranties as important decision-making factors and full warranties added more value to a product than limited warranties [33].

4.3.3 Comprehension of Warranty Terms

Generally, studies completed soon after the Magnuson–Moss Warranty Act was passed concluded that warranties were still difficult to understand [29, 30]. It is unlikely that warranties will be written in language that is simpler and easier to understand as society has continued down a path of a more legalistic interpretation of all product information.

4.3.4 Overall Effectiveness of the Magnuson–Moss Warranty Act

Although individual provisions of the Magnuson–Moss Warranty Act have been investigated, only one comprehensive study of the act has been conducted. This study concluded that consumers use a product's warranty as a signal of the product's reliability. However, in investigating the effectiveness of the act in meeting its objectives, the study found that warranty coverage had declined since the act was implemented in five of the eight product categories included in the study [34].

Based on the research conducted to date, it would appear that the Magnuson–Moss Warranty Act has had a small effect on consumer decision making. It seems that warranties are as difficult to read as before the act existed. In addition, consumers seem to lack awareness of the major provisions of the act nearly 20 years later [35].

4.4 IMPLICATIONS OF WARRANTY LEGISLATION FOR BUSINESS AND CONSUMERS

4.4.1 Implications for Business

The Uniform Commercial Code and Magnuson–Moss Warranty Act have major implications for business. These implications are more fully detailed in Part H—"Warranty and Management"—of this handbook. Given the increased emphasis on product quality and service as ways of achieving a competitive advantage, sellers must understand how their warranty policy relates to quality and service. The following are some suggestions for business when reconciling current warranty legislation and their warranty policy [23].

1.   Sellers that offer written warranties need to be aware of federal and state laws that govern their warranties. Although uniformity of requirements that govern consumer product warranties was achieved under the Magnuson–Moss Warranty Act of 1975, commercial warranties and implied warranties of merchantability and fitness are still covered under the various forms of the state-adopted Uniform Commercial Code. Failure to keep up to date with current state warranty laws, especially when the seller's products are distributed in many states, may lead to breach of warranty claims based on an implied warranty. For example, states differ in whether tort law or contract law governs product warranties. Courts that allow warranty claims under both tort and contract law must use contract law exclusively to determine if privity is required for the consumer to pursue a breach of implied warranty claim under the Magnuson–Moss Warranty Act [36].

2.   Manufacturers and others in the distribution chain should reach an understanding about the disclosure of warranty information in advertising and other forms of promotion. This is important not only because the Magnuson–Moss Warranty Act requires prepurchase disclosure of warranty terms on consumer products that retail for $15 or more, but it

helps to ensure that all forms of product promotion, including advertising and sales representations, conform to the terms of the written warranty. Advertising copy should be tested before it is released to make sure that it can be understood by the target audience. Failing to disclose warranty information is a violation of the Magnuson–Moss Warranty Act regardless of whether it is the fault of the trade or the manufacturer. For example, if a manufacturer engages in co-op advertising with a retail chain and does not proofread the advertising copy that contains claims about its products, the manufacturer may unknowingly establish a warranty if a consumer relies on an advertising claim as a promise or guarantee.

3. Sellers should encourage their customers to read and compare written warranties. This suggestion has two positive results. First, a competitive advantage may be achieved if a seller's warranty is "superior" to the industry standard warranty. Second, it is in a seller's best interest for the consumer to understand what is covered by the warranty. Fewer claims due to misinterpretation of the warranty should reduce the seller's opportunity costs due to loss of future sales because of customer confusion over what is covered in its warranty. In addition, the enhanced reputation of the seller who stands behind the warranty should build a positive business image and customer loyalty.

4. Sellers should establish informal mechanisms to resolve warranty disputes. Customer satisfaction may be enhanced by disclosing and then adhering to a set of rules that govern the resolution of complaints that inevitably rise if a product fails when it is still under warranty. Furthermore, problem areas may be identified in service and sales that could be corrected to avoid future problems. For example, court costs have escalated in recent years to the point where it may be less expensive to use an arbitrator to resolve warranty complaints. Such a system is used in the U.S. automobile industry to resolve customer complaints.

5. Sellers should disclose the warranty terms in simple, understandable language. The Magnuson–Moss Warranty Act requires disclosure of warranty terms on consumer products. Disclosing warranty terms, even when it is not required, may lead to fewer dissatisfied customers that result from misunderstanding a warranty and overestimating the warranty coverage. These fewer dissatisfied customers will have a positive economic effect because it is estimated that it costs less to retain existing customers than to attract new customers.

6. Sellers of products with written warranties should review the warranty literature for studies that potentially affect enforcement of warranty laws. Studies are constantly being done in the fields of economics, law, and business related to warranties and warranty laws. Keeping abreast of current research may allow a seller to maintain conformity

with federal and state laws governing warranties. For example, research evidence indicates that warranties may serve as signals of product reliability. Knowing this may suggest to a seller that additional government regulation and policing may be in order if consumers are misinterpreting the warranty signal.

4.4.2 Implications for Consumers

The implications warranty legislation has for consumers is provided in greater detail in Chapters 27 and 28 of this handbook. Here, it may be concluded that the Magnuson–Moss Warranty Act has resulted in some improvement of how consumers can use warranties in the process they use to make a purchase decision. Although product warranties may still be written in a complex manner, making them difficult to read, they must be made available prior to the sale of products with a retail price of $15 or more. This allows the consumer more time to weigh the warranty information carefully before buying a product. In addition, the clarification of terminology used in express warranties and disclosure of remedies provided under the Magnuson–Moss Warranty Act should reduce consumer confusion when evaluating different product warranties.

4.5 FUTURE RESEARCH OF WARRANTY LEGISLATION

One area needing additional research is business and consumer response to warranty legislation. Specifically, the question of what consumers would like included in a warranty needs to be investigated. What are the implications for a business that consults its consumers regarding what information they would find helpful in making a purchase decision and how that information should be presented to them? Related to this question is whether consumers understand the terms "full" and "limited." If they do not understand the terms, what public policy initiatives are needed to increase the understanding of the terms?

Another question that deserves some study is whether the Magnuson–Moss Warranty Act has resulted in increased competition and better warranties. These are two basic objectives of the act. Preliminary evidence has been mixed on this question. A related question is whether sellers have opted for more limited warranties and, therefore, have decreased the amount of protection the warranty provides the consumer.

Finally, a question remains regarding how conformance to warranty legislation may influence the pricing, promotion, and distribution of products. Do sellers consider these other business decisions when designing their warranty programs? If so, do sellers periodically review their warranty policy to react to changes in the consumer and business environments?

REFERENCES

1. Baxter v. Ford Motor Company 12 P.2d 409 (1939).
2. White, J. J., and Summers, R. S. (1972). *Uniform Commercial Code*, West Publishing Co., St. Paul, MN.
3. Epstein, D. G., and Nickles, S. H. (1981). *Consumer Law in a Nutshell*, West Publishing Co., St. Paul, MN.
4. Magnuson–Moss Warranty and Federal Trade Commission Improvement Act of 1975, Pub. Law No. 93–637 (15 U.S.C. 2301).
5. Bixby, M. B. (1984). Judicial interpretation of the Magnuson–Moss Warranty Act, *American Business Law Journal*, **22,** 125–163.
6. Task Force on Appliance Warranties and Service. (1969). Report on Appliance Warranties and Service, Federal Trade Commission, 100–108.
7. 15 U.S.C.A. 1992. Section 2301(6).
8. 16 C.F.R. 1992. Section 701.3(a).
9. 16 C.F.R. 1992. Sections 701.3(a)(1)–701.3(a)(8).
10. 15 U.S.C.A. 1992. Sections 2302 and 2303.
11. 16 C.F.R. 1992. Section 702.3(a)(1).
12. 16 C.F.R. 1992. Section 702.3(c).
13. 16 C.F.R. 1992. Section 702(d).
14. 15 U.S.C.A. 1992. Sections 2303, 2304, and 2308.
15. Hughes v. Segal Enterprises, Inc. 627 F. Supp. 1231 (1986).
16. Bush v. American Motors Sales Corp. 575 F. Supp. 1581 (1984).
17. 15 U.S.C.A. 1992. Section 2310(a).
18. 15 U.S.C.A. 1992 Section 2310(d).
19. Steele, E. H. (1975). Fraud, dispute, and the consumer: Responding to consumer complaints, *University of Pennsylvania Law Review*, **123,** 1107–1186.
20. McGuire, E. P. (1980). *Industrial Product Warranties: Policies and Practices*, The Conference Board, Inc., New York.
21. Najran Co. for General Contracting and Trading v. Fleetwood Enterprises, Inc. 659 F. Supp. 1081 (1986).
22. Black v. Don Schmid Motor, Inc. 657 P.2d 517 (1983).
23. Office of Consumer Affairs. U.S. Department of Commerce (1980). *Product Warranties and Servicing: Responsive Business Approaches to Consumer Needs*. Government Printing Office, Washington, DC.
24. 15 U.S.C.A. 1992. Section 2311(c).
25. Litker, M. (1988). *International Dimensions of the Legal Environment of Business*, PWS–Kent Publishing Co., Boston, MA.
26. Lehman, J. C., Gentry, J. C., and Ellis, H. W. (1983). The readability of warranties: Did they improve after the Magnuson–Moss Act and are they more complex than other product-related communications? *Proceedings of the Southwestern Marketing Association*, 19–23.

27. Schmitt, J., Kanter, L., and Miller, R. (1979). *Impact of the Magnuson–Moss Act: A Comparison of 40 Major Consumer Product Warranties from Before and After the Act*, Bureau of Consumer Protection, Staff Report to the Federal Trade Commission.

28. Shuptrine, F. K. (1976). A comparison of two readability measures—fog index and the Dale–Chall Method—on the comprehensibility of warranties, *Proceedings of the Southern Marketing Association*, 9–10.

29. Shuptrine, F. K., and Moore, E. M. (1980). Even after the Magnuson–Moss act of 1975, warranties are not easy to understand, *Journal of Consumer Affairs*, **14**, 394–404.

30. Wisdom, M. J. (1979). An empirical study of the Magnuson–Moss Warranty Act, *Stanford Law Review*, **31**, 1117–1146.

31. Wilkes, R. E., and Wilcox, J. B. (1981). Limited versus full warranties: The retail perspective, *Journal of Consumer Affairs*, **15**, 65–77.

32. Kelley, C. A., and Swartz, T. A. (1984). An empirical test of warranty label and duration effects on consumer decision making: An information processing perspective, *1984 AMA Summer Educators' Proceedings*, 309–313.

33. Kangun, N., Cox, K. K., and Staples, W. A. (1976). An exploratory investigation of selected aspects of the Magnuson–Moss Warranty Act, *Consumerism: New Challenges for Marketing*, 3–11.

34. Wiener, J. L. (1988). An Evaluation of the Magnuson–Moss Warranty and Federal Trade Commission Improvement Act of 1975, *Journal of Public Policy*, **7**, 65–82.

35. Kelley, C. A. (1986). Consumer product warranties under the Magnuson–Moss Warranty Act: A review of the literature, *1986 AMA Summer Educators' Proceedings*, 369–373.

36. Walsh v. Ford Motor Co. 588 F. Supp. 1513 (1984).

5

Warranty and the Courts

Rachel S. Kowal

New York University
New York, New York

5.1 INTRODUCTION

Chapter 4 discussed how warranties have become regulated by various forms of state and federal legislation. This chapter explores some of the legal issues that have arisen when warranties have been litigated under these statutes in both the federal and state courts. Also included is a brief discussion of current judicial trends in the interpretation of both federal- and state-created warranty laws. These legal issues are examined from the points of view of the consumer, warrantor, and manufacturer.

5.2 WARRANTY LITIGATION UNDER THE MAGNUSON–MOSS WARRANTY ACT

5.2.1 Scope and Nature of Consumer Litigation

As indicated in Chapter 4, The Magnuson–Moss Warranty and Federal Trade Commission Improvement Act of 1975 (hereafter referred to as "the act") was enacted in order to provide new opportunities for consumers in enforcing written product warranties. From a litigation point of view, these new consumer opportunities include the ability to dispute product warranties in the federal court system, to mount class action lawsuits, and to recover attorneys' fees and costs. In view of this, one would have expected the act to have generated extensive consumer litigation; however, during the first few years after the passage of the act, very few cases were reported [1].

It is only since 1981 that the use of the act by consumers, plaintiff's counsel, and corporate counsel in consumer product cases has increased [2]. This slow but steady increase in the number of reported state cases is reflected by an increase from 14 reported cases in 1986 to 30 cases in 1989. Twenty-seven state cases were reported from November 1990 through December 1993.* Despite this increase, as of November 1990, fewer than 180 appellate cases have been decided since the act was enacted in 1975. From November 1990 through December 1993, there were only 19 reported federal cases decided under the act. Seven out of 19 of these federal cases were appellate cases.† These meager statistics indicate that the act is still not being utilized to its fullest potential by both the legal profession and consumers [3].

The act's burdensome federal jurisdictional requirements have made it difficult to either attain federal jurisdiction or mount a successful class action. As a result, most of the federal cases have been dismissed for lack of federal jurisdiction prior to adjudication of the Magnuson–Moss warranty claims [4]. In addition, the lack of knowledge possessed by the general consumer public regarding the usefulness of the act has contributed to the relatively small number of reported Magnuson–Moss cases despite its potentially large legal significance.

5.2.2 Jurisdiction of Federal and State Courts Over Individual and Class Action Lawsuits

The jurisdictional requirements of the act vary depending on whether the lawsuit is commenced in state court or in federal court. In state court actions, consumers may file a warranty action in any state court, but such lawsuits are subject to each state's specific jurisdictional requirements.

In contrast, lawsuits commenced in federal court must meet all three following federal jurisdictional requirements:

---

* Statistics derived from LEXIS All States Mega database.
† Statistics derived from LEXIS All Fed Mega database.

1. Each plaintiff's claim must equal at least $25 [5].
2. The total amount of all claims in controversy must be at least $50,000 [6].
3. If the action is brought as a class action, there must be at least 100 named plaintiffs [7].

Because the federal courts have been strict in enforcing the above three federal jurisdictional requirements, the vast majority of Magnuson–Moss cases have been brought in state, rather than federal, courts. These stringent federal jurisdictional requirements are in keeping with the legislative purpose of the act which is to prevent the commencement of trivial warranty actions in the federal courts [8]. By imposing stiff jurisdictional standards for federal actions, Congress has reserved for the federal courts only the most significant warranty abuse cases.

This aim has been realized. Very few cases satisfy federal jurisdictional standards. The net result is that the federal warranty rights created by the act are mostly litigated and enforced in state, rather than federal, courts. This is consistent with the aims of the Senate drafters [9].

The $25 per Claim Requirement

It is generally not difficult to fulfill the $25 minimum allowable individual claim requirement for federal jurisdiction due to the kinds of products covered by the act. The act covers "consumer products" which are defined as "any tangible personal property which is normally used for personal, family or household purposes" [10]. Examples of covered "consumer products," as set forth in the act's House report, include some large expensive household items, such as dishwashers and ranges, as well as mostly smaller, inexpensive products, such as blenders, coffee makers, and blankets [11]. Automobiles are, by far, the most commonly litigated "consumer product" under the act.

The $50,000 in Controversy Requirement

One main deterrent to obtaining federal jurisdiction is the requirement that the amount in dispute total at least $50,000, exclusive of claims for costs, interest, and fees [12]. Attorneys' fees claims are excludable as a "cost" under 15 U.S.C. §2310(d) (3) [13]. Federal courts have routinely dismissed Magnuson–Moss cases for failure to meet the $50,000 amount in controversy requirement. Many state courts acquire jurisdiction over breach of warranty cases due to the plaintiff's failure to meet the $50,000 federal jurisdictional requirement.

For example, in *Barnette v. Chrysler Corp.* [14], the plaintiff sued Chrysler, pursuant to the act, to recover the $7,000 value of a defective Chrysler automobile. The federal district court dismissed plaintiff's suit based on lack of federal jurisdiction because the plaintiff had alleged that less than $50,000 was in dispute. The *Barnette* court suggested, as an alternative, that the plaintiff attempt to satisfy state jurisdictional requirements and bring this same action in state court. Similarly, in *Fleming v. Apollo Motor Homes, Inc.* [15],* a breach of warranty action brought under the act by a motor-home buyer against the seller for contract rescission or $33,000 in damages was dismissed for lack of federal jurisdiction because less than $50,000 was in dispute.

Some plaintiffs have added punitive damages claims to their breach of warranty claims in order to meet the act's $50,000 in dispute federal jurisdictional requirement. Neither the act nor its legislative history indicate whether punitive damages should or should not be included in calculating the $50,000 jurisdictional amount. Although, the Seventh Circuit United States Court of Appeals in *In re General Motors Engine Interchange Litigation* [16] has indicated that punitive damages may be recoverable under the act, this issue has not yet been authoritatively decided by the federal courts.

Some federal courts have resolved this issue by deferring to the state law regarding punitive damages. In *Novosel v. Northway Motor Car Corp.* [17], a Buick purchaser sued the dealer and GM under the act for $9,638 in compensatory damages and $50,000 in punitive damages due to automobile defects. The *Novosel* court dismissed the case for lack of federal jurisdiction because the $50,000 punitive damages claim could not be aggregated with the $9,638 in compensatory damages to satisfy the $50,000 federal jurisdictional requirement. They reasoned that punitive damages are awarded in New York contract actions only when willful intent has been shown and the *Novosel* plaintiff had not shown such willful intent. In contrast, a motion to dismiss for lack of jurisdiction was denied in *Schafer v. Chrysler Corp.* [18],† even though the plaintiff only alleged $8,295 in compensatory damages. The *Schafer* court ruled that the plaintiff met the $50,000 requirement because the plaintiff's additional allegation of $65,000 in punitive damages due to intentional fraud was supported by Indiana case law. Recent case law shows that federal courts will refer to the relevant state law on punitive damages when determining whether

---

* See also *Reiff v. Don Rosen Cadillac–BMW Inc.*, 501 F.Supp.77 (E.D.Pa. 1980).

† Accord, Barnette v. Chrysler Corp., 434 F.Supp. 1167 (D.Neb. 1977).

punitive damages claims can be added to breach of warranty claims for satisfying the $50,000 federal jurisdictional requirement.

The 100 Named Plaintiffs Requirement

Consumers may more successfully opt for class action lawsuits rather than individual actions, in order to meet the $50,000 in dispute requirement. However, only a few cases have met the act's class action requirements because the act requires that 100 plaintiffs be named in order to mount a class action lawsuit.

This requirement, like the $50,000 in dispute requirement, has been strictly construed by the courts. The phrase "named plaintiffs" is not defined in either the act or in its legislative history; therefore, most courts have applied the literal meaning of the word "named" by dismissing actions where each of the 100 plaintiffs were not actually identified by name [19]. One commentator [20] has noted that this literal judicial interpretation has defeated the congressional purpose of giving consumers an effective class action mechanism that constitutes "a fair and efficient procedure for handling claims where the claims or defenses of the represented parties are typical of the claims or defenses of the class . . . " [21].

In *Barr v. General Motors Corp.* [22], plaintiff filed a Magnuson–Moss class action in federal court alleging that the paint on her car and on every other car manufactured by General Motors that year was defective. Plaintiff's complaint alleged that the members in plaintiff's class probably exceeded several thousand persons. The *Barr* court rejected the plaintiff's class action suit because the plaintiff did not actually name 100 plaintiffs in her complaint. The court noted that the 100 plaintiffs must at least be named in order to certify a class action lawsuit and that a group of unidentified potential class members is not sufficient [23].* The 100 named plaintiffs jurisdictional requirement was met in *In re General Motors Corp. Engine Interchange Litigation* [24] because the Illinois Attorney General, on behalf of the state, consolidated its case and several individual and small class actions into one major class action which named more than 100 plaintiffs.

In addition to the practical difficulties of naming 100 plaintiffs, ethical considerations prohibit a plaintiff's attorney from soliciting the required 99 additional plaintiff-clients. The Code of Responsibility of the American

---

* See also *Stuessy v. Micrsoft Corp.*, 1993 U.S. Dist. LEXIS 16368 (Penn. 1993) where plaintiff's class action suit against Microsoft Corp. for breach of warranty in computer software was dismissed because there was only one named plaintiff.

Bar Association defines guidelines and standards of conduct for attorneys in its Ethical Considerations and Disciplinary Rules [25]. The Ethical Considerations encourage lawyers to assist laypersons in recognizing their legal problems provided such assistance is not motivated by monetary gain [26]. The Disciplinary Rules prohibit lawyers from soliciting clients for themselves [27]. These two provisions conflict: Lawyers are encouraged to give advice but discouraged from soliciting employment. These ethical considerations prevent a lawyer from obtaining the requisite 100 named plaintiffs because lawyers are prohibited from soliciting clients and it is unlikely that a lawyer will know of 100 plaintiffs when a lawsuit is brought to his or her office.

5.2.3 Class Action Lawsuits

A major advantage of the act is the availability of class action suits as a means for similarly situated consumers to enforce the act's substantive provisions regarding written warranties. Class actions may be brought in either federal or state court [28]. As discussed above, the federal class action remedy has not been frequently used by consumers due to the difficulty in naming a class of 100 plaintiffs and in aggregating a total of $50,000 in dispute [29].

Class actions brought in federal court are also subject to the notice requirements of Rule 23 of the Federal Rules of Civil Procedure. The U.S. Supreme Court held, in *Eisen v. Carlisle & Jacquelin* [30], that pursuant to Rule 23 the plaintiff must individually notify every class member who can be reasonably identified and bear the costs of such notification. One legal commentator makes the following argument:

> If courts impose the *Eisen* Court's strict interpretation of rule 23 (c) (2) notice requirements on Magnuson–Moss Act class actions, they will cripple the Act's principal enforcement mechanism by making many Magnuson–Moss Act class actions too expensive for consumers to pursue. Such a result would frustrate the purpose of the statute as well as the policy that underlies rule 23 (b) (3) itself, that is, securing a "fair and efficient" determination of every controversy. [31]

This commentator advocates that each court be permitted to design less expensive methods of notice, depending on the facts of each case, in order to make federal class actions more affordable for consumers [32].

It appears that lawsuits which allege product design defects are the most amenable to class action suits under the act [33]. For example, four early class action suits based on product design defects were successfully brought under the act. One, *General Motors Engine Interchange Litigation* [34], was based on an undisclosed "engine-switch" of Chevrolet engines in Oldsmobile cars. A second case, *Feinstein v. Firestone Tire & Rubber Co.* [35], involved the Firestone 500 tire. A third case, *Skelton v. General Motors Corp.*, concerned GM's undisclosed substitution of THM 200 transmissions instead of THM 300 transmissions in various GM models manufactured between 1976 and 1979 [36]. In a fourth case, *In Re Cadillac V8-6-4 Class Action*, the New Jersey appellate court certified a statewide class of approximately 7500 purchasers of 1981 Cadillac automobiles who alleged design defects in their V8-6-4 engines [37].

5.2.4 Recovery of Attorneys' Fees and Costs

Traditionally, in the United States, parties to a lawsuit must each pay the cost of their own litigation expenses, regardless of the outcome of the litigation. This is known as the "American Rule." In contrast, British law has always awarded the fees and costs of the litigation to the prevailing party. The American Rule continues to govern most cases in the United States, except for certain statutes where Congress has made statutory fee-shifting arrangements.* The act represents one of these statutory fee-shifting exceptions to the American Rule.

Section 110 (d) (2) provides that the prevailing plaintiff/consumer, in any action brought under the act, may recover their attorneys' fees and costs "based on actual time expended" and "reasonably incurred" by the plaintiff, unless the court determines that "such an award of attorneys' fees would be inappropriate" [38]. The purpose of the act's fee-shifting provision is to vindicate the rights of consumers who are injured by warrantors and to encourage consumers to pursue their legal remedies by providing them with ready access to legal assistance. Courts have acknowledged that the act's fee-shifting provisions are a legislative attempt to encourage consumers to initiate breach of warranty actions by providing the successful plaintiff/consumer with adequate attorneys' awards.

Successful defendant suppliers, manufacturers, warrantors, or retailers are excluded by the act's plain language from recovering their attorneys' fees and expenses incurred in defending or appealing Magnuson–Moss warranty actions that are brought against them. A successful

---

* See, for example, 15 U.S.C. §§78i (e), 78r (a) (Securities Exchange Act of 1934); 15 U.S.C. §1640 (a) (Truth In Lending Act); 15 U.S.C. §2310 (d) (2) (Magnuson-Moss Act); 42 U.S.C. §7604 (d) (Clean Air Act); 42 U.S.C. §2000a-3b (Civil Rights Act of 1964, Title II); 42 U.S.C. §2000e-5 (k) (Civil Rights Act of 1964, Title VII); 42 U.S.C. §3612 (c) (Fair Housing Act).

defendant manufacturer was fined in *State Farm Fire v. Miller Electric Company* [39] for bringing a frivolous petition to recover their attorneys' fees and costs because this is specifically not permitted by the act.

The attorneys' fee provision is one of the most significant consumer-oriented provisions of the act because it makes warranty litigation affordable for consumers and eliminates consumer worry about the expense of hiring an attorney. It is also significant from the plaintiff attorney's point of view because it enables attorneys to accept consumer cases which they previously could not have afforded to accept. Many commentators feel that the attorney's fee provision is *the most important* section of the act because consumers who suffer minimal warranty damages of $500 to $1,000, but who cannot afford to pay the legal costs of enforcing their rights, can both secure adequate legal representation and guarantee their successful attorney payment of their reasonable legal fees and actual costs [40]. Additionally, the act's provisions for attorneys' fees may make this federal remedy more popular than bringing a parallel breach of warranty action under state law.

In the majority of cases litigated under the act, courts have awarded attorneys' fees and costs to the successful plaintiff. There are many examples of this. In *Ballentine* [41], the court awarded, in addition to the damages award of $23,000, $3200 in attorney's fees and $600 in costs. After a jury found that the defendant had breached a written warranty in the purchase of a recreational vehicle for $72,201, the court in *Dunlap v. Jimmy GMC of Tucson Inc.* [42] awarded the plaintiff $20,500 in attorneys' fees under the act (in addition to awarding the plaintiffs compensatory damages of $45,000 and $35,000 against the dealer and manufacturer, respectively). In *Clarkson v. Overseas Imported Cars,** an award of $2000 in automobile damages as well as $600 in attorneys' fees and costs (plus $110 in interest) was granted. In *Royal Lincoln–Mercury Sales, Inc. v. Wallace* [43], the purchaser of a new 1978 Continental received a damage award in the amount of the purchase price ($13,812), as well as $2162.50 in attorneys' fees and expenses.

Ventura v. Ford Motor Co. [44] shows how this provision can be quite favorable to attorneys. In *Ventura*, plaintiff's attorney had agreed to take the case for a $350 retainer plus whatever the court awarded. The *Ventura* plaintiff won rescission of the car contract and a damages award of $7,000. The court awarded $5,165 in attorneys' fees (consisting of $75 per hour for 34.2 hours of office time and $100 per hour for 26 hours of

---

* Clarkson v. Overseas Imported Cars, No. Cir. 77-25194 (Michigan, 1978).

court time) even though the court had been apprised of the contingency arrangement between the parties.

It is entirely within the court's judicial discretion to award attorneys' fees and to determine the appropriate amount; however, appellate courts usually reverse lower courts that do not grant adequate attorneys' fee awards [45]. In *Gardner v. Nimnicht Chevrolet Co.* [46], the appellate court reversed the lower court's denial of attorneys' fees. In *Vieweg v. Ford Motor Corp.* [47] and *Maserati Autos, Inc. v. Caplan* [48], the appellate court remanded the case for an upward redetermination of the amount of attorneys' fees allowable due to the trial judge's "abuse of discretion" [49].

The successful plaintiff's request for attorneys' fees has been denied or reduced in only a few cases. In *Trost v. Porreco Motors* [50], it was not an abuse of discretion to deny attorneys' fees because the successful plaintiff was relieved, due to breach of the implied warranty of merchantability, of his duty to pay the $4,777 balance due for the purchase of a $8,876 jeep. In *Hibbs v. Jeep Corp.* [51], the trial court did not abuse its discretion by disallowing plaintiff's attorneys' fees where the attorney's itemized list of hours spent and subject matter covered showed that most of his time was spent on claims against defendants other than Jeep Corp., did not show what portion was spent on claims against this defendant seller, and did not clarify which lawyers worked on the case nor their specific charges. Although, the attorney in *Hanks v. Pandolfo* [52] requested fees of $2,825, the court only awarded attorneys' fees of $450 in a case involving a damage award of $900 for breach of a refrigerator warranty since this was a $900 case and plaintiff's attorney was a legal services attorney [53].

Courts consider the following factors when determining the amount of attorneys' fees to be awarded: the actual time reasonably expended multiplied by a reasonable hourly rate; the nature of the services rendered; the amount in dispute; the results obtained; the novelty or difficulty of the issues involved; and other relevant issues, such as whether the attorney took the case on a contingency fee basis. In addition, many courts, in determining their fee allowance, also consider the act's key policy of encouraging consumers to bring breach of warranty actions [54].

Most courts determine what constitutes a reasonable fee by multiplying the number of hours reasonably expended on the litigation by a reasonable hourly rate. Travel and lodging expenses incurred during the conduct of the trial by the successful plaintiff's attorney are included in this sum. In *Troutman v. Pierce* [55], the appellate court affirmed the trial court's award of attorneys' fees to a consumer who successfully revoked their acceptance of a mobile home due to a breach of warranty. The *Troutman* court awarded attorneys' fees based on the number of hours spent by plaintiff's attorney minus the hours spent by plaintiff's attorney on the issues that he did not prevail upon. In addition, the appellate court believed that the act was sufficiently broad enough to include the travel and lodging expenses of plaintiff's attorney plus the additional costs incurred in defending itself on appeal because the court did not believe that the "prevailing consumer's attorney-fee award under the Magnuson–Moss Act at the trial level should be dissipated by uncompensated costs, expenses and attorney's fees in successfully defending a judgement on appeal" [56]. Consumers who are forced on appeal to defend their award of attorneys' fees will most likely recover their appellate costs in defending a successful appeal.

A court abuses its discretion if they flatly limit their attorneys' fees award to a maximum amount as dictated by the plaintiff's damages recovery. The plaintiff's damages recovery is not the only factor to be considered in determining the attorneys' fee award [57]. For example, in *Croft v. P&W Foreign Car Service* [58],* the trial court limited the amount of attorneys' fees to $1,000 based on the plaintiff's $3,000 damage award. The trial court reasoned that, although plaintiff's attorney had submitted a fee request of $7,539 based on 89 hours of trial preparation, it was not appropriate to award attorneys' fees that were two and one-half times the amount of the jury award. On appeal, the appellate court reversed the trial court, indicating that the amount of the damage award is but one factor among many to consider in determining a reasonable award of attorneys' fees and it should not be used as an overall cap.

5.2.5 Consumer Complaints and Remedies

The act's remedies are only available to "consumers" who purchase "consumer products." A "consumer" is defined by the act as anyone who buys a consumer product "for purposes other than resale" [59]. A consumer product is defined by the act as "any tangible personal property which is distributed in commerce and which is normally used for personal, family or household purposes" [60]. Examples of "consumer products" as set forth in the federal regulations include "boats, photographic film and chemicals, clothing, appliances, jewelry, furniture, typewriters, motor homes, automobiles, mobile homes, vehicle parts and accessories,

---

* See also, Drouin v. Fleetwood Enterprises, 163 Cal.App.3d 486, 209 Cal. Rptr. 623 (1985).

stereos, carpeting, small aircraft, toys, and food" [61].* The act's legislative history gives additional "consumer product" examples: washing machines and dryers, dishwashers, food waste disposers, freezers, ranges, refrigerators, water heaters, bed coverings, blenders, broilers, can openers, coffee makers, corn poppers, floor polishers, frypans, hair dryers, irons, toasters, vacuum cleaners, waffle and sandwich grills, air conditioners, fans, radios, televisions, and tape recorders [62]. The courts have held that "consumer products" also includes "used" consumer products, specifically, the sale of used cars [63].

What is not a consumer product was addressed in one recent federal case, *Kemp v. Pfizer*, wherein the district court held that a surgically implanted prosthetic heart valve did not constitute a "consumer product" covered by the act [64]. According to *Kemp*, surgically implanted heart valves or other specially made medical devices are not consumer goods under the act because they are not customarily used by the general public. Similarly, in *Reynolds v. S&D Foods, Inc.*, dog food marketed in 10-pound tubes to kennels and breeders were not consumer products because they were not intended to be used as private pet food in consumer's homes [65]. In contrast, the recent case of *Muchisky v. Frederic Roofing Co.* expanded the act's definition of "consumer products" to include a contract to reroof a home [66]. The *Michisky* court held that building materials purchased by consumers for the improvement, modification, and/or repair of their homes are consumer goods, if at the "time of sale" (upon creation of the sales contract), the goods have not yet been integrated into the real estate. Materials that at the time of sale have already been integrated into the real estate are not consumer goods.

The "consumer" may choose among four possible federal causes of action under the act. The first is a cause of action for damage caused by a supplier, warrantor, or service contractor who fails to comply with the act. Included within this cause of action would be enforcement of consumers rights regarding the proper disclosure of warranty terms, labeling of the warranty as "full" or "limited," presale availability of warranties, and providing for informal dispute settlement procedures (as more fully explained in Chapter 4, Section 4.2.2). Although only actual damages are recoverable, state laws may provide for additional minimum or statutory penalties [67]. The second is a cause of action for breach of any written

---

* For cases holding that the Act applies to "mobile homes," see Brigadier Homes v. Thompson, 551 So.2d 1031 (Ala. 1989) and Liberty Homes, Inc. v. Epperson, 581 So.2d 449 (Ala. 1991).

"full" or "limited" warranty [68]. The third is a cause of action for the breach of or exclusion of either the implied warranty of merchantability or implied warranty of fitness for a particular purpose that arise pursuant to each state's Uniform Commercial Code (as previously defined in Chapter 4, Section 4.2.1). For example, the *In re General Motors Engine Interchange Litigation* case, which involved the undisclosed use of Chevrolet engines in Oldsmobiles, included a cause of action for breach of the implied warranty of merchantability [69]. The fourth is a cause of action for breach of any service contract.

The consumer's remedies are counterbalanced by the warrantor's right to elect from among the act's following remedies:

(A) repair, (B) replacement, or (C) refund; except that the warrantor may not elect refund unless (I) the warrantor is unable to provide replacement and repair is not commercially practicable or cannot be timely made, or (II) the consumer is willing to accept such refund. [70]

Three types of damages due to breach of warranty are commonly sought by consumers: claims for personal injury, economic loss, and punitive damages.

Personal Injury Claims

To date, the reported decisions have uniformly held that the act does not create a federal cause of action for personal injury claims arising out of a breach of warranty.* The recovery of personal injury claims arising out of defective products is generally a matter of state law.† A few courts have allowed recovery of personal injury claims if any of the following substantive provisions of the act have been violated: Section 2304(a) (2) prohibiting warrantors from limiting the duration of implied warranty coverage; Section 2304(a) (3) requiring that limitations on liability for consequential damages be conspicuously displayed on the warranty; or Section 2308 forbidding disclaimers of implied warranties.‡

---

* See Boelens v. Redman Homes, Inc., 748 F.2d 1058 (5th Cir. 1984); Bush v. American Motors Sales Corp., 575 F.Supp. 1581 (D. Colo. 1984); Gorman v. Saf-T-Mate, Inc., 513 F.Supp.1028 (N.D. Ind. 1981).

† See Washington v. Otasco, 603 F.Supp. 1295 (D.C. Miss. 1985) wherein the plaintiff could not recover under the act for personal injuries sustained on a bicycle manufactured by defendant because this remained a matter of state law.

‡ See Hughes v. Segal Enterprises, Inc., 627 F.Supp. 1231 (W.D.Ark. 1986).

Economic Loss Injuries

"Economic loss" has been defined as the diminution in the value of the product because it is inferior in quality and does not work for the general purposes for which it was manufactured and sold. In *Hughes v. Segal Enterprises, Inc.*, plaintiff sought $15,000 in economic loss for the loss in value of his mobile home and replacement housing. The *Hughes* court affirmed the well-settled rule that damages for economic losses are recoverable under the act.*

Punitive Damages

In *Hughes*, the plaintiff also sought $45,000 in punitive damages based on their economic loss. When Congress passed the act it intended the warranty statute to be based on contract, rather than on tort law [71]. In a breach of contract action, the main purpose of damages awards is to place the plaintiff in the same position he would have been in had the contract not been breached.† Punitive damages are usually not awarded in breach of contract actions because contract remedies are designed to compensate the injured party rather than to punish the breaching party.

It is a well-established judicial rule that punitive damages are not recoverable under the act. Punitive damages may be recovered only if the applicable state law permits such a recovery. The allowance of punitive damages takes different forms and requires different showings in different states. For example, some states allow recovery of punitive damages in a breach of contract action if the breach of contract amounts to an independent tort or is accompanied by fraudulent conduct [72]. In *Hughes*, although punitive damages are ordinarily not recoverable for breach of contract, the plaintiff alleged sufficient willful conduct to support a claim for punitive damages under Arkansas state law [73].

5.2.6 Alternative Dispute Settlement

As mentioned in Chapter 4, the act encourages consumers and warrantors to resolve their disputes without litigation by use of "Informal Dispute Settlement Mechanisms" established by the warrantor pursuant to Federal Trade Commission (FTC) promulgated regulations. Section 2310(a) (1) provides that if the warrantor has issued a written warranty that estab-

---

* Id. at 1237. See also In re General Motors Corp. Engine Interchange Litigation, 594 F.2d 1106 (7th Cir. 1979).

† See, for example, 5A Corbin, Corbin on Contracts §992 at 5 (1964).

lishes an informal dispute settlement mechanism, in accordance with FTC rule 16 C.F.R §703, the consumer must first utilize these settlement procedures prior to initiating a formal lawsuit against the warrantor.* The decision to establish such a mechanism is at the discretion of the warrantor; but if a mechanism is established, it must meet the FTC's minimum procedural requirements.

The Federal Trade Commission regulations provide that the decisions of dispute settlement mechanisms are not legally binding on any person [74]. A dissatisfied consumer may choose to disregard the dispute settlement decision and to pursue legal remedies [75]. The warrantor's establishment of an informal dispute settlement mechanism does not prejudice a consumer from eventually seeking legal redress. Because the decision of a dispute settlement mechanism is not final, their overall impact is questionable; however, an informal dispute settlement mechanism's decision is admissible as evidence in any subsequent legal proceeding.

The informal dispute settlement mechanisms required by the act may be disadvantageous to the consumer. These settlement mechanisms may not be readily understood by consumers and can be easily rigged in favor of the warrantors. In addition, they are often inconveniently located at a distance from the consumer's residence and are usually conducted without the benefit of an attorney [76]. On the other hand, some of the advantages to consumers include a quick and inexpensive way of settling warranty disputes that saves court costs, reduces attorneys' fees, and shortens the length of litigation.

Of course, a consumer may choose to totally disregard any settlement procedure and initiate an action entirely under state law. However, if a consumer chooses to only pursue a state law remedy, then he may not opt for the relief afforded by the act, including the right to recover attorneys' fees and costs (unless otherwise provided by state law), because he has not complied with the act's requirement of submitting to settlement procedures prior to resorting to court action [77].

5.2.7   Retailer, Manufacturer, Supplier, or Warrantor Defenses

There are several key defenses that the defendant supplier or warrantor may raise when confronted with a Magnuson–Moss warranty lawsuit. Section 2301(4) of the act defines a "Supplier" as "any person engaged in the business of making a consumer product directly or indirectly avail-

---

* See Wolf v. Ford Motor Co., 829 F.2d 1277 (4th Cir. 1987), where plaintiffs were required to follow the informal dispute settlement procedures set forth in a written automobile warranty prior to commencing a lawsuit under the act.

able to consumers.'' Section 2301 (5) of the act defines a ''Warrantor'' as including ''any supplier or other person who gives or offers to give a written warranty or who is or may be obligated under an implied warranty.'' The successful assertion, by either the supplier or warrantor, of any of the following defenses may cause dismissal of the consumer's breach of warranty action.

The Right to Cure Defense

Section 2310(e) of the act provides that ''No action . . . may be brought . . . for failure to comply with any obligation under any written or implied warranty . . . unless the person obligated under the warranty . . . is afforded a reasonable opportunity to cure such failure to comply.'' The act requires that, prior to instituting a breach of warranty action, the consumer first afford the seller a reasonable opportunity to either repair the product, replace the product, or refund the actual purchase price less reasonable depreciation.

The seller, in *Tucker v. Aqua Yacht Harbor Corp.*, was not afforded reasonable opportunity to cure the defects in a boat engine where the buyer revoked his acceptance and sued the seller while the seller was in the process of making the final engine repairs and less than 20 days had elapsed between the seller's receipt of the new engine parts and completion of the final repairs [78]. The consumer, in *Champion Ford Sales, Inc. v. Levine*, prevailed against the seller under the state's Uniform Commercial Code (UCC) for breach of implied warranty and revocation of acceptance because, prior to instituting their action, the consumer had afforded the seller a reasonable opportunity to cure the nonconformity, that is, to replace the engine or the car [79]. The *Champion* case also demonstrates a similarity between state-based Uniform Commercial Codes and the act. The UCC, like the act, makes the consumer's remedy subject to the seller's right to cure the defect within a reasonable amount of time after receiving notification of the product defect.

The Written Warranty Defense

A common defense raised by suppliers or warrantors is that the act is inapplicable because no written warranty has been given. The act's various remedies are only applicable when a ''written warranty,'' as defined in Chapter 4, has been given to a consumer with respect to a particular product. If a written warranty has not been given, then the consumer has no cause of action under the act [80].

The key legal issue is to determine whether or not a written warranty has been given.* The most important case in this area is *Skelton v. General Motors Corp.*, wherein purchasers of GM automobiles brought suit against GM because of the automaker's undisclosed transmission switch [81]. The plaintiffs argued that brochures, manuals, consumer advertising, and other forms of communication distributed by GM constituted a "written warranty" under the act. The *Skelton* court held that communications such as these are not written warranties under the act because they are subject to general public circulation and they do not directly relate to a specific contract for the sale of goods.

Similarly, in *Lytle v. Roto Lincoln Mercury*, the court found that neither the language inside the car dealer's "New Car Get Ready" form (stating that the car dealer prepared the car for delivery) nor inside the Subaru manufacturer's "1984 Warranty and Service Booklet" constituted a written warranty because the dealer did not agree in his sales contract to perform all terms and conditions of the Subaru owner's service policy [82]. Sufficient evidence of a written warranty was found where the defendant's warranty was written on the back of a one-page sales contract in the *Carr v. Hashman* case [83] or on an inspection form completed and signed by the service manager of a motor-home dealer in the *Marine Midland Bank, N.A. v. Carroll* case [84].†

Written service contracts frequently accompany a consumer's purchase of goods. Section 2301(8) of the act defines "service contracts" as "a contract in writing to perform, over a fixed period of time or for a specified duration, services relating to the maintenance or repair (or both) of a consumer product." As a general rule, these service contracts also constitute written warranties under the act.

5.2.8 Warrantor's Disclosure Requirements

A consumer who knows, prior to making a purchase, the warranty information concerning a consumer good may make a more informed purchasing decision from among a variety of similar consumer products. For this

---

* The act broadly defines a written warranty as: "Any undertaking in writing in connection with the sale by a supplier of a consumer product to refund, repair, replace, or take other remedial action with respect to such product in the event that such product fails to meet the specifications set forth in the undertaking, which written affirmation, promise, or undertaking becomes part of the basis of the bargain between a supplier and a buyer for purposes other than resale of such product." 15 U.S.C. §2301(6)(B).

† For a similar holding see also *Murphy v. Mallard Coach Company*, 179 A.D.2d 187, 582 N.Y.S.2d 528 (App.Div. 1992).

reason, if a seller chooses to offer a written product warranty, then the act requires that the written warranty terms be made available to consumers prior to the sale. As explained in Chapter 4 (Section 4.2.2), the act requires that any warrantor who gives a written warranty disclose "fully and conspicuously" in "simple and readily understood language" the terms and conditions of the warranty prior to the sale of the consumer product.* The Federal Trade Commission (FTC) has been delegated the responsibility by the Act to create presale warranty availability disclosure rules which the FTC has promulgated in rule 16 C.F.R. §702.3 [85].

Chapter 4 delineates the four FTC alternatives for making warranty information available to consumers prior to sale: displaying the warranty text next to the warranted product; maintaining binders containing copies of each warranty in the department where the product is sold; showing the package of the product on which the warranty text is disclosed nearby the product being sold; or placing a sign disclosing the warranty terms in close proximity to the warranted product. Failure to make the required disclosures constitutes a violation of both FTC rules and of section 2305(a)(1) of the act.

A car dealer in one early case, *In re Bob Rice Ford, Inc.*, was required to make his posted signs more conspicuous, in terms of their height, width, language, and placement, in order to ensure proper notice to consumers [86]. The *Montgomery Ward & Co., Inc., v. FTC* case [87] contains the most definitive judicial interpretation of the FTC's presale disclosure rules. Montgomery Ward and Co., a large retail store chain, picked the FTC's binder option as their method for warranty disclosure. In *Montgomery*, the Ninth Circuit Court of Appeals held that the binder option required large retailers to maintain one set of warranty binders on each sales floor but did not also require large retailers to post signs in each selling department indicating the location of the warranty binders. The FTC was unsuccessful in compelling Montgomery Ward to post signs in each selling department regarding warranty binder availability. The net result of the *Montgomery* decision is to limit the effectiveness of the FTC's presale disclosure rules because only the most diligent or aggressive consumer will seek out warranty binders which could be located anywhere on the selling floor of a large retail store.

---

* 15 U.S.C. §2302 (a). This section of the act provides that: "Any warrantor warranting to a consumer by means of a written warranty a consumer product actually costing the consumer more than $15.00 shall clearly and conspicuously disclose in a single document in simple and readily understood language, the following items of information: . . ." pertaining to the terms and conditions of the warranty.

5.2.9 Privity of Contract in Warranty Cases

It is the traditional common-law rule that privity of contract is a prerequisite to bringing any breach of warranty action because such actions are based on contract law principles* As modern chains of product distribution have become increasingly complex creating a more distant relationship between consumers and their manufacturers and consumer protection has become a federal and state priority, modern courts and legislatures have eroded the traditional common-law privity of contract rule. Now, the privity of contract requirement varies from state to state. The most liberal states have completely abolished the privity requirement in lawsuits between consumers and manufacturers, whereas other more conservative states still maintain the traditional privity requirements.†

In relationship to the act, the privity issue concerns whether a manufacturer's written warranties allow consumers to sue remote manufacturers for breach of the written warranty and/or breach of the simultaneously created UCC implied warranties. Several commentators have suggested that the act's broad definition of a ''consumer,'' which includes both the immediate buyer plus any transferee who subsequently acquires the goods from the buyer, effectively abolishes the privity requirement whenever a written warranty is given.‡

One judicial approach toward privity has not yet been formulated. Key decisions rendered by the Illinois Supreme Court in *Szajna v. General Motors Corp.* [88] and the Second Circuit Court of Appeals in *Abraham v. Volkswagen of America, Inc.* [89] provide guidance in this area. In *Szajna*, the plaintiff sued an automobile manufacturer for economic losses based on breach of the UCC implied warranty of merchantability. The *Szajna* trial court dismissed plaintiff's complaint because the privity of contract requirements, pursuant to Illinois state law, had not been met. On appeal, the Illinois Supreme Court held that once a written warranty is given, a privity relationship is established between the consumer and

---

* Privity of contract means that a direct contractual relationship must exist between the plaintiff and the defendant.

† A minority of states still require privity as a condition precedent to bringing a breach of warranty action. Those states are Arizona, Connecticut, Georgia, Indiana, Kentucky, Oregon, Washington and Wisconsin. See, Harris, O., Jr., and Squillante A. (1989). *Warranty Law in Tort and Contract Actions*, J. Wiley & Sons, New York, p. 346.

‡ See A. Devience, Jr. (1990). Magnuson–Moss Act: Substitution for UCC warranty protection?, *Commercial Law Journal*, **95**, 323, 324; Comment (1975). Consumer product warranties under the Magnuson–Moss Act and the Uniform Commercial Code, *Cornell Law Review*, **62**, 738, 755–59.

the manufacturer with respect to any implied warranties. According to *Szajna*, once a manufacturer has extended a written warranty to a consumer, then the consumer may sue the manufacturer for breach of implied warranty even though there is no privity between them as required by state law. If no written warranty has been given, then the state law privity requirements prevail. If state law requires privity of contract in order to bring an action for a breach of implied warranty and no written warranty has been given, then no action may be brought. The *Abraham* decision was in accord with the *Szajna* holding.

These decisions are consistent with other sections of the act which provide that consumers with written warranties also enjoy the continued protection of any implied warranties. For example, Section 2308(a) provides that implied warranties may not be disclaimed if either a full or limited written warranty is given. If a full warranty is given, then the implied warranties may not be limited in duration. If a limited warranty is given, then the implied warranties may be limited to the duration of the limited warranty, provided that such limitation is not unreasonable or unconscionable [90].*

The act enlarges the scope of the traditional common-law rule regarding privity of contract by encompassing any implied warranties created by state law, extending the warrantor's written warranties to any transferee of the product and restricting limitations placed upon implied warranty disclaimers. This interpretation makes sense in terms of the consumer policies underlying the act. Otherwise, manufacturers could assert lack of privity as a means of avoiding implied warranties which would leave consumers with worthless watered-down written warranties and without the protection of the implied warranties.

One commentator gives an example of how the act significantly increases a consumer's right of recovery from nonprivity warrantors: "In the event a new car purchaser, within a 5 year/50,000 mile warranty, sells the car to a second purchaser, the second purchaser will be able to seek recovery for economic loss against the remote manufacturer, which action would be barred under Sections 2-371 and 2-318 of the Uniform Commercial Code. Indeed, Magnuson–Moss does fill a deep consumer void" [91].

---

* See also Ismael v. Goodman Toyota, 106 N.C. App. 421, 417 S.E.2d 290 (N.C. 1992), where the act prohibited the car dealer from disclaiming the implied warranty of merchantability in an "as is" sale of a car where at the time of the sale the car dealer had entered into a "service contract" with the consumer with respect to the car. Where a warrantor enters into a written service contract with respect to a product, then the warrantor may not disclaim any implied warranty with respect to that product even in an "as is" sale.

5.3 WARRANTY LITIGATION UNDER THE UNIFORM COMMERCIAL CODE AND OTHER STATE LAWS

Consumer protection has traditionally been regulated by state law. The subsequent enactment of federal warranty laws runs parallel to other state consumer protection laws and does not restrict "any right or remedy of any consumer under State law or any other Federal law" [92].* The act, for the most part, does not invalidate any rights or remedies otherwise provided to consumers. Except, the act does limit implied warranty disclaimers, abolish state privity of contract requirements, and impose certain duties on warrantors who give "full" warranties. The act also gives consumers additional rights and remedies that might not exist under state law. For example, the UCC, unlike the act, does not contain a provision awarding attorneys' fees to the successful warranty plaintiff.

The most common and important source of state-created consumer relief are the Uniform Commercial Code (UCC) and state "lemon laws." In addition, the Uniform Deceptive Trade Practices Acts (UDTPA) of 1964 and 1966 have been adopted in 12 states† and 4 states have adopted the Uniform Consumer Sales Practices Act of 1970.‡

5.3.1 The Uniform Commercial Code

The Uniform Commercial Code, a state statute, which has been enacted in every state except Louisiana, sets forth the creation of product warranties and consumer remedies for breach of warranty. As discussed in Chapter 4, the UCC establishes two main types of warranties: express warranties and implied warranties of merchantability and fitness for a particular use. Consumers are provided with three main remedies for breach of these warranties: rejection of the goods; recovery of damages for breach of warranty; or revocation of acceptance of the goods.

A consumer who chooses the UCC remedy of rejection should be careful to not use the goods because they may forfeit their right of rejection from such use. A consumer who seeks damages in a breach of warranty suit may recover the difference between the value of the goods as received

---

* See also Schroeder (1978) Private actions under the Magnuson–Moss Warranty Act, *California Law Review*, **66**, 1. (1978).

† The Uniform Deceptive Trade Practices Act of 1964, 7A U.L.A. 299 (1985), has been adopted in Delaware, Illinois, Maine, and Oklahoma. The Uniform Deceptive Trade Practices Act of 1966, 7A U.L.A. 265 (1985), has been adopted in Colorado, Georgia, Hawaii, Minnesota, Nebraska, New Mexico, Ohio, and Oregon.

‡ The Uniform Consumer Sales Practices Act, 7A U.L.A. 231 (1985), has been adopted (with some variations) in Kansas, Ohio, Texas, and Utah.

and their value as originally warranted, unless, the warrantor limits their remedy to repair or replacement of the defect. A consumer who tries to revoke their acceptance may do so only if the defect substantially impairs the value of the goods and revocation is done within a reasonable time after the defect was discovered or should have been discovered.

5.3.2 Warranty Disclaimers and Warranties That Fail of Their Essential Purpose: UCC Sections 2-316 and 2-719

The UCC permits sellers to disclaim all warranties, express and implied, as long as the requirements of Section 2-316 are followed. For example, the implied warranty of merchantability may only be disclaimed if the seller does so conspicuously and uses the word ''merchantable.'' The implied warranty of fitness for a particular use may only be disclaimed if the seller does so conspicuously and in writing. An effective warranty disclaimer excludes the disclaimed warranty from the contract. As noted in Section 5.2.9 above, whenever a written warranty had been given, the Magnuson–Moss Warranty Act places some limitations on the disclaimer of UCC implied warranties.

The UCC permits a seller to limit or modify the buyer's remedies for breach of warranty. A limitation of remedies clause allows a cause of action for warranty breach but limits a buyer's remedies to those specified in the contract. If a seller's limitation of remedies clause is either ''unconscionable'' or causes the contract to fail of its essential purpose, then it will be unenforceable.

A limitation of remedies clause is ''unconscionable'' if, in light of the general commercial background of the case and the status of the parties, it is oppressive, overly one-sided, unfair, and the result of grossly unequal bargaining power between the parties. For example, courts have generally held that it is unconscionable for a seller to exclude their liability for consequential damages for personal injury to a consumer. A limitation of remedies clause also would cause a contract to fail of its essential purpose if the proposed remedy does not restore the main function of the goods to the consumer. A Delaware court concluded, in *Norman Gershman's v. Mercedes-Benz* [93], that a limitation of remedies clause that only allowed repair or replacement of defective automobiles failed of its essential purpose when the warrantor refused to repair the vehicle, it was not repaired within a reasonable time, and it was not repaired after a reasonable number of attempts had been made.

5.3.3 State "Lemon Laws"

State-created lemon laws make automobile manufacturers accountable if their newly manufactured automobiles fail to substantially conform to their express written warranties [94]. Prior to the enactment of lemon laws, most automobile manufacturers offered limited, rather than full, warranties due to the stringent requirements imposed by the Magnuson–Moss Warranty Act upon full written warranties. Typically, a car manufacturer's limited warranty did not provide for replacement or refund of the automobile's purchase price. Warranty litigation under the UCC provided limited consumer relief due to its high cost and lengthiness.

Most states have adopted lemon laws in response to the prevalence of limited manufacturer warranties and the complexities of UCC litigation in order to provide automobile purchasers with more easily accessible remedies. Motor vehicles were the chosen lemon law subject matter because, other than the purchase of a new home, a motor vehicle purchase is one of the most expensive purchases made by consumers. As of 1989, 44 states and the District of Columbia had enacted lemon laws [95]. Most courts have held that the state-created lemon laws are constitutionally permissible [96]. Consumers who have already sought warranty relief under the federal act are not precluded from also seeking relief under state lemon laws.*

A detailed discussion of local state lemon law statutes is beyond the scope of this chapter. For specific details of local lemon laws, the reader is advised to check the particular statutes of his or her jurisdiction.† For

---

* For example, a Texas consumer was not precluded from using his administrative remedy before the Texas Motor Vehicle Commission by virtue of having sought relief from his manufacturer under the act, Chrysler Corp. v. Texas Motor Vehicle Com., 755 F.2d 1192 (5th Cir. 1985), reh. den, en banc 761 F.2d 695. Similarly, the act did not preempt New York State Lemon Law, Motor Vehicle Mfrs. Assn. v. Abrams, 899 F.2d 1315 (2nd Cir. 1990), nor Minnesota laws requiring automobile manufacturers to establish informal dispute resolution forums, Automobile Importers of America v. Minnesota, 871 F.2d 717 (8th Cir. 1989) cert. den. 110 S.Ct. 201.

† [115] Nearly all of the states and the District of Columbia have enacted their own versions of lemon laws. See Alaska Stat. §§45.45.300–360 (1990); Ariz. Rev. Stat. Ann. §§44-1261 to -1265 (1987 & Supp. 1990); Cal. Civ. Code §1793.2 (Deering 1990); Colo. Rev. Stat. §§42-12-101 to -107 (1984 & Supp. 1990); Conn. Gen. Stat. Ann. §42-179 (West Supp. 1990); Del. Code Ann. tit. 6 §§5001–5009 (Supp. 1990); D.C. Code Ann. §§40-1301 to -1309 (1990); Fla. Stat. §§681.10-.111 (1990); Haw. Rev. Stat. §490:2-313.1 (Supp. 1989); Ill. Rev. Stat. ch.121 1/2, para. 1201–1208 (1988); Iowa Code §322E.1 (1989); Kan. Stat. Ann. §8-2419 (1985); Ky. Rev. Stat. Ann. §§367.840-.846 (Baldwin 1987); La. Rev. Stat. Ann. §§51:1941–1948 (West 1987); Me. Rev. Stat. Ann. tit. 10, §§1161–1169 (Supp. 1990); Md. Com. Law Code Ann. §§14-1501 to 1504 (1990); Mass. Ann. Laws ch.90, §7N1/2 (Law. Co-op 1985 & Supp. 1990); Mich. Comp. Laws §§257.1401-.1410 (1990); Minn. Stat. §325F.665 (1990); Mont. Code Ann. §§61-4-501 to .533 (1989); Neb. Rev. Stat. Ann. §§357-D:1 to :8 (1984 & Supp. 1990); N.M. Stat. Ann. §§57-16A-1 to -9 (1987); N.Y. Gen. Bus. Law §198-a (McKinney 1988 & Supp. 1991); N.D. Cent. Code §§51-07-16 to -22 (1989); Okla. Stat. Ann. tit. 15, §901 (West

example, the Texas "lemon law" requires that the new motor vehicle owner report substantial warranty nonconformities either during the warranty period or within 1 year of the sale, whichever is earlier [97]. The New York law is similar to the Texas law but extends protection to the consumer for as long as 2 years or 18,000 miles [98].

There are characteristic features common to most lemon laws. Lemon laws only apply to new passenger (noncommercial) vehicles. Owners of used cars or commercial vehicles are not entitled to lemon law protection. The vehicle must carry with it an express written warranty that is fully disclosed in the manufacturer's warranty or owner's manual. If the car contains a defect that substantially impairs the use, value, or safety of the car and the vehicle cannot be repaired after a reasonable number of attempts, the consumer is entitled to either replacement of the vehicle or full refund of the purchase price (less a reasonable allowance for the consumer's use of the vehicle). What constitutes a reasonable attempt to repair varies from state to state. For example, in Texas, if the manufacturer has made four attempts to repair the same problem during the warranty period or within 1 year after the vehicle's delivery or the vehicle is out of service for 30 business days, then a reasonable attempt to repair has been made and the vehicle is considered a "lemon" [99]. These statutory presumptions provide more certainty than the UCC, which does not specify how long or how many attempts to cure must be made before a warranty fails of its essential purpose.

Before bringing a lemon law action, the consumer must first resort to any informal dispute settlement procedures established by the manufacturer. Initially, it was intended that these alternative dispute resolution programs be voluntarily created and operated by car manufacturers pursuant to FTC regulations. As of 1990, only eight manufacturers, Ford, General Motors, Nissan, Honda, Volkswagen, Toyota, Saab, and Rolls Royce, had created informal dispute settlement mechanisms that met FTC requirements. Also, these manufacturer-run dispute settlement programs were criticized for the following shortcomings: Arbitration panels were biased against consumers because they often contained arbitrators with

---

Supp. 1991); Or. Rev. Stat. §§646.315–.375 (1985); Pa. Stat. Ann. tit. 73, §§1951–1963 (Purdon Supp. 1990); R.I. Gen. Laws §§31-5.2-1 to -13 (Supp.1990); Tenn. Code Ann. §§55-24-201 to -211 (1988); Tex. Rev. Civ. Stat. Ann. art. 4413 (36) (Vernon Supp. 1991); Utah Code Ann. §§13-20-1 to -7 (1986 & Supp. 1990); Vt. Stat. Ann. tit. 9, §§4170–4181 (1984 & Supp.1990); Va. Code Ann. §§59.1-207.9 to -14 (1987 & Supp. 1990); Wash. Rev. Code §§19.118.005–.902 (Supp. 1988); W. Va. Code §§46A-6A-1 to -9 (1986 & Supp. 1990); Wis. Stat. §218.015 (1988); Wyo. Stat. §40-17-101 (Supp. 1990).

close automobile industry ties or who were car manufacturer personnel; arbitration proceedings were not monitored in order to ensure FTC compliance; arbitrators were poorly trained and unfamiliar with lemon laws; and manufacturers tried to divert consumers away from external arbitration boards and toward their own internal settlement procedures.

Dissatisfaction with manufacturer-run arbitration programs prompted several states to establish mandatory state-run programs for the arbitration of lemon law disputes.* For example, in 1985, Massachusetts became one of the first states to provide its consumers with mandatory state-run arbitration of lemon law disputes. Since its inception, the Massachusetts state-run arbitration program has awarded consumers more than $3.2 million in refunds and replacements, decided in favor of consumers in nearly two-thirds of the cases heard through 1987 and doubled (in 1987) the percentage of cases settled prior to arbitration [100].†

Advocates of these mandatory state-run lemon law arbitration programs allege that they provide consumers with a forum that is more expedient, accessible, fair, and uniform than manufacturer-run settlement mechanisms [101]. Members of the state-run arbitration panels are generally appointed by the governor or attorney general. Prospective arbitrators are carefully screened to ensure against possible prejudice. They receive formal training regarding state lemon law statutes and arbitration procedures. A higher level of professionalism and fairness in arbitration proceedings has been achieved in states that devote time and effort to arbitrator recruitment, selection, and training.

Lemon law statutes provide manufacturers with the following defenses: the nonconformities were caused by consumer abuse, neglect, alteration, odometer tampering, or modification; the defects do not sub-

---

* Connecticut, the District of Columbia, Massachusetts, New York, Texas, Vermont and Washington all have state-run arbitration programs.

† In Connecticut, as of March 31, 1990, 1698 cases had been resolved with an estimated value of $15,406,285. Connecticut Department of Consumer Protection, *Semi-Annual Statistical Report: October 1, 1984–March 31, 1990.* In Florida, after the January 1, 1989 effective date of their lemon law, Florida consumers prevailed in 65% of state arbitration board decisions. [Ingalsbe, R. G. (1990). Florida's new car lemon law: An effective tool for the consumer, *The Florida Bar Journal*, 61–64 (October)]. The estimated value of these Florida awards was $1,917,943 and over 95% of the customers who prevailed elected a refund rather than a replacement of the vehicle. In 1990, consumers prevailed in 66.5% of the cases heard by Florida state-run arbitration boards and 73.1% of the cases approved for arbitration resulted in settlements in favor of the consumers resulting in consumer awards totalling $11,058,810 in 1990 [Adams, R. J. (1992). Florida's Motor Vehicle Arbitration Board—A two year review, *Arbitration Journal*, 36–43 (March)].

stantially impair the use, value, or safety of the car; or the claims were not filed in good faith. Also, only car manufacturers, and not car dealers, can be sued under lemon laws [102].

Some commentators argue that a better consumer remedy for nonconforming motor vehicles, as well as for any other consumer product, is provided by the UCC [103]. Section 2-608 of the UCC permits consumers to revoke their acceptance of new cars if the nonconformity substantially impairs the car and the revocation is done within a reasonable time after the consumer discovers or should have discovered the defect. The UCC provides car purchasers with two distinct advantages. First, if the lemon law warranty period of limitations has expired (i.e., a typical lemon law warranty period might expire after either 2 years or 24,000 miles), then the consumer may still be able to revoke his acceptance under the UCC if this is done in a timely manner. Second, the UCC does not require that consumers first resort to any manufacturer-created or state-run alternative dispute resolution procedures prior to commencing their lawsuit.

There are other disadvantages of lemon laws. Lemon laws do not always provide for recovery of attorneys' fees; therefore, consumers should always include in their complaint a request for attorneys' fees and costs pursuant to the Magnuson–Moss Warranty Act. Consumers who use lemon laws are sometimes required to give up their right to proceed for breach of warranty under the UCC. Lemon laws typically contain short statutes of limitations for asserting a breach of warranty. Lemon law causes of action typically occur when the same repeated defect, rather than numerous different defects, goes unremedied. This makes it an undesirable remedy for cars with many different problems.

5.3.4 Uniform Deceptive Trade Practices Acts and Uniform Consumer Sales Practices Acts

State Consumer Sales Practices Acts create consumer remedies beyond those available under either the Magnuson–Moss Warranty Act, the Uniform Commercial Code, lemon laws, or other state statutes. The Uniform Consumer Sales Practices Act, promulgated by the National Conference on Uniform State Laws and approved by the American Bar Association, prohibits unfair, deceptive, or unconscionable acts by suppliers in consumer transactions. The main purpose of the uniform act is to supplement traditional consumer remedies by simplifying, clarifying, and modernizing the laws governing consumer sales practices.

One example of a uniform act is the Ohio Consumer Sales Practices Act which is modeled after the Uniform Consumer Sales Practices Act. One portion of the Ohio Act creates liability for unfair and deceptive acts when a supplier fails to honor its express and implied product warranties [104]. It is also an unfair and deceptive act to make a statement in connection with a consumer transaction that could create a false warranty in the mind of a reasonable consumer as to any characteristic of the goods.

The contents of the Ohio act reveals some of the advantages of this type of consumer protection statute. The Ohio act offers breach of warranty remedies that are unavailable under the UCC or the Magnuson–Moss Warranty Act. For example, a new car purchaser who enters into a contract with a limitation of remedies clause and seeks to recover under the UCC must establish that the remedies limitation either causes the contract to fail of its essential purpose or is unconscionable. Revocation of acceptance under the UCC is only possible if the condition of the vehicle has not substantially changed from the date of purchase.

The Ohio act gives consumers two basic remedies—rescission or damages. If the damages option is elected, consumers may recover either three times the amount of their actual damages (treble actual damages) or $200, whichever is greater. The treble actual damages provision encourages compliance with the Ohio act by making consumer recoveries lucrative and penalizing suppliers for their unfair, deceptive, or unconscionable acts. Consumers may also find it easier to obtain rescission because these acts are not limited by the common-law or UCC rules regarding disclaimers, privity, limitation of remedies, or bad faith.

Equitable remedies are also available. Consumers may seek a declaratory judgment, an injunction, or revocation/suspension of the supplier's license or permit. These additional remedies give consumers some bargaining power in settlement discussions.

The Ohio act does have disadvantages. Not every state has a consumer sales practices act. In the states that have adopted such a statute, there is a low level of public awareness of the substantive provisions and remedies that these acts provide. As a result, many consumers fail to seek redress under the Ohio act. Because the Ohio act offers a variety of remedies and is not hampered by common-law privity, disclaimer, limitation of remedy, or bad faith restrictions, an allegation of unfair or deceptive conduct should always be asserted by a consumer in their breach of warranty complaint.

5.3.5 The Statute of Limitations in Warranty Cases

Warranty actions may only be brought within a specified time period known as a statute of limitation. The UCC has a 4-year statute of limitations. UCC Section 2-725 provides that the buyer has 4 years from the date of the breach of warranty to bring an action. The warranty is considered to be breached when delivery is made, unless the warranty extends to the future performance of the goods. If the warranty extends to future performance, then the breach occurs at the time of such future performance. Most warranties do not contain promises as to the future performance of goods. In most instances, lawsuits filed more than 4 years after delivery of the goods are time-barred. A promise to repair or replace defective goods is a common warranty term. Most courts have held that a promise to repair or replace does not make the warranty extend to future performance.

The UCC warranty principles apply to the Magnuson–Moss Warranty Act. The act prohibits disclaimers of implied warranties when a written warranty is given. Implied warranties may be limited to the duration of a limited warranty provided that such limitation is reasonable. Most commentators feel that the act's reference to the duration of the implied warranties refers to the UCC statute of limitations on actions for breach of warranty [105].

5.4 CONCLUSION

5.4.1 Significance and Uniformity of Warranty Litigation

To date, there is still not a large body of federal case law interpreting the meaning of federal warranty law under the Magnuson–Moss Warranty Act. The paucity of federal judicial decisions is largely due to the fact that few warranty lawsuits meet the onerous federal jurisdictional requirements. As a result, judicial interpretation of the federal warranty statute has become a matter of state court interpretation as reflected in many diverse and individual state court cases. It is still too early to establish a uniform federal or state judicial interpretation of the Act.

One commentator argues that in attempting to determine the meaning of the act, it would be more accurate to study the wording and purpose of the act's statutory provisions rather than the isolated decisions or several similar decisions of one or several state courts [106]. This commentator notes: "The decisions of individual state courts, however, are ordinarily not controlling, because they represent only one voice of more than fifty" [107]. As a general rule, the state courts should follow the federal courts in the interpretation of federal statutes, not vice versa.

There is more uniformity within the states in their interpretation of their own state-created consumer protection statutes—their Uniform Commercial Codes, lemon laws, and Uniform Consumer Sales Practices

Acts—as they can determine the statutory meaning for themselves through their own judicial precedent.

5.4.2 Future Implications of Warranty Litigation for Consumers

Chapter 4 (Section 4.4.2) outlines some of the major implications of warranty legislation in relationship to consumers. In recent years, legal liability arising from warranty litigation has become an integral component of business transactions among consumers, retailers, and manufacturers. As warranty concepts have been developed, they have become an integral component in the sale of consumer goods. There is a growing body of warranty lawsuits brought by consumers against warrantors. A variety of statutes, including the Magnuson–Moss Warranty Act, the Uniform Commercial Code, lemon laws, and other state consumer protection statutes have codified certain aspects of warranty law. Now it is the warrantor, rather than the consumer, who has the duty to beware.

5.4.3 Future Implications of Warranty Litigation for Manufacturers and Business

Chapter 4 (Section 4.4.1) also outlines some of the major implications of warranty legislation in relationship to business. Many of these same concerns are apparent with respect to the impact of warranty litigation on manufacturers. Current federal warranty litigation indicates that manufacturers have several advantages over consumers. First, it is extremely hard to mount either an individual or class action in the federal court system due to the difficulty in attaining federal jurisdiction. Thus, manufacturers do not need to fear that they will be constantly called into federal court whenever one of their written warranties has been breached. Second, manufacturers can also reduce their chances of becoming involved in federal warranty litigation by setting up their own informal dispute settlement mechanisms which must be resorted to prior to the initiation of any federal lawsuit. Third, manufacturers also have the right to cure any defects before they are sued under federal warranty laws. However, manufacturers should be aware that federal warranty litigation can be quite costly to business because the successful federal consumer plaintiff will be able to recover their attorneys' fees and costs. In addition, in order to circumvent federal warranty litigation, manufacturers must be careful to comply with all of the presale warranty disclosure requirements set forth by the FTC.

State, rather than federal, warranty litigation appears to have a greater potential impact on business. First, it is much easier for consumers to attain state court jurisdiction over manufacturers in state warranty litigation, especially due to the modern expansion of traditional privity of contract rules to encompass even remote manufacturers. Second, under state-enacted uniform commercial codes, many state warranties are automatically implied by operation of law whenever a product is sold, regardless of whether or not a warranty has been written.

Finally, due to the current trend favoring consumer protection, manufacturers need to be more aware of the myriad of individual state statues regulating warranties in the particular states where they do business. There are many individual state warranty statutes, such as uniform commercial codes, lemon laws, and deceptive trade practices acts, which provide consumers with various diverse forms of relief and contain a variety of litigation requirements. These individual state laws coexist with the federal warranty laws and their interpretation is a matter of individual state court decision. Manufacturers need to be aware of the vast body of state consumer protection laws and local judicial trends in interpreting these laws with respect to the products that they sell to the public.

REFERENCES

1. Bixby, M. B. (1964). Judicial interpretation of the Magnuson–Moss Warranty Act, *American Business Law Journal*, **22**, 125–163.
2. *National Law Journal*, July 4, 1983, at 1, col. 1.
3. Riegert, R. A. (1990). An overview of the Magnuson–Moss Warranty Act and the successful consumer plaintiff's right to attorney's fees, *Commercial Law Journal*, **95**, 468–485.
4. Comment, Magnuson-Moss federal court class actions–Federal right without a federal forum, *Cumb. Law Review*, **11**, 133–154.
5. 15 U.S.C. §2310(d) (3) (A).
6. 15 U.S.C. §2310(d) (3) (B).
7. 15 U.S.C. §2310(d) (3) (C).
8. H. R. Rep. No. 93-1107, 93d Cong., 2d Sess. 42 (1974).
9. S. Rep. No. 93-151, 93d Cong. 1st Sess. 23 (1973).
10. 15 U.S.C. §2301.
11. H. R. Rep. No. 1107, 93d Cong., 2d Sess. 22–23 (1974), reprinted in 1974 U.S. Code Cong. & Ad. News 7702, 7703–7704.
12. Henderson v. Murphy, 1993 U.S. Dist. Lexis 13189 (Michigan 1993) and 15 U.S.C.A. §2310(d) (3).
13. Boelens v. Redman Homes, Inc., 748 F2d. 1058 (5th Cir. 1984), reh. denied on other grounds, 759 F.2d 504 (5th Cir. 1985).
14. Barnette v. Chrysler Corp., 434 F.Supp. 1167 (D. Neb. 1977).
15. Fleming v. Apollo Motor Homes, Inc., 87 F.R.D. 408 (M.D.N.C. 1980).
16. In re General Motors Engine Interchange Litigation, 594 F.2d 1106, 1133 n.44 (7th Cir. 1979), cert. denied, 444 U.S. 870 (1979).
17. Novosel v. Northway Motor Car Corp., 460 F.Supp. 541 (N.D.N.Y. 1978).

18. Schafer v. Chrysler, 544 F.Supp. 182.
19. Watts v. Volkswagen Artiengesellschaft, 488 F.Supp. 1233 (W.D. Ark. 1980).
20. Comment, supra note 4, at 137.
21. Fischer v. Wolfinbarger, 55 F.R.D. 129, 132, 433 F.2d 117, (W.D. Ky. 1971).
22. Barr v. General Motors Corp., 80 F.R.D. 136 (S.D. Ohio 1978).
23. Id. at 138.
24. GM, supra note 16, at 1114–15 nn.2-4.
25. ABA Code of Professional Responsibility (Preamble) (1978).
26. ABA Code of Professional Responsibility, Ethical Consideration 2-3.
27. Id. at 2-2.
28. 15 U.S.C. §§2310(d)-10(e) (1976).
29. Miller & Kantor (1980). Litigation under Magnuson–Moss: New opportunities in private actions, 13 U.C.C.L.J., **10**, 18–21.
30. Eisen v. Carlisle & Jacquelin, 417 U.S. 156, 173–77 (1974).
31. Safer, R. (1982), The Magnuson–Moss Act class action provisions: Consumers' remedy or an empty promise?, *The Georgetown Law Journal*, **70**, 1399, 1419.
32. Id. at 1415.
33. Miller and Kantor, supra note 29, at 18.
34. GM, supra note 16.
35. Feinstein v. Firestone Tire & Rubber Co., 535 F.Supp. 595 (S.D.N.Y. 1978).
36. Skelton v. General Motors Corp., 500 F.Supp.1181 (D.C. Ill. 1980), rev'd 660 F.2d 311 (7th Cir., 1981), cert. denied, 456 U.S. 974 (1982).
37. In Re Cadillac V8-6-4 Class Action, 461 A.2d 736 (N.J. 1983).
38. 15 U.S.C. §2310(d) (2).
39. State Farm Fire v. Miller Electric Company, 231 Ill. App. 3d 355, 596 N.E. 2d 169 (Ill. App. 1992).
40. Bixby, supra note 1, at 154.
41. Ballentine v. Rollin Homes Corp., 386 So.2d 727 (Ala. 1979).
42. Dunlap v. Jimmy GMC of Tucson Inc., 136 Ariz. 338, 666 P.2d 83 (1983).
43. Royal Lincoln–Mercury Sales, Inc. v. Wallace, 415 So.2d 1024 (Miss. 1982).
44. Ventura v. Ford Motor Co., 173 N.J. Super. 501, 414 A.2d 611 (1980), aff'd 180 N.J.Super. 45, 433 A.2d 801 (1981).
45. Riegert, supra note 3, at 478.
46. Gardner v. Nimnicht Chevolet Co., 532 So.2d 26, 27 (Fla.Dist.Ct.App. 1988).
47. Vieweg v. Ford Motor Co., 173 Ill.App.3d 471, 526 N.E. 2d 364, 368 (1988).
48. Maserati Autos, Inc. v. Caplan, 522 So.2d 993, 997 (Fla.Dist.Ct.App. 1988).
49. Hibbs v. Jeep Corp., 666 S.W.2d 792, 799 (Mo.Ct.App.), cert. denied, 469 U.S. 853 (1984).
50. Trost v. Porreco Motors, 443 A.2d 1179 (Pa.Super. 1982).
51. Hibbs v. Jeep Corp., 666 S.W.2d 792 (Mo.App. 1984).
52. Hanks v. Pandolfo, 38 Conn. Supp. 447, 450 A.2d 1167 (1982).
53. Id. at 1170.

54. GMAC v. Jankowitz, 230 N.J.Super 555, 553 A.2d 1380 (1989).
55. Troutman v. Pierce, 402 N.W.2d 920 (N.D. 1987).
56. Id. at 925.
57. GMC v. Jankowitz, 555 A.2d 1380, 1384 (N.J.Super. A.D. 1989).
58. Croft v. P&W Foreign Car Service, 557 A.2d 18 (Pa. Super. 1989).
59. 15 U.S.C. §2301(3).
60. 15 U.S.C. §2310(1).
61. Magnuson–Moss Warranty Act: Implementation Enforcement Policy, 40 *Federal Register*, **40**, 25,721, 25,722 (1975).
62. H.R. Rep. No. 1107, 93d Cong., 2d Sess. 4, reprinted in 1974 U.S.C.C.A.N. 7702, 7705–06.
63. Ismael v. Goodman Toyota, 106 N.C. App. 421, 417 S.E. 2d. 290 (N.C. App. 1992).
64. Kemp v. Pfizer, 835 F.Supp. 1015 (E.D. Mich. 1993).
65. Reynolds v. S&D Foods, 1993 U.S. Dist. Lexis 3040.
66. Muchisky v. Frederic Roofing Company, 838 S.W.2d 74 (Mo. App. 1992).
67. 15 U.S.C. §2310(d) (1).
68. 15 U.S.C. §2301 (6).
69. GM, supra note 16.
70. Section 2301 (10).
71. H.R. Rep. No. 1107, 93d Cong., 2d Sess. 6, reprinted in 1974 U.S. Code Cong. & Ad. News 7702, 7707.
72. Walsh v. Ford Motor Co., 627 F.Supp. 1519 (DC Dist. Col., 1986).
73. Id. at 1238.
74. 16 C.F.R. Pt.703.5(j).
75. 16 C.F.R. Pt.703.5(g) (1).
76. Riegert, supra note 3, at 477.
77. Id. at 474.
78. Tucker v. Aqua Yacht Harbor Corp., 749 F.Supp. 142 (N.D. Miss., 1990).
79. Champion Ford Sales, Inc. v. Levine, 433 A.2d 1218 (1981, Md.).
80. Harmon v. Concord Volkswagen, Inc., 598 A.2d 696 (Del. Super. 1991).
81. Skelton v. General Motors Corp., 500 F.Supp. 1181 (D.C.Ill. 1980), rev'd 660 F.2d 311 (7th Cir., 1981). cert. denied, 456 U.S. 974 (1982).
82. Lytle v. Roto Lincoln Mercury, 167 Ill.App. 3d 508, 521 N.E.2d 201 (Ill.App., 1988).
83. Carr v. Hashman, No. 88AP-165 (Ohio App. July 28, 1988) (WESTLAW 79078).
84. Marine Midland Bank, N.A. v. Carroll, 471 N.Y.S.2d 409, (App. Div. 1984).
85. 15 U.S.C. §2302(b) (1) (A).
86. In re Bob Rice Ford, Inc., 96 F.T.C. 18 (1980).
87. Montgomery Ward & Co., Inc. v. FTC, 691 F.2d 1322 (9th Cir. 1982).
88. Szajna v. General Motors Corporation, 115 Ill.2d 294, 503 N.E.2d 760 (1986).
89. Abraham v. Volkswagen of America, Inc., 795 F.2d 238 (CA 2d 1986).
90. 15 U.S.C. §§2304(a) (2) and 2308(b).

91. Devience, supra footnote 21, at 336–337.
92. Quoting 15 U.S.C. §2311(b) (1).
93. Norman Gershman's v. Mercedes-Benz, 558 A.2d 1066 (Del.Super. 1989).
94. 17 Am.Jur.2d.
95. 39 Am.Jur.Trials 17.
96. Chrysler Corp. v. Texas Motor Vehicle Comm'n, 755 F.2d 1192 (5th Cir., 1985); State by Abrams v. Ford Motor Co., 74 N.Y.2d 495, 549 N.Y.S.2d 368, 548 N.E.2d 906 (1989); Ford Motor Co. v. Barrett, 800 P.2d 367 (1990); Motor Vehicle Mfrs. Ass'n v. O'Neill, 212 Conn. 83, 561 A.2d 917 (1989); Muzzy v. Chevrolet Div., GMC, 153 Vt 179, 571 A.2d 609 (1989), Motor Vehicles Mfrs. Ass'n v. State, 75 N.Y.2d 175, 551 N.Y.S.2d 470, 550 N.E.2d 919 (1990).
97. Texas RS Art 4413 (36) §6.07 (b).
98. NY CLS Gen Bus L §198-a (b).
99. Texas RS Art 4413 (36) §6.07 (d).
100. Gold, *Massachusetts Lemon Law Arbitration Program*: *1987 Report*, Arb. J., Sept. 1988, note 55, at 53.
101. Clark, Jr., D. L. (1991), Alternative dispute resolution under Ohio's lemon laws: A critical analysis, *Journal on Dispute Resolution*, **6**(2), 333.
102. Harmon v. Concord Volkswagen, Inc., 598 A.2d 696 (Del.Super. 1991).
103. Hanson, R. K. (1992). Lemon laws and the UCC: Advising clients with lemon cars, *The North Carolina State Bar Quarterly*, 24–27 (Summer).
104. Ohio Rev. Code Ann. §1345.02 (B).
105. Saxe and Blejwas (1976). The Federal Warranty Act: Progress and pitfalls, *N.Y.L.Sch.L.Rev.* **1**, 21–22; Smith, The Magnuson–Moss Warranty Act: Turning the tables on caveat emptor, Cal. W.L. Rev., **13**, 391, 409.
106. Riegert, supra note 3, at 479.
107. Id. at 479.

Part C

Mathematical and Statistical Techniques for Analysis of Warranty

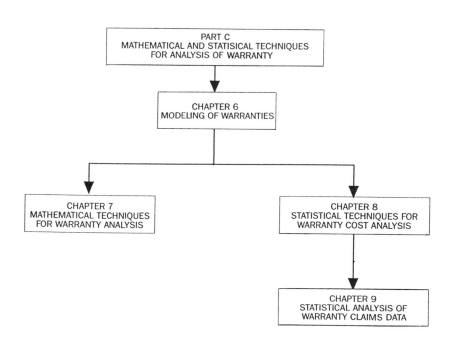

Introduction

From a cost analysis point of view, the warranty process is a complex one, involving many difficult mathematical and statistical issues. In Part C of the *Handbook* the authors present some important mathematical techniques used for modeling and analysis of warranties, as well as important statistical techniques used in the analysis of warranty data.

This part begins with a presentation by D. N. P. Murthy in Chapter 6 of some of the important tools used to model the warranty process. Beginning with the concept of a mathematical and systems approach, Murthy develops the simplified and complex representations introduced in Chapter 3 from the points of view of manufacturer, consumer, and society. Many aspects of the process are modeled, including many failure models for items under warranty, rectification actions, warranty costing, consumer behavior, manufacturer behavior, market outcome, and others.

In Chapter 7, J. J. Hunter deals with the mathematical tools needed for analysis of the models. A key element in assessing and predicting the cost of a warranty is the time to failure of the item in question. Since this is an unpredictable quantity, the major mathematical tools are probabilistic models. Hunter looks at warranty as a point process. A number of such processes are considered, including Poisson, one- and two-dimensional renewal processes, among others. Although this is a chapter on mathematics and most of the results are theoretical, attention is also given to applications; useful techniques such as approximations and numerical (computer) methods are given as well.

Chapter 8, by W. R. Blischke, is concerned with some basic statistical techniques for analysis of data in the context of warranty cost estimation. Basic probability distributions commonly used in modeling product lifetimes and associated renewal functions are presented. Methods of estimation of the parameters of these distributions, including confidence intervals, are given. The use of these results in estimation of renewal functions and their application in warranty analysis are discussed. Many numerical examples are included.

The final topic in this part, analysis of warranty claims data, is dealt with by J. Kalbfliesch and J. F. Lawless in Chapter 9. Most manufacturers collect large amounts of claims data of various types. These data are used to monitor warranty costs, to predict future costs, and often for many other purposes. The authors present some broadly applicable statistical techniques for analysis of such data. They show how to adjust for reporting delays, how to adjust for uncertainty in actual sales volume, and how to deal with calendar time effects. Some methods for estimation and prediction of warranty costs and of operational product reliability are also given.

6

Modeling for the Study of Warranties

D. N. Prabhakar Murthy

The University of Queensland
Brisbane, Queensland, Australia

6.1 INTRODUCTION

The systems approach [1] involves the following four steps:

Step 1: System characterization
Step 2: Mathematical model building
Step 3: Analysis of the mathematical model
Step 4: Interpretation of the analysis

Alternate systems characterizations for the study of warranties were discussed in Chapter 3, Section 3.3. If the aim is to study the warranty process in a limited qualitative sense, then one only needs to execute Step 1. However, if the aim is to obtain quantitative insights into the warranty process, then one needs to execute all four steps. In this chapter, attention is focused on Step 2, that is, building of mathematical models for studying the different aspects of warranty. The outline of the chapter is as follows: In Section 6.2, the model building process is discussed briefly. This essentially involves translating the system characterization into a mathematical description. As a result, the model depends on the system characterization, and different models will result based on different system characterizations. This chapter deals with model building based on the three characterizations given in Chapter 3, namely, characterizations from manufacturer, consumer, and public policy perspectives. Section 6.3 examines the model building for study from consumer and manufacturer perspectives and based on the simplified characterization. Here, the models are used to evaluate costs to manufacturer and consumer. Section 6.4 considers the modeling issues for study from the public policy perspective. Finally, Section 6.5 deals with the modeling of a few specific issues involving elements from the detailed system characterization discussed in Chapter 3.

6.2 MATHEMATICAL MODEL BUILDING

The model building process is a three-step iterative process. The three steps are as follows:

Step 1: Selection of mathematical formulation
Step 2: Estimation of model parameters
Step 3: Validation of model

Each of these is discussed briefly in this section and the remainder of the chapter deals mainly with Step 1.

6.2.1 Selection of Mathematical Formulation

As stated in Chapter 3, the system characterization is a simplified representation of the real world relevant to the study of interest. It involves identifying the variables and relationships among variables that are of importance.

A mathematical model is a translation of a system characterization into a mathematical description suitable for analysis. In a sense, it can be viewed as linking the variables and relationships of an abstract mathematical formulation to the physical variables of a system characterization and their relationships (or vice versa). Without such a linking, the abstract mathematical formulation makes no sense outside mathematics and is of no use for the study.

The type of mathematical formulation appropriate for model building depends on the system characterization. If time plays no part, then a static formulation suffices. On the other hand, if the variables of the system characterization change with time, then dynamic formulations are needed. If the variables of the system characterization change only at set time instants, then the independent variable of the formulation (representing time) assumes only discrete values and, hence, the modeling would involve discrete time formulations (e.g., time series). In contrast, if changes can occur over a time continuum, then one would need formulations with continuous time formulations (e.g., differential equations) for modeling the process. Finally, depending on whether the variables of the system characterization change in a deterministic or uncertain manner, one would need either a deterministic framework or a stochastic framework for modeling.

The complexity of the model depends on the framework (deterministic or stochastic) and the degree of detail incorporated into the system characterization. In general, deterministic models are simpler than their stochastic counterparts.

6.2.2 Estimation of Model Parameters

Model parameters refer to the parameters (or coefficients) of the underlying mathematical formulation used in the modeling. Often, the parameters are related to physical parameters of the system characterization. In order to carry out any quantitative study, one needs to assign specific values to model parameters. Often this is done using historical data from the real world. The process of assigning values to parameters based on such data is called parameter estimation, and many different methods have been developed for this purpose. The appropriateness of a particular method depends on the nature of data available. This topic is discussed in Chapters 8 and 9.

6.2.3 Validation of Model

For a mathematical model to be of use in the study, it is essential to establish that it represents the real world relevant to the study in an adequate manner. A model which is satisfactory for the study is defined as an adequate model. Not every model built is an adequate model. A model may be inadequate because (1) the system characterization is inadequate due to oversimplification, (2) the mathematical formulation used in the modeling is inappropriate, or (3) the parameters have been assigned wrong values. It is necessary to validate a model before any analysis is performed. The process is called validation and requires a testing procedure in which model behavior is compared with the real world to ensure that they are in agreement according to some specified criterion.

6.2.4 Data-Related Issues

For both model parameter estimation and model validation, one needs real data. Often the data available are not the most suitable for the problem at hand (e.g., having been collected for reasons very different from the current objective). Furthermore, data can be corrupted due to various causes—improper sampling, data missing, recording inaccuracies, and so on. A variety of methods have been developed for both parameter estimation and model validation for different types of models and data structures. Some of these issues are discussed in Chapters 8 and 9.

6.2.5 Iterative Process

The building of an adequate model involves, in general, an iterative process. If the model built is inadequate, one needs to repeat the sequence using either a modified system characterization and/or a different mathematical formulation.

6.3 MODELING BASED ON A SIMPLIFIED SYSTEM CHARACTERIZATION (CONSUMER AND MANUFACTURER PERSPECTIVES)

The simplified system characterization described in Chapter 3 is suited for cost analysis for which the interest is in the warranty cost per item over the warranty period or the cost per item over the product life cycle from consumer and manufacturer perspectives.

Here, the appropriate framework for system characterization is a stochastic framework with time treated as a continuum. Because warranty claims result from random item failures, changes to variables occur at points along the time continuum in a random manner. We need to differentiate the first failure from subsequent failures because the latter may depend on the type of rectification action used after the first failure. Modeling of the first failure can be done using basic probability theory, whereas the modeling of subsequent failures requires formulations belonging to the theory of stochastic processes—in particular, the theory of point processes. These are discussed in Chapter 7.

In order to carry out a cost analysis, it is necessary to model many other issues as well. Some of the more important of these are alternative rectification actions, rectification costs, and warranty service strategies (repair versus replace).

6.3.1 Modeling First Failures (1-D Formulation)

The case of a single-component system is considered first, and later the multicomponent case is discussed. Often, by viewing the multicomponent system as a single entity, one can model it as if it is a single-component system.

Let X_1 denote the age of an item at its first failure. (This is also called the time to first failure.) One can model X_1 in two different ways. The first, called "black-box" modeling, models X_1 directly as a random variable with a distribution function based on the modeler's intuitive judgment or on historical data. The second, called "physically based" modeling, models the physical mechanism of the item failure and then derives the distribution function for X_1 from that model. Thus, the black-box approach is based on a simple system characterization where an item is characterized through two states—working or failed—and the physically based model involves a more detailed system characterization of the physics of the failure.

Black-Box Approach

Let $F(x)$ [or more completely, $F(x;\theta)$ with θ the parameter] denote the distribution function for the first time to failure, that is,

$$F(x) = P\{X_1 \leq x\}$$

The probability that the first failure does not occur prior to x is given by

$$\overline{F}(x) = 1 - F(x) = P\{X_1 > x\}$$

If $F(x)$ is differentiable almost everywhere, then $f(x) = dF(x)/dx$ exists almost everywhere and is called the density function associated with $F(x)$. Another important associated concept is the failure rate $r(x)$ [associated with a distribution function $F(x)$] and defined as $r(x) = f(x)/\overline{F}(x)$. $r(x)\delta x$ is interpreted as the probability that the item will fail in $[x, x + \delta x]$ given that it has not failed prior to x. In other words, it characterizes the effect of age on item failure more explicitly than the failure distribution or density function. The failure rate $r(x)$, density function $f(x)$, and distribution $F(x)$ are related to each other as follows:

$$F(x) = 1 - \exp\left\{ -\int_0^x r(t)\, dt \right\}$$

and

$$f(x) = r(x) \exp\left\{ -\int_0^x r(t)\, dt \right\}$$

One can classify the distribution function $F(x)$ into many categories based on the failure rate $r(x)$. The three most important such categories are the following:

Definition: $F(x)$ is said to have an *increasing failure rate* (IFR) if $r(x)$ is increasing in $x \geq 0$.

Definition: $F(x)$ is said to have a *decreasing failure rate* (DFR) if $r(x)$ is decreasing in $x \geq 0$.

Definition: $F(x)$ is said to have a *constant failure rate* if $r(x)$ is constant for all $x \geq 0$.

Some additional categories are the following:

Definition: $F(x)$ is said to be *new worse than used* (NWU) if for all x, $y \geq 0$

$$\overline{F}(x + y) \geq \overline{F}(x)\overline{F}(y)$$

Definition: $F(x)$ is said to be *new better than used* (NBU) if for all x, $y \geq 0$

$$\overline{F}(x + y) \leq \overline{F}(x)\overline{F}(y)$$

Definition: $F(x)$ is said to be *new worse than used in expectation* (NWUE) if

   1. $F(x)$ has finite mean μ

   2. $\int_x^\infty \overline{F}(t)\, dt \geq \mu\overline{F}(x)$ for all $x \geq 0$

Definition: $F(x)$ is said to be *new better than used in expectation* (NBUE) if

   1. $F(x)$ has finite mean μ

   2. $\int_x^\infty \overline{F}(t)\, dt \leq \mu\overline{F}(x)$ for all $x \geq 0$

Definition: $F(x)$ is said to have *increasing failure rate average* (IFRA) if $-(1/x) \log \overline{F}(x)$ is increasing in $x \geq 0$.

Definition: $F(x)$ is said to have *decreasing failure rate average* (DFRA) if $-(1/x) \log \overline{F}(x)$ is decreasing in $x \geq 0$.

For further details and a chain of implications that link these different concepts, see Ref. 2.

Example 6.1 (Exponential Distribution)

The density function and the failure rate are given by

$$f(x) = \lambda \exp(-\lambda x), \quad 0 \leq x < \infty, \quad \lambda > 0$$

and

$$r(x) = \lambda$$

Example 6.2   (Gamma Distribution)
   The density function and the failure rate are given by

$$f(x) = f(x; \lambda, \beta) = \frac{\lambda^\beta x^{\beta-1} e^{-\lambda x}}{\Gamma(\beta)}, \quad 0 \le x < \infty, \quad \beta \text{ and } \lambda > 0$$

and

$$r(x) = \left\{ \int_x^\infty \left(\frac{t}{x}\right)^{\beta-1} e^{-\lambda(t-x)} \, dt \right\}^{-1}$$

Figure 6.1 shows $r(x)$ for different values of β. $\beta = 1$ corresponds to constant failure rate; $\beta < 1$ implies a decreasing failure rate (DFR) and $\beta > 1$ implies an increasing failure rate (IFR).

Example 6.3   (Weibull Distribution)
   The distribution function and the failure rate are as follows:

$$F(x) = F(x; \lambda, \beta) = 1 - \exp[-(\lambda x)^\beta], \quad 0 \le x < \infty, \quad \beta \text{ and } \lambda > 0$$

and

$$r(x) = \beta\lambda(\lambda x)^{\beta-1}$$

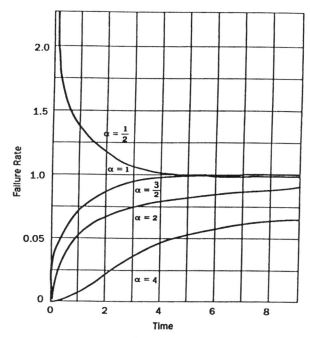

Figure 6.1   Failure rate for the Gamma distribution as a function of β.

Figure 6.2 shows $r(x)$ for different values of β. $\beta = 1$ corresponds to constant failure rate; $\beta < 1$ implies DFR and $\beta > 1$ implies IFR.

The Gamma and Weibull distributions both reduce to the exponential distribution when $\beta = 1$. They are the most commonly used distributions for characterizing time to failure in the black-box approach. In addition, many physical failure mechanisms lead to these distributions as well. Several additional failure distributions are given in Chapter 8. For a comprehensive list of distributions, see Ref. 3.

Many electrical and mechanical products exhibit a failure rate which has a "bathtub" shape (see Refs. 4 and 5). It is characterized by a decreasing failure rate from zero to some point x_1, a nearly constant failure rate over a range x_1 to x_2 and an increasing failure rate beyond x_2, as shown in Figure 6.3. Failures during the initial period are mainly due to defective material and/or poor manufacturing processes. In the case of repairable items, such failures are called "teething problems" and may often be fixed (i.e., discovered and repaired) through some form of testing program. Failures over the middle period are due purely to chance and, hence, are not influenced by age. Finally, failures over the last period reflect a true

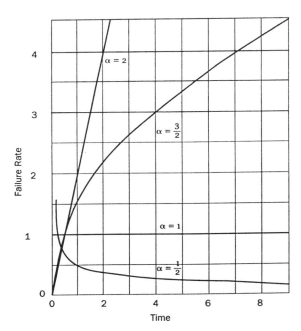

Figure 6.2   Failure rate for the Weibull distribution as a function of β.

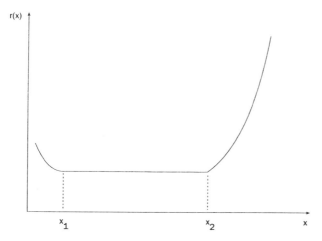

Figure 6.3 Bathtub failure rate.

aging process which results in the failure rate increasing with age. Note that in some instances x_1 can be zero and/or equal to x_2.

Physically Based Models

In the physically based modeling approach, one models explicitly the mechanism which causes item failure. See Ref. 6 for more on such models. Two illustrative examples are given below.

Example 6.4   (Shock Damage Model)

Suppose that an item is subjected to shocks which occur randomly over time. Suppose further that the magnitude of the shock is also a random variable and results in a short-duration stress on the item with the magnitude of the stress related to the intensity of the shock. Failure is assumed to occur at the first instant the stress exceeds a critical value. In other words, the item fails at the first time instant the shock magnitude exceeds some critical level. As a result, the time to failure is a random variable. The distribution function for this random variable can be obtained using a model involving a marked point process, where the point process characterizes the occurrence of shocks and the associated mark characterizes the magnitude of the shock.

A typical example of a failure phenomenon of this type is an electronic component failing due to current surges in a network.

Example 6.5 (Cumulative Damage Model)

Here the item is subjected to shocks as in the previous model. Each shock does a certain amount of damage to the item and the damage is cumulative. The item fails at the first time instant the cumulative damage exceeds some critical value. Here again, the time to failure is a random variable. The distribution function for the time to failure can be obtained using a cumulative point process.

Typical examples of item failure due to cumulative damage are crack growth in metals and tears in a conveyor belt.

Multicomponent Items

Most items and systems are made of more than one component. One approach to modeling multicomponent item failure is to model each component failure separately either as a "black box" or based on the physical mechanism causing the failure and then to relate component failures to item failure. The distribution for item failure would depend on how the components are interconnected and the effect of component failures on item failure.

When component failures are statistically dependent, one cannot model each separately. In the black-box approach, one needs to model component failures by a multidimensional distribution function and from this obtain the distribution of time to item failure. The analysis of such formulations is, in general, fairly involved and difficult.

Often when a component of a complex item or system fails, it can either induce failure of one or more components or cause damage so as to weaken them and hence accelerate their failure. Such types of failures are termed "failure interactions" [7].

Complex items and systems are built with modular structure, with a module being a collection of components. For such items, failure of a component results in a module failing [8]. Thus, one needs to relate component failure to module failure first and then module failure to item failure in order to obtain the distribution function for item failure. These models are even more complex and usually require computer analysis or simulation to obtain even approximate results.

Modeling Failure of Items Used Intermittently

Some products are used continuously—clocks, pumps in industrial operations, air conditioning in large, closed buildings, and so forth. Many other products are used intermittently or only occasionally (e.g., a dishwasher in a home, the compressor in a refrigerator, an elevator in a building, special equipment such as an emergency generator in a hospital. Here, usage and idle periods for the item alternate and the failure rate during usage will ordinarily be different from that when idle. In this case, in order to obtain the distribution function for the time to failure, it is necessary to model the usage pattern. In this case, the usage and idle periods are of random duration and can be modeled by a two-state, continuous-time Markov chain (a special type of stochastic process). For further details, see Ref. 9. When the duration of each usage is very small in relation to the time interval between usages, one can model usages as random points along a time continuum [10].

Justification for the Black-Box Approach

Although modeling items in terms of components and component failure in terms of the mechanisms of failure results in models which may be more realistic, this approach is not appropriate for study of warranty for two reasons:

1. The resulting models become extremely complex and very difficult to analyze.
2. The validation of such detailed models requires a large amount of data and, in general, it is very difficult, if not impossible, to obtain sufficient data of the types needed for model validation.

Consequently, most of warranty models for analysis are based on the black-box approach. The starting point is the characterization of the first time to failure through a distribution function selected either on an intuitive basis or the basis of on an analysis of available failure data. This is an approximation to the real world and is in the spirit of modeling, for model selection must be based on a sensible trade-off between the sometimes conflicting factors of realism, complexity, solvability, and verifiability.

6.3.2 Modeling First Failure (2-D Formulations)

As in the previous section, one can either take a "black box" or a "physically based" approach to modeling the first failure. The discussion will be confined to the former case for reasons discussed in the previous subsection.

Black-Box Approach

Let (X_1, Y_1) denote the age of an item and its usage at first failure. Because failure occurs in an uncertain manner and the item usage is also uncertain,

X_1 and Y_1 are non-negative random variables. We can model (X_1, Y_1) by a two-dimensional distribution $F(x, y)$, defined by

$$F(x, y) = P\{X_1 \leq x, Y_1 \leq y)$$

Analogous to the one-dimensional case, under appropriate assumptions on $F(x, y)$, one can characterize the age and usage at first failure by a density function $f(x, y)$ which is related to $F(x, y)$ by the relation

$$f(x, y) = \frac{\partial^2 F(x, y)}{\partial x\, \partial y}$$

or through a instantaneous failure rate $r(x, y)$ given by

$$r(x, y) = \frac{f(x, y)}{\overline{F}(x, y)}$$

where

$$\overline{F}(x, y) = P\{X_1 > x, Y_1 > y\}$$

$r(x, y)\, \delta x\, \delta y$ is essentially the probability that the first failure will occur with $(X_1, Y_1) \in [x, x + \delta x) \times [y, y + \delta y)$ given that $X_1 > x$ and $Y_1 > y$.

Alternately, one can model the time to first failure, X_1, by a one-dimensional distribution function $F_1(x)$ as discussed in Section 6.3.1. The item usage at first failure is modeled by a conditional distribution function $F_2(y|x)$ with

$$F_2(y|x) = P\{Y_1 \leq y|X_1 = x\}$$

The product of $F_1(x)$ and $F_2(y|x)$ is the two-dimensional distribution function $F(x, y)$.

As Y_1 represents the item usage at failure, it is reasonable to assume that $E(Y_1|X_1)$ is an increasing function of X_1. As a result, one must choose distribution functions $F(x, y)$ which have this property. The following are three such distribution functions.

Example 6.6   (Beta Stacey Distribution)
        The density function $f(x, y)$ for (X_1, Y_1) is given by

$$f(x, y) = \frac{bx^{\alpha b - \theta_1 - \theta_2}\, (y/\phi)^{\theta_1 - 1}\, (x - y/\phi)^{\theta_2 - 1}\, \exp(-t/a)^b}{\Gamma(\alpha)B(\theta_1, \theta_2)a^{\phi b}\phi}$$

where $x > 0$, $0 < y < \phi x$, and α, b, a, ϕ, θ_1, $\theta_2 > 0$. This is a slightly modified version of the Beta Stacey distribution proposed by Mihram and

Hultquist (see Ref. 11). The conditional expectation $E(Y_1|X_1)$ is given by

$$E[Y_1|X_1 = x] = \left(\frac{\theta_1\phi}{\theta_1 + \theta_2}\right)x$$

Example 6.7 (Multivariate Pareto Distribution)
The density function $f(x, y)$ for (X_1, Y_1) is given by

$$f(x, y) = a(a + 1) (\theta_1\theta_2)^{a+1} (\theta_2 x + \theta_1 y - \theta_1\theta_2)^{-(a+2)}$$

with $a > 1$, $x > \theta_1 > 0$, and $y > \theta_2 > 0$. The conditional expectation $E(Y_1|X_1)$ is given by

$$E[Y_1|X_1 = x] = \theta_2\{1 + x(\theta_1 a)^{-1}\}$$

Example 6.8 (Multivariate Pareto Distribution of the Second Kind)
The density function $f(x, y)$ for (X_1, Y_1) is given by

$$f(x, y) = \frac{a_1 a_2}{xy(1 - \rho^2)} \left[\left(\frac{\theta_1}{x}\right)^{a_1} \left(\frac{\theta_2}{y}\right)^{a_2}\right]^{1/(1-\rho^2)}$$

$$\times I_0\left(\frac{2\rho\sqrt{a_1 a_2}\{\log(x/\theta_1)\log(y/\theta_2)\}}{1 - \rho^2}\right)$$

with $x > \theta_1$, $y > \theta_2$, a_1 and $a_2 > 2$, $\rho^2 < 1$, and $I_0(\cdot)$ a modified Bessel function of order zero [12].
The expected usage conditioned on the time to failure, $E[Y_1|X_1]$, is given by

$$E[Y_1|X_1 = x] = \frac{a_2\theta_2}{a_2 - 1 + \rho^2} \left(\frac{x}{a_1}\right)^{a_1\rho^2/(a_1-1+\rho^2)}$$

Note that in contrast to the earlier two models, this conditional expectation is no longer a linear function of x. Different choice of parameters allows one to model a range of nonlinear relationships for the expected conditional usage.
Johnson and Kotz [11] give details of many other distributions useful for modeling failures in two dimensions.

6.3.3 Modeling Rectification Actions

Whenever a repairable item fails under warranty, the manufacturer has the option of either repairing the failed item or replacing it by a new item.

For nonrepairable items, the only option is to replace a failed item by a new one. This section is concerned with modeling various issues related to rectification actions. Only one-dimensional warranties will be considered. The extension to the two-dimensional case is straightforward.

Types of Repair

In the case of repairable items, a failed item can be made operational by subjecting it to repair. The behavior of the item after repair depends on the nature of repair carried out. In this context, one can define five types of repair actions:

1. *Repaired Good as New*: Here, after each repair, the condition of the repaired item is assumed to be as good as that of a new item. In other words, the failure distribution of repaired items is the same as that of a new item. In real life, this is seldom the case.

2. *Minimal Repair*: When a failed item is subjected to a minimal repair [13,14], the failure rate of the item after repair is the same as the failure rate of the item just before item failure.

This type of rectification model is appropriate for repair of multicomponent items where item failure occurs due to a component failure. When the failed component is replaced by a new working one, the item becomes operational. Because all other components are of age X_1, the repaired item as a whole is effectively a working item of age X_1 and hence the failure rate after repair is the same as that just before failure.

3. *Repaired Items Are Different from New* (*I*): Often when an item fails, not only are all the failed components replaced but also components which have deteriorated sufficiently. In other words, the item is subjected to a major overhaul which results in the failure distribution of all repaired items being $F_1(x)$, say, which is different from the failure distribution, $F(x)$, for new items. Because repaired items are assumed to be inferior to new ones, the mean time to failure for a repaired item is taken to be smaller than that for a new item.

4. *Repaired Items Are Different from New* (*II*): In type 3, the failure distribution for repaired item is different from that of a new item but is independent of the number of times the item has been subjected to repair. In some instances, the failure distribution of a repaired item is a function of the number of times the item has been repaired. This can be modeled by assuming that the failure distribution of an item after the jth repair ($j \geq 1$) is given by $F_j(x)$ with mean μ_j. μ_j is assumed to be a decreasing sequence in j, implying that an item repaired j times is inferior to an item repaired $j - 1$ times.

5. *Imperfect Repair*: Minimal repair implies no change to the failure rate, whereas repair action type 3 results in a predictable failure rate associated with the distribution function $F_1(x)$. Often, however, the failure rate of a repaired item after repair is uncertain. This is called "imperfect repair" and can be modeled in many different ways. Figure 6.4 shows two different imperfect repair actions: (a) corresponds to the failure rate after repair being lower than that before failure and (b) corresponds to the reverse situation. The change in the failure rate is a random variable in both cases.

Another form of imperfect repair is one where the item becomes operational with probability p after it is subjected to a repair action and continues to be in a failed state with probability $1 - p$. This implies that the item would need to be subjected to repair more than once before it becomes operational or perhaps would have to be replaced.

Rectification Duration

The time duration needed to carry out a rectification action is important in the context of warranty. When warranty terms include a penalty for downtime, it is in the manufacturer's interest to reduce this duration to the minimum possible. The duration is also of interest to buyers, as an item that is out of action deprives the buyer of its use and, in some instances, of the revenue that may be generated by the item.

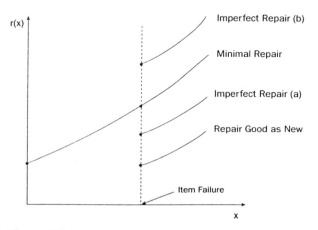

Figure 6.4 Failure rate function under different types of repair.

The total time involved in rectification action consists of the following:

1. Processing time of warranty
2. Investigation time
3. Repair/replace time
4. Testing time
5. Time to return the item to the buyer

Processing time consists of the time needed (at the retail-outlet level) for handling the claim for rectification under warranty, the time involved in transporting the failed item to the manufacturer (or workshop), and waiting time at the workshop. Investigation time is the time needed to locate the fault and decide on an appropriate action. Repair time includes the time needed to carry out the actual repair and the waiting times that can result due to lack of spares or because of other failed items awaiting rectification actions. This time is dependent on the inventory of spares and the manning of the repair facility. For the case where the failed item is replaced by a new one, the replacement time is nearly zero as long as there is a new unit in storage. Testing time is important where the success of a rectification action demands that the item be subjected to considerable testing before it is returned to the buyer.

Some of these times can be predicted precisely, whereas others (e.g., repair time) can be highly variable depending on the type of product. The easiest approach is to aggregate all the above-mentioned times into a single time called "service" time, $\tilde{X}$, modeled as a random variable with a distribution function $G(x) = P\{\tilde{X} \leq x\}$. We assume that $G(x)$ is differentiable and let $g(x) = dG(x)/dx$ denote the density function and $\overline{G}(x)$ the probability that the service time will exceed x, that is,

$$\overline{G}(x) = 1 - G(x) = P\{\tilde{X} > x\}$$

Analogous to the concept of failure rate function, we can define a "service" rate function $\rho(x)$ given by

$$\rho(x) = \frac{g(x)}{\overline{G}(x)}$$

$\rho(x)\,\delta x$ is interpreted as the probability that the service activity will be completed in $[x, x + \delta x)$, given that it has not been completed in $[0, x)$. In general, $\rho(x)$ would be a decreasing function of x, indicating that the probability of the service being completed in a short time increases with the duration of the service. In other words, $\rho(x)$ has a "decreasing service rate," a concept similar to decreasing failure rate.

If the variability in the service time is small in relation to the mean time for service, then one can approximate the service time as being deterministic. If the mean service time is very small in comparison to time between failures, then we can view service time as being nearly zero. This point of view allows a much simpler characterization of failures over time, as will be demonstrated in Chapter 7.

Repair Cost

When an item is returned under warranty, the manufacturer incurs a variety of costs. These are as follows:

1. Administration cost
2. Transportation cost
3. Repair/replacement cost—comprising material cost and labor cost
4. Transportation cost
5. Handling costs of retailer
6. Spare parts inventory costs

One can aggregate all of these costs into a single cost termed "service" cost for each warranty claim. Because some of the costs are uncertain (e.g., repair cost being dependent on the type of repair), the service cost is a random variable and needs to be modeled by a suitable distribution function. If the variability in the service cost is small, it can be treated approximately as a deterministic quantity. Often, for simplicity, these costs are aggregated, and in this case, a single variable characterizes the total repair cost.

Replace Versus Repair Decision

When a failed item is returned under warranty, the manufacturer has to make a decision whether to repair the failed item or replace it with a new one. The aim of the manufacturer is to minimize the cost of warranty servicing and hence the choice must be decided on economic considerations. This topic is discussed in Chapter 24.

6.3.4 Modeling Subsequent Failures

Thus far, the discussion has focused on the modeling of the first failure and the different rectification actions available to the manufacturer. Subsequent failures of an item often may depend on the nature of the rectification action after the first failure.

In the case of one-dimensional warranties, the time instants of item failures, and of rectified items being returned to the buyer, can be viewed as events occurring randomly along a time continuum. As such, they can be modeled by a stochastic point process, the form of which would depend on the nature of the rectification action. In Chapter 7, a variety of point process formulations and their use in modeling item failures over the warranty period are discussed.

In the case of two-dimensional warranties, the failures and the return of rectified items to the buyer can be viewed as points on a two-dimensional plane, with the horizontal axis representing time or age and the vertical axis representing item usage. The modeling would involve two-dimensional point processes and these are discussed further in Chapter 7.

6.3.5   Cost Analysis

In the context of warranty study, the following costs are of importance to both consumers and manufacturers:

1.   Expected warranty cost per unit sale
2.   Expected cost of operating a unit over its lifetime
3.   Expected cost of operation over the product life cycle

The last two costs are often called life cycle costs (LCC). As such, we shall denote them as LCC-I and LCC-II, respectively. In this section, some issues related to obtaining these cost measures are discussed.

Warranty Cost per Unit Sale

Whenever an item is returned for rectification action under warranty, the manufacturer incurs various costs as outlined in Section 6.3. These costs, too, can be random variables. The total warranty cost (i.e., the cost of servicing all warranty claims for an item over the warranty period) is thus a random sum of such individual costs, because the number of claims over the warranty period is also a random variable.

Life Cycle Costing: LCC-I

The total cost of operating a single item over its life consists of the following cost elements:

1.   Acquisition (or purchase) cost (C_A)
2.   Maintenance and repair cost of operating the item beyond the warranty period (C_M)
3.   Operating cost (energy, labor, etc.) (C_O)

4. Incidental ownership costs (C_I)
5. Disposal cost (C_S)

C_A is a fixed cost. C_M is a function of the time period beyond warranty for which the item is in use. This time period usually depends on the warranty period and will be denoted by $g(W)$. The remaining costs are functions of $W + g(W)$, the life of the unit.

Life Cycle Costing: LCC-II

This cost depends on the life cycle of the product, that is, the time interval over which consumers buy the product. After this time period, sales of the product cease, often because of the introduction of a new and better replacement. Let L denote the product life cycle. We assume that the consumer continues repeat purchases over this period. The number of repeat purchases is a random variable. The total life cycle cost is given by the product of this random variable and the life cycle cost per item, LCC-I. As a result, the total cost over the product life cycle is also random.

6.3.6 Repeat Purchase Sales

For evaluating the life cycle cost LCC-II, one needs to model repeat purchases over the product life cycle. We define a repeat purchase as any purchase, other than the first purchase, where the consumer pays the full price. As such, replacements bought at reduced price and under warranty are not considered to be repeat purchases.

The time intervals between the first purchase and the first repeat purchase, and between successive repeat purchases, are random variables. The distribution functions for these random variables are dependent on many factors: product reliability, type of warranty policy, and consumers' reactions to the product and to the warranty services offered.

In the case of nonrepairable products, a potential repeat purchase situation occurs at the first time instant the item fails outside the warranty period. Whether a purchase eventuates or not depends on factors such as advertising, competition, the general economy, buyer satisfaction with regard to the product and its performance, and the warranty service provided by the manufacturer. This uncertainty can be modeled by a binary-valued random variable assuming the value 1 (implying a repeat purchase) with probability p and 0 (implying no repeat purchase sale) with probability $1 - p$.

In the case of a repairable product, when an item fails outside warranty, the buyer has the option of either repairing it or replacing it with a new item. In this case, the probability of a repeat purchase is a function of the buyers economic evaluation of the choice between repair and replacement.

In addition, for both repairable and nonrepairable products, a buyer may sometimes decide to replace a working item by a new one. In this case, we have an additional feature, namely, the market for second-hand items. A typical example is the market for autos. When a change of ownership occurs before the warranty has expired, the original warranty terms are, in many cases, invalidated. (See Chapter 4 for relevant legislation concerning this action.)

Models based on the simplified system characterizations to study warranty costs for one-dimensional warranties can be found in Chapters 10–12 and for two-dimensional warranties in Chapters 13 and 14.

6.4   MODELING FOR STUDY FROM PUBLIC POLICY PERSPECTIVE

Models to study the effect of warranties on market behavior and the resulting social welfare implications are of importance to public policy makers for reasons discussed earlier. This section deals with modeling of issues of relevance to public policy makers.

The study of such issues involves aggregate behavior of consumers and manufacturers and the interactions between the two groups that determines the outcome of the market. The starting point is modeling of the behaviors of consumers and manufacturers. Modeling of the interaction depends on the role assigned to warranties. Market outcomes have been studied for the following four roles for warranties:

1.   Warranties as insurance
2.   Warranties as signals
3.   Warranties as incentives
4.   Warranties as marketing devices

The literature on this topic is vast. See Chapter 14 of Ref. 15 for some of the relevant references.

Finally, one needs to model the judicial side, which deals with dispute resolution, to make the model complete.

6.4.1   Modeling Consumer Behavior

If consumers are all assumed to be similar in their reaction to risk, one can model aggregate consumer behavior through a characterization of an individual's behavior. This is best done through a utility function which characterizes the behavior of the consumer to risk. If the population is heterogeneous—for example, if it consists of two groups, one risk-averse and the other risk-neutral—then one needs a different utility function for each group to characterize consumer behavior. Consumer actions consist of either buying or not buying from a set of similar products with identical or different warranty terms. For each action (except the trivial case of not buying), the outcome is random due to uncertainty in the revenue generated and/or the total cost involved. The cost depends on the failure characteristics of the product, warranty terms, and life of the product. A rational consumer chooses the action to maximize expected utility.

Further discussion on this topic can be found in Chapters 16 and 17.

6.4.2 Modeling Manufacturer Behavior

Two slightly different cases are considered, the monopoly case and the oligopoly case. In the former case, there is only one manufacturer and the behavior can be modeled through profit maximization as the underlying basis for the actions. In the latter case, there is more than one manufacturer and, here again, each is driven by profit maximization. However, one needs to differentiate two scenarios—collusion versus competition. The simplest case is one in which there is competition and all the manufacturers are similar in terms of their operations and the products produced. In this case, the only difference is in the price and/or warranty terms.

6.4.3 Modeling Market Outcome

One can best model this in a game-theoretic framework (see, for example, Ref. 16 or 17). Different scenarios lead to different market outcomes. One commonly used is the leader–follower game formulation where the manufacturer is the leader and the consumer is the follower. Here, the consumer chooses actions to maximize expected utility, given the set of manufacturer's decisions. The manufacturer chooses decision variables to maximize expected profit, taking into account the optimal actions of the consumer. As a result, the optimal actions are given by the equilibrium points of a Bertrand Stackelberg game (see Ref. 16) formulation and these characterize the outcome of the market. Based on the market outcome, one can evaluate the social welfare implications in terms of the costs and benefits to consumers and manufacturers.

The modeling effort becomes more complex when one tries to incorporate other factors, such as consumer search, quality of products, consumer maintenance effort, and so on. The modeling of each of these is discussed briefly.

The consumer search becomes important in the context of warranties as signals. To incorporate consumer search, one needs to model the search pattern. In the simplest case, one can model it by an integer-valued random variable indicating the number of retail outlets visited by a consumer during the search phase. The price and warranty terms obtained at each outlet can also be modeled in many different ways, depending on its distribution across retail outlets. The consumer then chooses the best action based on the information obtained. The information collected increases as more outlets are visited, but this needs to be traded against the time and effort involved.

The quality of a product is an important factor in the context of warranties as insurance. The reasoning behind this is that, for most products, the consumer is unable to differentiate good products from bad products before purchase and warranties serve as an insurance to protect consumers. The simplest way of modeling quality in this context is to characterize good products through a distribution function $F(t)$ and bad products with a different distribution function $G(t)$, with the mean time to failure for bad product being much smaller than that for good product. For more on this, see Chapters 23 and 29.

Consumer maintenance has a significant impact on item failures during the warranty period and may significantly affect the number of claims resulting from it. One way of modeling the effect of maintenance actions on warranty claims is to make the parameters of the failure distribution dependent on the maintenance effort such that the mean time to failure increases with greater maintenance effort.

One other factor which affects the market outcome is the information that consumers and manufacturers have. When consumers cannot evaluate the quality and manufacturers cannot evaluate the maintenance effort, then a situation for a double moral hazard exists and this can affect the market outcome in a dramatic way. As such, one needs to model the informational aspects very carefully to capture the real world of warranty in an effective manner. For more on this, see Chapter 25.

6.4.4 Modeling Dispute Resolution

A dispute arises when the manufacturer refuses to service a warranty claim. This could arise due to the claim being valid and the manufacturer not discharging his obligation or the claim being invalid (due to warranty terms being violated by the consumer) and the consumer refusing to accept the manufacturer's action not to accept the claim. This is easily modeled through three different discrete probabilities. The first characterizes the probability that a claim is valid or not. The second characterizes the probability that the manufacturer accepts or rejects the claim given that it is a valid claim. The third is similar to the second but is conditional on the claim being not valid.

Once the manufacturer refuses a claim, the consumer has the option to seek help from a third party, for example, a small-claims tribunal or the legal courts. Not every dissatisfied consumer chooses this option. One can model this by a probability which characterizes the consumer choosing further action or no action. Once the dispute proceeds to a third-party resolution, one needs a probabilistic model to characterize the outcome. This is because the third party (e.g., a judge or referee) has only limited information (and knowledge) and the actions are not only constrained by this but also by time and cost factors. See Chapters 4 and 5 for more on warranty and the law.

6.5 MODELING BASED ON A DETAILED SYSTEM CHARACTERIZATION

The detailed system characterization allows one to study many different aspects of warranty of interest to both consumer and manufacturer. There is no single model which incorporates all the factors of the detailed system characterization. Even if such a model were to be developed, it would be so complex as to be of limited use. However, a variety of models have been developed to study specific issues or topics of the warranty process and incorporating one or more of the factors of the detailed system characterization. For further details in the context of warranty cost analysis, see Chapters 10–14; in the context of warranty accounting, see Chapter 26; in the context of engineering, see Chapters 21, 23, and 24 (these cover design, manufacturing, and servicing); and in the context of marketing, see Chapter 15.

REFERENCES

1. Murthy, D. N. P., Page, N. W., and Rodin, E. Y. (1990). *Mathematical Modelling*, Pergamon Press, Oxford.
2. Barlow, R. E., and Proschan, F. (1965). *Mathematical Theory of Reliability*, John Wiley and Sons, New York.
3. Johnson, N. L., and Kotz, S. (1970). *Continuous Univariate Distributions I and II*, Houghton Mifflin, Boston.
4. Hjorth, U. (1980). A reliabilty distribution with increasing, decreasing, constant and bathtub-shaped failure rates, *Technometrics*, **22**, 99–107.
5. Glaser, R. E. (1980). Bathtub and related failure rate characterizations, *Journal of the American Statistical Association*, **75**, 667–672.
6. Gertsbakh, I. B., and Kordonsky, Kh. B. (1969). *Models of Failure*, Springer-Verlag, Berlin.
7. Murthy, D. N. P., and Nguyen, D. G. (1985). Study of two component system with failure interaction, *Naval Research Logistics Quarterly*, **32**, 239–248.

8. Murthy, D. N. P. (1983). Analysis and design of unreliable multicomponent system with modular structure, *Large Scale Systems*, **5**, 245–254.
9. Murthy, D. N. P. (1992). A usage dependent model for warranty costing, *European Journal of Operations Research*, **57**, 89–99.
10. Murthy, D. N. P. (1991). A new warranty costing model, *Computer and Mathematical Modelling*, **13**, 59–69.
11. Johnson, N. L., and Kotz, S. (1972). *Distributions in Statistics: Continuous Multivariate Distributions*, John Wiley and Sons, New York.
12. Abromowitz, M., and Stegun, I. A. (1964). *Handbook of Mathematical Functions*, Applied Mathematics Series No. 55, National Bureau of Standards, Washington, DC.
13. Murthy, D. N. P. (1991). A note on minimal repair, *IEEE Transactions on Reliability*, **R-40**, 245–246.
14. Nakagawa, T., and Kowada, M. (1983). Analysis of a system with minimal repair and its application to a replacement policy, *European Journal of Operations Research*, **12**, 176–182.
15. Blischke, W. R., and Murthy, D. N. P. (1993). *Warranty Cost Analysis*, Marcel Dekker, Inc., New York.
16. Eatwell, J., Milgate, M., and Newman, P. (eds.) (1989). *Game Theory*, W. W. Norton, New York.
17. Moorthy, K. S. (1985). Using game theory to model competition, *Journal of Marketing Research*, **22**, 262–282.

7

Mathematical Techniques for Warranty Analysis

Jeffrey J. Hunter

Massey University
Palmerston North, New Zealand

7.1 INTRODUCTION

The mathematical basis for the study of warranties lies very much in the area of applied probability modeling—the application of probabilistic methods to the solution of real-life problems. As these problems typically contain an element of randomness, such methods appear eminently suited for their solution.

We first explore ways that we can model warranty claims as a point process, a class of probabilistic models. This is followed by the classification and analysis of three applicable processes: Poisson processes and their variants, and one- and two-dimensional renewal processes.

In a final section, some more general models, alternating renewal processes and Markov renewal processes, are briefly considered.

7.2 WARRANTY CLAIMS MODELED AS POINT PROCESSES

In any applied probability modeling situation the first step is to describe the process of interest in simplistic terms to capture the essential characteristics. With this descriptive view of the overall setup as a basis, the next step is to formulate a mathematical model that reproduces and mimics the behavior of the situation. (See Chapter 6.) In those situations where uncertainty is present, it is desirable to introduce a probabilistic model in an attempt to describe the randomness present.

Let us consider a simplistic model for warranty claims. At some initial instant of time, a certain piece of equipment, subject to wear and/or breakdown, is installed. After a period of time, this item will fail and either be repaired or replaced with a similar item. This second item, which typically has similar operational characteristics to the first item, will also be subject to failure after a period of time. The process continues with successive items failing and being either repaired or replaced.

If we extract the essential modeling characteristics from this description, we see that it is possible to construct a probabilistic model for this situation in a variety of ways.

Let us describe an item failure as the occurrence of an "event." One way of formulating a probability model for the occurrence of these events would be to specify the number of events that occur in the time interval $(0, t]$, for each value of t, a *counting specification*. Another way would be to give a specification in terms of the intervals between successive events, an *interval specification*. A further procedure would be to specify the chance that events would or would not occur in any small interval $(t, t + \delta t)$, for each value of t, the *intensity specification*.

These three approaches have all been used in modeling the occurrence of events, regarded as "points" occurring along a time axis. A rich theory of *point processes* has been developed and there are many potential fields of applications; see Ref. 1. For warranty claims modeling, we need to restrict attention to certain applicable models.

In order to unify the above specifications, we introduce some notation and relationships. Let $N(t)$ represent the number of events that occur in the interval $(0, t]$, that is, the number of events that occur up to and including time t. We shall require that, for each t, $N(t)$ is a random variable. This will ensure that the point process $\{N(t), t \ge 0\}$ is a one-dimensional continuous-time point process.

Definition: $\{N(t), t \ge 0\}$ is said to be a *counting process* if

1. $N(t) \ge 0$.
2. $N(t)$ is nondecreasing, that is, if $s < t$, then $N(s) \le N(t)$.
3. $N(t)$ increases by jumps only, that is, $N(t)$ takes values in N, the non-negative integers.
4. $N(t)$ is right continuous, that is, $N(t+) = N(t)$.

With the above requirements, if $s < t$, then $N(t) - N(s)$ is the number of events in the interval $(s, t]$.

In the applications of interest, we shall also assume that events occur singly and, consequently, $N(t)$ has jumps of unit magnitude.

[It is also convenient to assume that an event occurs at the origin, at time $0-$ but that this event is not counted. This implies that $N(0) = 0$.]

Let

X_1 = time to the occurrence of the first event
X_n = time between the nth and $(n - 1)$th events, $n \geq 2$
$S_n = \sum_{i=1}^{n} X_i$ = time to the occurrence of the nth event, $n \geq 1$

so that

$$X_n = S_n - S_{n-1}, \quad n \geq 1, \text{ with } S_0 \equiv 0$$

In a probabilistic formulation, each of these quantities will be random variables. To ensure that we do not have multiple occurrences of events at any time point, we shall require that $\{X_n, n \geq 1\}$ is a family of positive random variables. We refer to X_n as an "interevent" random variable.

A key relationship connecting the counting and interval specifications is the fact that

$$\{N(t) \geq n\} \equiv \{S_n \leq t\}$$

Under the intensity specification we can make assumptions regarding the probabilities of the occurrence of events in the interval $(t, t + \delta t]$. Typically, we specify that

$$P\{N(t + \delta t) - N(t) = 1\} = \lambda(t)\delta t + o(\delta t)$$

[where the remainder term $o(\delta t)$, when divided by δt, approaches 0 as δt approaches 0 through positive values]. The function $\lambda(t)$ is called the *intensity function* of the process.

We shall tie together these specifications for the relevant applicable models in Section 7.3.

The one-dimensional point process formulations in terms of $\{N(t), t \geq 0\}$ or $\{X_n, n \geq 1\}$ will have relevance to those warranty models where decisions are based solely on the times items have been in operation. In some more general situations, warranties are sometimes expressed both in terms of time of operation and in terms of usage. For such situations, we need to examine appropriate two-dimensional point process models where events occur in the positive quadrant of the two-dimensional plane with one axis representing time and the other axis representing usage.

Not all generalizations to spatial processes are relevant, as we only wish to consider those models where both time and usage are directional. We consider an appropriate two-dimensional model in Section 7.3.3.

7.3 CLASSIFICATION OF POINT PROCESS FORMULATIONS

7.3.1 Poisson Processes

There are a variety of equivalent ways that we can specify the behavior of a Poisson process—each based on the different specifications of a point process. (See Chapter 4 of Ref. 2.) The definitions which follow apply to "stationary" Poisson processes where the intensity function $\lambda(t) = \lambda$ is constant.

Definition 1:   A counting process $\{N(t), t \geq 0\}$ is a *Poisson process* if

1. $N(0) = 0$.
2. Each jump of $N(t)$ is of unit magnitude.
3. The process has *independent increments*, that is, if for all choices $0 \leq t_1 < t_2 < \cdots < t_n$, the $n - 1$ random variables $N(t_2) - N(t_1), N(t_3) - N(t_2), \ldots, N(t_n) - N(t_{n-1})$ are independent.
4. The process has *stationary increments*, that is, the distribution function of the random variable $N(t + s) - N(t)$ does not depend on t.

Note that this definition is entirely qualitative. It is surprising that this set of axioms completely specifies the distribution of $N(t)$.

Definition 2:   The counting process $\{N(t), t \geq 0\}$ is a *Poisson process* if the sequence of interevent random variables $\{X_n, n \geq 1\}$ is a sequence of independent and identically distributed exponential random variables, that is, the X_n have a probability density function given by $f(x) = \lambda e^{-\lambda x}$.

The expected value of X_n, $E[X_n] = 1/\lambda$. We say that the Poisson process so generated with this particular parameter value has intensity, or rate, λ.

Definition 3:   The counting process $\{N(t), t \geq 0\}$ is a *Poisson process* with rate λ if

1. $N(0) = 0$.
2. The process $\{N(t), t \geq 0\}$ has independent increments.
3. $P\{N(t + \delta t) - N(t) = 1\} = \lambda \delta t + o(\delta t)$.
4. $P\{N(t + \delta t) - N(t) \geq 2\} = o(\delta t)$.

The equivalence of these three definitions can be established (see Section 7.4).

In particular, it can be shown that these definitions all imply that

$$P\{N(t) = n\} = \frac{e^{-\lambda t}(\lambda t)^n}{n!}, \quad n = 0, 1, 2, \ldots \tag{7.1}$$

showing that $N(t)$ is distributed as a Poisson random variable with mean, $E[N(t)] = \lambda t$, and variance, $\text{var}[N(t)] = \lambda t$.

Various generalizations are possible. By making changes to one or more of the axioms in the aforementioned definitions, we can widen the class of possible models. For example in Axiom 2 of Definition 3 we can permit the intensity to vary with time.

Definition: A counting process $\{N(t), t \geq 0\}$ is a *nonstationary* (or *nonhomogeneous*) *Poisson process*, with intensity $\lambda(t)$, if

1. $N(0) = 0$.
2. The process $\{N(t), t \geq 0\}$ has independent increments.
3. $P\{N(t + \delta t) - N(t) = 1\} = \lambda(t)\delta t + o(\delta t)$.
4. $P\{N(t + \delta t) - N(t) \geq 2\} = o(\delta t)$.

This leads to the result that

$$P\{N(t) = n\} = \frac{e^{-\Lambda(t)}\{\Lambda(t)\}^n}{n!}, \quad n = 0, 1, 2, \ldots$$

where $\Lambda(t) = \int_0^t \lambda(s) \, ds$, showing that $N(t)$ is distributed as a Poisson random variable with mean $E[N(t)] = \Lambda(t)$.

Another generalization arises when we permit the intensity to be a random variable Λ with a distribution $G(\cdot)$.

Definition: A counting process $\{N(t), t \geq 0\}$ is a *conditional Poisson process*, if conditional on the event $\Lambda = \lambda$, $\{N(t), t \geq 0\}$ is a (stationary) Poisson process with intensity λ.

This implies that

$$P\{N(t) = n\} = \int_0^\infty \frac{e^{-\lambda t}(\lambda t)^n}{n!} \, dG(\lambda)$$

This does not yield a Poisson process except for the trivial case when $G(\lambda)$ is the distribution function of a constant random variable.

The Poisson process is used in warranty analysis when item failures can be assumed to occur "at random." Each failed item is replaced with a new item and returned to the same environment. Should the underlying environment change over time, this can be reflected in the intensity function and a nonstationary Poisson process can be employed. With different items in different environments with failures due to chance, the conditional Poisson process may be appropriate.

7.3.2 One-Dimensional Renewal Processes

Definition 2 of a Poisson process lead to the characterization of the interevent times as independent and identically distributed exponential random variables. A natural generalization is to permit the exponential distribution to be replaced by a more general distribution.

Definition: A *renewal process* is a sequence of independent and identically distributed non-negative random variables $\{X_n, n \geq 1\}$.

$\{N(t), t \geq 0\}$ is the associated *renewal counting process*.

If $S_n = t$, we say that the nth event or nth "renewal" occurs at time t.

If each X_n has distribution function $F(x)$ and probability density function $f(x)$, then the renewal counting process associated with the case with $F(x) = 1 - e^{-\lambda x}$, $x \geq 0$, or equivalently with $f(x) = \lambda e^{-\lambda x}$, $x > 0$, is a stationary Poisson process in accordance with Definition 2 of Section 7.3.1.

Because we assume that the random variables X_n are continuous, they assume zero values with probability 0 and, consequently, with probability 1, multiple occurrences do not occur at any renewal point.

In some instances, it is desirable to permit the distribution of X_1 to be different from that of X_n ($n \geq 2$). For example, the lifetimes of repaired items which may be installed at each component breakdown may have a different characteristic behavior than that of the initial new item. The following model permits such an extension.

Definition: A *delayed renewal process* is a sequence of independent random variables $\{X_n, n \geq 1\}$ with X_1 having a distribution function $F(x)$ and each X_n, $n \geq 2$, having a distribution function $G(x)$, where $F(x)$ and $G(x)$ are not necessarily the same.

The associated counting process $\{N(t), t \geq 0\}$ is called the *delayed renewal counting process*. Alternative terms for such a renewal process are *general* or *modified* renewal process. When $F(x) = G(x)$, we refer to the renewal process as an *ordinary renewal process*.

These types of renewal processes lead to very useful classes of models which have been used extensively in the modeling of warranty times. If we relax the identical distribution or the independence assumptions, the mathematical analysis becomes very difficult and, in most cases, impractical in a modeling situation.

For free-replacement policies, when items that fail need to be replaced by new ones because the items are nonrepairable, then, under the assumption that replacement times are negligibly small, failures over time can be modeled as an ordinary renewal process. If, however, repairable items are sold with a nonrenewing free-replacement policy, all failed items are repaired and the lifetimes of these may have a different distribution $G(x)$ from that of the initial new item, $F(x)$. In this case, the "item failures" or "renewals" would be more appropriately modeled as a delayed renewal process.

7.3.3 Two-Dimensional Renewal Processes

The extension of ordinary one-dimensional renewal processes to two-dimensional processes follows in a natural manner.

Definition: A *two-dimensional renewal process* is a sequence of independent and identically distributed non-negative random variables $\{X_n, Y_n\}$.

With $S_0^{(1)} \equiv 0$, $S_n^{(1)} = \sum_{i=1}^n X_i$, $n \geq 1$, and $S_0^{(2)} \equiv 0$, $S_n^{(2)} = \sum_{i=1}^n Y_i$, $n \geq 1$, let

$$N_x^{(1)} = \max\{n: S_n^{(1)} \leq x\}$$
$$N_y^{(2)} = \max\{n: S_n^{(2)} \leq y\}$$
$$N(x, y) = \max\{n: S_n^{(1)} \leq x, S_n^{(2)} \leq y\}$$

so that $N(x, y)$ gives the number of occurrences of the events (renewals) that occur in the region $(0, x] \times (0, y]$ of $\mathbf{R}_+^2$, the positive quadrant in two dimensions.

$\{N(x, y), (x, y) \in \mathbf{R}_+^2\}$ is called the *two-dimensional renewal counting process*.

Let the common joint distribution function for the (X_n, Y_n) be given by $F(x, y)$. Thus, $\{X_n, n \geq 1\}$ is a sequence of independent and identically distributed random variables with common distribution function $F_1(x) = F(x, \infty)$, the marginal distribution of $F(x, y)$. Consequently, $\{X_n, n \geq 1\}$ is a one-dimensional renewal process. Similarly, the sequence $\{Y_n, n \geq 1\}$ is a one-dimensional renewal process with interevent distribution given by $F_2(y) = F(\infty, y)$.

As for one-dimensional renewal processes, we can define a two-dimensional modified renewal process by assuming that (X_1, Y_1) has a joint distribution function $F(x, y)$ and that (X_n, Y_n), for each $n \geq 2$, have a different joint distribution function $G(x, y)$.

A typical warranty application of two-dimensional renewal processes occurs in the situation where the manufacturer agrees to repair or provide a replacement for failed items free of charge up to a time W or up to a usage U, whichever occurs first, from the time of the initial purchase. Under the replacement option, the ordinary two-dimensional renewal process gives an appropriate model, whereas the repair option leads to an analysis utilizing the modified two-dimensional renewal process.

7.4 ANALYSIS OF POISSON PROCESSES

The standard mathematical techniques utilized in establishing the equivalence of the definitions given in Section 7.3.1 for a Poisson process is the solution of differential-difference equations for $p_n(t)$, where

$$p_n(t) = P\{N(t) = n\}, \quad n = 0, 1, 2, \ldots \quad t \geq 0$$

It can be shown (see Ref. 2) that for $n \geq 1$, $t > 0$,

$$\frac{dp_n(t)}{dt} = -\lambda p_n(t) + \lambda p_{n-1}(t)$$

and   (7.2)

$$\frac{dp_0(t)}{dt} = -\lambda p_0(t)$$

Equations (7.2) can be solved, by standard techniques, to deduce the expressions given by (7.1).

In particular, if we use Laplace transforms, with

$$\hat{p}_n(s) \equiv \int_0^\infty e^{-st} p_n(t) \, dt, \quad n = 0, 1, 2, \ldots$$

Equations (7.2) imply that

$$s\hat{p}_n(s) - \hat{p}_n(0) = -\lambda \hat{p}_n(s) + \lambda \hat{p}_{n-1}(s), \quad n \geq 1$$

with $\hat{p}_0(s) = 1/(s + \lambda)$.

An inductive argument shows that

$$\hat{p}_n(s) = \frac{\lambda^n}{(s + \lambda)^{n+1}}, \quad n \geq 0$$

which can be inverted, using tables of Laplace transforms, to give (7.1).

Observe that $M(t) \equiv E[N(t)] = \lambda t$ and that $\text{var}[N(t)] = \lambda t$.

To derive the interval specification of a Poisson process, it is easy to show that

$$P\{X_1 > x\} = P\{N(x) = 0\} = e^{-\lambda x}$$

implying $F(x) = P\{X_1 \leq x\} = 1 - e^{-\lambda x}$, $x \geq 0$, the distribution of an exponential random variable.

Note also that, for $n \geq 0$,

$$P\{X_{n+1} > x\} = P\{S_{n+1} - S_n > x\} = P\{N(S_n + x) - N(S_n)$$
$$= 0\} = P\{N(x) = 0\} = e^{-\lambda x}$$

leading to an exponential distribution for X_{n+1}.

The independence of the random variables $\{X_n, n \geq 1\}$ can also be established (see Ref. 2).

Without going into all the relevant details, we have provided sufficient background to enable the equivalence of the three definitions of a Poisson process to be verified.

The establishment of the distribution for $N(t)$ for a nonstationary Poisson process follows along similar lines to that used for a stationary Poisson process by setting up a system of differential equations. In particular, for $t \geq 0$, $s > 0$, and $n \geq 0$, it can be shown that

$$P\{N(t + s) - N(t) = n\} = \exp\{-[\Lambda(t + s) - \Lambda(t)]\}$$
$$\times \frac{[\Lambda(t + s) - \Lambda(t)]^n}{n!}$$

where $\Lambda(t) = \int_0^t \lambda(s)\, ds$.

For the conditional Poisson process,

$$E[N(t)] = \sum_{n=0}^{\infty} nP[N(t) = n] = \sum_{n=0}^{\infty} n \int_0^{\infty} e^{-\lambda t} \frac{(\lambda t)^n}{n!}\, dG(\lambda)$$

$$= \int_0^{\infty} \lambda t\, dG(\lambda) = tE[\lambda].$$

7.5 ANALYSIS OF ONE-DIMENSIONAL RENEWAL PROCESSES

Renewal processes and especially the renewal function, $M(t) \equiv E[N(t)]$, play an important role in the analysis of warranty claims. In Section 7.5.1, we give an overview of the general theory for such processes which we follow in Section 7.5.2 with discussions concerning techniques for computing, approximating, and simulating the renewal function $M(t)$. Section 7.5.3 contains some relevant related results. Key references for background results in renewal theory are Refs. 3–5.

7.5.1 General Theory

Let $F^{(n)}(x)$ be the distribution function of $S_n = \sum_{i=1}^{n} X_i$, $n \geq 1$, with $F^{(0)}(x) \equiv 1$, $x \geq 0$.

Because $N(t) \geq n$ if and only if $S_n \leq t$, we have that

$$
\begin{aligned}
P\{N(t) = n\} &= P\{N(t) \geq n\} - P\{N(t) \geq n + 1\} \\
&= P\{S_n \leq t\} - P\{S_{n+1} \leq t\} \\
&= F^{(n)}(t) - F^{(n+1)}(t), \quad n \geq 0
\end{aligned}
\tag{7.3}
$$

From Equation (7.3), we can find an expression for the expected number of renewals in $(0, t]$, $M(t)$, the *renewal function*

$$
\begin{aligned}
M(t) = E[N(t)] &= \sum_{n=0}^{\infty} nP\{N(t) = n\} \\
&= \sum_{n=1}^{\infty} n\{F^{(n)}(t) - F^{(n+1)}(t)\} = \sum_{n=1}^{\infty} F^{(n)}(t)
\end{aligned}
\tag{7.4}
$$

The above results for the distribution of $N(t)$ and the renewal function hold for ordinary and delayed renewal processes. In the more general setting, X_1 has a distribution function $F(x)$ and the X_n, $n \geq 2$, each have a distribution function $G(x)$.

Observe that $S_1 = X_1$ and, for $n \geq 2$, $S_n = S_{n-1} + X_n$, the sum of two independent non-negative random variables. Thus, for $n \geq 2$,

$$
F^{(n)}(t) = \int_0^{\infty} F^{(n-1)}(t - x) \, dG(x)
$$

with $F^{(1)}(t) = F(t)$.

Note that because the random variables are all non-negative, the actual range of integration in the above Stieltjes integral is actually $[0, t]$. Furthermore, as the random variables are (absolutely) continuous, $F^{(n)}(t)$ will have a probability density function $f^{(n)}(t)$. Thus, for $n \geq 2$,

$$
F^{(n)}(t) = \int_0^{\infty} F^{(n-1)}(t - x)g(x) \, dx \equiv F^{(n-1)} * g(t)
$$

and
$\qquad\qquad\qquad\qquad\qquad\qquad\qquad\qquad\qquad\qquad\qquad$ (7.5)

$$
f^{(n)}(t) = \int_0^{\infty} f^{(n-1)}(t - x)g(x) \, dx = f^{(n-1)} * g(t)
$$

using the convolution notation, with $f^{(1)}(t) = f(t)$.

A possible way of evaluating convolutions is to use Laplace transforms. Using integration by parts, it is easy to show that

$$
\hat{F}_n(s) \equiv \int_0^{\infty} e^{-st}F_n(t) \, dt = \frac{1}{s}\hat{f}_n(s)
\tag{7.6}
$$

Furthermore, for integrable functions which are zero for negative arguments, if

$$h(t) = f*g(t) = \int_0^t f(t - x)g(x)\, dx$$

then

$$\hat{h}(s) = \hat{f}(s)\hat{g}(s) \qquad (7.7)$$

Using induction, it is easily seen that

$$\hat{F}_n(s) = \begin{cases} \dfrac{1}{s}\,\hat{f}(s), & n = 1 \\[2mm] \dfrac{1}{s}\,\hat{f}(s)\,[\hat{g}(s)]^{n-1}, & n \geq 2 \end{cases} \qquad (7.8)$$

Let us denote $M(t)$ as the renewal function for the ordinary renewal process associated with $F(x)$, $M_d(t)$ as the renewal function associated with delayed renewal process, and $M_g(t)$ as the renewal function for the ordinary renewal process associated with $G(x)$.

Taking Laplace transforms of (7.4) and using (7.8), it is easy to see that

$$\hat{M}_d(s) = \frac{\hat{f}(s)}{s[1 - \hat{g}(s)]} \qquad (7.9)$$

and hence

$$\hat{M}(s) = \frac{\hat{f}(s)}{s[1 - \hat{f}(s)]} \qquad (7.10)$$

In special cases, expressions for $M(t)$ and $M_d(t)$ can be obtained using inverse tables of Laplace transforms (e.g., Ref. 6).

Alternatively, from (7.4) and (7.5) and reexpressing using the convolution operation,

$$M_d(t) = F^{(1)}(t) + \sum_{n=2}^{\infty} F^{(n)}(t) = F(t) + \sum_{n=2}^{\infty} F^{(n-1)}*g(t)$$

$$= F(t) + M_d*g(t)$$

yielding

$$M_d(t) = F(t) + \int_0^t M_d(t - x)g(x)\, dx \qquad (7.11)$$

For ordinary renewal processes,

$$M(t) = F(t) + \int_0^t M(t - x)f(x)\, dx \tag{7.12}$$

Equation (7.12) is known as the *Renewal Integral Equation* and is the starting point for many numerical and computational procedures (see Section 7.5.2).

Conditioning arguments can also be used to derive these integral equations. Conditioning on X_1, the time to the first renewal,

$$M_d(t) = E[N(t)] = \int_0^\infty E[N(t) \mid X_1 = x]f(x)\, dx$$

where

$$E[N(t) \mid X_1 = x] = \begin{cases} 0 & \text{if } x > t \\ 1 + M_g(t - x) & \text{if } x \leq t \end{cases}$$

because if the first event occurs at $x \leq t$, then over the interval $(x, t]$ events are generated by an ordinary renewal process associated with the distribution $G(x)$.

Thus,

$$\begin{aligned} M_d(t) &= \int_0^t [1 + M_g(t - x)]f(x)\, dx \\ &= F(t) + \int_0^t M_g(t - x)f(x)\, dx \end{aligned} \tag{7.13}$$

which, in the case of an ordinary renewal process, reduces to Equation (7.12). Equation (7.13) differs from Equation (7.11), but the equivalence can be established using Laplace transforms and the appropriate form for $\hat{M}_g(s)$ from Equation (7.10) with $F(x)$ replaced by $G(x)$.

Note that, for an ordinary renewal process, the renewal function $M(t)$ implies complete knowledge of all aspects of the renewal process. This reason is contained in Equation (7.10) which shows that $F(t)$ is immediately deducible from $M(t)$.

As a further illustration, $M(t)$ contains sufficient information to determine all the higher moments of $N(t)$. Consider the derivation of $M_2(t) \equiv E[N(t)]^2$. From Equation (7.3), it can be shown that

$$M_2(t) = \sum_{n=1}^\infty (2n - 1)F^{(n)}(t)$$

from which we can deduce, via Laplace transforms, that

$$M_2(t) = M(t) + 2\int_0^t M(t - x)\, dM(x)$$

The variance of $N(t)$ is given by

$$\text{var}[N(t)] = M_2(t) - [M(t)]^2$$

Another function of interest is the *renewal density function*, $m(t)$, given by

$$m(t) = \frac{dM(t)}{dt}$$

For ordinary renewal processes,

$$m(t) = f(t) + \int_0^t m(t - x)f(x)\, dx$$

with the Laplace transform given by

$$\hat{m}(s) = \frac{\hat{f}(s)}{1 - \hat{f}(s)}$$

The renewal density function has an interpretation akin to the intensity function of a nonstationary Poisson process. To order $o(\delta t)$, $m(t)\delta t$ is the probability that a renewal event occurs in the interval $(t, t + \delta t]$.

Two other quantities are often of interest. Consider the item currently in operation at time t. The last replacement would have occurred at time $S_{N(t)}$ and the next failure will occur at time $S_{N(t)+1}$. Thus,

$$A(t) \equiv t - S_{N(t)}$$

and

$$B(t) \equiv S_{N(t)+1} - t$$

are respectively the *age* and the *excess* (or *residual*) *life* of the item in use at time t. [$A(t)$ is sometimes referred to as the *backward recurrence time* at t and $B(t)$ as the *forward recurrence time* at t.]

It can be shown (see Ref. 2) that for ordinary renewal process

$$P\{A(t) \le x\} = \begin{cases} F(t) - \int_0^{t-x} [1 - F(t - y)]\, dM(y), & 0 \le x \le t \\ 1, & x > t \end{cases}$$

$$(7.14)$$

$$P\{B(t) \le x\} = F(t + x) - \int_0^t [1 - F(t + x - y)]\, dM(y), \quad x \ge 0$$

$$(7.15)$$

These distributions are, in general, difficult to derive explicitly.

7.5.2   The Renewal Function

The renewal function plays an important role in warranty analysis but, unfortunately, there are only a few special situations where analytical expressions can be obtained for $M(t)$. As a consequence, invariably some computational procedure is required. We survey, successively, asymptotic results, bounds, approximations, numerical integration, and simulation procedures that have been used to give useful estimates of $M(t)$. Unless otherwise specified, we list results for the renewal function associated with an ordinary renewal process.

Analytical Results

We list examples, relevant to warranty analysis, where specific mathematical expressions have been obtained for $M(t)$. Invariably, the procedures used to derive such results have been based on the Laplace transform $\hat{M}(s)$, given by Equation (7.10).

Example 7.1   (Exponential Distribution)
 If $f(t) = \lambda e^{-\lambda t}$, $t > 0$, then $\hat{f}(s) = \lambda/(\lambda + s)$, $\hat{M}(s) = \lambda/s^2$ implying that

$$M(t) = \lambda t, \quad t > 0$$

as to be expected, as the renewal process is a stationary Poisson process, with intensity λ.

Example 7.2   (Erlang Distribution with Parameters k and λ)
 If $f(t) = \lambda^k t^{k-1} e^{-\lambda t}/(k - 1)!$, $t > 0$, then $\hat{f}(s) = [\lambda/(\lambda + s)]^k$, $\hat{M}(s) = \lambda^k/s[(s + \lambda)^k - \lambda^k]$, yielding

$$M(t) = \frac{\lambda t}{k} + \frac{1}{k} \sum_{j=1}^{k-1} \left\{ \frac{\theta^j}{1 - \theta^j} \right\} \{1 - \exp[-\lambda t(1 - \theta^j)]\}$$

where $\theta = \exp(2\pi i/k)$ with $i = \sqrt{-1}$ [7].

Example 7.3   (Uniform Distribution)
 If $f(t) = 1$, $\theta \leq t \leq \theta + 1$, then

$$M(t) = \sum_{j=1}^{[t/\theta]} \sum_{i=0}^{\min(j, [t-j\theta])} \frac{(-1)^i \{t - (i + j\theta)\}^j}{i! \, (j - i)!}$$

with the convention that the first sum is zero whenever the lower summation limit exceeds the upper (i.e., $t < \theta$) [8–10]
 In the special case when $\theta = 0$ ([5], p. 385),

$$M(t) = e^t \sum_{i=0}^{m} \left(\frac{i - t}{e} \right)^i \frac{1}{i!} - 1, \quad m \leq t \leq tm + 1 \ (m = 0, 1, 2, \ldots)$$

Explicit expressions for $M(t)$ can also be derived for the cases (1) $X_1 = Z_\alpha Y^{1/\alpha}$, where Z_α has a positive stable distribution, Y is exponentially distributed, and Z_α and Y are independent [11], (2) the shifted exponential ([2], p. 288), (3) the truncated exponential [12], (4) mixed exponentials [13], and (5) renewal processes of phase type [14].

Asymptotic Results

The *Elementary Renewal Theorem* [15] states that

$$\lim_{t \to \infty} \frac{M(t)}{t} = \frac{1}{\mu_1}$$

where $\mu_1 = E[X_1] \leq \infty$ with the limit μ_1^{-1} interpreted as zero if $\mu_1 = \infty$.

This result was refined by Smith [16] who showed that if $\mu_2 = E[X_1^2] < \infty$,

$$\lim_{t \to \infty} \left(M(t) - \frac{t}{\mu_1} \right) = \frac{\mu_2}{2\mu_1^2} - 1$$

or, equivalently, because $\text{var}(X_1) = \sigma^2 = \mu_2 - \mu_1^2$, as $t \to \infty$,

$$M(t) = \frac{t}{\mu_1} + \left(\frac{\sigma^2 - \mu_1^2}{2\mu_1^2} \right) + o(1)$$

Murthy [17], generalizing Smith's result, showed that for delayed renewal processes, if $v_r = E[X_1^r]$ $(r = 1, 2, \ldots)$ and $\mu_r = E[X_2^r]$ $(r = 1, 2, \ldots)$, as $t \to \infty$,

$$M_d(t) = \frac{t}{\mu_1} + \left(\frac{\mu_2}{2\mu_1^2} - \frac{v_1}{\mu_1} \right) + o(1)$$

Blischke and Scheuer [18] showed that if we approximate $M(t)$ by its asymptotic form $M_a(t)$ where

$$M_a(t) \equiv \frac{t}{\mu_1} + \frac{\sigma^2 - \mu_1^2}{2\mu_1^2}$$

then, for selected Gamma and Weibull distributions, this provides an approximation to within 2%, and mostly well below 1%, of the renewal function estimated by simulation of the renewal processes. $M_a(t)$ is exact when the distribution is exponential.

Bounds

The observation that $M_a(t)$ closely approximates $M(t)$ has provided interest in the determination of bounds for $M(t) - t/\mu_1$ with a view to assessing the accuracy of the approximation. In particular, bounds of the form

$$c_1 \leq M(t) - \frac{t}{\mu_1} \leq c_2$$

for all t, have been obtained. The most general results available are the following.

For all renewal processes:

$$c_1 = -1 \quad ([7], \text{ p. } 53)$$

$$c_2 = \frac{\sigma^2}{\mu_1^2} \quad [19]$$

More specifically, if A is the set of all $t \geq 0$ with $\overline{F}(t) > 0$,

$$c_1 = \inf_{t \in A} \frac{\{F(t) - F_e(t)\}}{\overline{F}(t)}$$

$$c_2 = \sup_{t \in A} \frac{\{F(t) - F_e(t)\}}{\overline{F}(t)}$$

where $\overline{F}(t) \equiv 1 - F(t)$ and [20,21]

$$F_e(t) \equiv \frac{1}{\mu_1} \int_0^t \overline{F}(x) \, dx$$

Under specific restrictions on $F(x)$, further refinements to these bounds are possible. (See Section 6.3 for definitions of the relevant terms.)
If $F(x)$ is NBUE:

$$c_2 = 0 \quad ([22], \text{ p. } 171)$$

[Note that this result is also true under the unnecessarily strong assumption that $F(x)$ is IFR, (see Ref. 7, p. 54), and hence also if $F(x)$ is IFRA or NBU.]
If $F(x)$ is NWUE:

$$c_1 = 0 \quad ([22], \text{ p. } 171)$$

As above, this result is also true under the stronger assumptions that $F(x)$ is DFR and hence also if $F(x)$ is DFRA or NWU. (See Ref. 7, p. 54).
More specifically,

$$c_1 = -F_e(t) \quad \text{and} \quad c_2 = F(t) - F_e(t) \quad [23]$$

If $F(x)$ is DFR:

$$c_2 = \frac{\mu_2}{2\mu_1^2} - 1 \quad [24]$$

Approximations

Although the asymptotic nature of the approximation $M_a(t)$ and the global nature of the bounds, as given in the previous subsection, are useful, it is often desirable to be able to approximate $M(t)$ for a particular finite t value. A variety of approximations have been derived, all of them using as their basis the integral equation of renewal theory, Equation (7.12).

For the Weibull distribution, Leadbetter [25] (see also Ref. 26) proposed approximating the distribution function $F(t) = 1 - e^{-t^\beta}$, by a truncated power series, as

$$\hat{F}(t) = \sum_{r=1}^{K} \frac{(-1)^{r-1}}{\Gamma(\beta r + 1)} c_r t^{\beta r}$$

With $M(t)$ approximated as

$$M_\ell(t) = \sum_{r=1}^{K} \frac{(-1)^{r-1}}{\Gamma(\beta r + 1)} A_r t^{\beta r}$$

the coefficients can be obtained recursively by $A_1 = C_1$, $A_j = C_j - \sum_{r=1}^{j-1} A_r C_{j-r}, j \geq 2$, with $C_j = \Gamma(\beta j + 1)/j!, j \geq 1$. The accuracy of the approximation depends on the truncation parameter, K.

In order to obtain an analytic form for an approximation, Jaquette [27] approximates the renewal function $M_d(t)$ for a modified gamma renewal process with different exponential distributions for X_1 and X_2, under the assumption that rate parameters of the underlying delayed renewal process have uncertain Bayesian parameters.

A number of studies have advocated evaluating the renewal function by direct numerical solution of the approximate integral. In particular, the following three approximations to $M(t)$ have been proposed.

1. Bartholomew [28]:

$$M_b(t) = F(t) + \frac{1}{\mu_1} \int_0^t \left[\frac{F^2(x)}{F_e(x)} \right] dx$$

2. Ozbaykal [29]:

$$M_o(t) = \frac{t}{\mu_1} - F_e(t) + \frac{1}{\mu_1} \int_0^t [1 - F_e(t - x)] \, dx$$

3. Deligönül [30]:

$$M_{de}(t) = \frac{t}{\mu_1} - F_e(t) + \int_0^t \left(1 - F_e(t - x)][f(x)\right.$$

$$\left. + \frac{F^2(x)}{\mu_1 F_e(x)}\right) dx$$

Deligönül provides some numerical comparisons, involving IFR and DFR distributions, showing that $M_{de}(t)$ is superior to both $M_b(t)$ and $M_o(t)$.

Numerical Procedures

A variety of computational procedures for $M(t)$ have been considered, based either on its evaluation as a sum of n-fold convolutions [Equation (7.4)] or as the solution of the integral equation of renewal theory, or a variant thereof [Equation (7.14)].

For example, Baxter et al. [31] use the cubic splining algorithm of Cleroux and McConalogue [32] to generate piecewise continuous polynomial approximations to $F^{(n)}(t)$, and Equation (7.4), with an appropriate convergence criterion. They produce comprehensive tables for $M(t)$ for five distributions—Gamma, inverse Gaussian, Lognormal, Truncated Normal, and Weibull; see also Ref. 33. A limited version of this can be found in Ref. 34. A Fortran computer code can be found in Appendix C2 of Ref. 34.

A similar approach using numerical evaluation of convolutions was carried out by Soland [35,36] to obtain $M(t)$ for Weibull and Gamma distributions.

An alternative method was given by Deligönül and Bilgen [37] by approximating $M(t)$ by a linear combination of cubic splines and calculating the coefficients from a system of linear equations obtained via the computation of inner products with both sides of a variant of Equation (7.12).

More recently Xie [38] considered evaluating $M(t)$ based on direct numerical Riemann–Stieltjes integration. A comparison with the procedure of Deligönül and Bilgen for Gamma distributions shows the technique is very fast and very accurate, especially when the interevent distribution function, rather than just its probability density function, is known.

Kao [39] uses a randomization procedure for computing the renewal functions associated with phase-type distributions and discusses three approaches to approximating interevent distributions with phase-type distributions in order to use this computational method for determining $M(t)$.

Simulation Procedures

Simulation procedures are based on computer generation of sample paths, or time histories, of the underlying renewal process. This is achieved by using random number generators to produce typical observed values X_1, $X_2, \ldots$ from the distribution function $F(x)$. From this observed pattern, an estimator $\hat{M}(t)$ of $M(t)$ is computed. The simulation is repeated a large number of times and statistics produced to compare the accuracy and efficiency of the particular estimator $\hat{M}(t)$.

Blischke and Scheuer [18] considered the simple estimator based on the average number of renewals in $(0, t]$ for selected t values and compared the results for a variety of Gamma and Weibull distributions with the asymptotic approximation $M_a(t)$. Brown et al. [40] constructed another five estimators and Ross [41] proposed a further estimator. A numerical comparison of all 7 (unbiased) estimators was carried out by Iskandar et al. [42] based on 1000 independent simulation runs for 5 different Weibull distributions and a similar number of Gamma distributions. The simple estimator $\hat{M}_1(t) = N(t)$ is shown to be the worst, whereas the estimator $\hat{M}_2(t) = \{S_{N(t)+1}\}/\mu_1 - 1$ has the smallest standard deviation on average and is recommended as the best simulation estimator.

7.5.3 Related Limit Results

Whereas the renewal function $M(t)$ has a major role in the analysis of warranties, other properties of the random variable $N(t)$ are also important.

In particular, it is often useful to have some knowledge of the limiting distribution of $N(t)$. For any real x,

$$\lim_{t \to \infty} P \left\{ \frac{N(t) - t/\mu_1}{\sigma \sqrt{t/\mu_1^3}} \leq x \right\} = \Phi(x)$$

where $\Phi(x)$ is the distribution function of a standard normal random variable with mean zero and variance unity (see Ref. 5). Roginsky [43], correcting a result given by Ahmad [44], shows that, under certain regularity conditions, as $t \to \infty$,

$$\left| P \left\{ \frac{N(t) - t/\mu_1}{\sigma \sqrt{t/\mu_1^3}} \leq x \right\} - \Phi(x) \right| = O \left(\frac{1}{\sqrt{t}} \right).$$

Although the above result suggests that var $N(t) \sim \sigma^2 t/\mu_1^3$, Smith [45] was able to obtain a more accurate approximation. If $\mu_n = E[X_1^n]$,

$n \geq 1$, and $\mu_3 < \infty$, then as $t \to \infty$

$$\text{var } N(t) = \frac{\sigma^2 t}{\mu_1^3} + \left\{ \frac{5\mu_2^2}{4\mu_1^4} - \frac{2\mu_3}{3\mu_1^3} - \frac{\mu_2}{2\mu_1^2} \right\} + o(1)$$

For the modified renewal process, Murthy [17] derived a general expression for var $N(t)$. (See Ref. 46).

The behavior of the excess (or residual) life at time t, $B(t)$, is often of interest, especially at the time of expiration of a warranty. From Equation (7.15), it can be shown [47] that

$$E[B(t)] = \mu_1[1 + M(t)] - t$$

so that expressions, or approximations, for $M(t)$ can be used to evaluate this expected value. For example,

$$\lim_{t \to \infty} E[B(t)] = \frac{\mu_2}{2\mu_1}$$

Çinlar [2] shows that

$$\lim_{t \to \infty} P\{B(t) \leq x\} = F_e(x)$$

7.6   ANALYSIS OF TWO-DIMENSIONAL RENEWAL PROCESSES

7.6.1   General Theory

We develop a general theory for ordinary two-dimensional renewal processes as given by Definition 3 in Section 7.3.1. The extension to the modified case follows in an analogous manner to that given for one-dimensional renewal processes in Section 7.5.1. For further background results and details, the reader should refer to Ref. 48.

From the definition of $N(x, y)$, the number of renewals in $(0, x] \times (0, y]$, it is easily seen that

$$N(x, y) = \min\{N_x^{(1)}, N_y^{(2)}\} \tag{7.16}$$

As a result, for $n \geq 1$,

$$\begin{aligned}
P\{N(x, y) \geq n\} &= P\{(N_x^{(1)} \geq n) \cap (N_y^{(2)} \geq n)\} \\
&= P\{S_n^{(1)} \leq x, S_n^{(2)} \leq y\} = F^{(n)}(x, y)
\end{aligned}$$

where $F^{(n)}(x, y)$ is the joint distribution function of $(S_n^{(1)}, S_n^{(2)})$.

Consequently, for $x \geq 0$, $y \geq 0$, and $n \geq 0$,

$$P\{N(x, y) = n\} = F^{(n)}(x, y) - F^{(n+1)}(x, y)$$

with $F^{(0)}(x, y) \equiv 1$, $x \geq 0$, and $y \geq 0$.

For $n \geq 0$,

$$F^{(n+1)}(x, y) = F^{(n)} ** F(x, y)$$

where $**$ is the bivariate convolution operation given by

$$F ** G(x, y) \equiv \int_0^x \int_0^y F(x - u, y - v) \, dG(u, v)$$

for any Stieltjes integrable functions of two variables. Note that the operation is commutative and the order of integration is immaterial.

In analogy with the univariate theory, we define the *two-dimensional renewal function*, $M(x, y) \equiv E[N(x, y)]$.

It is easily seen that, for all $x \geq 0$, $y \geq 0$,

$$
\begin{aligned}
M(x, y) &= \sum_{n=0}^{\infty} P\{N(x, y) = n\} = \sum_{n=1}^{\infty} P\{N(x, y) \geq n\} \\
&= \sum_{n=1}^{\infty} F^{(n)}(x, y)
\end{aligned}
\tag{7.17}
$$

Because $F^{(n)}(x, \infty) = F_1^{(n)}(x)$, the marginal distribution of $S_n^{(1)}$, $M(x, \infty)$ is the renewal function of the one-dimensional ordinary renewal process $\{X_n, n \geq 1\}$, where each X_n has a distribution function $F_1(x) = F(x, \infty)$. Similarly, $M(\infty, y)$ is the renewal function associated with the Y renewals.

Let $\hat{F}(s_1, s_2) \equiv \int_0^{\infty} \int_0^{\infty} \exp(-s_1 x - s_2 y) F(x, y) \, dx \, dy$ be the bivariate Laplace transform of the distribution function $F(x, y)$. In an analogous way to that used for one-dimensional renewal processes,

$$\hat{F}^{(n)}(s_1, s_2) = \frac{1}{s_1 s_2} [\hat{f}(s_1, s_2)]^n, \quad n \geq 1 \tag{7.18}$$

where $\hat{f}(s_1, s_2)$ is the bivariate Laplace transform of the joint probability density function of the joint random variables (X_1, Y_1), $f(x, y)$, where

$$f(x, y) = \frac{\partial^2 F(x, y)}{\partial x \, \partial y}$$

Thus, from Equations (7.16) and (7.17), the bivariate Laplace transform of $M(x, y)$ is given by

$$\hat{M}(s_1, s_2) = \frac{\hat{f}(s_1, s_2)}{s_1 s_2 [1 - \hat{f}(s_1, s_2)]} \tag{7.19}$$

From tables of bivariate Laplace transforms (e.g., Ref. 49), expressions for $M(x, y)$ can be deduced upon inverting the expression given by Equation (7.19).

An alternative method is to consider numerical techniques for solving for $M(x, y)$ from the *integral equation of two-dimensional renewal theory*

$$M(x, y) = F(x, y) + \int_0^x \int_0^y M(x - u), (y - v) \, dF(u, v) \qquad (7.20)$$

which is derived from Equation (7.17) by observing that

$$M**F(x, y) = \sum_{n=1}^{\infty} F^{(n+1)}(x, y) = M(x, y) - F(x, y)$$

Alternatively, (7.20) can be derived by using conditional expectations. Because

$$M(x, y \mid X_1 = u, Y_1 = v) = 1 + M(x - u, y - v)$$

if $u \leq x$, $v \leq y$, and 0 otherwise.

$$M(x, y) = E[M(x, y) \mid X_1 = u, Y_1 = v]$$

$$= \int_0^x \int_0^y [1 + M(x - u, y - v)] f(u, v) \, du \, dv$$

$$= F(x, y) + \int_0^x \int_0^y M(x - u, y - v) f(u, v) \, du \, dv$$

In analogy with the univariate theory, we define the *two-dimensional renewal density*

$$m(x, y) = \frac{\partial^2 M(x, y)}{\partial x \, \partial y} = \sum_{n=1}^{\infty} f^{(n)}(x, y)$$

where $f^{(n)}(x, y)$ is the joint probability density function of $(S_n^{(1)}, S_n^{(2)})$.

From (7.20), we obtain the *two-dimensional renewal density integral equation*

$$m(x, y) = f(x, y) + \int_0^x \int_0^y m(x - u, y - v) \, f(u, v) \, du \, dv$$

which, upon taking bivariate Laplace transforms, yields

$$\hat{m}(s_1, s_2) = \frac{\hat{f}(s_1, s_2)}{1 - \hat{f}(s_1, s_2)} \qquad (7.21)$$

The univariate renewal density functions of the associated marginal renewal processes can be obtained as

$$m_1(x) = \int_0^\infty m(x, y)\, dy \quad \text{and} \quad m_2(y) = \int_0^\infty m(x, y)\, dx$$

Note also that

$$M(x, y) = \int_0^x \int_0^y m(u, v)\, du\, dv \tag{7.22}$$

We should remark that $\{N_x^{(1)}\}$ and $\{N_y^{(2)}\}$ are both renewal counting processes. Consequently, for each x and y, $(N_x^{(1)}, N_y^{(2)})$ will have a joint distribution, the properties of which are explored in Ref. 48. The behavior of $N(x, y)$ can be explored, through relationship (7.16), as the minimum of two correlated random variables. Of interest is the relationship between $N_x^{(1)}$ and $N_y^{(2)}$ and, in particular, $K(x, y) = \text{cov}(N_x^{(1)}, N_y^{(2)})$, as a measure of the dependence between the random variables. The bivariate Laplace transform for $K(x, y)$ is given by

$$\hat{K}(s_1, s_2) = \frac{\hat{f}(s_1, s_2) - \hat{f}(s_1, 0)\hat{f}(0, s_2)}{s_1 s_2 \,[1 - \hat{f}(s_1, s_2)]\, [1 - \hat{f}(s_1, 0)]\, [1 - \hat{f}(0, s_2)]}$$

$$\tag{7.23}$$

Another counting process, of relevance in warranty analysis, is

$$\tilde{N}(x, y) = \max\{N_x^{(1)}, N_y^{(2)}\}$$

which gives the number of renewals in the region formed as the union of the two strips $(0, x] \times \mathbf{R}_1^+$ and $\mathbf{R}_1^+ \times (0, y]$ in the positive quadrant $\mathbf{R}_2^+$. As $\{\tilde{N}(x, y) \geq n\} = \{N_x^{(1)} \geq n\} \cup \{N_y^{(2)} \geq n\}$,

$$P\{\tilde{N}(x, y) \geq n\} = F^{(n)}(x, \infty) + F^{(n)}(\infty, y) - F^{(n)}(x, y)$$

Also, because $\tilde{N}(x, y) = N_x^{(1)} + N_y^{(2)} - N(x, y)$, the associated renewal function for this counting process is given by

$$E[\tilde{N}(x, y)] = M(x, \infty) + M(\infty, y) - M(x, y)$$

7.6.2 The Two-Dimensional Renewal Function

The two-dimensional renewal function plays an important role in the analysis of two-dimensional warranty policies. Unfortunately, it is extremely difficult to obtain analytic expressions for $M(x, y)$ and computational procedures are generally required. Alternatively, approximations and bounds serve a useful role in any parametric study.

Analytical Results

To our knowledge, there is only one example applicable to warranty analysis that yields an analytical expression for $M(x, y)$. The following example, Downton's bivariate exponential distribution is discussed in detail in Ref. 48. Suppose (X_1, Y_1) has joint probability density function given by

$$f(x, y) = \frac{\lambda_1 \lambda_2}{1 - \rho} \exp\left[-\frac{\lambda_1 x + \lambda_2 y}{1 - \rho} \right] I_0\left[\frac{2(\rho \lambda_1 \lambda_2 xy)^{1/2}}{1 - \rho} \right]$$

where $I_n(\cdot)$ is the modified Bessel function of the first kind of nth order, (See, for example, Ref. 50 Chapter 9, for the definition of Bessel functions.)

It can be shown that X_1 and Y_1 each have marginal exponential distributions with means $\mu_1 = 1/\lambda_1$ and $\nu_1 = 1/\lambda_2$ and correlation, $\text{corr}(X_1, Y_1) = \rho$ $(0 \le \rho < 1)$.

The bivariate Laplace transform of $f(x, y)$ is given by

$$\hat{f}(s_1, s_2) = \frac{1}{(1 + \mu_1 s_1)(1 + \nu_1 s_2) - \rho \mu_1 \nu_1 s_1 s_2} \tag{7.24}$$

The joint distribution of $(S_n^{(1)}, S_n^{(2)})$ is Kibble's bivariate Gamma distribution with joint density function given by

$$f^{(n)}(x, y) = \frac{(\lambda_1 \lambda_2)^n}{(1 - \rho) \Gamma(n)} \left(\frac{xy}{\rho \lambda_1 \lambda_2} \right)^{(n-1)/2} \exp\left[-\frac{\lambda_1 x + \lambda_2 y}{1 - \rho} \right]$$

$$\times I_{n-1}\left[\frac{(\rho \lambda_1 \lambda_2 xy)^{1/2}}{1 - \rho} \right]$$

When $\rho = 0$,

$$F^{(n)}(x, y) = P_n(\lambda_1 x) P_n(\lambda_2 y) \tag{7.25}$$

where $P_n(x) = \int_0^x u^{n-1} e^{-u} du/\Gamma(n)$, the incomplete gamma function.

As $N_x^{(1)}$ and $N_y^{(2)}$ are each marginally distributed as Poisson (x/μ_1) and Poisson (y/ν_1) random variables, respectively, it is easy to see that

$$M(x, \infty) = \frac{x}{\mu_1} \quad \text{and} \quad M(\infty, y) = \frac{y}{\nu_1}$$

This would lead us to conjecture that $M(x, y)$ has a nice simple form. However, that is not the case!

Using equations (7.19) and (7.24),

$$\hat{M}(s_1, s_2) = \frac{1}{s_1 s_2 [\mu_1 s_1 + \nu_1 s_2 + (1 - \rho) \mu_1 \nu_1 s_1 s_2]} \tag{7.26}$$

A direct inversion of this bivariate Laplace transform is not easy. However, because, from (7.21) and (7.24),

$$\hat{m}(s_1, s_2) = \frac{1}{\mu_1 s_1 + \nu_1 s_2 + (1 - \rho)\mu_1 \nu_1 s_1 s_2}$$

using the tables of Voelker and Doetsch ([49] p. 209),

$$m(x, y) = \frac{1}{\mu_1 \nu_1 (1 - \rho)} \exp\left[-\frac{(x/\mu_1) + (y/\nu_1)}{1 - \rho}\right] I_0$$

$$\left[\frac{2}{1 - \rho}\left(\frac{xy}{\mu_1 \nu_1}\right)^{1/2}\right]$$

If $M_\rho(x, y)$ represents the renewal function for the two-dimensional renewal process with $\rho = corr(X_1, Y_1)$, it can be shown that

$$M_\rho(x, y) = (1 - \rho) M_0 \left(\frac{x}{1 - \rho}, \frac{y}{1 - \rho}\right)$$

From Equations (7.17) and (7.25),

$$M_0(x, y) = \sum_{n=1}^{\infty} P_n\left(\frac{x}{\mu_1}\right) P_n\left(\frac{y}{\nu_1}\right)$$

If we restrict attention to the line in the plane $x/\mu_1 = y/\nu_1 (= t$, say), we can show that

$$M_\rho(\mu_1 t, \nu_1 t) = t - t \exp\left[-\frac{2t}{1 - \rho}\right]\left[I_0\left(\frac{2t}{1 - \rho}\right) + I_1\left(\frac{2t}{1 - \rho}\right)\right]$$

$$(7.27)$$

In this particular case, the correlation between these two Poisson processes can be found, for each x and y, as

$$corr(N_x^{(1)}, N_y^{(2)}) = \rho \left(\frac{\mu_1 \nu_1}{xy}\right)^{1/2} M_\rho(x, y)$$

Asymptotic Expressions

One of the earliest limiting results in two-dimensional renewal theory, the *elementary renewal theorem for the plane* was given by Bickel and Yahav [51]. If (X_1, Y_1) are non-negative, nonarithmetic random variables with $E[X_1] = \mu_1 < \infty$ and $E[Y_1] = \nu_1 < \infty$, then

$$\lim_{t \to \infty} \frac{M(t, t)}{t} = \frac{1}{\max(\mu_1, \nu_1)}$$

From this result, Hunter [52] showed that

$$\lim_{t \to \infty} \frac{M(\mu_1 t, \nu_1 t)}{t} = 1 \tag{7.28}$$

Furthermore, under the assumption $\mathrm{var}(X_1) = \sigma_1^2 < \infty$, $\mathrm{var}(Y_1) = \sigma_2^2$, and $\mathrm{corr}(X_1, Y_1) = \rho$, Hunter [52] refined result (7.28) to show that

$$M(\mu_1 t, \nu_1 t) = t - D \left(\frac{t}{2\pi} \right)^{1/2} + o(t^{1/2}) \tag{7.29}$$

where

$$D = \left(\frac{\sigma_1}{\mu_1} \right)^2 + \left(\frac{\sigma_2}{\nu_1} \right)^2 - 2\rho \left(\frac{\sigma_1 \sigma_2}{\mu_1 \nu_1} \right)$$

In the same article [52], it was conjectured that

$$M(x, y) \sim \min \left(\frac{x}{\mu_1}, \frac{y}{\nu_1} \right) \tag{7.30}$$

even though it was known that the right-hand expression gives an upper bound for $M(x, y)$.

Utilizing the development of Abelian and Tauberian theorems for bivariate Laplace transforms, Omey and Willekens [53] were, in fact, able to refine the conjecture and establish that

$$\lim_{t \to \infty} \frac{M(tx, ty)}{t} = \min \left(\frac{x}{\mu_1}, \frac{y}{\nu_1} \right) \tag{7.31}$$

In fact, if $\mu_2 = E[X_1^2]$ and $\nu_2 = E[Y_1^2]$ and $\mu_2 + \nu_2 < \infty$, Omey and Willekens [53] refined (7.31) further to establish that

$$\lim_{t \to \infty} \left[M(tx, ty) - t \min \left(\frac{x}{\mu_1}, \frac{y}{\nu_1} \right) \right] = \begin{cases} \dfrac{\mu_2}{2\mu_1^2} & \text{if } y > \dfrac{\nu_1}{\mu_1} x \\[2mm] \dfrac{\nu_2}{2\nu_1^2} & \text{if } y < \dfrac{\nu_1}{\mu_1} x \end{cases} \tag{7.32}$$

Also, if $Y_1 \equiv (\nu_1/\mu_1) X_1$, then, for all $x, y > 0$,

$$\lim_{t \to \infty} \left[M(tx, ty) - t \min \left(\frac{x}{\mu_1}, \frac{y}{\nu_1} \right) \right] = \frac{\mu_2}{2\mu_1^2}$$

Finally, along the line of expectation, Omey and Willekens [53] give a refinement to (7.30), under the assumption $\rho \neq 1$, as $t \to \infty$,

$$M(\mu_1 t, t v_1) - t = \frac{\mu_2}{4\mu_1^2} + \frac{v_2}{4v_1^2} - D\left(\frac{t}{2\pi}\right)^{1/2} + o(t^{1/2}) \tag{7.33}$$

Results (7.32) and (7.33) give the most general approximations for $M(x, y)$, for large values of x and y, currently available.

Bounds

In Ref. 54, utilizing the Fréchet bounds on joint distributions in terms of the marginal distributions, a series of bounds for $M(x, y)$ were developed.

Let $F_1(x)$ and $F_2(y)$ be the marginal distribution functions of (X_1, Y_1) with joint distribution function $F(x, y)$. Then

$$A(x, y) \leq F(x, y) \leq B(x, y) \quad \text{for all } x, y \tag{7.34}$$

where the joint distribution functions A and B are given by

$$A(x, y) = \max\{F_1(x) + F_2(y) - 1, 0\}$$

$$B(x, y) = \min\{F_1(x), F_2(y)\}$$

From result (7.34), it can easily be seen that, for all $x, y \geq 0$,

$$A_n(x, y) \leq A^{(n)}(x, y) \leq F^{(n)}(x, y) \leq B^{(n)}(x, y) \leq B_n(x, y) \tag{7.35}$$

where

$$A_n(x, y) = \max\{F_1^{(n)}(x) + F_2^{(n)}(y) - 1, 0\}$$

$$B_n(x, y) = \min\{F_1^{(n)}(x), F_2^{(n)}(y)\}$$

with $A^{(n)}(x, y)$ and $B^{(n)}(x, y)$ as the n-fold convolutions of $A(x, y)$ and $B(x, y)$, respectively.

If $M_A(x, y)$ and $M_B(x, y)$ are the two-dimensional renewal functions associated respectively with joint distribution functions $A(x, y)$ and $B(x, y)$, then

$$M_-(x, y) \leq M_A(x, y) \leq M(x, y) \leq M_B(x, y) \leq M_+(x, y) \tag{7.36}$$

where $M_-(x, y) = \sum_{n=1}^{\infty} A_n(x, y)$ and $M_+(x, y) = \sum_{n=1}^{\infty} B_n(x, y)$.

Furthermore,

$$M_+(x, y) \leq \min(M_1(x), M_2(y))$$

where $M_1(x)$ and $M_2(y)$ are the renewal functions associated with the one-dimensional renewal process with interevent distributions $F_1(x)$ and $F_2(y)$, respectively.

In Ref. 54, conditions under which the equalities in (7.36) hold were also explored and the reader is referred to that article for derivations and further results. These bounds were evaluated in the special case of the family of bivariate Poisson processes.

An alternative set of bounds, utilizing results from one-dimensional renewal theory, were also derived in Ref. 54. In particular, let $Z_i \equiv aX_i + bY_i$, $i \geq 1$, with $a, b \geq 0$, where (X_i, Y_i), $i \geq 1$, is a two-dimensional renewal process. Define $N_z = \max\{n: aS_n^1 + bS_n^2 \leq z\}$ so that N_z is the renewal counting process for $\{Z_n\}$. If $\overline{M}(z)$ denotes the renewal function associated with N_z, then for all $x, y > 0$, and $a, b \geq 0$ $(a + b > 0)$,

$$\min(\overline{M}(ax), \overline{M}(by)) \leq M(x, y) \leq \overline{M}(ax + by)$$

Furthermore,

$$M(x, y) \leq \min_{\substack{a+b>0 \\ a\geq 0,\, b\geq 0}} \overline{M}(ax + by) \leq \min(M_1(x), M_2(y))$$

Computational Procedures

Murthy et al. [55] introduced three bivariate distribution functions that have direct relevance to two-dimensional renewal theory in warranty analysis. With X_i and Y_i referring respectively to the lifetime and usage of the ith item, any suitable joint distribution for (X_i, Y_i) would require $E(X_i \mid Y_i = y)$ to be an increasing function of y, because the greater the time between failures, the greater the usage. The following examples satisfy such a restriction. In each case, $f(x, y)$ is the joint probability density function of (X_i, Y_i), $i \geq 1$.

Example 7.4 (Beta–Stacy Distribution)

$$f(x, y) = \frac{bx^{\alpha b - \theta_1 - \theta_2}(y/\phi)^{\theta_1 - 1} (x - y/\phi)^{\theta_2 - 1} \exp(-x/a)^b}{\Gamma(\alpha)B(\theta_1, \theta_2)a^{\phi b}\phi}$$

where $x > 0$, $0 < y < \phi x$, and $\alpha, b, a, \phi, \theta_1, \theta_2 > 0$.

Example 7.5 (Bivariate Pareto Distribution)

$$f(x, y) = a(a + 1)(\theta_1 \theta_2)^{a+1}(\theta_2 x + \theta_1 y - \theta_1 \theta_2)^{-(a+2)}$$

with $x > \theta_1 > 0$, $y > \theta_2 > 0$, and $a > 1$.

Example 7.6 (Bivariate Pareto Distribution of the Second Kind)

$$f(x, y) = \frac{a_1 a_2}{xy(1 - \rho^2)} \left[\left(\frac{\theta_1}{x}\right)^{a_1} \left(\frac{\theta_2}{y}\right)^{a_2} \right]^{1/(1 - \rho^2)}$$

$$\times I_0 \left(\frac{2\rho \, \{a_1 a_2 \, \log(x/\theta_1) \, \log(y/\theta_2)\}^{1/2}}{1 - \rho^2} \right)$$

where $x > \theta_1$, $y > \theta_2$, $a_1 > 2$, $a_2 > 2$, and $\rho^2 < 1$.

Iskander [56] developed a numerical procedure based on the integral equation of two-dimensional renewal theory, Equation (7.20), using quadrature rules to evaluate the double integral numerically. His report gives details of the computer code (in FORTRAN 77) and computes the renewal function for Example 7.4 and Downton's bivariate exponential distribution. In this latter case, a comparison between the analytical solution (7.27) and the computed solution is given.

Simulation Procedures

Using techniques based on the simulation procedures developed for one-dimensional renewal processes, Iskander et al. [42] present two unbiased estimators for $M(x, y)$ and give results of simulation studies, for three different sets of parameters, for the Beta–Stacy distribution.

7.6.3 Further Limit Results

Although the renewal function $M(x, y)$ is of importance, the actual distribution of $N(x, y)$ has a role to play in assessing the accuracy of any warranty procedure. Consequently, some knowledge of the distribution of $N(x, y)$ is of interest. The only relevant results to date concern the limiting distribution of $N(x, y)$ with x and y constrained to lie on the line of expectation.

Using the result that, under a suitable limiting operation on the plane, $(N_x^{(1)}, N_y^{(2)})$ has a bivariate normal distribution, Hunter [52] derived the following result. If $\operatorname{var}(X_1) = \sigma_1^2 < \infty$, $\operatorname{var}(Y_1) = \sigma_2^2 < \infty$ and $\operatorname{corr}(X_1, Y_1) = \rho$, then, for any fixed c,

$$\lim_{t \to \infty} P \left\{ \frac{N_{\mu_1 t, \nu_1 t} - t}{\sqrt{t}} \le c \right\} = 1 - \Phi_\rho \left(\frac{-\mu_1 c}{\sigma_1}, \frac{-\nu_1 c}{\sigma_2} \right)$$

where $\Phi_\rho(x, y)$ is the joint distribution function of the bivariate normal distribution with zero means, unit variances, and correlation ρ.

Griniuviené [57] extended this result to show that, as $x \to \infty$ and $y \to \infty$,

$$P\{N_{x,y} = n\} = \frac{\mu_1}{\sigma_1 \sqrt{x}} \phi\left(\frac{x - n\mu_1}{\sigma_1 \sqrt{x}}\right) \Phi\left(\frac{y - n v_1}{\sigma_2 \sqrt{y}}\right)$$
$$+ \frac{v_1}{\sigma_2 \sqrt{y}} \phi\left(\frac{y - n v_1}{\sigma_2 \sqrt{y}}\right) \Phi\left(\frac{x - n\mu_1}{\sigma_1 \sqrt{x}}\right)$$
$$+ o\left(\frac{1}{\sqrt{x}}\right) + o\left(\frac{1}{\sqrt{y}}\right)$$

where $\phi(x)$ and $\Phi(x)$ are respectively the probability density function and distribution function of the standard univariate normal distribution.

7.7   OTHER PROCESS FORMULATIONS RELEVANT TO WARRANTY ANALYSIS

The following processes are presented to illustrate that additional features and more complex behavior in a warranty analysis situation can be modeled mathematically.

7.7.1   Alternating Renewal Processes

Suppose that when a repairable item, sold with a free-replacement policy, fails, it requires a random time, with distribution function $G(x)$, to be repaired before it is restored to new condition and returned to operation. Suppose the failure times have distribution function $F(x)$. Under the assumption that the warranty clock continues to run during repair, an evolution of the process consists of a sequence of independent random variables $X_1, Y_1, X_2, Y_2, X_3, \ldots$, where $\{X_i, i \geq 1\}$, the successive lifetimes, have a distribution function $F(x)$ and $\{Y_i, i \geq 1\}$, the successive repair times have a distribution function $G(x)$. Of key interest is the number of failures by time t. The theory for this process, an alternating renewal process is discussed in a variety of texts on stochastic processes. Refer to Refs. 2, 4, and 5.

7.7.2   Markov Renewal Processes

A Markov renewal process $\{(X_n, T_n), n \geq 1\}$ is generated very simply: The specifications require a Markov chain $\{X_n\}$, where the future "state" of the process at time $n + 1$, X_{n+1}, depends on the past evolution X_n, $X_{n-1}, \ldots, X_1, X_0$, only on the "present state" X_n, a finite state space $S = \{1, 2, \ldots, m\}$, a transition matrix $P = [p_{ij}]$, where $p_{ij} = P\{X_{n+1}$

$= j | X_n = i\}$, and a collection of distribution functions $F_{ij}(t)$. The process evolves by generating the sequence of states $\{X_n\}$ through which the process will pass. Once it is known that the process will next make a transition from state i to state j, the time that the process remains in state i until it makes this transition is determined by sampling a time interval from the distribution $F_{ij}(t)$, which depends on both of the states i and j. The transitions, or change of state, take place at the times T_n.

Observe that a renewal process is a very simple Markov renewal process with a single state, so that $S = \{1\}$ and $F_{11}(t) = F(t)$. An alternating renewal process is also a special Markov renewal process with $S = \{1, 2\}$, where state $1 \equiv$ item is operating, state $2 \equiv$ item is being repaired, $p_{12} = p_{21} = 1$, with $F_{12}(t) = F(t)$ and $F_{21}(t) = G(t)$.

In more complex warranty situations, the item may consist of a variety of different components, each subject to failure. The failure of some of the components may lead to overall system failure, whereas the failure of other items may not degrade the system as a whole. In another context, the manufacturer may wish to make a decision as to whether to repair or replace any failed item with such a decision being dependent on the failure time of the item. Situations such as these can be suitably modeled by a Markov renewal process. The theory of such processes is now well developed; see Refs. 2 and 47.

7.8 CONCLUSIONS

The results of this chapter are used in Part D of this *Handbook* for evaluating warranty costs for a variety of one- and two-dimensional warranties; in Part F in the context of engineering of the product and in Part H for cost management.

This chapter has focused on a variety of probabilistic models that have been utilized in the study of warranties. There is still, however, much scope for further development and refinement of appropriate mathematical models.

For a variety of warranty-related problems, the analytical approach is intractable. Some of these problems are discussed in Chapter 11 of Ref. 34. For such problems, the simulation approach is the most effective approach. Further relevant details can be found in the above-cited reference.

In particular, further computational techniques are required in the computation of two-dimensional renewal functions. Associated with this is also a study of appropriate two-dimensional distributions for two-dimensional warranties. The comprehensive study by Hutchinson and Lai [58] may prove to be useful.

The application of Markov renewal processes offers a rich class of stochastic models that have not been specifically used in warranty analysis, apart from special models. As models are developed in an attempt to reflect more of the actual behavior being observed, such models, because of their well-developed theory, will play an important role.

REFERENCES

1. Cox, D. R., and Isham, V. (1980). *Point Processes*, Chapman & Hall, London.
2. Çinlar, E. (1975). *Introduction to Stochastic Processes*, Prentice-Hall, Englewood Cliffs, NJ.
3. Smith, W. L. (1958). Renewal theory and its ramifications, *Journal of the Royal Statistical Society, Series B*, **20**, 243–302.
4. Cox, D. R. (1960). *Renewal Theory*, Methuen, London.
5. Feller, W. (1971). *An Introduction to Probability Theory and Its Applications, Volume 2*, 2nd ed., John Wiley & Sons, New York.
6. Erdélyi, A. (ed.) (1954). *Tables of Integral Transforms, Volume 1*, McGraw-Hill, New York.
7. Barlow, R. E., and Proschan F. (1965). *Mathematical Theory of Reliability*, John Wiley & Sons, New York.
8. Munford, A. C. (1979). The uniform renewal process, *International Journal of Mathematical Education in Science and Technology*, **10**, 361–364.
9. Russell, K. G. (1983). On the number of uniform random variables which must be added to exceed a given level, *Journal of Applied Probability*, **20**, 172–177.
10. Jensen, U. (1984). Some remarks on the renewal function of the uniform distribution, *Advances in Applied Probability*, **16**, 214–215.
11. Anderson, K. K., and Athreya, K. B. (1987). A renewal theorem in the infinite mean case, *Annals of Probability*, **15**, 388–393.
12. Stadje, W. (1987). Note on the renewal process for truncated exponential random variables, *Statistical and Probability Letters*, **6**, 61–66.
13. Uche, P. I. (1984). Models for some stochastic systems, *International Journal of Mathematical Education in Science and Technology*, **15**, 737–743.
14. Neuts, M. F. (1978). Renewal processes of phase type, *Naval Research Logistics Quarterly*, **25**, 445–454.
15. Feller, W. (1941). On the integral equation of renewal theory, *Annals of Mathematical Statistics*, **12**, 243–267.
16. Smith, W. L. (1954). Asymptotic renewal theorems, *Proceedings of the Royal Society Edinburgh A*, **64**, 9–48.
17. Murthy, V. K. (1961). *On the General Renewal Process*, Institute of Statistics Mimeo Series No 293, University of North Carolina Press, Chapel Hill, NC.
18. Blische, W. R., and Scheuer, E. M. (1981). Applications of renewal theory in analysis of the free-replacement warranty, *Naval Research Logistics Quarterly*, **28**, 193–205.

19. Lorden, G. (1970). On excess over the boundary, *Annals of Mathematical Statistics*, **41**, 520–527.
20. Marshall, K. T. (1973). Linear bounds on the renewal function, SIAM *Journal of Applied Mathematics*, **25**, 245–250.
21. Stoyan, D. (1983). *Comparison Methods for Queues and Other Stochastic Models*, D. J. Daley (ed.), John Wiley & Sons, New York.
22. Barlow, R. E., and Proschan, F. (1981). *Statistical Theory of Reliability and Life Testing*, To Begin With, Silver Spring, MD.
23. Xie, M. (1989). Some results on renewal equations, *Communications in Statistics—Theory and Methods*, **18**, 1159–1171.
24. Brown, M. (1980). Bounds, inequalities, and monotonicity properties for some specialised renewal processes, *Annals of Probability*, **8**, 227–240.
25. Leadbetter, M. R. (1963). On series expansion for the renewal moments, *Biometrika*, **50**, 75–80.
26. Smith, W. L., and Leadbetter, M. R. (1963). On the renewal function for the Weibull distribution, *Technometrics*, **5**, 393–396.
27. Jaquette, D. L. (1972). Approximations to the renewal function $m(t)$, *Operations Research*, **20**, 722–727.
28. Bartholomew, D. J. (1963). An approximate solution to the integral equation of renewal theory, *Journal of the Royal Statistical Society B*, **25**, 432–441.
29. Ozbaykal, T. (1971). *Bounds and Approximations for the Renewal Function*, M. S. Thesis, Naval Postgraduate School, Monterey, CA.
30. Deligönül, Z. S. (1985). An approximate solution of the integral equation of renewal theory, *Journal of Applied Probability*, **22**, 926–931.
31. Baxter, L. A., Scheuer, L. M., Blischke, W. R., and McConalogue, D. J. (1982). On the tabulation of the renewal function, *Technometrics*, **24**, 151–156.
32. Cleroux, R., and McConalogue, D. J. (1976). A numerical algorithm for recursively-defined convolution integrals involving distribution functions, *Management Science*, **22**, 1138–1146.
33. Baxter, L. A., Scheuer, L. M., McConalogue, D. J., and Blischke, W. R. (1981). *Renewal Tables: Tables of Functions Arising in Renewal Theory*, Technical Report, Decision Systems Dept., University of Southern California, Los Angeles, CA.
34. Blischke, W. R., and Murthy D. N. P. (1993). *Warranty Cost Analysis*, Marcel Dekker, Inc, New York.
35. Soland, R. M. (1968). A renewal theoretic approach to the estimation of future demand for replacement parts, *Operations Research*, **16**, 36–51.
36. Soland, R. M. (1969). Availability of renewal functions for Gamma and Weibull distributions with increasing hazard rate, *Operations Research*, **17**, 536–543.
37. Deligönül, Z. S., and Bilgen, S. (1984). Solution of the Volterra equation of renewal theory with Galerkin technique using cubic splines, *Journal of Statistical Computation and Simulation*, **20**, 37–45.
38. Xie, X. (1989). On the solution of renewal-type integral equations, *Communications in Statistics—Simulation and Computation* **18**, 281–293.

39. Kao, E. P. C. (1988). Computing the phase-type renewal and related functions, *Technometrics*, **30**, 87–93.
40. Brown, M., Solomon, H., and Stephens, M. A. (1981). Monte Carlo simulation of the renewal function, *Journal of Applied Probability*, **18**, 426–434.
41. Ross, S. M. (1989). Estimating the mean number of renewals by simulation, *Probability in the Engineering and Informational Sciences*, **3**, 319–321.
42. Iskander, B. P., Murthy, D. N. P., and Wilson, R. J. (1992). *Simulation Approaches to Solving One- and Two-Dimensional Renewal Integral Equations*, Preprint, Dept. of Mechanical Engineering, The University of Queensland, St. Lucia.
43. Roginsky, A. L. (1992). A central limit theorem with remainder term for renewal processes, *Advances in Applied Probability*, **24**, 267–287.
44. Ahmad, I. A. (1981). The exact order of normal approximation in bivariate renewal theory, *Advances in Applied Probability*, **13**, 113–128.
45. Smith, W. L. (1959). On the cumulants of renewal processes, *Biometrika*, **46**, 1–29.
46. Hunter, J. J. (1969). On the moments of Markov renewal processes, *Advances in Applied Probability*, **1**, 188–210.
47. Ross, S. M. (1970). *Applied Probability Models with Optimization Applications*, Holden-Day, San Francisco.
48. Hunter, J. J. (1974). Renewal theory in two-dimensions: Basic results, *Advances in Applied Probability*, **6**, 376–391.
49. Voelker, D., and Doetsch, G. (1950). *Die Zweidimensionale Laplace-transformation*, Verlag Birkhäuser, Basel.
50. Abramowitz, M. and Stegun, I. A., (1965). *Handbook of Mathematical Functions with Formulas, Graphs, and Mathematical Tables*, Dover, New York.
51. Bickel, P. L., and Yahav, J. A. (1965). Renewal theory in the plane, *Annals of Mathematical Statistics*, **36**, 946–955.
52. Hunter, J. J. (1974). Renewal theory in two-dimensions: Asymptotic results, *Advances in Applied Probability*, **6**, 546–562.
53. Omey, E., and Willekens, E. (1989). Abelian and Tauberian theorems for the Laplace transform of functions in several variables, *Journal of Multivariate Analysis*, **30**, 292–306.
54. Hunter, J. J. (1977). Renewal theory in two-dimensions: Bounds on the renewal function, *Advances in Applied Probability*, **9**, 527–541.
55. Murthy, D. N. P., Iskander, B. P., and Wilson, R. J. (1990). Two-dimensional warranty policies: A mathematical study, *Proceedings of Tenth ASOR Conference*, 89–104.
56. Iskander, B. P. (1991). *Two-Dimensional Renewal Function Solver*, Research Report No. 4/91, Department of Mechanical Engineering, The University of Queensland.
57. Griniuvienė, L. (1980). On the convergence of a renewal process (Russian), *Litovskii Matematicheskii Sbornik*, **20**, 41–49.
58. Hutchinson, T. P., and Lai, C. D. (1990). *Continuous Bivariate Distributions, Emphasising Applications*, Rumsby Scientific Publishing, Adelaide, Australia.

8

Statistical Techniques for Warranty Cost Analysis

Wallace R. Blischke

University of Southern California
Los Angeles, California

8.1 INTRODUCTION

Determining the cost of warranty is an essential aspect of the proper management of a warranty program. There are many approaches to this problem, depending on the type and amount of data and other information available at any given time and the nature of the models utilized for determining warranty costs. Prior to the introduction of a product, this cost determination is a prediction, usually based on relatively sparse and sometimes somewhat unreliable information. Later, costs can be estimated and future costs predicted on the basis of more detailed information on product performance obtained from additional testing and from data on warranty claims. Ultimately, it is usually necessary to monitor warranty costs carefully as soon as a product is introduced into the marketplace. This involves periodic updating of results as data become available over time.

    The types of information that a company may have include engineering specifications, preliminary test data, engineering judgment regarding failure rates and means of failure, detailed laboratory test data, historical data on similar products, data on product performance in test markets, mathematical cost models, and, ultimately, warranty claims data. In this chapter, we are primarily concerned with statistical information (i.e., *data*) and methods for analysis of such information. Here, the emphasis will be on relatively well-structured experimental data, although some comments on other types of data and references to methods of analysis will be included as well. Proper experimentation leads to reliable and valid data, which are essential in the analysis of warranty costs.

Claims data present some special problems; for example, they will usually be incomplete and unreliable and can seldom be considered to be a random sample, which is the theoretical basis for most standard statistical analyses. The data will be incomplete because failures outside the warranty period will rarely be reported and some failures within the warranty period will not be observed because claims are not made. They will be unreliable because of false claims, failure to report the actual time of failure accurately, and for many other reasons associated with the collection and compilation of data of this type. These problems are discussed is Chapter 9.

Analysis of warranty data, then, may be done at many levels, depending on the type of data available. It is often done primarily through summarization, using descriptive statistics such as means, standard deviations, time histories, histograms, and so forth, with little formal statistical analysis of the results. In other cases, well-developed prediction models are used and the estimated costs and predictions are updated at regular intervals as new information becomes available. This usually requires the use of considerably more sophisticated statistical methods. The purpose of this chapter is to present some of the techniques that can be used in analyses of this type.

The usual statistical approach to the analysis of data is based on certain key assumptions concerning the structure of the data and the probability distributions of the observed random variables. The most important variable in determining warranty costs is the time to failure (or lifetime) X of the product. If this quantity exceeds W, the length of the warranty period, the seller incurs no warranty cost; if $X < W$, a cost is incurred. The average amount of this cost depends on the warranty policy (free replacement, pro rata, etc.) covering the product (see Chapter 1) as well as on the probability of occurrence of the event $X < W$, written $P(X < W)$.

To determine $P(X < W)$, it is necessary to specify the probability distribution of time to failure (called both the "failure distribution" and the "life distribution") of the item. A number of important failure distributions will be given in Section 8.2. In analyzing data, it will be assumed that the form of this distribution is known. If it is not known, data can be used to select a failure distribution by "fitting" alternative distributions. (See "goodness-of-fit" in any standard statistics text, e.g., Refs. 1 and 2. Graphical methods for fitting are discussed by Nelson [3].)

It is assumed that a sample of n independent observations $X_1, \ldots,$ X_n is available and that the X_i are *identically distributed*, that is, all X_i have the same (known) failure distribution $f(x) = f(x; \Theta)$ [also written $f_X(x) = f_X(x; \Theta)$ *], where Θ represents the set of *parameters* of the distribution. The initial objective of the statistical analysis is to *estimate* Θ, that is, assign numerical values to the components of Θ using the information in the sample $X_1, \ldots, X_n$. This will be discussed in Section 8.3, which is devoted to point estimation, and Section 8.4, which deals with confidence intervals.

A number of the warranty cost models involve an important function that is derived from the failure distribution, called the *renewal function*. (See Chapter 7.) In the context of warranty, a "renewal" is the (assumed instantaneous) replacement of a failed item by an identical new item. The renewal function $M(t)$ is the expected (or average) number of renewals in the interval $[0, t)$. Estimation of renewal functions poses some special problems. Some of these will be addressed in Section 8.5.

Once the parameters of a failure distribution are estimated, the results can be used to estimate many additional quantities of interest, for example, the "reliability function," $R(x) = P(X > x)$. Of particular interest in warranty analysis are cost models that have been developed for evaluating the various warranty policies. Many models of this type are presented in this handbook, particularly in Parts D, E, F, and H. These will typically involve the parameters of the assumed life distribution. It follows that costs can be estimated using the parameter estimates that will be given in the next section. Some comments on how this is done are given in Section 8.6.

Finally, some comments on applications, recommendations, and conclusions for the practitioner and some suggestions for further research for the theoretician will be given in Section 8.7.

---

* Here we use the standard convention in probability theory whereby a capital letter indicates a random variable (e.g., a random lifetime) and a lowercase letter indicates a particular value of the random variable (e.g., 2.7 years).

8.2 PROBABILITY DISTRIBUTIONS AND RELATED CONCEPTS

In this section, six life distributions will be listed along with certain important characteristics of each. Because of its importance in statistical inference (to be discussed in the next section), the normal distribution will also be defined. Finally, some comments on the calculation of the renewal functions associated with the six life distributions will be included.

8.2.1 Standard Life Distributions

By far, the most commonly used distribution in reliability and related disciplines is the exponential distribution. It is appropriate as long as the items fail at a constant rate (i.e., the probability of a failure in a given small interval is the same regardless of the age of the item). The advantages of the exponential distribution are that it is applicable to many types of items, at least during a significant portion of the item lifetime, and that it is analytically tractable. However, many items do not have a constant failure rate, so other life distributions must be considered as well.*

Two common alternatives to the exponential are the Weibull and Gamma distributions, both of which can be used to model increasing failure rates (IFR) and decreasing failure rates (DFR) by appropriate choices of parameter values. The exponential distribution is a special case of both of these.

Another alternative, one that is always DFR, is the mixed exponential distribution. This occurs if a process produces two kinds of items, each of which has an exponential life distribution but with different failure rates. Two other alternatives, with variable failure rates, are the lognormal and inverse Gaussian distributions.

Formulas for these six distributions as well as for the mean (μ) and variance (σ^2) for each are given below. Illustrative graphs for selected parameter values will also be given. Many additional details about these and other important distributions may be found in Ref. 4. Their application in the context of failure data is discussed in many standard texts on reliability and life testing, for example, Refs. 5–7.

In defining probability distributions, the parameters of each distribution will be identified. Three types of parameters are encountered—scale parameters, whose values change when units are changed, shape parameters, whose values determine in some sense the shape of the distribution, and shift (or location) parameters, whose values change when the data

---

* The failure rate $r(t)$ is formally defined as $r(t) = f(t)/[1 - F(t)]$, where $F(t) = \int_0^t f(x)\, dx$ is the *cumulative distribution function* (CDF). (See Chapter 6.)

are shifted by adding a constant to each value. Because we are interested in data on lifetimes, the values of the variable will necessarily be non-negative, so that the shift parameter will typically be omitted in the formulations to follow. For versions of these distributions that include this additional parameter, see Ref. 4.

The Exponential Distribution

The exponential distribution with parameter λ (>0) is given by

$$f(x; \lambda) = \begin{cases} 0 & \text{if } x < 0 \\ \lambda e^{-\lambda x} & \text{if } x \geq 0 \end{cases} \tag{8.1}$$

The mean and variance respectively are

$$\mu = \frac{1}{\lambda} \tag{8.2}$$

$$\sigma^2 = \frac{1}{\lambda^2} \tag{8.3}$$

Plots of the exponential distribution for $\lambda = 0.5$, 1, and 2 (i.e., $\mu = 2$, 1, and 0.5, respectively) are given in Figure 8.1.

As noted, the exponential distribution is appropriate if it is reasonable to assume that the failure rate of the item is constant.

The Weibull Distribution

The Weibull distribution with scale parameter λ (>0) and shape parameter β (>0) is given by

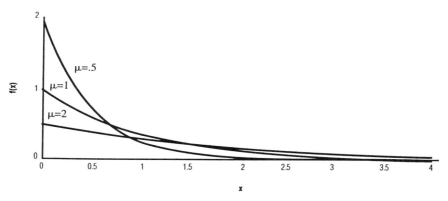

Figure 8.1 Exponential distribution; $\mu = 0.50, 1.00, 2.00$.

$$f(x; \lambda, \beta) = \begin{cases} 0 & \text{if } x < 0 \\ \lambda^\beta x^{\beta-1} e^{-(\lambda x)^\beta} & \text{if } x \geq 0 \end{cases} \qquad (8.4)$$

The mean and variance respectively are

$$\mu = \lambda^{-1} \Gamma\left(1 + \frac{1}{\beta}\right) \qquad (8.5)$$

$$\sigma^2 = \lambda^{-2} \left[\Gamma\left(1 + \frac{2}{\beta}\right) - \Gamma^2\left(1 + \frac{1}{\beta}\right) \right] \qquad (8.6)$$

where $\Gamma(x)$ is the Gamma function (See Ref. [8, Chap. 6].)

The Weibull distribution is DFR if $\beta < 1$, and IFR if $\beta > 1$. It reduces to the exponential distribution if $\beta = 1$. The distribution is plotted for $\beta = 0.5, 2$, and 4 in Figure 8.2, with λ chosen in each case so that the means are 2, 1, and 0.5, that is, the same as for the exponential distribution discussed previously.

The Weibull distribution was used initially in the context of breaking strength of materials [9]. It is used more generally as an extension of the exponential distribution in applications where the failure rate may not be constant.

The Gamma Distribution

The Gamma distribution with scale parameter λ (> 0) and shape parameter β (> 0) is given by

$$f(x; \lambda, \beta) = \begin{cases} 0 & \text{if } x < 0 \\ [\Gamma(\beta)]^{-1} \lambda^\beta x^{\beta-1} e^{-\lambda x} & \text{if } x \geq 0 \end{cases} \qquad (8.7)$$

The Gamma distribution reduces to the exponential when $\beta = 1$. It is DFR if $\beta < 1$, and IFR if $B > 1$. If β is a positive integer, the distribution is also called the *Erlang* or *Erlangian* distribution. If $\beta = v/2$, where v is a positive integer, and $\lambda = 2$, the distribution is a *chi-square* (χ^2) *distribution* with v *degrees of freedom*. The mean and variance are

$$\mu = \frac{\beta}{\lambda} \qquad (8.8)$$

$$\sigma^2 = \frac{\beta}{\lambda^2} \qquad (8.9)$$

The Gamma distribution is plotted in Figure 8.3 with $\beta = 0.5, 1.5$, and 2 and λ values selected so that $\mu = 0.5, 1$, and 2. This distribution is also used as an alternative to the exponential in life-testing applications;

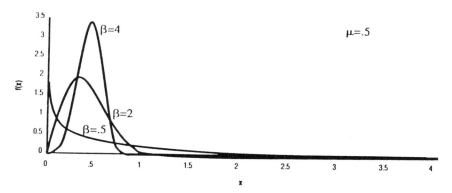

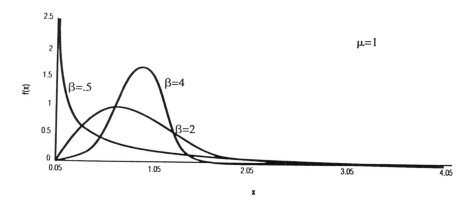

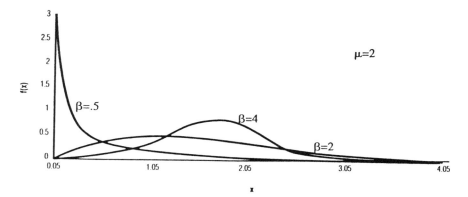

Figure 8.2 Weibull distribution; β = 0.50, 2.00, 4.00. (a) μ = 0.50; (b) μ = 1.00; (c) μ = 2.00.

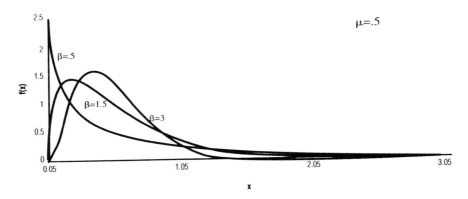

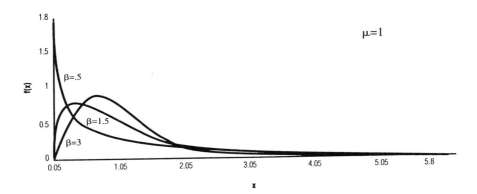

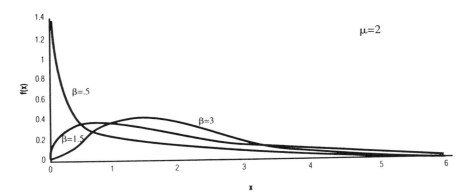

Figure 8.3   Gamma distribution; β = 0.50, 1.50, 3.00. (a) μ = 0.50; (b) μ = 1.00; (c) μ = 2.00.

its shape is somewhat different from that of the Weibull distribution. For more on applications, see Ref. 4.

The Mixed Exponential Distribution

The mixed exponential distribution is a weighted sum of exponentials with distinct parameters λ_1 and λ_2 and weights p $(0 < p < 1)$ and $q = 1 - p$. The formula is

$$f(x; \lambda_1, \lambda_2, p) = \begin{cases} 0 & \text{if } x < 0 \\ p\lambda_1 e^{-\lambda_1 x} + q\lambda_2 e^{-\lambda_2 x} & \text{if } x \geq 0 \end{cases} \tag{8.10}$$

The mean and variance are, respectively,

$$\mu = p\lambda_1^{-1} + q\lambda_2^{-1} \tag{8.11}$$

$$\sigma^2 = p\lambda_1^{-2} + q\lambda_2^{-2} + pq(\lambda_1^{-1} - \lambda_2^{-1})^2 \tag{8.12}$$

Plots of the mixed exponential distribution with $p = 0.05$ and $\lambda_1 = 4$ ($\mu_1 = 0.25$) and λ_2 selected so that $\mu = 0.5, 1,$ and 2, are given in Figure 8.4.

The mixed exponential distribution is used to model situations where a process produces two kinds of items, each of which has a constant failure rate. Examples are two machines or assembly lines or a single operation that may be in or out of control.

The Lognormal Distribution

The lognormal distribution with parameters η $(-\infty < \eta < \infty)$ and θ (> 0) is given by

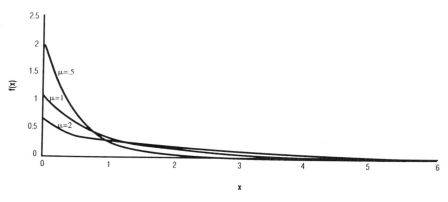

Figure 8.4 Mixed exponential distribution; $p = 0.05$, $\lambda_1 = 4$.

$$f(x; \eta, \theta) = \begin{cases} 0 & \text{if } x < 0 \\ \dfrac{1}{x\theta\sqrt{2\pi}} \exp\left(-\dfrac{(\log x - \eta)^2}{2\theta^2}\right) & \text{if } x \geq 0 \end{cases} \qquad (8.13)$$

The mean and variance are, respectively,

$$\mu = e^{\eta + \theta^2/2} \qquad (8.14)$$

$$\sigma^2 = e^{2\eta + \theta^2}(e^{\theta^2} - 1) \qquad (8.15)$$

If X is lognormally distributed, then e^X is normally distributed with mean η and variance θ^2. (See Section 8.2.2.) Plots of the lognormal distribution with $\theta = 0.25, 0.50,$ and 1 and η chosen so that $\mu = 0.5, 1,$ and 2 are given in Figure 8.5.

Applications in reliability and quality control for manufactured products are discussed in Ref. 4 and in many of the references cited. The distribution is appropriate when failures are caused by material fracture or breakage.

The Inverse Gaussian Distribution

The inverse Gaussian distribution with parameters $\eta \, (> 0)$ and $\theta \, (> 0)$ is given by

$$f(x; \eta, \theta) = \begin{cases} 0 & \text{if } x < 0 \\ \left(\dfrac{\theta}{2\pi x^3}\right)^{1/2} \exp\left(-\dfrac{\theta(x - \eta)^2}{2\eta^2 x}\right) & \text{if } x \geq 0 \end{cases} \qquad (8.16)$$

Here, η is a location parameter and $1/\theta$ is a scale parameter. The mean and variance are, respectively,

$$\mu = \eta \qquad (8.17)$$

$$\sigma^2 = \frac{\eta^3}{\theta} \qquad (8.18)$$

The inverse Gaussian distribution is plotted in Figure 8.6 with $\eta = 0.5, 1,$ and 2 and $\theta = 1$. The distribution is also called the Wald distribution [4] because of applications in sequential analysis. It also has important applications in diffusion processes and has been used as another alternative life distribution in reliability applications where the previous distributions do not provide an adequate fit to data.

8.2.2 The Normal Distribution

The normal distribution with mean $\mu \, (-\infty < \mu < \infty)$ and variance $\sigma^2 \, (> 0)$, which are also the parameters of the distribution, is given by

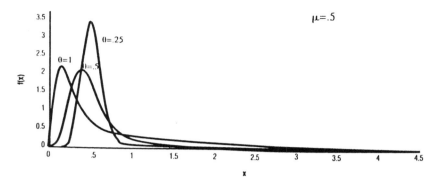

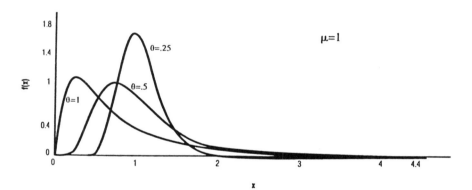

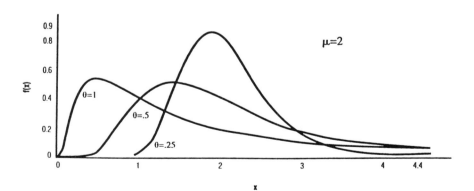

Figure 8.5 Lognormal distribution; $\theta = 0.25, 0.50, 1.00$. (a) $\mu = 0.50$; (b) $\mu = 1.00$; (c) $\mu = 2.00$.

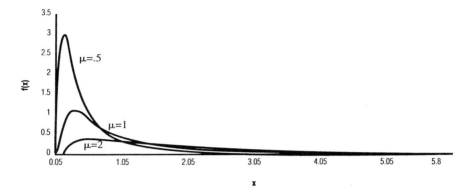

Figure 8.6 Inverse Gaussian distribution; $\theta = 1.00$.

$$f(x; \mu, \sigma^2) = \frac{1}{\sqrt{2\pi}\sigma} \exp\left(-\frac{(x - \mu)^2}{2\sigma^2}\right), \quad -\infty < x < \infty \qquad (8.19)$$

The normal distribution plots as the familiar bell-shaped curve. Tables of normal probabilities may be found in almost every statistics text. In Section 8.3, percentiles of the normal distribution will be used in the calculation of approximate confidence intervals for the parameters of some of the distributions given in the previous section.

8.2.3 Renewal Functions

Renewal functions are defined and discussed in detail in Chapter 7. If a failed item is always replaced by a new one and time to replace is zero, the renewal function $M(t)$ is defined to be the expected number of replacements in the interval $[0, t)$. $M(t)$ can be expressed in terms of the convolutions of $F(\cdot)$ with itself (i.e., distributions of sums of random variables) or as the solution of an integral equation. With few exceptions, closed-form analytical expressions for renewal functions do not exist, and they must be evaluated numerically.

Renewal functions for the distributions of Section 8.2.1 for which closed-form expressions exist are as follows:

Exponential distribution

$$M(t) = \lambda t$$

Erlang distribution

$$M(t) = \frac{\lambda t}{\beta} + \frac{1}{\beta} \sum_{j=1}^{\beta-1} \frac{a^j}{1 - a^j} \{1 - \exp[-\lambda(1 - a^j)]\}$$

where $a = \exp(2\pi i/\beta)$, $i = \sqrt{-1}$, $\beta = 2,3, \ldots$.

For other distributions, $M(t)$ may be approximated or evaluated by table look-up or by computer. An asymptotic approximation is

$$M(t) \approx \frac{t}{\mu} + \frac{\sigma^2}{2\mu^2} - \frac{1}{2} \qquad (8.20)$$

valid for large t. This is not useful for many warranty applications, however, as relatively small values of t are usually of interest. Tables of renewal functions have been prepared by Baxter et al. [10], Giblin [11], Soland [12], and White [13]. Together, these provide coverage of all of the distributions of Section 8.2.1 with the exception of the mixed exponential. Renewal functions for all of these distributions may be evaluated by computer using, for example, the algorithm discussed in Ref. 14 or the FORTRAN program given by Blischke and Murthy [15]. (Also see Ref. 15 for additional discussion regarding renewal functions, their evaluation, and their importance in warranty analysis.)

8.3 PARAMETER ESTIMATION

Each of the life distributions given in Section 8.2 involves one or more parameters. In order to use these distributions in the calculation of probabilities, renewal functions, or related quantities, it is necessary to determine numerical values (called *point estimates*) for the parameters of whichever distribution is appropriate in a given application. Ordinarily, these parameters are unknown and must be estimated from available data. In the remainder of this chapter, we assume that a sample of size n, X_1, $\ldots$, X_n, is available for this purpose. Data of this type are called *complete*.

There are many other types of data that may be obtained in warranty and related studies. In life testing, for example, data are often *censored*, that is, truncated in some way. Data of this type are *incomplete* in this sense.

There are two types of censoring that are commonly encountered. Type I censored data are data having the property that for some of the observations the exact time of failure is not known, only that it exceeds some (known) quantity. Failure data from laboratory tests that are terminated after a fixed period of time are of this type. Warranty claims data are Type I censored in that failures after the end of the warranty period are not reported.

Type II censored data are data for which testing or data collection stops as soon as a fixed number of failures have been observed. Lab tests often give rise to data of this type as well. One use of both types of censoring is to limit the amount of time required to complete the testing process.

In this section, the statistical methodology for estimation of parameters will be briefly discussed, following which procedures for each of the seven distributions (including the normal) given in Section 8.2 will be listed. As noted, here and in the remainder of the chapter we consider only complete data. For analyses appropriate for incomplete data, see Refs. 3 and 5–7.

8.3.1 Methods of Point Estimation

In estimating parameters, we distinguish between an *estimator*, which is the procedure or formula that is used for estimation, and an *estimate* or *point estimate*, which is a specific numerical value of an estimator obtained by substitution of sample values into the formula. Much of statistical theory is devoted to the derivation of estimators that are "best" in some sense. Criteria of "goodness" in this context include unbiasedness, efficiency (in the sense of minimum variance or minimum mean square error, etc.), sufficiency, consistency, admissibility, and so forth. These terms are defined precisely and discussed in some detail in texts on theoretical statistics, for example, Refs. 2 and 16. Efficiency is often considered to be of particular importance because an inefficient estimator requires larger sample sizes, thus increasing the cost of experimentation. The objective in theoretical investigations of estimators is to choose an estimator that uses all of the sample information in an efficient and effective manner.

In this section, two methods of estimation will be discussed. The first, the *method of moments*, is often not the best, according to many of the accepted criteria mentioned in the previous paragraph. In particular, moment estimators are often not efficient. They are, however, useful for a number of purposes: (1) as initial estimates in cases where they are relatively easily calculated, (2) as starting points in the numerical solution of estimation equations for other procedures, and (3) in cases where data are plentiful and efficiency is not an important consideration.

The second method of estimation, *maximum likelihood*, is optimal in many respects. In particular, it is efficient, at least for large samples (called *asymptotically efficient*), under fairly general conditions. In many cases, however, maximum likelihood estimation leads to computational difficulties.

Additional information on these and other methods of estimation may be found in Refs. [2, 4–6, and 16]. More on applications in warranty analysis is given in Chapter 13 of Ref. 15.

The Method of Moments

The expected value $E(X)$ of a (continuous) random variable is defined to be

$$\mu = E(X) = \int_{-\infty}^{\infty} xf(x)\, dx \tag{8.21}$$

The rth moment about zero of a random variable is given by

$$\mu_r' = E(X^r) = \int_{-\infty}^{\infty} x^r f(x)\, dx \tag{8.22}$$

The rth moment about the mean (also called the rth *central moment*) is

$$\mu_r = E[(X - \mu)^r] = \int_{-\infty}^{\infty} (x - \mu)^r f(x)\, dx \tag{8.23}$$

Note that $\mu = \mu_1'$ is the population mean and $\mu_2 = \sigma^2$ is the population variance.

The method of moments involves expressing a selected set of population moments in terms of the parameters of the assumed probability distribution, solving the resulting system of equations for the parameters, and replacing the population moments by the corresponding sample quantities in the solution. If the distribution in question involves k parameters, usually the first k moments are used for this purpose. (They may be ordinary moments, moments about 0, or other moments. The choice is usually made on the basis of ease of solution of the resulting equations.)

Moment estimators for the six life distributions given in Section 8.2.1 are given in Section 8.3.2.

The Method of Maximum Likelihood

The likelihood function is defined to be

$$L(X_1, \ldots, X_n; \Theta) = \prod_{i=1}^{n} f(X_i; \Theta) \tag{8.24}$$

The *maximum likelihood* (ML) *estimator* of Θ is the set of values of the components of Θ that maximize L. For the parameters of the life distributions in question, the ML estimators are obtained by differentiating the likelihood function, equating the derivatives to zero, and solving the resulting equations.

Procedures for obtaining maximum likelihood estimators of the parameters of the life distributions of Section 8.2.1 are given in the next section.

Other Methods of Estimation

As noted previously, moment estimators are unbiased or asymptotically unbiased but are often not efficient. Maximum likelihood estimators are, under fairly general conditions, asymptotically unbiased, consistent, and asymptotically efficient. They are also asymptotically normally distributed.[*] There is a whole class of estimators that share these properties, called BAN (best asymptotically normal) estimators. In addition, there are many other approaches to estimation. (See Refs. 2 and 6.)

Moment and ML estimators for the parameters of the distributions previously listed will be given below. These suffice for nearly all applications. For more on moment, ML, and other estimators for these and many other distributions and for both complete and censored data, see Refs. 4 and 6.

8.3.2 Moment Estimators of the Parameters of Life Distributions

The life distributions discussed in Section 8.2 involve up to three parameters. In defining moment estimators for these distribution, we use the first sample moment for a single parameter and the first two sample moments, namely, the sample mean and variance, for two parameters. The sample mean and variance respectively are

$$\bar{x} = \frac{1}{n} \sum_{i=1}^{n} x_i \tag{8.25}$$

and

$$s^2 = \frac{1}{n-1} \sum_{i=1}^{n} (x_i - \bar{x})^2 \tag{8.26}$$

where $x_1, \ldots, x_n$ are the sample values.

---

[*] Strictly speaking, the ML estimator must be properly normalized, for example, by multiplication by $\sqrt{n}$, so that the distribution of the estimator converges to the normal distribution as $n \to \infty$.

For the mixed exponential, the only three-parameter distribution given in the previous section, we use the first three sample moments about zero, given by

$$m'_r = \frac{1}{n} \sum_{i=1}^{n} x_i^r \tag{8.27}$$

for $r = 1, 2, 3$. (Note that $m'_1 = \bar{x}$.)

The moment estimates and estimators are designated by a "hat," for example the moment estimator of a parameter θ is denoted $\hat{\theta}$. The moment estimates of the parameters of the distributions listed in Section 8.2 are as follows:

1. Exponential distribution

$$\hat{\lambda} = \frac{1}{\bar{x}} \tag{8.28}$$

2. Weibull distribution
    The estimate of the shape parameter is obtained by solving

$$1 + \frac{s^2}{\bar{x}^2} = \frac{\Gamma(1 + 2/\hat{\beta})}{\Gamma^2(1 + 1/\hat{\beta})} \tag{8.29}$$

for $\hat{\beta}$. Tables to facilitate the solution are given in Ref. 17. The scale parameter is estimated by

$$\hat{\lambda} = \frac{\Gamma(1 + 1/\hat{\beta})}{\bar{x}} \tag{8.30}$$

3. Gamma distribution

$$\hat{\lambda} = \frac{\bar{x}}{s^2} \tag{8.31}$$

$$\hat{\beta} = \frac{\bar{x}^2}{s^2} \tag{8.32}$$

4. Mixed exponential distribution
    The moment estimates are obtained by solving

$$\bar{x} = \frac{\hat{p}}{\hat{\lambda}_1} + \frac{1 - \hat{p}}{\hat{\lambda}_2} \tag{8.33}$$

$$\frac{m'_2}{2} = \frac{\hat{p}}{\hat{\lambda}_1^2} + \frac{1 - \hat{p}}{\hat{\lambda}_2^2} \tag{8.34}$$

$$\frac{m'_3}{6} = \frac{\hat{p}}{\hat{\lambda}_1^3} + \frac{1 - \hat{p}}{\hat{\lambda}_2^3} \tag{8.35}$$

An algebraic solution for the moment estimates of the three parameters is given by Rider [18]. Computer solution is relatively straightforward.

If p is known (e.g., if it is known that a specified proportion of items is produced by one of two machines), the moment estimates of λ_1 and λ_2 are obtained as follows: $1/\hat{\lambda}_2$ is the admissible solution (if any) of the quadratic equation

$$y^2 - 2\bar{x}y + \frac{2\bar{x}^2 - m'_2 p}{2q} = 0 \tag{8.36}$$

where $q = 1 - p$. $\hat{\lambda}_1$ is then calculated as

$$\hat{\lambda}_1 = \frac{(\bar{x} - q\hat{\lambda}_2)}{p} \tag{8.37}$$

An approach to the three-parameter problem, easily implemented on a computer, is to solve (8.36) and (8.37) for a range of p-values and then take as the overall solution the set of p, λ_1, λ_2-values that satisfy (8.34).

5. Lognormal distribution
    Instead of solving (8.14) and (8.15) for η and θ, transform the data to $y_i = \ln(x_i)$. The moment estimates are the sample mean and standard deviation of the y_i's, namely,

$$\hat{\eta} = \frac{1}{n} \sum_{i=1}^{n} y_i = \bar{y} \tag{8.38}$$

$$\hat{\theta} = \left(\frac{1}{n-1} \sum_{i=1}^{n} (y_i - \bar{y})^2 \right)^{1/2} = s_y \tag{8.39}$$

6. Inverse Gaussian distribution

$$\hat{\eta} = \bar{x} \tag{8.40}$$

$$\hat{\theta} = \frac{\bar{x}^3}{s^2} \tag{8.41}$$

7. Normal distribution

$$\hat{\mu} = \bar{x} \tag{8.42}$$

$$\hat{\sigma} = s \tag{8.43}$$

Example Data Sets

To illustrate the computations of the moment estimates as well as many of the procedures to follow, we use two sample data sets. The first is a set of simulated data that appear to exhibit a decreasing failure rate such as might be encountered in certain electronic components or systems. The data, in years of service time, are given in Table 8.1. Summary statistics for these data are $n = 127$, $\bar{x} = 3.2194$ years, $m_2' = 25.9175$, $s^2 = 15.6737$, and $m_3' = 375.493$.

The second set of data is given in Table 8.2. These data are also simulated. Here, the appearance is of an increasing failure rate, as is the case for many mechanical items. The unit is hours of operation until the first failure. The summary statistics are $n = 95$, $\bar{x} = 2963$ hours, $m_2' = 8,779,369$, $s^2 = 2,883,204$, and $m_3' = 54,259,109,888$.

Example 8.1   (Moment Estimates for Data Set 1)

Moment estimates under the various distributional assumptions (omitting the normal distribution because it is rarely appropriate for failure data) are as follows:

1. Exponential distribution: From (8.28), the estimated failure rate is $\hat{\lambda} = 1/\bar{x} = 1/3.2194 = 0.3106$.
2. Weibull distribution: We have $1 + s^2/\bar{x}^2 = 2.5126$. The value of $\hat{\beta}$ that solves (8.29) is found (by interpolation in tables of the Gamma function or the methods of Ref. 17) to be $\hat{\beta} = 0.818$. Note that this does correspond to a decreasing failure rate. Similarly, the estimate of the scale parameter is $\hat{\lambda} = \Gamma(1 + 1/\hat{\beta})\bar{x}^{-1} = 1.116/3.219 = 0.3467$.

Table 8.1   Data Set 1—Failure Times of an Electronic Component

4.38	4.94	1.79	1.23	0.57	1.56	0.92	0.14	13.54
0.96	5.70	1.63	2.02	2.35	1.48	0.82	0.11	1.55
7.77	6.57	0.10	2.20	3.58	2.20	7.83	7.62	6.97
0.23	2.63	4.57	0.03	13.54	0.61	0.31	2.44	1.74
1.38	6.18	30.10	1.56	2.25	0.40	1.43	1.96	0.31
5.38	4.61	1.97	7.98	0.52	0.71	3.63	0.71	0.01
0.01	0.17	0.41	2.48	3.68	2.87	0.05	14.62	0.05
0.24	0.06	4.17	3.28	2.81	0.83	0.77	0.30	0.28
0.01	0.24	1.82	0.20	6.24	1.98	3.67	2.14	0.28
0.04	0.27	1.85	2.36	0.52	0.78	1.71	0.23	1.33
0.03	0.04	0.66	2.92	1.60	0.12	0.46	0.17	0.79
0.04	4.90	3.39	3.28	4.42	10.14	1.65	7.65	3.94
8.05	5.06	2.93	10.55	2.11	6.37	7.87	5.13	5.09
3.22	6.02	11.67	7.58	1.61	5.43	4.87	6.93	6.04
5.69								

Table 8.2   Data Set 2—Failure Times of a Mechanical Product

3843	6066	3024	2887	3776	3211	2660	6451	3216	3254	2048
3029	3613	6712	3466	5396	228	2737	3371	5227	1603	2469
1114	1224	2813	7883	4711	4523	6907	3931	2204	2688	2097
5346	1757	6356	3140	600	2011	1900	3033	5385	1160	2703
1850	3354	2627	2738	2389	1116	3719	1341	2729	2510	455
880	3194	5736	2802	2288	1170	953	4291	2628	2012	722
162	804	719	4408	2041	1350	929	1669	4958	6116	3971
1265	2338	3756	2941	4392	4160	3123	5186	2706	4255	1396
4351	3839	1054	3186	1802	308	1007				

3. Gamma distribution: From (8.31) and (8.32), the estimates are $\hat{\lambda} = 3.2194/3.959^2 = 0.2054$ and $\hat{\beta} = 3.219^2/15.6737 = 0.6611$, which also corresponds to a decreasing failure rate.

4. Mixed exponential distribution: The moment estimates, obtained by solving (8.36) and evaluating (8.37) for various values of p and selecting the result that satisfies (8.35), are $\hat{p} = 0.277$, $\hat{\lambda}_1 = 0.1718$, and $\hat{\lambda}_2 = 0.4500$. This corresponds to a process in which 27.7% of the items have an average lifetime of $1/0.1718 = 5.82$ and the remaining 72.3% have an average lifetime of $1/0.45 = 2.22$.

5. Lognormal distribution: The sample mean and variance of $y_i = \ln(x_i)$ are $\bar{y} = 0.2708$ and $s_y^2 = 2.9036$. From (8.38) and (8.39), the moment estimates are $\hat{\eta} = 0.2708$ and $\hat{\theta} = \sqrt{2.9036} = 1.704$. For this distribution, it can be seen from a plot of $r(t)$ that the failure rate is also decreasing.

6. Inverse Gaussian distribution: From (8.40) and (8.41), we obtain $\hat{\eta} = 3.219$ and $\hat{\theta} = 3.219^3/15.6737 = 2.1281$.

Comment: Plots of the six estimated distributions show similar patterns—essentially exponentially decreasing forms. They differ somewhat in the tail of the distribution (i.e., as x increases), and this can significantly influence the values of the corresponding reliability and renewal functions.

Example 8.2   (Moment Estimates for Data Set 2)

Moment estimates, calculated as indicated in Example 8.1, are as follows:

1. Exponential distribution

$\hat{\lambda} = 0.0003375$

2. Weibull distribution

$$\hat{\beta} = 1.81, \quad \hat{\lambda} = 0.0003001$$

3. Gamma distribution

$$\hat{\beta} = 3.0450, \quad \hat{\lambda} = 0.001028$$

Both the Weibull and Gamma shape parameters indicate IFR distributions.

4. Mixed exponential distribution: There are no real roots of (8.35) for any value of p with $0 < p < 1$, and hence no admissible solution to the moment equations. This is to be expected because the mixed exponential is always DFR.
5. Lognormal distribution: Here, $\bar{y} = 7.780$ and $s_y^2 = 0.5577$, so $\hat{\eta} = 7.780$ and $\hat{\theta} = 0.747$.

6. Inverse Gaussian distribution

$$\hat{\eta} = 2963, \quad \hat{\theta} = 9022.3$$

Comment: Plots of Distributions 2, 3, 5, and 6 show similar forms but differ in certain essential features. The exponential distribution looks quite different and appears to provide a poor fit to the data.

8.3.3 Maximum Likelihood Estimators of the Parameters of Life Distributions

The likelihood function is defined in (8.24). Maximum likelihood estimators for the distributions of Section 8.2.1 are obtained by maximizing (8.24) with respect to the parameters, with $f(x)$ the specified distribution. This is usually done by differentiating L with respect to the parameters,* equating the derivatives to zero, and solving the resulting system of equations. The ML estimates or estimators will be denoted by an asterisk; for example, the ML estimator of θ is θ^*. For the distributions considered previously, the ML estimates or the equations that must be solved to determine the estimates are as follows:

1. Exponential distribution

$$\lambda^* = \frac{1}{\bar{x}} \tag{8.44}$$

---

* An equivalent, and usually easier, approach is the maximize $\ln(L)$.

2. Weibull distribution: Solve

$$\frac{1}{n} \sum_{i=1}^{n} \ln(x_i) = \frac{\sum_{i=1}^{n} [x_i^{\beta^*} \ln(x_i)]}{\sum_{i=1}^{n} x_i^{\beta^*}} - \frac{1}{\beta^*} \tag{8.45}$$

for β^*. Computer programs (e.g., based on a simple search) are relatively straightforward. Once β^* is determined, λ^* is calculated as

$$\lambda^* = \left(\frac{n}{\sum_{i=1}^{n} x_i^{\beta^*}} \right)^{1/\beta^*} \tag{8.46}$$

3. Gamma distribution: Let

$$k = \ln(\bar{x}) - \frac{1}{n} \sum_{i=1}^{n} \ln(x_i) \tag{8.47}$$

β^* is the solution of

$$\ln(\beta^*) - \psi(\beta^*) = k \tag{8.48}$$

where $\psi(u)$ is the *digamma function*, given by

$$\psi(u) = \frac{d \ln[\Gamma(u)]}{du} = \frac{\Gamma'(u)}{\Gamma(u)} \tag{8.49}$$

$\psi(u)$ is tabulated in Chapter 6 of Ref. 8, where various series expressions and other aids to calculation are also given. λ^* is calculated as

$$\lambda^* = \frac{\beta^*}{\bar{x}} \tag{8.50}$$

The solution of (8.48) may be closely approximated [4, p. 189] by

$$\beta^* \approx$$

$$\begin{cases} k^{-1}(0.5000876 + 0.1648852k - 0.0544274k^2) & \text{if } 0 < k \le 0.5772 \\ \dfrac{8.898919 + 9.059950k + 0.9775373k^2}{k(17.79728 + 11.968477k + k^2)} & \text{if } 0.5772 < k \le 17 \end{cases}$$

$$\tag{8.51}$$

4. Mixed exponential distribution: The ML equations are difficult to solve and can only be dealt with by computer methods. As is the case for mixed distributions generally, the solution will often be unstable unless n is very large. If that is the case, the moment estimators will generally be adequate.

5. Lognormal distribution: The ML estimates are based on $y_i = \ln(x_i)$. The formulas are

$$\eta^* = \frac{1}{n} \sum_{i=1}^{n} \ln(x_i) = \bar{y} \tag{8.52}$$

$$\theta^* = \left(\frac{1}{n} \sum_{i=1}^{n} (y_i - \bar{y})^2 \right)^{1/2} \tag{8.53}$$

6. Inverse Gaussian distribution: The ML estimates are

$$\eta^* = \bar{x} \tag{8.54}$$

$$\theta^* = n \left[\sum_{i=1}^{n} \left(\frac{1}{x_i} - \frac{1}{\bar{x}} \right) \right]^{-1} \tag{8.55}$$

7. Normal distribution

$$\mu^* = \bar{x} \tag{8.56}$$

$$\sigma^* = \left(\frac{1}{n} \sum_{i=1}^{n} (x_i - \bar{x})^2 \right)^{1/2} \tag{8.57}$$

Example 8.3 (Maximum Likelihood Estimates for Data Set 1)

ML estimates for Data Set 1 for the parameters of the distributions other than normal and mixed exponential are as follows:

1. Exponential distribution: The ML estimate is the same as the moment estimate, namely, $\lambda^* = 0.3106$.
2. Weibull distribution: A simple search using a microcomputer found β^* satisfying (8.45) to be $\beta^* = 0.781$, giving, from (8.46), $\lambda^* = (127/285.02)^{1/0.781} = 0.3552$.
3. Gamma distribution: $\sum \ln(x_i) = 34.396$, so $k = \ln(3.219) - 34.396/127 = 0.8982$. From (8.51), $\beta^* = 0.676$. From (8.50), $\lambda^* = 0.676/3.219. = 0.2100$.
5. Lognormal distribution: From (8.52) and (8.53), $\eta^* = 34.396/127 = 0.2708$ and $\theta^* = 1.697$.
6. Inverse Gaussian distribution: From (8.54) and (8.55), $\eta^* = 3.219$ and $\theta^* = 0.2108$.

Example 8.4   (Maximum Likelihood Estimates for Data Set 2)
ML estimates for Data Set 2, obtained as for Data Set 1, are as follows:

1.   Exponential distribution

$\lambda^* = 0.0003375$

2.   Weibull distribution

$\lambda^* = 0.0003007, \qquad \beta^* = 1.795$

3.   Gamma distribution

$k = \ln(2963) - 7.780 = 0.214$, from which we obtain
$\beta^* = 3.024$ and $\lambda^* = 0.001021$.

5.   Lognormal distribution

$\eta^* = 7.780, \qquad \theta^* = 0.743$

6.   Inverse Gaussian distribution

$\eta^* = 2963, \qquad \theta^* = 3699.5$

Comment: For these data sets, the moment and ML estimates agree quite well with the exception of $\hat{\theta}$ and θ^* for the inverse Gaussian distribution. For very large samples, good agreement would be expected in all cases because both estimators are consistent.

8.4   CONFIDENCE INTERVAL ESTIMATION

A *confidence interval* or *confidence interval estimator* of a parameter is an interval defined by two limits, L_1 and L_2, which have the property that the true parameter value lies between L_1 and L_2 a specified proportion γ of the time. γ is called the *confidence coefficient* and the interval is described as either a confidence interval with confidence coefficient γ (e.g., 0.95) or as a $100\gamma\%$ (e.g., 95%) confidence interval.

Theoretical methods for deriving confidence intervals are discussed in Refs. 2 and 16. Procedures are given in Ref. 4 for many distributions. Applications in warranty analysis are discussed in Chapter 12 of Ref. 15.

Ordinarily, the limits L_1 and L_2 are constructed as functions of point estimators such as the moment or ML estimators of the previous sections. The calculations require fractiles of the distribution of the estimator or of an approximating asymptotic (large-sample) distribution. Because the latter is usually a normal distribution, we begin with the formula for a confidence interval for the mean of a normal distribution. Procedures will then be given for confidence intervals for one or more parameters of most of the life distributions considered previously.

8.4.1 Confidence Interval for the Mean of a Normal Distribution

Known Standard Deviation

If the standard deviation σ is known, a $100\gamma\%$ confidence interval for μ is given by

$$\bar{x} \pm \frac{z_p \sigma}{\sqrt{n}} \tag{8.58}$$

where z_p is the p-fractile of the standard normal distribution, with $p = 1 - (1 - \gamma)/2$. Values of z_p for selected choices of γ are given in Table 8.3.

Example 8.5 (Confidence Interval for μ, Normal Distribution)
     Sheet metal is produced for use in an application in which the reliability of a manufactured item depends critically on the thickness of the product. Data on thickness (mm) of finished sheet metal are recorded for a random sample selected from a day's production. It is known from past experience that the measurements are normally distributed with $\sigma = 0.0021$. A sample of size 15 results in the following observations:

$$1.000, \; 1.000, \; 0.997, \; 0.998, \; 1.002, \; 1.002, \; 0.999,$$
$$0.998, \; 0.999, \; 1.002, \; 1.001, \; 1.003, \; 0.999, \; 1.001, \; 1.002$$

For these data, $\bar{x} = 1.0002$. For a 95% confidence interval, the tabulated value of z_p is found from Table 8.3 to be 1.9600, A 95% confidence interval for the true mean thickness is $1.0002 \pm 1.96\,(0.0021)/\sqrt{15} = 1.0002 \pm 0.0009$, or $(0.9991, 1.0013)$.

Table 8.3 Fractiles of the Standard Normal Distribution for Calculating Confidence Intervals for a Population Mean with Confidence Coefficient γ

γ	Fractile
0.80	1.282
0.90	1.645
0.95	1.960
0.98	2.326
0.99	2.576
0.995	2.807

The procedure for the mean of a normal distribution is also used to calculate a confidence interval for μ in the case of non-normal distributions if n is large. This is appropriate because of the Central Limit Theorem, which states that for samples of size n from nearly any distribution, the sample mean $\overline{X}$ is asymptotically normally distributed with mean μ and standard deviation $\sigma/\sqrt{n}$. Because of this result, (8.58) can be used to calculate approximate large-sample confidence intervals for μ as well as for many other parameters of non-normal distributions as long as n is large enough. (How large n must be depends on the skewness of the true distribution of the estimator. Sample sizes of 50 to 100 will usually be adequate if the distribution is moderately skewed. Highly skewed distributions may require ns of hundreds or thousands.)

Example 8.6   (Asymptotic Confidence Interval for μ, Nonnormal Distribution)

Data Set 2 exhibits a moderate amount of skewness; $n = 95$ is considered adequate for use of the normal approximation. The approximate 90% confidence interval, using the sample standard deviation as an estimate of σ and $z_p = 1.645$ (from Table 8.3), is found to be $2963 \pm 1.645(1698)/\sqrt{95}$, or (2676, 3250).

Unknown Standard Deviation

Example 8.6 illustrates the technique that may be used if σ is not known and the sample size is large: Substitute the sample standard deviation for σ in (8.58). For small samples from normal distributions, a further modification is necessary. In place of the fractiles of the standard normal distribution (z_p) given in Table 8.3, use the corresponding fractile of the *Student's* t distribution with $n - 1$ *degrees of freedom** (df). Tables of the Student's distribution may be found in nearly any standard statistical text, for example, Ref. 1.

8.4.2   Confidence Intervals for the Parameters of Life Distributions

In presenting confidence intervals for the parameters of the life distributions discussed previously, we give exact results where possible and approximate results in other cases.† The approximate results are based on the fact that under fairly general conditions both the moment and ML

---

* "Degrees of freedom" is a term indicating the value of a parameter of the Student's t and a number of other statistical distributions.

† By "exact results" is meant that it can be proven mathematically that the confidence coefficient is exactly γ.

estimators are asymptotically normally distributed. In order to use these theoretical results, it is necessary to determine and be able to estimate the standard deviation of this asymptotic distribution. There are a number of ways of approaching this problem (see Ref. 2). The asymptotic results that follow are derived by determining the variance of the large-sample distribution of the estimator and then estimating this variance by substituting estimates for parameters in the resulting expressions.

The confidence intervals are as follows.

The Exponential Distribution

For the exponential distribution with complete data (i.e., no censoring), an exact $100\gamma\%$ confidence interval for λ is given by

$$\left(\frac{\chi^2_{1-p,2n}}{2n\bar{x}}, \frac{\chi^2_{p,2n}}{2n\bar{x}} \right) \tag{8.59}$$

where $p = 1 - (1 - \gamma)/2$ and $\chi^2_{p,2n}$ is the p-fractile of the chi-square (χ^2) distribution with $n - 1$ degrees of freedom. Table 8.4 gives values of $\chi^2_{p,2n}$ and $\chi^2_{1-p,2n}$ for selected values of γ and n. More extensive tables of the χ^2 distribution may be found in most statistical texts.

From this confidence interval, it is easy to determine a confidence interval for the μ, the mean time to failure (MTTF), as well. The result

Table 8.4 Values of $\chi_p{}^2$ for Calculating Confidence Intervals for μ, Exponential Distribution

| n | $\gamma = 0.90$ | | $\gamma = 0.95$ | | $\gamma = 0.99$ | |
	$\chi^2_{0.95,2n}$	$\chi^2_{0.05,2n}$	$\chi^2_{0.975,2n}$	$\chi^2_{0.025,2n}$	$\chi^2_{0.995,2n}$	$\chi^2_{0.005,2n}$
5	18.31	3.94	20.48	3.25	25.19	2.16
10	31.41	10.58	34.17	9.59	40.00	7.43
15	43.77	18.49	46.98	16.79	53.67	13.79
20	55.76	26.51	59.35	24.42	66.79	20.67
25	67.50	34.76	71.42	32.35	79.51	27.96
30	79.08	43.19	83.30	40.47	91.97	35.51
35	90.53	51.74	95.03	48.75	104.23	43.25
40	101.88	60.39	106.63	57.15	116.33	51.15
45	113.15	69.12	118.14	65.64	128.31	59.18
50	124.34	77.93	129.57	74.22	140.18	67.31
55	135.48	86.79	140.92	82.86	151.96	75.54
60	146.57	95.70	152.22	91.57	163.65	83.84

is

$$\left(\frac{2n\bar{x}}{\chi_{p,2n}^2}, \frac{2n\bar{x}}{\chi_{1-p,2n}^2} \right) \tag{8.60}$$

For confidence intervals based on censored data, see Refs. 4–6.

Example 8.7 [Confidence Interval for λ, Exponential Distribution]

Fifty items, assumed to have constant failure rate, are put on test. The observed MTTF is $\bar{x} = 1422$ hours. The exponential assumption is appropriate, and we find the estimated failure rate to be $\lambda^* = 1/1422 = .000703$. For a 95% confidence interval and $n = 50$, the tabulated χ^2 values are 77.93 and 124.34. The 95% confidence interval for λ is

$$\left(\frac{77.93}{2(50)(1422)}, \frac{124.34}{2(50)(1422)} \right) = (0.000548, 0.000874)$$

The Weibull Distribution

For large samples, confidence intervals may be based on the ML estimators, λ^* and β^*, using asymptotic results to approximate the variances of these estimators and the fact that they are asymptotically normally distributed. The large-sample approximations to the variances [4, p. 256] are

$$V(\lambda^*) \approx \frac{1.087\lambda^2}{n\beta^2} \tag{8.61}$$

and

$$V(\beta^*) \approx \frac{0.608\beta^2}{n} \tag{8.62}$$

These variances are estimated by substituting estimates of λ and β in (8.61) and (8.62). The resulting approximate confidence intervals are

$$\lambda^* \pm \frac{1.0426 z_p \lambda^*}{\beta^* \sqrt{n}} \tag{8.63}$$

and

$$\beta^* \pm \frac{0.7797 z_p \beta^*}{\sqrt{n}} \tag{8.64}$$

where z_p is as previously defined.

Example 8.8 (Asymptotic Confidence Intervals for λ, β Weibull Distribution)

If the data of the previous examples are assumed to be samples from Weibull distributions, the following confidence intervals are obtained:

Data Set 1: We calculate 90% confidence intervals. From Table 8.3, the appropriate tabulated z-value is found to be 1.645. From Example 8.1, the ML estimates are $\lambda^* = 0.3552$ and $\beta^* = 0.781$. The estimated variances are $\hat{V}(\lambda^*) = 1.087(0.3552)^2/127(0.781)^2 = 0.00177$ and $\hat{V}(\beta^*) = 0.608(0.781)^2/127 = 0.00292$. The 90% confidence interval for λ is

$$0.355 \pm 1.645\sqrt{0.00177} = 0.355 \pm 0.069$$

or $(0.286, 0.424)$. The 90% confidence interval for β is

$$0.781 \pm 1.645\sqrt{0.00292} = 0.781 \pm 0.089.$$

Data Set 2: Similarly, 99% confidence intervals based on this data set are found to be

λ: 0.0003007 ± 0.0000458

β: 1.81 ± 0.37

The Gamma Distribution

The variances of the asymptotic distributions of the ML estimators are [4, p. 190]

$$V(\beta^*) \approx \frac{\beta}{n[\beta\psi'(\beta) - 1]} \tag{8.65}$$

and

$$V(\lambda^*) \approx \frac{\lambda^2\psi'(\beta)}{n[\beta\psi'(\beta) - 1]} \tag{8.66}$$

where $\psi'(u) = d\psi(u)/du$ is the *trigamma function*. Tables, recursion relationships and other aids for calculation of the trigamma function are given in Chapter 6 of Ref. 8. The approximate confidence intervals are

$$\beta^* \pm z_p \left(\frac{\beta^*}{n[\beta^*\psi'(\beta^*) - 1]}\right)^{1/2} \tag{8.67}$$

and

$$\lambda^* \pm z_p \left(\frac{\lambda^{*2}\psi'(\beta^*)}{n[\beta^*\psi'(\beta^*) - 1]}\right) \tag{6.68}$$

More on estimation, particularly for small samples and incomplete data, can be found in Chapter 5 of Ref. 6 and Chapter 17 of Ref. 4.

Example 8.9 (Asymptotic Confidence Intervals for λ, β Gamma Distribution)

Data Set 1: We require $\psi(\beta^*) = \psi(0.676)$. From Ref. 8, this is found to be 2.996. The estimated variances are

$$\hat{V}(\beta^*) = \frac{0.676}{127[0.676(2.996 - 1]} = 0.00519$$

and

$$\hat{V}(\lambda^*) = \frac{0.21^2(2.996)}{127[0.676(2.996 - 1]} = 0.001015$$

95% confidence intervals for the parameters are

β: $0.676 \pm 1.96(0.072) = 0.676 \pm 0.141$

λ: $0.210 \pm 1.96(0.032) = 0.210 \pm 0.062$

Data Set 2: $\psi'(3.024) = 0.3912$. Ninety-five percent confidence intervals are

β: 3.024 ± 0.817

λ: 0.001021 ± 0.000300

The Lognormal Distribution

Normal theory is directly applicable on the log scale. Equation (8.58) with $\bar{x}$ replaced by $\bar{y}$, given in (8.38) and σ replaced by s if σ is unknown, may be used to obtain a confidence interval for μ for large samples. For small samples, σ is replaced by s in (8.58) and z_p is replaced by the corresponding fractile of the Student's t distribution with $(n - 1)$ df. The confidence interval for σ is based on the χ^2 distribution. The formula is

$$\left(\left(\frac{n - 1}{\chi^2_{p,n-1}} \right)^{1/2} s_y, \left(\frac{n - 1}{\chi^2_{1-p,n-1}} \right)^{1/2} s_y \right) \tag{8.69}$$

where s_y is given by (8.39) and $p = 1 - (1 - \gamma)/2$.

Example 8.10 (Confidence Intervals for μ, σ Lognormal Distribution)

We illustrate the computations with Data Set 2, for which $\bar{y} = 7.780$ and $s_y = 0.7468$. The 98% confidence interval for μ is [from Table 8.3 and Equation (8.58)]

$$7.780 \pm \frac{2.326(0.7468)}{\sqrt{95}} = 7.780 \pm 0.0178.$$

From (8.67), the 95% confidence interval for σ (with $\chi^2_{0.975,94} = 122.72$ and $\chi^2_{0.025,94} = 69.07$ obtained from an extended χ^2 table, a statistical program package, or by interpolation in Table 8.4) is

$$\left(\left(\frac{94}{122.72} \right)^{1/2} (0.7468), \left(\frac{94}{69.07} \right)^{1/2} (0.7468) \right) = (0.6536, 0.8712)$$

The Inverse Gaussian Distribution

Because the ML estimator of η is simply the sample mean $\bar{x}$, (8.58) may be used directly to obtain a large-sample confidence interval for η. Further, $\theta n/\theta^*$ is distributed as χ^2 with $(n - 1)$ df [4, Chap. 15]. This results in the confidence interval for θ given by

$$\left(\frac{\theta^* \chi^2_{p,n-1}}{n} \quad \frac{\theta^* \chi^2_{1-p,n-1}}{n} \right) \tag{8.70}$$

Example 8.11 (Confidence Intervals for η, θ, Inverse Gaussian Distribution)

For Data Set 2, $\bar{x} = 2963$ and $s = 1698$. The following 95% confidence intervals (using the tabulated values given in Example 8.10) are obtained:

$$\eta: \quad 2963 \pm \frac{1.96(1698)}{\sqrt{95}} = 2963 \pm 341$$

$$\theta: \quad \left(\frac{3699.5(69.07)}{95}, \frac{3699.5(122.72)}{95} \right) = (2690, 4779)$$

8.5 ESTIMATION OF RENEWAL FUNCTIONS

The renewal function $M(t)$ was defined in Section 8.2. This function is used in the cost analysis of warranty policies, particularly the free-replacement warranty, where $M(W)$ represents the number of free replacements that must be provided during a warranty period of length W. Ordinarily, the renewal function is not known and must be estimated on the basis of test data in order to predict future warranty costs.

There are a number of ways of estimating $M(t)$. (See Chapter 12 of Ref. 15 and Refs. 19–21.) The simplest approach is to estimate the param-

Table 8.5   Estimated Renewal Functions $M(t)$ Under Various Distributional Assumptions, Data Set 1

	Distribution[a]					
t	1	2	3	4	5	6
1	0.310	0.468	0.439	0.366	0.603	0.274
2	0.621	0.831	0.773	0.719	1.016	0.663
3	0.931	1.171	1.089	1.064	1.366	1.025
4	1.241	1.501	1.396	1.400	1.681	1.371
5	1.552	1.824	1.698	1.731	1.972	1.708

[a] 1 = exponential; 2 = Weibull; 3 = Gamma; 4 = mixed exponential; 5 = lognormal; 6 = inverse Gaussian.

eters of the life distribution as in the previous section and evaluate the renewal function at these parameter values. This is usually a reasonable approach (but see Refs. 19 and 20) and it is the approach that will be taken in the examples to follow.

The variance of the number of renewals, $V[N(t)]$, can also be estimated in this way. (See Refs. 9, 15, and 19.) Confidence intervals for $M(t)$ and related quantities are usually more difficult. An exception is the exponential distribution, for which $M(t) = \lambda t$, so a confidence is obtained directly from that for λ simply by multiplying the confidence limits by t. In other cases, the nonparametric method developed by Frees [20] is suggested.

We illustrate the basic estimation procedure for the two data sets considered previously in the examples which follow. In each case, $M(t)$ is estimated under each of the distributional assumptions discussed in the previous examples. $M(t)$ was calculated in these examples by use of the FORTRAN program given in Appendix C of Ref. 15.

Example 8.12   (Estimated Renewal Functions for Data Set 1)

The mean for Data Set 1 is $\bar{x} = 3.219$ years. In investigating the renewal functions under the various distributional assumptions, we look at $M(t)$ for $t = 1, \ldots, 5$. The results are given in Table 8.5. These values were calculated using the ML estimates, except for the mixed exponential distribution, for which the ML estimates were not calculated, and the inverse Gaussian distribution, for which the ML estimator appears to be unstable.* In the latter two cases, moment estimates were used in the calculation of $M(t)$.

---

* Values calculated using the ML estimates differed substantially from all other values in Table 8.5.

Table 8.6 Estimated Renewal Functions $M(t)$ Under Various Distributional Assumptions, Data Set 2

t	Distribution[a]				
	1	2	3	4	5
1000	0.338	0.112	0.082	0.121	0.036
2000	0.675	0.360	0.346	0.439	0.331
3000	1.013	0.675	0.676	0.770	0.676
4000	1.350	1.013	1.015	1.098	1.014
5000	1.688	1.355	1.353	1.424	1.351

[a] 1 = exponential; 2 = Weibull; 3 = Gamma; 4 = lognormal; 5 = inverse Gaussian.

Note from the results that $M(t)$, the expected number of replacements in the interval $[0,t)$, can vary quite significantly, depending on the assumed failure distribution. These data indicate a decreasing failure rate. Thus, as expected, the exponential renewal function significantly underestimates the expected number of failures.

Example 8.13   (Estimated Renewal Functions for Data Set 2)
    For Data Set 2, $\bar{x} = 2963$. $M(t)$ is tabulated for $t = 1000, 2000, \ldots ,$ 5000 in Table 8.6 for all except the mixed exponential distribution, for which no solution exists for a set of data exhibiting the characteristics of an IFR distribution. Again, moment estimates were used for the inverse Gaussian distribution and ML estimates for the remaining distributions.
    In this case, the exponential distribution provides a very conservative estimate of the expected number of failures. For small values of t, $M(t)$ can vary somewhat, depending on the distributional assumption made. As t increases, several of the distributions give very similar results for this set of data. As noted, however, in warranty analysis, relatively small values of t (relative to the MTTF) are of interest.

8.6   APPLICATIONS IN WARRANTY ANALYSIS

The cost of a warranty to both buyer and seller depends on the structure of the warranty, that is, the warranty terms, as well as on the life distribution of the items sold under warranty. In this section, estimation of warranty costs will be illustrated for a few relatively simple but commonly used warranties, the *free-replacement* warranty (FRW) and the *pro-rata* warranty (PRW). In each case, W is used to indicate the length of the warranty period.

Assuming that the distribution of lifetimes of the items is known, costs may be estimated by (1) performing a mathematical analysis to derive an expression for these costs as functions of the assumed life distribution and/or its parameters, (2) obtaining valid and reliable data on lifetimes of the item, (3) using these, as discussed in the previous sections, to estimate the parameters of the life distribution, and (4) using these parameter estimates to estimate the cost model.

Depending on the type of warranty and the assumed distribution of times to failure, the cost models may involve parameters directly or through functions of varying complexity. Examples of such functions, which must be estimated in order to evaluate the cost models, are the CDF $F(t)$, the MTTF μ, the *partial expectation* μ_W, given by

$$\mu_W = \int_0^W xf(x)\ dx \tag{8.71}$$

as well as the ordinary renewal function, defined previously, any of several other renewal functions, and solutions of complex, renewal-type integral equations.

For relatively straightforward (explicitly stated) functions, expressions for approximate variances can be given as functions of the parameters of the asymptotic distribution of the estimators. This is discussed in Chapter 12 of Ref. 15 and Ref. 16. For the more complex models, such as those for which no analytical solution exists, this is not possible. In the remainder of this section, results will be restricted to point estimation of a few simple cost models. In these models, c_b denotes the selling price (cost to buyer), c_s denotes the average cost to the seller (manufacturer) of supplying an item to the buyer, and $E(C_s)$ is the average total cost to the seller of selling an item with warranty.

8.6.1 Estimating Simple Cost Models: Free-Replacement Warranties

Cost models for the FRW are discussed in Chapter 10. Here we consider a few selected models that are relatively easily estimated.

Rebate FRW ("Money-Back Guarantee")

Under this policy, the buyer is given a full refund if the item fails prior to W after time of purchase. The expected cost to the seller is

$$E(C_s) = c_s + c_b F(W) \tag{8.72}$$

where $F(W)$ is the CDF of time to failure of the item. This is estimated by substitution of an estimate $\hat{F}(W)$ for $F(W)$ in (8.72). The resulting estimate, denoted $\hat{E}(C_s)$, is

$$\hat{E}(C_s) = c_s + c_b\hat{F}(W) \tag{8.73}$$

There are a number of approaches to the problem of estimating $F(W)$. We use that of the previous section, namely, substitution of estimates of the parameters into the formula for $F(W)$. [Note that $F(W)$ can be written explicitly for only a few of the life distributions of Section 8.2. Many computer routines are available, however, for calculation of the CDF in the other cases.]

Nonrenewing FRW

Under this policy, the seller supplies free-replacement items for all items that fail prior to W. A replacement at time $X < W$ is warrantied only for the remainder of the original warranty period, that is, the length of the FRW on the replacement item is $W - X$. The per-item cost to the seller for this warranty is estimated by

$$\hat{E}(C_s) = c_s[1 + \hat{M}(W)] \tag{8.74}$$

where $\hat{M}(W)$ is the estimated ordinary renewal function associated with the CDF $F(\cdot)$.

Renewing FRW

Under this policy, free replacements are given full-warranty coverage W. Thus, free replacements are provided until the first time an item has a lifetime of at least W. The estimated average cost to the seller per item sold is

$$\hat{E}(C_s) = c_s\left[1 + \frac{\hat{F}(W)}{1 - \hat{F}(W)}\right] \tag{8.75}$$

Example 8.14. (Estimated FRW Cost Models; Data Set 2)

Suppose that the three forms of FRW were being considered as possible warranty policies on a new item, with Data Set 2 being test data on the item, and that warranty periods of $W = 1000$ h and $W = 2000$ h were being considered. Estimated average costs per item to the seller for each of these policies under each of three possible distributional assumptions—Weibull, Gamma, and lognormal—are given in Table 8.7. In calculating the entries for this table, the ML estimates from Example 8.4 were used in evaluating $\hat{F}(W)$ in (8.73) and (8.75) and $\hat{M}(W)$ in (8.74). (See Table 8.6.)

It is instructive to compare both policies and distributions. For example, if the warranty period were taken to be 1000 h, the selling price c_b = \$100, and the cost to the seller c_s = \$60, then the average cost under

Table 8.7   Estimated Warranty Cost Models, Free-Replacement Warranty;
Data Set 2

W	Distribution	FRW Policy		
		Rebate	Nonrenewing	Renewing
1000	Weibull	$c_s + 0.1092c_b$	$1.112c_s$	$1.123c_s$
	Gamma	$c_s + 0.0816c_b$	$1.082c_s$	$1.089c_s$
	Lognormal	$c_s + 0.1202c_b$	$1.121c_s$	$1.137c_s$
2000	Weibull	$c_s + 0.3306c_b$	$1.360c_s$	$1.494c_s$
	Gamma	$c_s + 0.3291c_b$	$1.346c_s$	$1.491c_s$
	Lognormal	$c_s + 0.4048c_b$	$1.439c_s$	$1.680c_s$

the rebate FRW, including expected warranty costs, is $60 + 0.1092(100)$
= $70.92 under the assumption of a Weibull distribution, $68.16 assuming
Gamma-distributed lifetimes, and $72.02 assuming a lognormal distribu-
tion. Under the ordinary nonrenewing FRW, these costs are $66.72 under
the Weibull assumption, $64.92 assuming the Gamma distribution, and
$67.26 if lognormal. Under the renewing FRW, these costs are $67.38,
$65.34, and $68.22, respectively.

Note that a warranty period of 1000 h is just under one-third the
MTTF. If the warranty period were 2000 h, the costs would be substantially higher.

8.6.2   Estimated Simple Cost Models; Pro-Rata Warranty

Cost models for the PRW are given in Chapter 11. We consider the seller's
expected cost for an item sold under renewing PRW with linear proration.
Under this policy, the buyer receives a replacement upon failure of an
item covered under warranty at a cost proportional to the service time
achieved to that point in the warranty period (e.g., if the failure occurs
at $x = 0.8W$, the replacement item is provided at 80% of the purchase
price). The replacement item is covered under warranty terms identical
to those of the original item. For this warranty, the estimated average
cost to the seller is given by

$$E(C_s) = c_s + c_b \left(\hat{F}(W) - \frac{\mu_W}{W} \right) \qquad (8.76)$$

where μ_W is given by (8.71).

Except for the exponential distribution, μ_W presents some computational difficulties. (See Chapter 5 of Ref. 15.) For the Weibull distribution,

μ_w may be calculated as [15]

$$\mu_w = \frac{\beta}{\lambda(\beta + 1)} (\lambda\beta)^{\beta + 1} e^{-(\lambda W)^\beta} M\left(1, 1 + \frac{\beta}{\beta + 1}, \lambda^\beta W^\beta\right) \quad (8.77)$$

where

$$M(1, 1 + a, x) = 1 + \frac{x}{a + 1} + \frac{x^2}{(a + 1)(a + 2)}$$
$$+ \frac{x^3}{(a + 1)(a + 2)(a + 3)} + \cdots \quad (8.78)$$

(See Section 6.6 of Ref. 8.) The series converges reasonably rapidly unless β is small. In estimating warranty costs, these expressions are evaluated at the ML estimates, β^* and λ^*.

Example 8.15 (Estimated PRW Cost Model; Data Set 2, Weibull Distribution)

For Data Set 2, the ML estimates for the Weibull distribution (Example 8.4) are $\lambda^* = 0.0003007$ and $\beta^* = 1.795$. For $W = 1000$, (8.78) must be evaluated at $a = (\beta^* + 1)/\beta^* = 1.5571$ and $x = (1000\lambda^*)^{\beta^*} = 0.3007^{1.795} = 0.11568$. The result is $M(1, 2.557, 0.1157) = 1.0467$, from which it follows that

$$\mu_w = \frac{0.11568^{1.5571}}{0.0003007(1.5571)} e^{-0.11567} (1.0467) = 69.27$$

For $W = 2000$, we have $x = 0.40142$, from which we obtain $\mu_w = 406.04$. The estimated costs, from (8.76), are $c_s + (0.1092 - 0.0693)c_b = c_s + 0.0399c_b$ for $W = 1000$ and $c_s + 0.1276c_b$ for $W = 2000$. Note that, as would be expected, these costs are less than those of any version of the FRW.

8.7 RECOMMENDATIONS AND CONCLUSIONS

8.7.1 Data Analysis

In this chapter, we have been concerned with analysis of experimental data for purposes of estimating warranty models. In analyzing such data, the following precautions should be taken:

1. Experimental conditions must reasonably approximate actual consumer use of the product.
2. A large amount of valid, reliable data will usually be required in order to obtain sufficiently precise statistical results.
3. It is important that the correct failure distribution be used in

analysis of the data, particularly for IFR situations (which are the most common). The estimated costs may vary significantly for different distributions. The appropriate failure distribution may sometimes be deduced by the nature of the item and its failure modes. Some guidelines are given in Section 8.2 and in the references cited.

4. If the life distribution is not known, goodness-of-fit tests may be used (see Refs. 1–3) in an attempt to determine the appropriate distribution. These analyses should be updated periodically as additional data become available.

5. Alternatively, nonparametric procedures may be appropriate in some cases. (For example, see Ref. 21 for nonparametric estimation of renewal functions.)

6. If the data are incomplete (censored or truncated), the methods of this chapter must be modified. (See Refs. 4–6.)

7. Analysis of warranty claims data presents special problems. (See Chapter 9.)

8. Some of the models given in the ensuing chapters cannot easily be estimated by the methods of this and the following chapters. In these cases, simulation is often a feasible alternative. (See Chapter 11 of Ref. 15)

Comment: An overall managerial conclusion, demonstrated by the numerical examples of this chapter, is that both the warranty terms and the duration of the warranty period must be carefully selected. Both can dramatically affect the cost of the warranty.

8.7.2 Future Research

There are many unsolved problems in estimation of warranty cost models. In addition, there are a number of statistical methodologies that have not been applied in this area.

Ordinarily, a warranty manager or analyst will have available many types of data and other information, of varying degrees of relevance. These include the following:

- True experimental test data
- Data on lifetimes of similar products
- Data on lifetimes of some components or parts
- "Engineering judgment"
- Claims data

and so forth. Ideally, what is needed are models, methods of analysis, and information systems that would enable the user to

- Aggregate data of various types
- Provide cost predictions with confidence intervals for various alternative warranty policies
- Update the analyses and predictions as additional data become available

Note that the statistical methods of this chapter would form only a part of such a system.

Statistical research needed includes the following:

- Development of methods for aggregating and analyzing the different types of information
- Development of Bayesian approaches to warranty cost analysis
- Derivation of confidence interval procedures for many of the warranty cost models
- Investigation of methods of estimating implicit functions such as solutions of integral equations (see Chapters 10 and 11 of this Handbook and Chapters 5 and 6 of Ref. 15) and of the properties of the estimators
- Derivation of optimal estimators of partial moments
- Application of bootstrap and other nonparametric methods to estimation of warranty cost models

REFERENCES

1. Mansfield, E. (1991). *Statistics for Business and Economics*, 4th ed., W. W. Norton and Co., New York.
2. Stuart, A., and Ord, J. K. (1991). *Kendall's Advanced Theory of Statistics*, Vol. 2, 5th ed., Oxford University Press, New York.
3. Nelson, W. (1990). *Accelerated Testing*, John Wiley & Sons, New York.
4. Johnson, N. L., and Kotz, S. (1970). *Distributions in Statistics. Continuous Univariate Distributions—1*, John Wiley & Sons, New York.
5. Mann, N. R., Schafer, R. E., and Singpurwalla, N. (1974). *Methods for Statistical Analysis of Reliability and Life Data*, John Wiley & Sons, New York.
6. Lawless, J. F. (1982). *Statistical Models and Methods for Lifetime Data*, John Wiley & Sons, New York.
7. Kapur, K. C., and Lamberson, L. R. (1977). *Reliability in Engineering Design*, John Wiley & Sons, New York.
8. Abramowitz, M., and Stegun, I. A. (1964). *Handbook of Mathematical Functions*, National Bureau of Standards, Washington, DC.
9. Weibull, W. (1951). A statistical distribution function of wide applicability, *Journal of Applied Mechanics*, **18**, 293–297.
10. Baxter, L. A., Scheuer, E. M., Blischke, W. R., and McConalogue, D. J. (1981). *Renewal Tables: Tables of Functions Arising in Renewal Theory*,

Technical Report, Department of Information and Operations Management, University of Southern California, Los Angeles.

11. Giblin, M. T. (1983). *Tables of Renewal Functions Using a Generating Function Algorithm*, Technical Report, Postgraduate School of Studies in Industrial Technology, University of Bradford, England.

12. Soland, R. M. (1968). *Renewal Functions for Gamma and Weibull Distributions with Increasing Hazard Rate*, Technical Paper RAC-TP-329, Research Analysis Corp., McLean, VA.

13. White, J. S. (1964). Weibull renewal analysis, *Proc. Aerospace Reliability and Maintainability Conference*, Society of Automotive Engineers, Washington, DC, pp. 639–657.

14. Baxter, L. A., Scheuer, E. M., McConalogue, D. J., and Blischke, W. R. (1982). On the tabulation of the renewal function, *Technometrics*, **24**, 151–156.

15. Blischke, W. R., and Murthy, D. N. P. (1993). *Warranty Cost Analysis*, Marcel Dekker, Inc., New York.

16. Mood, A. M., Graybill, F. A., and Boes, D. C. (1974). *Introduction to the Theory of Statistics*, third edition, McGraw-Hill Book Co., New York.

17. Blischke, W. R., and Scheuer, E. M. (1986). Tabular aids for fitting Weibull moment estimates, *Naval Research Logistics Quarterly*, **33**, 145–153.

18. Rider, P. R. (1961). The method of moments applied to a mixture of two exponential distributions, *Annals of Mathematical Statistics*, **32**, 143–147.

19. Frees, E. W. (1986). Warranty analysis and renewal function estimation, *Naval Research Logistics Quarterly*, **33**, 361–372.

20. Frees, E. W. (1986). Estimating the cost of a warranty, *Journal of Business and Economic Statistics*, **6**, 79–86.

21. Frees, E. W. (1986). Nonparametric renewal function estimation, *Annals of Statistics*, **14**, 1366–1378.

9

Statistical Analysis of Warranty Claims Data

J. D. Kalbfleisch and J. F. Lawless

University of Waterloo
Waterloo, Ontario, Canada

9.1 INTRODUCTION

Manufacturers generally collect fairly detailed data on warranty claims. Besides providing a record of claims and their costs, such data may be used to predict future claims, to compare claims experience for different components or product lines, and to study variations in claims relative to variables such as time and place of manufacture or usage environment. These data also provide information on the field reliability of products and so may influence design and manufacturing decisions. To take full advantage of such possibilities, it is important to obtain good data and to employ sound analysis. This chapter presents some statistical methods of quite broad applicability.

We suppose that units of a product are sold over time and that information becomes available on the time and type of all warranty claims for each unit. We assume that when a claim is recorded, the date of sale of the unit is noted; this is typically necessary in order to verify that the unit is under warranty. In some sections, we also assume that the manufacturer knows the dates of sale for all the units that have been sold. This is the typical situation for expensive items such as automobiles but not for many other less expensive products.

Our objectives are to analyze warranty claims and, to a lesser extent, costs. Important aspects include comparisons, the assessment of trends and explanatory variables, and prediction. The emphasis is on aggregate claims for the population of units sold rather than on the detailed examination of claims patterns for individual units. We do, however, discuss individual claims and field reliability estimation briefly in Section 9.10. We consider only analysis in terms of calendar time and age of units (i.e., time since the unit was sold.) Situations involving usage time analysis are not considered but are mentioned briefly in Section 9.10.

Section 9.2 presents methods of age-specific claims analysis when complete data on both claims and units sold are available. Section 9.3 discusses adjustments for delays in the reporting of claims, and Section 9.4 provides an illustration based on automobile warranty data. Section 9.5 deals with adjustments when data on sales are incomplete. Section 9.6 deals with prediction, Section 9.7 with covariates and regression analysis, and Section 9.8 with calendar time effects on claims. The main focus of this chapter is on numbers of claims, but Section 9.9 briefly discusses how similar methods apply to the analysis of costs. Section 9.10 discusses the utilization of warranty claims data for reliability estimation, and Section 9.11 concludes with a few additional remarks and a summary of the Chapter's methodology.

9.2 SIMPLE AGE-SPECIFIC CLAIMS ANALYSIS

9.2.1 Notation and Assumptions

We assume to start that time is measured in discrete units and that data are available on the numbers of units sold and numbers of claims in each time period. For convenience of exposition we will take the time to be days. We assume that units are sold on calendar days $0, 1, \ldots, \tau$, with

$$N(d) = \text{number of units sold on day } d, \quad d = 0, 1, \ldots, \tau$$

We consider claims data that have accrued up to day $T \geq \tau$ and assume that we can determine

$$n^*(t, a) = \text{number of claims on day } t \text{ for units sold on day } t - a,$$
$$0 \leq a \leq t \leq T$$

Here, the variable a is the age of the unit at the time t of a claim.

Sometimes it is useful to classify claims according to the day of sale. For this purpose, we define $n(d, a)$ to be the number of claims at age a, for units sold on day d. Clearly,

$$n(d, a) = n^*(d + a, a), \quad a \geq 0, d \geq 0, a + d \leq T$$

In some cases, claims may be grouped or aggregated according to one or more of the following: age, time period, or date of sale. To deal with this later (see Section 9.2.3), we use n^* and n with capital letter arguments:

$$n^*(P, A) = \sum_{t \in P} \sum_{a \in A} n^*(t, a)$$

$$n(D, A) = \sum_{d \in D} \sum_{a \in A} n(d, a)$$

where D, P, and A represent sets of sales days, time period days, and age days, respectively.

In what follows, we suppose that $N(d)$, the number of units sold on day d, is known, $d = 0, 1, \ldots, \tau$. To develop methods of analysis, we assume that claims occur according to some random process such that, given $N(0), \ldots, N(\tau)$, $n^*(t, a)$ and $n(d, a)$ have (conditional) expected values

$$\mu^*(t, a) = E\{n^*(t, a)\}, \qquad \mu(d, a) = E\{n(d, a)\}$$

We similarly write $\mu^*(P, A)$ and $\mu(D, A)$ for grouped data.

The notation above refers to one specific type of claim for a unit. More generally, we may want to consider claims of several types simultaneously. In such cases, we subscript the quantities above to designate the type. For example, $n_j^*(t, a)$ refers to the number of type j claims on day t, for units of age a. Generally, different claim types can be analyzed separately, so we will proceed as though only one type is being considered.

9.2.2 Age-Specific Analysis

First, we consider situations in which the average or expected number of claims per unit per day may depend on the age of the unit but is roughly independent of other factors. This leads us to define age-specific expected claim frequencies:

$\lambda(a) = $ expected number of claims per unit while at age a

$$\Lambda(a) = \sum_{u=0}^{a} \lambda(u) = \text{expected number of claims per unit up to age } a$$

Note that $\lambda(a)$ is typically not the expected number of failures per unit at age a, as some units sold may no longer be in use or under warranty at age a. In addition, we note that with warranties under which a failed unit may be replaced, the "age" with respect to the definition above is the number of days since the unit that originated the warranty coverage was sold.

It follows that the expected values of $n^*(t, a)$ and $n(d, a)$ are

$$\mu^*(t, a) = N(t - a)\lambda(a), \qquad \mu(d, a) = N(d)\lambda(a) \tag{9.1}$$

When the $N(d)$'s are known, we can estimate the $\lambda(a)$'s from the warranty claim counts $n^*(t, a)$ up to day T (i.e., with $0 \le a \le t \le T$) as

$$\hat{\lambda}(a) = \frac{n_T(a)}{R_T(a)}, \quad a = 0, 1, \ldots, T \tag{9.2}$$

where

$$n_T(a) = \sum_{t=a}^{T} n^*(t, a) = \sum_{d=0}^{T-a} n(d, a)$$

is the total number of age a claims occurring up to day T, and

$$R_T(a) = \sum_{d=0}^{T-a} N(d) \tag{9.3}$$

is the total number of units that have reached age a on or before day T. The estimates (9.2) are maximum likelihood estimates (see Chapter 8) when the $n(d, a)$'s are independent Poisson random variables but are valid quite generally as long as (9.1) is true. These estimates and the corresponding estimates of cumulative claims

$$\hat{\Lambda}(a) = \sum_{u=0}^{a} \hat{\lambda}(u) \tag{9.4}$$

are very useful. An illustration based on real data is given in Section 9.4, but let us consider a simple artificial example here.

Example 9.1

Suppose that a product has a 1-year warranty period and that $N(d)$ = 100 units are sold on each day $d = 0, 1, \ldots, 364$ of a 1-year period. In a real situation, we would, of course, not have the same number of units sold on each day, and days when businesses are closed would preclude sales or claims being made on those days; but for simplicity of illustration, we ignore such features. We assume also for simplicity that the $N(d)$'s are known as each day passes, as are the numbers of warranty claims each day. We discuss how to handle uncertainty about sales or delays in the reporting of sales and claims in Sections 9.3 and 9.5.

Table 9.1 shows fictitious claims data for the 700 units sold over days 0, 1, . . . , 6. For example, on day 2 there was one claim on a unit of age 1 day (i.e., that unit had been sold on day 1) and no claims on units of age 0 days (i.e., units sold on day 2) or 2 days (units sold on day 0). The column totals in Table 9.1 show the total age a claims $n_6(a)$ up to day 6 for each a. The estimates $\hat{\lambda}(a)$ given by (9.2) are displayed in Table 9.2, along with the values $n_6(a)$ and $R_6(a)$ [see (9.3)] needed to compute them. The corresponding estimates of $\hat{\Lambda}(a)$ [see (9.4)] are also shown.

In the following, we will for convenience suppress the T in $n_T(a)$ and $R_T(a)$, writing just $n(a)$ and $R(a)$. Variance estimates for $\hat{\lambda}(a)$ and $\hat{\Lambda}(a)$ can be obtained under various assumptions and derivations are outlined in the appendix. If the $n(d, a)$'s are independent Poisson variables, then $\text{Var}\{n(a)\} = R(a)\lambda(a)$ and the variance of $\hat{\Lambda}(a)$ is estimated by

$$\hat{V}_p(a) = \widehat{\text{Var}}\{\hat{\Lambda}(a)\} = \sum_{u=0}^{a} \frac{\hat{\lambda}(u)}{R(u)} \tag{9.5}$$

This estimate is reasonable if units generate claims randomly and in an identical fashion. However, even though (9.1) may be acceptable, there is often extra-Poisson variation evident in the claim frequencies and sometimes a degree of correlation. Because (9.5) can in those cases substantially underestimate $\text{Var}\{\hat{\Lambda}(a)\}$, we will note two approaches that allow for extra-Poisson variation. A third approach is mentioned in the appendix.

A simple approach is to assume that the $n(d, a)$'s [or equivalently the $n^*(t, a)$'s] are independent random variables with means $\mu(d, a) =$

Table 9.1 Numbers of Claims $n^*(t, a)$ up to Day $T = 6$

Day (t)	Age (a)						
	0	1	2	3	4	5	6
0	1						
1	0	0					
2	0	1	0				
3	0	1	1	1			
4	1	2	1	0	1		
5	1	0	0	2	0	1	
6	0	1	1	1	0	1	1
Totals	3	5	3	4	1	2	1

Table 9.2 Calculation of Estimates of $\lambda(a)$ and $\Lambda(a)$

	Age (a)						
	0	1	2	3	4	5	6
$R_6(a)$	700	600	500	400	300	200	100
$n_6(a)$	3	5	3	4	1	2	1
$\hat{\lambda}(a)$	0.0043	0.0083	0.0060	0.0010	0.0033	0.0010	0.0100
$\hat{\Lambda}(a)$	0.0043	0.0126	0.0186	0.0196	0.0229	0.0239	0.0339

$N(d)\lambda(a)$ and variances

$$v(d, a) = \text{Var}\{n(d, a)\} = \sigma^2 \, \mu(d, a) \tag{9.6}$$

where $\sigma > 0$ is a dispersion parameter. The case $\sigma = 1$ gives the Poisson variance function and $\sigma > 1$ gives extra-Poisson variation. The model (9.6) often provides a reasonable approximation to reality. The parameter σ^2 may be estimated in several ways, by equating expressions for observed and expected second moments. For example, the fact that $E\{[n(d, a) - \mu(d, a)]^2/[\sigma^2 \, \mu(d, a)]\} = 1$ suggests the estimators

$$\hat{\sigma}^2 = \frac{1}{df} \sum_d \sum_a \frac{[n(d, a) - \hat{\mu}(d, a)]^2}{\hat{\mu}(d, a)} \tag{9.7}$$

and

$$\hat{\sigma}^2 = \frac{\sum_d \sum_a [n(d, a) - \hat{\mu}(d, a)]^2}{\sum_d \sum_a \hat{\mu}(d, a)}$$

where df equals the number of terms in the sum less the number of $\lambda(a)$'s that are estimated. In the case where we observe all $n(d, a)$'s with $0 \le d + a \le T$, we have $df = T(T + 1)/2$. The adequacy of (9.1) and (9.6) and the independence of the $n(d, a)$'s can be assessed by examining the residuals

$$r(d, a) = \frac{n(d, a) - \hat{\mu}(d, a)}{\hat{\sigma}\hat{\mu}(d, a)^{1/2}} \tag{9.8}$$

If the overall model is adequate, the $r(d, a)$'s should look roughly like independent standard normal variables, provided the $\mu(d, a)$'s are not too small. If the $\mu(d, a)$'s and $n(d, a)$'s are small (this will depend on the time units being used to record the counts as well as the claim rates and number of units sold), then it is better to use larger age and time intervals in defining (9.7) and (9.8), or to consider other types of residuals; Ref. 1 (Chapters 2 and 12) may be consulted for more information about residual analysis.

Under the model (9.6) and independent counts, the estimated variance of $\hat{\Lambda}(t)$ is simply $\hat{\sigma}^2$ times the Poisson estimate (9.5):

$$\hat{V}(a) = \hat{\sigma}^2 \sum_{u=0}^{a} \frac{\hat{\lambda}(u)}{R(u)} \tag{9.9}$$

Extra-Poisson variation may arise from several sources, including inherent variation in the robustness of units, variations in usage environment, and non-Poisson claim patterns for individual units. Kalbfleisch et al. [2] consider an approach that is motivated by assuming that there is an unobservable random variable α_i associated with each unit sold. The α_i's are assumed to be i.i.d., and the ith unit is assumed to generate claims according to a Poisson model with age-specific expected claims $\alpha_i \lambda(a)$. This leads to both extra-Poisson variation and a degree of correlation in claim counts corresponding to the same day of sale; we have

$$\text{Var}\{n(d, a)\} = \mu(d, a)[1 + \sigma_\alpha^2 \lambda(a)]$$

$$\text{cov}\{n(d, a_1), n(d, a_2)\} = \sigma_\alpha^2 N(d)\lambda(a_1)\lambda(a_2), \quad a_1 \neq a_2 \tag{9.10}$$

$$\text{cov}\{n(d_1, a_1), n(d_2, a_2)\} = 0, \quad d_1 \neq d_2$$

where σ_α^2 is the variance of the α_i's. The parameter σ_α^2 may be estimated by equations analogous to (9.9), for example,

$$\sum_d \sum_a \frac{[n(d, a) - \hat{\mu}(d, a)]^2}{\hat{\mu}(d, a)[1 + \sigma_\alpha^2 \hat{\lambda}(a)]} = df \tag{9.11}$$

where $\hat{\mu}(d, a)$, $\hat{\lambda}(a)$, and df are the same as for (9.7). Under this model, $\text{Var}\{\hat{\Lambda}(a)\}$ is estimated by

$$\hat{V}(a) = \hat{V}_p(a) + \hat{\sigma}_\alpha^2 \sum_{u=0}^{a} \frac{\hat{\lambda}(u)^2}{R(u)} + 2\hat{\sigma}_\alpha^2 \sum_{u=1}^{a} \hat{\lambda}(u)\hat{V}_p(u - 1) \tag{9.12}$$

where $\hat{V}_p(a)$ is the Poisson estimate (9.5). A derivation of this is sketched in the appendix.

As with the previous variance specification, assumptions (9.1) and (9.10) can be assessed by examining residuals, now defined as

$$r(d, a) = \frac{n(d, a) - \hat{\mu}(d, a)}{\{\hat{\mu}(d, a)[1 + \hat{\sigma}_\alpha^2 \hat{\lambda}(a)]\}^{1/2}} \tag{9.13}$$

In many situations, either of the variance estimation methods will be satisfactory; however, (9.6) is slightly easier to use. An additional check on (9.9) and (9.12) is mentioned in the appendix of this chapter. More refined variance function modeling is beyond the scope of this chapter; in any event, with most types of warranty data, detailed modeling of the variance function is not warranted because there will generally be departures from the mean specification (9.1).

Approximate confidence limits for $\Lambda(a)$ may be obtained by using the fact that if (9.1) and the variance specification (9.6) or (9.10) used are correct, then $[\hat{\Lambda}(a) - \Lambda(a)]/\hat{V}(a)^{1/2}$ has an approximate standard normal distribution. For example, an approximate 0.95 confidence interval for $\Lambda(a)$ is given by

$$\hat{\Lambda}(a) \pm 1.96\hat{V}(a)^{1/2}$$

Confidence limits are illustrated in the automobile warranty claims example of Section 9.4.

9.2.3 Grouped Data

Frequently, data are grouped into total claims for units in various age groups, over specific time periods. For example, the number of claims on units aged 0–30 days, 31–60 days, and so on might be given on a monthly basis. If the age groups and reporting periods are of the same fixed length, then one possibility is to use the procedures in the previous section with the periods taken as the units of time. However, if the age groups and reporting periods are of different lengths or if the periods are fairly long relative to the time units in which the warranty period is specified, some adjustments are necessary.

Suppose that claims are grouped into age intervals $A_i = [a_{i-1}, a_i)$ with $a_0 = 0 < a_1 < a_2 < \cdots$. In this case, we seek to estimate the expected number of claims for that interval,

$$\Lambda(A_i) = \sum_{a=a_{i-1}}^{a_i-1} \lambda(a) \tag{9.14}$$

A natural estimate is

$$\hat{\Lambda}(A_i) = \frac{n(A_i)}{R(A_i)} \tag{9.15}$$

where, extending the notation used for (9.2) and suppressing T,

$$n(A_i) = \sum_t n^*(t, A_i) = \text{number of claims on units of age } a \in A_i$$

$$R(A_i) = \frac{1}{a_i - a_{i-1}} \sum_{a=a_{i-1}}^{a_i-1} R(a) \tag{9.16}$$

where $R(a) = \sum_{d=0}^{T-a} N(d)$, as before. To motivate (9.15), we note that

$$E\{n(A_i)\} = \sum_{a=a_{i-1}}^{a_i-1} \sum_{d=0}^{T-a} E\{n(d, a)\} = \sum_{a=a_{i-1}}^{a_i-1} \lambda(a)R(a)$$

and assume (with little practical consequence if $a_i - a_{i-1}$ is not too large) that the $\lambda(a)$'s are constant for $a_{i-1} \leq a < a_i$. Thus, $E\{n(A_i)\} = \Lambda(A_i)R(A_i)$, which motivates (9.15).

If sales data are only available in aggregate for different time periods, then (9.16) has to be estimated. The simplest way to do this is to estimate the daily sales figures $N(d)$ from the available period data, thus providing estimates of the $R(a)$'s and of (9.16); the $N(d)$'s can be estimated using plausible assumptions about sales patterns. Provided that the sales periods are not too long relative to the age periods, errors due to estimation of the $R(A_i)$'s should not be substantial. Further discussion of estimated sales data is given in Section 9.4.

The easiest approach to variance estimation is to assume independence of the $n^*(t, a)$'s along with (9.6). Then $\text{Var}\{n(A_i)\} = \sigma^2\Lambda(A_i)R(A_i)$, where we continue to assume that the $\lambda(a)$'s are constant over A_i. Thus, by (9.15), $\text{Var}\{\hat{\Lambda}(A_i)\}$ is estimated by

$$\hat{V}(A_i) = \hat{\sigma}^2 \frac{\hat{\Lambda}(A_i)}{R(A_i)} \tag{9.17}$$

A suitable estimate for σ^2 is

$$\hat{\sigma}^2 = \frac{1}{df} \sum_j \sum_i \frac{\{n^*(P_j, A_i) - \hat{\mu}(P_j, A_i)\}^2}{\hat{\mu}(P_j, A_i)} \tag{9.18}$$

where the sum ranges over all of the time periods (P_j) and age intervals (A_i) in the data, df equals the number of terms in the sum minus the number of age intervals, and $\mu(P_j, A_i) = E\{n^*(P_j, A_i)\}$ is estimated by

$$\hat{\mu}(P_j, A_i) = \frac{\hat{\Lambda}(A_i)}{a_i - a_{i-1}} \sum_{a=a_{i-1}}^{a_i-1} \sum_{t \in P_j} N(t - a) \tag{9.19}$$

As before, the $N(d)$'s will need to be estimated if sales data are available only in aggregate form.

Confidence limits for $\Lambda(A_i)$'s or sums of $\Lambda(A_i)$'s may be obtained by treating the $\hat{\Lambda}(A_i)$'s as independent and approximately normally distributed with means $\Lambda(A_i)$ and variances $\hat{V}(A_i)$.

9.3   ADJUSTMENTS FOR REPORTING DELAYS

Frequently, there are delays in the reporting or ratification of warranty claims and hence delays in their being recorded in the database used for analysis. Recent claim counts in that case are an underestimate of the number of claims actually made. In this case, one approach to analysis is to exclude recently reported claims: for example, if reporting delays are up to 60 days duration we could, at any point in time, use only data on claims that had occurred at least 60 days before. Another approach, which is important when prompt reviews of warranty claims are needed, is to adjust recent claim counts for underreporting. We describe the techniques presented by Kalbfleisch et al. [2].

We assume a stationary reporting delay distribution and let

$$f(r) = Pr(\text{a claim is reported exactly } r \text{ days after it occurs}),$$

$$r = 0, 1, \ldots$$

and $F(r) = f(0) + \cdots + f(r)$. The term "reported" here refers to the point at which the claim is entered into the database used for analysis. The $f(r)$'s can be estimated from previous claims data, and we will thus consider them known; Kalbfleisch et al. [2] and Kalbfleisch and Lawless [3] discuss the estimation of the $f(r)$'s. The procedure for estimating the $\lambda(a)$'s is now to consider counts $n(d, a, r) = $ the number of age a claims that had a reporting delay of r days, for units that were sold on day d. Then, assuming that $E\{n(d, a, r)\} = N(d)\lambda(a)f(r)$, we find that $\lambda(a)$ can be estimated by the same kind of expression as (9.2),

$$\hat{\lambda}(a) = \frac{n_T(a)}{R_T(a)} \tag{9.20}$$

where now [temporarily reintroducing the T in $n(a)$ and $R(a)$ for clarity]

$$n_T(a) = \text{total number of age } a \text{ claims reported up to day } T$$

$$= \sum_{d=0}^{T-a} \sum_{r=0}^{T-d-a} n(d, a, r)$$

$$R_T(a) = \sum_{d=0}^{T-a} N(d)F(T - a - d) \tag{9.21}$$

Thus, as in (9.2), $n_T(a)$ is the number of age a claims reported, but $R_T(a)$ is now a discounted number of units reaching age a, the discounting factor being the probability that a claim occurring on day $t = a + d$ would be reported by day T and, hence, be in the data.

A variance estimate for $\hat{\Lambda}(a) = \sum_{u=0}^{a} \hat{\lambda}(u)$ based on the Poisson model is once again given by (9.5), with $R(a)$ now given by (9.21). It is sensible, however, to allow for extra-Poisson variation. A simple approach is to use the model (9.6) with independent counts, which yields the variance estimate (9.9). In this case, σ^2 may be estimated by moment methods, as with (9.7). One reasonable estimator, analogous to the second estimator in (9.7), is

$$\hat{\sigma}^2 = \frac{\sum\limits_{d=0}^{T} \sum\limits_{a=0}^{T-d} [n(d, a, \cdot) - \hat{\mu}(d, a, \cdot)]^2}{\sum\limits_{d=0}^{T} \sum\limits_{a=0}^{T-d} \hat{\mu}(d, a, \cdot)} \tag{9.22}$$

where $\hat{\mu}(d, a, \cdot) = N(d)\hat{\lambda}(a)F(T - d - a)$. Another way to estimate σ^2 is to use the total observed and expected claims for each product unit; this gives

$$\hat{\sigma}^2 = \frac{\sum\limits_{i=1}^{m} (y_i - \hat{\mu}_i)^2}{\sum\limits_{i=1}^{m} \hat{\mu}_i} \tag{9.23}$$

where y_i is the total number of claims for unit i ($i = 1, \ldots, m$) and $\hat{\mu}_i = \sum_{a=0}^{T-d_i} F(T - d_i - a)\hat{\lambda}(a)$, where d_i is the day unit i was sold, estimates $E(y_i)$.

Kalbfleisch et al. [2] discuss another model for extra-Poisson variation. They also discuss adjustments when claims or sales data are aggregated, as in Section 9.3. In particular, (9.15) and (9.17) still hold, with $R(a)$ in (9.16) replaced by (9.21).

9.4 EXAMPLES

Example 9.2

We start with the same artificial illustration of age-specific analysis as in Example 9.1. Consider a product with a 1-year warranty period and suppose that $N(d) = 100$ units are sold on each day $d = 0, 1, \ldots,$ 364 of a 1-year period. For this example, we generated artificial data by assuming that the numbers of claims at age a for any unit are independent

Poisson random variables with constant age-specific claim rates $\lambda(a) =$ 0.002 for $a = 0, 1, \ldots, 364$. We assume that reporting delays are distributed over 0 to 59 days, with probabilities $f(r) = 1/30$ for $r = 20, \ldots,$ 39 and $f(r) = 1/120$ for $r = 0, \ldots, 19$ and $40, \ldots, 59$.

Figure 9.1 shows estimates $\hat{\Lambda}(a)$, $a = 0, 1, \ldots, 364$ based on (9.20) and (9.21) and data on claims reported up to 1 year after the first units were sold (i.e., $T = 364$). Table 9.3 shows part of the calculations involved: the first three columns show for selected ages a the value of $R_T(a)$ of (9.21) and $\hat{\Lambda}(a) = \sum_{u=0}^{a} \hat{\lambda}(u)$. Figure 9.1 also shows (lower curve) the biased estimate of $\Lambda(a)$ that results if we ignore the reporting delays and incorrectly use (9.2) for $\hat{\lambda}(a)$ with (9.3) for $R_T(a)$. The underestimation of $\Lambda(a)$ is especially severe for a close to 364 because only a small fraction of claims made close to 364 days (and none past day 345 in this example) have been reported by $T = 364$ days.

If claims data are grouped into larger age classes, we use (9.15) to estimate the expected claims per unit for each class. The last five columns

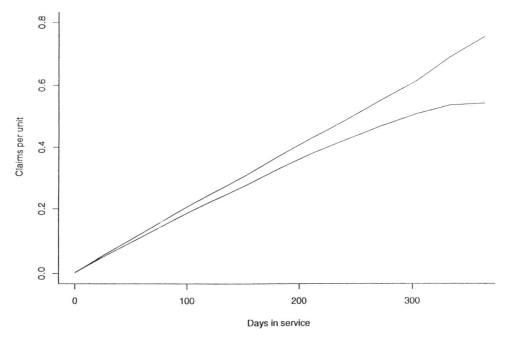

Figure 9.1   Estimated average cumulative claims per unit, $\hat{\Lambda}(a)$, based on artificial data. (The lower curve ignores reporting delays and so is biased.)

Table 9.3 Estimation of $\Lambda(a)$ from Artificial Data at $T = 364$ Days

a	$R_T(a)$	$\hat{\Lambda}(a)$	$A_i = (a_{i-1}, a_i)$	$n(A_i)$	$R(A_i)$	$\hat{\Lambda}(A_i)$	$\hat{\Lambda}(a_i)$
0	33,550	0.002					
30	30,550	0.064	(0, 30)	2064	32,050	0.0644	0.064
60	27,550	0.124	(31, 60)	1726	29,000	0.0595	0.124
90	24,550	0.185	(61, 90)	1586	26,000	0.0610	0.185
121	21,450	0.247	(91, 121)	1416	22,950	0.0617	0.247
151	18,450	0.304	(122, 151)	1132	19,900	0.0569	0.304
181	15,450	0.366	(152, 181)	1053	16,900	0.0623	0.366
211	12,450	0.428	(182, 211)	870	13,900	0.0626	0.428
242	9,450	0.488	(212, 242)	648	10,850	0.0597	0.488
272	6,352	0.548	(243, 272)	470	7,800	0.0603	0.548
303	3,250	0.610	(273, 303)	294	4,750	0.0619	0.610
333	635	0.684	(304, 333)	136	1808.5	0.0752	0.686
364	0.83	0.750	(334, 364)	10	169.7	0.0589	0.744

of Table 9.3 show results when age at claim is assigned to classes 0–30 days, 31–60 days, and so on, as shown in column 4. Column 5 shows the total claims $n(A_i)$ [see (9.16)] for each age class A_i, and column 6 shows $R(A_i)$ as in (9.16) but with $R(a)$ defined as in (9.21) to allow for reporting delays. Column 7 shows the estimated average claims per unit $\hat{\Lambda}(A_i)$ for each age class, and column 8 the cumulative estimates $\hat{\Lambda}(a_i) = \sum_{j=1}^{i} \hat{\Lambda}(A_j)$. It is noted that the $\hat{\Lambda}(a_i)$'s are virtually identical to those obtained from the raw data (column 3).

Example 9.3

We consider some real warranty data for a system on a particular car model and year. As of the final database update, 36,683 cars had been sold and 5760 claims had been reported. For the discussion here, day 0 is defined as the day on which the first sales occurred.

Figure 9.2 shows the pattern of sales by week. We estimated claim reporting delay probabilities from the data; see Kalbfleisch et al. [2, Sect. 1] and Kalbfleisch and Lawless [3] for a discussion of this. The estimated cumulative probability function $F(r)$ is summarized in Table 9.4. Figure 9.3 shows the estimated average cumulative claims per vehicle, $\hat{\Lambda}(a)$; for illustration, we have shown the curves that result from the sales and claims reported up to each of $T = 91, 182, 273, 365, 456,$ and 547 days, respectively; that is, these are the estimates that we would obtain approximately 3, 6, 9, 12, 15, and 18 months, respectively, after the first cars of that

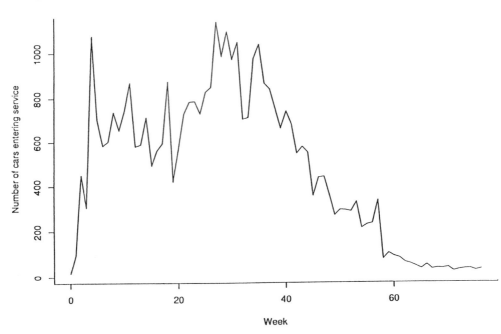

Figure 9.2   Number of cars entering service (sold), by week.

Table 9.4   Estimated Probability $F(r)$
That Reporting Delay Is $\leq r$ Days

r	$F(r)$
10	0.058
20	0.553
30	0.811
40	0.903
50	0.934
60	0.960
80	0.981
100	0.989
120	0.992

model year were sold. We observe that the estimates at 9–18 months agree well, but those for $T = 91$ and 182 are somewhat lower, suggesting that cars sold early in the model year had a somewhat lower frequency of early claims than cars sold later. Kalbfleisch et al. [2] discuss and suggest explanations for this. One point to note is that the estimates $\hat{\Lambda}(a)$ for $T = 91$ have rather large standard errors.

Plots like Figure 9.3 are very useful for tracking warranty claims experience as time progresses. They may also be used to compare claim rates for different time periods or groups of products. In Figure 9.4, we show estimates of $\Lambda(a)$ for cars that were manufactured in each of six 2-month production periods, going from mid-July to mid-September (Period 1) to mid-May to mid-July the following year (Period 6). The plots are based on all claims reported up to $T = 547$ days, of which there were 5701. They reveal the striking fact that average claims per vehicle appear to be very similar for all periods except Period 3 (November–January),

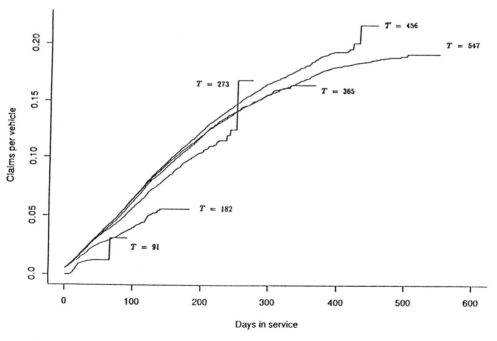

Figure 9.3 Estimated average cumulative claims per car, $\hat{\Lambda}(a)$, based on claims reported on various days T.

Warranty Claim Frequencies

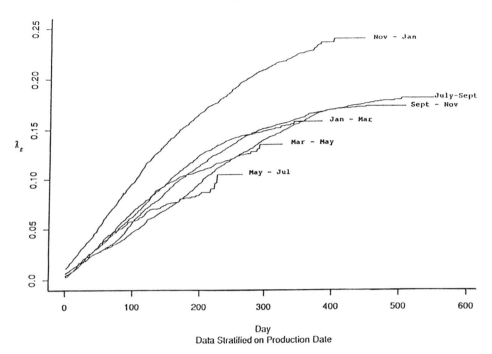

Figure 9.4   Estimated average cumulative claims per car, $\hat{\Lambda}(a)$, stratified according to production period.

for which claims are much higher. It would be interesting to determine the source of this difference.

It is a good idea to compute variance estimates or standard errors for $\hat{\Lambda}(a)$'s so that we may be aware of the uncertainty about $\Lambda(a)$. For example, if we use the variance function (9.6) and estimate σ^2 separately for the six production period groups using (9.23), we obtain $\hat{\sigma}^2 = 1.86$, 1.67, 1.87, 1.59, 1.65, and 1.30, respectively. These may be used in (9.9) to estimate $\text{Var}\{\hat{\Lambda}(a)\}$ and to obtain approximate confidence limits for $\Lambda(a)$ as described at the end of Section 9.2.2. As an illustration, let us consider groups 1 and 3 and the average claims per vehicle up to age 1 year [i.e., $\Lambda(364)$]. Using (9.9) and the estimates of σ^2 we find that the standard errors $\hat{V}(364)^{1/2}$ for $\hat{\Lambda}(364)$ are approximately 0.006 and 0.0085 for groups 1 and 3, respectively. Approximate 0.95 confidence limits [i.e.,

$\hat{\Lambda}(364) \pm 1.96\hat{V}(364)^{1/2}]$ for the two periods are roughly 0.15–0.17 and 0.21–0.25. It seems clear that the two average claim values are different. In Section 9.7, we test the equality of the $\Lambda(a)$ functions for all six groups and find that there is very strong evidence against equality.

We conclude this example with a few additional remarks. We note first that the rather large jumps at the right-hand ends of some of the plots in Figures 9.3 and 9.4 are due to the fact that for ages a close to T, $R_T(a)$ is small and the variability in $\hat{\lambda}(a)$ is consequently large. For the point estimation of $\Lambda(a)$, one may want to smooth the last portion of the plots; standard errors for $\hat{\Lambda}(a)$ in such cases make it clear that there is considerable uncertainty about $\Lambda(a)$, based solely on the data. We remark also that the warranty in this example is 1 year and has a 12,000-mile limit. Consequently, one should not see any claims at ages above 1 year: Figures 9.3 and 9.4 show that there are indeed a few claims allowed, presumably on a courtesy basis or due to data recording errors. Note too that the fact that some vehicles reach the 12,000-mile limit in less than 1 year and cannot have claims after that does not affect the analysis presented here, because $\lambda(a)$ is defined as the average number of age a claims per vehicle sold, taking into account that some vehicles drop out of the warranty coverage before 1 year. A third point is that we have assumed reporting delay probabilities to be constant over the entire period of the data. This may be checked by examining the delays for claims reported in different time periods. There is no evidence here that delay probabilities are changing, but if there were, it would be important to adjust the analysis to deal with it; this is readily done. Finally, there are some days on which claims cannot be made or reported, in particular weekend days and holidays. It is usually not worth incorporating this additional complexity into the models, as there is little effect on estimates of $\lambda(a)$ or $\Lambda(a)$, and we have not done it here.

9.5 ADJUSTMENTS WHEN SALES ARE ESTIMATED

Frequently, the exact number of product units sold in different time periods is unknown to the manufacturer. If sales up to time T are known to a fairly close approximation, then close approximations to $R_T(a)$ in (9.2) are available and we may proceed as though these values were known. If only imprecise estimates of sales are available, however, it may be desirable to reflect this by including an extra term in variance estimates for $\hat{\lambda}(a)$ or $\hat{\Lambda}(a)$ or by a sensitivity analysis to determine the effect of uncertainty about sales.

It may be shown that if $\hat{R}(a)$'s are approximately unbiased estimates of $R(a)$'s, then the variance of $\hat{\Lambda}(a)$ is estimated by

$$\hat{V}_1(a) = \hat{V}(a) + \hat{\mathrm{Var}}\left\{\sum_{u=0}^{a} \lambda(u) \frac{\hat{R}(u)}{R(u)}\right\} \qquad (9.24)$$

where $\hat{V}(a)$ is the estimated variance for $\hat{\Lambda}(a)$ without the adjustment; this is given by (9.5) for the Poisson model or, more generally, (9.9) for the model (9.6) that incorporates extra-Poisson variation. The second term on the right-hand side of (9.24) accounts for the fact that the $R(a)$'s are estimated. In order to evaluate this term, we need to have an estimated covariance matrix for the $\hat{R}(a)/R(a)$ values, from which an algebraic expression can be obtained. However, the simplest approach is to approximate the distribution of the estimated sales data $\hat{N}(d)$ by some distribution, and to estimate the right-hand part of (9.24) by simulation.

An alternative procedure that is adequate for most practical situations is to ignore the additional variance term in (9.24) but to carry out a check on the sensitivity of estimates $\hat{\Lambda}(a)$ and $\hat{V}(a)$ to the estimated $R(a)$ values. By varying the $R(a)$'s in ways that are considered plausible, we can see to what extent $\hat{\Lambda}(a)$ and $\hat{V}(a)$ vary, and use this to modify confidence limits for $\Lambda(a)$ in an informal way.

9.6   PREDICTION

We have so far focused on the estimation of expected claim counts for a hypothetical infinite population of units, of which those sold are considered a random sample. Such estimates are very useful for summarizing claims experience and for comparing different groups of units. However, we should remember that the population of units that is eventually sold is finite. Thus, if the total number of units sold over the time period $(0, \tau)$ is $N. = \sum_{d=0}^{\tau} N(d)$, then the actual (finite population) average number of claims per unit at age a is

$$m(a) = \frac{\sum_{d=0}^{\tau} n(d, a)}{N.}, \qquad a = 0, 1, 2, \dots$$

and

$$M(a) = \sum_{u=0}^{a} m(u)$$

is the average number of claims per unit up to age a. If data on claims up to time T are available, then the $m(a)$'s are partially known, and the problem of estimating them reduces to one of predicting the $n(d, a)$'s that are still unobserved. This problem is discussed in detail by Kalbfleisch et al. [2], and we summarize the main points.

Sensible estimates of $m(a)$ and $M(a)$ based on data to time T are clearly

$$\hat{m}(a) = \hat{\lambda}(a), \qquad \hat{M}(a) = \hat{\Lambda}(a) \tag{9.25}$$

where $\hat{\lambda}(a)$ and $\hat{\Lambda}(a)$ are given by (9.2) and (9.4), respectively. The difference that consideration of the finite population makes is in the variance estimates; under the extra-Poisson dispersion model with variance function (2.6), we have (assuming $\tau = T$, so that $N. = N$)

$$\text{Var}\{\hat{m}(a) - m(a)\} = \sigma^2 \left\{ \frac{N - R(a)}{NR(a)} \right\} \lambda(a) \tag{9.26}$$

In addition, the $\hat{m}(a)$'s are independent, and

$$V_{\text{Pred}}(a) = \text{Var}\{\hat{M}(a) - M(a)\} = \sigma^2 \sum_{u=0}^{a} \left\{ \frac{N - R(u)}{NR(u)} \right\} \lambda(u) \tag{9.27}$$

The overdispersion parameters σ^2 and the $\lambda(a)$'s are estimated as in Section 9.2; setting $\sigma^2 = 1$ gives the Poisson distribution results. Note that if $R(u) = N$ for $0 \le u \le a$ (which implies that all units sold over $0, 1, \ldots,$ T have already reached age a by time T), then $V_{\text{Pred}}(a) = 0$. This is as it should be, because in that case, $M(a) = \hat{M}(a)$ is fully known.

Grouped data are handled similarly, according to the procedures in Section 9.3. In particular, quantities $M(A_i) = \sum_{a=a_{i-1}}^{a_i - 1} m(a)$ [see (9.14)] are estimated as $\hat{M}(A_i) = \hat{\Lambda}(A_i)$, from (9.15), and $\text{Var}\{\hat{M}(A_i) - M(A_i)\}$ is estimated by [compare (9.17)]

$$\hat{V}_{\text{Pred}}(A_i) = \hat{\sigma}^2 \left\{ \frac{N - R(A_i)}{NR(A_i)} \right\} \hat{\Lambda}(A_i) \tag{9.28}$$

where $R(A_i)$ is given by (9.16). Calculations are outlined in the appendix.

Reporting delays as discussed in Section 9.3 create no additional problems; we merely replace $R(a) = \sum_{d=0}^{T-a} N(d)$ with (9.21) in any formulas.

One may wish to predict the total numbers of claims, as opposed to the average claims per unit; this is also easily done. For example, suppose that we wish to predict $n^*(t, a)$, the number of age a claims in a future time period t. To do this, we consider $n^*(t, a) - \hat{\mu}^*(t, a)$, where $\hat{\mu}^*$ $(t, a) = N(t - a)\hat{\lambda}(a)$ is a point estimate of $n^*(t, a)$ or of $\mu^*(t, a)$. It is clear that

$$\text{Var}\{n^*(t, a) - \hat{\mu}^*(t, a)\} = \sigma^2\mu^*(t, a) + N(t - a)^2\frac{\sigma^2\lambda(a)}{R(a)}$$

$$= \sigma^2N(t - a)\lambda(a)\left\{1 + \frac{N(t - a)}{R(a)}\right\} \qquad (9.29)$$

where we continue to use the variance model (9.6).

Prediction intervals for quantities such as $m(a)$, $M(a)$, or $n^*(t, a)$ are obtained by treating $m(a) - \hat{m}(a)$, $M(a) - \hat{M}(a)$, or $n^*(t, a) - \hat{\mu}^*(t, a)$, respectively, as approximately normal with mean 0 and variance estimated from (9.26), (9.27), or (9.29), respectively. As earlier, such approximate intervals work best when the quantity being predicted is not too small.

9.7 COVARIATES AND REGRESSION ANALYSIS

The methods described so far are appropriate when the expected number of age a claims per unit does not vary substantially with respect to calendar time or other factors. Sometimes expected claims may depend on factors such as manufacturing conditions or the environment in which the unit is used, however. The car data at the end of Section 9.4 provide a graphic example.

In situations where a few separate conditions can be identified as of possible importance, the simplest procedure is to analyze the data separately under the different conditions; this was the approach taken at the end of Section 9.4 where we examined the claims experience of cars manufactured in six distinct production periods. Another approach that is especially useful when conditions thought to affect claims cannot be categorized with a few groups is regression analysis. In this case, covariates are associated with units or groups of units, and a model for the way that the covariates affect the expected claims is postulated. Regression models for count data such as claims have been widely studied, and standard methods and software are available for analysis (e.g., Ref. 1, especially Chapter 6). We will merely outline a few points, as a thorough treatment of regression is outside the scope of this chapter.

We suppose that there is a vector of covariates z_i associated with each individual unit, i, that enters service. The expected number of claims at age a for unit i, entering service on day d, can be conveniently modeled in the log linear form

$$E\{n_i(d, a)\} = \mu_i(d, a) = \lambda(a)e^{z_i'\beta} \qquad (9.30)$$

where β is a vector of regression parameters. The covariate z_i can specify, for example, manufacturing characteristics, model lines, and, if they are available, even time-varying factors such as environmental variables. In this latter case, z_i could vary according to the current calendar time $t = d + a$.

The models (9.30), with variance function $\text{Var}\{n_i(d, a)\} = \sigma^2 \mu_i(d, a)$ to allow for extra-Poisson variation, can be fitted by Poisson maximum likelihood or quasi-likelihood methods [1, Chap. 6]. Most general statistical analysis systems (e.g., SAS, BMDP, GLIM, SPSS, S) perform the necessary calculations. Checks on the form of (9.30) may be made by examination of residuals, as described in Ref. 1 and other texts.

Example 9.4

As a simple example, we consider the comparison of different production periods, described in Example 9.3, via regression analysis.

We define a covariate vector $z_i = (z_{i1}, \ldots, z_{i5})'$ where $z_{ij} = 1$ if vehicle i was produced in period j and 0 if not ($j = 1, \ldots, 5$), and use the regression model (9.30). We can test whether the $\Lambda(a)$ functions for the six periods are equal by considering the hypothesis H: $\beta = 0$. A test of H may be carried out by obtaining $\hat{\beta}$ and its estimated asymptotic variance (e.g., see Chapter 6 of Ref. 1 or Ref. 4) $V(\hat{\beta})$. If H is true, $\hat{\beta}'V(\hat{\beta})^{-1}\hat{\beta}$ is approximately distributed as $\chi^2_{(5)}$. Lawless and Nadeau [5] find $\hat{\beta}'V(\hat{\beta})^{-1}\hat{\beta} = 85.6$ with one particular choice of variance estimate. The 99.9 percentile of $\chi^2_{(5)}$ is only 20.52, so there is clearly very strong evidence against H, as seemed clear from Figure 9.4 and the standard errors referred to in Section 9.4.

9.8 CALENDAR TIME EFFECTS

Sometimes there are systematic calendar time effects on rates of claims. For example, a monotonic trend over time might occur because of an improvement or deterioration in the reliability of the units being produced over time, whereas products that experience a pronounced seasonal usage pattern typically show a similar pattern in warranty claims.

Often calendar time effects can be examined simply by estimating age-specific expected claims via the models (9.1), separately for claims occurring in different time periods. When calendar time is an important (perhaps predominant) factor in the occurrence of claims, however, it is best to focus on the counts $n^*(t, a)$ and their expected values $\mu^*(t, a)$ described in Section 9.2.1. If $\mu^*(t, a) = E\{n^*(t, a)\}$ is the expected number of claims in time period t due to units of age a (i.e., units sold in period $t - a$), then

$$\mu^*(t, a) = N(t - a)g(t, a) \tag{9.31}$$

where $g(t, a)$ is the expected number of claims per unit for a unit of age a at calendar time t. Models for $g(t, a)$ may be fitted and examined using the log linear regression software mentioned in Section 9.7. The obvious place to start is with a "main effects" model with

$$g(t, a) = h(t)\lambda(a) \tag{9.32}$$

where the $h(t)$'s and $\lambda(a)$'s are non-negative parameters. As (9.32) stands, the parameters are identifiable only up to a multiplicative constant. Identifiability may be achieved by placing a single constraint on the parameters, for example, $h(0) = 1$ or $\lambda(0) = 1$. An examination of residuals $r^*(t, a)$ $= n^*(t, a) - \hat{\mu}^*(t, a)$ will usually indicate whether (9.32) is suitable. Tests may also be carried out for hypotheses of interest; for example, if the $\lambda(a)$'s in (9.32) are equal, there is no age effect (i.e., the expected number of claims per unit does not vary with age, although it may vary with calendar time), so we may wish to test that $\lambda(0) = \lambda(1) = \cdots = \lambda(T)$. Such tests are easily carried out with the log linear framework; see Chapter 6 of Ref. 1. Note that our age-specific models of Sections 9.2–9.5 are given by (9.32) with $h(t) = $ a constant.

Finally, we could also fit parametric models for $h(t)$ or $\lambda(a)$ of (9.32), or more generally for $g(t, a)$ in (9.31). If, for example, we suspected a monotonic trend in calendar time, we might consider

$$\mu^*(t, a) = N(t - a)\lambda(a)e^{\beta t} \tag{9.33}$$

where β and the $\lambda(a)$'s are unknown parameters to be estimated.

9.9 ANALYSIS OF COSTS

Warranty costs may also be analyzed using the methods described in this chapter. To do this, we assume that claim costs are indexed by $c = 1$, $2, \ldots, r$ and $k(c)$ is the cost of a claim in group c. The amount of grouping used will depend on the application, but there is no difficulty with r being large, as, for example, if dollar amounts were used. We will consider just the case of age-specific analysis and let $\lambda_c(a)$ be the expected number of claims of cost $k(c)$ for a unit at age a. Under the assumptions of Section 9.2, $\lambda_c(a)$ is estimated by

$$\hat{\lambda}_c(a) = \frac{n_c(a)}{R(a)} \tag{9.34}$$

where, in an obvious notation, $n_c(a) = \sum_{d=0}^{T-a} n_c(d, a)$ is the number of claims of age a and cost $k(c)$ up to time T. Examination of the estimated cumulative expected frequencies

$$\hat{\Lambda}_c(a) = \sum_{u=0}^{a} \hat{\lambda}_c(a) \tag{9.35}$$

is often useful. Cumulative total expected cost per unit up to age a is estimated by

$$\hat{K}(a) = \sum_{c=1}^{r} k(c)\hat{\Lambda}_c(a) \tag{9.36}$$

Predictions and associated variance estimates may also be readily obtained. Let

$$M_c(a) = \frac{1}{N} \sum_{u=0}^{a} \sum_{d=0}^{T} n_c(d, u)$$

be the actual average number of cost $k(c)$ repairs per unit, up to age a, for units sold over $0, 1, \ldots, T$ and

$$K(a) = \sum_{c=1}^{r} k(c)M_c(a)$$

the average total cost per unit. Clearly, $M_c(a)$ and $K(a)$ are estimated by (9.35) and (9.36), respectively. In addition,

$$\text{Var}\{\hat{K}(a) - K(a)\} = \sum_{c=1}^{r} k^2(c) \, \text{Var}\{M_c(a) - \hat{M}_c(a)\}$$

where $\text{Var}\{M_c(a) - \hat{M}_c(a)\}$ is given by (9.27), with $\lambda_c(u)$ in place of $\lambda(u)$. Prediction limits for the average total cost $K(a)$ per unit up to age a are obtained as in Section 9.6. In particular, $\hat{K}(a) \pm 1.96 \, \hat{V}\text{ar}\{\hat{K}(a) - K(a)\}^{1/2}$ is an approximate 0.95 prediction interval for $K(a)$.

9.10 ESTIMATION OF FIELD RELIABILITY

Warranty data are valuable sources of information about the reliability of manufactured products. However, for reliability assessments, it is desirable to study failure or repair processes for individual units. Therefore, the type of warranty must be taken into consideration, as it affects how long units stay under observation (i.e., under warranty coverage). In addition, the amount of data kept on each unit will affect how well certain quantities may be estimated. It is beyond the scope of this chapter to provide a comprehensive treatment in this area, but we will summarize a few important points.

    1. Suppose that a particular type of failure is of interest and that when such failures occur under warranty, they result in a claim. If we

wish to estimate the expected number of failures $\lambda^F(a)$ per unit at age a and the cumulative expected number $\Lambda^F(a) = \sum_{u=0}^{a} \lambda^F(u)$, then we must adjust the $\hat{\lambda}(a)$ values discussed earlier in this chapter for the probability $p(a)$ that a unit failing at age a is under warranty (and that the claim is reported, if this is not always done). Because $\lambda(a) = \lambda^F(a)p(a)$, we can estimate $\lambda^F(a)$ as

$$\hat{\lambda}^F(a) = \frac{\hat{\lambda}(a)}{\hat{p}(a)} \tag{9.37}$$

provided that we are able to obtain an estimate $\hat{p}(a)$ of $p(a)$. With motor vehicles, for example, warranties typically have both age and mileage limits. In that case, we could base $\hat{p}(a)$ on the warranty limits (e.g., 2 years or 20,000 miles) and data on the rates at which vehicles in the population accumulate mileage. Vehicle manufacturers usually collect some such data by means of surveys. Note, however, that (9.37) is in this case appropriate only for failures that depend primarily on age, not on mileage. See points 4 and 5 for additional remarks.

  2. Sometimes the distribution of time to the first failure of some type is of interest. If the dates of sale (or of replacement for units replaced under warranty) are available for each unit and if we have only age-based or calendar-time-based warranty limits, then estimation of the time-to-failure distribution can be based on well-known methods for censored lifetime data (e.g., Refs. 6–8). If, however, dates of sale are not available until a warranty claim is made, then the estimation either has to be based on data conditional on failure having occurred by a certain age or on assumptions regarding sales.

  To be specific, suppose that t_i represents the age at failure of the ith unit and that t_i has a distribution with density function $f(t; \theta)$ and cumulative distribution function $F(t; \theta) = \int_0^t f(u; \theta)\, du$, where θ is an unknown parameter. Suppose that we consider warranty data for units sold over the calendar time period $(0, T)$ and that the warranty age limit is A. If the ith unit is sold at d_i, then we will observe its failure time t_i only if $t_i \le \tau_i = \min(T - d_i, A)$. If the sales dates d_i are available for all of the N units sold over $(0, T)$, then the τ_i's may be calculated for all $i = 1, \ldots, N$. Then, if the n units $i = 1, \ldots, n$ experience failures and the remainder do not, we know that $t_i > \tau_i$ for items $i = n + 1, \ldots, N$, and the probability of the data gives the so-called censored data likelihood

$$L_1 = \prod_{i=1}^{n} f(t_i; \theta) \prod_{i=n+1}^{N} \{1 - F(\tau_i; \theta)\} \tag{9.38}$$

If, however, the d_i's only become available when a claim is made, then the τ_i's cannot be calculated for $i = n + 1, \ldots, N$. In this case, we still have a likelihood based on $t_1, \ldots, t_n$, given that $t_i \leq \tau_i$ in each case; this is

$$L_2 = \prod_{i=1}^{n} \frac{f(t_i; \theta)}{F(\tau_i; \theta)} \tag{9.39}$$

Kalbfleisch and Lawless [3] discuss truncated data in some detail.

Usually, L_2 is a lot less informative about θ than L_1 (e.g., Ref. 9). It is, therefore, desirable to have dates of sale for all items; but if these are not available, it is worth trying to estimate them from sales information. In many cases, it is possible to get sufficiently accurate estimates of N and $\tau_{n+1}, \ldots, \tau_N$ [e.g., from estimates of the numbers of items sold in different days over $(0, T)$] to allow us to use (9.38).

3.   A distinction should be made between average failure rates (or other characteristics) across the population of units in service and the form of the failure processes for individual units. Aggregate warranty data as discussed in earlier sections allow us to estimate population average characteristics such as $\lambda^F(a)$ in point 1, but not to ascertain the nature of individual processes; for this, we need data on individual units. For repeated events, we would, in particular, need to keep track of the events occurring for each unit. For example, for engineering purposes, it is of much interest whether a certain average failure rate arises from each unit having roughly the same rate or from a situation in which most units experience no failures but a few experience many.

4.   Sometimes there is an easily measurable "usage" time and it is wished to estimate failure rates or distributions in terms of that. For example, with motor vehicles, it is common to evaluate certain reliability characteristics in terms of mileage. Assuming that mileage is recorded for units experiencing failures under warranty, the main difficulty is similar to that in point 2: the usage (mileage) at which units are censored are usually unknown and, consequently, even the likelihood (9.39) may not be available. To overcome this, it is necessary to postulate a model for mileage accumulation; Lawless and Kalbfleisch [10, Sect. 3.2] indicate how this can be done.

5.   When failures or other events depend on both age and other factors (including usage), we must have ways of observing or modeling these other factors. In particular, if the factors are observed only for units that experience warranty claims, then a model has to be postulated for how these factors are distributed in the population of units in service. One can learn about this by supplementing the warranty data with random samples from the population; see Refs. 9, 11, and 12 for methods based on this approach.

9.11 CONCLUDING REMARKS

There seem to be few sources that deal comprehensively with the analysis of warranty claims data. Robinson and McDonald [13] discuss several general issues in field reliability and warranty analysis, which supplement the present chapter nicely.

The methods presented in this chapter are designed to allow us to portray warranty claims data clearly and to extract as much information as possible from them. The main focus was on analyzing aggregate claims data; this is important for understanding total claims behavior and factors affecting it either for specific types of claims or for all types combined. The methods also allow claim numbers or costs to be predicted from past data. Section 9.10 dealt briefly with another issue, namely, the use of warranty data for reliability estimation. For a thorough analysis, it is usually necessary to supplement the warranty data with additional information.

In summary, the basic steps and procedures in the methods we have presented for analyzing claims data are as follows:

1. Obtain actual or estimated numbers $N(d)$ of product units sold in each time period d (Section 9.2).
2. Obtain the number of claims $n^*(t, a)$ in each time period, broken down according to age of the unit at the time of the claim (Section 9.2).
3. Calculate estimates of the expected number of claims per unit at age a, $\hat{\lambda}(a)$, and of the expected cumulative number per unit, $\hat{\Lambda}(a)$ (Section 9.2).
4. Plots of $\hat{\Lambda}(a)$ versus a, perhaps with product units grouped in various ways, are highly informative. If desired, variance estimates can be calculated for $\hat{\Lambda}(a)$, leading to confidence intervals for $\Lambda(a)$ (Sections 9.2 and 9.4).
5. The methods are readily extended to deal with aggregated data (Section 9.2.3), delays in the reporting of claims (Section 9.3), or uncertainty regarding sales (Section 9.5). Warranty costs may be analyzed using the same methods (Section 9.9).
6. Slightly more involved procedures such as prediction (Section 9.6) and regression analysis (Sections 9.8) may also be implemented easily.

Finally, we should stress that the foundation of good statistical analysis is good data. If warranty data are to be unambiguous and useful, it is important that claims and sales information be obtained in a timely and accurate way and that it be accessible for analysis.

APPENDIX

A1. Variance Estimates for Age-Specific Expected Claims Estimates

Under the Poisson model or the model with variance function (9.6), the $n(d, a)$'s are independent, with $\text{Var}\{n(d, a)\}$ given by (9.5). Thus, the $n_T(a)$'s in (9.2) are independent for $a = 0, 1, \ldots, T$, with variance $\sigma^2 \sum_{d=0}^{T-a} \mu(d, a) = \sigma^2 R_T(a)\lambda(a)$, so the $\hat{\lambda}(a)$'s are independent, with $\text{Var}\{\hat{\lambda}(a)\} = \sigma^2 \lambda(a)/R_T(a)$. This leads to (9.5) and (9.9), the former being given by the case when $\sigma = 1$.

For the model (9.10), the terms in $n_T(a) = \sum_{d=0}^{T-a} n(d, a)$ are independent, so $\text{Var}\{n_T(a)\} = \sum_{d=0}^{T-a} \mu(d, a)[1 + \sigma^2 \lambda(a)] = R_T(a)\lambda(a)[1 + \sigma^2 \lambda(a)]$, and so

$$\text{Var}\{\hat{\lambda}(a)\} = \frac{\lambda(a)[1 + \sigma^2 \lambda(a)]}{R(a)}$$

Because of correlation between $n(d, a_1)$ and $n(d, a_2)$, $\hat{\lambda}(a_1)$ and $\hat{\lambda}(a_2)$ are correlated. We have, for $a_1 \neq a_2$,

$$\text{cov}\{n_T(a_1), n_T(a_2)\} = \sum_{d=0}^{\min(T-a_1,\ T-a_2)} \text{cov}\{n(d, a_1), n(d, a_2)\}$$

$$= \sigma^2 \lambda(a_1)\lambda(a_2)R_T\{\max(a_1, a_2)\}$$

$$\text{cov}\{\hat{\lambda}(a_1), \hat{\lambda}(a_2)\} = \frac{\sigma^2 \lambda(a_1)\lambda(a_2)}{R_T\{\min(a_1, a_2)\}}$$

Then

$$\text{Var}\{\hat{\Lambda}(a)\} = \sum_{u=0}^{a} \text{Var}\{\hat{\lambda}(u)\} + 2 \sum_{u<} \sum_{v} \text{cov}\{\hat{\lambda}(u), \hat{\lambda}(v)\}$$

$$= V_P(a) + \sigma^2 \sum_{u=0}^{a} \frac{\lambda(u)^2}{R(u)} + 2\sigma^2 \sum_{u=1}^{a} \lambda(u)V_P(u - 1)$$

giving (9.12).

A robust variance estimate for $\hat{\Lambda}(a)$ can also be obtained [5], provided that data on individual units is available. In this case, let m be the number of products sold by time T, and for unit i, let d_i be the day of sale and $n_i(u)$ the number of age u claims. Then the robust estimate for $\text{Var}\{\hat{\Lambda}(a)\}$ is

$$\hat{V}_R(a) = \sum_{i=1}^{m} \left\{ \sum_{u=0}^{\min(a,\ T-d_i)} \frac{1}{R(u)} [n_i(u) - \hat{\lambda}(u)] \right\}^2 \tag{9.A1}$$

Variance estimates (9.9) and (9.12), which are based on specific assumptions about variance, may be checked against (9.A1), which is valid under quite general conditions.

A2. Variance Estimates (9.28) for Prediction with Grouped Data

The calculations follow the same lines as in Kalbfleisch et al. [2, Appendix B]. We note that (ignoring reporting delays for simplicity)

$$
M(A_i) - \hat{M}(A_i) = \frac{1}{N} \sum_{a=a_{i-1}}^{a_i-1} \left\{ \sum_{d=0}^{T-a} n(d, a) + \sum_{d=T-a+1}^{T} n(d, a) \right\}
$$
$$
- \sum_{a=a_{i-1}}^{a_i-1} \sum_{d=0}^{T-a} \frac{n(d, a)}{R(A_i)}
$$

where

$$
R(A_i) = \frac{1}{a_i - a_{i-1}} \sum_{a=a_{i-1}}^{a_i-1} R(a)
$$

as described in Section 9.3. Because the $n(d, a)$'s are mutually independent with variance function (9.6), we find that

$$
\mathrm{Var}\{M(A_i) - \hat{M}(A_i)\} = \left(\frac{N - R(A_i)}{NR(A_i)} \right)^2 \sum_{a} \sum_{d=0}^{T-a} \sigma^2 N(d)\lambda(a)
$$
$$
+ \frac{1}{N^2} \sum_{a} \sum_{d=T-a+1}^{T} \sigma^2 N(d)\lambda(a)
$$
$$
= \left(\frac{N - R(A_i)}{NR(A_i)} \right)^2 \sigma^2 \Lambda(A_i)R(A_i)
$$
$$
+ \frac{1}{N^2} \sigma^2 \Lambda(A_i)[N - R(A_i)]
$$
$$
= \left(\frac{N - R(A_i)}{NR(A_i)} \right) \sigma^2 \Lambda(A_i)
$$

ACKNOWLEDGMENTS

This research was supported in part by grants from the Manufacturing Research Corporation of Ontario and the Natural Sciences and Engineering Research Council of Canada.

The authors thank Claude Nadeau for suggestions and computational assistance, and Diane Gibbons and Jeff Robinson for helpful discussions.

REFERENCES

1. McCullagh, P. and Nelder, J. A. (1989). *Generalized Linear Models*, Chapman & Hall, London.
2. Kalbfleisch, J. D., Lawless, J. F., and Robinson, J. A. (1991). Methods for the analysis and prediction of warranty claims. *Technometrics*, **33**, 273–285.
3. Kalbfleisch, J. D., and Lawless, J. F. (1992). Some useful statistical methods for truncated data. *Journal of Quality Technology*, **24**, 145–152.
4. Breslow, N. (1990). Tests of hypotheses in overdispersed Poisson regression and other quasi-likelihood models. *Journal of the American Statistical Association*, **85**, 565–571.
5. Lawless, J. F., and Nadeau, J. C. (1995). Some simple robust methods for the analysis of recurrent events. *Technometrics*, **37**, 158–168.
6. Kalbfleisch, J. D., and Prentice, R. L. (1980). *The Statistical Analysis of Failure Time Data*, John Wiley & Sons, New York.
7. Lawless, J. F. (1982). *Statistical Models and Methods for Lifetime Data*, John Wiley & Sons, New York.
8. Nelson, W. B. (1982). *Applied Life Data Analysis*, John Wiley & Sons, New York.
9. Kalbfleisch, J. D., and Lawless, J. F. (1988). Estimation of reliability in field performance studies (with discussion). *Technometrics*, **30**, 365–388.
10. Lawless, J. F., and Kalbfleisch, J. D. (1992). Some issues in the collection and analysis of field reliability data, in *Survival Analysis: State of the Art* (J. P. Klein and P. K. Goel, eds.), Kluwer, Amsterdam, pp. 141–152.
11. Suzuki, K. (1985). Estimation of lifetime parameters from incomplete field data. *Technometrics*, **27**, 263–271.
12. Hu, X. J., and Lawless, J. F. (1994). *Analysis of Truncated Recurrent Event Data, with Application to Warranty Claims*, IIQP Technical Report 94-04, University of Waterloo.
13. Robinson, J. A., and McDonald, G. C. (1991). Issues related to field reliability and warranty data, in *Data Quality Control: Theory and Pragmatics* (G. E. Liepins and V. R. R. Uppuluri, eds.), Marcel Dekker, Inc., New York, pp. 60–90.

Part D

Warranty Cost Models

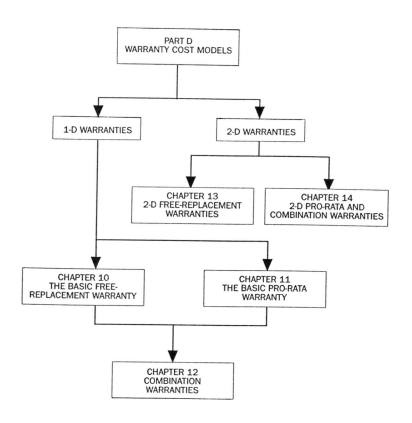

Introduction

Part D is concerned with cost analysis of the most commonly used consumer product warranties, including the basic free-replacement, pro-rata, and combination free-replacement/pro-rata warranties. Also included are the two-dimensional versions of these, such as warranties based both on calendar time and usage. This is a relatively new area of warranty analysis and much additional research remains to be done.

The cost of a warranty depends crucially on the structure of the warranty and on the life distribution of the product. The warranty structures given in Chapter 1 and the models discussed in Part C play an important role in cost analysis. The authors of the chapters in Part D give many numerical examples, discuss applications, and provide suggestions for the practitioner.

We begin, in Chapter 10, by W. R. Blischke, with analysis of the basic free-replacement warranty and several policies under which a cash rebate is given rather than a replacement product. Both manufacturers' and buyers' warranty costs are analyzed under several assumed life distributions. Present value and life-cycle cost models are given. A variable price structure (with or without warranty), under which buyer and seller would be indifferent as to whether or not a warranty were offered, is discussed. Statistical estimation of the cost models and a number of examples and applications are included.

In Chapter 11, J. G. Patankar and A. Mitra deal with the basic pro-rata warranty. They discuss the value of a warranty and various warranty theories and then turn to the pro-rata warranty under which a rebate or a discount on next purchase is a linear function of time to item failure. An extensive review of cost models for the pro-rata warranty is given, along with methods for estimating warranty costs. Examples and suggestions for use of the models are included.

Chapter 12, also by W. R. Blischke, extends some of the results of Chapters 10 and 11 to analysis of various types of combination warranties and develops some new approaches for use in this context. Again, sellers' and buyers' points of view are considered. Present value and life-cycle cost models are analyzed and the results are illustrated by many examples. Sensitivity of the results to assumptions regarding the life distribution of the item is explored.

H. Moskowitz and Y. H. Chun turn to two-dimensional versions of the free-replacement warranty in Chapter 13. The most common example is the x-year, y-mile warranty offered on automobiles. As can be seen from Chapters 1 and 7, both the conceptual and mathematical problems become considerably more complex in this case. Here the authors look at some specific two-dimensional policies and suggest some approaches to modeling and cost analysis. They present a decision model for the manufacturer and discuss the sensitivity of the optimal warranty price. Implementation issues are also addressed.

The final chapter of Part D, by R. J. Wilson and D. N. P. Murthy, is concerned with cost analysis of two-dimensional versions of the pro-rata and combination warranties. There are many possibilities under the two-dimensional structure, particularly for combination warranties. These are discussed and several specific policies are analyzed in detail. Analysis requires the bivariate distribution of the two characteristics on which the warranty is based. Cost analysis is performed for two such distributions. The results are illustrated by means of numerical examples.

10

The Basic Free-Replacement Warranty and Related Rebate Warranties

Wallace R. Blischke

University of Southern California
Los Angeles, California

10.1 INTRODUCTION

The free-replacement warranty, under which items that fail under warranty are replaced at no charge to the buyer, is one of the most commonly employed warranties for consumer goods. There are a number of versions of this warranty and a number of related policies, including cash refunds rather than replacements, refunds of a portion of the purchase price ("rebate warranties"), and refunds that depend on the amount of service (i.e., the length of time since purchase). In this chapter, we present a number of models for costing warranties of this type. The emphasis is primarily from the point of view of the seller. A few results from the buyer's point of view will be given as well.

The free-replacement warranty has also been called the "failure-free" warranty, the "standard" warranty, "full" warranty, and the "lump-sum" or "lump-sum rebate" warranty. Here, the generic term "free-replacement warranty" and the abbreviation FRW will be used.

The FRW is usually not renewing; that is, the warranty period for a replacement item is ordinarily not W, the length of the warranty item on the original item, but the remaining time in this period, which can be expressed as $W - S_n$, where $S_n = \sum_{i=1}^{n} X_i$, X_i is the life or service time of the ith item, and n is the number of items that have failed up to that point of time in the warranty period. Thus, for a nonrenewing FRW, warranty coverage ends with the $(n - 1)$th replacement provided that $S_{n-1} < W$ and $S_n \geq W$, and the item then in service is no longer covered by warranty.

Occasionally, items are sold under renewing FRW. Under this policy, all replacement items would be warrantied anew for the same period W. Thus, the warranty process associated with a single sale ends only when a single item with a lifetime of at least W is encountered. The renewing FRW is often called an "unlimited" warranty or "unlimited free-replacement" warranty.

Rebate warranties are, by their very nature, nonrenewing.

In this chapter, cost models for the basic free-replacement warranty and several related warranties will be presented. Specifically, the following warranty policies are covered:

Warranty 1. *Free-Replacement Warranty* (*nonrenewing*): The seller agrees to repair or replace, at no cost to the consumer, any failed item up to time W (the length of the warranty period) from the time of purchase. Repaired or replaced items are warrantied only for the time remaining in the warranty period.

Warranty 2. *Free-Replacement Warranty* (*renewing*): The seller agrees to repair or replace, at no cost to the consumer, any failed item up to time W (the length of the warranty period) from the time of purchase. Repaired or replaced items are warrantied for a period W from the time of repair or replacement.

Warranty 3. *Rebate Warranty* ("*money-back guarantee*"): The seller agrees to refund the full purchase price of the warranty if the item fails prior to time W from the time of purchase.

Warranty 4. *Partial Rebate Warranty*: The seller agrees to refund a portion of the purchase price if the item fails prior to time W from the time of purchase.

These are Warranties 1, 14, 3, and 2, respectively, in Chapter 1. They are renumbered here for convenience.

Note that Warranties 3 and 4 are not true free-replacement warranties because actual replacements are not provided, although under Warranty 3, the buyer could, at his or her option, purchase a replacement with the rebate. Warranty 4 provides only a partial rebate of the purchase price. The amount not refunded may be considered compensation to the seller for the service obtained by the buyer prior to failure of the item. It is convenient to include this warranty here because it is easily analyzed by a simple modification of the cost model for Warranty 3.

Applications of the basic FRW and the other warranties listed range from small, relatively inexpensive items such as photographic film, basic household goods, and cheap tools, to very expensive items such as luxury automobiles and watches. In applications, Warranty 1 is the most widely used of these four warranties.

Warranty costs for the seller and benefits to the buyer depend fundamentally on two elements—the terms of the warranty policy and the life distribution(s) of the items, including the life distribution of the original items sold under warranty and that of replacement or repaired items in the event of a failure. (Replacement items are assumed to have the same life distribution as that of the item originally purchased.) Together, these determine the magnitude of the costs that will be incurred, the time and frequency of their occurrence, and the length of time over which the warranty coverage will be in effect. The warranty terms to be considered in this chapter are those of Warranties 1–4, given in the previous paragraphs. Life distributions that may be appropriate for various types of items are discussed in Chapter 8. A number of these will be briefly reviewed and used to illustrate the cost models in this chapter.

Models for per-unit cost for these warranties will be given along with some comments on the calculation of life-cycle costs. In this context, per-unit cost is taken to mean the total cost associated with the sale or purchase of an item, including warranty costs. In life-cycle costing, repeat purchases after warranty expiration are assumed up to some future time L and costs are calculated over this entire time period.

The organization of the remainder of this chapter is as follows: In Section 10.2, a list of the distributions used to illustrate the cost models and the choices of parameter values for these and for the warranty policies will be given. Models of sellers' costs for each of the four policies listed previously will be given in Section 10.3. Sections 10.4–10.6 will discuss buyer's cost models, present value cost models, and models of life-cycle costs, respectively. In Section 10.7, we discuss a differential pricing structures for warrantied versus unwarrantied items under which either buyer's or seller's long-run costs would be the same whether items were sold with or without warranty. Statistical estimation of warranty cost models will be discussed briefly in Section 10.8, and some comments and suggestions with regard to application of the cost models will be provided in Section 10.9

10.2   ASSUMED LIFE DISTRIBUTIONS AND WARRANTY PARAMETERS

In illustrating the cost models for Warranties 1 through 4, we use warranty periods of $W = 0.5$ and $W = 1$. The distributions used for this purpose are the exponential, Weibull, Gamma, lognormal, and inverse Gaussian. Formulas for these distributions will be given in this section. (The distributions are discussed in more detail in Chapters 6–9.)

   Warranty costs will be calculated for mean times to failure (MTTFs) of $\mu = 1.0$, 1.5, 2.0, and 2.5. These are equivalent to MTTFs equal to the warranty period up to 2.5 times the length of the warranty period for $W = 1$ and 5 times the length for $W = 5$. Distribution parameters that will give these values of μ will be used in each case. The remaining parameters of the distributions, if any, will be chosen to provide a variety of shapes for the life distributions. This will not only provide illustrative examples, but it will give an indication to the user, on a small scale, of the sensitivity of the results to the distributional assumptions made.

   The distributions and their characteristics are as follows.

10.2.1   The Exponential Distribution

The formula for the exponential distribution is

$$f(x; \lambda) = \begin{cases} 0, & x < 0 \\ \lambda e^{-\lambda x}, & x \geq 0 \end{cases} \tag{10.1}$$

The *cumulative distribution function* (CDF), $F(x) = P(X \leq x)$, is given by

$$F(x) = \begin{cases} 0, & x < 0 \\ 1 - e^{-\lambda x}, & x \geq 0 \end{cases} \tag{10.2}$$

The parameter is λ. The mean is $\mu = 1/\lambda$; the variance is $\sigma^2 = 1/\lambda^2$. In the illustrations, λ will be taken to be 1, 0.6667, 0.5, and 0.4. The distribution is appropriate only for items having constant failure rate. (See Chapter 8.)

10.2.2   The Weibull Distribution

The formula for the Weibull distribution is

$$f(x; \lambda, \beta) = \begin{cases} 0, & x < 0 \\ \lambda^\beta x^{\beta - 1} e^{-\lambda x^\beta}, & x \geq 0 \end{cases} \tag{10.3}$$

The CDF is

$$F(x; \lambda, \beta) = \begin{cases} 0, & x < 0 \\ 1 - e^{-\lambda x^\beta}, & x \geq 0 \end{cases} \tag{10.4}$$

Here, λ is a scale parameter and β is a shape parameter. The mean is $\mu = \lambda^{-1}\Gamma(1 + 1/\beta)$; the variance is $\lambda^{-2}[\Gamma(1 + 2/\beta) - \Gamma^2(1 + 1/\beta)]$, where $\Gamma(\cdot)$ is the Gamma function. (See Chapter 8.) We use $\beta = 0.5$, which is a decreasing failure rate distribution and $\beta = 2$, for which the failure rate in increasing in x. The corresponding values of $\Gamma(1 + 1/\beta)$ are $\Gamma(3) = 2$ and $\Gamma(1.5) = 0.88623$. Values of λ used are those that provide the desired values of μ, namely, $\lambda = 2/\mu$ for $\beta = 0.5$ and $\lambda = 0.88623/\mu$ for $\beta = 2$. The resulting values of the variance are $\sigma^2 = (\mu/2)^2[4! - (2!)^2] = 5\mu^2$ for $\beta = 0.5$ and $\sigma^2 = 0.27323\mu^2$ for $\beta = 2$.

10.2.3 The Gamma distribution

The formula for the Gamma distribution is

$$f(x; \lambda, \beta) = \begin{cases} 0, & x < 0 \\ [\Gamma(\beta)]^{-1}\lambda^\beta x^{\beta-1}e^{-\lambda x}, & x \geq 0 \end{cases} \tag{10.5}$$

As for the Weibull distribution, here λ is a scale parameter and β is a shape parameter. Unless β is a positive integer, there is no closed-form expression for the CDF. The mean is $\mu = \beta/\lambda$; the variance is β/λ^2. For the examples of the next section, values of β and λ that provide the same mean and variance as in the Weibull distribution will be used. For the Gamma distribution, the β-values corresponding to 0.5 and 2 in the Weibull distribution are $\beta = 0.2$ and $\beta = 3.6599$. The corresponding values of λ for the Gamma distribution are $\lambda = 0.2/\mu$ and $\lambda = 3.6599/\mu$.

The values of β used for the Gamma distribution again correspond to decreasing and increasing failure rates, respectively.

10.2.4 The Lognormal Distribution

The formula for the lognormal distribution is

$$f(x; \eta, \theta) = \begin{cases} 0, & x < 0 \\ (x\theta\sqrt{2\pi})^{-1}\exp\left(-\frac{[\ln(x) - \eta]^2}{2\theta^2}\right), & x \geq 0 \end{cases} \tag{10.6}$$

There is no closed-form expression for the CDF. The mean is $\mu = \exp(\eta + \theta^2/2)$; the variance is $\sigma^2 = \exp(2\eta + \theta^2)[\exp(\theta^2) - 1]$. For the numerical examples, values of η and θ which give the desired values of μ and the same values of σ^2 used in the Weibull and Gamma cases are chosen. Here the Weibull β-values of 0.5 and 2 lead to θ-values of 1.3386 and 0.49148, respectively. The corresponding values of η are obtained as $\eta = \ln(\mu) - \theta^2/2$. The values are given in Table 10.1.

10.2.5   The Inverse Gaussian Distribution

The formula for the inverse Gaussian distribution is

$$f(x; \eta, \theta) = \begin{cases} 0, & x < 0 \\ \left(\dfrac{\theta}{2\pi x^3}\right)^{1/2} \exp\left(-\dfrac{\theta(x - \eta)^2}{2\eta^2 x}\right), & x \geq 0 \end{cases} \tag{10.7}$$

The CDF is given (see Ref. 1, p. 141) by

$$F(x; \eta, \theta) = \begin{cases} 0, & x < 0 \\ \Phi\left((x - 1)\sqrt{\dfrac{\theta}{\eta x}}\right) + \exp\left(\dfrac{2\theta}{\eta}\right)\Phi\left(-(x + 1)\sqrt{\dfrac{\theta}{\eta x}}\right), & x \geq 0 \end{cases} \tag{10.8}$$

where $\Phi(x)$ is the CDF of the standard normal distribution, given by

$$\Phi(x) = \int_{-\infty}^{\infty} \frac{1}{\sqrt{2\pi}} e^{-u^2/2} \, du \tag{10.9}$$

The mean of the inverse Gaussian distribution is $\mu = \eta$; the variance is $\sigma^2 = \eta^3/\theta = \mu^3/\theta$. In the examples, the same values of μ will be used as in the previous cases, with θ chosen so that $\theta = \mu^3/\sigma^2$, where σ^2 is the Weibull variance for these cases. The result is $\theta = 0.2\mu$ when $\beta = 0.5$, and $\theta = 3.6599\mu$ when $\beta = 2$.

Table 10.1   Values of Lognormal Parameters Corresponding to Weibull Parameters of $\beta = 0.5$ ($\theta = 1.3386$) and $\beta = 2$ ($\theta = 0.49148$)

	Values of η	
μ	$\theta = 1.3386$	$\theta = 0.49148$
1.0	-0.89588	-0.12078
1.5	-0.49041	0.28469
2.0	-0.20273	0.57237
2.5	0.02041	0.79551

10.3 MODELING SELLER'S PER-UNIT WARRANTY COST

The cost to the seller of a warrantied item will be modeled as the total expected cost of the sale, say $E_s[C(W)]$, where W is the warranty period, including the cost of supplying the original item and the costs of all replacements or repairs necessary under warranty. On a per-item basis, these costs are taken to include manufacturing costs, distribution costs, and all other costs associated with providing the item to the consumer, as well as marketing and all other costs associated with the sale, exclusive of warranty costs, amortized over all items provided to the buyer. The average cost to the seller of supplying an item is denoted by c_s. Note that c_s may be interpreted as the cost per item to the seller if items were sold without warranty. The selling price, that is, the cost to the buyer of the original item purchased, is denoted by c_b. It is assumed that both of these costs are constant throughout the period of interest.

Actual costs to seller and buyer of a single item sold under warranty are denoted $C_s(W)$ and $C_b(W)$, respectively. Both of these are random variables whose distributions depend on the distributions of the lifetimes of the items and the type of warranty offered and can depend on many other factors as well (e.g., usage rate, maintenance, and environment). Cost models deal with the average or expected costs, $E[C_s(W)]$ and $E[C_b(W)]$.

A number of additional simplifying assumptions are made. Unless otherwise stated, it will be assumed that all failures before time W from the time of purchase result in a warranty claim, that is, that all legitimate claims under warranty are pursued,* that no false claims are made, that replacement of a failed item by a repaired or new item is instantaneous,† and that the lifetimes X_1, X_2, ... of the original and all replacement items are independent and identically distributed.

The cost models to be discussed in this chapter allow for the possibility of multiple failures within the warranty period. In this connection, we define $N = N(W) =$ number of repairs or replacements in the warranty period. Under the assumption that the X_i are independent and identically distributed, the expected value of $N(W)$ is given by $E[N(W)] = M(W)$, where $M(\cdot)$ is the *ordinary renewal function* associated with the distribution function $F(\cdot)$. (See Chapters 6–8.)

The seller's expected cost for Warranties 1–4 will now be given. Both nonrepairable and repairable items will be considered.

---

* See Chapter 17 for modifications of models that account for other claim patterns.

† In practice, this means that the time to repair or replace an item is small relative to the expected time to failure.

10.3.1   Cost Models for the Nonrenewing Free-Replacement Warranty

Nonrepairable Items

In this section, it is assumed that items are sold under nonrenewing free-replacement warranty (Warranty 1) and that the items are not repairable (or that it is less costly to replace a failed item than to repair it). In this case, warranty claims are settled by supplying an identical new item at an average cost c_s to the seller. If the warranty period is of length W from the time of purchase and the original item fails at time X_1, the replacement is warrantied for a period of length $W - X_1$ from the time of replacement. If the lifetime of the replacement item is X_2, with $X_2 < W - X_1$, a second replacement is provided. This item is warrantied for a period $W - X_1 - X_2$. The process continues until a total service time of at least W is attained.

The expected total cost to the seller of an item sold under this type of warranty is given (see Chapter 4 of Ref. 2) by

$$E[C_s(W)] = c_s[1 + M(W)] \tag{10.10}$$

Here, the expected cost is expressed as the sum of the cost of supplying the original item, c_s, and the expected cost of supplying replacements under warranty, which involves the ordinary renewal function, $M(W)$. In the following example, we look at this total expected cost for warranties of length $W = 0.5$ and $W = 1$, with MTTFs of $\mu = 1$, 1.5, 2, and 2.5, for the five distributions given in the previous section.

Example 10.1

Calculation of expected costs using Equation (10.10) requires evaluation of the ordinary renewal functions corresponding to the distributions given in Section 10.2. There are a number of tables available for this purpose. (See Chapter 8.) In addition, a FORTRAN computer program (available from the editors) that runs on an IBM compatible microcomputer or any of a number of other programs can be used to calculate the renewal functions.

The FORTRAN program was used to compute the values of $M(W)$ for $W = 0.5$ and $W = 1$ for the exponential, Weibull, Gamma, lognormal, and inverse Gaussian distributions for the values of the MTTF previously listed, all except the exponential having the same values for the variance. Expected costs were then calculated using Equation (10.10). The results, tabulated as $E[C_s(W)]/c_s$, are given in Table 10.2.

Thus, for example, if item lifetimes follow a Weibull distribution with $\mu = 1.5$ years and $\beta = 2$, and a FRW of length $W = 0.5$ years is given, the expected cost to the seller is $1.085c_s$; that is, the warranty cost

Table 10.2 Expected Cost to Seller, Nonrenewing FRW, for Various Life Distributions (Table of $E[C_s W]/c_s$)

μ	Exponential	Weibull		Gamma		Lognormal		Inv. Gaussian	
		β = 0.5	β = 2	β = 0.2	β = 3.66	θ = 1.339	θ = 0.4915	θ = 0.2μ	θ = 3.66μ
				W = 0.5					
1.0	1.500	2.308	1.184	1.791	1.159	1.835	1.122	2.060	1.125
1.5	1.333	2.018	1.085	1.702	1.056	1.571	1.023	1.746	1.021
2.0	1.250	1.857	1.048	1.645	1.025	1.430	1.005	1.570	1.003
2.5	1.200	1.752	1.031	1.605	1.012	1.342	1.001	1.455	1.001
				W = 1					
1.0	2.000	3.048	1.624	2.091	1.631	2.541	1.635	2.859	1.637
1.5	1.667	2.570	1.313	1.960	1.301	2.081	1.283	2.345	1.290
2.0	1.500	2.308	1.184	1.877	1.159	1.835	1.122	2.060	1.125
2.5	1.400	2.139	1.121	1.818	1.091	1.679	1.053	1.876	1.052

is 8.5% of the cost of producing an item and getting it into the hands of the consumer (including per item cost for marketing, administration, distribution, etc.) The comparable cost under a lognormal distribution with the same mean and variance would be $1.023c_s$. If the warranty were for 1 year, this cost would increase to 31.3% under the Weibull assumption. These costs would be slightly less under the other IFR (increasing failure rate) distributions tabulated; they would be substantially more under the exponential assumption (33.3% for $W = 0.5$ and 66.7% for $W = 1$ year).

Note from the table that the Weibull and Gamma distributions with decreasing failure rates and the corresponding lognormal and inverse Gaussian distributions (i.e., those with the same means and variances) always give higher expected costs than does the exponential distribution. The distributions with increasing failure rates always lead to lower costs than the exponential.

Note also that the distributional assumptions made can have a very significant effect on the seller's expected cost, particularly if the items have distribution with decreasing failure rate. For example, if the warranty period is $W = 1$ and the MTTF is $\mu = 1.5$, the expected cost to the seller is $1.67c_s$ if the distribution is exponential, that is, if the failure rate is constant through time. On the other hand, this cost may range from $1.96c_s$ to $2.57c_s$ if the failure rate is decreasing and time to failure follows one of the four distributions considered, all of which have the same mean and variance.

This difference, as well as the warranty costs, are much less if the item has a decreasing failure rate distribution. In practical terms, however, the difference can still be very significant. For example, if $W = 0.5$ and $\mu = 1$, the warranty cost can range from 12% to 18% of the total cost to the manufacturer of providing an item to the consumer.

Repairable Items

If the items sold under warranty are repairable, two factors may affect warranty costs: (1) the cost of repair versus the cost of replacing a failed item by a new item and (2) the life distribution of a repaired item as opposed to that of a new item. (See Chapter 4 of Ref. 2.) Cost models for repairable items express the total cost to the seller as a function of the average cost of repair, denoted c_r.

The expected number of repairs, and hence the total expected cost to the seller, depends on the nature of the repair. The simplest model is that under which a repaired item is "good-as-new." This might be appropriate, for example, if failures are almost always due to failure of a single key component and repair consists of replacement of that component by a new one. In this case, the failure distribution of a repaired item may be assumed to be the same as that of a new item, namely, $F(x)$, and the cost model becomes simply

$$E[C_s(W)] = c_s + c_r M(W) \tag{10.11}$$

Example 10.2

For the distribution and parameter combinations previously considered, the results of Table 10.2 may be used with slight modifications to evaluate cost model (10.11): If x is a tabulated value (in Table 10.2), the expected cost for a repairable item is calculated as $E[C_s(W)] = c_s + (x - 1)c_r$. Suppose, for example, that the cost of repair (including all cost associated with servicing the warranty) is 40% of the cost of supplying the original item, that is, $c_r = 0.4c_s$. Then under the Weibull assumption, with $\mu = 1.5$ and $\beta = 2$, the expected cost to the seller would be $c_s + 0.085(0.4)c_s = 1.034c_s$, that is, for a repairable item, the cost of warranty is 3.4% of the total cost to the seller.

If failure times were lognormally distributed, the total cost to the seller would be $c_s + 0.023(0.4)c_s = 1.009c_s$, that is, the warranty cost would be less than 1% of total cost.

There are many other assumptions regarding the effect of repairs that might be considered. (See Chapter 4 of Ref. 2.) These lead to repaired items having life distributions different from $F(x)$ and to renewal functions different from the ordinary renewal function $M(t)$. For example, if all repaired items have the same distribution $G(x)$, different from $F(x)$, then the expected number of renewals is given by the "delayed" renewal function, $M_d(t)$, obtained, in theory, as the solution to

$$M_d(t) = F(t) + \int_0^t M_G(t - x)f(x)\,dx \tag{10.12}$$

where $M_G(\cdot)$ is the ordinary renewal function associated with $G(x)$. Explicit expressions for $M_d(t)$ can be obtained for only a few distributions and numerical evaluation or simulation are usually required for solution.

The resulting cost model is

$$E[C_s(W)] = c_s + c_r M_d(W) \tag{10.13}$$

Another type of repair is "minimal repair," whereby a repaired item is returned to a state identical to that immediately prior to failure. This is appropriate for more complex items having many critical components. If repair consists of replacement of a single component at time t after beginning of item operation, the item should have about the same failure distribution as at time t, because all other components are of age t. In this case, the number of failures and repairs over the warranty period occur according to a nonhomogeneous Poisson process with intensity function equal to the failure rate function $r(t) = 1 - F(t)$. (See Chapter 7.) The expected cost to the seller becomes

$$E[C_s(W)] = c_s + c_r \int_0^W r(t)\,dt \tag{10.14}$$

Still another model is one in which the distribution (or the MTTF) changes after each repair. Expected numbers of replacements and cost models for situations such as these are discussed by Nguyen and Murthy [3].

10.3.2 Cost Models for the Renewing Free-Replacement Warranty

Nonrepairable Items

Under the renewing FRW (Warranty 2 of Section 10.1), items are replaced free of charge until the first time an item has a lifetime of at least W. The number of replacements required to achieve this has a *geometric distribution* with mean $F(W)/[1 - F(W)]$. (See Section 4.4 of Ref. 4.) From this, it follows that for nonrepairable items the expected cost to the seller (under the conditions and assumptions stated previously) is

$$E[C_s(W)] = c_s[1 - F(W)]^{-1} \tag{10.15}$$

Repairable Items

For repairable items, with lifetimes of all repaired items, regardless of the number of repairs, distributed as $G(x)$, with $G(x)$ different from $F(x)$, the seller's expected cost is

$$E[C_s(W)] = c_s + c_r F(W)[1 - G(W)]^{-1} \qquad (10.16)$$

Example 10.3

For the distributions and parameter values discussed previously and used in the previous examples, the seller's expected cost, again expressed as $E[C_s(W)]/c_s$, is given in Table 10.3. As before, the results overall show highest costs for DFR (decreasing failure rate) distributions and lowest for IFR, with the exponential between these two extremes. The specific influence of the distributional assumptions on expected cost for Warranty 2, however, differs somewhat from that for Warranty 1.

Note the curious result for the inverse Gaussian distribution—warranty cost remains constant as the MTTF varies. This is due to the nature of the relationship between the mean and variance of the Weibull distribution and the parameters of the inverse Gaussian for the particular values used in the examples.

10.3.3   Cost Models for the Rebate Warranty

Nonrepairable Items

Warranty 3, the full rebate warranty or money-back guarantee, is analyzed by means of cost models of the same structure as those for the nonrenewing FRW. The primary difference is that the seller's cost of warranty service is a function of the buyer's cost c_b (i.e., the selling price) rather than of c_s. For Warranty 3, the seller's expected per-unit cost is given by

$$E[C_s(W)] = c_s + c_b F(W) \qquad (10.17)$$

Repairable Items

For rebate warranties, whether or not the item is repairable is irrelevant because a replacement item is not supplied under the terms of the warranty. An exception would be a situation in which failed items are returned to the manufacturer and can be reworked or repaired and sold as new items or sold on the secondhand market. The models of the previous two paragraphs can be adjusted to account for this by reducing c_b by the average net value of a returned item.

If returned items have some scrap value, this would also affect the seller's net cost. This can also be accounted for by suitable reduction of c_b in the cost models.

Table 10.3 Expected Cost to Seller, Renewing FRW, for Various Life Distributions (Table of $E[C_s W)]/c_s$)

		Weibull		Gamma		Lognormal		Inv. Gaussian	
								Life Distribution	
μ	Exponential	$\beta = 0.5$	$\beta = 2$	$\beta = 0.2$	$\beta = 3.66$	$\theta = 1.339$	$\theta = 0.4915$	$\theta = 0.2\mu$	$\theta = 3.66\mu$
				$W = 0.5$					
1.0	1.649	2.718	1.217	3.087	1.187	2.274	1.139	2.715	1.134
1.5	1.396	2.263	1.091	2.679	1.059	1.785	1.024	2.715	1.134
2.0	1.284	2.028	1.050	2.489	1.025	1.555	1.005	2.715	1.134
2.5	1.221	1.882	1.032	2.317	1.013	1.423	1.001	2.715	1.134
				$W = 1$					
1.0	2.718	4.113	2.193	4.245	2.323	3.974	2.482	4.480	2.536
1.5	1.948	3.173	1.418	3.476	1.413	2.801	1.391	4.480	2.536
2.0	1.649	2.718	1.217	3.087	1.187	2.274	1.139	4.480	2.536
2.5	1.492	2.446	1.134	2.846	1.100	1.976	1.056	4.480	2.536

Example 10.4

For the distributions and parameter values used previously, values of $E[C_s(W)]$ are easily obtained from the results given in Table 10.3. For example, if $W = 0.5$ and a Weibull distribution with $\mu = 1.5$ and $\beta = 2$ is assumed, then the expected cost to the seller is $c_s + c_b(1 - 1/1.0912)$ $= c_s + 0.084c_b$. Thus, if the seller's average cost per item is \$60, exclusive of warranty costs, and the item sells for \$100, the cost to the seller of a warrantied item is \$68.40, that is, the cost of warranty is \$8.40 per item.

Note that for the nonrenewing FRW, the average cost of a warrantied item (from Table 10.2) is $1.085c_s$. In the numerical example, this amount is \$65.10, still a fairly substantial cost of warranty but considerably less than that for the full rebate warranty. In applications, the actual warranty cost of Warranty 3 depends critically on the markup of the item, that is, on the relationship between c_s and c_b.

10.3.4 Cost Models for the Partial Rebate Warranty

Under Warranty 4, failure of an item prior to W from the time of purchase results in a refund of a fraction, say α, of the original purchase price. The expected per-unit cost to the seller is

$$E[C_s(W)] = c_s + \alpha c_b F(W) \tag{10.18}$$

Repairability affects this cost only as indicated in the previous section.

Example 10.5

Suppose that an item is sold at c_b = \$100 and that the item cost to the seller is c_s = \$60, exclusive of warrant costs (the gross profit of \$40 covering the cost of warranty as well as profits to manufacturer, retailer, etc.). Suppose that the warranty period is either 6 months (W = 0.5) or one year (W = 1), the MTTF is μ = 1.5 years and the item has either a constant or an increasing failure rate. Table 10.4 gives $E[C_s(W)]$ for Warranties 1 through 4 assuming exponential time to failure, a Weibull distribution with β = 2, and the corresponding Gamma, lognormal, and inverse Gaussian distributions. For Warranty 4, α is taken to be 0.75, that is, a refund of 75% of the purchase price of the item is given if it fails within 1 year of purchase.

In interpreting these results, recall that the basic cost to the manufacturer is \$60 per unit. The remainder of the costs given in Table 10.4 is due to warranty.

Note the wide variability in expected costs, depending on the life distribution of the item (all of which have the same mean and all but the exponential the same variance). For example, a 6-month warranty on an item with a constant failure rate would increase cost to the seller by one-third (and decrease profits by 50%). This is a very substantial warranty cost for an item sold with a warranty that is one-third the length of the average lifetime of the item. On the other hand, for IFR items, a 6-month

Table 10.4 Expected Per-Unit Cost to Manufacturer Under Warranty Policies 1–4 with c_s = \$60, c_b = \$100, W = 0.5 and 1 year, and μ = 1.5 years for Various Life Distributions

Warranty policy	Life Distribution				
	Exponental	Weibull	Gamma	Lognormal	Inv. Gaussian
			W = 0.5		
1	\$80.00	\$65.09	\$63.35	\$61.40	\$61.27
2	83.74	65.47	63.55	61.43	68.05
3	88.35	68.36	65.58	62.33	71.82
4	81.26	66.27	64.19	61.75	68.87
			W = 1		
1	\$100.00	\$78.77	\$78.07	\$76.97	\$77.39
2	116.86	85.06	84.80	83.47	152.17
3	108.66	89.46	89.25	88.12	120.57
4	96.49	82.10	81.94	81.11	105.43

warranty would lead to reasonably acceptable costs under many of the distributions tabulated.

It is also seen from these results that any 1-year FRW or rebate warranty would be very costly and in some cases would lead to a loss on the item due solely to warranty costs.

10.4 MODELING BUYER'S EXPECTED COST OF WARRANTIED ITEMS

Not as much research has been devoted to modeling the cost of warranty to the buyer, for two reasons. First, few buyers have the required expertise, resources, or information necessary to develop and apply such models. (This is certainly true of the individual consumer. In almost all cases, it is also true of institutional buyers, at least as far as the required information is concerned.) Second, the models generally deal with expected costs; even if the information and models were available, this is of limited usefulness to the purchaser of an individual item.

Some results with regard to buyer's expected costs are available, however, and a few will be presented in this section. Most are basically modifications of the sellers' cost models. The same basic assumptions as in that case will be made: instantaneous replacement or repair, independent lifetimes of items and replacements, and known failure distributions. Here, the analysis will be further restricted to only nonrepairable items or items that are good-as-new after repair.

10.4.1 Buyer's Cost Models for the Nonrenewing FRW

For the buyer, the basic cost is c_b, the selling price of the item. Three ways of defining unit cost in this context will be considered.

Average Cost per Item in the Warranty Period

Calculation of the expected cost to the buyer over the period during which the warranty is in effect is useful in that it enables the buyer to compare the average cost of the item under warranty, c_b, with the average cost of an unwarrantied item purchased at the same price, that is, to assess the value of the warranty. The approximate expected cost* to the buyer is

$$E[C_b(W)] \approx \frac{c_b}{1 + M(W)} \tag{10.19}$$

---

* This is an asymptotic result. The exact result is $c_b E[1/\{1 + N(W)\}]$. For large W, this is approximately $c_b/\{1 + E[N(W)]\} = c_b/[1 + M(W)]$.

Average Cost to Buyer per Unit of Time

To calculate the average cost per unit of time of an item purchased under nonrenewing FRW, the ratio of c_b to the total expected service time of the original item and its replacements under warranty is determined. The total service time is $W + \gamma(W)$, where $\gamma(W)$ is the remaining life of the item in service at time W (which may be the original item or a replacement provided under warranty). The approximate average cost per unit of time, $E_T[C_b(W)]$, is given by

$$E_T[C_b(W)] \approx \frac{c_b}{W + E[\gamma(W)]} \tag{10.20}$$

The expected excess life, $E[\gamma(W)]$, is difficult to determine, except for the exponential distribution, for which $E[\gamma(W)] = \mu$, the mean time to failure of the items. In other cases, the distribution of $\gamma(W)$ is complicated and its expectation is not easily determined.* For additional results, see Refs. 1, 4, and 5.

Example 10.6

For $W = 0.5$ and $W = 1$ and the distributions and parameters considered in the previous examples, the values of $M(W)$ are given in Table 10.2. From these, the buyer's approximate expected unit costs are easily calculated. For example, with $W = 1$ and $\mu = 2$, under a Weibull distribution with $\beta = 2$, the unit cost to the buyer would be $c_b/1.184 = 0.845c_b$, or about 84.5% the cost of an unwarrantied item sold at the same price. For the exponential distribution with $\mu = 2$, this cost would be 66.7% of the cost of an unwarrantied item.

The approximate average cost per unit of time can be calculated for the exponential distribution because in this case $E_T[C_b(W)] = c_b/(W + \mu)$. With $W = 1$ and $\mu = 2$, measured in years, say, the expected cost per year of a warrantied item is $0.33c_b$. Note that the expected cost per year of an unwarrantied item would be $c_b/\mu = 0.5c_b$.

Total Cost of an Item Purchased Under Warranty

The models given in the previous paragraphs consider only the purchase price of the item. In fact, a buyer incurs many additional costs associated with the purchase and use of that item. These include acquisition cost (the cost of shopping, brand comparison, etc.), operating costs, maintenance

---

* It can be shown [4] that asymptotically $E[\gamma(t)] \to \frac{1}{2}(\mu + \sigma^2/\mu)$, where μ and σ^2 are the mean and variance of the life distribution of the items.

costs, and many incidental costs (e.g., installation, if appropriate, disposal, etc.).

The *total* cost of ownership of an item over its lifetime, $C_b(W)$, may be expressed as

$$C_b(W) = C_a + c_b + C_0(W + \gamma(W)) \tag{10.21}$$
$$+ C_m(W, W + \gamma(W)) + C_I(W + \gamma(W))$$

where C_a is the cost of acquisition, c_b is the fixed purchase price, $C_0(W + \gamma(W))$ is the total operating cost over the warranty period and the remaining lifetime of the item in service at time W after purchase, $C_m(W, W + \gamma(W))$ is the cost of maintaining the item for its remaining life after warranty expiration, and $C_I(W + \gamma(W))$ includes all incidental costs of ownership. Note that all cost elements except c_b are random.

10.4.2 Buyer's Cost Models for the Renewing FRW

Only the per-unit cost will be considered. This is again calculated as the ratio of the purchase price to the expected number of items supplied. The result is

$$E[C_b(W)] = c_b[1 - F(W)] \tag{10.22}$$

The average cost to the buyer per unit of time is considerably more complex. See Ref. 6.

Example 10.7

The buyer's average cost per item may be obtained for the distributions of the previous examples as c_b divided by the factor given in Table 10.3. For example, if $W = 1$, $\mu = 2$, and a Weibull distribution with $\beta = 2$ is assumed, the buyer's expected cost is $c_s/2.193 = 0.4560c_s$.

10.4.3 Buyer's Cost Models for the Rebate and Partial Rebate Warranties

For these warranties, the buyer's expected cost is c_b reduced by the expected amount of the rebate. For the full rebate warranty, this is

$$E[C_b(W)] = c_b[1 - F(W)] \tag{10.23}$$

Note that this is exactly the same as the expected cost per item of items purchased under renewing FRW.

For the partial rebate warranty, the buyer's average cost per item is

$$E[C_b(W)] = c_b[1 - \alpha F(W)] \tag{10.24}$$

Examples illustrating these costs may be obtained as in Example 10.7.

10.5   PRESENT VALUE COST MODELS FOR THE FRW

The models of the previous sections assume that seller's costs are constant through time. If this is not the case, the models can be modified to account for changing costs by replacing c_s by a specified function $c_s(t)$ in the previous expressions. (See Ref. 2.)

Time-varying cost models can also be used to discount both buyer's and seller's future costs to their present value. The most common discounting function used for this purpose is the exponential function, given by

$$c_s(t) = c_s e^{-\delta t} \tag{10.25}$$

where δ is the discount rate. Let $C_s(W; \delta)$ be the seller's cost of supplying an item sold under nonrenewing FRW, discounted to present value at rate δ. For nonrepairable items, the seller's discounted expected cost [given in (10.10) without discounting] is given by

$$E[C_s(W; \delta)] = c_s \left(1 + \int_0^W e^{-\delta t} \, dM(t) \right) \tag{10.26}$$

For repairable items with good-as-new repair, the seller's discounted expected cost [analogous to (10.11)] is given by

$$E[C_s(W; \delta)] = c_s + c_r \int_0^W e^{-\delta t} \, dM(t) \tag{10.27}$$

For other types of repair, similar expressions can be obtained. (See Chapter 4 of Ref. 2.)

For the partial rebate warranty, assuming nonrepairable items, the present value of seller's expected cost is given by

$$E[C_s(W; \delta)] = c_s + \alpha c_b \int_0^W e^{-\delta t} \, dF(t) \tag{10.28}$$

Discounted expected seller's cost for the full rebate warranty is given by (10.28) with $\alpha = 1.0$.

The buyer's expected costs can also be discounted to present value by the same modifications of the undiscounted cost models. Furthermore, discounting can also be employed in the cost models given in the sections which follow. See Chapter 4 of Ref. 2 for details and examples.

10.6 LIFE-CYCLE COST MODELS FOR THE FRW

In calculating the life-cycle cost of items purchased under warranty, it is assumed that an identical replacement item is purchased at the same price c_b immediately upon failure of the item in service at the expiration of the warranty period. This replacement item is assumed to be covered by an identical warranty, and this process continues over a life cycle of length L. L is assumed to be a known, fixed quantity.

Examples of applications of life-cycle analyses are replacement tires, batteries, and other automotive components over the lifetime of an automobile; replacement tires, engines, windshields, electronic equipment, or any of a large number of other components on jet aircraft; or any consumer or commercial items that are replaced on failure over an extended period of time.

10.6.1 Renewal Processes for Life-Cycle Cost Models

Over the life cycle of the item, the seller will incur a cost each time an item fails. The buyer, on the other hand, will incur a cost (and hence the seller will realize income) only upon failure of items not covered under warranty. Thus, two renewal processes are involved in the analysis. The first is the renewal process associated with the sequence $X_1, X_2, \ldots$ of lifetimes of the items. The second is the renewal process associated with the sequence of buyer's purchase times, say $Y_1, Y_2, \ldots$. (The warranty process for renewing and nonrenewing FRW is illustrated in Figures 4.1 and 4.2 of Ref. 2.)

Because purchases occur upon failure of the item in service at the end of the warranty coverage, the Y_i's are of the form $Y = W + \gamma(W)$ for the nonrenewing FRW, where $\gamma(W)$ is the remaining life of the item in use at time W from purchase. The distribution of $\gamma(W)$ (see Chapter 3 of Ref. 2) is given by

$$F_\gamma(t) = F(W + t) - \int_0^W [1 - F(W + t - x)] \, dM(t) \qquad (10.29)$$

The renewal function associated with the distribution of Y can be expressed in terms of $F_\gamma(t)$.

In the models which follow, it is assumed that repairs and replacements (including purchased replacements) occur instantaneously. This is reasonable as long as the actual times until repair, replacement, or repurchase are small relative to W and L.

10.6.2 Modeling Buyer's Life-Cycle Cost

It is convenient to begin with buyer's cost. This depends only on the renewal function associated with Y. The results of this section apply to the nonrenewing FRW, assuming nonrepairable items. Most of the results can be modified to cover repairable items as well, assuming that repair costs are constant and that repair is good-as-new.

Lifetimes $X_1, X_2, \ldots$ are assumed to be independent and identically distributed with distribution function $F(\cdot)$. In this case, the buyer's purchase intervals, $Y_1, Y_2, \ldots$, are also independent and identically distributed. Let $M_Y(\cdot)$ be the renewal function associated with the Y_i. The expected number of purchases in the interval $[0, L]$, that is, during the life cycle of the item, is $1 + M_Y(L)$. Let $C_b(L, W)$ denote the buyer's life-cycle cost. The buyer's undiscounted expected total cost of purchasing the item over its life cycle is then

$$E[C_b(L, W)] = c_b[1 + M_Y(L)] \tag{10.30}$$

$M_Y(\cdot)$ can be obtained [7] as the solution to

$$M_Y(t) =$$
$$\begin{cases} 0, & 0 \le t \le W \\ F_\gamma(t - W), & W < t \le 2W \\ F_\gamma(t - W) + \int_0^{t-2W} M_Y(t - W - x)\, dF_\gamma(x), & 2W < t < \infty \end{cases}$$

$$\tag{10.31}$$

It is not possible to obtain explicit expressions for $M_Y(t)$ except for a few simple cases. In addition, tables of this renewal function are not available for any distribution $F(\cdot)$. Thus, it is usually necessary to use numerical computation, bounds, simulation, or analytical approximations to evaluate $M_Y(\cdot)$.

If L is large relative to W and $E(Y)$, approximations based on asymptotic results for renewal functions may be used. The simplest of these is based on the Elementary Renewal Theorem (see Chapter 7), which states that $M_Y(t)/t \to 1/E(Y)$ as $t \to \infty$. $E(Y)$ is given by $E(Y) = \mu[1 + M(W)]$, where $\mu = E(X)$ and $M(\cdot)$ is the ordinary renewal function associated with $F(\cdot)$. For large L, the resulting approximation is

$$M_Y(L) \approx \frac{L}{\mu[1 + M(W)]} \tag{10.32}$$

An improvement on this result is the second-order approximation of this type, given [4] by

$$M_Y(L) \approx \frac{L}{E(Y)} + \frac{V(Y)}{2E^2(Y)} - \frac{1}{2} \tag{10.33}$$

where $V(Y)$ is the variance of the random variable Y. An expression for $V(Y)$ is given by Nguyen and Murthy [7]. The result is

$$V(Y) = [\sigma^2 - \mu^2 M(W)][1 + M(W)]$$
$$+ 2\mu \left[WM(W) - \int_0^W M(t)\, dt \right] \tag{10.34}$$

where σ^2 is the variance of X. The integral of the renewal function, which appears in the last term of this expression, has been tabulated for several important life distributions. (See, e.g., Ref. 8.) These tables may be used to obtain numerical results.

For other approximations, exact results for the exponential distribution, present value calculations, additional discussion, and numerical examples, see Section 4.5.1 of Ref. 2.

10.6.3 Modeling Seller's Life-Cycle Profit

The seller's life-cycle profit may be expressed as the difference between the buyer's expected life-cycle cost, given in the previous section, and the seller's expected life-cycle cost.

Over the life cycle L of the item, the expected number of sales is $1 + M_Y(L)$. The expected cost to the seller of each sale is $c_s[1 + M(W)]$. Let $P_{FRW}(L, W)$ be the profit on a sequence of items sold under warranty of length W over a life cycle of length L to a single buyer, including the initial purchase item and all purchased replacements. The seller's expected profit is given by

$$E[P_{FRW}(L, W)] = \{c_b - c_s[1 + M(W)]\}[1 + M_Y(L)] \tag{10.35}$$

10.7 INDIFFERENCE PRICES FOR THE NONRENEWING FRW

An approach to determining the value of a warranty is to consider the proportional price differential that a buyer should be willing to pay or that a seller would have to charge for a warrantied as opposed to an unwarrantied item. From the buyer's point of view, this is equivalent to finding an "indifference" price c_b^*, below which a warrantied item is cheaper in the long run and above which an unwarrantied item, purchased at a lower price c_u, say, is cheaper in the long run. Thus, c_b^* is such that if the buyer were offered the option of purchasing an unwarrantied item at price c_u or a warrantied item at price c_b, the warrantied item would be preferred if

$c_b < c_b^*$, the unwarrantied item would be preferred if $c_b > c_b^*$, and the buyer would be indifferent with regard to the warranty if $c_b = c_b^*$.

From the seller's point of view, the equivalent result is an indifference price c_b^{**}, below which the seller would prefer to sell the item without warranty at price c_u, above which selling the item with warranty is preferable, and the buyer is indifferent if $c_b = c_b^{**}$.

Buyer's and seller's indifference prices for the nonrenewing FRW over a life cycle of length L are given in the following sections.

10.7.1 Buyer's Indifference Price

For unwarrantied items, sold at cost c_u to the buyer, the total expected cost over a life cycle of length L is $c_u[1 + M(L)]$, because the buyer must purchase a replacement at full price each time an item fails in this period. For warrantied items, the buyer's total expected cost is $c_b[1 + M_y(L)]$. The indifference price is obtained by equating these two long-run costs and solving for c_b. The result is

$$c_b^* = \frac{c_u[1 + M(L)]}{1 + M_Y(L)} \tag{10.36}$$

For large L, this may be approximated by

$$c_b^* \approx \frac{c_u\mu[1 + M(L)][1 + M(W)]}{\mu[1 + M(W)] + L} \tag{10.37}$$

This approximation is discussed by Blischke and Scheuer [9]. The approximation is good for large values of L but overestimates c_b^* if L is not large. As $L \to \infty$, $M(L)$ is approximately L/μ, and (10.37) becomes

$$c_b^* \approx c_u[1 + M(W)] \tag{10.38}$$

If future payments are discounted to their present value at discount rate δ, the indifference price as $L \to \infty$ becomes

$$c_b^*(\delta) = \frac{c_u(1 - \exp\{-\delta\mu[1 + M(W)]\})}{1 - \exp(-\delta\mu)} \tag{10.39}$$

Numerical Examples

Values of $1 + M(W)$ are given in Table 10.2 for the six life distributions considered in the previous examples. For selected values of L, $M(L)$ may be obtained by table look-up or computer calculation as indicated previously. These may be substituted into Equations (10.37) and (10.39) to obtain numerical illustrations of indifference prices.

10.7.2 Seller's Indifference Price

To analyze indifference prices from the seller's point of view, it is necessary to consider the differential pricing structure such that the seller's long-run profit would be the same if items were sold with nonrenewing FRW at price c_b or without warranty at price c_u. Without warranty, each sale results in a profit of $c_u - c_s$, so that the seller's total expected profit is $(c_u - c_s)[1 + M(L)]$. With warranty, if the administrative cost of servicing the warranty is ignored, the expected long-run profit to the seller is expected income of $c_b[1 + M_Y(L)]$ minus expected cost of $c_s[1 + M(L)]$. Thus, $c_b^{**} = c_b^*$, that is, the seller's and buyer's indifference prices are the same.

If the per-item administrative cost of servicing the warranty, say c_a, is not included in c_s, then the indifference price to the seller is increased. The result in this case is

$$c_b^{**} = \frac{c_u[1 + M(L)] + [M(L) - M_Y(L)]c_a}{1 + M_Y(L)} \qquad (10.40)$$

Under this model, the seller prefers to sell an item with warranty at price c_b to selling an item without warranty at price c_u if $c_b > c_b^{**}$, the buyer prefers an item with warranty if $c_b < c_b^*$, and *both* prefer an unwarrantied item if $c_b^* < c_b < c_b^{**}$.

These expressions can be evaluated as indicated previously. Numerical examples are given in Chapter 4 of Ref. 2.

10.8 STATISTICAL ESTIMATION OF COST MODELS

For the four warranties considered in this chapter, the cost models involve primarily the distribution function $F(\cdot)$ of X, its moments, μ and σ^2, the ordinary renewal function $M(\cdot)$, and some related distributions and renewal functions. Statistical estimation of these requires an appropriate database.

The simplest situation is that in which a sample of test data consisting of n independent lifetimes $X_1, X_2, \ldots, X_n$ is available and the form of the distribution $F(\cdot)$ is known (e.g., exponential, Weibull, etc.). In this case, the most straightforward analysis is to use the data to estimate the parameters of $F(\cdot)$, then use tables or a computer algorithm to determine estimates of $F(\cdot)$, $M(\cdot)$, and the related quantities required to estimate the cost models. Methods for estimating parameters for the life distributions used in this chapter for purposes of illustration are discussed in Chapter 8 and in Chapter 12 of Ref. 2.

If $F(\cdot)$ is not known, goodness-of-fit techniques may be used to select a distribution that "best" fits the data. (See Chapters 2 and 3 of Ref. 10 for graphical methods and Chapter 9 of Ref. 11 for analytical methods for fitting life distributions.) The distributions of Section 10.2 are reasonable candidate distributions for this purpose. Alternatively, nonparametric methods may be used to estimate both $F(\cdot)$ and $M(\cdot)$. See Chapter 12 of Ref. 2 and Ref. 12 for details.

Alternative approaches that may be preferred for certain distributions, particularly if data are limited, are also discussed in Chapter 8, in Chapter 12 of Ref. 2, and in the references cited. Several methods of estimating renewal functions are discussed as well. For analysis of other types of data (e.g., claims data), see Chapter 9.

10.9 CONCLUDING REMARKS

10.9.1 Recommendations for the Practitioner

Analyses of Benefit to the Seller

As noted, the seller is usually in a more advantageous position than the buyer with regard to analysis of warranty costs, as much more information is available to the seller. The following analyses are suggested:

1. Consider various warranty policies and terms and compare expected costs. In addition to the FRW and rebate policies discussed in this chapter, the pro-rata warranty, combination warranties, and other alternatives should be considered as well. Cost models for these will be given in the chapters which follow.
2. Costs can be quite sensitive to the assumptions made regarding the life distribution of the item under warranty. This is apparent in some of the numerical examples of this chapter. Whether items have increasing, decreasing, or a constant failure rate is especially important in this regard. It is instructive to consider various alternatives in calculating the cost models.
3. It is also useful to investigate the relationship between the length of the warranty period W and the mean time to failure μ. In some cases, it may be possible to increase W with only a small increase in costs, and thereby gain a competitive advantage.
4. Careful testing and data collection are essential for statistical estimation of warranty costs. Claims data are usually much messier, and, in any case, will probably be acquired too late to be of much use in formulating a warranty policy, except possibly for future similar products.

5. The cost models that have been developed (those discussed in this chapter and elsewhere) are based on many assumptions, some of which may have been used for mathematical convenience and may or may not be realistic. The possible limitations of the models should be carefully considered in any application.

Suggestions for the Buyer

Buyers are seldom in a position to obtain the necessary information to evaluate a product adequately. Usually, however, some information is available. Our suggestions are as follows:

1. Information is often available from consumer's groups and publications. Although an individual consumer cannot economically test and evaluate products, testing by such organizations is often feasible.
2. Comparison shopping for warranties on otherwise comparable products is often instructive. A longer warranty, or warranty terms that are better in some other respect, may signal a more reliable product. On the other hand, this must be balanced against the fact that such an item will generally have a higher purchase price.
3. For large organizations such as large corporations or government agencies, it may be possible to compel the potential seller to provide the necessary information for product evaluation. Alternatively, user information on product performance for similar items may have been collected. In such cases, the buyer has the opportunity to analyze and predict costs as well. Any available information of this type will be useful in negotiating warranty terms.

10.9.2 Topics for Future Research

As is apparent in many of the comments of this chapter regarding cost estimation or prediction for the FRW, many unsolved problems remain. In the following, a few, but by no means all of these, are listed:

1. The models given in this chapter (and elsewhere) either assume constant costs or use average or expected costs. In reality, costs are random and both average costs and the distributions of costs change through time.
2. Neither repair nor replacement times are instantaneous. Models which reflect random repair and replacement times

and costs would be more realistic in assessing warranty cost.

3. Many models are analytically intractable and many more simulation studies are needed to investigate warranty costs in realistic situations.

4. Extensions of models such as those given in Chapter 24 to account for the fact that not all potential warranty claims are exercised are needed.

5. Extensions to account for invalid or rejected claims are also needed. (Some results along these lines as well as analysis of some related problems are given in Ref. 13.)

6. Some very difficult problems remain in estimating warranty costs on the basis of sample data in realistic contexts. These include the following:

    i. Additional realistic models for claims data, building on the results of Chapter 9

    ii. Estimation of parameters of life distributions based on aggregated data, including test data, claims data, data on previous similar products, and so forth

    iii. Estimation of renewal functions and other important elements of the cost models based on aggregate data

    iv. Development of confidence intervals for warranty cost models

    v. Application of nonparametric methods to warranty cost estimation

7. Many additional sensitivity studies are needed to investigate the robustness of the models to distributional and other assumptions.

REFERENCES

1. Johnson, N. L., and Kotz, S. (1970). *Distributions in Statistics: Continuous Univariate Distributions—1*, John Wiley & Sons, New York.

2. Blischke, W. R., and Murthy, D. N. P. (1993). *Warranty Cost Analysis*, Marcel Dekker, Inc., New York.

3. Nguyen, D. G., and Murthy, D. N. P. (1984). A general model for estimating warranty costs for repairable products. *IIE Transactions*, **16**, 379–386.

4. Cox, D. R. (1960). *Renewal Theory*, Metheun, London.

5. Ross, S. M. (1970). *Applied Probability Theory with Optimization Applications*, Holden-Day, San Francisco.

6. Mamer, J. W. (1987). Discounted and per unit costs of product warranty, *Management Science*, **33**, 916–930.

7. Nguyen, D. G., and Murthy, D. N. P. (1988). Failure free warranty policies

for nonrepairable products: A review and some extensions, *RAIRO Operational Research*, **22**, 205–220.

8. Baxter, L. A., Scheuer, E. M., Blischke, W. R., and McConalogue, D. J. (1981). *Renewal Tables: Tables of Functions Arising in Renewal Theory.* Technical Report, Department of Information and Operations Management, University of Southern California, Los Angeles.

9. Blischke, W. R., and Scheuer, E. M. (1981). Applications of renewal theory in analysis of the free-replacement warranty, *Naval Research Logistics Quarterly*, **28**, 193–205.

10. Nelson, W. (1990). *Accelerated Testing. Statistical Models, Test Plans, and Data Analysis*, John Wiley & Sons, New York.

11. Lawless, J. F. (1982). *Statistical Models and Methods for Lifetime Data*, John Wiley & Sons, New York.

12. Frees, E. W. (1986). Warranty analysis and renewal function estimation, *Naval Logistics Research Quarterly*, **33**, 361–372.

13. Palfrey, T., and Romer, T. (1983). Warranties, performance, and the resolution of buyer-seller disputes, *Bell Journal of Economics*, **14**, 97–117.

11

The Basic Pro-Rata Warranty

Jayprakash G. Patankar

The University of Akron
Akron, Ohio

Amitava Mitra

Auburn University
Auburn, Alabama

11.1 INTRODUCTION

As discussed in detail in Chapter 1, a warranty is an obligation to a manufacturer of products or to a vendor of services for a certain duration of time known as the warranty period. Its purpose is to establish liability in the event that the product does not perform effectively within the warranty period or if services rendered are unacceptable to the customer. The Magnuson–Moss Warranty Act [1] mandates that manufacturers must offer a warranty for all consumer products sold for $15 or more (see Chapters 4 and 5).

A taxonomy of the different types of warranty policies may be found in Chapter 2 of the work of Blischke and Murthy [2]. Considering warranty policies that do not involve product development after sale, policies exist for a single item or for a group of items. With our focus on single items, policies may be subdivided into the two categories of nonrenewing and renewing. In a renewing policy, if an item fails within the warranty time, it is replaced by a new item with a new warranty. In effect, warranty begins anew with each replacement. On the other hand, for a nonrenewing policy, replacement of a failed item does not alter the original warranty. Within each of these two categories, policies may be subcategorized as simple or combination. Examples of a simple policy are those that incorporate replacement or repair of the product, either free or on a pro-rata basis. The proportion of the warranty time that the product was operational is typically used as a basis for determining the cost to the customer for a pro-rata warranty. As an example, consider a linear pro-rata warranty for 12 months for a product which was sold at $60. Suppose the product failed after 5 months. Then, on replacement of the product, the customer would pay only $25, or in effect would receive a rebate of $35. To the manufacturer, it then becomes important to estimate the expected warranty reserves that it would have to set aside to meet warranty obligations. A combination warranty policy is a simple policy with additional features or a policy that combines the terms of two or more simple policies. This chapter will consider the simple pro-rata policy.

Pro-rata policies may involve a refund of a fraction of the selling price or a replacement of the product. Let us consider the first situation, where the customer is not obligated to buy a replacement item. Three subcategories of a pro-rata policy involving refund are as follows: notation described below:

(a) The refund or rebate to the customer, if $t_1 < W$, is a linear function of the remaining time in the warranty period:

$$C(t_1) = \left(\frac{(W - t_1)}{W}\right) c$$

(b) The rebate is a proportional linear function of the remaining time in the warranty period:

$$C(t_1) = \left(\frac{\rho(W - t_1)}{W}\right) c$$

where $0 < \rho < 1$.

(c) The rebate is a nonlinear function, for example, quadratic, of the time remaining in the warranty period:

$$C(t_1) = \left(\frac{W - t_1}{W}\right)^2 c$$

In the above, W is the warranty period c is the unit product price, t_1 is the age of the product at failure, and $C(t_1)$ is the refund or rebate at time t_1.

Examples of pro-rata policies involving refund are relatively inexpensive nonrepairable products such as automobile batteries.

In the class of renewing policies, a pro-rata form also exists. Under this situation, if the product fails within the warranty period, the manufacturer agrees to provide a replacement item at prorated cost. This applies to the original item as well as any replacements made within the warranty period. Proration can be either a linear or a nonlinear function of $W - t_1$, the time remaining in the warranty period. The difference between this class of policies and the previous class is that a refund is unconditional with the previous class. Here, it is only provided as a discount on the purchase of a replacement item. Several nonrepairable items are sold under such a policy. For example, automobile tires are sold under a renewing pro-rata warranty, where the consumer is offered a replacement for a failed item at a reduced price. In this chapter, our focus is on the linear pro-rata warranty involving refund.

11.1.1   Linear Pro-rata Warranty

A basic linear pro-rata warranty is considered in this chapter. The concept of this warranty is that the rebate paid to the customer decreases linearly from the purchase price of the product, c, to zero, during the time period of 0 to the warranty period, W. Common examples of pro-rata warranties are automobile tires, automobile batteries, and so on.

Let $C(t)$ denote the rebate paid at time t. Under the linear pro-rata rebate plan, $C(t)$ is given by

$$C(t) = c\left(1 - \frac{t}{W}\right), \quad 0 < t < W \tag{11.1}$$

where c denotes the unit product price, W denotes the warranty period, and t denotes the time of product failure.

11.1.2   Warranty Theories

We may group the problems studied in this area into several categories, such as promotional and qualitative issues, cost estimation for single products, cost estimation for renewable policies, and multiobjective models. In this chapter, we focus on single objective models.

In the area of studying the impact of warranty on marketing and consumer perception of quality, the study by Udell and Anderson [3] investigated the effect of warranty on sales. They found that warranties could be used to attract market segments from competitors by using it as a promotional tool. Further, by incorporating an acceptable warranty policy, a company could retain and improve its market share, thus using it as a protectional marketing tool. Five hypotheses were investigated.

Hypothesis 1: The effectiveness of promotional warranties and, consequently, management's use of them should vary directly with the price of the product. The empirical evidence confirms this decision rule. Promotional warranties were used in 86% of cases in which product price exceeded $2500 and were used only in 50% of cases in which the unit price was $2 or less.

Hypothesis 2: The potency of a promotional warranty increases with product complexity. Empirical evidence shows that among "highly technical and complex" products, 46% of them used promotional warranties, whereas for the "non-technical and not complex" product category, only 23% of them used promotional warranties. This hypothesis was supported at a significance level of 0.001.

Hypothesis 3: The effectiveness of a promotional warranty increases with the life of the product. Empirical evidence suggested that of products purchased least often (every 15 years or more), 79% of the products offered promotional warranties. Of products purchased most often (once a month), only 44% offered promotional warranties. A test of hypothesis supported this rule at a level of significance of 0.02.

Hypothesis 4: The effectiveness of a promotional warranty varies inversely with the purchaser's knowledge concerning the product and its want-satisfying power. Evidence shows that 42% of the products sold to knowledgeable customers used promotional warranties. Twenty-five percent of those selling to buyers with a low degree of knowledge used promotional warranties. This hypothesis was not supported at a level of significance of 0.05.

Hypothesis 5: The utilization of promotional warranties varies inversely with the share of market. Evidence shows that 42% of firms with less than 5% of market share used promotional warranties, whereas of firms with 60% of market share, only 14% of them used promotional warranties. This hypothesis was supported at a level of significance of 0.01.

They also found that when all variables were considered together, a statistically significant relationship exists between the use of promotional warranties and market share, product complexity, and length of the repurchase cycle. They concluded that the promotional warranties will be most effective when the following occurs:

1. The retail price of the product is high.
2. The product purchase cycle is long.
3. The product is perceived to be complex.
4. The knowledge of the buyer is low.
5. The firm's market share is small

Shimp and Bearden [4] studied the effect of warranty quality, warranty reputation, and price on the consumer's perception of financial and performance risks. They believe that the risk reduction capacity of warranties is realized in conjunction with other factors such as warrantor reputation and price. A series of experiments was conducted to study how these extrinsic cues such as warrantor reputation and price, separately and interactively, influence the consumer's perceptions of financial risks (F-risk) and performance risk (P-risk), associated with purchase decisions. They tested five hypotheses:

H1: The higher the perceived warranty quality, the less the perceived F- and P-risks associated with an innovative product. Empirical evidence shows that warranty quality does not reduce P-risk, whereas warranty quality does reduce F-risk significantly.

H2: The higher the price, the less the perceived P-risk associated with an innovative product.

H3: The higher the price, the greater the perceived F-risk associated with an innovative product.

Experimental results showed that price did not significantly reduce P-risk or increase F-risk at a level of significance of 0.05.

H4: The more favorable a warrantor's reputation is perceived to be, the less the F- and P-risks associated with an innovative product.

Evidence did not support the hypothesis that the warrantor's reputation reduced P- and F-risks.

H5: There is a significant interaction between warranty quality (WQ) and warrantor reputation (WR) in influencing F- and P-risks. Risks should decrease with warranty quality if warrantor's reputation is high, but risk will not decrease with warranty quality if warrantor is perceived to be less reputable.

The evidence does not support this hypothesis.

Wiener [5] concludes that a product's warranty is an accurate signal of a product's quality. He conducted an empirical study to test the above hypothesis and tested consumer durables such as air conditioners, televisions, washing machines, and dryers, and also tested automobiles and pickup trucks from 1977 to 1979. Consumer durables were tested using the Kolmagorov–Smirnov two-sample test to see if a product's warranty is an accurate signal of product quality. The conclusion was that it was so. For motor vehicles, he tested whether high-priced vehicles are more reliable than low-priced vehicles, as consumers believe that a higher price may reflect better quality. In addition, there is a perception that foreign cars are better than domestic cars; that hypothesis was tested. Finally, the hypothesis that vehicles with a higher signal are more reliable than those with a low signal is also tested. The conclusions were that when all vehicles are included, the price and foreign manufacturer were found to be accurate signals of reliability. It was found that even after prescreening for price and foreign manufacturer, warranty proved to be an accurate signal of reliability.

11.2   MODELING THE PRO-RATA WARRANTY

Menke [6] estimated expected warranty cost for a linear and lump-sum rebate policies. He assumed an exponential failure distribution and showed that the warranty cost per dollar sale is given by

$$\frac{r}{c} = 1 - \frac{m}{W}(1 - e^{-m/W})$$

where r is the expected warranty cost per unit product and m is the mean life of a product. If c' is the unit price, then, c, selling price which includes warranty cost, is given by

$$c = c' + r, \quad c' = c\left(1 - \frac{r}{c}\right)$$

For a lump-sum rebate policy, Menke showed that

$$\frac{r}{c} = \frac{L}{c}(1 - e^{-W/m})$$

where L is the lump-sum rebate. Menke compares the two policies in order to make strategic decisions.

The main drawback in Menke's [6] estimation was that he ignored the time value of money. Because warranty costs are incurred in the future, it is important that we discount them using a proper discounting rate. Amato and Anderson [7] estimated discounted warranty reserve per dollar sale (r^*/c^*) as follows:

$$\frac{r^*}{c^*} = (1 + m \ln(1 + \theta + \phi))^{-1}$$

$$\times \left(1 - \frac{m}{W}(1 + m \ln(1 + \theta + \phi))^{-1}[1 - (1 + \theta + \phi)^{-W}e^{-W/m}]\right)$$

where r^* is the expected discounted warranty reserve per unit product, $c^* = c' + r^*$, c' is the product unit price excluding warranty cost, θ is the discount rate, and ϕ is the inflation rate. Amato and Anderson showed that r^*/c^* is less than r/c as estimated by Menke and concluded that by setting smaller warranty reserves, the saved amount can be invested to increase the firm's profits. Discounting may cause the product price to fall in real terms and be more competitive in the marketplace.

Whereas these models estimate costs based on a selected warranty period, the formulation by Lowerre [8] was motivated by the manufacturer's concern. His model determined the warranty time assuming that a selected percentage of revenue is used to meet warranty claims, thus providing the manufacturer with some guidelines for selecting the warranty period. Heschel [9] developed his model with the consumer in mind. He found the expected repair cost to the consumer over the life of the product. His model assumed a full rebate for an initial period (up to time t_1), followed by a pro-rata rebate. The expected repair cost over a useful life (t_3) to the consumer was found as

$$E(\text{cost}) = C\left(\frac{e^{-\lambda t_1} - e^{-\lambda t_2}}{\lambda(t_2 - t_1)} - e^{-\lambda t_3}\right)$$

where C is the cost to repair, λ is the constant average failure rate in an exponential distribution, t_1 is the time until full rebate, t_2 is the time when pro-rata rebate expires, and t_3 is the useful life of the product. The expected repair cost during the useful life on a product without warranty will be $\lambda t_3 C$ for identical equipment. Thus, the cost savings due to warranty will be $[\lambda t_3 C - \lambda t_3 E(\text{cost})] = \lambda t_3[C - E(\text{cost})]$.

Patankar and Worm [10] developed prediction intervals for warranty reserves and cash flows associated with linear pro-rata and lump-sum rebate plans. They developed confidence intervals on total warranty reserves and cash flows. Upper bounds on cash flows and warranty reserves are determined to analyze the uncertainty or risk involved. If the amounts paid on each item, $r_i(t)$, are independent and identically distributed, then $R = \sum_{i=1}^{N} r_i(t)$ is normally distributed with mean $NE(r_i(t))$ and variance

$NV(r_i(t))$ where $V(r_i(t))$ represents the variance of $r_i(t)$. Thus, after estimating $E(r_i(t))$ and $V(r_i(t)$, prediction intervals are determined using standard normal formulas as follows.

The present value of expected warranty reserve per unit is

$$PV(r) = \frac{cs}{m}\left(1 - \frac{s}{W}(1 - e^{-W/s})\right)$$

where $s = \theta + \Phi + 1/m$. The variance of the present value of expected warranty reserve is given by

$$V(PV(r)) = \frac{c^2x}{m}\left(1 - \frac{2x}{W}\left[1 - \frac{x}{W}(1 - e^{-W/x})\right]\right) - (PV(r))^2$$

where $x = 1/(2\theta + \Phi + 1/m)$. The mean and variance of total warranty reserves (R) is found from

$$E(R) = NPV(r)$$

$$V(R) = NV(PV(r))$$

So, a $100(1 - \alpha)\%$ prediction interval on R is given by

$$E(R) \pm z_{\alpha/2}\sqrt{V(R)}$$

where $z_{\alpha/2}$ is the standard normal variate corresponding to an upper tail area of $\alpha/2$. A similar procedure is used to determine prediction intervals on cash flows. The upper bound on cash flow is used for planning purposes so that the firm is able to meet its warranty liability for that particular period or year.

Whereas Refs. 6–10 dealt with nonrenewable warranties, Blischke and Sheuer [11] applied renewal theory to estimate warranty costs for renewable policies. They assumed that the buyer purchases an identical replacement when an item in service fails at the end of the warranty period. The customer is provided free replacements during the warranty period. The customer pays an initial price C and then again if the product fails beyond the warranty time, until the useful life, L (life cycle), of the product expires. Purchases occur at random with a cycle length of $Y = W + r(W)$, where $r(W)$ is the remaining lifetime of the item W time units after the beginning of the cycle. The expected number of payments $= M_Y(L)$, where M_Y is the renewal function of Y. They considered the exponential, uniform, Gamma, and Weibull failure distributions. The values of $M_Y(L)/L$ were estimated using a simulation approach.

An analysis from an accounting perspective was considered by Balachandran et al. [12]. They used a Markovian approach to estimate repair and replacement costs of a warranty program during the warranty period and during an accounting period. They assumed an exponential failture distribution because its failure rate, λ, is constant and fits the assumption that during the useful period of the product the failure rate is constant. For illustration, they used a product with three components which have independent failure distributions. They assume that the components are repaired on the first two failures and are replaced on the third failure. Unit repair and replacement costs are assumed. If the time to failure is exponentially distributed with a failure rate of λ, then the number of failures are Poisson distributed with rate λ. They developed a Markovian model to study breakdowns. State $i(0, 0, 0)$ means no breakdown in any of the three parts. State $j(1, 0, 0)$ means one breakdown on part 1 but no breakdowns on parts 2 and 3. They developed transition probabilities for all states under the claim that because failures are Poisson distributed, the transition probabilities do not depend on time and therefore the process is stationary. A cost is computed for transition. It depends on whether the product is repaired or replaced. An expression for warranty cost using a Markov process was developed as

$$C_i^{(n)} = \sum_{j=1}^{k} [c_{ij} + C_j^{(n-1)}P_{ij}]$$

where $C_i^{(n)}$ is the expected cost over an n-week period given that the current state is i, c_{ij} is the transition cost from state i to state j, and P_{ij} is the transition probability from state i to state j. We may write

$$C_i^{(1)} = \sum_{j=1}^{k} c_{ij}P_{ij}$$

So,

$$C_i^{(n)} = C_i^{(1)} + \sum_{j=1}^{k} C_j^{(n-1)} P_{ij}$$

This equation is known as the Chapman–Kolmogorov equation. The problem is solved recursively. To estimate warranty cost in an accounting period, the following approach is used. Suppose five units are under warranty and all in state $(0, 0, 1)$ and 2 weeks are left in the accounting period. The contribution to total expected warranty cost due to these five units is $5C_{(0,0,1)}^{(2)}$. Suppose eight new items are sold. They are all in state $(0, 0, 0)$, and 3 weeks are left in accounting period, and warranty extends beyond this period. Then, the contribution of these eight units toward the total expected warranty cost for this accounting period is $8C_{(0,0,0)}^{(3)}$.

Mamer [13] estimated short-run total costs and long-run average costs of products under warranty. He studied pro-rata and free-replacement warranty policies under different failure distributions (such as new-better-than-used). Mamer showed that the expected warranty cost depends on the mean of the product life time distribution and its failure rate.

Mamer looked at pro-rata and free-replacement warranties. He assumed that a customer purchases a new product instanteously during the product life cycle. This provides a good assumption for using renewal theory. Cost of the replacement depends on the type of rebate offered. In the short run, under pro-rata warranty, consumer's expected cost is given by

$$E[\mathrm{TC_{PR}}(T)] = q_2 + q_2\left(F(T) - F(W) + \frac{\mu_W}{W}\right)$$

$$+ q_2 \int_0^{T-W} F(T-x)\, dm(x) + \int_{T-W}^T \int_0^{T-x} \left(\frac{q_2 z}{w}\right) dF(z)\, dm(x)$$

where T is the time up to which owner will replace the product, $F(T)$ is the failure distribution function at time T, q_2 is the cost to the owner of replacing the product, $m(x)$ is the renewal function associated with the distribution function $F = \sum_{n=1}^{\infty} F^{(n)}(x)$, where $F^{(n)}(x)$ represents the n-fold convolution of F evaluated at the point x, and $\mu_W = \int_0^W x\, dF(x)$.

For free replacement, the total expected cost during the period $[0, T]$ is given by

$$E(\mathrm{TC_{FR}}(T, W)) = q_2 + q_2(F(T) - F(W))$$

$$+ \int_W^T E[C_{\mathrm{FR}}(T-x, W)]\, dF(x) + \int_0^W E[C_{\mathrm{FR}}(T-x, W-x)]\, dF(x)$$

where $\mathrm{TC_{FR}}(T, W)$ is the total cost of ownership during the period $[0, T]$ under a free-replacement warranty and $C_{\mathrm{FR}}(T, W)$ is the cost of ownership of a product with warranty length W for a length of time T, exclusive of initial purchase cost.

Mamer also estimated long-run average costs for both warranty policies. For the no-warranty case, average long-run cost ($\mathrm{AC_{NW}}$) is given by

$$\mathrm{AC_{NW}} = \frac{q_2}{E(X)}$$

where $E(X)$ represents the expected life of the product. Under pro-rata warranty,

$$\text{AC}_{\text{PR}}(W) = \left(\frac{q_2}{W}\right) \int_0^W \frac{e^{-H(x)} \, dx}{E(X)}$$

where $H(x)$ represents the cumulative failure rate up to time x and

$$H(x) = \int_0^x h(u) \, du$$

where

$$h(x) = \frac{f(x)}{1 - F(x)}$$

Under a free-replacement warranty policy,

$$\text{AC}_{\text{FR}}(W) = \frac{q_2}{E(Y)} = \frac{q_2}{(m(W) + 1)E(X)}$$

He also gives expressions for profit to the producer, under both plans, for short-run as well as long-run situations.

Mamer [14] later expanded his previous research with present value analysis and analyzed the trade-off between warranty and quality control. He considered three warranty policies: ordinary replacement (OR), free replacement (FR) and pro-rata replacement (PR). In OR, an item is replaced and warranty is offered only up to the original warranty period. In FR, a new warranty is offered when an item is replaced. The new warranty is the same as the original warranty. In PR, the consumer is charged based on time of failure, and a new warranty, the same as the original, is offered.

He also discusses the trade-off between warranty and quality control and attempted to find a warranty length such that the expected discounted cost of ownership of a sequence of products under warranty, with certain reliability, is equal to some target level of ownership costs. The relationship between warranty length and product reliability is illustrated. As reliability drops, so does the warranty length. Mamer concludes that producers will offer ordinary free replacements where damage rate is unknown or hard to estimate. If damage rate is known or can be estimated, they will offer pro-rata or unlimited replacement warranties.

Blacer and Sahin [15] estimated warranty costs under free-replacement and pro-rata warranty policies using renewal theory. They used the concept of a renewal process to estimate warranty costs over the product life cycle $(0, T)$ for pro-rata and free-replacement warranties and computed first and second cost moments using Gamma, exponential, and mixed exponential failure distributions. Because they computed the mean as well as the variance of warranty costs under various assumptions, they provide some estimation of risk involved with warranty cost estimation. They estimated the reliability of items over the warranty interval. Because the failure distribution influences reliability, Blacer and Sahin also computed warranty cost assuming conditions where reliability was the same for the failure distributions considered.

Frees and Nam [16] found expressions for expected warranty costs in terms of distribution functions. They considered the free-replacement and pro-rata policies to estimate warranty costs, where warranties are renewed under certain conditions. They concluded that it is very difficult to estimate renewal functions and, therefore, warranty costs mathematically and suggest a couple of approximations. They used NBU (new better than used) distribution and SLA (straight-line approximation) methods. They found that SLA gives a very good approximation to the estimations provided by Nguyen and Murthy [17]. The drawback of Nguyen and Murthy's model was that they assumed a monotonic failure rate for lifetime distributions.

Frees [18] showed that estimating warranty costs is similar to estimations renewal functions in renewal theory. He also estimated the variability of warranty costs using parametric and nonparamatric estimations. Let $X_1, X_2, X_3, \ldots$ be a sequence of i.i.d. (independent identically distributed) random variables that represent successive failure times of a product, where C_1 is the initial fixed cost per unit. Then, expected warranty cost for free replacement is given by

$$EC_{FR}(W) = C_1 H(W)$$

where

$$H(W) = E(N(W)) = \sum_{k>1} P(X_1 + \cdots + X_k \leq W)$$

is a renewal function at time W. The variance is given by

$$V(W) = \text{Var}(N(W))$$
$$= \sum_{k>1} (2k - 1)P(X_1 + \cdots + X_k \leq W) - (H(W))^2$$

The renewal function, $H(W)$, is estimated by using parametric as well as nonparametric estimators.

Let us consider the renewing pro-rata warranty (PRW) policy. In estimating costs for this situation, renewal-type equations are used. Initial purchase of an item under PRW leads to a whole sequence of purchases, which stops either when the buyer chooses not to repurchase the item or when an item achieves a lifetime of at least W. We assume that the process does not stop for the first reason. Under these conditions, Blischke and Murthy [2] list the expected total cost to the buyer as

$$E[C_b(W)] = c_b \left(1 + \frac{\mu_W F(W)}{W F(W)} \right)$$

where μ_W is the partial mean life

$$\mu_W = \int_0^W t \, dF(t)$$

$$\overline{F}(W) = 1 - F(W)$$

and c_b is the cost of the initial item. It is observed that the expected cost is an increasing function of W. As W increases, failures are more likely and, therefore, additional purchases will be required. One should note that the cost to the buyer is equivalent to the income to the seller.

Another form of analyses would be to look at life-cycle cost models, for which the total cost to the buyer and seller are considered in a period $[0, L]$, where L is relatively long compared to W [2]. An expression for the expected long-run life-cycle cost to the buyer is given by

$$E[C_b(L, W)] = H(L) + \int_0^L H(L - x) \, dM(x)$$

where $M(\cdot)$ is the ordinary renewal function associated with $F(\cdot)$ and

$$H(L) = c_b \left(\frac{\mu_W}{W} + F(L) - F(W) \right)$$

The expected long-run cost to the seller may be found from

$$E[C_S(L, W)] = c_S[1 + M(L)]$$

where c_S is the average total cost of supplying a single item.

11.3   ESTIMATION OF EXPECTED WARRANTY COSTS

Manufacturers must have adequate resources to meet warranty costs that occur due to product failure within the warranty period. This section presents a model to estimate the expected warranty costs for a linear pro-rata warranty that is nonrenewing, based on a certain product failure distribution. Furthermore, because warranty claims are paid in the future, it is appropriate to estimate the present value of expected warranty costs assuming a chosen discount rate and an inflation rate.

11.3.1   Product Failure Distributions

A variety of failure distributions is considered in order to model general classes of products. If a product life-cycle curve is considered, for most products it is the traditional "bathtub" curve with three phases [2]. The first phase is known as the burn-in or infant-mortality phase and involves debugging. During this period, the failure of marginal parts occur resulting in a decrease in the failure rate. The chance failure phase follows, wherein the failure rate remains constant with failures assumed to occur in a random manner. The assumption of a constant failure rate is considered reasonable for the phase of a product's useful life. The third phase represents the period of wearout which is characterized by an increase in the failure rate. In order to model situations where the product failure rate decreases, remains constant, or increases, several failure distributions are considered.

Exponential Distribution

The exponential distribution has a constant failure rate. It may be used to model the chance failure phase of a product. The density function is given by

$$f(t) = \lambda e^{-\lambda t}, \quad 0 < t < \infty \tag{11.2}$$

where λ represents the failure rate. Its mean is given by $1/\lambda$ and its variance by $1/\lambda^2$.

Truncated Exponential Distribution

The truncated exponential distribution is similar to the exponential, with the exception that time to failure is truncated at some chosen value, T. This may apply to products that undergo preventive maintenance at a certain selected time period. Its density function is given by

$$f(t) = \frac{1}{1 - e^{-\lambda T}} \lambda e^{-\lambda t}, \quad 0 < t < T \tag{11.3}$$

Gamma Distribution

The density function for the Gamma distribution depends on two parameters, λ and α. The Gamma density takes on a variety of shapes depending on the values of these two parameters. Its density function is given by

$$f(t) = \frac{\lambda^\alpha}{\Gamma(\alpha)} t^{\alpha - 1} e^{-\lambda t}, \quad 0 < t < \infty \tag{11.4}$$

Its mean and variance are given by α/λ and α/λ^2, respectively. When $\alpha = 1$, the Gamma distribution reduces to the exponential distribution.

Weibull Distribution

The two-parameter Weibull distribution, like the Gamma, maybe used to model a wide variety of situations. Based on a shape parameter, β, and a scale parameter, α, its density function is expressed as

$$f(t) = \alpha\beta t^{\beta-1} \exp(-\alpha t^\beta), \quad \alpha > 0, \beta \geq 1 \tag{11.5}$$

Its mean and variance are given by $\alpha^{-1/\beta}\Gamma(1 + 1/\beta)$ and $\alpha^{-2/\beta}[\Gamma(2/\beta + 1) - (\Gamma(1/\beta + 1))^2]$, respectively. Depending on the values of the parameters α and β, the Weibull density function may be similar to an exponential distribution, or a normal distribution, among others. The failure rate of such a distribution may be constant, decreasing, or increasing with time and is influenced by the parameter values, which is also true of the Gamma distribution. As the value of β increases from 1, the failure rate increases.

Lognormal Distribution

A random variable, T, is said to have a lognormal distribution if $\ln(T)$ has a normal distribution, with, say, mean μ and variance σ^2. The density function of T, time to product failure, is given by

$$f(t) = \frac{1}{\sqrt{2\pi}\sigma t} \exp\left(\frac{-(\ln(t) - \mu)^2}{2\sigma^2}\right) \tag{11.6}$$

The mean and variance of T are given by

$$\mu_T = E(T) = \exp\left(\mu + \frac{\sigma^2}{2}\right) \tag{11.7}$$

$$\sigma_T^2 = \text{Var}(T) = [\exp(\sigma^2) - 1]\,[\exp(2\mu + \sigma^2)] \tag{11.8}$$

respectively.

11.3.2 Present Value of Expected Warranty Costs

The problem of estimation of warranty costs that are to be borne by a company is of major significance to the manufacturing sector. First and foremost are the economic implications associated with the problem. Manufacturers need to offer warranty plans and, hence, need to estimate the amount of reserves that they should set aside for meeting obligations arising out of product failure within the warranty period. If warranty costs are not included in the financial statement, then income will be overstated as will be the rate of return on investment. Because the manufacturer, typically, passes the burden of increased warranty costs on to the consumer by increasing the selling price, it is of obvious importance to accurately estimate these costs.

The following notation is used:

c:	unit product price
$C(t)$:	rebate at time t
W:	warranty period
$f(t)$:	failure density function
δ:	firm's discount rate
ϕ:	inflation rate
R:	total warranty cost
r:	expected warranty cost per unit product $= E[PV(R)]/N$
N:	production quantity

The expected present values of warranty costs to the manufacturer is given by

$$E[PV(R)] = N \int_0^W C(t)f(t) \exp[-(\delta + \phi)t] \, dt \qquad (11.9)$$

Depending on the product failure distribution, the expression for $f(t)$ with appropriate parameter values is to be used in (11.9). In situations where a closed-form expression for the integral given by (11.9) is not feasible, numerical integration methods may be used.

In addition to the total warranty costs, management might be more interested in the warranty cost per unit product price, r/c. This value provides some sort of a standardized measure for unit warranty costs. It depicts the proportion of the product price that should be set aside for warranty reserves. For example, if r/c is 0.2, it means that 20% of the unit product price should be allocated for warranty costs.

11.4   RESULTS

For each of the failure distributions discussed previously, the expected present value of warranty costs per unit price, r/c, is calculated. Several values of the parameter are considered to encompass a variety of situations. The warranty period, W, is changed from 0.1 to 1. The combined discount and inflation rate $(\delta + \phi)$ is assumed to be 0.125.

11.4.1 Exponential Failure Distribution

The failure rates λ were chosen to be 1.5, 2, and 2.5, respectively. The corresponding values of mean time to failure (MTTF) are 0.667, 0.5, and 0.4, respectively. Table 11.1 gives values of r/c for W ranging between 0.1 and 1. For a given warranty period, as the failure rate increases, obviously, unit warranty reserves costs increase. However, the effect of the parameter λ may be observed from the values in Table 11.1. Figure 11.1 shows warranty reserve costs per unit dollar sales (r/c) as a function of the warranty period for different values of λ. One observes that for small values of W, the rate of increase in r/c is much more compared to larger warranty periods. For instance, when $\lambda = 1.5$, unit warranty reserve costs increase by about 90% as the warranty period increases from 0.1 to 0.2. On the other hand, for the same value of λ, a marginal increase of about 7% is observed when the warranty period increases from 0.9 to 1.0. The reason for this is that most of the failures have already occurred within the long warranty period (say, 0.9) and so the extension of warranty leads to a minimal increase in r/c. From Figure 11.1, one finds that the values of r/c increase at a linear rate of about 10% for $W \geq 5$. This may provide some guidelines to the manufacturer for selecting a warranty period given the budget constraints for warranty costs.

11.4.2 Truncated Exponential Distribution

If the product is replaced at time T, regardless of whether it failed or not, warranty reserve costs will be higher. This practice may be adopted to

Table 11.1 Values of r/c for Exponential Failure Distribution

Warranty period W	r/c		
	$\lambda = 1.5$	$\lambda = 2.0$	$\lambda = 2.5$
0.1	0.0711	0.0933	0.1147
0.2	0.1350	0.1744	0.2114
0.3	0.1925	0.2452	0.2932
0.4	0.2443	0.3072	0.3628
0.5	0.2991	0.3615	0.4221
0.6	0.3334	0.4093	0.4729
0.7	0.3718	0.4514	0.5166
0.8	0.4065	0.4887	0.5544
0.9	0.4381	0.5217	0.5872
1.0	0.4669	0.5512	0.6159

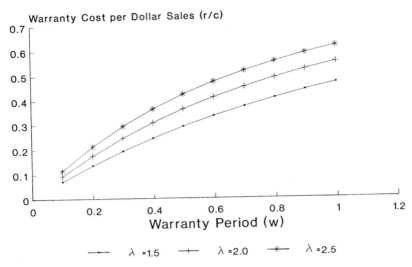

Figure 11.1   Warranty reserve costs per unit dollar sales using the exponential failure distribution.

reduce downtime or delays in repairing a product. Table 11.2 shows values of r/c for $\lambda = 1.5$ and values of T of 0.5, 0.6, and 0.7, when the warranty period is changed from 0.1 to 1. A similar trend is observed as in the exponential distribution case. Figure 11.2 depicts the warranty costs per unit dollar sales (r/c) for values of T of 0.5, 0.6, and 0.7. The rate of increase in r/c decreases with W. As T increases, there is a decrease in the unit warranty costs, as expected. For large values of T compared to MTTF, the values of r/c should approach those for the exponential distribution. Comparing the values of r/c with those for the exponential distribution, if $W = 0.2$ and $T = 0.7$, the increase in r/c is about 54%.

11.4.3 Gamma Distribution

For the Gamma distribution, the selected parameter values are $\lambda = 2.5$ and $\alpha = 2, 2.5$, and 3.0. The corresponding values of MTTF are 0.8, 1.0, and 1.2, respectively. Table 11.3 shows values of r/c for these parameter combinations with W ranging from 0.1 to 1. As W increases, the rate of increase of r/c decreases. Figure 11.3 shows the values of warranty costs for $\alpha = 2, 2.5$, and 3, when $\lambda = 2.5$ for the Gamma distribution. The shape of the graph is concave up, compared to Figures 11.1 and 11.2,

Table 11.2 Values of *r/c* for Truncated Exponential Failure Distribution with λ = 1.5

Warranty period *W*	*T* = 0.5	*T* = 0.6	*T* = 0.7
0.1	0.1347	0.1198	0.1094
0.2	0.2558	0.2275	0.2077
0.3	0.3648	0.3244	0.2961
0.4	0.4631	0.4117	0.3758
0.5	0.5517	0.4906	0.4478
0.6	0.6319	0.5619	0.5129
0.7	0.7046	0.6265	0.5719
0.8	0.7705	0.6851	0.6254
0.9	0.8304	0.7383	0.6740
1.0	0.8849	0.7868	0.7182

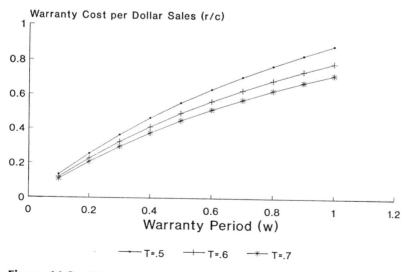

Figure 11.2 Warranty reserve costs per unit dollar sales using the truncated exponential distribution.

Table 11.3 Values of *r/c* for Gamma Failure Distribution with λ = 2.5

Warranty period W	*r/c*		
	α = 2	α = 2.5	α = 3.0
0.1	0.0092	0.0023	0.0006
0.2	0.0323	0.0114	0.0038
0.3	0.0642	0.0274	0.0111
0.4	0.1013	0.0492	0.0227
0.5	0.1410	0.0753	0.0384
0.6	0.1814	0.1045	0.0575
0.7	0.2214	0.1355	0.0795
0.8	0.2600	0.1674	0.1035
0.9	0.2969	0.1994	0.1290
1.0	0.3317	0.2310	0.1554

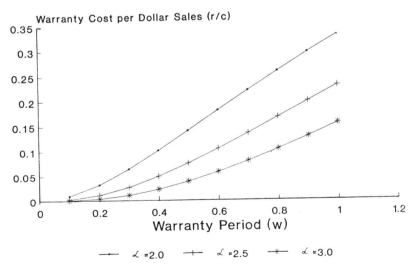

Figure 11.3 Warranty reserve costs per unit dollar sales using the Gamma distribution.

which were convex in nature. For values of $W \geq 0.4$, a steep increase is observed in the r/c values. For example, as W changes from 0.5 to 0.6, for $\alpha = 2$, r/c increases by about 29%. This rate of increase becomes fairly constant as W increases beyond 0.6. The impact of the parameter α may be observed from Figure 11.3. As α increases, the unit warranty costs decrease.

A company could estimate the parameters of the failure distribution from historical data. These values could then be used to estimate values of r/c. Knowing the unit price of the product and annual sales, such information would aid the company in determining annual warranty costs as well as the corresponding warranty period to be offered. If the company estimated λ to be 2.5 and α to be 2.5 and decided that they could set aside about 10 cents per sales dollar for warranty reserves, the selected warranty period would be about 0.6, from Figure 11.3.

11.4.4 Weibull Distribution

The Weibull distribution with parameter values of $\alpha = 1.5$ and $\beta = 2$, 2.5, and 3 is selected for analysis. Table 11.4 shows values of r/c for these parameter combinations with W ranging from 0.1 to 1. As β increases, the unit warranty costs decrease. The trend of the r/c values may be observed from Figure 11.4, with the graph being concave upward. In the considered region of W, the rate of increase of r/c increases with W. For

Table 11.4 Values of r/c for Weibull Failure Distribution with $\alpha = 1.5$

	r/c		
Warranty period W	$\beta = 2$	$\beta = 2.5$	$\beta = 3.0$
0.1	0.0049	0.0013	0.0004
0.2	0.0194	0.0075	0.0029
0.3	0.0424	0.0203	0.0098
0.4	0.0728	0.0404	0.0227
0.5	0.1088	0.0679	0.0429
0.6	0.1489	0.1021	0.0709
0.7	0.1913	0.1416	0.1063
0.8	0.2347	0.1847	0.1478
0.9	0.2776	0.2296	0.1934
1.0	0.3191	0.2746	0.2408

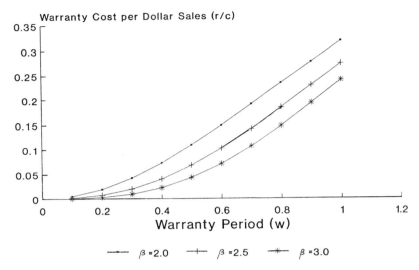

Figure 11.4   Warranty reserve costs per unit dollar sales using the Weibull distribution.

values of $W \geq 0.6$, a constant differential seems to be maintained among the r/c values for different values of β.

11.4.5   Lognormal Distribution

Two lognormal distributions are used in the analysis. One corresponds to the parameter values being $\mu = 0.2$ and $\sigma = 0.5$, which leads to $\mu_T = 1.384$ and $\sigma_T = 0.7376$. The second lognormal distribution has parameter values of $\mu = 0.2$ and $\sigma = 0.7$, which leads to $\mu_T = 1.5605$ and $\sigma_T = 1.2409$. Table 11.5 shows values of r/c for these two distributions when W ranges from 0.1 to 1. Figure 11.5 shows the graphs of r/c versus W for the two lognormal distributions.

Note that the r/c curves for the two lognormal distributions become divergent as W increases. The first lognormal distribution has a larger value of r/c. Observe that the second distribution has a larger mean (μ_T) and a larger standard deviation (σ_T) compared to the first. So, for a given warranty period, fewer units may fail compared to the first distribution and, hence, the smaller values of r/c. The effect of an increase in the mean of T (compared to the first) may more than offset the increase in the

Table 11.5   Values of r/c for Lognormal Failure Distribution

Warranty Period W	r/c	
	$\mu = 0.2, \sigma = 0.5$ $\mu_T = 1.384, \sigma_T = 0.7376$	$\mu = 0.2, \sigma = 0.7$ $\mu_T = 1.5605, \sigma_T = 1.2409$
0.1	0.	0.0002
0.2	0.	0.0015
0.3	0.	0.0039
0.4	0.0001	0.0072
0.5	0.0004	0.0113
0.6	0.0009	0.0160
0.7	0.0017	0.0210
0.8	0.0029	0.0265
0.9	0.0044	0.0321
1.0	0.0063	0.0379

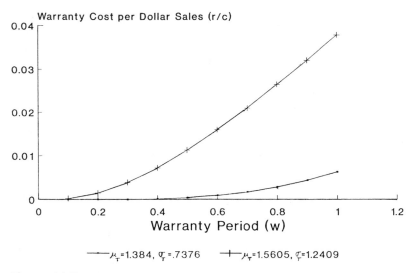

Figure 11.5   Warranty reserve costs per unit dollar sales using the lognormal distribution.

variability of T. With the average life of the product having increased, the unit warranty reserve costs are lower, relative to the first distribution.

11.5 CONCLUSIONS

This chapter discusses the basic linear pro-rata rebate plan, where it is assumed that the rebate to the consumer decreases linearly with the time of failure. Companies are often interested in estimating their warranty expenditures, for a selected warranty period, or in determining the warranty period for a selected constraint on the amount of warranty expenditures that they can budget. The analysis and results of this chapter will help management address both of these issues.

One of the difficulties in estimating warranty costs is due to the lack of knowledge of the failure distribution of the product. This chapter considers a variety of product failure distributions and demonstrates the impact of such distributions on the warranty reserve costs. In practice, a company will, therefore, need to estimate the failure distribution and its relevant parameters for the product under consideration. The methods of parameter estimation described in Chapter 10 of Ref. 2 may be used. Some procedures are the method of moments and the maximum likelihood method of estimation using sample data on product failure. As an example, if from historical data, the failure rate is found to be constant, the exponential distribution may be used as an approximation of the failure distribution. With the exponential distribution being a single parameter distribution, one need only estimate the constant failure rate. Suppose sample data yields a product mean life as $\overline{X}$. Then, an estimate of the failure rate is obtained as $\lambda = 1/\overline{X}$. For the two-parameter Weibull distribution, the first two sample moments may be equated to the population moments to obtain estimates of the scale and shape parameter. However, moment estimators are not always efficient. The maximum likelihood estimators, under certain conditions, are optimal, at least asymptotically.

Once the parameters of the failure distribution have been estimated, using the procedures described in this chapter, the manufacturer may estimate expected warranty costs associated with a pro-rata policy for a selected warranty time. Alternatively, if management has an idea of the amount to be allocated for expected warranty costs, a determination can be made of the warranty period. Consider the example where sample information yields a constant failure rate. The mean time to failure has been estimated as 0.5. The unit product price is $50 and 10,000 items are sold in a unit time period. Management has decided to set aside no more than $150,000 for warranty costs during the time period. What is the maximum warranty that they should offer for the product? Assuming that a constant failure rate justifies use of the exponential failure distribution, the failure rate is estimated as $1/0.5 = 2$. In order to stay within the constraints of warranty costs set by management, expected warranty cost per dollar sales should not exceed 0.3 [= 150,000/(50)(10,000)]. Now, using Figure 11.1, one estimates a warranty period of approximately 0.4, which will satisfy the specified criteria. This demonstrates an application of the determination of the parameters of a warranty policy. Alternatively, for a selected warranty policy, an estimation of warranty costs may be obtained. Such estimates should help in providing a more accurate picture of the rate of return on investment, as warranty costs are included as part of the expenditures.

Whereas this chapter has focused on the linear pro-rata warranty for the nonrenewing case, a comparison of the unit costs to the seller for the nonrenewing and renewing pro-rata policies may be found in Ref. 2, Table 5.18. Blischke and Murthy [2] also compare pro-rata warranties and free-replacement warranties. In particular, although the pro-rata warranties are usually thought to be more costly to the seller than the free-replacement warranty, it is shown that such is not always the case. They conclude that the effect of varying the warranty period on the warranty costs of the various policies could not be determined, as it was held constant in the comparison. For the practitioner, they recommend a simulation approach to estimate warranty costs and, thereby, select an acceptable policy.

REFERENCES

1. *U.S. Federal Trade Commission Improvement Act* (1975). 88 Stat 2183, 101–112.
2. Blischke, W. R., and Murthy, D. N. P. (1994). *Warranty Cost Analysis*, Marcel Dekker, Inc., New York.
3. Udell, J. G., and Anderson, E. E. (1968). The product warranty as an element of competitive strategy, *Journal of Marketing*, **32**, 1–8.
4. Shimp, T. A., and Bearden, W. O. (1982). Warranty and other extrinsic cue effects on consumers' risk perceptions, *Journal of Consumer Research*, **9**, 38–46.
5. Wiener, J. L. (1985). Are warranties accurate signals of product liability? *Journal of Consumer Research*, **12**(9), 245–250.
6. Menke, W. W. (1969). Determination of warranty reserves, *Management Science*, **15**(10), 542–549.
7. Amato, H. N., and Anderson, E. E. (1976). Determination of warranty reserves: An extension, *Management Science*, **22**(12), 1391–1394.
8. Lowerre, J. M. (1968). On warranties, *Journal of Industrial Engineering*, **19**(3), 359–360.

9. Heschel, M. S. (1971). How much is a guarantee worth? *Industrial Engineering*, **3**(5), 14–15.
10. Patankar, J. G., and Worm, G. H. (1981). Prediction intervals for warranty reserves and cash flows, *Management Science*, **27**(2), 237–241.
11. Blischke, W. R., and Scheuer, E. M. (1981). Applications of renewal theory in analysis of the free-replacement warranty, *Naval Research Logistics Quarterly*, **28**, 193–205.
12. Balachandran, K. R., Maschmeyer, R. A., and Livingstone, J. L. (1981). Product warranty period: A Markovian approach to estimation and analysis of repair and replacement costs, *The Accounting Review*, **56**(1), 115–124.
13. Mamer, J. W. (1982). Cost analysis of pro rata and free-replacement warranties, *Naval Research Logistics Quarterly*, **29**(2), 345–356.
14. Mamer, J. W. (1987). Discounted and per unit costs of product warranty, *Management Science*, **33**(7), 916–930.
15. Blacer, U., and Sahin, I. (1986). Replacement costs under warranty: Cost moments and time variability, *Operations Research*, **34**(4), 554–559.
16. Frees, E. W., and Nam, S. H. (1988). Approximating expected warranty cost, *Management Science*, **43**(12), 1441–1449.
17. Nguyen, D. G., and Murthy, D. N. P. (1984). Cost analysis of warranty policies, *Naval Research Logistics Quarterly*, **31**, 525–541.
18. Frees, E. W. (1988). Estimating the cost of a warranty, *Journal of Business and Economic Statistics*, **4**(1), 79–86.

12

Combination Warranties

Wallace R. Blischke

University of Southern California
Los Angeles, California

12.1 INTRODUCTION

Chapters 10 and 11 were concerned with cost models for the two most common basic warranty structures, the free-replacement and pro-rata warranties (hereafter denoted FRW and PRW). These, in fact, form the basis of virtually all consumer warranties and most commercial warranties as well, in the sense that more complex warranty structures are typically structured as combinations of the FRW and/or the PRW, with either rebates or replacements serving as the form of rectification. A more detailed description and a number of examples will be given in the next two subsections. The final item in this section will be an outline of the remainder of the chapter.

12.1.1 Structure of Combination Warranties

The combination warranty results when warranty terms change at one or more points in time during the course of the warranty period, which is taken to be the time interval $[0, W]$. Warranty coverage in each subinterval may be free replacement, rebate, proportional rebate, or prorata (with any reasonable proration function), subject only to the logical constraint that compensation be a nonincreasing function of time throughout the warranty period. The most common combination in nearly all types of applications is an initial period of FRW coverage, followed by a (usually longer) period under PRW. In fact, in applications this is usually what is meant by "combination warranty." (Here, the term is used more generally, as indicated in the above definition.)

A combination warranty may be renewing or nonrenewing* in each subinterval, that is, whether or not the warranty renews may also change at one or more of the times when the coverage changes.

A combination warranty, then, is any multiperiod warranty with rebate, free-replacement, or pro-rata terms, and renewing or nonrenewing, in any combination. A great many such combination warranties are possible. As examples, Warranty Policies 8–12, 16–18, 27, and 32 of Chapter 1 are combination warranties of various types.

The rationale for combination warranties is that they are a compromise between the free-replacement warranty, which tends to favor the buyer, particularly in the latter portion of the warranty (where failures are usually more likely to occur), and the pro-rata warranty, which tends to favor the seller, because failures late in the warranty period entail very little compensation to the buyer and, hence, very little cost to the seller. Thus, the combination warranty gives the buyer full protection against potentially costly early failures while also protecting the seller against full liability for later failures, where the buyer has received nearly the full amount of service that was guaranteed under the warranty. In short, this type of warranty is very common because it has a significant promotional value to the seller while at the same time providing adequate control over costs for both buyer and seller in most applications. Specific applications illustrating the various types of combination warranties and the types of products on which they are offered are given in Chapter 2 of Ref. 1.

12.1.2 Examples of Combination Warranties

To illustrate the richness of this class of warranties, ten relatively simple combination warranties are listed below. The first two are combination

---

* A renewing warranty is one under which all items replaced under warranty are provided full-warranty coverage under the original warranty terms. Under a nonrenewing warranty, the buyer is either given a rebate or replacement items are covered only for the remaining time in the original warranty period. (See Chapter 1.)

rebate warranties. The remaining eight are combination FRW/PRW warranties or combinations of these that also include rebate features, with various renewing options. In the PRW combinations, the proration is usually taken to be linear in applications. (See Chapter 11.) The notation used is W = warranty period; W_1, W_2, and so forth are times after purchase at which the warranty coverage changes; and c_b is the cost to the buyer (i.e., the purchase price) of the initial item. Other symbols will be defined as they are introduced.

The ten warranties are as follows:

Warranty 1 (*Combination Lump-Sum Rebate Warranty*): A rebate in the amount $\alpha_1 c_b$ is given for any item that fails prior to time W_1 from the time of purchase; the rebate is $\alpha_2 c_b$ for items that fail in the interval $(W_1, W_2]$, $\alpha_3 c_b$ for items that fail in the interval $(W_2, W_3]$, and so forth, up to a final interval $(W_{k-1}, W]$, in which the rebate is $\alpha_k c_b$, with $1 \geq \alpha_1 > \alpha_2 > \cdots > \alpha_k > 0$ and $0 < W_1 < W_2 < \cdots < W_{k-1} < W$.

An example of a policy of this type (with W = 25 years!) is given in Chapter 2 of Ref. 1.

Warranty 2 (*Combination Full Rebate/Pro-Rata Rebate*): Under this policy, the seller agrees to provide a full refund of the original purchase price up to time W_1 from the time of initial purchase; any failure in the interval from W_1 to W ($>W_1$) results in a pro-rata refund.

Note that the proration may be linear, in which case the rebate would be $\alpha(W - X)/(W - W_1)$, for X, the lifetime of the item, in the interval $[W_1, W]$, and $0 < \alpha \leq 1$, or the proration may be nonlinear. (See Chapter 5 of Ref. 1.)

Warranty 3 (*Combination Full Rebate/Nonrenewing PRW*): Under this policy, the seller agrees to provide a full refund of the original purchase price up to time W_1 from the time of initial purchase; for any failure in the interval from W_1 to W ($>W_1$), a replacement will be provided at pro-rata cost to the buyer. Replacements are covered under warranty terms identical to those in force at the time of failure.

Note that this warranty is entirely nonrenewing because in no case is coverage provided beyond the original warranty period. Proration may again be linear or nonlinear.

Warranty 4 (*Combination Nonrenewing FRW/PRW*): The seller will replace any item that fails prior to W_1 from the time of purchase with an identical new item at no cost to the buyer; from time W_1 to W, any item that fails is replaced with an identical new item at pro-rata cost to the buyer. All replacement items assume the remaining time and terms of the original warranty.

The next three warranties are combination FRW/PRWs featuring the three different possibilities with regard to renewing.

Warranty 5 (*Combination Nonrenewing FRW/Renewing PRW*): The seller will replace any item that fails prior to W_1 from the time of purchase with an identical new item at no cost to the buyer. Replacement items in this period are covered for the remaining time and terms of the original warranty. From W_1 to W, failed items are replaced at pro-rata cost to the buyer and the warranty begins anew, that is, all replacement items in this period are covered under terms identical to those of the original warranty.

Warranty 6 (*Combination Renewing FRW/Nonrenewing PRW*): The seller will replace any item that fails prior to W_1 from the time of purchase with an identical new item at no cost to the buyer and the warranty begins anew. From time W_1 to W, any item that fails is replaced with an identical new item at pro-rata cost to the buyer. Items replaced during this period are covered under the pro-rata terms and for the remaining time of coverage of the item they replace.

Warranty 7 (*Fully Renewing Combination FRW/PRW*): The seller will replace any item that fails prior to W_1 from the time of purchase with an identical new item at no cost to the buyer; from time W_1 to W, any item that fails is replaced with an identical new item at pro-rata cost to the buyer. Upon failure of any item, the warranty begins anew.

Extensions to three or more warranty periods (Warranty 1 is an example) offer an even larger class of warranty policies. The final three warranties illustrate a few of the many additional possibilities. All three involve an initial FRW period followed by a period during which a replacement is provided at a discount, and a final PRW period. The warranties differ with regard to renewing.

Warranty 8 (*Fully Renewing Three-Period Combination Warranty*): The seller will replace any item that fails prior to W_1 from the time of purchase with an identical new item at no cost to the buyer. From time W_1 to W_2, any item that fails is replaced at cost αc_b ($\alpha < 1$) to the buyer. From time W_2 to W after purchase, any item that fails is replaced with an identical new item at pro-rata cost to the buyer. Upon failure of any item, the warranty begins anew.

Warranty 9 (*Partially Renewing Three-Period Combination Warranty*): The seller will replace any item that fails prior to W_1 from the time of purchase with an identical new item at no cost to the buyer. Replacement items during this period assume the remaining time and coverage of the original item. From time W_1 to W_2 after purchase, any item that fails is replaced at cost αc_b ($\alpha < 1$) to the buyer, and the warranty begins anew. From time W_2 to W after the most recent purchase (i.e., the original item or any item partially paid for by the buyer), any item that fails is replaced with an identical new item at pro-rata cost to the buyer, and the original warranty coverage is provided.

Warranty 10 (*Partially Renewing Three-Period Combination Warranty*): The seller will replace any item that fails prior to W_1 from the time of purchase with an identical new item at no cost to the buyer. From time W_1 to W_2 after purchase, any item that fails is replaced at cost αc_b ($\alpha < 1$) to the buyer. Replacement items during these periods assume the remaining time and coverage of the original item. From time W_2 to W after purchase, any item that fails is replaced with an identical new item at pro-rata cost to the buyer, and the original warranty coverage is provided.

Warranty 8 renews upon failure of any item prior to time W from the time of original purchase, Warranty 9 renews in the second and third periods, and Warranty 10 only in the last. Many other permutations are possible.

12.1.3 Chapter Outline

In this chapter, the primary objective is to provide cost models that may be used to assess the value to both buyer and seller of several of the policies given in Section 12.1.2. The cost analysis is not intended to be exhaustive of all of the possible combination warranties or even of those listed in Chapter 1 and the previous section. Instead, a few of the more widely used combination warranties will be analyzed in order to illustrate the methodology and provide cost models for some important applications. Other combination warranties may be analyzed by the methods illustrated here. For additional details concerning the methodology and additional analyses, see Chapter 6 of Ref. 1.

It will be assumed in the analyses of this chapter that the items covered under warranty are nonrepairable, although most models are easily modified to include simple repair regimes such as good-as-new and minimal repair.

As noted, the concept of renewability introduces still further latitude in specifying warranty terms, as this, too, may vary from period to period. Cost models for renewable warranties will be given only for the standard two-period warranty, that is, FRW followed by PRW. In this case, the warranty may renew only in the PRW period or it may be set up so that it is always renewing.

The organization of the remainder of this chapter is as follows: In Sections 12.2 and 12.3, per-unit cost models from the seller's and buyer's points of view will be given for Warranties 1–3. Costs discounted to present value will be discussed in Section 12.4. Models for life-cycle costs of Warranty 7 will be given in Section 12.5. Some comments on estimating cost models based on test data will be given in Section 12.6. Finally, some recommendations for practitioners and areas for future research will be presented in Section 12.7.

12.2 MODELING SELLER'S PER-UNIT WARRANTY COST

In this section, Warranties 1–3 will be analyzed in detail. These are the most commonly used combination warranties, although Warranties 4–7 are occasionally encountered.

Analysis of Warranty 4 results in a model involving renewal functions for the FRW portion (see Chapter 10) and integral equations such as those of Chapter 5 of Ref. 1 for analysis of the PRW portion. This warranty has not been analyzed. (It is not often used in practice, as it is quite unfavorable to the buyer in the latter part of the warranty period because replacements are purchased at nearly full price but have warranties of very limited duration.)

Warranties 5–7 are analyzed in Ref. 2 and in Chapter 6 of Ref. 1, where some examples using the exponential and Weibull distributions are given. Some results for Warranty 7 are basically the same as for Warranty 2; see Chapter 6 of Ref. 1. Warranties 8–10 have not been analyzed.

Other combination warranties can be analyzed similarly. One possibility that can be useful in certain applications is a three-period warranty with a full-refund period followed by a period in which a smaller lump-sum rebate is given, and concluding with a pro-rata rebate period. A modification of this would be a proportional pro-rated rebate in the final period. Many other such combinations are possible. Analyses of the nonrenewing, rebate versions of such warranties, using the methods of this chapter, are straightforward.

As in previous chapters, the notation used is c_b is the cost to buyer (selling price), c_s is the average cost to seller of supplying a single item (including all incidental costs), and C_b and C_s are the (random) net costs to buyer and seller of an item sold under a specified warranty policy. The seller's average cost c_s may be interpreted as the average cost of supplying an unwarrantied item.

12.2.1 Seller's Cost per Unit Sold Under Warranty 1

Cost analyses of Warranty 1 are most easily approached by means of a rebate function, denoted $q(\cdot)$, which specifies the amount of compensation in the event of an item failure as a function of the item lifetime X. For Warranty 1, this function is given by

$$q(X) = \begin{cases} \alpha_i c_b, & W_{i-1} < X \le W_i, i = 1, \ldots, k \\ 0, & \text{otherwise} \end{cases} \tag{12.1}$$

where $W_0 = 0$, $W_k = W$, and the α_i are as indicated in the policy statement.

Expected costs to both buyer and seller can easily be determined as functions of the expected rebate. This is given by

$$E[q(X)] = \int_0^W q(x) \, dF(x) = c_b \sum_{i=1}^k \alpha_i [F(W_i) - F(W_{i-1})] \tag{12.2}$$

where $F(x)$ is the CDF (cumulative distribution function) of X.

We use the vector notation $\alpha = (\alpha_1, \ldots, \alpha_k)'$ and $W = (W_1, \ldots, W_k)'$. The seller's cost under Policy 1 is given by $C_s(\alpha, W) = c_s + q(X)$. The expected per-unit cost to the seller is

$$E[C_s(\alpha, W)] = c_s + c_b \sum_{i=1}^k \alpha_i [F(W_i) - F(W_{i-1})] \tag{12.3}$$

Example 12.1

We consider three different three-stage rebate policies: (1) a 6-month warranty with full refund on failure during the first month, 75% refund of the initial purchase price during the second month, and 50% refund during the remainder of the warranty period; (2) a 1-year warranty with full refund if a failure occurs in the first month, a rebate of 75% of the initial purchase price if a failure occurs in the second month, and a 50% rebate if a failure occurs in the remaining 10 months of the warranty period; and (3) a 1-year warranty with full refund in the first 2 months, 75% in the third and fourth months, and 50% during the remaining 8 months.

The warranty parameters are $k = 3$, $\alpha_1 = 1.0$, $\alpha_2 = 0.75$, and $\alpha_3 = 0.50$, and the warranty periods for the three cases are (1) $W_1 = 0.0833$, $W_2 = 0.1667$, and $W = W_3 = 0.5$ years, (2) $W_1 = 0.0833$, $W_2 = 0.1667$, and $W = 1.0$ year, and (3) $W_1 = 0.1667$, $W_2 = 0.3333$, and $W = 1$ year.

We consider $\mu = 1.0, 1.5, 2.0$, and 2.5 years and assume exponential, Weibull, Gamma, and lognormal distributions for time to failure. (See Chapter 8 for formulas and discussion of these distributions.) Parameter values for the distributions are as follows: The exponential parameter is determined by the assumed mean, because the parameter is $\lambda = 1/\mu$. For the Weibull distribution, we use values of the shape parameter of $\beta = 0.5$, a decreasing failure rate (DFR) distribution, and $\beta = 2.0$, an increasing failure rate (IFR) distribution. Parameter values for the Gamma and lognormal distributions are chosen to give the same means and variances as those of the corresponding DFR and IFR Weibull distributions. (See Chapter 10.) This leads to Gamma shape parameters of 0.2 and 3.66, and lognormal shape parameters of $\theta = 1.339$ and 0.4915. In Table 12.1, these distributions and designated Weib1, Weib2, and so on.

Substitution into Equation (12.3) will provide the seller's expected cost for this warranty. The results are given in Table 12.1. The factor tabulated is the multiplier of c_b in the cost model of Equation (12.3). Thus, for Case 1, assuming the exponential distribution with $\mu = 1.5$, the expected cost to the seller is $c_s + 0.182c_b$. The expected cost of an item that is made for $60 and sold for $100 under Warranty 1 is $60 + 0.182(100) = \$78.20$. For the DFR Weibull distribution of the example, the cost is $60 + 100(0.444) = \$104.40$; for the IFR Weibull, it is $64.50. For the

Table 12.1   Factors for Calculating Seller's Average Cost of Items Sold Under Warranty 1

	μ	Distribution						
		Exp.	Weib1	Weib2	Gamma1	Gamma2	Logn1	Logn2
Case 1	1.0	0.255	0.510	0.096	0.595	0.081	0.372	0.061
	1.5	0.182	0.444	0.045	0.551	0.028	0.278	0.012
	2.0	0.141	0.400	0.026	0.521	0.012	0.219	0.0025
	2.5	0.115	0.368	0.017	0.499	0.006	0.178	0.0006
Case 2	1.0	0.374	0.572	0.279	0.639	0.287	0.466	0.299
	1.5	0.283	0.507	0.150	0.593	0.146	0.380	0.141
	2.0	0.227	0.463	0.091	0.562	0.079	0.321	0.061
	2.5	0.189	0.429	0.061	0.539	0.045	0.277	0.026
Case 3	1.0	0.425	0.628	0.298	0.676	0.300	0.547	0.304
	1.5	0.319	0.558	0.159	0.628	0.151	0.444	0.141
	2.0	0.255	0.510	0.096	0.595	0.081	0.372	0.061
	2.5	0.212	0.473	0.064	0.570	0.046	0.319	0.026

DFR Gamma, the cost is \$115.10; for the IFR Gamma, it is \$62.80. For the two lognormal examples, the costs are \$87.80 and \$61.20. Note the importance of the distributional assumptions in assessing average costs. (The DFR distributions have the same mean and variance, as do the IFR distributions.)

12.2.2 Seller's Cost per Unit Sold Under Warranty 2

Combination warranties which feature free replacement or full refund up to time W_1 from purchase, followed by linear pro-rata coverage from W_1 until some later time, $W_2 = W$, are the most common combination warranties. In this section, we consider Warranty 2, which is the rebate form of this warranty.

For this warranty, with linear proration, the rebate function is

$$q(t) = \begin{cases} c_b, & 0 \leq t < W_1 \\ \dfrac{c_b(W - t)}{W - W_1}, & W_1 \leq t < W \\ 0, & \text{otherwise} \end{cases} \tag{12.4}$$

Note, incidentally, that this combination reduces to the FRW if $W_1 = W$, and to the linear PRW if $W_1 = 0$.

The expected cost to the seller, $E[C_s(W_1, W)]$, is obtained, as in the previous section, as the fixed cost of supplying the item plus the expected rebate. This gives

$$E[C_s(W_1, W)] = c_s + \int_0^\infty q(t) \, dF(t)$$

$$= c_s + \frac{c_b[WF(W) - W_1F(W_1) - \mu_W + \mu_{W_1}]}{W - W_1} \tag{12.5}$$

where μ_W and μ_{W_1} are *partial expectations* of X, given by

$$\mu_t = \int_0^t x \, dF(x) \tag{12.6}$$

For the distributions considered in Example 12.1, the partial expectations are as follows:

Exponential Distribution:

$$\mu_t = \frac{1}{\lambda} [1 - (1 + \lambda t)e^{-\lambda t}] \tag{12.7}$$

Weibull Distribution:

$$\mu_t = \frac{1}{\lambda} \gamma \left(1 + \frac{1}{\beta}, (\lambda t)^\beta \right) \tag{12.8}$$

where $\gamma(a, x)/\Gamma(a)$ is the incomplete Gamma function (see Ref. 3), which, for warranty periods of interest, may be evaluated by use of the series

$$\gamma(a, x) = a^{-1}x^a e^{-x}$$

$$\times \left(1 + \frac{x}{a + 1} + \frac{x^2}{(a + 1)(a + 2)} + \frac{x^3}{(a + 1)(a + 2)(a + 3)} + \cdots \right) \tag{12.9}$$

Gamma Distribution:

$$\mu_t = \frac{1}{\lambda\Gamma(\beta)} \gamma(1 + \beta, \lambda t) \tag{12.10}$$

Lognormal Distribution:

$$\mu_t = \exp\left(\eta + \frac{\theta^2}{2} \right) \Phi \left(\frac{\ln W - (\eta + \theta^2)}{\theta} \right) \tag{12.11}$$

where $\Phi(x)$ is the CDF of the standard normal distribution.

The CDFs for these distributions are given in Chapter 8.

Example 12.2

Three cases are considered: (1) full rebate for 1 month ($W_1 = 0.08333$ years) and pro-rata rebate for the remaining 5 months of a 6-month warranty ($W = 0.5$ years); (2) $W_1 = 0.08333$ and $W = 1$; and (3) $W_1 = 0.25$ and $W = 1$ year. The same distributions and parameter values as those considered in Example 12.1 are used. Seller's costs per unit for items sold under Warranty 2 were calculated by evaluation of the CDFs and partial expectations given above. The results are given in Table 12.2. As in Table 12.1, the tabulated values here are the multiple of c_b in the seller's cost model given, in this case, by Equation (12.5).

To illustrate these results, suppose that the seller's average cost of supplying an item is $c_s = \$60$ and the item is sold for $c_b = \$100$. Suppose that the warranty offered is Case 2 and that the distribution of time to failure is exponential with $\mu = 1.5$. Then, from Equation (12.5) and Table 12.2, the seller's total cost per unit, including warranty costs, is $\$60 + 0.292(100) = \$89.20$. If the time to failure is distributed as the Gamma

Table 12.2 Factors for Calculating Seller's Average Cost of Items Sold Under Warranty 2

	μ	Exp.	Weib1	Weib2	Gamma1	Gamma2	Logn1	Logn2
					Distribution			
Case 1	1.0	0.248	0.519	0.074	0.601	0.053	0.229	0.022
	1.5	0.174	0.451	0.034	0.556	0.017	0.212	0.0075
	2.0	0.134	0.406	0.019	0.526	0.007	0.192	0.0021
	2.5	0.109	0.372	0.013	0.504	0.004	0.172	0.0006
Case 2	1.0	0.398	0.619	0.229	0.669	0.223	0.386	0.184
	1.5	0.292	0.548	0.115	0.620	0.097	0.354	0.105
	2.0	0.231	0.499	0.067	0.588	0.048	0.326	0.053
	2.5	0.190	0.462	0.045	0.563	0.026	0.301	0.026
Case 3	1.0	0.452	0.660	0.275	0.696	0.270	0.446	0.225
	1.5	0.334	0.587	0.138	0.646	0.118	0.408	0.128
	2.0	0.264	0.536	0.081	0.612	0.058	0.376	0.065
	2.5	0.218	0.497	0.054	0.587	0.032	0.348	0.031

distribution with decreasing failure rate, this cost is $60 + 0.620(100) = $122, that is, a situation in which the total of production and warranty costs exceed the selling price, resulting in a net loss on the item. (In such cases, it is necessary to take action to weed out early failures—see Chapter 23.) For the Gamma distribution with increasing failure rate, seller's expected total cost per unit is $69.70.

Note from Table 12.2 that costs increase from Case 1 to Case 3, as expected, as these warranties involve increased coverage. Note also that again costs may vary considerably, depending on the distributional assumptions made.

Although the results are not strictly comparable because of the rather different structure of the warranties, the results given in Table 12.2 follow a pattern quite similar to those of Table 12.1.

12.2.3 Seller's Cost per Unit Sold Under Warranty 3

Warranty 3 is similar in structure to Warranty 2, the difference being that the buyer must purchase a new item at pro-rata cost on failure under warranty rather than being given a pro-rata refund. As a result, the seller's cost is a function only of c_s and not of the purchase price c_b.

The resulting cost model is

$$E[C_s(W_1, W)] = c_s \left(1 + \frac{WF(W) - W_1 F(W_1) - \mu_W + \mu_{W_1}}{W - W_1}\right)$$

(12.12)

This expected cost can be calculated easily by use of the relationship between Equations (12.5) and (12.12).

Example 12.3

For the three cases and the distributions considered in Example 12.2, the seller's expected cost of Warranty 3 can be calculated by use of the factors given in Table 12.2.

Suppose, for example that $c_s = \$60$ per item and that the warranty is Warranty 3 with parameters and assumptions as in Case 2 of Example 12.2. For the exponential distribution with $\mu = 1.5$, seller's expected cost is $\$60(1 + 0.292) = \$77.52$ (as opposed to $\$89.20$ under Warranty 2). For the Gamma distributions of Example 12.2, the costs would be $\$60(1.620) = \$97.20$ and $\$60(1.097) = \$65.82$ as opposed to $\$122.00$ and $\$60.70$, respectively.

12.3   PER-UNIT COST TO BUYER

The expected cost per unit to the buyer is easily calculated based on the results of the previous section. It is simply the price paid for the item, c_b, minus the expected rebate. The buyer's cost models for Warranties 1–3 are given in this section.

12.3.1   Buyer's Cost per Unit for Items Purchased Under Warranty 1

For this warranty, which features decreasing lump-sum rebates as compensation for failed items, the expected rebate is given by Equation (12.2). The average cost per item to the buyer, denoted $E[C_b(\alpha, \mathbf{W})]$, is given by

$$E[C_b(\alpha, \mathbf{W})] = c_b \left\{1 - \sum_{i=1}^{k} \alpha_i [F(W_i) - F(W_{i-1})]\right\}$$

(12.13)

Example 12.4

In Example 12.1, Warranty 1 was considered for 3 three-stage lump-sum rebate policies. Each compensated the buyer with full refund it the item failed in the first portion of the warranty period, provided a 75% rebate in the second portion, and a 50% rebate in the final portion of the warranty period. Resulting costs to the seller for various warranty periods, mean times to failure, and various failure distributions are given in Table 12.1. The values tabulated are the weighted sums in expression (12.13). Thus, the buyer's expected cost can be calculated from these results as c_b times (1 − tabulated value).

Suppose, for example, that an item is warrantied for a period of 6 months, with full rebate for failures during the first month, 75% rebate during the second month, and 50% rebate for the remaining 4 months of the warranty period, and that the item is produced for $60 and sold for $100. If the lifetime of the item is exponentially distributed with mean time to failure of 1.5 years, the expected cost to the buyer is $100(1 − 0.182) = $81.80, a substantial saving over the purchase price (but with the potential annoyance of an early failure and the resulting effort required to obtain the rebate).

If the lifetime of the item has a Gamma distribution with β = 3.66 [the increasing failure rate Gamma distribution of Example 12.1 (Gamma2 in Table 12.1)], the buyer's expected cost per item is $100(1 − 0.028) = $97.20.

12.3.2 Buyer's Cost per Unit for Items Purchased Under Warranty 2

Under Warranty 2, the expected cost per item for the buyer is gotten by modification of Equation (12.5). The result is

$$E[C_b(W_1, W)] = c_b \left(1 - \frac{WF(W) - W_1 F(W_1) - \mu_W + \mu_{W_1}}{W - W_1} \right)$$

$$(12.14)$$

Example 12.5

Here we consider the same warranties, distributions, and parameter choices as in previous examples. Factors for calculating the seller's expected costs under Warranty 2 are given in Table 12.2. The buyer's expected costs are calculated as in Example 12.4.

Suppose, for example, that item lifetimes are exponentially distributed with μ = 1.5, that the seller's average cost per item is $60, and that the item sells for $100. If a full refund is given for items that fail within the first month of purchase and a pro-rata rebate is given for items that fail during the next 5 months of a 6-month warranty, the buyer's average cost per item is $100(1 − 0.174) = $82.60. Here the warranty has a significant value to the buyer, but note that relatively frequent replacements are necessary.

The buyer's expected costs under other distributional assumptions and parameter choices may be calculated similarly from Table 12.2 for selected parameter values or by use of Equations (12.5) and (12.14).

12.3.3 Buyer's Cost per Unit for Items Purchased Under Warranty 3

Under this warranty, the buyer purchases replacements at reduced cost, with the replacement item being warrantied for the remaining time in the warranty period, that is, up to time W from the time of original purchase. The buyer's total cost during the warranty period is, therefore, c_b, the initial purchase price, plus the discounted cost of each additional item purchased under warranty. The per-unit cost is this cost divided by the number of units purchased. This is a complicated expression, involving the ratio of two random quantities. The random cost, $C_b(t)$, where t is the time to failure, is given by

$$C_b(t) = \begin{cases} 0 & \text{if } 0 \le t < W_1 \\ c_b \left(1 - \dfrac{W - t}{W - W_1} \right) & \text{if } W_1 \le t \le W \end{cases} \qquad (12.15)$$

In analyzing the expected cost per unit up to time W from purchase (at which time the warranty expires), it is necessary to consider the possibility that more that one replacement will be needed (each purchased at a higher price than the preceding). The resulting cost model involves an integral equation similar to that used in analysis of the renewing PRW and combination PRW/FRW. Results for these warranties are given in Chapters 5 and 6 of Ref. 1.

There are also additional difficulties in analysis of Warranty 3 from the buyer's point of view. As a result, this warranty has not been analyzed in detail from this perspective. In any case, the costs over the warranty period $[0, W]$ would not be exactly comparable to those of Warranties 1 and 2, as those warranties terminated upon failure of an item. Long-run average costs for this warranty are discussed in Chapter 6 of Ref. 1 and in the references cited.

12.4 PRESENT VALUE COST MODELS

In analyzing costs over time, it is often useful to discount costs to their present value. Many discount functions are used for this purpose, the most common being an exponential function. Under this mode of discounting, the present value of an amount x at a future time t is given by

$$x\, d(t) = x e^{-\delta t} \qquad (12.16)$$

where $d(t)$ is the discount function and δ is the discount rate.

With discounting, the expected rebate $E[q_D(X)]$ is given by

$$E[q_D(X)] = E[d(X)q(X)] = \int_0^\infty d(x)q(x)\,dF(x) \tag{12.17}$$

In the following sections, the discount function given in Equation (12.16) will be used to determine present values for the nonrenewing warranties, Warranties 1–3, discussed previously. In principle, the results can easily be modified for use of other discount functions. The analysis can also be done for renewing combination warranties, but the mathematics becomes increasingly difficult and computer simulation would be required.

Here we look at seller's discounted costs. This is particularly relevant because the seller is generally dealing in a large volume of sales, whereas the buyer typically buys a single item. If it is of interest, the buyer's discounted cost per item can be determined by use of a similar analysis.

12.4.1 Discounted Costs for Warranty 1

The seller's discounted expected cost, denoted $E[C_s(\boldsymbol{\alpha}, \mathbf{W}; \delta)]$, is c_s plus the expected rebate paid out under warranty. The rebate function for Warranty 1 is given in Equation (12.2). With discounting as given by Equation (12.16), the seller's expected rebate per unit becomes

$$E[q_D(X)] = \int_0^W q(x)e^{-\delta x}\,dF(x)$$

$$= c_b\left[\alpha_1\int_0^{W_1} e^{-\delta x}\,dF(x) + \alpha_2\int_{W_1}^{W_2} e^{-\delta x}\,dF(x) + \cdots\right.$$

$$\left. + \alpha_k\int_{W_{k-1}}^{W_k} e^{-\delta x}\,dF(x)\right] \tag{12.18}$$

This is easily evaluated for the exponential distribution. If lifetimes of the items follow this distribution, the seller's discounted expected cost per item is

$$E[C_s(\boldsymbol{\alpha}, \mathbf{W}; \delta)] = c_s + \frac{\lambda}{\lambda + \delta}c_b\sum_{i=1}^k \alpha_i\{\exp[-(\delta + \lambda)W_i]$$

$$- \exp[-(\delta + \lambda)W_{i-1}]\} \tag{12.19}$$

A similar result can be obtained for the Gamma distribution, as the introduction of the discount factor also amounts to a change of scale.

Table 12.3 Factors for Calculating Seller's Discounted Average Cost of Warranty for Items Sold Under Warranty 1, Exponential Distribution

		Case 1			Case 2			Case 3	
μ	$\delta =$	0	0.05	0.10	0	0.05	0.10	0	0.05	0.10
1.0		0.255	0.253	0.250	0.374	0.368	0.361	0.425	0.418	0.411
1.5		0.182	0.180	0.178	0.283	0.278	0.272	0.319	0.314	0.308
2.0		0.141	0.139	0.138	0.227	0.222	0.218	0.255	0.250	0.246
2.5		0.115	0.114	0.113	0.189	0.185	0.182	0.212	0.208	0.204

For the Weibull and lognormal distributions, however, the result is more complex and numerical methods would be needed for evaluation of the integrals.

Example 12.6

We consider the exponential distribution with means 1, 1.5, 2, and 2.5 ($\lambda = 1, 0.6667, 0.5$, and 0.4) and the three warranty policies of Example 12.1. These had $\alpha = (1, 0.75, 0.5)$ and warranty periods **W** (in years) of (0.0833, 0.1667, 0.5), (0.0833, 0.1667, 1), and (0.1667, 0.3333, 1). The warranty cost factors, discounted at rates $\delta = 0, 0.05$, and 0.10 are given in Table 12.3. [These are the multiples of c_b is Equation (12.19).] Note that the value $\delta = 0$ is equivalent to no discounting to present value and is included for purposes of comparison of the effect of discounting. As expected, discounting to present value provides a reduced, more realistic assessment of the cost of warranty.

12.4.2 Discounted Costs for Warranties 2 and 3.

Discounting can be introduced into the analysis of seller's costs under other warranties as well. The same principle is involved; the mathematics is more difficult. For Warranties 2 and 3, the results are obtained from Equation (12.17), using $q(t)$ as given in Equation (12.4). Analytical results can be obtained for the exponential and Gamma distributions as in the previous section. For the Weibull and lognormal distributions, simulation or numerical methods would be required.

12.5 LIFE-CYCLE COST MODELS

12.5.1 Assumptions

When repeated purchases of an item are made over a long period of time, it is of interest to look at the long-run cost of the item and its replacements, some of which may be provided to the buyer free of charge, some of which may be purchased at reduced cost, and the remainder of which will be purchased at full price. Models which deal with such long-run costs are called *life-cycle* cost models.

Life-cycle costs are usually of most interest to the buyer. Furthermore, they are generally only useful if it assumed that repeat purchases are made on failure of an item, whether this occurs within the warranty period or outside of it. This effectively makes the warranty renewing, whether that is so stated or not, because any rebate may be applied to the purchase of the replacement item. Thus, we look at the buyer's expected life-cycle cost over a period L (the life cycle, usually considerably larger than the warranty period W). In the analysis, it is assumed that the buyer always replaces a failed item during this period with an identical new item (no brand switching) and that the time to replacement is sufficiently small that replacement can be considered to be instantaneous.

Analysis of renewing combination warranties is quite complex, involving renewal-type integral equations that typically cannot be evaluated analytically for any of the common life distributions. (See Chapter 6 of Ref. 1 and Refs. 2 and 4.) Here, we restrict attention to Warranty 7 of Section 12.1. This is the fully renewing combination FRW/PRW. For some results on the partially renewing versions of this combination warranty, see Refs. 1 and 2.

12.5.2 Buyer's Life-Cycle Cost for Items Purchased Under Warranty 7

The analysis of buyer's LCC (life-cycle cost) under Warranty 7 is given in Section 6.2.5 of Ref. 1. The analysis is based on the expected value of the replacement cost to the buyer of items purchased over the life cycle, that is, on the expected value of $C_b(L, W_1, W)$, the buyer's (random) life-cycle cost under this warranty. The result is

$$
\begin{aligned}
E[C_b(L, W_1, W)] = c_b \Bigg(&1 + \frac{1}{W - W_1} \int_{W_1}^{W} (x - W_1)\, dF(x) \\
&+ F(L) - F(W_1) + \frac{1}{W - W_1} \\
&\times \int_{W_1}^{W} (x - W_1) M(L - x)\, dF(x) \\
&+ \int_{W}^{L} M(L - x)\, dF(x) \Bigg)
\end{aligned}
\tag{12.20}
$$

Alternatively, this may be written as

$$E[C_b(L, W_1, W)] = c_b \left\{ 1 + M(L) - \int_0^{W_1} M(L - x) \, dF(x) \right.$$

$$- \frac{1}{W - W_1} \left[\int_0^{W - W_1} F(x + W_1) d - \right.$$

$$\int_0^{W - W_1} (x - W + W_1) M(L - W_1 - x)$$

$$\left. \left. dF(x + W_1) \right] \right\}$$

(12.21)

Closed-form solutions to these integral equations do not exist for most life distributions of interest. Nguyen and Murthy [2] provide bounds and approximations. These may be expressed in terms of either the ordinary renewal function $M(\cdot)$ or the renewal function $M_Y(\cdot)$ associated with the random variable indicating the purchase interval (i.e., the time between failures after expiration of the warranty).

Bounds based on the former, which are usually found to be easier to evaluate, are given by

$$K(W_1, W) [1 + M(L)] \le E[C_b(L, W_1, W)] \le c_b + K(W_1, W)M(L)$$

(12.22)

where

$$K(W_1, W) = c_b \left(1 - F(W) - \frac{W_1[F(W) - F(W_1)]}{W - W_1} + \frac{(\mu_W - \mu_{W_1})}{W - W_1} \right)$$

(12.23)

This leads to the approximation

$$E[C_b(L, W_1, W)] \approx K(W_1, W)[0.5 + M(L)] + 0.5c_b$$

(12.24)

Alternative bounds and some numerical examples are given in Chapter 6 of Ref. 1.

For the exponential distribution, exact expressions for buyer's life-cycle costs can be determined explicitly. The result is

$$E[C_b(L, W_1, W)] = c_b + c_b e^{-\lambda W_1}$$

$$\times \left\{ e^{-\lambda(W - W_1)} + \frac{1}{\lambda(W - W_1)} [1 - e^{-\lambda(W - W_1)}][\lambda(L - W_1) - 1] \right\}$$

(12.25)

Bounds and approximations of costs under the assumption of a Weibull distribution are also discussed in Chapter 6 of Ref. 1. Other distributions have not been analyzed. Computer simulation is suggested as an appropriate approach to these analyses.

Buyer's long-run average cost per purchased item and long-run average cost per unit of time can also be analyzed by these methods. Because the buyer's cost in these models is equivalent to income to the seller, the seller's long-run profit per buyer with life cycle L can also be determined. See Ref. 1 for details.

12.6 ESTIMATING COST MODELS

Estimation of warranty cost models based on experimental data is discussed in Chapter 8. The use of claims data for this purpose is discussed in Chapters 9 and 31. Here, we consider briefly the use of experimental data in estimating the models of the previous sections.

All of the quantities in the cost models for the warranties analyzed (as well as the other warranties listed) are functions of the assumed life distribution of the items and the parameters of this distribution. The most straightforward method of estimating the models is to estimate the parameters of the assumed life distribution using an efficient procedure such as maximum likelihood (see Chapter 8) and then substitute the estimated values into the expressions for the cost models.

Other approaches are discussed in Chapters 8 and 9 and in Chapter 12 of Ref. 1. In particular, there are a number of alternatives for estimating renewal functions. Direct estimation of partial moments has not been investigated.

Once the individual elements of the models have been estimated, expected costs can be estimated in a straightforward manner by the methods illustrated in the examples given in the previous sections. Evaluation of the ordinary renewal function can be accomplished by use of tables (e.g., Ref. 5) or by use of a computer program available from the *Handbook* editors.

Example 12.7

Data were obtained on time to failure of 241 airborne radar transmitters in military aircraft. The data are summarized in Ref. 6. Plots and a statistical test suggest that the data follow an exponential distribution. The mean time to failure is estimated to be 265 flight hours, so the estimate of the parameter of the exponential distribution is $\hat{\lambda} = 0.003774$ for these data.

Suppose that the unit is warrantied under Warranty 3 with $W_1 = 50$ flight hours and $W = 200$ flight hours. Estimated cost to the seller can be obtained by substitution of these results into Equations (12.7) and

(12.12). This gives

$$E[C_s(50, 200)] = c_s \left(1 + \frac{200(0.530) - 50(0.172) - 46.39 + 4.16}{150} \right)$$

$$= 1.37c_s$$

Hence, the warranty cost to the seller is estimated to be 37% of the production cost of the transmitter.

Note, however, that in this application, the unit would probably be repairable. This may be taken into consideration either by interpreting c_s to be the average cost of supplying a unit, averaging over repaired and new units, or by rewriting Equation (12.12) as

$$E[C_s(W_1, W)] = c_s + \frac{c_r[WF(W) - W_1F(W_1) - \mu_W + \mu_{w_1}]}{W - W_1}$$

(12.26)

where c_r is the average repair cost per failed unit.

In this example, the methods of Chapter 8 may be used to obtain a confidence interval for λ. Confidence intervals for costs can be obtained as well by substitution of the upper and lower confidence limits into the above results.

12.7 RECOMMENDATIONS AND CONCLUSIONS

12.7.1 Applications

As is apparent from some of the examples of this and other chapters on cost modeling, the results can depend quite critically on the distributional assumptions made. It is essential in applying these results that (1) high-quality experimental data on item lifetimes are available and (2) the data are available in sufficient quantity so that the most appropriate life distribution can be determined with reasonable certainty. This cannot be done with small samples; samples of at least a few hundred observations are usually necessary.

In some applications, only field data are available in sufficient quantity to be useful in parameter estimation. Example 12.7 is an illustration of this. This may be more realistic in a sense, because the context reflects true operational conditions rather than controlled laboratory conditions. The drawbacks are that factors causing failure may be very difficult to determine and the data are often very much less reliable. Particular care must be taken to assure that valid and reliable data are obtained.

For the more complex combination warranties (e.g., those defined in Section 12.1 and any of the many other possibilities), the mathematical analysis of cost models is complex and will seldom lead to tractable results. In these cases, simulation is a useful approach. Determine the life distribution (or select several candidate distributions), estimate the parameters, select warranty policies for consideration, and estimate warranty costs by simulation methods. (See Chapter 11 of Ref. 1 and Ref. 7.)

12.7.2   Future Research

The scope for additional research on combination warranties is endless. The complexity of the mathematical results leads to considerable difficulties in pursuing a purely theoretical approach. Beyond this, a tie of theory to real applications is very much needed. Some suggestions for future research are as follows:

1. Realistic computer simulation models are needed. Input should include a choice of distributions, parameter values, a selection from various warranty policies for analysis, warranty parameters, and type of analysis (cost per item, life-cycle cost, etc.). Output should be the estimated warranty cost (with confidence intervals, if possible) for a variety of warranty alternatives.
2. Many results in this and other chapters indicate that estimated costs are quite sensitive to distributional assumptions. A systematic study of this would be very useful to the practitioner.
3. Comparisons of theoretical with experienced costs would be most enlightening. Obtaining adequate data would not be easy, but empirical studies of this type would provide important information both to the analyst and the practitioner.
4. Models that more nearly reflect reality are needed. These must take into account the fact that many legitimate claims under warranty are not made and that many claims made are not legitimate. (See Chapter 17.) In addition, discounting to present value requires further analysis.
5. In the analyses discussed in this chapter, it was usually assumed that items were nonrepairable. Modifications of the models to take into account repairability are needed. In some cases, this is not difficult (for instance, in Example 12.7). In other cases, this is a significant additional complication.

REFERENCES

1. Blischke, W. R., and Murthy, D. N. P. (1994). *Warranty Cost Analysis*, Marcel Dekker, Inc., New York.
2. Nguyen, D. G., and Murthy, D. N. P. (1984). Cost analysis of warranty policies, *Naval Research Logistics Quarterly*, **31**, 525–543.

3. Abramowitz, M., and Stegun, I. (eds.) (1964). *Handbook of Mathematical Functions with Formulas, Graphs, and Mathematical Tables*, National Bureau of Standards Applied Mathematics Series No. 55, U.S. Government Printing Office, Washington, DC.
4. Biedenweg, F. M. (1981). *Warranty Policies: Consumer Value vs. Manufacturer Cost*, Doctoral Dissertation, Department of Operations Research and Statistics, Stanford University.
5. Baxter, L. A., Scheuer, E. M., Blischke, W. R., and McConologue, D. J. (1981). *Renewal Tables: Tables of Functions Arising in Renewal Theory*, Technical Report, Department of Information and Operations Management, University of Southern California, Los Angeles.
6. Lakey, M. J. (1991). Statistical analysis of field data for aircraft warranties, *Proc. Annual Reliability and Maintainability Symposium*, 340–344.
7. Hill, V. L., Beall, C. W., and Blischke, W. R. (1991). A simulation model for warranty analysis, *International Journal of Production Economics*, **22**, 131–140.

13

Two-Dimensional Free-Replacement Warranties

Herbert Moskowitz

Purdue University
West Lafayette, Indiana

Young Hak Chun

Louisiana State University
Baton Rouge, Louisiana

13.1 INTRODUCTION

As defined in Chapter 1 of this Handbook, a warranty is a producer's promise that a product (or certain of its performance characteristics) is free from defects in materials and workmanship, and should any product prove to be defective during the warranty duration, it will be repaired or replaced according to the warranty terms. Thus, a warranty can be thought of as a contractual *obligation* [1] or a potential *liability* to the producer [2]. However, the limit of a producer's responsibility is usually specified at the time of product sale; the producer is liable only for the product failures satisfying the prespecified warranty criteria.

The *warranty period*, measured from the time of product sale to a certain point in the product lifetime, is one of the most popular criteria in warranty analysis. Some of the reasons for this popularity are that product failures are usually dependent on product age, and the product age can be measured objectively so that conflicts between a producer and a customer on the time of failure occurrence are avoided [3].

In some cases, however, warranty eligibility is measured by more than one criterion. The most common example of these types of warranty policies is the x-year, y-mile protection plan in the automobile industry. Under the *two-dimensional warranty policy*, a producer provides the warranty service only when a defective automobile satisfies the two warranty criteria simultaneously; namely, the age and the mileage should be less than x years and y miles, respectively. In this case, product failures are considered to be dependent not only on the product age but also on its mileage. We can further imagine a *multidimensional warranty policy* in which a producer identifies more than two factors which cause the product failures, thus judging warranty eligibility by several warranty criteria.

Notwithstanding the practical popularity of the two-dimensional warranty policy in practice, little research has been conducted on the theoretical foundation of the two-dimensional warranty policy. Singpurwalla [4] discussed a strategy for choosing an optimal warranty policy based on the principle of maximization of expected utilities involving both profit and costs. His approach incorporated a bivariate probability model involving time and usage as warranty criteria. Recently, Moskowitz and Chun [3] considered a Poisson regression model to determine the warranty cost for two-dimensional warranty policies, assuming that the total number of product failures is distributed as Poisson and its parameter can be expressed as a regression function of the age and usage of a product. Murthy et al. [5] discussed several types of bivariate probability distributions in modeling product failures on the two-dimensional plane.

In this chapter, we propose a decision model that determines the optimal warranty price under the two-dimensional warranty policy and compare it with the classical one-dimensional warranty policy. In Section 13.2, we formally define the two-dimensional warranty policy considered in this chapter. In Section 13.3, we explain major assumptions made in this chapter, which make our warranty models mathematically tractable. Two different approaches to modeling the two-dimensional warranties are discussed in Section 13.4 and, based on one of the approaches, we propose in Section 13.5 a mathematical model for the two-dimensional warranty policy, which can be naturally extended to multidimensional cases. The two-dimensional warranty model is then compared with the single-dimensional model, which is obtained by releasing a restriction on one of the two warranty variables in the two-dimensional warranty model. In Section 13.6, we analyze the sensitivity of the optimal warranty price to several design factors such as the producer's risk aversion and the product life pattern. Issues regarding the implementation of a two-dimensional warranty policy are discussed in Section 13.7. The glossary of symbols used and the derivation of the mathematical models are given in appendices.

13.2 TWO-DIMENSIONAL WARRANTY POLICIES

Under the typical situation of a warranty transaction, a customer pays the *warranty price* at the time of product purchase and a producer provides some kind of repair or replacement service for any product failures during some specified *period of time*. Therefore, two of the major concerns in warranty analysis are how to determine the warranty price and/or the warranty period.

Based on how much a customer should pay to receive the warranty service, we can categorize warranty policies as follows: (1) free-replacement warranty, (2) pro-rata warranty, and (3) lump-sum warranty. Based on how long a producer should provide the warranty service, the warranty policies are divided into (1) a fixed-period warranty and (2) renewable warranty policy. Furthermore, according to the number of criteria used in determining the warranty eligibility, warranty policies can be classified as (1) a single-dimensional and (2) multidimensional warranty policy. See Chapter 1 of the *Handbook* or Blischke and Murthy's [6] excellent review article for more details about the classifications of warranty policies.

The warranty model we consider in this chapter is for the free-replacement, fixed-period, two-dimensional warranty policy; a model for the pro-rata, two-dimensional warranty policy is discussed in Chapter 14 by Wilson and Murthy. According to the taxonomy in Chapter 1 of the *Handbook,* our free-replacement, two-dimensional warranty policy, which is one of the most common types of two-dimensional warranty policies in practice, is classified as Policy 19 (two-dimensional free-replacement warranty).

In the two-dimensional warranty policy, let X and Y denote the two warranty variables and x and y be their specific values. For example, in a 5-year 50,000-mile automobile protection plan, X and Y denote the product age and the mileage, whereas x and y are 5 years and 50,000 miles, respectively. Under the two-dimensional warranty policy denoted by $\Omega(x, y)$, a producer provides as many repairs or replacements as necessary free of charge, in exchange for a customer's initial payment of the warranty price W for any product failures satisfying the two warranty criteria, x and y. Original warranty terms are not renewable after repairs or replacements.

13.3 ASSUMPTIONS

In developing a warranty model for the free-replacement, fixed-period, two-dimensional warranty policy, it is necessary to make certain simplifying assumptions so as to render the mathematics tractable. First, we assume that the product failure rate is constant over time; that is, the failure process follows an exponential distribution [7–10]. The constant failure rate assumption is reasonable, especially for sophisticated products composed of a large number of components; a repair or replacement of a component has little effect on the overall failure rate of the whole product.

Consider, for example, an integrated circuit (IC) in a computer. After the IC is replaced upon failure, the computer as a whole will be about as prone to failure after the replacement as before the IC failure [11]. Furthermore, the constant failure rate assumption is valid if, in the classic *bathtub product failure pattern* shown in Figure 13.1, the *early failure* or *infant-mortality* stage is effectively eliminated by a refund policy and the warranty period is set at some point within the constant failure (chance-failure) stage, the wearout stage being excluded.

Second, we assume that a producer's repair cost is constant regardless of the type and time of product failures. The constant repair cost assumption is also an idealization which will strictly apply only rarely in reality. However, it is a valid approximation in cases where the labor

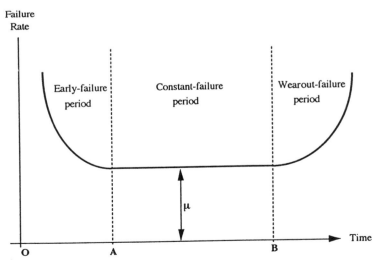

Figure 13.1 Bathtub product failure pattern.

costs of handling and diagnosis (which tend to be constant) dominate the cost of failed components [8].

Third, it is assumed that the duration of time needed to repair or replace a failed component is negligible when compared to the whole warranty period (years) or usage (mileage), so that the failure process is continuous and is not interrupted during the product "down time" [9].

In a two-dimensional warranty policy, product failures can be treated as a family of bivariate failure models indexed by time and usage. Thus, as a mathematical model for a two-dimensional warranty policy, a bivariate Poisson failure model, for example, may be a reasonable approximation for representing the product failure process on the X-Y plane [12–14]. In Chapter 14 of this Handbook, for example, Wilson and Murthy describe two-dimensional warranty policies using two types of bivariate distribution functions: Beta–Stacy distribution and multivariate Pareto distribution of the second kind. (Two different approaches to modeling two-dimensional warranties are reviewed in the next section in more details.)

In this chapter, however, we model the product failure as a function of the age (X) only, assuming that the usage (Y) has a functional relationship with the product age (X). As a result, the modeling of product failures over the two-dimensional plane is done effectively in terms of a single-dimensional point process formulation. This representation, which is computationally more tractable than bivariate failure models, is made possible due to the assumption that the product usage Y can be expressed as a *linear* function of the product age X; that is, $Y = SX$, where the slope S is a random variable having a probability density function $g(s)$, $0 \leq S < \infty$.

This assumption can be explained easily with the *life curve* of a product on the X-Y plane. Figure 13.2 illustrates the product life curves and the corresponding probability density function (p.d.f.). On the X-Y plane, the vertical and horizontal lines x and y represent the limits of warranty duration; only the product failures occurring within $[0, x)$ and $[0, y)$ will be covered by the warranty. The line starting from the origin represents the life pattern of an individual product. All of the product life curves increase individually according to their unique life patterns.

The life curve of Product 1, for example, increases slowly because its warranty variable X is expanding faster than the variable Y, and finally crosses the line, x. The life curve of Product 2, on the other hand, soars sharply due to the fast consumption of Y and eventually encounters the line, y. Under the limitations of x and y, Product 1 and Product 2 finally lose their warranty eligibility by the restriction on X and Y, respectively.

If the product is being used in a constant rate or, in other words, the ratio Y:X is constant, then the product life curve increases constantly.

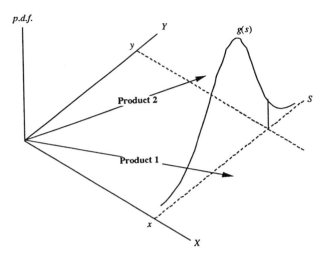

Figure 13.2 Product life pattern.

Because we assumed that the product life curve is not interrupted by product failures, the product life curve can be expressed as a *linearly* increasing line with slope S, $0 < S < \infty$, and intercept 0. In reality, no two products have the same life pattern. Consider, for example, a traveling salesman's and a grandmother's cars, which have different slopes S. Thus, we assume that the slope S is distributed as a p.d.f., $g(S)$.

In the two-dimensional warranty model, the total number of product failures can be expressed as a function of X and Y and we further assume, for simplicity, that the overall failure process is composed of two independent failure processes, each of which is due to X and Y. Based on these assumptions, we develop a decision model that determines the optimal warranty price W for any pair of x and y in the free-replacement, fixed-period, two-dimensional warranty policy $\Omega(x, y)$.

13.4 MODELING TWO-DIMENSIONAL WARRANTIES

In most two-dimensional warranty policies, the age X and the usage Y of a product are the most influential factors on the product failures. Thus, the total number of product failures under a two-dimensional warranty policy can be expressed as a function of the two variables, X and Y, and bivariate probability density functions seem to be the natural choices in describing the product failure process on the two-dimensional plane. In this chapter, however, we express the product failures as an univariate Poisson process, assuming in Section 13.3 that the product usage Y is a linear function of X. This formulation method is called the *one-dimensional* approach [5].

In the one-dimensional approach, the relationship between X and Y is expressed by the product usage rate S, which is usually assumed to be a non-negative random variable of the continuous type. The distribution of the usage rate S could be modeled satisfactorily by the Gamma distribution, as its two parameters α and β provide a great deal of flexibility. Murthy and Wilson [15] consider other types of distribution such as the uniform distribution, the beta-weighted two-point distribution, and the exponential distribution.

One may describe the product failures as a random point process on the two-dimensional plane, using a bivariate probability density function of X and Y. This method is called the *two-dimensional* approach [5]. Iskandar et al. [16] use this approach for the analysis of pro-rata two-dimensional warranty policies, whereas Murthy et al. [5] use it for the analysis of free-replacement, two-dimensional policies. Iskandar et al. [16] also follow the two-dimensional approach for the warranty policy that is a combination of the free-replacement and the pro-rata two-dimensional policies.

As a bivariate probability density function in the two-dimensional approach, the Beta–Stacy distribution and the multivariate Pareto distribution of the second kind are widely used in Refs. 5, 15, and 16, as well as in Chapter 14 of the *Handbook*.

No research works have been reported in the open literature that compare the effectiveness of the two aforementioned approaches based on the real data. According to Murthy and Wilson [15], they are in the process of carrying out an empirical analysis of product failures for two brands of automobiles with more than 1000 cars for each brand. Once their work is completed, some questions will be answered about which methods are more effective in modeling the two-dimensional failure process and in estimating the parameters of warranty models.

In the next section, we formulate a mathematical model for the free-replacement two-dimensional warranty policy based on the one-dimensional approach.

13.5 COST ANALYSIS OF TWO-DIMENSIONAL FREE-REPLACEMENT WARRANTY POLICIES

Because we have assumed a constant failure rate over X and Y, the number of product failures due to X or Y follows a Poisson process with mean

$\lambda_x X$ or $\lambda_y Y$, where λ_x and λ_y are the corresponding failure rates attributed to X and Y, respectively. When the slope of a product life curve is S, let $N(x, y|S)$ be the total number of product failures experienced during $[0, x)$ and $[0, y)$ and let $f(N|S)$ represent a probability mass function conditional on S. Because we assume that the overall failure process is a convolution of two independent failure processes, each of which is attributed to X and Y, respectively, the total number of product failures can be considered as a sum of two independent Poisson random variables with respective means $\lambda_x X$ and $\lambda_y Y$. Thus, $N(x, y|S)$ is distributed as Poisson with the mean

$$\mu = \lambda_x X + \lambda_y Y \tag{13.1}$$

In practice, the parameters λ_x and λ_y can be estimated by multiple regression analysis using the actual observations of μ, X, and Y.

Given the specific warranty limits x and y, let $s = y/x$. Any product life curve whose slope S is less than s will cross the vertical line x, which means that its warranty eligibility will be lost by the restriction on X. On the other hand, any product with $S > s$ will lose its warranty eligibility by the restriction on Y. By exploiting this property, we can rewrite (13.1) as

$$\mu(S) = \begin{cases} \lambda_x x + \lambda_y xS & \text{if } S \leq s \\ \lambda_x \dfrac{y}{S} + \lambda_y y & \text{if } S \geq s \end{cases} \tag{13.2}$$

When S is equal to s, both equations in (13.2) become identical, resulting in $\mu(s) = \lambda_x x + \lambda_y y$. Because $y \geq xS$ for $S \leq s$ and $x > y/S$ for $S \geq s$, the mean $\mu(s)$ is the maximum among all $\mu(S)$.

As shown in (13.2), μ is a function of S. Thus, the probability mass function $f(N|S)$ can be expressed alternatively as $f(N|\mu)$, which is a Poisson distribution with mean μ. From (13.2), the unconditional probability mass function of the total number of product failures under the warranty policy $\Omega(x, y)$ is

$$P[N(x, y)] = \int_0^s f(N|\mu = \lambda_x x + \lambda_y xS)g(S) \, dS$$
$$+ \int_s^\infty f\left(N\Big|\mu = \frac{\lambda_x y}{S} + \lambda_y y\right)g(S) \, dS \tag{13.3}$$

where $g(S)$ is a p.d.f. of S and $f(N|\mu)$ is a Poisson distribution with mean μ.

According to the von Neumann–Morgenstern expected utility theory, a producer's (dis)utility function for cost Z can be represented by a monotonically decreasing function $u(Z)$ [17, p. 179]. Let r denote the repair or replacement cost for a product failure. Then a producer's warranty cost per product, W, can be determined such that it equals the *certainty equivalent* for the sum of total repair costs under the warranty policy $\Omega(x, y)$. Thus, the warranty cost per product under the two-dimensional warranty policy can be derived from

$$
u(W) = \sum_{N=0}^{\infty} u(rN)P[N(x, y)]
$$

$$
= \int_0^s \sum_{N=0}^{\infty} u(rN)f(N|\mu = \lambda_x x + \lambda_y xS)g(S)\, dS
$$

$$
+ \int_s^{\infty} \sum_{N=0}^{\infty} u(rN)f(N|\mu = \frac{\lambda_x y}{S} + \lambda_y y)g(S)\, dS \qquad (13.4)
$$

The classical single-dimensional warranty model can be viewed as a special case of the two-dimensional warranty model in which the restriction on one of the two warranty variables is released. The mathematical form of the single-dimensional warranty model obtained by unrestricting y (i.e., setting y to ∞) is

$$
u(W) = \int_0^{\infty} \sum_{N=0}^{\infty} u(rN)f(N|\mu = \lambda_x x + \lambda_y xS)g(S)\, dS \qquad (13.5)
$$

Based on these mathematical models with specific functional forms and parameter values, the two-dimensional warranty policy will be compared in the next section with the single-dimensional warranty policy for the cases in which a producer is risk neutral or risk averse. The isowarranty cost curve will also be developed and analyzed for the two-dimensional warranty policy.

13.6 IMPLICATIONS FOR THE MANUFACTURER

13.6.1 Numerical Example

Risk Preference

Pratt [18] proposed a measure, called *absolute risk aversion* $r(Z)$, which captures a decision maker's risk preference with respect to cost Z. If $u(Z)$ is a twice differentiable decreasing utility function, $r(Z)$ is expressed as

$$
r(Z) = \frac{u''(Z)}{u'(Z)} \qquad (13.6)
$$

where u' and u'' are respectively the first- and the second-order derivatives of the utility function u. The relationships between $r(Z)$ and a producer's risk attitude toward cost Z is summarized as

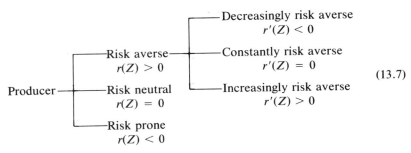

$$(13.7)$$

In the numerical analysis, we consider only the risk-neutral and the constantly risk-averse cases, excluding all other possible risk attitudes. The *risk-prone* case is not considered because it typically does not appear to characterize actual risk attitudes [17]. *Decreasing risk-averse* producers are also excluded because such a risk attitude implies that as their monetary losses increase, they would be willing to pay a smaller risk premium for a given risk. The *increasingly risk-averse* case is also excluded because the use of functions characterizing such behavior is computationally complex and the response surface is quite similar to that obtained using constant risk aversion [19].

To analyze the sensitivity of warranty cost to a producer's risk preference, we assume that a producer has a decreasing utility function $u(Z)$ for cost Z. According to the absolute risk-aversion factor, we use a linear utility function to represent a producer's risk neutrality for cost Z:

$$u(Z) = -Z \tag{13.8}$$

and the producer's (constant) risk aversion is represented by a exponential function:

$$u(Z) = -\exp(aZ) \tag{13.9}$$

where $a > 0$ is a risk parameter. A producer becomes more risk averse as the risk parameter a is increased. In our numerical study, the risk parameter, a, is set equal to 0.2 and is varied systematically from 0.01 to 0.68 to evaluate a wide spectrum of possible producers' risk attitudes.

Product Life Patterns

Because we assumed a constant failure rate over the warranty duration, the conditional probability mass function representing the total number of product failures can be expressed as a Poisson distribution with mean μ:

$$f[N = n|\mu] = \frac{\mu^n e^{-\mu}}{n!}, \quad n = 0, 1, \ldots, \infty \tag{13.10}$$

where

$$\mu = \begin{cases} \lambda_x x + \lambda_y xS & \text{if } S \le s \\ \lambda_x \dfrac{y}{S} + \lambda_y y & \text{if } S \ge s \end{cases}$$

In the numerical analysis, the default values of λ_x and λ_y are 1.

To numerically assess the sensitivity of the warranty cost to the product life patterns, we must specify the prior density function of the slope, S. The p.d.f. of the slope variable S has the following Gamma density function $g(S|\alpha, \beta)$ with parameters α and β; that is

$$g(S|\alpha, \beta) = \frac{1}{\Gamma(\alpha)\beta^\alpha} S^{\alpha-1} e^{-S/\beta}, \quad 0 \le S < \infty \tag{13.11}$$

where $E(S) = \alpha\beta$ and $\text{Var}(S) = \alpha\beta^2$. The default values of α and β are respectively 4 and 0.25 in the numerical analysis. The rationale for selecting the Gamma distribution as a p.d.f. is due to its two inherent properties: First, it is conjugate with the Poisson distribution $f(N|\mu)$ in (13.10), and second, it is rich in the sense that it can represent a wide variety and shapes of prior probability functions.

The sensitivity of the warranty cost to the product life patterns are analyzed by examining the effects of the mean and variance of $g(S)$ on the warranty cost. By manipulating the values of α and β in the Gamma distribution, we can systematically vary its mean and variance. The mean is first fixed arbitrarily at 4.00 while the variance is varied from 0.10 to 0.95. Then the variance is fixed at 0.25 with the mean varied from 0.3 to 2.5 to yield a broad range of Gamma density functions.

Decision Models

For our numerical analysis, we have applied the exponential, Poisson, and Gamma distributions to respectively represent the utility function $u(Z)$, the conditional mass function $f(N|\mu)$, and the probability density function $g(S)$. We have also provided the appropriate parameter values for each distribution. Because we compare the two-dimensional model with the single-dimensional warranty model for the risk-neutrality and risk-aversion cases, we can generate the following four decision models (in the model's names, "1" and "2" denote single and two-dimensional, respectively, and "N" and "A" denote risk neutral and risk averse).

Warranty policy	Producer's risk preference	
	Risk neutral	Risk averse
One-dimensional	Model N1	Model A1
Two-dimensional	Model N2	Model A2

The specific forms of these decision models that determine the optimal warranty cost W are given below and the mathematical derivations of Model N2 and Model A2 are presented in Appendix 13.2. Model N1 and Model A1, the single-dimensional warranty models for the risk-neutral and risk-averse cases, can be easily derived from Model N2 and Model A2 respectively by releasing a restriction on Y, that is, by setting y (or equivalently s) at infinity.

Model N2 (Two-Dimensional Warranty Model for a Risk-Neutral Producer)

$$W = r\left(\lambda_x x G(s|\alpha, \beta) + \lambda_y x\alpha\beta G(s|\alpha + 1, \beta)\right.$$

$$\left. + \frac{\lambda_x y[1 - G(s|\alpha - 1, \beta)]}{(\alpha - 1)\beta} + \lambda_y y[1 - G(s|\alpha, \beta)]\right) \qquad (13.12)$$

where

$$G(s|\alpha, \beta) = \int_0^s g(S|\alpha, \beta)\, dS$$

Model A2 (Two-Dimensional Warranty Model for a Risk-Averse Producer)

$$W = \frac{1}{a}\ln\left\{\exp(\lambda_x xe^{ar} - 1)(1 - \beta k_1)^{-\alpha}\, G\left(s\Big|\alpha, \frac{\beta}{1 - \beta k_1}\right)\right.$$

$$\left. + \exp(\lambda_y ye^{ar} - 1)\int_s^\infty e^{k_2/S} g(S)\, dS\right\} \qquad (13.13)$$

where

$$k_1 \equiv \lambda_y x(e^{ar} - 1), \qquad k_2 \equiv \lambda_x y(e^{ar} - 1), \qquad \frac{\beta}{1 - \beta k_1} > 0$$

Model N1 (One-Dimensional Warranty Model for a Risk-Neutral Producer)

$$W = r(\lambda_x x + \lambda_y x \alpha \beta) \tag{13.14}$$

Model A1 (One-Dimensional Warranty Model for a Risk-Averse Producer)

$$W = \frac{1}{a} \ln[\exp(\lambda_x x e^{ar} - 1)(1 - \beta k_1)^{-\alpha}] \tag{13.15}$$

where

$$k_1 \equiv \lambda_y x (e^{ar} - 1), \qquad \frac{\beta}{1 - \beta k_1} > 0$$

In the numerical analysis, we assume for illustrative purposes only and without loss of generality that x and y in warranty policies are 1. The default value of the repair cost r is $1.

Isowarranty Cost Curves

Under the two-dimensional warranty policy, a producer can provide a variety of warranty plans with the same warranty cost to customers. To illustrate this property, we draw isowarranty cost curves of $1.00 ($\times$ 1000) for Model N2 and Model A2. The closed forms for x or y are mathematically intractable in Model N2 and Model A2. Therefore, we first obtain pairs of x and y which yield a $1000 warranty cost, and then connect these points to draw isowarranty cost lines on the X-Y plane.

13.6.2 Results

Producer's Risk Preference

Figure 13.3 shows the effect of a producer's risk preference on the warranty cost. For the exponential risk function, risk neutrality is equivalent to constant risk aversion with risk parameter $a = 0.0$. Therefore, the warranty costs for Model N2 and Model N1 are the same as the respective starting points of Model A2 and Model A1 where a is 0.0, that is, $1.689 and $2.000. When a producer is risk neutral, the warranty cost is invariant to the risk parameter. As a consequence, the curves for Model N2 and Model N1 are shown to be horizontal. However, as a producer becomes more risk averse, the estimated warranty cost increases such as in the cases of Model A2 and Model A1.

The warranty cost of Model A1 is highly sensitive to the risk-aversion parameter and, given the same risk preference, the warranty cost of

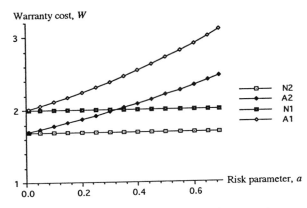

Figure 13.3 Effect of a producer's risk aversion.

Model A1 is always higher than the warranty cost of Model A2. Moreover, in the two-dimensional warranty policy, a producer is less sensitive to risk and thus estimates the warranty cost lower than in the single-dimensional warranty case. The reason is that, in the two-dimensional warranty policy, a producer is doubly protected from customers' excessive warranty claims by imposing one more restriction on the warranty terms. Therefore, we conclude that particularly when a producer is risk averse, the economic advantages of the two-dimensional warranty policy over the single-dimensional policy become increasingly significant.

Product Life Patterns

Figure 13.4 exhibits the sensitivity of the warranty cost to the change in variance of the prior distribution, $g(S)$. The warranty cost of Model N1 is horizontal at $2.00 because, as shown in (13.14), the warranty cost of Model N1 remains unchanged as long as the mean ($=\alpha\beta$) is fixed at some value (1.00 in the example). In other words, in the single-dimensional warranty policy, a risk-neutral producer is concerned with the mean only, ignoring the variance of the product life patterns. However, the warranty cost of Model A1 increases as the variance is increased. It shows that as a producer becomes more risk averse under the single-dimensional warranty policy, the producer is more sensitive to the diversity of the product life patterns.

Note that the warranty costs of Model N2 and Model A2 are decreasing as the variance of the product life patterns increases. This advanta-

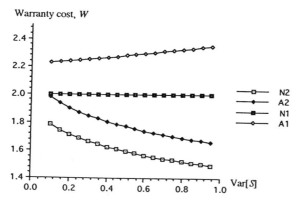

Figure 13.4 Effect of the variance on the warranty cost.

geous characteristic is attributed to the two-dimensional warranty model's *double protection property*: Customers' extraordinary warranty claims are prevented by a producer's restrictions imposed on two warranty variables. Therefore, we conclude that the heterogeneous or uncertain life patterns of a product are less sensitive to a producer's estimate of the warranty cost under the two-dimensional warranty policy than under the single-dimensional warranty policy.

Figure 13.5 shows the sensitivity of the warranty cost to the mean of product life patterns. As the mean of the slope increases, the expected

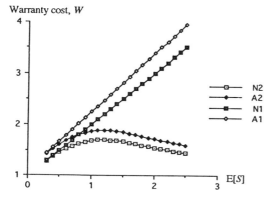

Figure 13.5 Effect of the mean on the warranty cost.

warranty costs of Model N1 and Model A1 increase significantly. This increase of the warranty cost is natural because the warranty variable Y is not restricted in the single-dimensional warranty model, which means that there is no horizontal restriction line in Figure 13.2. Therefore, as the slope S is increased in the direction of the unrestricted Y, the estimated cost for the warranty service is naturally increased.

The warranty costs of Model N2 and Model A2, however, increase and then slightly decrease after some maximum point around $s(s = 1$ in the example). This is because, as shown in Figure 13.2, the warranty eligibility of products are restricted more by the horizontal line y as the mean of the slope is increased. Therefore, again we see that a producer is more protected under the two-dimensional warranty policy from inaccurate estimation of the product life patterns than in the single-dimensional warranty policy.

When the mean of the slope is zero, the warranty costs of Model N2 and Model N1 and those of Model A2 and Model A1 are identical because the product life patterns are expressed only by the variable X.

Isowarranty Cost Curves

The isowarranty cost curves in Figure 13.6 represent the various combinations of x and y that can be offered for the same warranty cost, $1000, when a producer is risk neutral or risk averse. To explain the isowarranty curves, assume that the X axis represents product age (year) and the Y axis denotes product use (1000 miles). According to the isocost curve of Model N2, a risk-neutral producer is indifferent between any combination

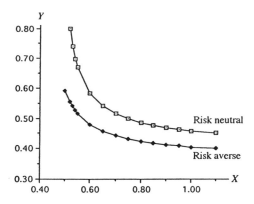

Figure 13.6   Isowarranty cost curves of $1 ($\times 1000$).

of x and y, say, {0.53, 0.74}, {0.70, 0.515}, and {1.0, 0.458}, as long as customers pay \$1000 for the warranty price.

On the other hand, the iso-cost curve of Model A2 exhibits a risk-averse producer's possible warranty plans for the \$1000 warranty price. The isowarranty cost curve of Model A2 is nearer to the origin than the iso-cost curve of Model N2. It shows that the warranty terms offered to customers for a given warranty price tends to be tighter as a producer becomes more risk averse.

Note that as y increases to infinity, x approaches a limiting value of 0.5 in Model N2. This means that the maximum warranty period with unlimited mileage is 0.5 for a risk-neutral producer. It can be also interpreted such that, under the single-dimensional warranty policy $\Omega(x, \infty)$, the maximum warranty period a risk-neutral producer can offer for the \$1000 warranty price is 0.5 or, for illustration, 0.5 year. However, under the two-dimensional (x and y) warranty policy, a producer can offer more than 6 months if customers are willing to give up the unlimited-mileage warranty service in return.

Meanwhile, the limiting value of y in Model N2 is 0.433, which means that the maximum mileage with the lifetime warranty service is 4330 miles. This mileage can be extended if customers prefer longer mileage with a shorter warranty period. Therefore, under the two-dimensional warranty policy, each customer can choose the most appropriate warranty plan suited to his/her own needs for the same warranty price.

13.7 CONCLUDING REMARKS

Based on the warranty model developed, we have compared the two-dimensional warranty policy with the classical single warranty policy and analyzed the iso-warranty cost curves of the two-warranty models. The numerical results show that the two-dimensional warranty policy may have significant advantages over the single-dimensional warranty policy both for the producer and customer.

A producer can be doubly protected from customers' excessive warranty claims by imposing one more restriction on the warranty variables. For example, under the single-dimensional 5-year warranty plan only, a producer is responsible for any failures for 5 years of, for example, a salesman's car, which presumably could have a considerable number of failures due to its excessive use.

Under the 5-year, 50,000 mile warranty policy, however, a producer provides warranty service only for the cars which belong to the 5-year or 50,000-mile warranty limits and, thus, can be protected from the consumers' excessive claims. This dual protection property becomes more significant particularly when the product life patterns are unpredictable or heterogeneous and a producer is risk averse with respect to total repair or replacement cost.

To a customer, the two-dimensional warranty policy provides a wide variety of warranty plans for the same warranty price. Thus, a customer can choose a warranty plan best suited to his/her own needs and tastes. For example, a traveling salesman may choose a 1-year, 100,000-mile protection plan for the same warranty price as the 5-year, 50,000-mile warranty plan, whereas a retired grandmother might select a 10-year, 10,000-mile plan.

Most existing warranty policies are concerned with a single warranty variable—the warranty period. However, according to our model and results, it appears desirable from both the producers' and customers' viewpoints to convert the single-dimensional warranty policy to the two-dimensional one. This conversion can be done by identifying one more warranty characteristic, such as product use, and limiting the warranty eligibility in terms of this new warranty characteristic. However, the new warranty characteristic should be easily and objectively measurable so as to avoid conflicts between a producer and a customer about the point of failure occurrence. One of the reasons for the popularity of the two-dimensional warranty policy in the automobile industry is attributed to the presence of a "tamper-proof" odometer, which shows the true mileage of a car easily and objectively.

In the two-dimensional warranty policy in the automobile industry, the set of two warranty terms has been usually determined arbitrarily or in response to a competitor's warranty strategy. However, a producer could diversify the warranty plans rather than sticking to the existing set of two warranty terms. As shown in the isowarranty cost curve, a producer can provide various sets of warranty plans for the same warranty cost. This strategy could give customers more flexible choices on the warranty plan and, thus, give a manufacturer a clear competitive advantage in terms of sales promotion.

APPENDIX

A.1. Glossary of Symbols

a risk parameter

α and β parameter in a Gamma distribution

$G(s|\alpha, \beta)$ cumulative Gamma distribution function with parameters α and β

$g(S|\alpha, \beta)$ Gamma probability density function of S with parameters α and β

λ_x product failure rate due to X

λ_y product failure rate due to Y

μ expected number of product failures during x and y

$\Omega(x, y)$ two-dimensional warranty policy with the limits $(0, x]$ and $(0, y]$

N total number of product failures during x and y

r repair or replacement cost per failure

S slope of product life pattern (variable)

s ratio of two warranty terms $(y:x)$ (constant)

$u(Z)$ producer's utility function

X, Y warranty attributes (variables)

x, y warranty terms (constants)

W warranty cost (or price)

A.2. Derivations of Model N2 and Model A2

Model N2:   Two-Dimensional Warranty Model for a Risk-Neutral Producer

From (13.4), the objective function for the two-dimensional warranty model is

$$u(W) = \int_0^s \sum_{N=0}^{\infty} u(rN)f(N|\mu = \lambda_x x + \lambda_y xS)g(S)\,dS$$

$$+ \int_s^{\infty} \sum_{N=0}^{\infty} u(rN)f\left(N\Big|\mu = \frac{\lambda_x y}{S} + \lambda_y y\right)g(S)\,dS \quad (A13.1)$$

In the numerical analysis, the following functions are supplied to the above objective function:

$$f[N = n|\mu] = \frac{\mu^n e^{-\mu}}{n!}, \quad n = 0, 1, \ldots, \infty$$

$$g(S|\alpha, \beta) = \frac{1}{\Gamma(\alpha)\beta^\alpha} S^{\alpha-1} e^{-S/\beta}, \quad 0 \le S < \infty$$

and, for the risk-neutral case, the following linear function is given as the utility function

$$u(Z) = -Z$$

Hence, the left-hand side term in (A13.1) becomes

$$u(W) = -W \quad (A13.2)$$

Note that the mixture of the linear utility function and Poisson distribution with mean μ is

$$\int_0^\infty \sum_{N=0}^\infty u(rN)f(N|\mu)g(S)\,dS = \int_0^\infty \sum_{N=0}^\infty (-rN)f(N|\mu)g(S)\,dS$$

$$= -r \int_0^\infty \sum_{N=0}^\infty Nf(N|\mu)g(S)\,dS = -r \int_0^\infty \mu g(S)\,dS \qquad \text{(A13.3)}$$

For the first term on the right-hand side in (A12.1), the range of S is between 0 and s, and μ is defined as

$$\mu = \lambda_x x + \lambda_y xS \qquad \text{(A13.4)}$$

Hence, from (A13.3) and (A13.4), the first term on the right-hand side in (A13.1) is expressed as

$$-r \int_0^s (\lambda_x x + \lambda_y xS)g(S)\,dS = -r\left(\int_0^s \lambda_x xg(S)\,dS + \int_0^s \lambda_y xSg(S)\,dS\right)$$

$$= -r[\lambda_x xG(s|\alpha, \beta) + \lambda_y x\alpha\beta G(s|\alpha + 1, \beta)] \qquad \text{(A13.5)}$$

For the second term on the right-hand side in (A13.1), the range of S is between s and ∞, and μ is defined as

$$\mu = \frac{\lambda_x y}{S} + \lambda_y y \qquad \text{(A13.6)}$$

Hence, from (A13.3) and (A13.6), the second term on the right-hand side in (A.1) is expressed as

$$-r \int_s^\infty \left(\frac{\lambda_x y}{S} + \lambda_y y\right)g(S)\,dS$$

$$= -r\left\{\lambda_x y \int_s^\infty S^{-1}g(S)\,dS + \lambda_y y[1 - G(s|\alpha, \beta)]\right\}$$

$$= -r\left(\frac{\lambda_x y[1 - G(s|\alpha - 1, \beta)]}{(\alpha - 1)\beta} + \lambda_y y[1 - G(s|\alpha, \beta)]\right) \qquad \text{(A13.7)}$$

Finally, from (A13.2), (A13.5), and (A13.7), Model N2 is derived as follows:

$$W = r\left(\lambda_x xG(s|\alpha, \beta) + \lambda_y x\alpha\beta G(s|\alpha + 1, \beta)\right.$$

$$\left. + \frac{\lambda_x y[1 - G(s|\alpha - 1, \beta)]}{(\alpha - 1)\beta} + \lambda_y y[1 - G(s|\alpha, \beta)]\right)$$

Model A2:   Two-Dimensional Warranty Model for a Risk-Averse Producer

In the numerical analysis, the utility function for Model A2 is given as

$$u(z) = -e^{az}, \quad a > 0$$

Hence, the left-hand side term in (A13.1) becomes

$$u(W) = -e^{aW}, \quad a > 0 \tag{A13.8}$$

Note that the mixture of the exponential utility function and Poisson distribution with mean μ is

$$\int_0^\infty \sum_{N=0}^\infty u(rN) f(N|\mu) g(S) \, dS = \int_0^\infty \sum_{N=0}^\infty -e^{arN} f(N|\mu) g(S) \, dS$$

$$= -\int_0^\infty \exp[\mu(e^{ar} - 1)] g(S) \, dS \tag{A13.9}$$

For the first term on the right-hand side in (A13.1), the range of S is between 0 and s, and μ is defined as

$$\mu = \lambda_x x + \lambda_y x S \tag{A13.10}$$

Hence, from (A13.9) and (A13.10), the first term on the right-hand side in (A13.1) becomes

$$-\int_0^\infty \exp[(\lambda_x x + \lambda_y x S)(e^{ar} - 1)] g(S) \, dS$$

$$= -\exp[\lambda_x x(e^{ar} - 1)] \int_0^s \exp[\lambda_y x S(e^{ar} - 1)] \, g(S) \, dS \tag{A13.11}$$

Let $k_1 \equiv \lambda_y x(e^{ar} - 1)$. Then, (A13.11) becomes

$$-\exp[\lambda_x x(e^{ar} - 1)] \int_0^s e^{k_1 S} g(S) \, dS$$

$$= -\exp[\lambda_x x(e^{ar} - 1)](1 - \beta k_1)^{-\alpha} G\left(s\Big|\alpha, \frac{\beta}{1 - \beta k_1}\right) \tag{A13.12}$$

For the second term on the right-hand side in (A13.1), the range of S is between s and ∞, and μ is defined as

$$\mu = \lambda_x \frac{y}{S} + \lambda_y y \tag{A13.13}$$

Hence, from (A13.9) and (A13.13), the second term on the right-hand side in (A13.1) is expressed as

$$-\int_s^\infty \exp\left[\left(\frac{\lambda_x y}{S} + \lambda_y y\right)(e^{ar} - 1)\right] g(S)\, dS$$

$$= -\exp[\lambda_y y(e^{ar} - 1)] \int_s^\infty \exp\left(\frac{\lambda_x y(e^{ar} - 1)}{S}\right) g(S)\, dS \quad \text{(A13.14)}$$

Let $k_2 \equiv \lambda_x y(e^{ar} - 1)$. Then, from (A13.8), (A13.12), and (A13.14), we have

$$-e^{aW} = -\exp[\lambda_x x(e^{ar} - 1)](1 - \beta k_1)^{-\alpha} G\left(S\Big|\alpha, \frac{\beta}{1 - \beta k_1}\right)$$

$$- \exp[\lambda_y y(e^{ar} - 1)] \int_s^\infty e^{k_2/S} g(S)\, dS$$

Hence, the warranty cost for Model A2 is derived as

$$W = \frac{1}{a} \ln\left[\exp[\lambda_x x(e^{ar} - 1)](1 - \beta k_1)^{-\alpha} G\left(s\Big|\alpha, \frac{\beta}{1 - \beta k_1}\right)\right.$$

$$\left. + \exp[\lambda_y y(e^{ar} - 1)] \int_s^\infty e^{k_2/S} g(S)\, dS\right]$$

REFERENCES

1. Nguyen, D. G., and Murthy, D. N. P. (1984). Cost analysis of warranty policies, *Naval Research Logistics Quarterly*, **31**, 525–541.
2. Ritchken, P., Dean, B. V., and Reisman, A. (1985). Establishing warranty reserves for a product which fails given imperfect information, *Applications of Management Science*, **4**, 177–193.
3. Moskowitz, H., and Chun, Y. H. (1994). A Poisson regression model for two-attribute warranty policies, *Naval Research Logistics* **41**, 355–376.
4. Singpurwalla, N. D. (1987). *A Strategy for Setting Optimal Warranties*, Report TR-87/4, Institute for Reliability and Risk Analysis, School of Engineering and Applied Science, George Washington University, Washington DC.
5. Murthy, D. N. P., Iskandar, B. P., and Wilson, R. J. (1995). Two dimensional failure free warranty policies: Two dimensional point process models, *Operations Research* **43**, 356–366.
6. Blischke, W. R., and Murthy, D. N. P. (1992). Product warranty management—I: A taxonomy for warranty policies, *European Journal of Operational Research*, **62**, 127–148.

7. Menke, W. W. (1969). Determination of warranty reserves, *Management Science*, **15**, B542–B549.

8. Glickman, T. S., and Berger, P. D. Optimal price and protection period decisions for a product under warranty, *Management Science*, **22**, 1381–1390.

9. Balachandran, K. R., Maschmeyer, R. A., and Livingstone, L. (1981). Product warranty period: A markovian approach to estimation and analysis of repair and replacement costs, *The Accounting Review*, **LVI**, 115–124.

10. Mitra, A., and Patankar, J. G. (1988). Warranty cost estimation: A goal programming approach, *Decision Sciences*, **19**, 409–423.

11. Nguyen, D. G., and Murthy, D. N. P. (1984). A general model for estimating warranty costs for reparable products, *IIE Transactions*, **16**, 379–386.

12. Holgate, P. (1964). Estimation for the bivariate Poisson distribution, *Biometrika*, **53**, 241–244.

13. Hunter, J. J. (1974). Renewal theory in two dimensions: Basic results, *Advances in Applied Probability*, **6**, 376–391.

14. Jorgenson, D. W. (1961). Multiple regression analysis of a Poisson process, *Journal of the American Statistics Association*, **56**, 235–245.

15. Murthy, D. N. P., and Wilson, R. J. (1991). Modelling two dimensional warranties, *Proceedings of the 5th International Symposium on Applied Stochastic Models and Data Analysis*, 481–492.

16. Iskandar, B. P., Wilson, R. J., and Murthy, D. N. P. Two-dimensional combination warranty policies, *RAIRO*, (in press).

17. Keeney, R. L., and Raiffa, H. (1976). *Decisions with Multiple Objectives: Preferences and Value Tradeoffs*, John Wiley & Sons, New York.

18. Pratt, J. W. (1964). Risk aversion in the small and in the large, *Econometrica*, **32**, 122–136.

19. Moskowitz, H., and Plante, R. (1984). Effect of risk aversion on single sample dimensional inspection plans, *Management Sciences*, **30**, 1226–1237.

14

Two-Dimensional Pro-Rata and Combination Warranties

Richard J. Wilson and D. N. Prabhakar Murthy

The University of Queensland
Brisbane, Queensland, Australia

14.1 INTRODUCTION

Under a free-replacement warranty (FRW), a free-replacement item is provided if the initial item or subsequent replacement items fail within the specified warranty region. This tends to favor the consumer, as the consumer may have benefited considerably from the use of the item before it failed and receives a new item when it fails. In contrast, a pro-rata warranty policy (PRW) provides either a partial refund or a new item at reduced cost if the initial item fails within the specified warranty region. This type of policy tends to favor the manufacturer. A third possibility is a combination warranty policy in which the warranty region is divided into two (or more) subregions, with different policies (FRW or PRW) applying over the different subregions. Such a policy can be used to balance the interests of both the consumer and the manufacturer.

Chapters 10–12 dealt with free-replacement, pro-rata, and combination policies for one-dimensional warranties, where the warranty applies over an *interval* of time (or usage). For a two-dimensional policy, the warranty applies over a two-dimensional region, Ω, where one dimension is usually age or time and the other dimension is usually usage. Different possibilities for Ω were given in Figure 1.2 of Chapter 1. Two-dimensional free-replacement warranty policies were considered in Chapter 13, and two-dimensional pro-rata warranties and combination warranties will be considered in this chapter. Attention is restricted to describing such policies for nonrepairable products (so that an item which fails under warranty needs to be replaced by a new item or a refund provided) and carrying out a cost analysis for them by obtaining the manufacturer's expected warranty costs.

Mathematical models used to analyze the manufacturer's expected warranty costs have been developed by Moskowitz and Chun [1], Singpurwalla [2], Murthy and Wilson [3], and Murthy et al. [4] for two-dimensional free-replacement warranty policies (see Chapter 13), by Iskandar et al. [5] for two-dimensional pro-rata (or rebate) warranty policies, and by Iskandar [6] and Iskandar et al. [7] for two-dimensional combination warranty policies.

The outline of the chapter is as follows: In Section 14.2, pro-rata warranty policies are described; combination warranty policies are described in Section 14.3. These include descriptions of a number of specific policies. The assumptions underlying item failures and approaches to modeling item failures is covered in Section 14.4. Cost analyses for pro-rata warranties and for combination warranties are carried out in Section 14.5 and Section 14.6, respectively, including some numerical examples for the specific policies of Sections 14.2 and 14.3. Various extensions are discussed in the final section.

14.2   PRO-RATA WARRANTY POLICIES

In a *pro-rata* warranty policy, the manufacturer agrees to provide the consumer with either a refund (full or partial) or with a replacement item at a reduced (pro-rated) price if the item fails during the warranty region (i.e., in Ω). Both the partial refund and the reduced price depend on the age and usage of the item at failure, as well as the sale price. If a replacement item is supplied, then it usually has a new warranty which is identical to that of the item sold initially; such a warranty is called a *renewing* policy. A policy which has a refund rather than a replacement item is called a *nonrenewing* warranty policy.

Under a nonrenewing pro-rata warranty, the buyer has no obligation to purchase a replacement item. Hence, this type of refund is sometimes called an unconditional refund. Nonrepairable products such as batteries, tires, ceramics, and so on, are often sold with this type of warranty. Under a renewing pro-rata policy, the reduction in price for the replacement item is called a rebate. This type of policy is also called a conditional refund policy because the rebate is conditional on a repeat purchase being made. Nonrepairable items can also be sold with this type of warranty.

As mentioned above, the partial refund or rebate (sale price minus the reduced price) is dependent on the age and usage of the item at failure. Let T and X denote the age and usage at failure, respectively. Let $\phi(t, x)$ denote the value of the refund/rebate when $(T, X) = (t, x)$. [The reduced price for the consumer for a replacement item is given by $C_b - \phi(t, x)$, if the rebate is conditional on a repeat purchase being made, where C_b is the sale price or cost to the buyer for the initial item.] The two possible pro-rata policies are the following:

Policy 1

> *NONRENEWING PRO-RATA WARRANTY*: Under this policy, if the item fails under warranty (i.e., if the failure occurs with $(T, X) \in \Omega$), then the manufacturer agrees to refund an amount $\phi(T, X)$ and the warranty ceases.

Policy 2

> *RENEWING PRO-RATA WARRANTY*: Under this policy, the manufacturer agrees to provide a replacement item with a new warranty at prorated cost $C_b - \phi(T, X)$ for any failure with $(T, X) \in \Omega$.

In general, $\phi(t, x)$ can be any nonincreasing positive function in t and x over Ω and taking the value zero outside Ω. Typically, $\phi(0, 0) = C_b$ and $\phi(t, x)$ decreases to zero as t and x increase. Different forms for $\phi(t, x)$ define a family of warranty policies. Some possibilities for $\phi(t, x)$ are given below, where α is a proportionality factor and $0 < \alpha \leq 1$. The first possibility can apply to any warranty region, whereas the second and third apply to a rectangular region as given in Figure 14.1 [see also Figure 1.2(a) in Chapter 1] and denoted here by $\Omega_R(W, U)$) and the fourth applies to an infinite strips region as given in Figure 14.2 [see also Figure 1.2(b) in Chapter 1] and denoted here by $\Omega_S(W, U)$. In the case of a nonrenewing policy with warranty region Ω_R, the warranty expires when either the time since purchase exceeds W or the usage exceeds U. As a result, the buyer obtains a *maximum* coverage of W units of time and U units of usage. In contrast, when the warranty region is Ω_S, the buyer is assured of an item which should last for a *minimum* coverage of W units of time and U units of usage. Note that Policy 24 and Policy 25 in Chapter 1 are

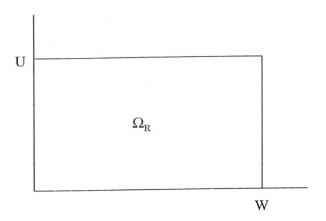

Figure 14.1   Rectangular warranty region.

examples of two-dimensional PRW with the second and fourth refund functions below, respectively.

Lump-sum rebate/refund:

$$\phi(t, x) = \begin{cases} \alpha C_b, & (t, x) \in \Omega \\ 0, & \text{otherwise} \end{cases}$$

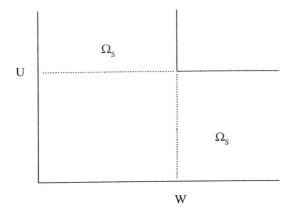

Figure 14.2   Infinite strips warranty region.

Proportional linear rebate/refund:

$$\phi(t, x) = \begin{cases} \alpha \left(1 - \dfrac{t}{W}\right)\left(1 - \dfrac{x}{U}\right) C_b, & (t, x) \in \Omega_R(W, \ U) \\ 0, & \text{otherwise} \end{cases}$$

Proportional quadratic rebate/refund:

$$\phi(t, x) = \begin{cases} \alpha \left(1 - \dfrac{t}{W}\right)^2 \left(1 - \dfrac{x}{U}\right)^2 C_b, & (t, x) \in \Omega_R(W, \ U) \\ 0, & \text{otherwise} \end{cases}$$

Proportional minimal rebate/refund:

$$\phi(t, x) = \begin{cases} \alpha \left[1 - \min \left(\dfrac{t}{W}, \dfrac{x}{U}\right)\right] C_b, & (t, x) \in \Omega_S(W, \ U) \\ 0, & \text{otherwise} \end{cases}$$

Obviously, there are many other possibilities, such as an exponential rebate function with exponent which is linear, quadratic, or of a similar form to the last rebate function above. Note too, that a renewing lump-sum rebate policy with $\alpha = 1$ is simply a renewing free-replacement warranty policy. A nonrenewing free-replacement warranty policy would be identical to a nonrenewing lump-sum rebate ($\alpha = 1$) *conditional* refund policy, with the replacement items covered by the initial warranty region.

14.3 COMBINATION WARRANTY POLICIES

In a combination warranty, the warranty region, Ω, is partitioned into two subregions, Ω_1 and Ω_2, and different simple warranties (subwarranties) applied on the different subregions. These subwarranties can be either free-replacement warranties or pro-rata warranties and either renewing or nonrenewing in nature. As indicated earlier, one advantage of the combination policy is to obtain a balance between the consumer's interests and the manufacturer's interests. This can be achieved by providing a policy which benefits the consumer if an early failure occurs but with less benefit to the consumer if a late failure occurs (relative to the warranty region).

Various partitions of the warranty region Ω are possible. The simplest form for the rectangular warranty region, $\Omega = \Omega_R(W, U) = [0, W) \times [0, U)$, is given in Figure 14.3 [see also Figure 1.2(e) in Chapter 1], where the subregions are $\Omega_1 = [0, W_1) \times [0, U_1)$ and $\Omega_2 = \Omega \cap \Omega_1{}^c$. Here, $0 \leq W_1 \leq W$, $0 \leq U_1 \leq U$, and $\Omega_1{}^c$ denotes the complement of Ω_1.

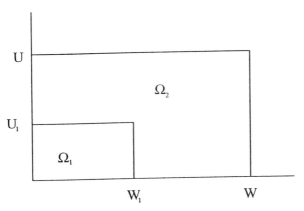

Figure 14.3   Rectangular combination warranty region.

Obviously, $\Omega_1 \cup \Omega_2 = \Omega$ and $\Omega_1 \cap \Omega_2 = \emptyset$. Other possibilities for Ω and Ω_1 (or Ω_2) define families of different warranty regions.

Five combination warranty policies are described below for the rectangular warranty region (Ω_R) with the partitioning given above. Some of these include renewing subwarranties for which the warranty (and, therefore, the warranty region) is renewed; that is, a new warranty, which is identical to the initial warranty, covers the replacement item. The descriptions include the costs of the refunds and replacement items required for failed items. These depend on the position in the warranty region where the failure occurs. In some instances, the position will be given by the age and usage of the item which failed (for example, for the first failure after an item is purchased or after a warranty has been renewed). On other occasions, it will be given by the total time and total usage since the initial purchase.

Suppose an item fails at age T and with usage X. Let T_T and X_T denote the total time and total usage, respectively, from the initial purchase until the item's failure. As before, let $\phi(t, x)$ denote the rebate or refund if a failure occurs at $(t, x) \in \Omega$.

Policy 3

NONRENEWING FRW AND NONRENEWING PRW: Under this policy, the manufacturer agrees to provide a free replacement for any failure with $(T_T, X_T) \in \Omega_1$ and to refund an amount $\phi(T_T, X_T)$ for any failures with $(T_T, X_T) \in \Omega_2$. Replacement items come with a reduced warranty region as the original warranty remains in place. (This policy is Policy 27 in Chapter 1.)

Policy 4

TWO NONRENEWING PRW: Under this policy, the first failure under warranty results in a refund and the warranty ceases, so that $(T_T, X_T) = (T, X)$. The manufacturer agrees to refund the sale price for any failure if $(T, X) \in \Omega_1$ and to refund an amount $\phi(T, X)$ if $(T, X) \in \Omega_2$.

Policy 5

RENEWING FRW AND RENEWING PRW: Under this policy, the manufacturer agrees to provide a free replacement with a new warranty for any failures with $(T, X) \in \Omega_1$ and to provide a replacement with a new warranty at prorated cost $C_b - \phi(T, X)$ for any failure with $(T, X) \in \Omega_2$.

Policy 6

RENEWING FRW AND NONRENEWING PRW: Under this policy the manufacturer agrees to provide a free replacement with a new warranty for any failures with $(T, X) \in \Omega_1$ and to refund an amount $\phi(T, X)$ for any failure with $(T, X) \in \Omega_2$.

Policy 7

NONRENEWING PRW AND RENEWING PRW: Under this policy, the manufacturer agrees to refund the sale price for any failure with $(T, X) \in \Omega_1$ and to provide a replacement with a new warranty at a prorated cost $C_b - \phi(T, X)$ for any failure with $(T, X) \in \Omega_2$.

Comments:

1. Policies 4 and 7 can be viewed as a money-back guarantee policy for items which fail in Ω_1.
2. For Policies 3, 4, and 6, whenever an item fails in Ω_2, the refund is unconditional, as the consumer is not required to buy a replacement item.
3. For Policies 5 and 7, whenever an item fails in Ω_2, the refund is conditional; that is, it is tied to a repeat purchase.
4. If $W_1 = W$ and $U_1 = U$ or if $W_1 = U_1 = 0$, the policies reduce to the appropriate simple policies.

For Policies 1–7, the total warranty servicing cost per item sold at full price is a random variable and its expected value depends on the various costs $[C_b, C_m, \phi(t, x)$, where C_m is the manufacturing cost per item], the parameters of the policy (Ω, Ω_1) and the modeling of the age and usage at failure for the product. The manufacturer's expected warranty cost for Policy j will be denoted by $E_j(\theta)$, where θ represents the set of policy parameters. If the expected profit per item sold at full price for Policy j is denoted by $P_j(\theta)$, then

$$P_j(\theta) = C_b - C_m - E_j(\theta)$$

and the expected profit as a fraction of the manufacturing cost is given by

$$\frac{P_j(\theta)}{C_m} = \frac{C_b}{C_m} - 1 - \frac{E_j(\theta)}{C_m} \tag{14.1}$$

The modeling aspects will be discussed in Section 14.4; cost analyses to obtain $E_j(\theta)$ for these policies are carried out in Sections 14.5 and 14.6. The results for the examples are presented in terms of $E_j(\theta)/C_m$, C_b/C_m, and $P_j(\theta)/C_m$.

14.4 MODEL FORMULATION

14.4.1 Assumptions

First, the following assumptions will be made:

1. The sale price per item (C_b) exceeds the manufacturing cost per item (C_m).
2. The warranty handling cost is small relative to C_m and C_b, and it is ignored.
3. All failures under warranty are claimed.
4. All claims are valid.
5. The claims are made immediately after failures.
6. The time to replace a failed item is small so that it can treated as zero.

14.4.2 Modeling Item Failures

As well as the above assumptions, the age and usage of the items at failure need to be modeled. Two approaches have been used. In the first approach, the usage is modeled as a function of time (or age) and the distribution of the age at failure is modeled using a conditional intensity function dependent on the usage rate. The simplest case is where usage is a linear function of time (or age), such as Usage $= R \times$ Time, where R represents the average usage per unit time. As this may vary from user to user, it can be taken to be some positive random variable whose probability distribution models the likelihood of the different possible values for the average usage rate over all the product's users. Two possible distributions for R are the uniform distribution over some fixed positive interval and the Gamma distribution. The time to failure is then modeled conditionally on a given usage rate, through a conditional intensity function, $\lambda(t|R = r)$. This may be a function of time, age, the usage rate, and the usage. Failures then occur according to a Poisson process with intensity function $\lambda(t|r)$. See Refs. 1, 3, and 6 for more on this approach.

In the second approach, which is taken here, the age and usage at failure are modeled using a two-dimensional distribution function. Let T_i and X_i be the age and usage, respectively, of the ith item at failure. Assume that (T_i, X_i) is a sequence of independent identically distributed bivariate random variables, with joint distribution function $F(t, x) = \Pr(T_i \le t, X_i \le x)$. The form of $F(t, x)$ must be such that the conditional expectation $E[X_i|T_i = t]$ is increasing in t. If $f(t, x)$ is the probability density function corresponding to $F(t, x)$, then two possible forms for $f(t, x)$ are the following:

Beta–Stacy distribution: The density function $f(t, x)$ is given by

$$f(t, x) = \frac{c \exp[-(t/a)^c]t^{\alpha c - \theta_1 - \theta_2}(x/\phi)^{\theta_1 - 1}[t - (x/\phi)]^{\theta_2 - 1}}{\phi a^{\alpha c}\Gamma(\alpha)B(\theta_1, \theta_2)}$$

(14.2)

where $t > 0$, $0 < x < \phi t$, and $\alpha, c, a, \phi, \theta_1, \theta_2 > 0$. The above is a slightly modified version of the Beta–Stacy distribution proposed by Mihram and Hultquist [8]. Expressions for the first and second moments are given in Ref. 9. Of particular interest is the conditional expectation, $E[X_i|T_i = t]$, which is given by

$$E[X_i|T_i = t] = \left(\frac{\phi\theta_1}{\theta_1 + \theta_2}\right)t$$

This is a linear function in t, implying that the usage increases with age in a linear manner in an expected sense.

Multivariate Pareto distribution of the second kind (type 2): The density function $f(t, x)$ is given by

$$f(t, x) = \frac{a_1 a_2[(\theta_1/t)^{a_1}(\theta_2/x)^{a_2}]^{(1 - \rho^2) - 1}}{tx(1 - \rho^2)} I_0\left(\frac{2\rho\sqrt{a_1 a_2 \ln(t/\theta_1)\ln(x/\theta_2)}}{1 - \rho^2}\right)$$

where $t > \theta_1$, $x > \theta_2$, $a_1, a_2 > 2$, $\rho^2 < 1$, ln denotes log to the base e, and I_0 is a modified Bessel function of order zero (see Ref. 10). Expressions for the first and second moments are given in Ref. 9. Of particular interest is the conditional expectation, $E[X_i|T_i = t]$, which is given by

$$E[X_i|T_i = t] = \frac{a_2\theta_2}{a_2 - 1 + \rho^2}\left(\frac{t}{a_1}\right)^{a_1\rho^2/(a_1 - 1 + \rho^2)}$$

In contrast to the earlier case, this is a nonlinear function in t.

Using the appropriate model (with any unknown parameters estimated), the expected warranty servicing cost per item sold at full price can be obtained and used by the manufacturer to compare different policies and to choose the policy which either minimizes the expected warranty cost per item or maximizes the expected profit per item sold. If the item is part of an expensive system which has a much longer life than the item life, then any items which fail outside of the warranty need to be replaced by new (purchased) items. The time period for which the system is kept in use is called the *life cycle* and the total expected warranty cost over the life cycle is called the *life-cycle* cost. This cost is also of interest to the manufacturer.

14.5   COST ANALYSIS: PRO-RATA POLICIES

In this section, the manufacturer's expected warranty cost per item sold will be obtained for Policies 1 and 2. Examples for these will be given later in this section.

14.5.1   Manufacturer's Expected Warranty Cost

Recall that T_i and X_i are the age and usage, respectively, of the ith item at failure and that (T_i, X_i) are independent identically distributed bivariate random variables, with joint distribution function $F(t, x) = \Pr(T_i \le t, X_i \le x)$. The expected warranty cost to the manufacturer for the two pro-rata policies are given below.

Policy 1: Nonrenewing Pro-Rata Warranty

For this policy, the manufacturer incurs a cost only if the first failure occurs under warranty [i.e., $(T_1, X_1) \in \Omega$] and the warranty ceases once the consumer is refunded $\phi(T_1, X_1)$. Consequently, the manufacturer's expected warranty cost is

$$E_1 = \int_\Omega \phi(t, x)\, dF(t, x) \tag{14.3}$$

Policy 2: Renewing Pro-Rata Warranty

In this case, the manufacturer's warranty cost is the sum of the (prorated) costs of the replacement items for failures under warranty, until the warranty ceases. The warranty ceases the first time an item does not fail before its age and usage exits from the warranty region Ω. Consequently, the number of replacement items provided is a random variable which has a geometric distribution with parameter $1 - F(\Omega)$, where $F(\Omega)$ is the probability that an item fails under warranty. If the ith item fails under warranty, then the prorated cost to the manufacturer is $C_m - C_b + \phi(T_i, X_i)$ (which may be negative; i.e., a profit). From this, it can be shown that the manufacturer's expected warranty cost is

$$E_2 = \frac{\int_\Omega (C_m - C_b + \phi(t, x)) \, dF(t, x)}{1 - F(\Omega)} \tag{14.4}$$

It is easy to see from (14.3) and (14.4) that the expected warranty cost for Policy 2 is less than that for Policy 1 ($E_2 < E_1$) if and only if $E_1 < C_b - C_m$, where $C_b - C_m$ is the initial profit. This implies that Policy 2 has a lower expected warranty cost than Policy 1 in all cases where Policy 1 has a positive expected final profit. Consequently, Policy 2 is always preferable from the manufacturer's viewpoint (unless a loss is expected for a product). A further point to note is that E_2 may be negative, so that an additional profit, rather than a loss, may result for the replacement items.

14.5.2 Examples—Pro-Rata Policies

In general, it is not possible to obtain analytical expressions for the expected total warranty costs—a computational scheme is required to obtain them. Two examples of nonrenewing PRW policies (the first with a rectangular warranty region and the second with an infinite strips warranty region) and one example of a renewing PRW policy (with a rectangular warranty region) will be given. For the policies with a rectangular warranty region, a linear rebate/refund function (with $\alpha = 1$) will be used; a minimal rebate/refund function (also with $\alpha = 1$) will be used for the policy with the infinite strips warranty region.

Assume that the item is a part of an automobile and that the units for age and usage are 1 year and 10,000 miles, respectively. Let $F(t, x)$ be given by a Beta–Stacy distribution [see (14.2)] with the following parameter values:

$$a = 0.6, \quad \alpha = 5, \quad c = 1, \quad \phi = 2.34, \quad \theta_1 = 18, \quad \theta_2 = 24$$

These were selected so that the mean age at failure ($E[T_i]$) equals 2 (years) and the mean usage at failure ($E[X_i]$) equals 2 ($\times 10,000$ miles). A range of values for W and U varying from 0.5 to 2 are considered, with $W_1 = 0.5W$ and $U_1 = 0.5U$.

Policy 1A: Nonrenewing PRW—Rectangular Region

Table 14.1 shows $E_1(\theta)/C_m$ for a rectangular warranty region (Ω_R) with different (W, U) and C_b/C_m combinations for Policy 1, where $E_1(\theta)$ is given by (14.3). As can be seen, for a given C_b/C_m, $E_1(\theta)/C_m$ increases with W and U, as expected because better warranty terms imply greater expected warranty costs. For a given (W, U) combination, $E_1(\theta)/C_m$ also increases with C_b/C_m. This is again to be expected, as a larger C_b/C_m implies greater payout under warranty.

Obviously, $P_1(\theta)/C_m$ [see (14.1)] depends on (W, U) and C_b/C_m. It is interesting to compare the following three cases:

1. $W = U = 1.0$ and $C_b/C_m = 1.1$: $P_1(\theta)/C_m = 0.09791$
2. $W = U = 1.5$ and $C_b/C_m = 1.4$: $P_1(\theta)/C_m = 0.38746$
3. $W = U = 2.0$ and $C_b/C_m = 1.8$: $P_1(\theta)/C_m = 0.75700$

They represent increasingly better warranty terms with increasing sale

Table 14.1 Ratio of Expected Warranty Cost to Manufacturing Cost for Policy 1A

C_b/C_m	W	U			
		0.50	1.00	1.50	2.00
1.1	0.50	0.00011	0.00022	0.00026	0.00028
	1.00	0.00030	0.00209	0.00344	0.00413
	1.50	0.00037	0.00401	0.00985	0.01411
	2.00	0.00041	0.00509	0.01537	0.02628
1.4	0.50	0.00014	0.00028	0.00033	0.00036
	1.00	0.00038	0.00266	0.00438	0.00525
	1.50	0.00047	0.00510	0.01254	0.01796
	2.00	0.00058	0.00648	0.01956	0.03345
1.8	0.50	0.00018	0.00037	0.00043	0.00046
	1.00	0.00049	0.00342	0.00563	0.00676
	1.50	0.00061	0.00656	0.01612	0.02309
	2.00	0.00067	0.00833	0.02515	0.04300

price. $P_1(\theta)/C_m$ is the smallest for the first case and largest for the last case. This follows as the increase in the expected warranty servicing cost (with better warranty terms) is more than offset by the increase in the revenue generated due to the increase in the sale price. As a result, from the manufacturer's viewpoint, the best strategy is the last case—that is, sell the product with $C_b/C_m = 1.8$ and with warranty terms $W = U = 2$. Alternatively, W and U could be increased with C_b/C_m being increased to have the same profit-to-cost ratio.

In general, as the warranty terms get better (i.e., as W and/or U increases), the total sales increase. This is due to two reasons: (1) manufacturers using better warranty terms as a promotional tool for marketing and (2) the consumers' perception that better warranty terms imply a better product. In contrast, as the sale price increases, the total sales decrease due to reasons such as value for money, budgetary constraints of consumers, and so on. Hence, to compare the three cases above in a more meaningful manner, a model needs to be developed for the total sales as a function of the sale price and the warranty terms (and other relevant variables such as, for example, advertising). A model dealing with this issue in the context of one-dimensional warranty policy can be found in Ref. 11.

Policy 1B: Nonrenewing PRW—Infinite Strips Region

Table 14.2 shows $E_1(\theta)/C_m$ for an infinite strips warranty region (Ω_s) with different (W, U) and C_b/C_m combinations for Policy 1, where $E_1(\theta)$ is given by (14.3). The results are very similar to that for Policy 1A. The expected costs for Policy 1B are greater than those for Policy 1A for all combinations of (W, U) and C_b/C_m. This is to be expected because the infinite strips region contains the rectangular region and, over the rectangular region, the linear rebate function is less than the minimal rebate function. Policy 1B is, therefore, more attractive to the buyer, whereas Policy 1A is more attractive to the manufacturer. As in Policy 1A, the manufacturer's best strategy is to sell the product with $C_b/C_m = 1.8$ and with warranty terms $W = U = 2$, in order to obtain the highest profit to manufacturing cost ratio.

Policy 2: Renewing PRW—Rectangular Region

Table 14.3 shows $E_2(\theta)/C_m$ for a rectangular warranty region (Ω_R) with different (W, U) and C_b/C_m combinations for Policy 2, where $E_2(\theta)$ is given by (14.4). Unlike the nonrenewing PRW policy, the expected warranty cost here is not always positive or always increasing with W, U, and C_b/C_m. The negative cost (i.e., profit) occurs because the replacement

Table 14.2   Ratio of Expected Warranty Cost to Manufacturing Cost
for Policy 1B

C_b/C_m	W	0.50	1.00	1.50	2.00
			U		
1.1	0.50	0.00058	0.00837	0.03385	0.07921
	1.00	0.00622	0.00949	0.03385	0.07921
	1.50	0.02728	0.02760	0.03834	0.07934
	2.00	0.06793	0.06802	0.06998	0.08945
1.4	0.50	0.00074	0.01065	0.04308	0.10081
	1.00	0.00791	0.01208	0.04308	0.10081
	1.50	0.03472	0.03513	0.04879	0.10098
	2.00	0.08645	0.08658	0.08907	0.11385
1.8	0.50	0.00095	0.01370	0.05539	0.12962
	1.00	0.01017	0.01553	0.05539	0.12962
	1.50	0.04464	0.04516	0.06273	0.12983
	2.00	0.11115	0.11131	0.11452	0.14638

items are being sold at a reduced price which, depending on the time and
usage at failure, may be greater than the manufacturing cost (i.e., $C_b -$
$\phi(t, x) > C_m$). Depending on the weight given by the probability distribution [$f(t, x)$] to the failure times and usages which result in a profit being
made for a replacement item, the expected cost may also be negative.

   Given that a replacement item may result in a profit rather than a
loss for the manufacturer, it then follows that the cost may decrease (i.e.,
the profit may increase) as the warranty parameters, (W, U), or the ratio
of the sale price to manufacturing cost, C_b/C_m, increase. This is seen to
occur for Policy 2, especially as W and U increase with $C_b/C_m = 1.4$ and
1.8. These remarks will also apply to combination warranties involving a
renewing subwarranty.

   For $C_b/C_m = 1.8$, the expected profit is the maximum for $W = U$
$= 2$. As a result, the manufacturer's strategy should be to sell the item
at a price with $C_b/C_m = 1.8$ and with $W = U = 2$. As W and U increase
beyond 2, the profits will initially increase, then start decreasing, and
finally become negative to imply losses rather than profits from the replacements under warranty.

   Because the rebate is conditional for the renewing policies, the buyer
is assumed to purchase a replacement item for each failure under war-

Table 14.3 Ratio of Expected Warranty Cost to Manufacturing Cost for Policy 2

C_b/C_m	W	U			
		0.50	1.00	1.50	2.00
1.1	0.50	−0.00003	0.00006	0.00010	0.00011
	1.00	0.00005	−0.00019	0.00070	0.00141
	1.50	0.00012	0.00055	0.00072	0.00367
	2.00	0.00016	0.00158	0.00352	0.00661
1.4	0.50	−0.00041	−0.00039	−0.00034	−0.00032
	1.00	−0.00061	−0.00659	−0.00683	−0.00592
	1.50	−0.00052	−0.00910	−0.02673	−0.02848
	2.00	−0.00048	−0.00807	−0.03371	−0.06436
1.8	0.50	−0.00091	−0.00098	−0.00092	−0.00089
	1.00	−0.00149	−0.01512	−0.01687	−0.01571
	1.50	−0.00138	−0.02198	−0.06332	−0.07134
	2.00	−0.00132	−0.02094	−0.08334	−0.15898

ranty. A more realistic model may be one in which the buyer may choose to purchase a replacement with certain probability; this leads to a type of combination policy.

14.6 COST ANALYSIS: COMBINATION POLICIES

In this section, the manufacturer's expected warranty cost per item sold will be obtained for Policies 3–7. These results are special cases of a more general result which can be found in Ref. 12. In this more general result, the buyer can choose among different warranty options if the item fails under warranty and is assumed to do so according to probabilities which vary depending on the amount of usage received and the time since purchase to failure. By specifying appropriate values for these probabilities over the subregions of the warranty region, each of Policies 3–7 can be obtained. Examples for Policies 4–7 will be given later in this section.

In order to determine the manufacturer's expected total warranty servicing cost for Policies 3–7, it is necessary to determine the manufacturer's warranty cost for an item failing after t units of time and x units of usage since the start of the (possibly renewed) warranty region. If this cost is denoted by $R(t, x)$, then it is given by

$$R(t, x) = \begin{cases} c_1 & (t, x) \in \Omega_1 \\ c_2 + \phi(t, x) & (t, x) \in \Omega_2 \\ 0 & \text{otherwise} \end{cases} \tag{14.5}$$

where c_1 and c_2 are constants which vary from policy to policy, and Ω_1 and Ω_2 are as before. Policies 3–7 and their warranty costs are as follows.

14.6.1 Manufacturer's Expected Warranty Cost

Policy 3: Nonrenewing FRW and Nonrenewing PRW

For Policy 3, a nonrenewing free replacement warranty applies over Ω_1 and a nonrenewing pro-rata warranty applies over Ω_2, with the warranty ceasing if an item fails in Ω_2 or when the total age and total usage exits from Ω. Consequently, the cost to the manufacturer for a single failure is given by (14.5), with $c_1 = C_m$ and $c_2 = 0$. It can be shown that the manufacturer's expected warranty cost for Policy 3 is given by

$$E_3(W_1, U_1, W, U)$$

$$= \int_{\Omega_1} [E_3(W_1 - t, U_1 - x, W - t, U - x) + C_m] \, dF(t, x)$$

$$+ \int_{\Omega_2} \phi(t, x) \, dF(t, x) \tag{14.6}$$

This is a two-dimensional renewal integral equation for which a computational approach usually would be required to obtain the solution.

Policy 4: Two Nonrenewing PRW

For Policy 4, a nonrenewing full-refund pro-rata warranty applies over Ω_1 and a nonrenewing partial refund pro-rata warranty applies over Ω_2, with the warranty ceasing when the first item fails (if in Ω) or when the age and usage for the purchased item exits from Ω. Hence, the cost to the manufacturer for a single failure is given by (14.5), with $c_1 = C_b$ and $c_2 = 0$ [with $\phi(t, x) < C_b$]. Because only one failure is possible under warranty, only the time and usage until the first failure are needed. It can be shown that the manufacturer's expected warranty cost for Policy 4 is given by

$$E_4(W_1, U_1, W, U) = C_b F(\Omega_1) + \int_{\Omega_2} \phi(t, x) \, dF(t, x) \tag{14.7}$$

where $F(\Omega_1)$ is the probability that an item fails in Ω_1.

Policy 5: Renewing FRW and Renewing PRW

For Policy 5, a renewing free-replacement warranty applies over Ω_1 and a renewing pro-rata warranty applies over Ω_2 with the warranty ceasing when an item does not fail under warranty for the first time. Consequently, the cost to the manufacturer for a single failure is given by (14.5), with $c_1 = C_m$ and $c_2 = C_m - C_b$. Replacement items always come with a new warranty in this policy. It can be shown that the manufacturer's expected warranty cost for Policy 5 is given by

$$E_5(W_1, U_1, W, U) = \frac{C_m F(\Omega_1) + \int_{\Omega_2} [C_m - C_b + \phi(t, x)] \, dF(t, x)}{1 - F(\Omega)}$$

(14.8)

where $F(\Omega)$ is the probability that an item fails in Ω.

Policy 6: Renewing FRW and Nonrenewing PRW

For Policy 6, a renewing free-replacement warranty applies over Ω_1 and a nonrenewing pro-rata warranty applies over Ω_2, with the warranty ceasing when an item does not fail in Ω_1 for the first time. Consequently, the cost to the manufacturer for a single failure is given by (14.5), with $c_1 = C_m$ and $c_2 = 0$. Again, replacement items always come with a new warranty in this policy. It can be shown that the manufacturer's expected warranty cost for Policy 6 is given by

$$E_6(W_1, U_1, W, U) = \frac{C_m F(\Omega_1) + \int_{\Omega_2} [\phi(t, x)] \, dF(t, x)}{1 - F(\Omega_1)}$$

(14.9)

Policy 7: Nonrenewing PRW and Renewing PRW

For Policy 7, a nonrenewing full-refund pro-rata warranty applies over Ω_1 and a renewing pro-rata warranty applies over Ω_2 with the warranty ceasing when an item fails in Ω_1 or does not fail in Ω for the first time. Consequently, the cost to the manufacturer for a single failure is given by (14.5), with $c_1 = C_b$ and $c_2 = C_m - C_b$. Replacement items always come with a new warranty here as well. It can be shown that the manufacturer's expected warranty cost for Policy 7 is given by:

$$E_7(W_1, U_1, W, U) = \frac{C_b F(\Omega_1) + \int_{\Omega_2} [C_m - C_b + \phi(t, x)] \, dF(t, x)}{1 - (F(\Omega) - F(\Omega_1))}$$

$$(14.10)$$

where $F(\Omega) - F(\Omega_1)$ is the probability that an item fails in Ω_2.

14.6.2 Examples—Combination Policies

As with the PRW examples, it is not possible to obtain analytical expressions for the expected total warranty costs—a computational scheme is required to obtain them. The expected warranty cost for Policy 3, given by (14.6), is obtained by solving a two-dimensional renewal-type equation. Efficient and accurate numerical methods for solving such equations are currently under consideration. The expected warranty costs for the remaining policies can be obtained using a standard two-dimensional numerical integration package. As such, attention here is confined to Policies 4–7.

Recall that the warranty region and its subregions are given in Figure 14.3 and consider the following form for $\phi(t, x)$:

$$\phi(t, x) = \begin{cases} C_b, & 0 \leq t \leq W_1, 0 \leq x \leq U_1 \\ \dfrac{C_b(W - t)}{W - W_1}, & W_1 \leq t \leq W, 0 \leq x \leq U_1 \\ \dfrac{C_b(U - x)}{U - U_1}, & 0 \leq t \leq W_1, U_1 \leq x \leq U \\ \dfrac{C_b(W - t)(U - x)}{(W - W_1)(U - U_1)}, & W_1 \leq t \leq W, U_1 \leq x \leq U \\ 0, & \text{otherwise} \end{cases}$$

$$(14.11)$$

It is easy to see that $\phi(t, x)$ is a continuous function for $t \geq 0$ and $x \geq 0$. Again, assume that the item is a part of an automobile and that the units for age and usage are 1 year and 10,000 miles, respectively, and let $F(t, x)$ be given by a Beta–Stacy distribution [see (14.2)], with the same parameter values as before:

$$a = 0.6, \quad \alpha = 5, \quad c = 1, \quad \phi = 2.34, \quad \theta_1 = 18, \quad \theta_2 = 24$$

The same range of values for W and U (from 0.5 to 2) are considered, with $W_1 = 0.5W$ and $U_1 = 0.5U$.

Policy 4: Two Nonrenewing PRW

Table 14.4 shows $E_4(\theta)/C_m$ for different (W, U) and C_b/C_m combinations for Policy 4, where $E_4(\theta)$ is given by (14.7). The results for this policy are very similar to those for Policy 1A and Policy 1B. For example, it can be seen that $E_4(\theta)/C_m$ increases with W and U for a given C_b/C_m and that $E_4(\theta)/C_m$ also increases with C_b/C_m for a given (W, U) combination. These are as expected, for the same reasons as before. From Table 14.1 and Table 14.4, it can be seen that the values for Policy 4 are greater than the corresponding values for Policy 1A. This is due to a full refund being given over Ω_1 for Policy 4, whereas only a partial refund is given for Policy 1A, and from a higher refund being given over Ω_2 for Policy 4 than for Policy 1A.

Similar comments as those made for Policy 1A and Policy 1B with regard to the ratio of the expected profit to manufacturing cost also apply here to $P_4(\theta)/C_m$. For the three cases mentioned earlier ($W = U = 1.0$, $C_b/C_m = 1.1$; $W = U = 1.5$, $C_b/C_m = 1.4$; $W = U = 2.0$, $C_b/C_m = 1.8$), the last again represents the manufacturer's best strategy.

Table 14.4 Ratio of Expected Warranty Cost to Manufacturing Cost for Policy 4 (with $W_1 = 0.5W$ and $U_1 = 0.5U$)

		U			
C_b/C_m	W	0.50	1.00	1.50	2.00
1.1	0.50	0.00039	0.00095	0.00101	0.00101
	1.00	0.00066	0.00743	0.01359	0.01566
	1.50	0.00067	0.01122	0.03418	0.05182
	2.00	0.00067	0.01199	0.04647	0.08875
1.4	0.50	0.00050	0.00121	0.00128	0.00128
	1.00	0.00085	0.00946	0.01730	0.01993
	1.50	0.00085	0.01428	0.04350	0.06595
	2.00	0.00085	0.01526	0.05914	0.11295
1.8	0.50	0.00064	0.00156	0.00165	0.00165
	1.00	0.00109	0.01216	0.02224	0.02562
	1.50	0.00109	0.01836	0.05592	0.08480
	2.00	0.00109	0.01962	0.07603	0.14522

Policy 5: Renewing FRW and Renewing PRW

Table 14.5 shows $E_5(\theta)/C_m$ for different (W, U) and C_b/C_m combinations for Policy 5, where $E_5(\theta)$ is given by (14.8). The results for this policy are similar to those for Policy 2. However, the corresponding values are greater due to a higher rebate (or refund) being given for Policy 5 than for Policy 2. For $C_b/C_m = 1.1$, the expected warranty cost is always positive and increases with W and/or U increasing, as with Policy 4. For $C_b/C_m = 1.4$, the expected warranty cost is negative for $W = U = 0.5$ and positive for the remaining combinations of (W, U). The reason for this happening is that, for some values of $(t, x) \in \Omega_2$, $\phi(t, x)$ is small and $R(t, x)$ is negative. This implies that the manufacturer makes a profit rather than a loss from replacing failed items under warranty, because the compensation under the warranty claim is less than the difference between the sale price and the manufacturing cost. This becomes more striking when $C_b/C_m = 1.8$, where the expected warranty costs are negative for all warranty combinations considered and initially decrease as W and U increase. As with Policy 2, the manufacturer is making a profit rather than loss by selling replacement items at a reduced price, with the best result being for the combination with $C_b/C_m = 1.8$ and with $W = U = 2$.

Table 14.5 Ratio of Expected Warranty Cost to Manufacturing Cost for Policy 5 (with $W_1 = 0.5W$ and $U_1 = 0.5U$)

		U			
C_b/C_m	W	0.50	1.00	1.50	2.00
1.1	0.50	0.00025	0.00071	0.00076	0.00076
	1.00	0.00050	0.00528	0.01049	0.01254
	1.50	0.00050	0.00871	0.02751	0.04508
	2.00	0.00050	0.00950	0.03996	0.08574
1.4	0.50	−0.00005	0.00022	0.00029	0.00029
	1.00	0.00017	0.00036	0.00354	0.00587
	1.50	0.00018	0.00336	0.00738	0.01918
	2.00	0.00018	0.00436	0.01771	0.03636
1.8	0.50	−0.00045	−0.00042	−0.00034	−0.00034
	1.00	−0.00026	−0.00618	−0.00573	−0.00302
	1.50	−0.00025	−0.00377	−0.01946	−0.01534
	2.00	−0.00025	−0.00249	−0.01195	−0.02949

Policy 6: Renewing FRW and Nonrenewing PRW

Table 14.6 shows $E_6(\theta)/C_m$ for different (W, U) and C_b/C_m combinations for Policy 6, where $E_6(\theta)$ is given by (14.9). The results are very similar to that for Policy 4. The expected costs for Policy 6 are smaller than those for Policy 4 for all combinations of (W, U) and C_b/C_m. This is to be expected because the cost to the manufacturer is C_b for any item failing in Ω_1 under Policy 4, whereas this cost is C_m ($< C_b$) under Policy 6. Although the policy is renewing for items failing in Ω_1, the probability of the original and the replaced item failing in Ω_1 is very small. Again by offering different warranty combinations to different types of user, the expected costs become smaller.

Policy 7: Nonrenewing PRW and Renewing PRW

Table 14.7 shows $E_7(\theta)/C_m$ for different (W, U) and C_b/C_m combinations for Policy 7, where $E_7(\theta)$ is given by (14.10). The results are similar to those for Policy 2 and Policy 5. However, the corresponding values are greater, due to the refund given over Ω_1 being higher for Policy 7 than for the other two policies. As well, it is interesting to note that the manufacturers profits are maximized when items are sold with $C_b/C_m = 1.8$

Table 14.6 Ratio of Expected Warranty Cost to Manufacturing Cost for Policy 6 (with $W_1 = 0.5W$ and $U_1 = 0.5U$)

C_b/C_m	W	U			
		0.50	1.00	1.50	2.00
1.1	0.50	0.00038	0.00094	0.00099	0.00100
	1.00	0.00066	0.00731	0.01339	0.01545
	1.50	0.00066	0.01107	0.03368	0.05127
	2.00	0.00066	0.01184	0.04598	0.08848
1.4	0.50	0.00047	0.00117	0.00123	0.00124
	1.00	0.00082	0.00893	0.01640	0.01899
	1.50	0.00082	0.01363	0.04082	0.06210
	2.00	0.00082	0.01461	0.05602	0.10626
1.8	0.50	0.00059	0.00147	0.00155	0.00156
	1.00	0.00103	0.01109	0.02041	0.02371
	1.50	0.00104	0.01704	0.05034	0.07654
	2.00	0.00104	0.01830	0.06940	0.12997

Table 14.7 Ratio of Expected Warranty Cost to Manufacturing Cost
for Policy 7 (with $W_1 = 0.5W$ and $U_1 = 0.5U$)

		U			
C_b/C_m	W	0.50	1.00	1.50	2.00
1.1	0.50	0.00026	0.00072	0.00077	0.00077
	1.00	0.00050	0.00541	0.01070	0.01276
	1.50	0.00051	0.00886	0.02810	0.04578
	2.00	0.00051	0.00965	0.04056	0.08614
1.4	0.50	−0.00002	0.00027	0.00033	0.00034
	1.00	0.00020	0.00092	0.00450	0.00688
	1.50	0.00021	0.00404	0.01057	0.02407
	2.00	0.00021	0.00505	0.02157	0.04665
1.8	0.50	−0.00040	−0.00033	−0.00025	−0.00024
	1.00	−0.00020	−0.00506	−0.00037	−0.00096
	1.50	−0.00019	−0.00238	−0.01279	−0.00487
	2.00	−0.00019	−0.00110	−0.00377	−0.00625

and $W = U = 1.5$. This is in contrast to Policy 2 and Policy 5, where
warranty terms for maximum profit are $W = U = 2$.

14.7 CONCLUSIONS

In this chapter, two-dimensional pro-rata and combination warranty policies have been described and a cost analysis carried out for a number of
specific cases. Numerical results have been presented for three different
pro-rata warranty policies and four different combination policies with a
rectangular warranty region in the case where the refund function is given
by (14.11) and item failures are modeled by the Beta–Stacy distribution.

As mentioned earlier, different rebate functions $\phi(t, x)$ (see Section
14.4) and different shapes for the warranty region Ω, such as the infinite
strips warranty region Ω_S given by Figure 14.2, define families of two-
dimensional pro-rata and combination warranty policies. The analysis for
policies based on other rebate functions and/or other shapes for the warranty region to those used here can be carried out as well (using the results
of Ref. 12). The form of the final expressions would be different.

In the policies studied here, the warranty region Ω is comprised of
two subregions with different warranty terms for each region. Warranty

policies can be defined with more than two subregions and the warranty terms varying from subregion to subregion. The cost analysis of such combination policies would be more complex, but the results would be similar to those stated in Section 14.6.

In the model formulation, it was assumed that all warranty claims are exercised and valid, which is often not the case. An additional option can be introduced to model the consumer's choice to claim for a failure. The probability that a claim is not exercised if the age and usage at failure is t and x, respectively, would then need to be modeled (see Chapter 17 and Ref. 13). The case where not all claims are valid poses additional problems. A resolution mechanism which models the outcome should a dispute arise then needs to be incorporated.

A further extension is to allow consumers to choose between different options for items which fail under warranty. The probabilities for each option would then need to be modeled. This extension can be obtained using the results of Wilson and Murthy [12].

Finally, as mentioned in Section 14.6, the optimal decisions with regard to warranty terms must take into account the total sales as a function of warranty terms and sale price. This is yet another open topic for further research.

REFERENCES

1. Moskowitz, H., and Chun, Y. H. (1988). *A Bayesian Approach to the Two-Attribute Warranty Policy*, Paper No 950, Krannert Graduate School of Management, Purdue University, West Lafayette, IN.
2. Singpurwalla, N. D. (1987). *A Strategy for Setting Optimal Warranties*, Report GWU/IRRA/Serial TR-87/4, George Washington University, Washington, DC.
3. Murthy, D. N. P., and Wilson, R. J. (1991). Modeling two dimensional warranties, *Proc. 5th. Int. Conf. Appl. Stochastic Models and Data Analysis*.
4. Murthy, D. N. P., Iskandar, B. P., and Wilson, R. J. (1990). Two-dimensional failure free warranties: two-dimensional point process models. *Operations Research* (in press).
5. Iskandar, B. P., Wilson, R. J., and Murthy, D. N. P. (1991). *Two Dimensional Rebate Warranty Policies: Modelling and Analysis*, Proc. 11th National Conference of ASOR.
6. Iskandar, B. P. (1993). *Modelling and Analysis of Two-Dimensional Warranty Policies*, Ph.D. Thesis, The University of Queensland, St. Lucia.
7. Iskandar, B. P., Wilson, R. J., and Murthy, D. N. P. (1994). Two-dimensional combination warranty policies, Récherche Opérationnelle, **28**, 57–75.
8. Mihram, G. A., and Hultquist, R. A. (1967). A bivariate warning-time/failure-time distribution. *Journal of the American Statistics Association*, **62**, 589–599.

9. Johnson, N. L., and Kotz, J. (1972). *Distributions in Statistics: Continuous Multivariate Distributions*, John Wiley & Sons, New York.
10. Abramowitz, M., and Stegun, I. A. (1964). *Handbook of Mathematical Functions*, Applied Mathematics Series No. 55, National Bureau of Standards, Washington, DC.
11. Murthy, D. N. P. (1990). Optimal reliability choice in product design, *Engineering Optimization*, **15,** 281–294.
12. Wilson, R. J., and Murthy, D. N. P. (1994). A general two-dimensional combination warranty policy. Research Report No. 41, The Centre for Statistics, The University of Queensland, St Lucia, Australia.
13. Patankar, J. G. (1978). *Estimation of Reserves and Cash Flows Associated with Different Warranty Policies*, Ph.D. Thesis, Clemson University, Clemson, SC.

Part E

Warranty and the Marketplace

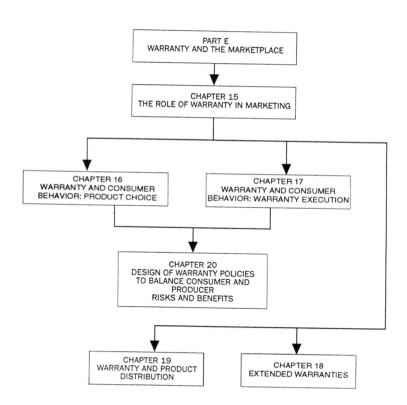

Introduction

Part E consists of six chapters dealing with topics that relate warranty to the marketplace. They address how consumers view warranties and their behavior (or actions) during pre- and postpurchase phases. An understanding of these issues is very important for the manufacturer in formulating marketing strategies when warranty is used as a marketing tool.

In Chapter 15, V. Padmanabhan examines the marketing of product warranty and develops a framework to put this function in a proper perspective. From a marketing viewpoint, consumers view warranty as serving a dual role. First, it influences the degree of perceived risk and, second, it provides information signaling product quality. The chapter briefly examines issues such as postpurchase behavior, warranty and distribution channels, and extended warranties. Although the focus is mainly on consumer goods, the chapter touches on industrial and packaged goods as well.

In Chapter 16, C. A. Kelly expands on the results of Chapter 15. The chapter examines in greater depth the role warranties play in impacting consumer purchase decisions and how warranties affect consumers after the purchase. In particular, it highlights why studies into the role of warranty in impacting purchase decisions have been inconclusive. Postpurchase behavior deals with issues such as claims behavior, impact of product failure, and repeat purchases.

Chapter 17, by J. G. Patankar and A. Mitra, deals with claims behavior and its impact on the expected warranty cost to the manufacturer. The authors focus their attention on nonrenewing linear pro-rata warranties and model consumer behavior through a warranty execution function. This function characterizes the probability that a failure results in a claim as a function of the the age at failure relative to the length of the warranty period.

Padmanabhan returns in Chapter 18 to discuss extended warranties. Extended warranty is an optional warranty provided in addition to the normal (base) warranty and is obtained by paying an additional premium. The chapter starts with a brief discussion on why such warranties have become popular from both consumer and manufacturer viewpoints. It then addresses the topic of designing extended warranties and consumer preferences among alternative extended warranties.

In Chapter 19, A. P. S. Loomba and K. R. Kumar study warranty as a marketing variable with the warranty period to be optimally selected. They discuss this in the context of channels of distribution by considering two cases. In the first (Centralized Distribution and Centralized Service Support), the manufacturer sells the product directly to the consumer and services the claims under warranty. In the second (Decentralized Distribution and Decentralized Service Support), the product is sold through a retailer and the warranty servicing is done by an external third party. They discuss the relative merits of the two approaches.

In the last chapter of Part E, Chapter 20, R. Marcellus and B. Pirojboot discuss the designing of warranties to incorporate both consumer and producer risks and benefits. The process involves selecting the parameters for a specific warranty policy (for example, warranty period) that achieve a balance between manufacturer and consumer risks.

15

Marketing and Warranty

V. Padmanabhan

Stanford University
Stanford, California

15.1 INTRODUCTION

The reliance on product warranty as a prominent factor in the marketing mix has increased substantially in recent times. Several companies have taken advantage of consumer uncertainty over postpurchase problems by making product warranty a very persuasive marketing variable. Long the province of fine print and obscurity, warranty facts and figures are prominently featured in advertisements and press releases nowadays.* The trends are visible across a wide variety of industries and services. Compaq Computer Corp. in December 1992 announced that every personal computer sold worldwide would have a 3-year warranty. CEO Eckhard Pfeiffer states this increase in base warranty from 1 year to 3 years "is an important part of the worldwide strategy." Eaton Corp., a major automo-

---

* In fact, the auto industry, despite being one of the most prominent spenders on advertisements, spent 80% more on warranties than they do on advertising [1].

tive drive train component builder, announced in early 1990 that the base warranty on transmissions, drive train line, and so on had been increased beyond the standard 5-year/500,000-mile limit. Roger Hobbie, General Service Manager at Eaton says, "Warranty is a marketing tool . . . they are a statement about the product . . . they foster a better understanding between Eaton and the end user." Electronic Realty Associates (ERA), one of the nations largest real estate franchisor, offers a unique buyer and seller protection plan to its clients. Curtis Mathes offers a 6-year all-inclusive parts and labor warranty on its product (televisions, stereos, videocassette recorders, etc.). SAS Service Partner, a worldwide airline catering service provider, guarantees its performance along three dimensions: on-time performance, meals to specification, and equipment to airline instruction. SAS feels that apart from the external benefits of these guarantees, they have a tremendous internal benefit on employee motivation and performance. Rank Xerox offers a "Total Satisfaction Guarantee" on all of its products for 3 years. Most observers expect that the trend toward increasing reliance on warranty as an important strategic marketing variable will continue in the near future. This chapter examines issues related to the marketing of product warranty.

15.2   THE CONVENTIONAL WISDOM

What accounts for the surge in the prominence of warranty? It was not long ago that warranty was viewed as a necessary evil: as a method of limiting the liability that the firm would otherwise have. There are a variety of reasons advanced in the literature for the prominence of product warranty. These have to do with the benefits that warranty provide to the consumer as well as the manufacturer. From the manufacturer's perspective, warranties represent an effective differentiating tool for firms wishing to make a definitive statement in an increasingly competitive world. Chrysler's move in the early eighties to a much higher base warranty on its products is an often-cited example of this.* Warranties make a statement about the product, its quality, and the firm's commitment. General Electric's "Satisfaction Guaranteed" program is a case in point. Warranties also help accelerate the product adoption process by reducing the severity of the perceived risk in the adoption of the product or service. L. L. Bean's "100% Guaranteed" program is an example of this. Finally, warranty is a persuasive element in the promotion of the product. In fact,

---

* Industry analysts estimate that Chrysler gained at least one market share percentage point as a result of its warranty advantage. See also Ref. 2.

firms that do not even produce durables have found it useful to offer product warranties. American Express and a host other issuers of credit cards double the manufacturer's warranty on products purchased with their card.

Changes in the characteristics of the consumer marketplace have also contributed to the prominence of product warranty by increasing the consumer reliance, evaluation, and demand for product warranty. Consumers are increasingly quality sensitive nowadays. They are more than willing to pay a higher price in return for better quality. The trend toward double-income households, single-parent households, and so on have all contributed to a premium on time. Most customers do not have the time and energy to deal with product failures. Furthermore, the increasing technological sophistication of products has made it very difficult for consumers to evaluate quality in any meaningful fashion. Consumers therefore look at the product warranty as a useful cue in product evaluation. The fact that more and more consumers are looking at warranty as a useful cue is reflected in the estimated sales response effects for product warranty. (See Ref. 3.)

To summarize, the conventional wisdom suggests that an effective warranty can help the producer in promoting the product by providing a persuasive element, differentiate the product, build a consumer franchise by lending credibility to the manufacturer, and, in general, facilitate the transaction by reducing the perceived cost. An effective warranty can also help the consumer by providing assurances of product quality and value, reducing the degree of perceived risk in the adoption of the product, and increasing satisfaction through reduction of dissonance. These are but a short list of the reasons culled from the popular press for the increasing prominence of warranty in the marketing mix.

15.3 THEORIES OF PRODUCT WARRANTY

There is a considerable literature in marketing that investigates the subject matter of product warranty. Warranty has long fascinated researchers in marketing. Surveys of new-product introduction have consistently shown that a majority of new-product introductions are failures.* An important reason for the failure of new products is the high degree of risk involved in the purchase decision. Most consumers are risk-averse and, consequently, the willingness to try new products is inversely related to the

---

* For example, see the Booz, Allen, and Hamilton [4,5] reports on new-product management. Other reports making similar observations are Refs. 6 and 7.

degree of perceived risk associated with the purchase of the product and the uncertainty about product quality or product performance. Although it is possible for manufacturers to use intrinsic cues or objective physical product characteristics to allay the risk or uncertainty associated with new-product adoption, this is not easy to implement for most experience goods (e.g., consumer durables), where consumers have to learn by experience and use. In these situations, manufacturers have to perforce rely on extrinsic cues to alleviate this risk or uncertainty. Product warranty is one such extrinsic cue that can be used and has thus become one of the most prominent weapons at the manufacturer's disposal to mitigate the risk and uncertainty associated with new-product adoption.

As might be anticipated, a large proportion of the marketing research on product warranty deals with the role of warranty in influencing the degree of perceived risk in the product adoption process. Craig Kelley presents a comprehensive review of the marketing literature on the role of warranty in mitigating risk as well as the other theories of product warranty in Chapter 16. The risk reduction view of warranty is also closely related to the economic literature on the insurance role of product warranty, which is reviewed comprehensively by Lutz in Chapter 25. A simple economic argument for product warranty is that with risk-averse consumers it is optimal for the manufacturer to offer warranty as a form of insurance. Although agreeing with this notion of risk and its relationship with warranty, researchers in marketing have theorized that the overall construct of perceived risk is, in reality, comprised of several dimensions that refer to more specific risks (Refs. 8 and 9). Bearden and Shimp [10,11] point out that the most prominent of these dimensions from the warranty perspective are the financial risk and performance risk associated with the product under consideration. Their experiments clearly demonstrate the pervasive influence of risk in product adoption and the role of extrinsic product cues such as product warranty in risk reduction. They further report that although an outstanding warranty is clearly capable of reducing consumers' perception of potential financial loss, a marginal warranty is no more superior than a poor warranty or no warranty at all, suggesting that there are threshold effects to warranty. The experimental results also suggest that warranty does not seem to play an important role in reducing consumers' perception of performance risk. This suggests that consumers are more likely to rely on word of mouth or some form of product usage experience in order to achieve desired levels of performance risk prior to purchase of a product.

There has also been considerable research in marketing on the consumer response to warranties from an information processing point of view. The focus here is on the types of arguments and counterarguments that warranties engender in consumers. For example, when faced with a claim that the manufacturer offers the longest warranty in the market, do consumers consider this to be a signal of the quality of the product or do they think that the manufacturer was forced to offer the longer warranty in order to compensate for poor quality? This is important because it sheds insight on possible consumer reactions to warranty claims. Dawar and Price [12] explore the idea that consumers' judgments may be driven not just by the warranty information but also by other information such as brand name. They develop and test hypothesis about conditions under which consumers' judgments of product quality are driven by brand name alone, by warranty information alone, and when they jointly and equally contribute to assessment of product quality.

The information processing view on product warranty investigates the arguments generated by product warranty and explores the different rationalizations reached by consumers. The signaling view of product warranty, on the other hand, suggests that consumers should rationally make only a single inference from the terms of the warranty. This theory argues that consumers should always expect that a longer warranty implies a better quality product. The intuition for the signaling role for warranty is based on the fact that only a firm that has a reliable product can afford to provide a comprehensive warranty. The interested reader is referred to Chapter 16 for a more detailed exposition of the signaling theory for product warranty and conditions under which it holds. The interesting issue then is how the market data reconciles with the signaling intuition. Using *Consumer Reports* data on warranty terms and reliability of consumer durables, Wiener [13] reports that all products with a superior warranty were likely to have above-average reliability. The test of the alternative hypothesis about the accuracy of price as a signal of product quality is not supported. Researchers have also looked at the impact that the Magnuson–Moss FTC Act (1975) has had on warranty provision and practice following its passage (see Chapters 4 and 27). Is it a fact that the act succeeded in its intent to make warranty a better signal of product quality and did it succeed in improving the levels of warranty provision provided to consumers by the manufacturers? Wiener [14] suggests that warranty availability and comprehensibility are slightly better in the postact period. However, there is no change in the number of consumers who read and comprehend the warranty and in the number who use them. Analysis of *Consumer Reports* data on durables suggests that in 54% of preact cases and 88% of the postact cases, products that offer more warranty coverage are associated with higher reliability. In terms of the pattern of coverage offered in the postact period, the author reports that significant declines in warranty coverage occurred for products that were not very reliable (e.g., console TVs, washing machines, air-conditioners, etc.). Overall, the data seem to suggest that the act had the expected impact in making the product warranty a better signal of product quality, although it was at the cost of reducing the level of warranty coverage in a number of markets.

15.4 WARRANTY PROVISION AND POSTPURCHASE BEHAVIOR

It is reasonable to expect that the provision of warranty does affect behavior of consumers. Does the provision of warranty coverage have a specific effect on incidence and reports of product failures? What effect does warranty have on replacement behavior for products? Patankar and Mitra in Chapter 17 discuss warranty and consumer behavior with regard to claims. They elaborate on the fact that different types of warranties create different patterns of consumer redemption and, hence, differences in the expected warranty costs and the amounts that need to be set aside as reserves. Kelley in Chapter 16 also discusses this phenomenon of postpurchase behavior of consumer with regard to claims. This issue is also related to the economics literature on consumer moral hazard which is discussed by Lutz in Chapter 25. The essential idea is that provision of warranty coverage reduces the incentive for consumers to invest in effort in the form or regular product care and maintenance. The reports of the Center for Policy Alternatives [15], Bryant and Gerner [16], and Padmanabhan and Rao [17] document this effect. Consumers who purchased warranty protection do invest in less maintenance than consumers who do not. This results in a greater incidence of repairs among the consumers who do purchase the additional warranty. This would also hold for situations in which the firm unilaterally increases the base warranty coverage provided on the product. Management has to be sensitive to this effect and take it into account in its calculation of the expected costs of the warranty program. It is, however, important to note that the theories discussed in the various chapters point out that even with the presence of these effects it is optimal for the manufacturer to offer menus of warranty coverage. In other words, although these effects do create certain costs for the manufacturer, they do not effect the optimality of provision of warranty.

Marcellus and Pirojboot in Chapter 20 discusses the design of warranty policies that balance the consumer and producer risks and benefits. Manufacturers provide warranties because of the benefits cited earlier. The provision of warranty, however, also creates costs for the manufacturer and include among other things the administrative costs, logistics, verification of legitimate claims, and so on. These costs have to be considered in the cost–benefit trade-off from the manufacturer perspective. The manufacturer has to design the warranty policy so that the net benefits outweigh the additional costs. It is shown in the chapter that for a given benefit, it is possible to design a warranty policy that is attractive to the consumer.

15.5 QUANTIFYING THE WARRANTY ELASTICITY

A summary statement of the results from the research on the role of warranty in the prepurchase and the adoption process is that it plays an important role and is effective in accelerating the adoption process. It then becomes important from the point of view of management to have a quantification of the exact impact of this effect. Researchers in marketing have been able to document the magnitude of this impact in terms of sales response and sales elasticity. Note that the role of warranty in accelerating the product adoption process through reduction in perceived risk is verified by the empirical validation of the anticipated sales response effect.

Menezes and Currim [3] posit a general demand function wherein the quantity q sold is a function of price P, warranty length W, advertising A, distribution D, quality R, and corresponding features F and corresponding variables (P', W', A', D', R', F') for the firm's competitors. This is formalized as

$$q = q(P, W, A, D, R, F, P', W', A', D', R', F') \tag{15.1}$$

The decision variables considered in the model are warranty length and price. The cost function posits that the unit cost of production C is a function of the manufacturing costs C_m and that the cost of the warranty C_w depends on warranty length (W), product failure rate (λ), and the proportion of defective items on which a claim is made (α). Formally,

$$C = f(C_m, W, \lambda, \alpha) \tag{15.2}$$

Using differential calculus, they then derive the optimal warranty length and price. The formal expressions are

$$W^* = \frac{C}{\partial C/\partial W}\left(\frac{\mu}{-\eta - 1}\right) \tag{15.3}$$

$$P^* = C\left(\frac{\eta}{\eta + 1}\right) \tag{15.4}$$

where μ and η are the warranty and price elasticity, respectively. This procedure is conducted for three types of product warranty schemes (replacement warranty, free-repair warranty, and pro-rata warranty) and two types of distributions (exponential and Weibull). They show how the procedure can be adapted for both long-run as well as short-run decision horizons.

The authors [3] illustrate the use of this procedure with historical data collected on automobiles. The data were yearly observations of American manufacturers' car sales by model for four categories (compact, intermediate, full size, and luxury) during the period 1981–1987. The unit of analysis was an automobile model. The sales model was estimated using OLS (ordinary least square) regression. All variables (except distribution) are significant at the $p < 0.10$ level. Signs of all variables except repair frequency are in the expected direction. Their analysis of competitive interactions reveal that firms respond very quickly to changes in warranty policy. The warranty elasticity for Chrysler was computed to be 0.143. Note that the direct elasticity was estimated to be 0.88; however, explicit consideration of competitive activities results in a total warranty elasticity of 0.143. As they report, this estimate of warranty elasticity is in the typical range of advertising elasticities reported in the marketing literature. The price elasticity was computed to be -2.49, which again is consistent with typical marketing estimates of the total price elasticity. Using these data, they show how it is possible to calculate the optimal warranty length of a product. For example, using the computed elasticity for warranty and price and typical values for variables such as C_r, C_m, α, β, and λ (in this instance, they chose $C_r/C_m = 0.125$, $\lambda = 1/120$ months, $\alpha = 0.95$, and $\beta = 1.7$), they calculate that the optimal warranty length for Chrysler is 80.2 months.

15.6 WARRANTY AND DISTRIBUTION CHANNEL STRUCTURE

An important issue for consideration in the implementation of any warranty program is the distribution channel implications for warranty policy. Loomba and Kumar discuss the role of product distribution on warranty policy in Chapter 19. There are various dimensions to product distribution. These include the structure of the distribution channel and its intensity, as well as the management of the distribution function. There are also considerable variations in the practice with regard to warranty policy in the channel of distribution. In certain categories, manufacturers take on the dominant role in the provision and administration of warranty, whereas in other product categories, the retailers take on a more dominant role. These factors play a significant role in the design and administration of the overall warranty policy. For example, a major concern of dealers is the effect of unilateral decisions by manufacturers on the terms and provisions of warranty coverages. Many automotive dealers are clearly worried at the effect that increases in the length of the manufacturer warranty will have on their profits on warranty. The authors discuss the various models that look at this subject matter and the analysis that has been conducted in this area. For instance, Loomba and Kumar [18] suggest that the preference for distribution and service support channel structure depends on the degree of substitutability among manufacturers[2] products. As substitutability increases, the channel structure preference for implementation of customer service support shifts from a purely centralized channel to different mixed distribution channels. Furthermore, for products that exhibit low substitutability (e.g., specialty products), manufacturer's should centralize the product distribution as well as customer support.

15.7 EXTENDED SERVICE CONTRACTS

An interesting trend in the marketplace is the increasing prominence of another item in the warranty portfolio: the extended service contract. Several firms offer optional service contracts of various types, lengths, and prices. Extended service contracts refer to the optional warranty coverage that provides protection against failure for some length of time beyond the manufacturer's warranty that can be purchased by the consumer for an added price. This service contract in some sense can be thought of as warranty on demand. Service contracts are a good example of how firms can cater to the different demands for warranty protection by different consumers. If there is one issue that all marketers agree on, it is the fact that no two consumers are alike. Padmanabhan discusses this issue and related details in Chapter 18, which is a review of the marketing literature on extended service contracts. There are a number of theories that have been advanced for the provision of service contracts by manufacturers, retailers, and other institutions. A common thread that ties the literature together is the notion that consumers are heterogeneous and that creates variations in their valuations of insurance. This heterogeneity in the consumer demand for warranty implies that a policy of offering uniform coverage to all consumers is neither desirable nor optimal. The firm can successfully satisfy the needs of all consumers by marketing a menu of warranties in the form of optional extended service contracts. The empirical support for these theories is also discussed there. The increasing prominence of independent providers of insurance in the after-market for insurance and its implications for the overall warranty policy of the manufacturer is also discussed here.

15.8   WARRANTY AND INDUSTRIAL GOODS MARKETS

The focus of most of the literature has been on warranty policies as they relate to typical consumer durables. It is useful to note that most of the essential ideas and, hence, the results apply equally well to another setting where product warranties play an important role: industrial goods. The Conference Board Report [19] discusses these issues in considerable detail. Although virtually all producers of industrial goods offer some sort of warranty, there is substantial variation in the kind of protection provided, the length of the warranties, and in the method of administration. This should sound familiar to the reader, as it identical to the nature of the warranty market in standard consumer durables product category. It is interesting to note that although the Magnuson–Moss Act is taken by observers as referring to consumer products, most manufacturers of industrial goods do believe that it applies to their products and services. The reason is that they believe that the definition of consumer goods is fairly general in the provisions of the Act and a reasonable argument could be made for its interpretation to include the products and service offered by these manufacturers. As a consequence, most producers in the industrial market think about warranty in very much the same fashion as producers of consumer durables. They all agree that warranty has an important role to play in preserving market share. That is the primary reason they cite for keeping informed about competitors' warranties and being in line with the market demands for warranty provision. In fact, what is even more interesting is that manufacturers have begun offering extended warranties on their products and services. This parallels the growth in the prominence of service contracts in the consumer durable market.

15.9   WARRANTY AND PACKAGED GOODS

An interesting aspect of the warranty problem is the apparent lack of popularity of product warranty in the typical consumer packaged-goods industry. McNeal and Hise [20] survey of the consumer packaged-goods industry found that only about a quarter of all the products in a standard supermarket carried some form of a warranty statement. Why do firms rely on product warranty in one product category and not in another? The issue is one of product-specific effects on warranty and its implications for decision making and behavior of the firm and the consumer. The conventional wisdom suggests that given the low dollar value of a standard packaged good, a warranty is unimportant to the consumer. The risk involved in the purchase is negligible and, consequently, the expected benefit from a warranty is very small. However, as these authors and Kendall and Russ [21] point out, these manufacturers may be incurring a significant opportunity cost by following the status quo. An explicit statement in simple and concise terms giving the terms and scope of the warranty and the procedure for its administration can have a significant benefit for the manufacturer. The literature in marketing on managing complaint behavior makes this very point. The firm actually gains a great deal by actively soliciting consumer feedback in the form of warranty claims and complaints. The point to remember is that only a fraction of the consumers who are dissatisfied with the product will bother to make a claim or complaint. Brand switching and/or store switching are the more frequent mode of consumer expression of dissatisfaction. The short-term advantage of avoiding warranty claims is not worth the price to be paid later in the form of lost market share. Furthermore, typical packaged goods, although inexpensive on a per-unit basis, account for a very large fraction of the total consumer expenditures on goods and services. Therefore, it is not clear that perceived risk or exposure is minimal as it is viewed conventionally. There are tremendous benefits to be gained by manufacturers that adopt a more proactive stance on warranty in this category. The fact that some of the more prominent marketers of consumer packaged goods have adopted such a posture is indicative of the wisdom of such a move. Fisk [22] and Kendall and Rusk [21] make these points in their analysis of warranty programs and recommendations for warranty strategy. For instance, Fisk [22] suggests that a more positive outlook at product warranty through explicit, truthful, and clear warranty statements, proper consumer education and training, appropriate pricing of repair services, recognition of legitimate claims can, in fact, stimulate warranty claims and increase positive word of mouth and consumer loyalty. Kendall and Rusk [21] argue that unless there are concrete reasons to believe that the cheating complaints will not increase at a faster rate than legitimate complaints, the warranty trade-off strictly favors a pro-complaint stance. A clear, simple, and obvious warranty policy in this situation adds value to the product through reduction in perceived risk and product differentiation. Furthermore, the data generated by warranty claims can supplement marketing research data in executive decision making.

15.10 SUMMARY DISCUSSION

The growing popularity of the marketing concept requires the firm to consciously understand the role of warranty in marketing strategy and make decisions on warranty policy with a view toward achieving sustainable competitive advantage. Menezes and Quelch [2] provide a useful framework for the design of overall warranty policy with that objective in mind. There are many aspects to the overall warranty program. Internally consistent and synergistic decisions need to be made on each of the following variables: warranty type (free-repair, replacement, or pro-rata warranty), warranty length, warranty breadth, product scope, market scope, coverage, and conditions. Each of these decisions require the consideration of an explicit trade-off analysis where the firm balances the cost implications with the benefit derived by the various consumers as well as the benefits derived by the firm from the adoption of a particular decision. Consumers evaluate a product and the firm not on the basis of a single warranty variable but on a holistic perspective. It is therefore very important for the firm to consider the synergy issue involved among the various dimensions of the warranty policy. The set of factors that influence the overall warranty program can be delineated as product–firm-specific factors (e.g., product and market characteristics, price, reliability, market structure, etc.) and consumer–market-specific factors (e.g., length of ownership, warranty elasticity, price elasticity, expected redemptions, etc.). Each of these factors impact on some or all of the elements of the warranty program. The marketing managers need to evaluate the consumer preferences for each of these elements as well as the cost–benefit of the decisions on those before determining the optimal warranty program.

   In the final analysis, the success of the warranty program depends a great deal on its execution. There is evidence that suggests that consumers are increasingly skeptical of warranty claims.* A great deal of this is attributable to shoddy implementation. A well-designed warranty program can lead to woefully inadequate results if its implementation is not done properly. It would behoove the firm to keep in mind that paying attention to the details for successful implementation can have great dividends to the firm in way of positive word of mouth, customer loyalty, and increased market share. It is extremely important that the warranty wording be simple, precise, and explicit: what is covered, what are the responsibilities of the consumer, and what are the responsibilities of the manufacturer. It would be useful to keep in mind that it is the potential consumer who will be reading the fine print on the warranty and not a lawyer. There is evidence that suggests that close to half the number of service calls made for warranty claims are due to lack of consumer education about proper use. As products get increasingly complex, the importance of consumer education cannot be overstressed. In its absence, both the consumer and the producer suffer the consequences. It is absolutely essential that a warranty program be promoted. Consumers who encoun-

---

* See, for example, Refs. 23 and 24.

ter problems need to be encouraged to bring them to the attention of the firms. This is absolutely essential to prevent consumers from switching to products offered by the competition. It is also a very effective way for the manufacturer to obtain useful diagnostic information on the performance of its products. Such information has been repeatedly proven to be very useful in product redesign and new product development. The warranty program can a be a live testing laboratory for the manufacturer with all its attendant advantages if the firm administers the program properly. Another benefit of the warranty program is that the data generated in this process can effectively supplement the marketing research data for executive decision making. The cost of generation for this data is minimal but the pay-offs can be tremendous. The data can be used for standard types of reliability analysis by product systems, parts, and so on, as well as for more innovative types of market assessment analysis for new-product development and new-product promotions. Another equally important benefit of a well-designed warranty program is its effect on postpurchase behavior of consumers. One of the most often cited benefits of product warranty from the firm's perspective is that it helps the firm stay in close touch with the consumer. This is particularly useful when it comes time for the consumer to replace or buy another product.

It has been suggested in many forums that the process of simply extending the length of warranty coverage is going to stop sooner or later. Most consumers do not feel that extending the length of the manufacturer warranty is by itself a major inducement for purchase. As Jim Warren, Director of Product Assurance at Rockwell International, says, "In the end, consumers are more interested in getting products that don't fail, not insurance policies." Why should a consumer be impressed with a longer warranty coverage on a product than his or her length of ownership? In fact, there is reason to believe that the major improvements in product warranty may not be as much in the scope and coverage as in the administration and implementation. The reason for this is that although the warranty may be comprehensive, the costs to the consumer of redeeming the warranty may be a bigger obstacle. Hence, any improvement in warranty provision without concomitant improvement in its implementation will have negligible impact on the firm's fortunes. This is not to say that there are no possible venues for improvement in warranty provisions. A major complaint of consumers is that they typically have to approach different manufacturers or different retailers for warranty complaints on different parts of the same product. A common warranty where possible would go a long way in reducing the transactions costs faced by consumers in the redemption of warranties. Another suggestion is to specify the warranty coverage in units of usage rather than units of time. For instance, it would be useful to consumers to have the warranty coverage specified as number of uses. The reason is that different consumers subject the product to different levels of use. Therefore, specifying the warranty as 12 months on a washer would mean different things to a family if five as opposed to a single individual. The single individual in this case probably feels that he or she is subsidizing the costs of warranty coverage for the heavier users of the product. The recent advances in microprocessor technology make it possible for the usage to be monitored in a reliable fashion, and that should help in any implementation of such a warranty. In fact, one could also theorize that these sorts of issues will lead to the eventual adoption of multidimensional warranties on more and more products. Automobiles are an example of one product category that already follows such a policy.

REFERENCES

1. Menezes, M. A. J. (1989). *Ford Motor Company: The product warranty program (A)*, Report 9-589-001, Harvard Business School.
2. Menezes, M. A. J., and Quelch, J. A. (1990). Leverage your warranty program, *Sloan Management Review*, 69–80.
3. Menezes, M. A. J., and Currim, I. S. (1992). An approach for determination of warranty length, *International Journal of Research in Marketing*, **9**, 177–195.
4. Booz, Allen, and Hamilton (1971), *Management of New Products*, Booz, Allen and Hamilton, Inc., New York.
5. Booz, Allen, and Hamilton (1982). *New Product Management for the 1980s*, Booz, Allen and Hamilton, Inc., New York.
6. Duerr, M. G. (1986). *The Commercial Development of New Products*, The Conference Board, New York.
7. Wind, J., Mahajan, V., and Bayless, J. L. (1990). *The Role of New Product Models in Supporting and Improving the New Product Development Process: Some Preliminary Results*, The Marketing Science Institute, Cambridge. MA.
8. Jacoby, J., Kaplan, L. B., and Szybillo, G. J. (1974). Components of perceived risk in product purchase, *Journal of Applied Psychology*, **59**, 287–291.
9. Tarpey, L. X., and Peter, J. P. (1975). A comparative analysis of three consumer decision strategies, *Journal of Consumer Research*, **2**, 29–37.
10. Bearden, W. O., and Shimp, T. A. (1982). The use of extrinsic cues to facilitate product adoption, *Journal of Marketing Research*, **9**, 229–239.
11. Shimp, T. A., and Bearden, W. O. (1982). Warranty and other extrinsic cue effects on consumers risk perception, *Journal of Marketing Research*, **9**, 38–46.
12. Dawar, N., and Price, L. J. (1994). *Brand name and Warranty Effects on Consumers Quality Judgments*, Working Paper, INSEAD, France.

13. Wiener, J. L. (1985). Are warranties accurate signals of product reliability, *Journal of Consumer Research*, **12**, 245–50.

14. Wiener, J. L. (1988). An evaluation of the Magnuson–Moss Warranty and Federal Trade Commission Improvement Act of 1975, *Journal of Product Policy and Marketing*, **7**, 65–82.

15. Center for Policy Alternatives at Massachusetts Institute of Technology (1978). *Consumer Durables, Warranties, Service Contracts and Alternatives*, Vol. I-IV, CPA-78-14.

16. Bryant, W. K., and Gerner, J. L. (1982). The demand for service contracts, *Journal of Business*, **55**(3), 345–366.

17. Padmanabhan, V., and Rao, R. C. (1993). Warranty policy and extended service contracts: Theory and an application to automobiles, *Marketing Science*, **12**(3), 230–247.

18. Loomba, A. P. S., and Ravi Kumar, K. (1994), *Optimal Product Sale and Service Support Policies in a Channel Environment under Duopoly*, Working Paper, University of Northern Iowa.

19. The Conference Board (1980), *Industrial Product Warranties: Policies and Practice*, The Conference Board, New York.

20. McNeal, J. U., and Hise, R. T. (1986). An examination of the absence of written warranties on routinely purchased supermarket items, *Akron Business and Economic Review*, **17**(3), 20–30.

21. Kendall, C. L., and Russ, F. A. (1975). Warranties and complaint policies: An opportunity for marketing management, *Journal of Marketing*, **39**, 36–43.

22. Fisk, G. (1970). Guidelines for warranty service after sale, *Journal of Marketing*, **34**, 63–67.

23. A shorter product warranty can be better, *The Marketing Navigator*, 1992, pp. 3, and 4.

24. More firms pledge guaranteed service, *Wall Street Journal*, July 17, 1991.

16

Warranty and Consumer Behavior: Product Choice

Craig A. Kelley

California State University, Sacramento
Sacramento, California

16.1 INTRODUCTION

A warranty is an often misunderstood product attribute. Differences of opinion have been expressed by business practitioners and consumers regarding the role that warranties play in influencing choices among products. Business practitioners have utilized product warranties in a variety of ways to influence consumer product choices. It is not uncommon for advertisements for some products (e.g., automobiles or major consumer appliances) to emphasize a warranty in an effort to promote the reliability of the product. In other cases, warranties are used to achieve price differentiation without changing the product's price or to convey certain rights to the consumer that are already contained in an implied warranty [1].

Consumers, on the other hand, may view warranties as legal documents that are designed to provide a way of securing redress in the event that the product is defective or happens to malfunction during the warranty period. In addition, they may use warranties as a way to choose among products.

This chapter is devoted to presenting the role warranties play in impacting consumer purchase decisions. The theoretical explanations for consumer warranties are discussed and how warranties affect consumers after the purchase are investigated. Finally, unanswered questions concerning the role warranties play in affecting consumer product choices are highlighted.

16.2 WARRANTIES AND PRODUCT ATTRIBUTES

A product is actually a bundle of "attributes" that collectively meet the needs of a consumer. A warranty is a product attribute that may take on one of many roles in the consumer decision-making process. The product's price, packaging, labeling, brand name, and materials used in its production also are examples of product attributes that serve a role in shaping consumer choice.

The search for understanding what role warranties play in influencing consumer product choices has attracted a great deal of research interest. But despite increased research on warranties and other product attributes in the consumer buying process, few concrete conclusions can be drawn to help business practitioners decide how warranties should fit into their product strategy. This condition exists for three reasons [2]. First, much of the warranty research is fragmented and explores a number of different warranty-related topics. Second, most warranty research has not been replicated. Therefore, caution must be used in drawing conclusions from the research. Finally, because many of the results of warranty research have been mixed, few generalizations about the influence of warranties on consumer product choices are possible.

As an example of the mixed results that have been reported, one study investigated the role express warranties, price, store image, and the presence of third-party certification play in affecting consumers' perceptions of the risk associated with a product's purchase [3].* The study tested the following hypotheses.

H_1: A high-quality warranty (i.e., full warranty) will result in less performance risk than a low-quality warranty (i.e., limited warranty).

H_2: A high-quality warranty will result in less financial risk than a low-quality warranty.

---

* Perceived risk will be discussed in greater detail in Section 16.2.1. Perceived risk is the amount of uncertainty that surrounds a particular act, such as the purchase of a product.

H₃: A more favorable store image will result in less performance risk than an unfavorable store image.

H₄: A more favorable store image will result in less financial risk than an unfavorable store image.

H₅: Certification by a reputable private testing organization will result in less performance risk than certification by a disreputable private testing organization.

H₆: Certification by a reputable private testing organization will result in less financial risk than certification by a disreputable private testing organization.

H₇: A high price will result in less performance risk than a low price.

H₈: A high price will result in more financial risk than a low price.

Only four of eight hypotheses were supported, either wholly or in part, in the study. Specifically, Hypotheses 1 and 3 were partially supported as there was a significant interaction between store image and warranty quality. A high-quality warranty and favorable store image reduced performance risk. Hypotheses 2 and 4 were fully supported; a high-quality warranty and a favorable store image each reduced financial risk. Hypotheses 5–8 were not supported.

16.2.1 Consumer Assessment of Warranties in Product Choice

Warranties have been acknowledged to have a significant impact on consumer product choices. The Federal Trade Commission found in a 1985 study that 75% of the consumers surveyed indicated that they would pay a premium for products with better warranty terms [4]. Investigations of the role warranties play in consumer choices of products has focused on the importance of warranties in the purchase decision and consumer knowledge of the difference between a full and limited warranty.

There is evidence that suggests warranties are important product choice attributes [5–8]. Specifically, studies have investigated whether consumers understand the difference between a full and limited warranty and, if so, whether the difference influences product choice. Consumer preference has been shown to exist for products with a more comprehensive full warranty [9].

16.3 EXPLANATIONS OF THE IMPACT OF WARRANTIES ON PRODUCT CHOICE

The perceived risk, information processing, and market signal theories have been advanced in an attempt to explain the impact that warranties have on consumer product choice. Each theory has been investigated and supported by empirical research. The following subsections describe each theory with its corresponding support.

16.3.1 Perceived Risk Theory

The basis of the perceived risk theory is that each purchase decision involves some element of risk [10]. The amount of perceived risk associated with the purchase of a product is thought to be a function of the amount of uncertainty and the magnitude of the consequences that surround the choice of a particular product [11]. Perceived risk may be manifested as financial, performance, social, psychological, or physical risk. The type of product will influence the relative importance of each type of risk that is present in the purchase decision. For example, social and performance risks can be the most important forms of risk present in a choice of a deodorant product, whereas performance and financial risks can be the most important risk elements when considering the purchase of an automobile or major appliance. Performance and financial risks are usually the most important in most purchase decisions and, therefore, are the variables that are studied in most of the investigations of the role that warranties play in the reduction of perceived risk [12].

Warranties may reduce perceived performance risk by providing protection against product defects and premature malfunction of the product during the time that the warranty is in force. Financial risk may be reduced by a warranty protecting the consumer against a large repair bill or having to replace the product during the warranty period.

Research supporting the expected effect of warranties on perceived performance and financial risk has been mixed. There is evidence that "better" warranty terms do have an effect on reducing perceived financial risk, whereas other studies have reported inconclusive results [3,13–15]. In many cases, the mixed results are due to the presence of interaction effects among the variables that are being studied [3,13,14]. Interactions occur when two or more variables together influence consumer perceptions of perceived risk. The effect of each of the variables that interact cannot be separated. For example, a high price and a full warranty combined may reduce performance risk. Yet the impact of each variable independently cannot be assessed.

16.3.2 Information Processing Theory

Information processing theory has been used to explain what information will be used in consumer purchase decisions. In order for information to be used, it must first attract the attention of the consumer. Second, the consumer must comprehend the information. According to the information processing theory, a warranty would only be used in a purchase decision if the consumer sought out the warranty information and then was able to understand it. If either one of these conditions is not met, the warranty could not be used in the decision to purchase a product.

The cognitive response methodology has been used to study how consumers process warranty information. Cognitive response research suggests that a consumer may respond to warranty information in one of three ways. Consumers may generate support arguments (agreement with the terms of the warranty), counterarguments (refutation of the warranty terms), or source derogations (negative images of the warrantor) [16]. In addition, different terms of the warranty, such as the words "full" or "limited," the length of the warranty period, and disclaimers may be processed as separate chunks of information. As some of this information may be easier to process and recall because it is required by law to be available prior to the purchase of consumer products retailing for $15 or more, the consumer may rely more heavily on this information when choosing between products.

The information processing theory predicts a consumer would view a full warranty with a long warranty period as providing a great deal of protection and, thus, generate support arguments that would reinforce the decision to purchase the product. Conversely, a product with a limited warranty with a short warranty period should lead a consumer to generate counterarguments and source derogations because this warranty does not provide as much protection. Research evidence exists which supports these predictions. Specifically, the length of warranty period and whether the warranty was full or limited were manipulated to test their effect on the number of support and counterarguments that a consumer generates [16]. The results of the study indicated that the duration of the warranty significantly impacted the number of support and counterarguments that were generated, whereas the terms full or limited warranty only influenced the number of support arguments. These findings suggest consumers may use more concrete warranty information, such as 3 months or 1 year, rather than the more abstract verbal warranty information (i.e., full versus limited warranty) when deciding which product to purchase.

16.3.3 Market Signal Theory

The market signal theory suggests that consumers use warranties as signals of a product's reliability. Products with warranties that provide better protection in terms of what is covered and for longer periods of time would be viewed as being highly reliable. Conversely, products that have warranties that offer little protection would be viewed by the consumer as less reliable. In essence, the consumer views the relationship between warranty terms and product reliability in terms of an investment. Warranty terms are costly to offer. Therefore, products with better warranty terms also must receive a corresponding investment in product quality. Otherwise, the cost of servicing better warranty terms on unreliable products would be prohibitive [17,18].

Conceptually, the relationship between product reliability and warranty terms can be written as

$$R = f(W, T)$$

where R is the perceived reliability, W is the warranty terms, and T is the length of warranty period.

The above equation suggests that the perceived reliability of the product is a function of its warranty terms (e.g., full versus limited warranties) and the length of time a warranty is in effect. Consumers would perceive a product as being less reliable if warranty terms are limited and the duration it is in effect is small. Conversely, longer-duration warranties with better terms would signal greater product reliability. What has not been tested in the market signal model is whether the consumer weighs warranty terms and length of warranty period differently. Evidence from the research utilizing information processing suggests the more concrete length of warranty information may be weighed more heavily.

Research support for the market signal theory has been mixed. When an aggregate measure of product reliability is used, the theory does not hold [18,19]. When brand-specific measures of reliability, such as a frequency of repair measure, the market signal theory does hold [20,21]. The results of brand-specific research is significant in that it suggests that a warrantor with warranty terms that are superior to the competition should aggressively promote their warranties to influence a consumer's purchase decision.

16.4   IMPACT OF WARRANTIES ON POSTPURCHASE BEHAVIOR

Not only do warranties affect prepurchase product choice but they have an impact on postpurchase behavior as well. Warranties may relieve postpurchase dissonance.* Warranties may reinforce the purchase decision if

---

* Dissonance is created when a product's performance does not meet a consumer's expectations about the product. If a consumer expects a product to be defect-free, then experiences a product defect, negative dissonance is created. However, if a consumer expects a product to have defects, then experiences a defect-free product, positive dissonance is created. In either situation, the consumer attempts to reduce the dissonance. This may be done by searching for additional product information to revise the consumer's expectations about the product.

they are viewed as "better" than the industry standard and the consumer does not have to use the warranty because the product's performance meets or exceeds the consumer's expectations.

Three areas where warranties impact postpurchase behavior, and hence subsequent product choice, are claims behavior, postpurchase product failure behavior, and brand loyalty.

16.4.1 Claims Behavior

Retaining existing consumers is the foundation of defensive marketing strategies that are becoming increasingly more important in highly competitive, low-growth markets. Part of any defensive marketing strategy is successfully satisfying customer complaints concerning warranty service [22]. Although an extensive discussion of warranty claims behavior can be found in Chapter 17, it is sufficient here to suggest how products under warranty that are being serviced may ultimately affect a consumer's subsequent choice of products. Often, customer grievances can be explained by ignorance or inflated expectations that do not result in the customer switching brands [23]. However, in situations where more serious differences exist, the inability to satisfactorily address a customer's warranty complaint may result in a loss of a significant amount of future sales. In these situations, the three theories discussed previously all support the link between the postpurchase evaluation of the choice of a product and the satisfactory service work completed under the terms of the warranty offered on the product.

The perceived risk theory suggests satisfactory completion of any warranty work would reinforce the perception that the warranty could be used to reduce a consumer's perception of the amount of risk that is associated with the purchase of a product with similar levels of risk. Information processing theory indicates that satisfactory completion of warranty work would cause support arguments to be generated. The support arguments would reinforce the purchase decision of that particular product and increase the importance of a warranty in subsequent similar purchase decisions. Finally, the market signal theory suggests that satisfactory completion of warranty work at first would appear to reduce the use of warranties as signals of reliability because the product did not perform beyond the warranty period. However, consumers would view the warranty as a safety device that would protect them from incurring extra costs because the warranty is part of the warrantor's investment in the product.

Just as satisfactory completion of servicing a product under its warranty would make the warranty more important to the consumer, unsatisfactory service would have the opposite effect. For example, if consumers are unduly inconvenienced by loss of time without the product or incurring additional costs not covered by the warranty, they may discount the warranty information in future, similar purchase decisions.

16.4.2 Postpurchase Product Failure Behavior

Whenever a product fails, it has a negative impact on a consumer's choice of that product. The magnitude of the impact will vary depending on when the product fails. Consumers do not expect products to last forever. If a product meets or exceeds the consumer's expectation of the life of the product before it fails, then the likelihood is that the failure would have a minimal impact on future purchase decisions. However, if the product fails relatively soon after the product is purchased, the magnitude of the negative view of the product would be significant. The warranty can help minimize the amount of postpurchase dissonance that a consumer may have by repairing or replacing the product such that the consumer does not view the purchase as being a mistake.

Estimates of the amount of product repairs made under warranty vary from industry to industry. Part of the reason for the variation is that warranties differ in terms of how a consumer would secure repairs under the warranty. For most consumer durable goods (e.g., automobiles or major appliances), the product is returned to a shop for repairs. The consumer is inconvenienced in terms of time, but little else.) In addition, the repairs on most consumer durable goods are relatively expensive. Therefore, the consumer has an economic interest in pursuing service satisfaction under the terms of the warranty. In these cases, the role of the warranty in the future choice of products would tend to be reinforced as an important product attribute.

Conversely, warranties on less expensive consumer goods typically have warranties that require the consumer to do far more when seeking service under the warranty. Often the product must be packaged and shipped to a location dictated by the warrantor. This leads to not only a loss of time but additional expense which the consumer may not view as being justified. Therefore, warranties on these products may cause the consumer to place less weight on the warranty in the future when purchasing similar products.

16.4.3 Brand Loyalty

Warranties may contribute to strengthening brand loyalty. When warranties are viewed as offering extensive coverage, consumers seem to impart that the product is equally reliable. If the product fails prematurely, the consumer may view the warranty as a form of protection. Satisfactory warranty service then is an essential component of ensuring that the warranty reinforces the purchase choice and, ultimately, brand loyalty.

16.5 DIRECTIONS FOR FUTURE RESEARCH

Practitioners are acutely aware of the need to know the role warranties play in influencing consumer product choices. Generally, the issues that are left unresolved have been classified below as those questions of a theoretical nature and questions involving the interaction of warranty terms and other product attributes which influence consumer product choice behavior.

16.5.1 Theoretical Questions Left Unanswered

Research evidence using perceived risk, information processing, and the market signal theory as a theoretical explanation has been mixed at best. Further development and testing of these and other theories that may be advanced is needed in order to predict with any degree of certainty the effect of offering warranties with specific warranty terms on consumer product choices.

Questions also remain concerning what role warranties play in the prepurchase decision as it relates to the Magnuson–Moss Warranty Improvement Act's provision to make warranty information available prior to the purchase. (See Chapter 4 for a complete discussion of this and other provisions of the Magnuson–Moss Act.)

Questions remain concerning the role that warranties play in post-purchase decisions. Some evidence exists on claims behavior [15], but few studies have examined whether warranties become more or less important in purchase decisions over time.

Unanswered questions remain in terms of which situations warranties do play an important role. Perceived risk, information processing, and market signal theories may explain that warranties are important in the purchase of expensive or technical products, but do warranties have the same level of importance in the purchase of common household products such as small appliances? Are there distinct consumer segments that weigh the importance of warranties more than other segments? If so, can these segments be identified and persuaded to choose products that have certain warranty terms.

16.5.2 Impact of Interaction of Warranties and Other Product Attributes on Consumer Product Choice Behavior

Very little evidence exists that gives insight into the interaction that warranties have with other product attributes. Is it possible to partition out the effect that warranties have on product choice relative to, say, promotional claims of quality and reliability? Is it possible that a store's image, the product's, brand name, and information from third-party sources (e.g., ratings from sources like the Consumers Union) can interact with the warranty on a product to communicate a perception of a good or bad choice? Future research needs to explore these and other questions to further understanding how a warranty fits into a consumer's decision-making process for choosing a product.

REFERENCES

1. Aggressive use of warranties is benefiting many concerns, *Wall Street Journal*, April 5, 1984, 33.
2. Kelley, C. A. (1986). Product warranties under the magnuson–Moss Act: A review of literature, in *AMA Summer Educators' Proceedings*, 369–373.
3. Kelley, C. A. (1987). The role of consumer product warranties and other extrinsic cues in reducing perceived risk, in *Developments in Marketing Science, Proceedings of the Academy of Marketing Science Conference*, 319–323.
4. Federal Trade Commission (1985). *An Evaluation of the Warranty Rules of the Magnuson–Moss Act*, U.S. Government Printing Office, Washington, DC.
5. Kangun, N., Cox, K. K., and Staples, W. A. (1976). An exploratory investigation of selected aspects of the magnuson–Moss Warranty Act, *Consumerism: New Challenges for Marketing*, **1**, 3–11.
6. Darden, W., and Rao, C. P. (1977). Satisfaction with repairs under warranty and perceived importance of warranties for appliances, *Consumer Satisfaction, Dissatisfaction, and Compelling Behavior*, **1**, 167–170.
7. Nordstrom, R. D., and Metzner, H. (1976). Warranties: How important as a marketing tool? In *Proceedings of the Southern Marketing Association*, 26–28.
8. Rao, C. P., and Weinrauch, J. D. (1976). Consumers and appliance warranties, *Baylor Business Studies*, **109**, 17–27.
9. Wilkes, R. E., and Wilcox, J. B. (1981). Limited versus full warranties: The retail perspective, *Journal of Consumer Affairs*, **7**, 65–77.
10. Bauer, R. A. (196). Consumer behavior and risk taking, in *Proceedings of the American Marketing Association*, 59–63.
11. Cox, D. F. (1967). *Risk Taking and Information Handling in Consumer Behavior*, Harvard University Press, Boston.
12. Ross, I. (1975). Perceived risk and consumer behavior: A critical review, in *Advances in Consumer Research*, 245–249.
13. Bearden W. O., and Shimp, T. A. (1982). The use of extrinsic cues to facilitate product adoption, *Journal of Marketing Research*, **19**, 229–239.
14. Shimp, T. A., and Bearden, W. D. (1982). Warranty and other extrinsic cue effects on consumers' risk perceptions, *Journal of Consumer Research*, **9**, 38–46.
15. Armstrong, G. C., Kendall C. L., and Russ, F. (1975). Applications of consumer information processing research to public policy issues, *Communication Research*, **3**, 232–245.

16. Kelley, C. A., and Swartz, T. A. (1984). An empirical test of warranty label and duration effects on consumer decision making: An information processing perspective, in *AMA Summer Educators' Proceedings*, 309–313.
17. Spence, M. (1977). Consumer misperceptions, product failure, and producer liability, *The Review of Economic Studies*, **44**, 561–572.
18. Priest, G. (1981). The theory of the consumer product warranty, *Yale Law Journal*, **90**, 1297–1352.
19. Gerner, J., and Bryant, W. K. (1981). Appliance warranties as a market signal? *Journal of Consumer Affairs*, **40**, 75–86.
20. Wiener, J. (1985). Are warranties accurate signals of product reliability? *Journal of Consumer Research*, **12**, 245–250.
21. Kelley, C. A. (1988). An investigation of consumer product warranties as market signals of product reliability, *Journal of the Academy of Marketing Science*, **16**, 72–78.
22. Fornell, C., and Wernerfelt, B. 1987. Defensive marketing strategy by customer complaint management: A theoretical analysis, *Journal of Marketing Research*, **24**, 337–346.
23. Steele, E. H. (1977). Two approaches to contemporary dispute behavior and consumer problems, *Law and Society*, **11**, 667–675.

17

Warranty and Consumer Behavior: Warranty Execution

Jayprakash G. Patankar

*The University of Akron
Akron, Ohio*

Amitava Mitra

*Auburn University
Auburn, Alabama*

17.1 INTRODUCTION

A survey of the business and trade publications reveals that warranty costs are very important to top management. U.S. Intec Inc. of Port Arthur, TX paid over $1 million and $1.2 million in warranty claims for the years 1990 and 1991, respectively. Plymouth Rubber Co. Inc. of Canton, MA had expenditures charged to warranty reserves of over $1.7 million in 1991. Ford Motor Co., citing problems that may occur with engine mounts on cars equipped with a 3.8-liter engine, extended warranty coverage on 550,000 1988–1991 models of Ford Taurus, Mercury Sable, and Lincoln Continental sedans [1]. Warranty claim procedures vary from manufacturer to manufacturer. In motor vehicle fleets [2], for example, the process between warranty filing and warranty reimbursement is one of the largest outlays of capital any fleet faces in a year.

Chapter 1 presents the fundamentals on the various types of warranties that are typically available. Examples of statements of warranties are provided. Such statements provide the user with information on the redemption or refund, repair, or replacement policy of the company. In Chapter 7, mathematical techniques that can be used for warranty analysis are discussed. Such methods will assist the management of the company in estimating the proportion of the product that will fail within the warranty period. This information can then serve as a basis for projecting financial obligations to meet warranty obligations.

An objective of this chapter is to consider the effect of warranty execution on warranty reserve costs. In the estimation of warranty reserves, it has historically been assumed [3–6] that the warranty is executed at the time of product failure if it is within the warranty time. This assumption, however, may not always be valid, as all customers may not exercise the warranty even if the product fails during the warranty period [7]. One can mention a number of instances when a warranty will not be exercised even though it is possible to do so. For example, a customer may develop a dissatisfaction for the product and prefer to switch brands rather than to exercise the warranty. In another instance, a customer may purchase some other product cheaper than the cost of repurchase of the same product using the warranty right. Instances such as lost warranties, customer relocation, the form of warranty reimbursement, the type of rebate plan offered, and so on, will also prevent or dissuade a customer from exercising the warranty.

In this chapter, customer behavior in exercising warranties is described by a warranty execution function. The shape of the warranty execution function is dependent on the product, the behavioral attitudes of the individual, economies of exercising warranty, change in product preference, form of reimbursement, and other such factors. Details on the warranty execution function are provided in Section 17.2 on model development. The implication of such warranty execution functions is more consistent with consumer behavior than the assumption that all customers will exercise their warranty in case of product failure within the warranty time. The mathematical equations describing the computation of the expected present value of the warranty reserve costs, based on a selected warranty execution function and a failure distribution, are also found in Section 17.2.

Section 17.3 provides applications of the models developed to illustrate their role in providing meaningful insights in decision making. Further, examples are provided for usage of the models on some common appliances. The applications provide an analysis of the impact of the warranty execution function as well as the sensitivity of the expected warranty reserve costs on some of the model parameters.

This chapter considers the linear pro-rata rebate plan. For a linear pro-rata plan, the amount paid to the consumer varies linearly from c, the selling price of the product, to zero, based on the time at which the product failure occurs. If failure of the product occurs after the warranty time (w), then the rebate paid by the manufacturer is zero. The failure distributions considered in the chapter are the exponential and Weibull distributions. Several different values of the parameters (α and β) of the Weibull distribution are considered, ensuring a broad spectrum of selected shapes of the distribution. Additionally, for the purpose of validation with some previous studies, the values of α and β were chosen to match those considered by Thomas [6].

17.2 MODEL DEVELOPMENT

Although it is true that the incorporation of selective execution of warranties by the customer, based on some of the reasons previously described in Section 17.1, will reduce the expected warranty reserve costs, it is of interest to determine the relative degree of reduction. Such reductions will be influenced by the behavior patterns of the consumer. To develop a model that incorporates consumer behavior through a warranty execution function, the following notation is introduced:

$$
\begin{aligned}
c &= \text{unit product price including warranty cost} \\
w &= \text{duration of warranty period} \\
t &= \text{time of product failure} \\
\lambda &= \text{product failure rate} \\
m &= \text{mean life of product} = 1/\lambda \\
N &= \text{production quantity for warranty reserve cost deter-} \\
  &\quad\ \text{mination (using, say, annual production)} \\
s(t) &= \text{rebate function at time } t \\
R &= \text{warranty reserve cost for production quantity } N \\
g(t) &= \text{warranty execution function at time } t \\
E[PV(R)] &= \text{expected present value of warranty reserve cost for} \\
  &\quad\ \text{the production quantity } N \\
r &= E[PV(R)]/N, \text{ the expected present value of warranty} \\
  &\quad\ \text{reserve cost per unit product} \\
\theta &= \text{the firm's discount rate} \\
\phi &= \text{inflation rate}
\end{aligned}
$$

17.2.1 Warranty Execution Function

As mentioned previously, there are a variety of reasons which may prevent the full execution of warranty. The form of the weight function, which describes warranty execution, could be influenced by a variety of factors. The warranty time, the type of rebate plan, warranty attrition due to costs of executing the warranty, the product class as to whether they are expensive, and the form of reimbursement as to whether the warranty provides cash, repair, or replacement, are examples of factors that influence the warranty execution function. We provide some insight as to how some of these factors would impact the weight function associated with warranty execution.

Figure 17.1 provides some examples of the warranty execution weight function. Four functions are described. The decision-maker will select an appropriate execution function based on factors that are relevant to his/her setting. Consider the first warranty execution function (WEF 1). The weight is 1 up to a certain time w_1 ($w_1 \leq w$), and then decreases linearly to 0. Note that if $w_1 = w$, we have full warranty execution. Such an execution function may be used with a linear pro-rata rebate plan to model warranty attrition, say, due to the small amount of rebate obtained for a linear pro-rata plan after an initial period ($w > w_1$). Also, the cost of exercising warranty along with reduced rebate may not provide enough of an incentive and the chance of execution diminishes as the time of

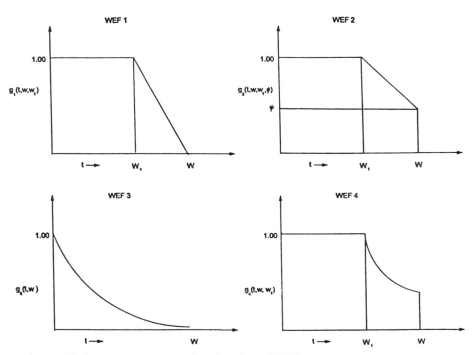

Figure 17.1   Warranty execution functions (WEF).

failure approaches the warranty time (w). The warranty execution function is given by

$$g_1(t, w_1, w) = \begin{cases} 1, & 0 \le t \le w_1 \\ 1 - \dfrac{(t - w_1)}{w - w_1}, & w_1 \le t \le w \\ 0, & t > w \end{cases} \tag{17.1}$$

The value of w_1 could be influenced by factors such as customer attitude, market characteristics that define present and future competitive offerings, amount of rebate to be obtained, and others. In the problems chosen for analysis, w_1 is selected as one-half of w. The values of w selected were 0.5, 1.0, 1.5, and 2.0.

The second form of warranty execution function is shown under WEF 2 in Figure 17.1. This is similar to WEF 1, in that the weight is 1.0 up to a certain time w_1. However, from the time period w_1 to w, the weight decreases linearly to some value, ψ, instead of zero. It implies that a nonzero proportion of the customers may not execute warranty when the product fails just prior to the warranty time. Reasons for such behavior could be the effort and cost associated with the execution of the warranty, dissatisfaction with the product, and the reduced utility in possessing the product due to new product innovations. The warranty execution function is given by

$$g_2(t, w_1, w, \psi) = \begin{cases} 1, & 0 \le t \le w_1 \\ \dfrac{w - \psi w_1}{w - w_1} - \dfrac{(1 - \psi)t}{w - w_1}, & w_1 \le t \le w \\ 0, & t > w \end{cases} \tag{17.2}$$

In the problems chosen for analysis, w_1 is selected as one-half of w, and the value of the parameter, ψ, is selected as 0.2. If a linear rebate plan were used, where just prior to the failure time being w when hardly any rebate would be received, the case may be stronger when dissatisfaction with the product may cause a certain segment of the customers to refrain from exercising warranty. Certainly, if warranty payment does not involve cash but rather repair of the existing product, product preference for a competitive offering could lead the customer to not exercise warranty.

The third form of the warranty execution function is shown under WEF 3 in Figure 17.1. In this case, the warranty weight function starts decreasing exponentially from time 0. It is different from WEF 1 and WEF 2 in that there is no minimum period defined by w_1, such that full execution takes place for product failure within the period of 0 to w_1, as in WEF 1 and

WEF 2. Dissatisfaction for the product or attractiveness to a competitive product develops right away. The longer the time to failure, the more the likelihood of the customer not exercising warranty and thereby switching brands, say. The warranty execution function is given by

$$g_3(t, w) = \begin{cases} e^{-t/\delta}, & 0 \le t \le w \\ 0, & t > w \end{cases} \tag{17.3}$$

The value of the parameter δ in (17.3) is influenced by factors such as the rate of warranty attrition due to some of the reasons discussed previously. The larger the value of δ, the slower the rate of attrition. In the examples considered, the values of δ were selected to be 0.5, 1.0, and 1.5. This allows us to gauge the effect of the parameter δ on the expected warranty reserve costs. The effect of the market characteristics, such as competitive offerings over a period of time, can be modeled by such an execution function. A manager could obtain data on the market share of the company versus that of a competitor over a period of time and estimate an attrition rate. This could subsequently be used to construct the warranty execution function.

The fourth form of the warranty execution function is shown under WEF 4 in Figure 17.1. Here, the warranty weight function has a value of 1.0 up to a certain time w_1. For this segment of the failure time ($t \le w_1$), it is similar to WEF 1 and WEF 2. For the period after w_1, the warranty execution function decreases exponentially, similar to WEF 3. The warranty execution function is given by

$$g_4(t, w_1, w) = \begin{cases} 1, & 0 \le t \le w_1 \\ \exp\left(-\dfrac{(t - w)}{\delta}\right), & w_1 \le t \le w \\ 0, & t > w \end{cases} \tag{17.4}$$

For some of the reasons described under WEF 1 and WEF 2, the justification for full execution up to time w_1 can be made. After time w_1, customers are more prone to switch brands or lose interest in the product, and the attrition rate increases. The value of δ will be influenced by similar factors as described under WEF 3. One particular situation that could be modeled by this execution function is the issue of product class and warranty attrition due to product dissatisfaction or preference to switch brands. For expensive products, which offer a linear pro-rata rebate plan, it is possible that for failure within the period $(0, w_1)$, full execution would take place. The utility associated with either getting a cash refund or having the product replaced or repaired far exceeds the discomforts and costs of executing the warranty. Now, after time w_1, because of the decreased cash rebate (if the plan offers that), the customer may be less inclined to exercise the warranty. A large value of δ would imply a slower rate of attrition. In the analysis, the value of w_1 is selected as one-half of w, and the values of δ are 0.5, 1.0, and 1.5, respectively.

It is possible for the form of reimbursement of the warranty to influence the warranty execution function. Depending on whether the warranty gives cash, repairs the product, or provides a replacement for the product, the execution function could vary. For instance, if the warranty provides cash, candidates WEF 1, WEF 2, and WEF 4 might be feasible execution functions. Here, more customers would tend to exercise their warranty in the early stages. Warranty functions WEF 2, WEF 3, and WEF 4 might be appropriate if the warranty calls for repair of the product. In this situation, some proportion of the customers would choose to not exercise their warranty because of product dissatisfaction and brand switching.

17.2.2 Product Failure Distributions

The exponential distribution is used as one of the failure distributions. It is used to describe the period of useful life of products when the failure rate is constant. For electronic components, this might be a feasible distribution. The failure density function for the exponential distribution is given by

$$f(t) = \lambda e^{-\lambda t}$$
$$= \left(\frac{1}{m}\right) e^{-t/m} \tag{17.5}$$

where λ denotes the failure rate and m is the mean life of the product. In the analysis, the values of λ are selected to be 0.5, 1.11, and 2. The value of λ = 1.11 yields a mean life of 0.9. This value was selected as a guideline for comparison purposes as used in other studies [6].

For mechanical components, however, the units may become more prone to failure with increasing time. This may happen because of deterioration and aging of the product. The failure rate for these products would increase with time. The Weibull distribution is used to model these situations. Its density function is given by

$$f(t) = \alpha\beta t^{\beta-1} \exp(-\alpha t^\beta), \quad \alpha > 0, 1 \le \beta < \infty \tag{17.6}$$

where β is a shape parameter and α is a scale parameter. The values of α and β can be varied to produce a variety of shapes of the failure distribution. As the value of β increases from 1, the failure rate increases also. In the study, in order to explore the effect of the failure distribution parameters on the expected warranty reserve costs, the values of α and β were each selected to be 1.0, 1.5, 2.0, and 3.0. Each combination of (α, β) yields a unique failure distribution. These combinations were chosen to provide a comparison of our results with the case when there is full execution.

17.2.3 Warranty Rebate Plan

The rebate function considered in this chapter is the linear pro-rata plan, which is given by

$$
s(t) = \begin{cases} c\left(1 - \dfrac{t}{w}\right), & 0 \le t \le w \\ 0, & t > w \end{cases} \tag{17.7}
$$

17.2.4 Expected Present Value of Warranty Reserve Costs

For a given product failure distribution and a warranty execution function, the expected present value of the warranty reserve costs are found. In determining present values, a continuous form of discounting is used. The expected present value of warranty reserve costs, in general, is calculated from the expression

$$
E[PV(R)] = N \int_0^w s(t)f(t)g(t)e^{-(\theta + \phi)t}\, dt \tag{17.8}
$$

The warranty reserve cost per unit (r) can be estimated by determining $E[PV(R)]/N$, from which the warranty reserve cost per unit product price, r/c, can be found.

We now provide expressions for $E[PV(R)]$ for some selected combinations of the failure distribution and the warranty execution function. For some of these, it is difficult to arrive at closed-form expressions. Thus, for the analysis section, numerical integration procedures are used. As a baseline case for comparison purposes, $E[PV(R)]$ is found for full execution, using $g(t) = 1$ for $0 \le t \le w$. The value of the firm's discount rate plus the inflation rate $(\theta + \phi)$ is selected to be 0.075.

Full Execution

When the failure distribution is exponential, under full execution, we have, after simplification [using 17.8],

$$
E[PV(R)] = \frac{Ncz}{m}\left(1 - \frac{z}{w}(1 - e^{-w/z})\right) \tag{17.9}
$$

where

$$\frac{1}{z} = \frac{1}{m} + \theta + \phi \tag{17.10}$$

Consequently, we get the warranty reserve cost per unit price as

$$\frac{r}{c} = \frac{z}{m}\left(1 - \frac{z}{w}(1 - e^{-w/z})\right) \tag{17.11}$$

If the failure distribution is assumed to be Weibull, as given by (17.6), closed-form expression for r/c under full execution is not feasible. Consequently, for this case, numerical integration methods are used.

Warranty Execution Function 1

Suppose the warranty execution function is WEF 1 given by (17.1). For a linear pro-rata rebate plan and exponential failure distribution, the expected present value of warranty reserve costs is found as

$$E[PV(R)] = \int_0^{w_1} Nc\left(1 - \frac{t}{w}\right)\left(\frac{1}{m}\right) e^{-t/m} e^{-(\theta + \phi)t} dt$$

$$+ \int_{w_1}^{w} Nc\left(1 - \frac{t}{w}\right)\left(\frac{1}{m}\right) e^{-t/m}\left(1 - \frac{t - w_1}{w - w_1}\right) e^{-(\theta + \phi)t} dt$$

$$= \frac{Ncz}{m}[1 - (z/w_1)(1 - e^{-w_1/z})]$$

$$+ \frac{Ncz}{wm(w - w_1)}[2z^2(e^{-w_1/z} - e^{-w/z})$$

$$+ (w - w_1)(w - w_1 - 2z)e^{-w_1/z}] \tag{17.12}$$

The value of the warranty reserve cost per unit price (r/c) is obtained by dividing the right-hand side of (17.12) by Nc.

When the failure distribution is Weibull, the expected present value of the warranty reserve costs is given by

$$E[PV(R)] = Nc\,\alpha\beta\left[\int_0^{w_1}\left(1 - \frac{t}{w}\right) t^{\beta - 1} e^{-(\theta + \phi)t - \alpha t^\beta} dt\right.$$

$$\left. + \int_{w_1}^{w}\left(1 - \frac{t}{w}\right)\left(1 - \frac{(t - w_1)}{w - w_1}\right) t^{\beta - 1} e^{-(\theta + \phi)t - \alpha t^\beta} dt\right] \tag{17.13}$$

Numerical integration methods are used to compete (17.13) and then determine r/c.

Warranty Execution Function 2

For an exponential failure distribution, using WEF 2, the expected present value of warranty reserve costs after simplification is given by

$$E[PV(R)] = \frac{Ncz}{m}\left[1 - \frac{1}{j} + \left(\frac{1}{j}\right)(1 + bd - 4bz)e^{-j} + \left(\frac{bd}{j}\right)e^{-k}\right]$$

(17.14)

where z is defined by (17.10), $j = w/z$, $k = w_1/z$, $b = (1 - \psi)(w - w_1)$, and $d = w_1 - w + 2z$.

When the failure distribution is Weibull, when using WEF 2, a closed-form expression for $E[PV(R)]$ is not feasible. Hence, numerical integration methods are used for computational purposes.

Warranty Execution Function 3

For an exponential failure distribution, using WEF 3, the expected present value of warranty reserve costs, after simplification, is given by

$$E[PV(R)] = \frac{Ncz}{m}\left(1 - \left(\frac{z_1}{w}\right)(1 - e^{-w/z})\right)$$

(17.15)

where

$$\frac{1}{z_1} = \frac{1}{m} + \frac{1}{\delta} + \theta + \phi$$

(17.16)

When the failure distribution is Weibull, when using WEF 3, a closed-form expression for $E[PV(R)]$ is not feasible. Numerical integration methods are used in computations.

Warranty Execution Function 4

For an exponential failure distribution, using WEF 4, the expected present value of warranty reserve costs, after simplification, is given by

$$E[PV(R)] = \frac{Ncz}{m}\left[\left(1 + \frac{z}{w}\right) - e^{-w_1/z}\left(1 - \frac{w_1}{w} - \frac{z}{w}\right)\right] + \frac{Ncz_1}{m}e^{-w_1/\delta}$$

$$\times \left[e^{-w_1/z_1}\left(\frac{1}{z_1} - \frac{w_1}{w} - \frac{z_1}{w}\right) - e^{-w/z_1}\left(\frac{1}{z_1} - 1 - \frac{z_1}{w}\right)\right]$$

(17.17)

where z and z_1 are given by (17.10) and (17.16), respectively.

When the failure distribution is Weibull and using WEF 4, a closed-form expression for $E[PV(R)]$ is not feasible. Numerical integration methods are used in computations.

17.3 ANALYSIS AND RESULTS

The expected present value of warranty reserve costs per unit price are calculated under the different failure distributions, using the different warranty execution functions.

For the selected values of the parameters (α, β) of the Weibull distribution, Table 17.1 shows the mean and the variance of the time to failure. Note that the mean time varies between 1.0 and 0.333 time units, thus covering a wide range. For a given value of β, an increase in α causes a decrease in the mean life. A wide range of the variance is also accomplished through the choice of α and β. Observe that for a given value of α, an increase in the value of β causes a decrease in the variability. From Table 17.1, we find that the range of variance of the product life is between 1.0 and 0.0506. Of interest to the consumer may be the coefficient of variation, which is the ratio of the mean to the standard deviation. This provides an idea of the average life of the product in units of standard deviation. For the parameter values shown in Table 17.1, the coefficient of variation varies between 1.0 and 2.75, which is a fairly broad spectrum.

17.3.1 Impact of Partial Execution of Warranty

Full-warranty execution has been studied in the past. Warranty cost models where full execution is assumed may be found in Chapters 10–12. Chapter 10 deals with the basic free-replacement warranty; Chapter 11 introduces the basic pro-rata warranty, and Chapter 12 exposes some combination warranties. This chapter introduces the notion of partial redemption of warranties and explores its impact on the expected present value of warranty reserve costs per unit price (r/c). Tables 17.2 and 17.3 show values of r/c for the various execution functions, for the discount and inflation rate $(\theta + \phi) = 0.075$; the failure distribution being Weibull with parameters $\alpha = 1.0, 1.5, 2.0,$ and 3.0 and $\beta = 1.0, 1.5, 2.0,$ and 3.0, and warranty time $(w) = 1$ and 2, respectively.

Obviously, the values of r/c are less with warranty execution function WEF 1 than with full execution. The degree of reduction, which is of interest, can be found on comparison of the corresponding r/c values. For example, from Table 17.2, for $(\alpha, \beta) = (1.0, 1.0)$, $w = 1.0$, and linear pro-rata rebate plan, for full execution, the value of r/c is 0.3602. When using WEF 1, under similar conditions, the value of r/c is 0.2644, indicating

Table 17.1 Mean and Variance of Weibull Failure Distribution for Various Parameter Values

Weibull parameters (α, β)	Mean	Variance
1.0, 1.0	1.0000	1.0000
1.0, 1.5	0.9027	0.3757
1.0, 2.0	0.8862	0.2146
1.0, 3.0	0.8930	0.1053
1.5, 1.0	0.6667	0.4444
1.5, 1.5	0.6889	0.2188
1.5, 2.0	0.7236	0.1431
1.5, 3.0	0.7801	0.0803
2.0, 1.0	0.5000	0.2500
2.0, 1.5	0.5687	0.1491
2.0, 2.0	0.6267	0.1072
2.0, 3.0	0.7088	0.0663
3.0, 1.0	0.3333	0.1111
3.0, 1.5	0.4340	0.0868
3.0, 2.0	0.5117	0.0715
3.0, 3.0	0.6192	0.0506

Table 17.2 Expected Present Value of Warranty Reserve Costs per Unit Price (r/c) for Various Execution Functions, Failure Distribution Weibull, $w = 1$, $\theta + \phi = 0.075$

Weibull parameters (α, β)	Full execution	WEF 1	WEF 2 $\psi = 0.2$	WEF 3 $\delta = 0.5$	WEF 3 $\delta = 1.0$	WEF 3 $\delta = 1.5$	WEF 4 $\delta = 0.5$	WEF 4 $\delta = 1.0$	WEF 4 $\delta = 1.5$
1.0, 1.0	0.3602	0.2644	0.3080	0.2243	0.2788	0.3023	0.2325	0.2831	0.3051
1.0, 1.5	0.2918	0.1841	0.2229	0.1509	0.2054	0.2297	0.1541	0.2070	0.2308
1.0, 2.0	0.2447	0.1353	0.1668	0.1089	0.1597	0.1832	0.1100	0.1603	0.1836
1.0, 3.0	0.1846	0.0819	0.1021	0.0655	0.1079	0.1285	0.0656	0.1079	0.1285
1.5, 1.0	0.4729	0.3582	0.4147	0.3055	0.3735	0.4024	0.3173	0.3796	0.4065
1.5, 1.5	0.3866	0.2528	0.3052	0.2076	0.2777	0.3086	0.2124	0.2802	0.3103
1.5, 2.0	0.3260	0.1871	0.2304	0.1504	0.2169	0.2473	0.1521	0.2178	0.2479
1.5, 3.0	0.2473	0.1143	0.1425	0.0908	0.1471	0.1742	0.0910	0.1472	0.1743
2.0, 1.0	0.5577	0.4345	0.5001	0.3724	0.4487	0.4806	0.3877	0.4566	0.4859
2.0, 1.5	0.4598	0.3103	0.3735	0.2555	0.3365	0.3718	0.2618	0.3392	0.3740
2.0, 2.0	0.3897	0.2313	0.2845	0.1859	0.2640	0.2993	0.1882	0.2652	0.3001
2.0, 3.0	0.2974	0.1423	0.1774	0.1125	0.1797	0.2117	0.1127	0.1798	0.2117
3.0, 1.0	0.6730	0.5493	0.6252	0.4754	0.5586	0.5926	0.4969	0.5697	0.6000
3.0, 1.5	0.5634	0.4009	0.4799	0.3321	0.4258	0.4657	0.3414	0.4306	0.4690
3.0, 2.0	0.4818	0.3027	0.3714	0.2437	0.3369	0.3781	0.2471	0.3387	0.3793
3.0, 3.0	0.3715	0.1890	0.2355	0.1483	0.2309	0.2696	0.1487	0.2312	0.2698

Table 17.3 Expected Present Value of Warranty Reserve Costs per Unit Price (r/c) for Various Execution Functions, Failure Distribution Weibull, $w = 2.0$, $\theta + \phi = 0.075$

Weibull parameters (α, β)	Full execution	WEF 1	WEF 2 $\psi = 0.2$	WEF 3			WEF 4		
				$\delta = 0.5$	$\delta = 1.0$	$\delta = 1.5$	$\delta = 0.5$	$\delta = 1.0$	$\delta = 1.5$
1.0, 1.0	0.5480	0.4286	0.4929	0.2724	0.3676	0.4144	0.3011	0.3829	0.4248
1.0, 1.5	0.5364	0.3802	0.4552	0.2061	0.3148	0.3708	0.2224	0.3235	0.3768
1.0, 2.0	0.5306	0.3498	0.4282	0.1676	0.2828	0.3443	0.1758	0.2872	0.3473
1.0, 3.0	0.5215	0.3158	0.3929	0.1274	0.2472	0.3139	0.1293	0.2482	0.3146
1.5, 1.0	0.6630	0.5429	0.6173	0.3609	0.4701	0.5216	0.4013	0.4915	0.5362
1.5, 1.5	0.6323	0.4748	0.5642	0.2716	0.3967	0.4583	0.2954	0.4094	0.4669
1.5, 2.0	0.6099	0.4291	0.5230	0.2183	0.3498	0.4167	0.2305	0.3563	0.4212
1.5, 3.0	0.5783	0.3741	0.4650	0.1610	0.2950	0.3663	0.1638	0.2964	0.3673
2.0, 1.0	0.7353	0.6237	0.7016	0.4306	0.5449	0.5970	0.4810	0.5716	0.6152
2.0, 1.5	0.6917	0.5420	0.6393	0.3237	0.4566	0.5197	0.3547	0.4732	0.5310
2.0, 2.0	0.6592	0.4851	0.5886	0.2585	0.3987	0.4675	0.2746	0.4074	0.4734
2.0, 3.0	0.6147	0.4151	0.5153	0.1874	0.3298	0.4032	0.1911	0.3317	0.4045
3.0, 1.0	0.8173	0.7286	0.8038	0.5329	0.6459	0.6947	0.5997	0.6813	0.7188
3.0, 1.5	0.7612	0.6316	0.7345	0.4025	0.5398	0.6015	0.4464	0.5632	0.6175
3.0, 2.0	0.7189	0.5607	0.6747	0.3199	0.4674	0.5362	0.3435	0.4801	0.5449
3.0, 3.0	0.6610	0.4713	0.5837	0.2277	0.3793	0.4541	0.2333	0.3823	0.4561

a reduction in the warranty reserve costs per unit price of about 27%. Such comparisons can be done for other values as well and they indicate the relative effect of the various warranty execution functions. Under the same conditions as before, the reduction in warranty reserve costs per unit price is about 14.5%, 38%, and 35% for WEF 2, WEF 3 (with $\delta = 0.5$), and WEF 4 (with $\delta = 0.5$), respectively, compared to full execution. As the warranty time increases, the impact on such reductions are dampened for WEF 1 and WEF 2. For example, from Table 17.3 where $w = 2$, under the same conditions where (α, β) = (1.0, 1.0), the corresponding reductions, compared to full execution, are about 21.8% and 10% for WEF 1 and WEF 2, respectively. For WEF 3 and WEF 4, however, based on the attrition rate, the reduction could increase. For instance, under similar conditions as before, the reductions in expected warranty reserve costs per unit price are about 50% and 45% for WEF 3 (with $\delta = 0.5$) and WEF 4 (with $\delta = 0.5$), respectively, compared to full execution. This implies that management could plan their budgets accordingly, if they had some knowledge of the attrition rate.

From Tables 17.2 and 17.3, the relative impact of the failure distribution on the warranty reserve costs can also be observed. As the parameter

α increases, the mean time to failure decreases and so do the expected warranty reserve costs. Note that for a linear pro-rata rebate plan, using WEF 1 and a warranty time of 1.0, as (α, β) changes from $(1.0, 1.0)$ to $(3.0, 1.0)$, the value of r/c increases by about 70%. For $w = 2$, for the same change in the values of (α, β), the value of r/c increases by about 108% under WEF 1. These values indicate the sensitivity of the failure distribution on r/c.

Tables 17.2 and 17.3 also provide us with information on the impact of the attrition rate on warranty reserve costs. When $w = 2$ and $(\alpha, \beta) = (2.0, 1.0)$, using WEF 3, the reduction in r/c compared to full execution is about 41%, 26%, and 19% for values of $\delta = 0.5$, 1.0, and 1.5, respectively. Under WEF 4, we would expect these reductions to be smaller. As pointed out previously, the parameter δ in WEF 3 and WEF 4 can be interpreted as a warranty attrition parameter. As δ increases, there is less attrition, and so the expected warranty reserve costs per unit price increase.

Another criterion which may help managers decide on the type of execution function to select is dependent on whether the warranty offers cash, repairs the product, or replaces it. If cash is offered, it may induce warranty execution function WEF 1 or WEF 2 or WEF 4, for a linear pro-rata rebate. If the product is repaired, an appropriate warranty execution form may be WEF 3. For example, the relative impact of the execution functions WEF 1 and WEF 3, based on the form of reimbursement being either cash or repair, respectively, can be observed from Tables 17.2 and 17.3. From Table 17.3, for $(\alpha, \beta) = (1.5, 1.0)$, $w = 2$ and $\delta = 0.5$, a decrease in r/c of about 33.5% is observed in going from WEF 1 to WEF 3. This may provide the manager with useful information on the amount of money to set aside for warranty expenses, depending on the form in which the rebate is offered. A possible procedure for a manager to estimate the attrition rate could be to obtain estimates of the proportion of failures and the proportion who file for warranty claims at various points in time. Parameter estimation concepts of Chapter 8 might be helpful.

17.3.2   Impact on Related Issues

The partial redemption of warranties, as modeled through the usage of various warranty execution functions, has a bearing on several issues. Although some insight on these has been gained through the previous discussion and analysis of results, we delineate some of these exclusively.

Bias Toward Longer Warranties

Because partial warranty execution causes a reduction in the warranty reserve costs for the same parameter values, compared to full execution, an alternative might be to determine the warranty time (w) that could be offered under partial redemption, with the expected warranty reserve costs per unit price remaining the same. If a linear pro-rata plan is used, the $E[PV(R)]$ for full execution would be found from (17.9). Suppose the warranty execution function used is WEF 1. Then, the value of w which would yield the same $E[PV(R)]$ as for full execution would be found from (17.12). Consider, for example, the situation when a linear pro-rata plan is used, failure distribution being Weibull with (α, β) = (1.0, 1.0), and under full execution the warranty time (w) is 1.0. From Table 17.3, under the same conditions, using WEF 3 (with δ = 1.0), an equivalent warranty time, while keeping the warranty reserve costs per unit price about the same, would be about 2. This indicates an increase of about 100% in the warranty time. On the other hand, under warranty execution plan WEF 3 with δ = 0.5, the warranty could be extended to more than 2.0, while maintaining the same amount of warranty reserve costs per unit. This represents an increase of over 100% in the warranty time. Thus, if the rate of attrition is high, the manufacturer could create a more lucrative warranty which would almost double the warranty time, at hardly any increased costs per unit price.

Impact of Attrition

The execution functions WEF 3 and WEF 4 can model warranty attrition by varying the parameter δ. Small values of δ imply high attrition. From Table 17.2, we find that for (α, β) = (2.0, 1.0) and w = 1.0, using WEF 3 as δ changes from 1.5 to 0.5, the warranty reserve costs per unit price decrease by about 22.5%. Such information will help manufacturers to allocate funds in a rational manner.

Impact of Product Class

The relative cost of the product, or the cost category in which it is placed, could have an impact on warranty execution and, consequently, on the warranty reserve costs. For instance, for very expensive products, where the chances of execution are initially high, warranty execution functions WEF 1, WEF 2, or WEF 4 might be feasible. For very expensive products, the attrition rate might be low. So, if using WEF 3, one with a large value of δ (say 1.5 rather than 0.5) might better model the situation. Thus, if a linear pro-rata rebate plan were to be considered for a Weibull failure distribution with (α, β) = (3.0, 1.0) and w = 1, as δ changes from 0.5 (for less expensive products) to 1.5 (for expensive products), the increase in warranty reserve costs per unit price is approximately 25%. This could serve as a motivational guide to further lower costs of production in order to keep the selling price competitive.

17.3.3 Applications to Some Commonly Used Appliances

The average product life (m) and the warranty period (w) is given for some commonly used appliances such as freezers, refrigerators, dryers, electric ranges, color televisions, washers, and automobiles by Priest [8]. We demonstrate the usage of the various warranty execution functions and their impact on the warranty reserve costs per unit price to these specific examples. As discussed previously, factors such as the product class, length of the warranty, and attrition rate influence the type of warranty execution function. In Table 17.4, the warranty execution function chosen for each product is shown. A linear pro-rata rebate plan and an exponential failure distribution are assumed. As only estimates of the mean time to failure are provided [8], as a first approximation one could assume the failure time to be exponentially distributed, with the failure rate obtained from the estimate of the mean time. Because the exponential distribution is a single parameter distribution, it is feasible to do so. With other distributions, such as the Weibull or Gamma distributions, involving more than one parameter, additional information would be needed to estimate the parameters. For further details, the reader is referred to Chapter 8 in which model selection and parameter estimation are discussed.

The expected present value of the warranty reserve cost per dollar sales is found for each situation and the results are shown in Table 17.4. The values in the table provide an idea of the amount of money that would

Table 17.4. Expected Present Value of Warranty Reserve Costs per Dollar Sales for Some Common Appliances, Failure Distribution Exponential, Linear Pro-Rata Rebate Plan, $\theta + \phi = 0.075$

Appliance	Product mean life in years m	Warranty time in years w	Warranty execution function type	r/c
Freezers	18	5	WEF 1	0.078
Refrigerator	15	1	WEF 1	0.021
Dryers	15	1	WEF 4 ($\delta = 1.5$)	0.023
Electric ranges	14	1	WEF 4 ($\delta = 1.5$)	0.024
Color televisions	12	1	WEF 3	0.032
Washers	12	2	WEF 4 ($\delta = 1.5$)	0.054
Automobiles	9.4	1	WEF 1	0.034

be set aside for the average warranty reserve costs for each sales dollar for these various appliances. For instance, for a freezer, which has a mean life of 18 years and the warranty is for 5 years, using WEF 1, on the average, for every dollar of sales about 7.8 cents should be set aside for warranty reserve costs. For a color television, with a mean life of 12 years and warranty of 1 year, using WEF 3 assuming attrition takes place soon after sales, about 3.2 cents, on the average, for every dollar of sales is needed of warranty reserve costs. Such information is of interest to management in planning and budgeting for costs.

17.3.4 Application of Warranty Cost Estimates in Pricing

The information on expected present value of warranty reserve costs per sales dollar can also be used to determine the price markup needed in the selling price to cover warranty expenses. If we define

c' = selling price per unit excluding warranty cost
a = proportion of selling price c, that will cover warranty cost, where c has already been defined as the selling price per unit including warranty cost, then

$$c = c' + ac \quad \text{or} \quad c = \frac{c'}{1 - a} \tag{17.18}$$

Thus, for a refrigerator, assuming a mean life of 15 years and a warranty time of 1 year, if we assume warranty execution function WEF 1 and the selling price excluding warranty costs (c') is \$700, then the selling price including warranty costs for a linear pro-rata rebate plan is given by

$$c = \frac{\$700}{1 - 0.021} = \$715.15$$

17.4 CONCLUSIONS

This chapter considers the estimation of warranty costs under a linear pro-rata rebate plan. Past research has assumed full execution of warranty, in case of product failure within the warranty time. This chapter extends the situation to consider a general warranty execution function which may provide a better way of modeling customer behavior, as all customers may not exercise the warranty even if the product fails within the warranty time. Reasons for nonexecution of warranty may be dissatisfaction for the product, preference to switch brands, cost of exercising warranty, and customer relocation, among others. The effect of the warranty execution function on the rebate plans in terms of expected warranty reserve costs under exponential and Weibull failure distributions is studied. Practical applications of the model are also discussed.

REFERENCES

1. Ford extends some warranties, *Wall Street Journal* April 7, 1992, 6.
2. Birkland, C. (1991). Fine tuning warranty recovery, *Fleet Equipment*, **17**, 47–49.
3. Menke, W. W. (1969). Determination of warranty reserves, *Management Science*, **15**(10), 542–549.
4. Amato, H. N., and Anderson, E. E. (1968). Determination of warranty reserves: An extension, *Management Science*, **22**(12), 1391–1394.
5. Mitra, A., and Patankar, J. G. (1988). Warranty cost estimation: A goal programming approach, *Decision Sciences*, **19**(2), 409–423.
6. Thomas, M. U. (1989). A prediction model for manufacturer warranty reserves, *Management Science*, **35**(12), 1515–1519.
7. Patankar, J. G., and Mitra, A. Effect of warranty execution on warranty reserve costs, *Management Science*, **41**(3), 395–400.
8. Priest, G. L. (1981). A theory of consumer product warranty, *The Yale Law Journal*, **90**(6), 1297–1352.

18

Extended Warranties

V. Padmanabhan

Stanford University
Stanford, California

18.1 INTRODUCTION

Extended warranties (hereafter referred to as EWs) refer to optional warranties that provide additional warranty coverage over the base (manufacturer) warranty on the product. They are obtained by paying an additional amount over the price of the product (either at the time of purchase or at some later time). The popularity of these contracts can be gauged from some sample statistics. In the category of consumer durables, Sears reports revenues in excess of $1 billion from EWs alone (*San Francisco Chronicle*, January 1992). In the case of home electronics, at least half of the profits at some major appliance store chains are due to sales of EWs (*Business Week*, January 14, 1991). The major automobile manufacturers report revenues in excess of $100 million from sales of EWs on new cars.

There is considerable variation in practice of the different manufacturers, retailers, and third-party companies with regard to the marketing of EWs. In the case of manufacturers, some attempt to sell EWs at the point of purchase (e.g., GM, Ford, Chrysler, etc.), others approach consumers through the mail after product purchase (e.g., General Electric, JVC, etc.), whereas still some others eschew the marketing of EWs altogether (e.g., Maytag, Curtis Mathes, American Honda, etc.). Similar variations are observed in the practices with regard to EWs at the retail level. Some retailers aggressively promote EWs at the point of sale (e.g., Sears, Circuit City, Electronic Realty Associates Inc., etc.), others take a much more subdued approach to the marketing of EWs (e.g., J. C. Penney's), and yet others approach consumers through direct mail at about the time of expiration of the base manufacturer warranty. A common phenomenon across most product markets has been the growth in the number of independent providers of EWs. These independent insurers market EWs for a variety of consumer durables as well as nontraditional products and services (e.g., optical services, dental care, home protection plans, veterinarian care, etc.).

What accounts for the popularity of EWs? Why do firms market EWs? Why do consumers buy EWs? How can one go about trying to design new EWs and how can we understand consumer preference for these alternative designs? This chapter will try to summarize the current wisdom on this subject and suggest some answers to the preceding set of questions.

The rest of the chapter is organized as follows. The conventional wisdom on EWs is recounted in Section 18.2 and is followed in Section 18.3 by a discussion of results of studies aimed at understanding consumer perceptions of EWs. Section 18.4 focuses on the basic theories of warranty and EWs and their implications for warranty policy. The warranty policy implications of increasing competition in the insurance aftermarket for products/services are discussed in Section 18.5. Section 18.6 discusses the use of market research methodology for design of EW policies. Section 18.7 concludes with a brief summary and suggestion of a research agenda on unresolved issues of interest.

18.2   THE CONVENTIONAL WISDOM

The literature in the popular press provides a number of qualitative insights into the benefits offered by EWs to the potential seller and buyer of these contracts. From the seller's perspective (and this is specific to the consumer durables product category), EWs provides a unique mechanism for building consumer loyalty and repeat purchase. Most consumer durables are characterized by long interpurchase cycles. Whereas most consumers go to the manufacturer or dealer for service and repair during the length of the base warranty, they typically go to the mass merchandiser or specialty store after the expiration of the base warranty. EWs help the manufacturer keep in touch with the consumer long after the expiration of the base warranty. The added benefits of such a situation is the market it creates for brand-authorized spare parts and allied services. EWs also provide valuable diagnostic data on the performance of their products over the life cycle of the product. Such data are particularly useful in R&D and Design activities. Finally, EWs generate profits to the manufacturer from the sales of insurance.

The most popular-press rationale for consumer adoption of EWs is that they provide consumers with "peace of mind." Having an EW implies that consumers need not worry about dealing with service people and/or contracting on their own in case of product failure. It also protects consumers against rising repair costs due to inflation. It is also likely that manufacturer-backed EWs are viewed by consumers as providing incentives for the firm to invest in quality because the reputation of the manufacturer is at stake. On the flip side, there is evidence that suggests that consumers do not view EWs totally positively. There is the widespread feeling that EWs are overpriced.* In summary, the popular press provides a number of reasons for the popularity of EWs both from the buyer and seller perspective. The literature also points out that there does exist some reservations in the mind of consumers and particularly the consumer activists and public policy agencies about the total worth of these contracts.

18.3 RESEARCH ON CONSUMER PERCEPTIONS OF EW

The academic research in this area is very recent in origin. The earliest work on this subject matter was conducted at The Center for Policy Alternatives at M.I.T. in 1971 [1]. The study of EWs was part of a larger study of product warranty. The study analyzed responses from 1317 households. Over 25% of the respondents owned EWs. Only half of them owned an EW at the time of the survey. There was a marked tendency for appliance owners not to renew their EWs. Consumers who owned an EW on one type of appliance were observed to be more likely to own one for another appliance. Consumers perception of repair far exceeded actual repair experience and this was particularly so for nonowners than for owners of appliances. EW ownership was also positively related to income, family size, urban residency, and with above-average repair experience of the respondent.

Day and Fox [2] report the results of a qualitative study aimed at understanding consumer perception and decision making with regard to

---

* *Consumer Reports* and other periodicals repeatedly speak to this issue.

EWs. Their findings suggest that consumers favorable evaluate EWs that provide insurance against catastrophic loss and EWs that offer periodic preventive maintenance. The major benefit of an EW to the consumer was not having the hassle of seeking out a repair person. Convenience was not a major benefit of EWs. Consumers perceive EWs as being over-priced and as being a way for the manufacturer to make money. Based on the results of the qualitative data they gathered, the authors hypothesize that a variety of factors may be related to the market potential and demand for EWs. They suggest that the demand for EWs is likely to be influenced by consumer risk-preference. Consumers who are risk-averse may perceive the need for greater protection offered by the EW. They suggest that larger families will derive greater benefit from the EW because they make more intense use of the product and, hence, will desire greater protection due to the higher probability of failure. They suggest that product experience (or to be more precise, the lack of product experience) is likely to significantly influence the demand for EWs. In other words, there is likely to be a strong market for EWs among newlyweds, singles, and younger persons. Finally, they suggest that the market for EWs is going be stronger for technologically complex products. The preceding research highlights variables that are likely determinants of consumer demand for warranty. Their implications for the warranty policy of the firm is the focus of the following section.

18.4   THEORIES OF PRODUCT WARRANTY AND EW

Padmanabhan and Rao [3] analytically characterize the optimal warranty policy for the manufacturer in a market characterized by consumer variation in risk-preference and consumer moral hazard using the agency theory framework. Warranties are viewed primarily as a form of insurance in this development. The interaction between the risk-neutral manufacturer and a risk-averse consumer is viewed as a game, where the consumer reacts to the warranty compensation package offered by the manufacturer by choosing the amount of effort to invest in product care and maintenance. This effort investment impacts the probability of failure of the product. However, the effort input cannot be observed or inferred by the firm. This is referred to as consumer moral hazard. The consumer's expected utility from the product is then determined by whether the product works or fails, the warranty compensation in case of failure, and the amount of effort in product maintenance and care. The unobservability of effort implies that (i) consumers will shirk on maintenance effort because it is costly to them and (ii) the firm cannot make warranties conditional on consumer effort. The moral hazard problem refers to the fact that the consumers can change their behavior with respect to effort investment in such a way that the firm's cost of servicing the warranty is increased. In this situation, the problem stems from the fact that it is not possible for the firm to verify or observe the consumer's effort investment. The reader can refer to Chapter 25 for a more complete description of this phenomenon.

The model is as follows. Consumers initially decide on the purchase of the product. Let p denote the price of the product and let y denote the initial wealth of the consumer. Assume that the production either work or fail. If it works, it confers a monetary benefit z, and failure creates a loss of k. The product is subject to probabilistic failure and the probability that the product works denoted by π depends on the consumer effort investment e whose cost to the consumer is $\psi(e)$. The expected utility from product adoption then is

$$\pi(e)U(y - p + z) + [1 - \pi(e)]U(y - p + z - k) - \psi(e)$$

$$(18.1)$$

Consumers are now required to make a decision on the amount of warranty protection to obtain. This is modeled as the selection of a degree or level of protection s where $0 \le s \le 1$. In case of failure, the firm repatriates an amount sk back to the consumer. Note that the consumer choice of level of coverage s translates to the choice of a deductible $k(1 - s)$ on the warranty. The expected utility from the choice of warranty then is

$$\pi(e)U(y - p + z) + [1 - \pi(e)]U(y - p + z - k(1 - s)) - \psi(e)$$

$$(18.2)$$

The consumer maximizes expected utility through choice of effort. The price that the consumer is willing to pay for the product and the optional warranty can be obtained by equating the expected utility to the initial wealth y. The monopolist's profit function can then expressed as

$$p(s) - p(0) - [1 - \pi(e)]sk \qquad (18.3)$$

The firm maximizes profit by the choice of an appropriate level of coverage s. The analysis provides several interesting insights. The optimal level of coverage offered by the firm increases monotonically as consumer risk-aversion increases. The profits from provision of insurance decreases monotonically as consumer risk-aversion increases. The manufacturer profits are maximized by the provision of a menu of warranty plans. The plan is implemented through a uniform base warranty on the product and the provision of an optional EW. The level of coverage provided by the base warranty is the optimal level of coverage for the least risk-averse segment of the population. The optional EWs overinsure the more risk-averse segment of the population. The level of overinsurance depends on the relative sizes of the segments in the market. The important point made in this chapter is that it is necessary to view the EW and the base warranty as representing items on a menu of warranty plans. The author empirically validates this explanation for the role of risk-preference in the choice of EW using data collected from a sample of new car buyers.

Padmanabhan [4] develops the segmentation role for warranty and extended warranty in the context of consumer heterogeneity in usage habits. The article argues that variation in consumer usage habits for the product creates variation in the consumer's demand for warranty. This heterogeneity leads to the optimality of a menu of warranty plans. Heavy users of the product will purchase the optional extended warranty. It is shown that the insurance profits decrease as the usage rate of consumers increase. As a result, if the size of the heavy-user segment is small in proportion to the light-user segment, as is often the case, then it is profitable for the manufacturer to exclude this segment from their potential market. This result helps provide a basis for understanding the exclusion of warranty privileges to certain segments of the market. For example, laundromat operators are not provided with the same warranty coverage as an average consumer. In fact, most manufacturer's expressly preclude warranty coverage to this segment.

Lutz and Padmanabhan [5] identify another source of consumer variation in the demand for warranty protection. This article motivates the role of income in influencing the demand for warranty protection. It is shown that high-income consumers have a higher willingness to pay for warranty coverage. This is because the high-income consumers have a lower marginal utility for wealth. Said differently, low-income consumers demand higher warranty coverages for an additional dollar in premium compared to high-income consumers. The high-income consumers will purchase the optional extended warranty. Profits from insurance increase as the consumer income increases.

Taken together, these articles provide a clear understanding of the role of base warranty and extended warranty in segmentation of typical consumer durable product markets. In economic terms, the analyses demonstrate clearly that base warranty and extended warranty are important in a theory of monopolistic price discrimination. Although both types of warranties provide insurance to consumers, they are different in that they are targeted at different segments of the population. In managerial terms, the theories clearly illustrate the role of base warranty and extended warranty in segmentation of the market on the basis of pertinent consumer characteristics (e.g., risk-preference, usage, income, etc.,). If a manager feels that there exists considerable consumer heterogeneity along any of these dimensions, then serious consideration should be given to the design of a total warranty package that addresses the warranty needs of the different market segments. The articles provide clear normative implications for the design of the various elements of the total warranty package (viz. the base warranty and extended warranty), its implications for consumer choice behavior, and its consequences in terms of consumer actions, especially those relating to product maintenance activities. Managers are often faced with a situation in which a competitor has made a change in the base warranty coverage on their product. In such a situation, managers need information on the consequences of such a change. One important effect of a change in base warranty is the change in sales of extended warranty. The estimates obtained from the empirical analyses are useful in predicting the magnitude of this effect. The elasticity of choice probability to the terms of the base warranty can easily be derived from the logit model of consumer choice. Padmanabhan and Rao [3] obtain an estimate of this elasticity of choice probability denoted by η as $\eta = -0.31$. This elasticity can be used to estimate the change in sales of extended warranty following changes in the base warranty. For instance, they predict that a change in the base warranty from 1 year to 3 years would result in the penetration figures of extended warranties dropping from 35% to about 13%. This figure is remarkably close to the sales-penetration figures of extended warranties at some major automobile manufacturers who moved to a 3-year comprehensive base warranty on their automobiles subsequently. This methodology is easily applicable to other product categories.

The models discussed up to this point in the chapter are all static models of product warranty. The omission of time dynamics in models of warranty can be a serious concern given the fact that consumer value the use of most durable goods for several time periods (i.e., not a single time period as is implied in static models of product warranty). Murthy and Padmanabhan [6] explicitly model the dynamic nature of product consumption and analyze its implications for manufacturers warranty policy as well as consumer behavior with regard to warranty choice and effort investment. The dynamic analysis provides interesting insights. For instance, they show that consumers invest in more effort during the warranty period than during the time period after the expiration of the warranty. This suggests that although warranty does create consumer moral hazard, the dynamics of consumer effort investment mitigate its detrimental effects on the manufacturer's warranty costs. The earlier insights about the optimality of a menu of warranty plans and its self-selection implications are shown to be robust to dynamic models.

The optimal policy with regard to base warranty and the menu of EWs does have its side effects on the consumer actions and, consequently, on the manufacturer bottom line. The consumer moral hazard problem implies that any provision of warranty coverage to the consumer reduces their incentive to invest in effort. Therefore, when consumers choose EWs, they will act in ways that increase the cost of servicing the EW. The data from the M.I.T. report is indicative of this moral hazard effect. The probability of at least one repair call and the average number of repair calls was higher for owners of EWs. Furthermore, consistent with Padmanabhan [4], the data indicated that owners of EWs were significantly higher in the average number of hours of use of their product (this was for televisions). The data from Padmanabhan and Rao [3] suggest that, on average, buyers of EWs invest in less regular maintenance of their product than nonbuyers of EWs. All of these numbers suggest that manufacturer's need to be sensitive to this effect. The manufacturer's can try to alleviate these effects in a couple of ways. The first is to make the honoring of claims contingent on the consumers taking certain actions in product maintenance that are verifiable. The case of automobiles where firms require proof of regular oil changes is a case in point. The data from Padmanabhan and Rao [3], in fact, show that buyers of EWs are more regular than nonbuyers with regard to oil changes, implying that such provisions can work. Another possibility is for the firm to try and reduce the cost to consumers of investing in maintenance activities. The firm could specify that it will pay fully or partly for those maintenance activities. Alternatively, it could set up systems that make it considerably easier for the consumer to access these services (e.g., get the product serviced at convenient outlets through licensed service centers other than the firm-specific outlets).

18.5 RECENT INSTITUTIONAL DEVELOPMENTS

The research discussed up to this point developed the motivation for warranty and extended warranty assuming that the firm is a monopolist in the insurance market. An intriguing development in the nineties has been the growth in the number of independent providers of insurance. It is now possible for a consumer to buy a Honda Accord and then obtain additional warranty coverage through an extended warranty underwritten by a company other than Honda.*

---

* The major domestic automobile manufacturers are fairly active in the sales of extended warranties on a variety of Japanese and European automobiles. Similar conditions hold in other consumer durable product markets.

The implications of an independent market for supplemental product insurance on the optimal warranty policy are not readily apparent. This is because the incentive effects of the provision of independent insurance are not clear at all. The provision of insurance by independent providers raises a whole new set of issues and questions; for instance: What determines the structure of the insurance market (i.e., monopoly versus competition)? What are the consumer-behavior implications of a competitive insurance market? What are the implications for the manufacturer warranty policy in the face of competition in the insurance market? These are some of the issues investigated in the following series of articles.

Lutz and Padmanabhan [7] investigate the issue of independent provision of insurance in the context where consumer investment in effort is significant in influencing the probability of product failure. The article makes several interesting points. First, it is shown that competition in the insurance market arises endogenously in the model. The reason is that it is cheaper for an independent agent to provide additional insurance in comparison to the manufacturer. The independent agent computes the cost of additional insurance based only on the amount of insurance coverage that it offers to the consumers. However, when the manufacturer offers the same level of coverage, his expected cost needs to consider the additional coverage as well as the basic coverage already provided to the consumer. Second, the provision of supplemental insurance has a significant negative externality on the manufacturer's cost, thereby reducing the profits by providing a base warranty. This is because consumers reduce their effort investment after they obtain additional warranty coverage from the insurer. This drives up the manufacturer's expected cost for the base warranty. As a result, it is optimal for the manufacturer to offer a minimal or zero-base warranty on the product. Third, by carefully choosing a price, the manufacturer can obtain the same profits as it would have in the absence of competition in the insurance market. Finally, the article demonstrates that extended warranties can be observed even in homogenous consumer markets. This motivation for extended warranty based on competition in the insurance market is different from the prior theories that rely on consumer heterogeneity. A study of the contact lens market provides a good illustration of this theory. As predicted by the theory, manufacturers of contact lenses provide no warranty on their products. Furthermore, there exists a thriving market for supplemental insurance on the contact lenses. A similar pattern is observed in the case of all-terrain vehicles and certain types of water sports equipment.

Lutz and Padmanabhan [8] investigate the implications of a competitive insurance market in which the manufacturer and insurer face identical costs for the provision of insurance. This is the case when there is no moral hazard. Consumer heterogeneity is sufficient to motivate the entry of competition in the insurance market: Competition in the insurance market results in the manufacturer becoming indifferent to the provision of a base warranty. Said differently, competition in the insurance market eliminates any role for base warranty in segmentation of the market. Lutz and Padmanabhan [9] investigate the implications of a competitive insurance market when product warranty serves as an incentive for the investment of quality.* When the manufacturer monopolizes the insurance market, the profit maximizing policy screens consumers by different warranty and quality combinations. Consumer heterogeneity is sufficient to motivate the entry of competition in the insurance market in this model as well. Screening is still optimal with competition in the insurance market. However, the presence of competition by reducing the manufacturer's insurance profits decreases the range of population distributions for which the manufacturer will serve the entire population. The manufacturer finds it profitable to concentrate on marketing extended warranties only to certain segments of the population, leaving the insurer to cater to the insurance demands of the other segments. All consumers are fully insured when there is competition, but only certain segments profit from the introduction of competition. By offering a low base warranty to certain segments of the population, the manufacturer is able to drive up the insurer' cost of providing insurance. This strategy helps the manufacturer limit the impact of competition on his insurance profits.

18.6 DESIGN OF ALTERNATIVE EW POLICIES

As mentioned earlier, studies of consumer perception of EWs seem to indicate that not all consumers view EWs on complimentary terms. Kelley and Conant [10] report that some consumers who purchased EWs believe that manufacturers who market EWs limited their express warranty coverage. Consumers were also of the opinion that manufacturers market EWs as a revenue-generating product. *Consumer Reports* suggests that EWs are overpriced. The M.I.T. report points out that consumers rarely renew their EWs. Collectively, these data seem to suggest that there does exist some strong reservation in the market about the worth of EWs.

Fox and Day [11] make the case that this could be very likely due to bad design of current EW policy and suggest the use of conjoint analysis for this purpose. Conjoint analysis is a measurement technique that has

---

* This is the case of producer moral hazard. In this situation, the manufacturer can increase the insurer's cost of insurance by changing the quality of the product.

been widely used in the market research community for new product development across many product and service categories.* It is easily applied in helping obtain data on consumer evaluation of various alternative designs of EWs. For instance, one could get consumers to evaluate EWs that vary along dimensions such as the amount of rebate offered, the amount of the deductible, the extent or length of coverage, the scope of coverage, the operationalization of the claims process, and so forth. The data obtained would be useful to the firms in ascertaining the relative importance of the various attributes of an EW policy (i.e., the relative importance of rebate versus deductible versus scope of coverage), the utility values associated by the consumers with various levels of the same attribute (i.e., different deductible amounts, different lengths of coverage), and in understanding what specific combination of attributes.

Fox and Day [11] assess ways in which the appeal and, thereby, consumer evaluations of the worth of EWs could be altered by redesigning the contractual terms of these policies. They specifically elicit consumer responses to two new types of EW: one with a rebate (a variation of the endowment life insurance or cash value life insurance policy) and one with a deductible. They demonstrate how it is possible to obtain consumer preference data and utility measurements of different EW with conjoint analysis. The study was conducted using washers as the product of interest. The rebate EW priced at $200 provided the standard EW coverage for 2 years. If no service calls (other than those for preventive maintenance) were made for the duration of the EW, a rebate of $100 was paid to the consumer. The deductible EW priced at $150 provided the same coverage but charged a deductible of $15 per service calls other than for preventive maintenance. The authors varied the price of the contracts, the amount of the rebate, and the amount of the deductible. The results suggest that consumers assess the rebate EW as providing them with greater utility than the deductible format. Both these formats performed better than the current format of the EW. The rebate format has the added advantage of reinforcing the purchase decision of the consumers in the event that they earn the rebate (this is the classical operant conditioning result) and, hence, could lead to higher renewal rates as well as higher sales of EW on other products. Note that both the rebate format as well as the deductible format have the added advantage of providing incentives for consumers to minimize the number of service calls and, hence, lessen the problem of consumer moral hazard. Obviously, it is necessary to carry

---

* See Green and Srinivasan [12] for a comprehensive discussion on conjoint analysis and Wittink and Cattin [13] for a survey of commercial applications of this technique.

out a full-fledged analysis of the financial implications of these alternative formats. But the important point to note is that there exist established methodology that can be used by firms to elicit consumer preferences and feedback for design of alternative formats of EWs.

18.7 CONCLUSION

In this chapter, we have focused on the subject of EWs. They refer to optional warranties that provide warranty coverage additional to the manufacturer warranty on a product/service. They do not come with the product and have to be purchased by paying an additional amount over the price of the product. Although there is considerable acknowledgement in the popular press of the popularity and importance of EWs, there is little understanding of the reasons for this prominence or of the role of EW in warranty policy and management.

The academic rationale for EWs is found in the diverse demands for product insurance created due to consumer heterogeneity along dimensions such as risk-preference, usage intensity, and income. The provision of optional EWs allows the firm to implement a segmented pricing strategy in these markets. This theory of EWs has been empirically examined and supported. The theory provides clear normative implications for the optimal warranty and EW policy. Another important issue highlighted by the research in the determination of the optimal warranty policy is the independent insurance market for the product/service. The research suggests that the different incentive effects of product warranty have very different implications for optimal warranty policy in the context of independent insurance markets. The firms decisions with regard to warranty and EW are critically influenced by these forces.

Given the emerging status of the literature on the topic of EWs, it is not surprising that there exist several directions for future research in this area. There is little understanding at present of the role of channel intermediaries in warranty policy. What specific roles do they play in the design of the warranty policy as well as its implementation? Their contribution in terms of information transmission, consumer self-selection, and overall efficiency are poorly understood. Another important area for research is the role of EWs in influencing customer satisfaction. A better understanding of this relationship will have important implications for the design of warranty contracts. Our understanding of this subject matter would also be considerably improved by a comprehensive study of the different warranty policies, practices, and consumer behaviors across a broad range of products/services.

REFERENCES

1. Center for Policy Alternatives at Massachusetts Institute of Technology (1978), *Consumer Durables, Warranties, Service Contracts and Alternatives.* Vols. *I–IV*, CPA-78-14.
2. Day, E., and Fox, R. J. (1985). Extended warranties, service contracts and maintenance agreements—A marketing opportunity? *Journal of Consumer Marketing*, 2(4), 77–86.
3. Padmanabhan, V., and Rao, R. C. (1993). Warranty policy and extended warranties: Theory and an application to automobiles, *Marketing Science*, 12(3), 230–247.
4. Padmanabhan, V. (1995). Usage Heterogeneity and Extended Warranties, *Journal of Economics and Management Strategy*, 4(1), 33–53.
5. Lutz, N. A., and Padmanabhan, V. (1994). Income variation and warranty policy, Working Paper, Graduate School of Business, Stanford University.
6. Murthy, D. N. P., and Padmanabhan, V. (1992). A continuous time model of product warranty, Working Paper, Graduate School of Business, Stanford University.
7. Lutz, N. A., and Padmanabhan, V. (1995). Why do we observe Minimal Warranty, *Marketing Science*, in press.
8. Lutz, N. A., and Padmanabhan, V. (1994). Competitive insurance market and manufacturer warranty policy, Working Paper, Graduate School of Business, Stanford University.
9. Lutz, N. A., and Padmanabhan, V. (1994). Warranty, extended warranty and product quality, Working Paper, Graduate School of Business, Stanford University.
10. Kelley, C. A., and Conant, J. S. (1991). Extended warranties: Consumer and manufacturer perceptions, *Journal of Consumer Affairs*, 25(1), 68–83.
11. Fox, R. J., and Day, E. (1988). Enhancing the appeal of service contracts: An empirical investigation of alternative offerings, *Journal of Retailing*, 64(3), 335–352.
12. Green, P. E., and Srinivasan, V. (1978). Conjoint analysis in consumer research: Issues and outlook, *Journal of Consumer Research*, 5, 103–123.
13. Wittink, D. R., and Cattin, P. (1989). Commercial use of conjoint analysis: An update, *Journal of Marketing*, 53, 91–96.

19

Warranty and Product Distribution

Arvinder P. S. Loomba

University of Northern Iowa
Cedar Falls, Iowa

K. Ravi Kumar

University of Southern California
Los Angeles, California

19.1 INTRODUCTION

Some of the critical issues confronting the management of a manufacturing firm in today's marketplace are concerns regarding product distribution and warranty/service support practices. Most consumers feel that when making a product purchase, the question of "where to buy the product" is equally important as "what measure of performance can be expected from the product and what will happen if problems arise after its sale."

According to a major study [1], manufacturer's primary rationale for offering product warranties in a distribution environment is the need to support their products. Also, the demand for warranty servicing in a distribution environment has steadily increased in recent times. This can be accounted for by several factors:

- Greater customer awareness of warranty rights*
- Products are now more complex and, therefore, more service sensitive
- Inflationary pressures creating greater sensitivity to repair costs
- Declines in product quality as well as instances of poor product design
- Inexperienced and unqualified personnel maintaining and servicing products
- Increased incidents of product misuse by the customer†

Many of these concerns translate directly to decisions regarding a firm's choice of channels used to distribute its product and provide for after-sale service support such as warranty, service contract, repairs, and so forth. Specifically, the manufacturing organization needs to address questions such as: What channel structure(s) is (are) best suitable for distributing and offering warranty support for their product in a particular industry? How should channels for product distribution and service support be selected and managed?

In most industrial distribution situations, there is a trade-off between a manufacturer's ability to control important channel functions and the financial resources required to exercise that control. The more channel intermediaries involved in getting a manufacturer's product to the end consumer, the less control the manufacturer can exercise over the flow of its product through the distribution channel. As the number of intermediaries in a distribution channel increases, the manufacturer's ability to influence product prices and service support attributes to the end user generally tend to decrease. On the other hand, reducing the length and the breadth of the distribution channel generally requires that the manufacturer allocate more financial resources for channel-related functions.

According to Armistead and Clark [2], the decision of how to distribute products and to offer after-sale service support such as warranty, service contract, repairs, and so forth is linked to the overall strategy of the organization on the products and the markets it is to serve. As an example, some manufacturers, such as IBM, have the financial resources needed to perform most or all channel functions, if they choose to do so.

---

* A detailed discussion of warranty rights and legislation is offered by Burton in Chapter 28 and by Kelley in Chapter 4.

† This translates into higher warranty costs to the warranty provider. Warranty cost models are discussed in greater detail by Blischke in Chapters 10 and 12, by Patankar and Mitra in Chapter 11, by Moskowitz and Chun in Chapter 13, and by Wilson and Murthy in Chapter 14.

On the other hand, other manufacturing units, operating with constrained financial resources,* lack the ability to perform many channel functions (including product distribution and warranty administration) directly, regardless of management's desire for direct control over these functions. In order to reach a decision, the following questions need to be addressed:

- How important is for a firm to distribute its product and/or to provide service support for the product? There may be legal, regulatory, and safety conditions to make this necessary.
- Do customers for the product expect that a firm will offer service support for its products rather than leave it to third-party servicers? Does this influence future sales?
- Will the product distribution and/or service support activities give sufficient returns to a firm for it to be financially feasible? Profitable?
- Can a firm allocate resources to distribute the product and/or to offer service support for its products in a particular market segment?
- Does a firm have necessary strengths or skill base to distribute its product and/or to provide service support for its products in a particular market segment?
- Are there enough opportunities in a particular market segment for a firm to consider distributing its product and/or providing service support for its products?

This chapter is organized as follows: Section 19.2 deals with current warranty/servicing trends and practices in today's market environment. Section 19.3 offers an overview of the product distribution, its evolution, and its organization in the current marketplace. Section 19.4 deals with the design of distribution and warranty/servicing channels. Role of channel members and prevalent channel structures are also addressed in this section. Specific notations, assumptions, and characterization of basic models are presented in Section 19.5. Also in that section, single-product monopoly models for optimizing channel member profits under two different distribution structures are proposed. In Section 19.6, model results are illustrated using a numerical example, which is supplemented with a sensitivity analysis characterizing the impact of model parameters on policies and profits under each distribution channel. Model inferences, future research agenda, and implications for business managers are discussed in the final section.

---

* This includes firms operating in markets in which a dispersed customer base simply increases the resources required for adequate market coverage.

19.2   WARRANTY/SERVICING PRACTICES IN DISTRIBUTION ENVIRONMENTS

Warranty is a service contract for which the warranty provider contractually agrees to repair or replace a product for the customer at no additional cost for a specified period of time after its sale. It is important to note that the cost of warranty is included in the price the customer pays for the product. Transmission of the warranty details to the customer is accomplished in a number of ways in today's market environment. Based on a major study [1], about half of the firms surveyed provide warranty along with their products, 30% send warranty details under separate cover, and 22% use both means. In addition, only one-third of all firms surveyed reported that they require warranty service be provided by company-authorized sources.

There are a number of factors that influence the type and extent of warranty coverage offered to the customer [3]. These factors include the following:

- *Marketing goals*, which determine what the organization wants to achieve in the marketplace in terms of positioning the product against its competitors (such as the warranty with a 7-year, 70,000-mile coverage offered on Chrysler automobiles). A warranty with longer duration would tend to suggest that the product is of higher quality or presents better value. However, a product with strong position in the marketplace may not demand a strong warranty because the product is so widely accepted by the potential consumers that the additional investment in an improved warranty would not significantly increase sales (such as the warranty on Honda automobiles with only a 3-year, 36,000-mile coverage in light of warranties offered by its competitors with longer and/or broader coverage).
- *Competition*, whose products and warranty/servicing strategy need to be critically evaluated and is often used as an input for determining extent and type of warranty coverage that should be offered on a firm's product.
- *Customer acceptance of product*, which will influence the extent and type of warranty coverage for that product. A new product from a well-established, reputable organization may not need a strong warranty. On the other hand, a manufacturer who has no track record in a particular product line or has a history of poor product quality may find it necessary to offer a strong warranty to build a perception of quality.

- *Position in product's life cycle*, which will influence the warranty requirement for that product. A stronger warranty can be used by a firm to (i) demonstrate its conviction in newly introduced product, (ii) differentiate a product (from competing products), which is at the saturation stage of its life cycle, and (iii) reestablish consumer confidence in a product, which is at the maturity/decline stage of its life cycle.
- *Initial cost of the product*, which will establish the need and type of warranty coverage for that product. Customers expect a better warranty coverage for an expensive product in order to derive full value from the investment. On the other hand, it will seem reasonable to offer minimal or no warranty on inexpensive products, which the customers perceive as "disposable items."

Traditionally, it is the manufacturing firm which assumes the cost of fulfilling its warranty obligation (including parts and/or labor). However, varying practices in warranty provision do exist in today's marketplace. Although, the cost of fulfilling the warranty obligation is borne by the manufacturer most of the time, it is shared by other market intermediaries in other instances. The latter is prevalent in industries where the products are sold through distributors, dealers, or retailers. In such industries, it is a common practice for the manufacturer to replace defective parts, whereas the cost of labor to diagnose the problem, install the replacement parts, and make any necessary adjustments is the responsibility of the distributor, dealer, or retailer.

Another trend in warranty provision is that a manufacturer may encourage or require a dealer or retailer to provide the specific warranty, even though the dealer or retailer must assume all or part of the cost of fulfilling that obligation. This practice is commonly witnessed in industries where products are manufactured by OEMs (original equipment manufacturers) and distributed/serviced by brand name distributors or retailers. There are numerous options available to the manufacturer for fulfilling warranty obligation in a distribution channel environment [3]. These include (i) reimbursing retailers or dealers at their full rates of labor and parts on warranty work performed, (ii) paying retailers or dealers a fixed dollar amount for labor on given repairs plus furnishing required replacement parts, (iii) manufacturer offering to provide the replacement parts only, with retailers or dealers bearing all the labor costs associated with warranty work, or (iv) all the warranty burden is on the retailers or dealers, who absorb the cost for all labor and parts in fulfilling the warranty.

Current management practices associated with offering extended service contracts by manufacturers, retailers, and independent servicers also vary widely in nature.* Manufacturers usually try to market service contracts either during the sale of the product (such as General Motors) or well after the product purchase (such as General Electric). Other manufacturers (such as Maytag) offer more comprehensive warranties in place of service contracts. Retailers also tend to market service contracts both in an aggressive (such as Sears, ERA, etc.) and dormant manner (such as J.C. Penney). The independent servicer, who is mainly in the business of offering service contracts, usually promotes them in an aggressive manner and at attractive terms because of the potential competition from manufacturers and retailers in the after-sale service business (such as ERA, American Homeowners Association, etc.) [4].

19.3 PRODUCT DISTRIBUTION OVERVIEW

"Distribution channels" can be viewed as interdependent organizations involved with the transfer of products from manufacturer to end consumer. The manufacturer is dependent on the distribution channel (and its intermediaries) for performing some of the product distribution, marketing, and servicing operations such as selling, distributing products efficiently, offering various product assortments to the consumer, transportation, storage, financial risk-sharing, and after-sale service support functions [5]. In getting its products to the end-user markets, a manufacturer must either assume all of these functions or shift some or all of them to channel intermediaries.

Recent industry trends indicate that a growing number of manufacturers are reaching consumers through distribution channel intermediaries such as distributors, wholesalers, independent retailers, and so forth. According to a major study of 200 U.S. companies [6], product manufacturers preferred distributing their products through distribution channel intermediaries over direct selling, to curb escalating distribution costs. Almost half of the surveyed firms (49%) used wholesalers, 42% used distributors, and 39% used independent retailers. More than 20% of the surveyed firms expected to use these channel intermediaries even more heavily in the future.

The specific breakdown of the distribution channels used by the manufacturers is presented in Table 19.1. The market trends indicate that costs are the driving force behind the major changes in sales and service support channels. Also, there seems to be a trend toward increased use

---

* A detailed discussion of extended service contracts is offered by Padmanabhan in Chapter 18.

Table 19.1 Distribution Channels Now Used by Product Manufacturers

Distribution channels and contact methods used	Percentage of reporting companies[a] (%)			
	Total companies	Industrial product producers	Consumer product producers	Both
Direct channels				
Company's sales force:				
Specialists	68	75	55	75
By type of product	54	63	38	63
By type of industry	30	37	13	46
By trade channel	23	16	28	42
By size of account	21	19	25	17
By type of user	18	23	8	25
Generalists	64	66	58	71
Company agents:				
Manufacturer's representatives	32	37	19	42
Independent agents	16	21	13	4
Brokers	14	9	23	17
Commission merchants	4	3	5	8
Independent telemarketers	1	1	3	—
Company-owned retail outlets	12	7	17	21
Company-owned wholesale outlets	9	9	5	25
Indirect channels				
Wholesalers	49	34	69	80
Industrial distributors	42	54	11	58
Independent retailers	39	15	70	80
Franchise outlets	16	12	23	13
Drop shippers	9	6	13	8
Other contact methods				
Telephone sales	25	25	20	38
Computer-to-computer shopping	12	10	19	9
Catalog sales	8	6	11	8
Number of companies	214	126	64	24

[a] Multiple responses.
Source: Ref. 6, p. 2 (reprinted with permission of *The Conference Board*).

of indirect (or decentralized) channels of distribution with an increased reliance on distributors and/or retailers.

19.3.1 Evolution of Product Distribution

According to McInnes [7], *markets* can be interpreted as "the gap which separates producer and consumer." Types of separation inherent in the interpretation of this gap include separation of space, time, perception, ownership, and values. Certain functional areas of the business, which are relevant in linking distinct markets by closing the gaps between producers and consumers, include the following:

- *Transportation*, which deals with spatial separation and includes all activities directly concerned with moving goods from place of production to place of sale
- *Inventory*, which deals with temporal separation and includes all activities directly concerned with holding goods between the time of production and the time of sale
- *Promotion*, which deals with separation of perception and includes all activities directly concerned with providing information regarding nature of goods and services and their relationships to the potential user's perceived needs
- *Transaction*, which deals with separation of ownership and values and includes all activities directly concerned with active buyer/seller negotiation and with the transfer of title of products

Distribution channels evolved as particular systems through which these functional areas are combined, integrated, and allocated among various business firms. These channels are viewed as effective linkages among various markets. Alderson [8] argued that the evolution of distribution channels (and its intermediaries) was primarily due to the following factors:

- Distribution channel intermediaries evolve from the product exchange process due to their potential to raise the efficiency of exchange by creating time, place, and possession utility
- Distribution channel intermediaries enable the adjustment of *discrepancy of assortments* by performing sorting and assorting functions
- Distribution channel arrangements are established to achieve routinization of transactions
- Distribution channels facilitate an effective consumer search process.

19.3.2 Organization of Product Distribution

Distribution channels are very difficult and time-consuming to build and, once in place, are not easily changed [9]. The nature of a distribution channel structure determines its performance in terms of risk sharing, allocative efficiency, and the distribution of profits among various channel members. Therefore, it is imperative that the manufacturing firm make judicious decisions regarding the selection of an appropriate channel structure for distributing and servicing its products.

Although most of the distribution channel literature has centered on the management of ongoing dyadic channel relationships, there is some research conducted on the nature of the distribution channel organization. Weigand [10] found a direct relationship between firm size and the type of distribution channel organization used. He concluded that large firms were more likely to be vertically integrated than smaller size firms. As illustrated in Figure 19.1, Bucklin [11] developed an elaborate theory of distribution channel structure or organization. He claimed that the objective of the distribution channel is to offer various service outputs (such as lot size, delivery waiting time, market decentralization, etc.) to the consumer. The consumer then makes his decision about what channel service outputs to purchase, which, in turn, determines the structure of distribution channel. The more channel service outputs that are demanded by the consumer, the more likely it is that the channel organization will include intermediaries. Figure 19.1 illustrates the determination of the distribution channel structure using channel service output criteria.

19.3.3 Distribution Channel Selection Decision

The decision of selecting a distribution channel largely depends on the specific characteristics of the product and its manufacturer [12]. Duncan [13] posited that the factors affecting the channel selection decision include the nature and the extent of the product market, existing distribution channels, channel choice of competitors, manufacturer's financial resources, and potential interaction with channel intermediaries. Other researchers add that selected target markets, the manufacturer's marketing mix (price, product, promotion, physical distribution), the industry type, and the economic context are also significant factors in this decision [14–16].

The distribution channel selection decision consists of numerous dimensions [17]. Out of these, three dimensions—distribution structure, distribution intensity, and distribution management, which are relevant to our discussion, are reviewed below.

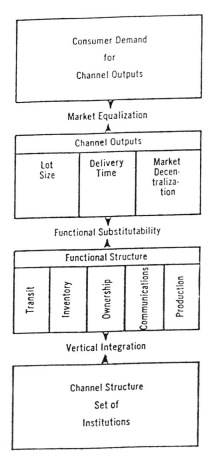

Figure 19.1   Determination of channel structure. (From Ref. 11, p. 16; reprinted with permission of *IBER Special Publications.*)

Distribution structure refers to management's decision of selecting a particular type of distribution channel to serve designated end-user markets. For instance, management alternatives, such as making use of a vertically integrated channel (where the product manufacturer performs retail and service activities) versus using a decentralized channel (which includes channel intermediaries, such as retailers and/or independent servicers), are evaluated in the area of distribution structure.

Research in the area of distribution structure was initiated by Artle and Berglund [18]. They analyzed a single channel decision choice between a direct sales force channel and using a retailer to distribute the products. A generalized extension of the above model was presented by Bucklin [19]. Weigand [20] emphasized the importance of choosing the right combination of markets, channels, and products. He claimed that under certain industry conditions it is better to use more than one market, channel, or product, which he refers to as *multimarketing*. Other researchers examined the channel structure decision from a distribution cost viewpoint, employing both theoretical [21] and empirical research approaches [22].

Distribution intensity refers to the decision concerning the number of channel intermediaries to be involved with the distribution channel. For example, should a product manufacturer deal directly with the end consumer (no intermediary) or have one or more intermediaries perform the distribution function?

In one of the earlier research efforts, Hartung and Fisher [23] used nonlinear programming techniques to determine the optimal number of retail outlets to be built by a company in one area. The level of distribution intensity (or length of the channel) depends on the number of consumers targeted, potential of consumer volume, availability of the channel intermediaries, significance of the purchase, level of channel control required, level of channel intermediary and consumer contact required, and concentration of geographic and industrial markets [24,25].

Distribution management deals with the decision to employ price and other marketing-mix variables to influence the performance of distribution channel components. It also represents the service output levels, such as inventory levels, product assortment, delivery time, spatial convenience, and so forth, offered to the consumer.

Miller et al. [26] proposed a value-exchange model to facilitate management of distribution channels. Cohen and Lee [27] furnished a strategic framework for an integrated production–distribution system addressing various service performances, such as customer service levels, finished-goods inventory levels, manufacturing cost, and distribution cost, under various distribution strategies.

19.4 CHANNEL DESIGN FOR DISTRIBUTION AND WARRANTY/SERVICING

In this section, we address design considerations of distribution and warranty/servicing channels. We accomplish this by discussing the role of various channel members and by examining the prevalent channel structures.

19.4.1   Role of Channel Members

In addition to the product manufacturer and the end consumer, the market channel environment may include intermediaries such as distributors, wholesalers, retailers, franchise outlets, drop shippers, and so on. In order to better understand specific policy decisions made by channel members under various channel environments, we limit our discussion to only three types of channel members: manufacturer, retailer, and servicer. Using these channel members, we present mathematical models to analyze channel policies and profits in different channel environments; each model considers these channel members operating in a specific channel structure with an objective of maximizing their respective profits.

> *Manufacturer*:   This channel member's primary function is to manufacture the product at a prespecified level of quality. The manufacturer may distribute the product to the consumer using a centralized (i.e., sell the product directly to the consumer) or decentralized strategy (i.e., sell the product to a retailer who, in turn, sells it to the consumer). Similarly, the manufacturer may choose to offer service support for the product using a centralized (i.e., provide service support to the consumer, either himself or through a retailer) or a decentralized strategy (i.e., allow an independent servicer to provide after-sale service support to the consumer). In addition, the manufacturer generally offers an initial warranty on the product for a specific time period, during which time consumers can get repairs for failed items free-of-charge from the warranty provider.
>
> *Retailer*:   This channel member will exist in a channel structure only when the manufacturer chooses to decentralize its product distribution. The primary function of the retailer is to buy the product from the manufacturer and sell it to the consumer for a profit. Generally, the retailer does not perform any value-added operation.* In addition, the product selling price charged by the retailer is a function of the wholesale product price charged by the manufacturer.
>
> *Servicer*:   This channel member will appear in a channel structure

---

* However, the retailer performs the value-added operation when, in addition to distributing the product, he also handles the service support function in a channel structure.

only when the manufacturer opts to decentralize the service support function. When present in the channel structure, the servicer's primary function is to perform the value-added operation of offering the after-sale service support for the product. This is accomplished by providing options such as repair service and service contracts to the consumers. In this case, the servicer is responsible for handling all product repairs over the effective product life span after the initial product warranty expires.

19.4.2 Distribution Channel Structures

We categorize distribution channel structures based on their product distribution and after-sale service support (warranty/servicing) functions. For example, channel structures can first be categorized based on whether or not a servicer is present in the channel structure. In a specific channel structure, where a servicer is present and is responsible for handling warranty/service support function, the channel structure is considered to be *horizontally decentralized*. Further, the channel structures can be classified based on whether or not a retailer is present in the channel structure. In channel structures, where the retailer is present and is responsible for handling the product distribution function, the channel structure is considered to be *vertically decentralized*.

In our discussion, we limit our analysis to examining the channel structure environments, to provide distribution and warranty/service support, that follow *pure* strategies only. These channel environments consists of (i) CC structure, which is a channel structure environment which is completely centralized in both product distribution and warranty/service functions and (ii) DD structure, which is a channel structure environment which is completely decentralized in both product distribution and warranty/service support functions. Both CC and DD channel structures are discussed in detail in the following section.

19.5 OPTIMAL DISTRIBUTION AND WARRANTY/SERVICING MODELS

In this section, we present models for two channel structure environments discussed earlier: Model CC and Model DD. To preserve brevity, only a limited discussion of models is presented here. Interested readers are referred to Ref. 28 for detailed discussions on these and other models, rationale for specific mathematical functions for consumer demand and production costs, and derivations of optimal policies and results.

19.5.1 Model Notation

Parameters

κ: Industry constant. Note that for positive product demand, $\kappa > 0$.

α: Product price sensitivity of the consumer.

β: Warranty duration sensitivity of the consumer.

δ: Repair charge sensitivity of the consumer.

γ: Service-contract fee sensitivity of the consumer.

λ: Product's repairable life-span sensitivity of the consumer.

R: Average cost of repairs per product breakdown to the servicer.

θ: Failure rate of the product (a prespecified measure of product quality).*

$\mathscr{C}(\theta)$: Per-unit manufacturing cost of the product (expressed as a function of its measure of quality, that is, the failure rate of the product, θ).

ψ: Proportion of the consumers who opt for repair service for a charge during every product breakdown.† For our analysis, this pre-specified proportion is assumed to be a constant.‡

Decision Variables

P: Product's per-unit final selling price charged to the end consumer.

W: Product's per-unit wholesale price paid by retailer to the manufacturer.

T: Warranty duration associated with the product.§

S: Repair service charge assessed for each product breakdown by the service support provider after the warranty on product (if any) expires.¶

F: One-time service-contract fee assessed at the time of product sale.**

---

* The failure rate of a product, θ, is defined as the average number of expected failures or breakdowns in a product over a given period of time. In our analysis, θ is not a function of time because we assume that product failure follows exponential distribution. Note that, $0 \le \theta \le 1$. Further, $\theta = 0$ represents a failure-free product.

† This implies that the proportion of consumers opting for service contracts is $1 - \psi$.

‡ The preference for service contracts over per-charge repair service depends on the individual risk-preference of consumers, an issue not considered here.

§ During the time the warranty is in force, from 0 to T, any expenses for repairing the product are borne by the warranty provider.

¶ This amount is charged only to consumers who do not purchase the service contract.

** This fee is assessed by the support provider to consumers who are *willing* to purchase the service contract. In return, the support provider promises to repair the product for its effective repairable life, L, at no cost to the consumer.

L: Effective repairable life span of the product.*

Q: Quantity (in number of units) of product demanded by the consumer.

M_M: Manufacturer's per-unit profit margin.

M_R: Retailer's per-unit profit margin.

M_S: Servicer's per-unit profit margin.

Π_M: Manufacturer's profits. Here, $\Pi_M = M_M \times Q$.

Π_R: Retailer's profits. Here, $\Pi_R = M_R \times Q$.

Π_S: Servicer's profits. Here, $\Pi_S = M_S \times Q$.

19.5.2 Model Assumptions

In this study, only one-product industrial market environments are considered.† In addition, we do not consider either synergies with other marketing variables (such as advertising, sales, promotions, etc.) or the demand effects of retailer's brand assortment. The manufacturer is considered to be the sole producer (i.e., no product competition). Also, in this chapter, we limit our analysis to the discussion of "ordinary free-replacement warranty" [29]. All channel members have individual profit-maximization objectives and play the game using Nash [30] strategies.‡ Further, all product sale and service support decision variables are implicit functions of product quality.§ Also, all consumers¶ are assumed to be homogenous in terms of product usage. Finally, we assume that the time to failure of the product can be closely approximated using the *exponential* distribution.**

---

* We define it as the duration for which the after-sale support services are available for the product after its initial sale. Beyond this time, the product is expected to breakdown or wear out at an increasing rate. We assume that after this duration L, once the product breaks down, the service support provider no longer offers repairs on the product and, therefore, the consumer has to replace it. Note that $L \geq T$.

† This assumption is essential because we can assume that the purchase of industrial products is generally guided by a rational decision making process. In reality, the quantity for consumer products demanded may also depend on issues such as intangible product attributes, impulse buying behavior, and so on, which are not captured by the current model.

‡ Here, optimal pricing and service support policies are derived assuming all channel members follow Nash [30] game strategy. This is true except in cases where a retailer is present in the channel; in those cases, the manufacturer follows a Stackelberg leader [31] game strategy to derive the wholesale price, W, and the product quality measure, θ.

§ In cases, in which a retailer also exists in the channel structure, decision variable values are also explicit functions of manufacturer's wholesale price of the product, W.

¶ The word "consumer" refers to the end user of industrial products being considered.

** This failure rate distribution assumption has proven effective in a number of industrial product failure applications [32,33].

19.5.3 Model Characterization

We posit a general formulation of a consumer-demand function wherein the quantity Q demanded by the consumer is assumed to depend on the final product price P, warranty duration T, repair service charge S, service-contract fee F, and repairable life span L. Mathematically speaking,

$$Q = \mathcal{F}(P, T, S, F, L)$$

where

$$\frac{\partial Q}{\partial P}, \frac{\partial Q}{\partial S}, \frac{\partial Q}{\partial F} < 0 \quad \text{and} \quad \frac{\partial Q}{\partial T}, \frac{\partial Q}{\partial L} > 0$$

Here, we utilize the following log-linear demand function:

$$Q = \kappa P^\alpha T^\beta S^\delta F^\gamma L^\lambda$$

where

$$\kappa > 0; \quad \alpha < -(1 + \beta + \lambda); \quad 0 < \beta, \lambda < -(1 + \gamma); \quad (19.1)$$
$$-(1 + \lambda) < \delta, \gamma < -1$$

Here, the consumer demand, Q, is assumed to be an increasing function of product distribution and service support prices and is a decreasing function of warranty and repairable product life span. Also, conditions $\alpha < -(1 + \beta + \lambda), 0 < \beta, \lambda < -(1 + \gamma)$, and $-(1 + \lambda) < \delta, \gamma < -1$ are necessary to obtain a well-behaved consumer-demand function.

In addition, $\mathcal{C}(\theta)$ is assumed to be a function of the level of production quality (such as product's failure rate, θ), which is presumed to be prespecified for the manufacturer. For instance, the specific functional form of $\mathcal{C}(\theta)$ could be written as

$$\mathcal{C}(\theta) = C^+(1 - \theta)^\epsilon, \quad \text{where } C^+ > 0; 0 < \theta < +1; \forall \epsilon. \quad (19.2)$$

19.5.4 Model CC

Model CC depicts a channel structure with Centralized Distribution and Centralized Service Support. As illustrated in Figure 19.2, this channel structure is comprised of only the manufacturer and the consumer. In this channel environment, the manufacturer centralizes both product distribution and service support functions and deals directly with the end consumer.

Here, for a prespecified level of product quality and for a given consumer-demand function, the manufacturer derives optimal values of product price, warranty duration, repair service charge, service-contract

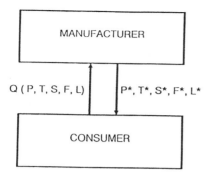

Figure 19.2 CC channel structure.

fee, and effective serviceable life span associated with the product. For example, companies such as Xerox Corporation follow the CC channel strategy. Xerox centralizes its product distribution by selling copiers directly to the consumers through its outlets and exclusive dealerships. It also centralizes the service support function by offering warranty, repairs, and service-contract provisions to the consumer through its service center network.

Here, the manufacturer wishes to maximize his profit function, Π_M, which depends on the per-unit profit margin, M_M, and the quantity demanded, Q. The per-unit profit margin can be expressed as follows: Per-unit profit margin = (Per-unit selling price) − (Per-unit manufacturing cost) + (Per-unit service support revenues) − (Per-unit cost of providing service support).

Mathematically speaking,

$$M_M = [P] - [\mathscr{C}(\theta)] + [\psi S\theta(L - T) + (1 - \psi)F] - [R\theta L]$$

Therefore, the profit maximization objective of the product manufacturer can be written as

$$\max_{P,T,S,F,L} \Pi_M = [P - \mathscr{C}(\theta) + \psi S\theta(L - T) + (1 - \psi)F - R\theta L]Q$$

$$(19.3)$$

where the value of Q is specified in Eq. (19.1). The following proposition outlines the optimal pricing and service support policies in a CC channel structure.

Proposition 19.1. *In a CC channel structure, the optimal pricing and service support policies are given by*

$$P^* = \frac{\alpha \mathscr{C}(\theta)}{\alpha + \beta + \gamma + \lambda + 1} \tag{19.4}$$

$$T^* = \frac{-\beta(\beta - \delta + \lambda)\mathscr{C}(\theta)}{R\theta(\beta - \delta)(\alpha + \beta + \gamma + \lambda + 1)} \tag{19.5}$$

$$S^* = \frac{R(\beta - \delta)}{\psi(\beta - \delta + \lambda)} \tag{19.6}$$

$$F^* = \frac{\gamma \mathscr{C}(\theta)}{(1 - \psi)(\alpha + \beta + \gamma + \lambda + 1)} \tag{19.7}$$

$$L^* = \frac{-(\beta - \delta + \lambda)\mathscr{C}(\theta)}{R\theta(\alpha + \beta + \gamma + \lambda + 1)} \tag{19.8}$$

Proof. See the Appendix.

19.5.5 Model DD

Model DD depicts a channel structure with "Decentralized Distribution and Decentralized Service Support." As illustrated in Figure 19.3, this channel structure is comprised of the manufacturer, the retailer, the servicer, and the consumer. In this channel environment, the manufacturer decentralizes both product distribution and service support functions. Whereas the retailer manages all product distribution activities, service

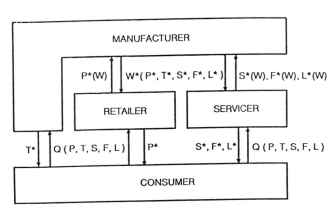

Figure 19.3 DD channel structure.

support operations is handled by the servicer who deals with the consumer in this regard. The manufacturer, however, is still responsible for offering basic warranty on the product to the consumer.

Here, for a prespecified level of product quality and for a given consumer-demand function, the retailer derives optimal value of product's retail price as a function of product's wholesale price, which is controlled by the manufacturer. In addition, the servicer derives optimal values of repair service charge, service-contract fee, and effective product life span, all as a function of product's wholesale price. The manufacturer, in turn, derives optimal value of the product's wholesale price and conveys it back to both retailer and servicer, who then optimize their respective profits. For example, companies such as Dell Computers follow a DD strategy. Although the company distributes its computer products through local dealers and retailers, it has contracted out its installation and service support to Xerox Corporation [34].

Now, the manufacturer's profit maximization objective is given by

$$\max_{W,T} \Pi_M = [W - \mathscr{C}(\theta) - R\theta T]Q \tag{19.9}$$

The retailer's profit maximization objective is given by

$$\max_{P} \Pi_R = (P - W)Q \tag{19.10}$$

Finally, the servicer's profit maximization objective is given by

$$\max_{S,F,L} \Pi_S = [\psi S\theta(L - T) + (1 - \psi)F - R\theta(L - T)]Q \tag{19.11}$$

Here, the value of Q is specified in Eq. (19.1). The following proposition outlines the optimal pricing and service support policies in a DD channel structure.

Proposition 19.2. *In a DD channel structure, the optimal pricing and service support policies are given by*

$$W^* = \frac{\alpha\mathscr{C}(\theta)}{(\alpha + \beta + \gamma + \lambda + 1)} \tag{19.12}$$

$$P^* = \frac{\alpha^2\mathscr{C}(\theta)}{(\alpha + 1)(\alpha + \beta + \gamma + \lambda + 1)} \tag{19.13}$$

$$T^* = \frac{-\beta(\beta + \gamma + \lambda + 1)\mathscr{C}(\theta)}{R\theta(\beta + 1)(\alpha + \beta + \gamma + \lambda + 1)} \tag{19.14}$$

$$S^* = \frac{\delta R}{\psi(\delta + \gamma + 1)} \tag{19.15}$$

$$F^* = \frac{\beta\gamma(\gamma + 1)(\beta + \gamma + \lambda + 1)\mathscr{C}(\theta)}{(1 - \psi)(\beta + 1)(\delta + \gamma + 1)(\gamma + \lambda + 1)(\alpha + \beta + \gamma + \lambda + 1)} \tag{19.16}$$

$$L^* = \frac{-\beta\lambda(\beta + \gamma + \lambda + 1)\mathscr{C}(\theta)}{R\theta(\beta + 1)(\gamma + \lambda + 1)(\alpha + \beta + \gamma + \lambda + 1)} \tag{19.17}$$

Proof. See the Appendix.

19.6 A NUMERICAL EXAMPLE

The model results can be best illustrated by employing a numerical example. In our example, we consider the following values of model parameters:

$$\alpha = -2.50, \quad \beta = +0.35, \quad \delta = -1.25, \quad \gamma = -1.25,$$

$$\lambda = +0.30, \quad \kappa = 1.00 \times 10^{16}, \quad R = \$200.00, \quad \psi = 0.75$$

Also, we assume $\mathscr{C}(\theta) = C^+(1 - \theta)^\epsilon$, where $C^+ = \$500.00$, $\theta = 0.10$ per year, and $\epsilon = +1.50$.

The optimal values of product distribution policies, service support policies, and channel members profits in each channel are presented in Table 19.2. From this example, we gather that, in Model CC, the consumers are offered lower product prices (P) but higher prices and durations of after-sale service support variables (i.e., T, S, F, and L). In contrast, in Model DD, the customer is offered higher product prices but lower prices and durations of after-sale service support variables.

Table 19.2 Optimal Decision Variable Values from Numerical Example

Decision variables	Model CC	Model DD
P^*	\$508.22	\$847.04
W^*	—	\$508.22
T^*	4.22 yr	1.05 yr
S^*	\$224.56	\$222.22
F^*	\$1,016.45	\$351.36
L^*	19.31 yr	6.32 yr
Q^{*a}	1.39	0.65
Channel margin	\$203.29	\$469.32
Channel profits[a]	\$281.66	\$305.03

[a] Here, quantity demanded and channel profits are in thousands.

In selecting between CC and DD channel structures, the preference largely depends on the trade-offs between sensitivities of product distribution and warranty/service support at tributes. A summary of the effects of the various model parameters on product price (P), warranty duration (T), per-breakdown repair charge (S), service-contract fee (F), effective product life span (L), consumer-demand quantity (Q), channel margins ($M_M + M_R + M_S$), and total channel profits ($\Pi_M + \Pi_R + \Pi_S$) is presented in Tables 19.3 and 19.4 for CC and DD channel structures, respectively.

Note that as a parameter value is increased, the resulting effect on each of the decision variables in both models is indicated by a "↑" for an increase, a "↓" for a decrease, "?" when the direction depends on the particular data, and a "*ne*" for no effect. As expected, Tables 19.3 and 19.4 indicate that product price sensitivity (α), per-breakdown repair charge sensitivity (δ), and service-contract fee sensitivity (γ) have direct negative impact, and warranty duration sensitivity (β) and the effective

Table 19.3 Sensitivity Analysis for CC Model Policies

Parameter	P	T	S	F	L	Q	Channel margin[a]	Channel profits[b]
κ	*ne*	*ne*	*ne*	*ne*	*ne*	↑	*ne*	↑
α	↓	↓	*ne*	↓	↓	↓	↓	↓
β	↑	↑	↑	↑	↑	↑	↑	↑
δ	*ne*	↓	↑	*ne*	↑	↓	↓	↓
γ	↓	↓	*ne*	↑	↓	↓	↓	↓
λ	↑	↑	↓	↑	↑	↑	↑	↑
θ	↓	↓	*ne*	↓	↓	↑	↓	?
ψ	*ne*	*ne*	?	?	*ne*	?	*ne*	?
ϵ	↓	↓	*ne*	↓	↓	↑	↓	?
R	*ne*	↓	↑	*ne*	↓	↓	*ne*	↓
C^+	↑	↑	*ne*	↑	↑	↓	↑	↓

[a] Here, channel margin = $M_M + M_R + M_S$.
[b] Here, channel profits = $\Pi_M + \Pi_R + \Pi_S$.

Legend:

"↑" represents an increase in decision variable value when parameter is increased.
"↓" represents a decrease in decision variable value when parameter is increased.
"?" represents that direction depends upon the particular parameter values.
"*ne*" represents no effect on decision variable value when parameter is increased.

Table 19.4   Sensitivity Analysis for DD Model Policies

	Effect on								
Parameter	P	W	T	S	F	L	Q	Channel margin[a]	Channel profits[b]
κ	*ne*	*ne*	*ne*	*ne*	*ne*	*ne*	↑	*ne*	↑
α	↓	↓	↓	*ne*	↓	↓	↓	↓	↓
β	↑	↑	↑	*ne*	↑	↑	↓	↑	↓
δ	*ne*	*ne*	*ne*	↑	↓	*ne*	↓	↓	↓
γ	↓	↓	↓	↓	↑	↑	↑	↑	↑
λ	↑	↑	↑	*ne*	↓	↓	↑	↓	↑
θ	↓	↓	↓	*ne*	↓	↓	↑	↓	?
ψ	*ne*	*ne*	*ne*	?	?	*ne*	?	*ne*	?
ε	↓	↓	↓	*ne*	↓	↓	↑	↓	?
R	*ne*	*ne*	↓	↑	*ne*	↓	↓	*ne*	↓
C^+	↑	↑	↑	*ne*	↑	↑	↓	↑	↓

[a] Here, channel margin $= M_M + M_R + M_S$.
[b] Here, channel profits $= \Pi_M + \Pi_R + \Pi_S$.

Legend: See Table 19.3

product life span sensitivity (λ) have direct positive impact on most decision variables.

It is interesting to note that although an increase in warranty duration sensitivity has a direct positive impact on total channel profits in case of CC channel structure, the impact is *negative* in case of DD channel structure. This makes intuitive sense because, in the case of DD channel structure, the manufacturer is responsible for offering basic warranty on the product to the consumer (which is a cost variable), whereas the remaining service support variables (including all variables that generate revenues) are controlled by the servicer. Here, an increase in warranty duration sensitivity will make a warranty with longer coverage more desirable to the consumer. This, in turn, will increase the total cost to the manufacturer, thereby reducing not only the manufacturer's profits but also the retailer's profits because such a cost increase will be reflected in manufacturer's wholesale price, which, in turn, dictates the final product price charged by the retailer to the consumer. On the other hand, the servicer's profits will increase because warranty duration sensitivity has a positive impact on the values of service contract fee and the effective product life span. However, the net effect of an increase in warranty duration sensitivity will be negative on total channel profits because major proportion of these profits belong of the manufacturer and retailer.

It can be observed that the changes in parameter values of product price sensitivity, average cost of per-breakdown repairs, proportion of consumers opting for per-breakdown repair service, measure of product quality, and product's manufacturing cost have no impact on channel structure choice. As an example in our numerical illustration, the DD channel structure is dominant (i.e., the DD channel structure provides the maximum total channel profits) and this channel structure will remain dominant in spite of any changes in the values of parameters indicated earlier. On the other hand, the preference for a channel structure, in terms of total channel profits, will depend on the sensitivities of after-sale service support-based product attributes, including warranty duration sensitivity, per-breakdown repair charge sensitivity, service-contract fee sensitivity, and the effective product life span sensitivity (i.e., the channel structure choice will change from DD channel structure to CC channel structure, or vice versa, depending on these parameter values).

It follows that in industries in which consumers are more sensitive to warranty duration and service-contract fee and less sensitive to per-breakdown repair charge and effective product life span, the preference should be to implement a CC channel structure in order to obtain the maximum total channel profits. On the other hand, in industries in which consumers are less sensitive to warranty duration and service-contract fee and more sensitive to per-breakdown repair charge and product life span, the preference should be to adopt a DD channel structure to realize the maximum total channel profits.

19.7 CONCLUSIONS

This chapter has given an overview of the interaction between the issue of product warranty and distribution channel structure design choices. The channel structure choice decision significantly impacts the product warranty decision and this chapter has demonstrated that it is imperative that firms consider both these choices simultaneously or else risk suboptimal decisions in today's business environment.

Our models of such environments include the choice of distribution as well as after-sale service support together with pricing and warranty decisions. Although simplistic in nature, the models are rich enough to illustrate design trade-offs between centralizing or decentralizing the distribution/service support operations. From the numerical example, optimal warranty time periods are much larger when the manufacturer centralizes distribution/service support operations. Also, it is clearly seen that superiority of centralization depends on product/firm/industry character-istics.

In general, our models conclude that in environments of products with "low substitutability," which are indicated by high sensitivity to both warranty duration and service-contract fee and low sensitivity to both per-breakdown repair charge and effective product life span, a CC channel structure will generate the maximum total channel profits. In con-trast, in environments of products with "high substitutability," which are indicated by low sensitivity to both warranty duration and service contract fee and high sensitivity to both per-breakdown repair charge and effective product life, a DD channel structure will generate the maximum total channel profits. This suggests that the nature of product and industry will have significant influence on the design and management of channels for product distribution and warranty/service support.

The managerial insight into product warranty/service support ele-ments and channel design choices, in the chapter, have been developed using monopoly models. It would be interesting to see if the same would be true under competitive conditions. Also, the model assumes a particu-lar type of product warranty (i.e., ordinary free-replacement warranty) offered by the manufacturer. Sensitivity to other types of warranty struc-tures would be useful to pursue.* Finally, some empirical analysis of prod-uct warranty/distribution channel environments in specific industries could reveal model improvements and, potentially, confirmation of the results from the analysis in this chapter.

APPENDIX

A.1 Policy Derivations for the CC Model

In this model, the manufacturer determines the optimal value of all product distribution and service support variables (i.e., P^*, T^*, S^*, F^*, and L^*). Here, the manufacturer's profit function is given by

$$\max_{P,T,S,F,L} \Pi_M = [P - \mathscr{C}(\theta) + \psi S\theta(L - T) + (1 - \psi)F - R\theta L]Q$$

The first-order conditions for profit maximization are

---

* Other types of warranty structures are discussed in greater detail by Blischke in Chapters 10 and 12, by Patankar and Mitra in Chapter 11, by Moskowitz and Chun in Chapter 13, and by Wilson and Murthy in Chapter 14.

$$\frac{\partial \Pi_M}{\partial P} = \frac{\alpha}{P} [P - \mathscr{C}(\theta) + \psi S\theta(L - T) + (1 - \psi)F$$
$$- R\theta L]Q + Q = 0$$

$$\frac{\partial \Pi_M}{\partial T} = \frac{\beta}{T} [P - \mathscr{C}(\theta) + \psi S\theta(L - T) + (1 - \psi)F$$
$$- R\theta L]Q - \psi S\theta Q = 0$$

$$\frac{\partial \Pi_M}{\partial S} = \frac{\delta}{S} [P - \mathscr{C}(\theta) + \psi S\theta(L - T) + (1 - \psi)F$$
$$- R\theta L]Q + \psi\theta(L - T)Q = 0$$

$$\frac{\partial \Pi_M}{\partial F} = \frac{\gamma}{F} [P - \mathscr{C}(\theta) + \psi S\theta(L - T) + (1 - \psi)F$$
$$- R\theta L]Q + (1 - \psi)Q = 0$$

$$\frac{\partial \Pi_M}{\partial L} = \frac{\lambda}{L} [P - \mathscr{C}(\theta) + \psi S\theta(L - T) + (1 - \psi)F$$
$$- R\theta L]Q + (\psi S - R)\theta Q = 0$$

Solving above equations for optimal values of manufacturer's decision variables, we get

$$P^* = \frac{\alpha \mathscr{C}(\theta)}{\alpha + \beta + \gamma + \lambda + 1}$$

$$T^* = \frac{-\beta(\beta - \delta + \lambda)\mathscr{C}(\theta)}{R\theta(\beta - \delta)(\alpha + \beta + \gamma + \lambda + 1)}$$

$$S^* = \frac{R(\beta - \delta)}{\psi(\beta - \delta + \lambda)}$$

$$F^* = \frac{\gamma \mathscr{C}(\theta)}{(1 - \psi)(\alpha + \beta + \gamma + \lambda + 1)}$$

$$L^* = \frac{-(\beta - \delta + \lambda)\mathscr{C}(\theta)}{R\theta(\alpha + \beta + \gamma + \lambda + 1)}$$

$\square$

A.2 Policy Derivations for the DD Model

In this model, the manufacturer is responsible for determining W^* and T^*. The retailer, on the other hand, is responsible for determining P^*. In addition, the servicer is responsible for determining S^*, F^*, and L^*. Here, the manufacturer wishes to maximize the following profit function:

$$\max_{W,T} \Pi_M = [W - \mathscr{C}(\theta) - R\theta T]Q$$

The retailer wants to maximize the following profit function:

$$\max_{P} \Pi_R = [P - W]Q$$

Finally, the servicer wants to maximize the following profit function:

$$\max_{S,F,L} \Pi_S = [\psi S\theta(L - T) + (1 - \psi)F - R\theta(L - T)]Q$$

The first-order conditions for profit maximization for manufacturer, retailer, and servicer are

$$\frac{\partial \Pi_M}{\partial T} = \frac{\beta}{T}[W - \mathscr{C}(\theta) - R\theta T]Q - R\theta Q = 0$$

$$\frac{\partial \Pi_R}{\partial P} = \frac{\alpha}{P}[P - W]Q + Q = 0$$

$$\frac{\partial \Pi_S}{\partial S} = \frac{\delta}{S}[\psi S\theta(L - T) + (1 - \psi)F - R\theta(L - T)]Q$$
$$+ \psi\theta(L - T)Q = 0$$

$$\frac{\partial \Pi_S}{\partial F} = \frac{\gamma}{F}[\psi S\theta(L - T) + (1 - \psi)F - R\theta(L - T)]Q$$
$$+ (1 - \psi)Q = 0$$

$$\frac{\partial \Pi_S}{\partial L} = \frac{\lambda}{L}[\psi S\theta(L - T) + (1 - \psi)F - R\theta(L - T)]Q$$
$$+ (\psi S - R)\theta Q = 0$$

Solving the above equations for optimal values of the decision variables, we get

$$P^* = \frac{\alpha W}{\alpha + 1}$$

$$T^* = \frac{\beta[W - \mathscr{C}(\theta)]}{R\theta(\beta + 1)}$$

$$S^* = \frac{\delta R}{\psi(\delta + \gamma + 1)}$$

$$F^* = \frac{-\beta\gamma(\gamma + 1)[W - \mathscr{C}(\theta)]}{(1 - \psi)(\beta + 1)(\delta + \gamma + 1)(\gamma + \lambda + 1)}$$

$$L^* = \frac{\beta\lambda[W - \mathscr{C}(\theta)]}{R\theta(\beta + 1)(\gamma + \lambda + 1)}$$

Here, it is interesting to note that optimal value of repair service charge, S^*, is a function of average repair cost, R, only. On the other hand, the optimal values of P^*, T^*, F^* and L^* are all functions of the selling price charged by the manufacturer, W, and product quality, θ. Therefore, after substituting the optimal values of decision variables back in the profit equation, we can rewrite the manufacturer's profit equation as a function of W and θ:

$$\Pi_M(W, \theta) = \left(\frac{W - \mathscr{C}(\theta)}{\beta + 1}\right) Q$$

The optimal selling price charged by the manufacturer, W, can be determined by solving the following first-order condition:

$$\frac{\partial \Pi_M}{\partial W} = \frac{\partial}{\partial W}\left(\frac{W - \mathscr{C}(\theta)}{\beta + 1}\right) Q = 0$$

$$\Rightarrow \left(\frac{\alpha}{P}\frac{\partial P}{\partial W} + \frac{\beta}{T}\frac{\partial T}{\partial W} + \frac{\gamma}{F}\frac{\partial F}{\partial W} + \frac{\lambda}{L}\frac{\partial L}{\partial W}\right)\left(\frac{W - \mathscr{C}(\theta)}{\beta + 1}\right) Q$$

$$+ \frac{\partial}{\partial W}\left(\frac{W - \mathscr{C}(\theta)}{\beta + 1}\right) Q = 0$$

After simplification, we obtain

$$W^* = \frac{\alpha\mathscr{C}(\theta)}{\alpha + \beta + \gamma + \lambda + 1}$$

Substituting, the value of W from the above equation, the optimal values of decision variables can be rewritten as function of θ:

$$P^* = \frac{\alpha^2\mathscr{C}(\theta)}{(\alpha + 1)(\alpha + \beta + \gamma + \lambda + 1)}$$

$$T^* = \frac{-\beta(\beta + \gamma + \lambda + 1)\mathscr{C}(\theta)}{R\theta(\beta + 1)(\alpha + \beta + \gamma + \lambda + 1)}$$

$$S^* = \frac{\delta R}{\psi(\delta + \gamma + 1)}$$

$$F^* = \frac{\beta\gamma(\gamma + 1)(\beta + \gamma + \lambda + 1)\mathcal{C}(\theta)}{(1 - \psi)(\beta + 1)(\delta + \gamma + 1)(\gamma + \lambda + 1)(\alpha + \beta + \gamma + \lambda + 1)}$$

$$L^* = \frac{-\beta\lambda(\beta + \gamma + \lambda + 1)\mathcal{C}(\theta)}{R\theta(\beta + 1)(\gamma + \lambda + 1)(\alpha + \beta + \gamma + \lambda + 1)}$$

□

REFERENCES

1. McGuire, E. P. (1980). *Industrial Product Warranties: Policies and Practices*, The Conference Board, New York.
2. Armistead, C., and Clark, G. (1994). *Outstanding Customer Service: Implementing the Best Ideas from Around the World*, Financial Times/Irwin Professional Publishing, Burr Ridge, IL.
3. Rizzo, M. R. (1987). *How To Make Profits with Service Contracts*, AMACOM, New York.
4. Day, E., and Fox, R. J. (1985). Extended warranties, service contracts, and maintenance agreements—A marketing opportunity, *Journal of Consumer Marketing*, 2(4), 77–86.
5. Kotler, P. (1988). *Marketing Management: Analysis, Planning, Implementation, and Control*, 6th ed., Prentice-Hall, Englewood Cliffs, NJ.
6. Sutton, H. (1986). *Rethinking the Company's Selling and Distribution Channels*, The Conference Board, New York.
7. McInnes, W. (1964). A conceptual approach to marketing, in *Theory in Marketing* (R. Cox, W. Alderson, and S. J. Shapiro, eds.), Richard D. Irwin, Homewood, IL, pp. 51–67.
8. Alderson, W. (1954). Factors governing the development of marketing channels, in *Marketing Channels for Manufactured Products* (R. M. Clewett, ed.), Richard D. Irwin, Homewood, IL, p. 263.
9. Corey, R. E. (1976). *Industrial Marketing: Cases and Concepts*, Prentice-Hall, Englewood Cliffs, NJ, p. 263.
10. Weigand, R. E. (1963). The marketing organization, channels, and firm size, *Journal of Business*, **36**, 228–236.
11. Bucklin, L. P. (1966). *A Theory of Distribution Channel Structure*, IBER Special Publications, University of California, Berkeley, CA, pp. 5, 16.
12. Aspinwall, L. (1958). The characteristics of goods and parallel systems theories, in *Managerial Marketing* (E. Kelly and W. Lazer, eds.), Richard D. Irwin, Homewood, IL, pp. 434–450.
13. Duncan, D. J. (1951). Selecting a channel of distribution, in *Marketing by Manufacturers*, rev. ed., (C. F. Phillips, ed.), Richard D. Irwin, Chicago, IL, pp. 173–238.
14. Mallen, B. E. (1973). Functional spin-off: A key to anticipating change in distribution structure, *Journal of Marketing*, 37(3), 18–25.
15. Stern, L. W., and El-Ansary, A. I. (1982). *Marketing Channels*, 2nd ed., Prentice-Hall, Englewood Cliffs, NJ.
16. Cespedes, F. V. (1988). Control vs. resources in channel design: Distribution differences in one industry, *Industrial Marketing Management*, **17**, 215–227.

17. Lilien, G. L., and Kotler, P. (1983). *Marketing Decision Making: A Model Building Approach*, Harper and Row, New York, pp. 433–488.

18. Artle, R., and Berglund, S. (1959). A note on manufacturer's choice on distribution channels, *Management Science*, **5**(4), 460–471.

19. Bucklin, L. P. (1970). Management of the channel, in *Managerial Analysis in Marketing*, (Sturdivant et al., eds.), Scott Foresman, Glenview, IL, pp. 620–662.

20. Weigand, R. E. (1977). Fit products and channels to your markets, *Harvard Business Review*, **55**, 95–105.

21. Burns, L. D., Hall, R. W., Blumenfeld, D. E., and Daganzo, C. F. (1985). Distribution strategies that minimize transportation and inventory costs, *Operations Research*, **33**(3), 469–490.

22. Anderson, E., and Coughlan, A. T. (1987). International market entry and expansion via independent or integrated channels of distribution, *Journal of Marketing*, **51**(1), 71–82.

23. Hartung, P. H., and Fisher, J. L. (1965). Brand switching and mathematical programming in market expansion, *Management Science*, **11**(10), B231–B243.

24. Jackson, D. M., Krampf, R. F., and Konopa, L. J. (1982). Factors that influence the length of industrial channels, *Industrial Marketing Management*, **11**, 263–268.

25. Powers, T. L. (1989). Industrial distribution options: Trade-offs to consider, *Industrial Marketing Management*, **18**, 155–161.

26. Miller, R. L., Lewis, W. F., and Merenski, J. P. (1985). A value exchange model for the channel of distribution: Implications for management and research, *Journal of the Academy of Marketing Science*, **13**(4), 1–17.

27. Cohen, M. A., and Lee, H. L. (1988). Strategic analysis of integrated production–distribution systems: Models and methods, *Operations Research*, **36**(2), 216–228.

28. Loomba, A. P. S., and Kumar, K. R. (1994). Product distribution and service support decisions in a monopoly environment, Working Paper, College of Business Administration, University of Northern Iowa.

29. Mamer, J. W. (1987). Discounted and per unit costs of product warranty, *Management Science*, **33**(7), 916–930.

30. Nash, J. F. (1950). Equilibrium points in *N*-person games, *Proceedings of the National Academy of Sciences*, **36**, 48–49.

31. Stackelberg, H. von (1934). *The Theory of Market Economy*, Oxford University Press, New York. [Reprinted 1952.]

32. Davis, D. J. (1952). An analysis of some failure data, *Journal of the American Statistical Association*, **47**(258), 113–150.

33. Epstein, B. (1958). The exponential distribution and its role in life-testing, *Industrial Quality Control*, **15**(6), 2–7.

34. Dell Computer Corp. (Changes Field Service, Installation Contractor from HoneyWell Inc. to Xerox Corp.), *Wall Street Journal*, March 6, 1989, A12(W) and A7(E).

20

Design of Warranty Policies to Balance Consumer and Producer Risks and Benefits

Richard Marcellus and Ba Pirojboot

Northern Illinois University
DeKalb, Illinois

20.1 INTRODUCTION

When a product is marketed with a warranty, the producer must choose a particular warranty policy and all the parameters that define it. Finding the best warranty is a nontrivial problem, as is evident from the variety of warranty policies displayed in Chapter 1. This chapter does not contain guidelines for inventing a warranty policy or even for choosing one of the warranty policies defined in Chapter 1. Rather, it is assumed that the producer has decided to use one of the three most common warranty policies. (These are defined in the next section.)

The contents of the chapter are as follows. First, the warranties considered are defined, followed by a discussion of warranty design, with emphasis on the approach that is presented in this chapter. Then, for each of three commonly used warranty policies, specific formulas are given that can be used to choose warranties acceptable to both consumer and producer.

20.2   THE WARRANTIES CONSIDERED

For the warranties in this chapter, it is assumed that the consumer has a continuing perpetual need for a product over a long time period. The product is nonrepairable and so must be replaced by a new, identical item upon each failure.

The consumer's acquisition cost for the initial item is $C_A(0)$ if the item is purchased without a warranty. If the initial item is purchased with a warranty of length w, then the consumer's acquisition cost is $C_A(w)$. If an item fails before its warranty expires, it is replaced with the consumer incurring either reduced cost or no cost, depending on the terms of the warranty. If an item fails after its warranty expires, it is replaced, with the consumer incurring the full cost. The cost of replacing the item is either $C_A(0)$ or $C_A(w)$, depending on whether or not the item is provided with a warranty. This chapter deals with nonrenewing and renewing free-replacement warranties (FRWs) and renewing pro-rata warranties (PRWs). These warranties are defined in Chapter 1 as follows:

Policy 1

> *NONRENEWING FREE-REPLACEMENT WARRANTY*: The producer agrees to provide replacements for failed items free of charge up to a time w from the time of the initial purchase. The warranty expires at time w after purchase.

Policy 14

> *RENEWING FREE-REPLACEMENT WARRANTY*: The producer agrees to provide replacements for failed items free of charge up to a time w from the initial purchase. Whenever there is a replacement, the failed item is replaced by a new item with a new warranty whose terms are identical to those of the original warranty.

Policy 15

> *RENEWING PRO-RATA WARRANTY*: The producer agrees to provide replacement items, at prorated cost, for failed items up to a time w from the initial purchase. Whenever there is a replacement, the failed item is replaced by a new item with a new warranty whose terms are identical to those of the original warranty. Specifically, it is assumed that the prorating is linear. If the length of the warranty is w and the product fails after being used for X time units, with X less than w, then the product is replaced with the consumer incurring cost $(X/w)C_A(w)$.

20.3 WARRANTY DESIGN

Many factors should be considered in warranty design. Among them are the distribution of the product's lifetime, the effect of the warranty on market share and total sales, the costs of servicing warranty claims, the costs of replacing or repairing the items that will fail while the warranty is in effect, and the probability of meeting a desired profit goal. Although several approaches to warranty design have been suggested, a unified and complete theory of warranty design does not exist.

Perhaps the most straightforward design strategy is to first choose the warranty type and length, and then estimate the total expected warranty reserve needed for the product. The product would then be priced to provide funds to cover the expected warranty reserve, perhaps with a built-in safety factor. The cost models of Lowerre [1] and Menke [2] and their many successors can be used in this way. Blischke and Murthy [3] have given an extensive account of these cost models.

Another approach is to set up an optimization model for the joint choice of price and warranty length. This has been done by Thomas [4], Glickman and Berger [5], Anderson [6], Murthy [7], and Mitra and Patankar [8]. Perhaps the most comprehensive framework is provided by Mitra and Patankar [8] who consider warranty design in the context of production planning, marketing, and financing.

Still another approach to warranty design is to first determine all the warranty–price combinations that are acceptable to the producer, and then from these, choose the one most attractive to the consumer. In this connection, Menke [2] has suggested choosing the most marketable warranty out of all those requiring the same warranty reserve, and several authors have suggested basing the design on "life-cycle costing."

Life-cycle costing is based on an idealized model of the relationship between the consumer and producer. In this model, the consumer requires the product over a certain time frame or "life cycle," and throughout the life cycle acquires a replacement from the same producer whenever the product fails (whether under warranty or not).

This creates an idealized sequence of transactions between the consumer and producer. At each of these transactions, the consumer will incur the total cost of the product if it has failed while not under warranty, and partial cost otherwise. Also, at each of these transactions, the producer will gain partial or complete profit, depending on whether or not the product failed under warranty. (This basic model may be extended by adding such things as inconvenience costs to the consumer or service costs for the producer or by assuming that the consumer may go to another producer when a replacement is required.)

The two sequences, one of costs and the other of profits, can be evaluated in several different ways. One is to calculate an equivalent present worth for a chosen minimal attractive rate of return (MARR). Another is to calculate the total cost or profit. Another is to calculate the cost or profit rate. Another is to first assign a disutility or utility to each cost or profit and then calculate the equivalent present worth of these disutilities or utilities. To actually calculate these different measures of the consumer's and producer's transactions is seldom a trivial undertaking. Many investigators have contributed to these calculations. Among them are Brown [9], Blischke and Scheuer [10,11], Mamer [12,13], Thomas [4], Ritchken [14], Ritchken and Tapiero [15], and Biedenweg [16].

Once the consumer's life-cycle cost and the producer's life-cycle profit have been calculated, they can be used to design warranties. In the indifference costing method, the consumer's cost for a single item without warranty is assumed known, and the producer chooses the consumer's cost for a single item with warranty in order to make the consumer's life-cycle cost with warranty identical to the consumer's life-cycle cost without warranty. The price determined in this method is called the consumer's indifference price.

The consumer's indifference price may or may not be attractive to the producer. If the producer has significant cost for servicing warranty claims, then offering a warranty at the consumer's indifference price will cause the producer's life-cycle profit with warranty to be less than the producer's life-cycle profit without warranty. The producer may still be compelled to offer a warranty or may wish to offer a warranty for promotional or protective purposes. In this case, the producer may consider the producer's indifference price, defined as the price that makes the producer's life-cycle profit with warranty identical to the producer's life-cycle profit without warranty. The producer's indifference price will be greater than the consumer's indifference price. The actual marketing price should be set somewhere between the consumer's indifference price and the producer's indifference price, depending on managerial judgment on the benefits of offering the warranty. Details of indifference pricing are given in Ref. 3 and the references therein.

The remainder of this chapter contains the details of a life-cycle costing method that allows explicit quantification of the consumer and producer benefits of warranties.

20.4 RISKS AND BENEFITS OF OFFERING WARRANTIES

20.4.1 Consumer Risks and Benefits

For the consumer, the primary benefit of a warranty is reduction of risk. In the case of a single purchase, a warranty ensures a minimal period of use without unreasonable replacement costs. In the case of multiple purchases over a long period of time, warranties decrease the consumer's overall expected cost. For these reasons, the consumer should be willing to pay more for a product with warranty than for a product without warranty, as long as the increased cost is not excessive. The increased cost would then be exchanged for decreased risk.

20.4.2 The Consumer's Acquisition Cost for a Single Item

It is assumed that the consumer must pay more for the product when it is provided with a warranty; that is, $C_A(w)$ is greater than $C_A(0)$. The producer must choose the value of $C_A(w)$. One possibility is to let $C_A(w) = C_A(0) + Aw$, where $A > 0$. Then the surcharge for a warranty is proportional to the length of the warranty. Another possibility is to let $C_A(w) = B(w)C_A(0)$, where $B(w) > 1$. Then the product cost with warranty magnifies the cost without a warranty. The magnification factor, $B(w)$, should increase as the length of the warranty increases, perhaps tapering off to a constant.

20.4.3 Producer Risks and Benefits

The producer's principal risk for offering a warranty is the potential cost for any necessary replacements that occur before warranties expire. There may also be significant administrative costs involved in maintaining a warranty program. The benefits include protection from litigation through definition and limitation of liability, acquisition of information about the consumers through collection of warranty cards, feedback about quality and production problems through warranty claims, and increased sales volume and market share through increased perceptions of quality and reliability. Issues of this sort are discussed by Udell and Anderson [17]. Thomas [4] has suggested a way to quantify the producer's benefits in life-cycle costing.

20.4.4 The Producer's Cost of Providing a Single Item

Life-cycle costing models require the producer's amortized cost per single item of manufacturing and marketing the product, excluding the costs due to replacement of items that fail while under warranty. This cost is denoted $C_M(w)$, with $C_M(0)$ denoting the similar cost for products offered without warranty.

If it is truly profitable to offer the warranty, then the producer's per-item cost of production, marketing, and servicing all warranties may very well be smaller when the product is offered with warranty than when the product is offered without warranty. This is because out of all the costs involved in offering the product—advertising, marketing, manufacturing, and possible future liabilities—some are fixed. If the sales volume is increased, the fixed cost per item will decrease. If there are economies of scale in the manufacture of the item, the variable cost per item will also decrease with increased sales volume. Offering the warranty may increase sales volume and therefore decrease the cost per item of selling the product. Glickman and Berger [5], Murthy [7], and Blischke and Murthy [3] have proposed models of total sales as a function of warranty length. Mitra and Patankar [8] have proposed a model for market share as a function of warranty length. In these models, more attractive warranties increase sales and market share.

If the producer's cost amortized per item is less when the product is provided with a warranty, then $C_M(w)$ is less than $C_M(0)$. In the examples, the value of $C_M(w)$ is set to $\gamma(w)C_M(0)$, with $0 < \gamma(w) \leq 1$. Then the product cost per item is decreased by a reduction factor $\gamma(w)$ that depends on the warranty length.

There is no well-recognized way to construct the function $\gamma(w)$. The actual values of $\gamma(w)$ should be set by managerial judgment. One possibility is to define $\gamma(w)$ in a piecewise fashion, assuming the following:

1. Short warranties, say from time 0 up to t_1, would not reduce the producer's per item cost at all.
2. Warranties of length t_1 would reduce the producer's per item cost by a certain factor, say γ_1.
3. The reduction factor would decrease linearly from γ_1 to γ_2 for warranties from length t_1 up to time t_2.
4. For all warranties greater than t_2 time units, the reduction factor would be γ_2.

Then $\gamma(w)$ can be defined as

$$\gamma(w) = \begin{cases} 1.0 & \text{for } 0 \leq w < t_1 \\ \gamma_1 - \dfrac{(w - t_1)(\gamma_1 - \gamma_2)}{t_2 - t_1} & \text{for } t_1 \leq w < t_2 \\ \gamma_2 & \text{for } t_2 \leq w \end{cases}$$

An example is shown in Figure 20.1. (This is the function used in Example 20.1.)

Another possibility is to use a backward-S-shaped curve such as

$$\gamma(w) = \gamma_{min} + (1 - \gamma_{min}) \exp\left(-\frac{w^2}{2}\right)$$

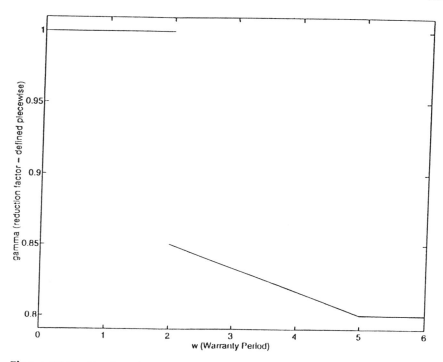

Figure 20.1 Producer's cost reduction factor defined piecewise.

This function satisfies $\gamma(0) = 1.0$ and decreases toward γ_{min} as w increases. Thus, the minimal reduction factor is γ_{min}. An example is shown in Figure 20.2. (This is the function used in Examples 20.4 and 20.6.)

Of course, if $\gamma(w) \equiv 1$, then the producer foresees no particular benefit from offering warranties.

20.4.5 Warranties Beneficial to Both Consumer and Producer

Sections 20.7 through 20.9 contain conditions that can be used to find warranties beneficial to both consumer and producer. The formulas in these conditions are based on the life-cycle model described in Section 20.3. It is assumed that the product is required over an essentially infinite life cycle, that the consumer will always purchase the same type of warranty from the same producer, that the producer will always be available to replace the product, that there is negligible cost involved in claiming

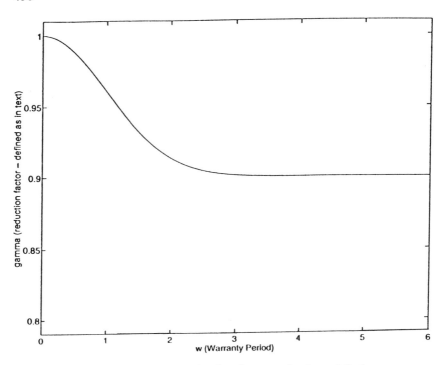

Figure 20.2   Producer's cost reduction factor as backward-S shape.

and collecting the warranty, and that there is negligible time involved in replacing the product.

If the product fails while the warranty is in force, the producer bears part or all of the cost of the replacement, thus reducing the profit for the individual transaction. For failures that occur after the warranty expires, the consumer bears the full cost of the replacement. In this case, the producer's profit is the consumer's cost minus the per item cost of offering the product with warranty.

Specifically, the consumer incurs a cost of $C_A(w)$ for the initial purchase and for each subsequent purchase resulting from a failure while the warranty is not in effect. Otherwise, for a failure that occurs while the warranty is in effect, the consumer incurs no cost for the FRW and cost $(X/w)C_A(w)$ for the PRW. The producer's profit is $C_A(w) - C_M(w)$ for a purchase that occurs while the warranty is not in effect. At each transaction that occurs while a warranty is in force, the producer's profit is reduced by $C_A(w)$ for the FRW and by $(1 - X/w)C_A(w)$ for the PRW.

A warranty is considered beneficial to the consumer if the life-cycle cost without warranty is greater than the life-cycle cost with the warranty. A warranty is considered beneficial to the producer if the life-cycle profit without warranty is less than the life-cycle profit with the warranty.

Although, in many cases, the decision to purchase with or without warranty is not available to the consumer, it is still advantageous for the producer to market a warranty attractive to the consumer. If consumers recognize a disadvantageous warranty, the product's reputation will be degraded, with possible reduction in sales volume. In a competitive market, the producer most successful at finding warranties attractive to consumers will gain a competitive advantage.

Two ways are presented for evaluating the cost or profit per consumer–producer relationship. They are the long-run-time-average cost or profit and the expected discounted value with minimal attractive rate of return ρ. Formally, the long-run-time-average cost or profit is defined as

$$\lim_{T \to \infty} \frac{1}{T} \sum_{j=0}^{N(T)} K_j$$

where K_0 is the cost or profit at the initial purchase, $N(T)$ is the number of replacements that occur from time zero to T, and $K_1, K_2, K_3, \ldots$ are the costs or profits that occur at the replacements.

The expected discounted value with minimal attractive rate of return ρ is defined formally as the expected value of

$$\sum_{j=0}^{\infty} K_j \exp(-\rho T_j)$$

where K_0 is the cost or profit at the initial purchase, T_0 is time zero, $K_1, K_2, K_3, \ldots$ are the costs or profits that occur at the replacements, and $T_1, T_2, T_3, \ldots$ are the times at which the replacements occur.

Sections 20.5 and 20.6 give formulas for these two quantities, under the assumption that warranties are not purchased. Each of Sections 20.7 through 20.9 gives formulas for a specific warranty policy, as well as conditions for warranties beneficial to both consumer and producer. The formulas have two sources. The formulas for long-run-time-average cost or profit are based on the formulas in Ref. 12. The formulas for expected discounted cost or profit are based on the formulas in Ref. 13.

20.4.6 Using the Inequalities to Design Warranties

Formulas (20.14), (20.15), (20.24), (20.25), (20.34), and (20.35) contain conditions for warranties beneficial to both consumer and producer. Each of these formulas contains two inequalities involving three expressions. The expression on the extreme right is associated with the consumer and is called the "consumer's curve." The expression on the extreme left is associated with the producer and is called the "producer's curve." The expression in the middle is called the "determining curve."

Warranties acceptable to both consumer and producer are those for which the determining curve lies between the consumer's curve and the producer's curve. To find these warranties, the curves can be plotted using readily available software. The desired warranties can then be estimated from the graphs, as in the examples. The graphs that accompany the examples were produced via MATLAB (see Ref. 18).

20.4.7 Summary of Notation

To use the formulas presented in the sections below, it is necessary to know or estimate the probability distribution of the product's lifetime. Methods for finding or choosing this distribution are presented in Chapter 6 of this volume and also in Chapter 2 of Ref. 3. Some of the formulas require relatively sophisticated functions associated with the probability distribution of the product's lifetime. These are its renewal function and its renewal density (the derivative of the renewal function). These are discussed in Chapter 7 of this volume and also in Chapter 3 of Ref. 3.

The following quantities are used in Sections 20.5 through 20.9:

$X \equiv$ the product lifetime (a random variable)
$f(x) \equiv$ the probability density function of the product lifetime
$F(x) \equiv$ the cumulative distribution function of the product lifetime
$E[X] \equiv$ the expected value of the product lifetime
$\hat{f}(x) \equiv$ the ordinary Laplace transform of the probability density function
$M(w) \equiv$ the renewal function corresponding to the distribution of the product lifetime
$m(w) \equiv$ the renewal density corresponding to the distribution of the product lifetime
$\rho \equiv$ the minimal attractive rate of return (MARR), assumed the same for both consumer and producer
$C_A(0) \equiv$ the consumer's acquisition cost for a single item without warranty
$C_A(w) \equiv$ the consumer's acquisition cost for a single item with a warranty of length w
$C_M(0) \equiv$ the producer's cost for providing a single item without warranty
$C_M(w) \equiv$ the producer's cost for providing a single item with a warranty of length w, excluding the cost due to replacement of items that fail while under warranty.

20.5 CONSUMER'S EXPECTED COSTS WITHOUT WARRANTIES

The formulas in this section are to be compared with the corresponding formulas in Sections 20.7 through 20.9.

The *consumer's long-run-time-average cost without warranties* is

$$\frac{C_A(0)}{E[X]} \tag{20.1}$$

The *consumer's expected discounted cost without warranties* is

$$\frac{C_A(0)}{1 - \hat{f}(\rho)} \tag{20.2}$$

20.6 PRODUCER'S EXPECTED PROFITS WITHOUT WARRANTIES

The formulas in this section are to be compared with the corresponding formulas in Sections 20.7 through 20.9.

The *producer's long-run-time-average profit without warranties* is

$$\frac{C_A(0) - C_M(0)}{E[X]} \tag{20.3}$$

The *producer's expected discounted profit without warranties* is

$$\frac{C_A(0) - C_M(0)}{1 - \hat{f}(\rho)} \tag{20.4}$$

20.7 NONRENEWING FREE-REPLACEMENT WARRANTY

20.7.1 Consumer's Expected Costs

The *consumer's long-run-time-average cost for nonrenewing free-replacement warranties* of length w is

$$\frac{C_A(w)}{[1 + M(w)]E[X]} \tag{20.5}$$

For warranties beneficial to the consumer, this should be less than formula (20.1). A few algebraic steps lead to the following condition for the long-run-time-average cost to be less than the corresponding cost for no warranties:

$$\frac{1}{1 + M(w)} < \frac{C_A(0)}{C_A(w)} \tag{20.6}$$

The *consumer's expected discounted cost for nonrenewing free-replacement warranties* is:

$$\frac{C_A(w)}{1 - G(w, \rho)} \tag{20.7}$$

where

$$G(w, \rho) = \int_w^\infty e^{-\rho x} f(x) \, dx + \int_0^w \int_{w-y}^\infty e^{-\rho(x+y)} f(x) \, dx \, m(y) \, dy \tag{20.8}$$

For warranties beneficial to the consumer, this should be less than formula (20.2). A few algebraic steps lead to the following condition for the expected discounted cost to be less than the corresponding cost for no warranties:

$$\frac{1 - \hat{f}(\rho)}{1 - G(w, \rho)} < \frac{C_A(0)}{C_A(w)} \tag{20.9}$$

20.7.2 Producer's Expected Profit

The *producer's long-run-time-average profit for nonrenewing free-replacement warranties* of length w is

$$\frac{C_A(w) - C_M(w)[1 + M(w)]}{[1 + M(w)]E[X]} \tag{20.10}$$

For warranties beneficial to the producer, this should be greater than formula (20.3). A few algebraic steps lead to the following condition for the long-run-time-average profit to be greater than the corresponding profit for no warranties:

$$\frac{C_A(0) - C_M(0) + C_M(w)}{C_A(w)} < \frac{1}{1 + M(w)} \tag{20.11}$$

The *producer's expected discounted profit for nonrenewing free-replacement warranties* is

$$\frac{C_A(w) - C_M(w)\left[1 + \int_0^w e^{-\rho x} m(x) \, dx\right]}{1 - G(w, \rho)} \tag{20.12}$$

For warranties beneficial to the producer, this should be greater than formula (20.4). A few algebraic steps lead to the following condition for the expected discounted profit to be greater than the corresponding profit for no warranties:

$$\frac{C_A(0) - C_M(0)}{C_A(w) - C_M(w)\left[1 + \int_0^w e^{-\rho x} m(x) \, dx\right]} < \frac{1 - \hat{f}(\rho)}{1 - G(w, \rho)} \tag{20.13}$$

20.7.3 Warranties Beneficial to Both Consumer and Producer

For the long-run-time-average cost and profit, a nonrenewing FRW is acceptable to both consumer and producer if

$$\frac{C_A(0) - C_M(0) + C_M(w)}{C_A(w)} < \frac{1}{1 + M(w)} < \frac{C_A(0)}{C_A(w)} \tag{20.14}$$

This condition combines formulas (20.6) and (20.11). Note that calculating $M(w)$ is discussed in Chapter 7.

For the discounted cost and profit, a nonrenewing FRW is acceptable to both consumer and producer if

$$\frac{C_A(0) - C_M(0)}{C_A(w) - C_M(w)\left[1 + \int_0^w e^{-\rho x} m(x)\, dx\right]} < \frac{1 - \hat{f}(\rho)}{1 - G(w, \rho)} < \frac{C_A(0)}{C_A(w)}$$

$$(20.15)$$

This condition combines formulas (20.9) and (20.13). Note that calculating $G(w, \rho)$ requires $m(w)$, which is defined in Chapter 7. For most product lifetime distributions, numerical integration is required.

Example 20.1 (Choosing a Warranty Length Given a Pricing Policy)
Suppose that $C_M(0) = \$70$, that $C_A(0) = \$100$, and that the pricing scheme $C_A(w) = C_A(0) + Aw$ is used, with $A = 20$. Also suppose that the producer believes the following:

1. Warranties of less than two time units would not reduce the producer's per item cost at all.
2. Warranties of length two time units would reduce the producer's per-item cost by 15%.
3. The reduction in the producer's per-item cost would increase linearly from 15% to 20% for warranties from length two time units up to warranties of length five time units.
4. All warranties greater than five time units would reduce the producer's per-item cost by 20%.

Then $\gamma(w)$ can be defined as

$$\gamma(w) = \begin{cases} 1.0 & \text{for } 0 \le w < 2 \\ 0.85 - \left(\dfrac{0.05}{3}\right)(w - 2) & \text{for } 2 \le w < 5 \\ 0.80 & \text{for } 5 \le w \le 10 \end{cases}$$

Suppose the producer believes that the product's lifetime has an exponential distribution with parameter $\lambda = 0.25$. For the exponential distribution, $M(w) = \lambda w$. Thus, for this example, $M(w) = 0.25w$.

To use the long-run-time-average cost and profit as performance measure, these values should be substituted into formula (20.14). This results in the following condition for a nonrenewing FRW to be acceptable to both consumer and producer:

$$\frac{100 - 70\,[1 - \gamma(w)]}{100 + 20w} < \frac{1}{1 + 0.25w} < \frac{100}{100 + 20w}$$

In this condition, the consumer's curve, the producer's curve, and the determining curve appear as functions of w. If all three of these are graphed, with the determining curve as a solid curve, the consumer's curve as dots, and the producer's curve as dashes, then acceptable warranty periods are easily identified as the values of w where the solid curve lies between the other two curves. This is shown in Figure 20.3. In the

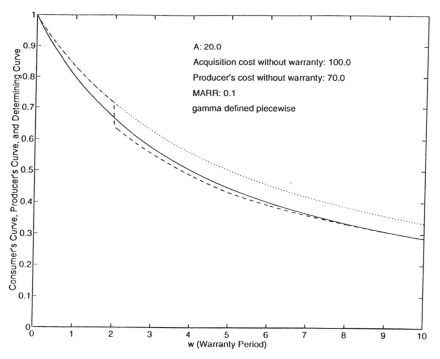

Figure 20.3 Choice of warranty period given cost parameters. (Nonrenewing FRW, expected discounted cost and profit.)

figure, the consumer's and producer's curves are identical up to $w = 2$. At that point, the proposed pricing of warranties becomes acceptable to both consumer and producer. Warranties of length 2 to about 8 are acceptable to both consumer and producer, with the shorter warranties more beneficial to the producer.

Example 20.2 (Finding Required Reduction in Producer's Per-Item Cost) Suppose that $C_M(0) = \$70$, that $C_A(0) = \$100$, and that the pricing scheme $C_A(w) = C_A(0) + Aw$ is used, with $A = 20$. Suppose the producer believes that the product's lifetime has an exponential distribution with parameter $\lambda = 0.25$. For the exponential distribution, $G(w, \rho) = \lambda \exp(-\rho w)/(\lambda + \rho)$, $1 + \int_0^w e^{-\rho x} m(x)\, dx = 1 + (\lambda/\rho)[1 - \exp(-\rho w)]$, and $\hat{f}(\rho) = \lambda/(\lambda + \rho)$. Suppose the MARR, ρ, is 0.10.

To use the expected discounted cost and profit as a performance measure, these values should be substituted into formula (20.15). This

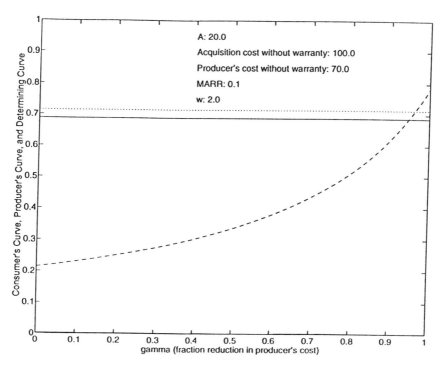

Figure 20.4 Determination of required cost reduction factor. (Nonrenewing FRW, long-run-time-average cost and profit.)

results in the following condition for a nonrenewing FRW to be acceptable to both consumer and producer:

$$\frac{30}{100 + 20w - 70\gamma(w)\{1 + 2.5[1 - \exp(-0.1w)]\}}$$
$$< \frac{1}{1 + 2.5[1 - \exp(-0.1w)]} < \frac{100}{100 + 20w}$$

Suppose the producer is contemplating a warranty length of two time units and wishes to know how small the reduction factor $\gamma(2)$ must be in order to make this warranty length attractive. Substituting $w = 2$ makes the consumer's curve and the determining curve constant. At the same time, the producer's curve becomes a function of $\gamma(2)$.

If all three of the curves are graphed as functions of $\gamma(2)$, with the determining curve as a solid curve, the consumer's curve as dots, and the producer's curve as dashes, then values of $\gamma(2)$ for which the solid curve lies between the other two curves would result in acceptable warranties. This is shown in Figure 20.4, where it is seen that all values of $\gamma(2)$ from zero up to about 0.95 would result in acceptable warranties. It is perhaps more important to note that at least a 5% per-single-item cost reduction is necessary to make a warranty of length 2 attractive to the producer.

20.8   RENEWING FREE-REPLACEMENT WARRANTY

20.8.1   Consumer's Expected Cost

The *consumer's long-run-time-average cost for renewing free-replacement warranties* is

$$\frac{C_A(w)[1 - F(w)]}{E[X]} \tag{20.16}$$

For warranties beneficial to the consumer, this should be less than formula (20.1). A few algebraic steps lead to the following condition for the long-run-time-average cost to be less than the corresponding cost for no warranties:

$$1 - F(w) < \frac{C_A(0)}{C_A(w)} \tag{20.17}$$

The *consumer's expected discounted cost for renewing free-replacement warranties* is

$$C_A(w) \left(\frac{1 - \int_0^w e^{-\rho x} f(x)\, dx}{1 - \hat{f}(\rho)} \right) \tag{20.18}$$

For warranties beneficial to the consumer, this should be less than formula (20.2). A few algebraic steps lead to the following condition for the expected discounted cost to be less than the corresponding cost for no warranties:

$$1 - \int_0^w e^{-\rho x} f(x) \, dx < \frac{C_A(0)}{C_A(w)} \tag{20.19}$$

20.8.2 Producer's Expected Profit

The *producer's long-run-time-average profit for renewing free-replacement warranties* is

$$\frac{[C_A(w) - C_M(w)][1 - F(w)] - C_M(w)F(w)}{E[X]} \tag{20.20}$$

For warranties beneficial to the producer, this should be greater than formula (20.3). A few algebraic steps lead to the following condition for the long-run-time-average profit to be greater than the corresponding profit for no warranties:

$$\frac{C_A(0) - C_M(0) + C_M(w)}{C_A(w)} < 1 - F(w) \tag{20.21}$$

The *producer's expected discounted profit for renewing free-replacement warranties* is

$$\frac{C_A(w)\left[1 - \int_0^w e^{-\rho x} f(x) \, dx\right] - C_M(w)}{1 - \hat{f}(\rho)} \tag{20.22}$$

For warranties beneficial to the producer, this should be greater than formula (20.4). A few algebraic steps lead to the following condition for the expected discounted profit to be greater than the corresponding profit for no warranties:

$$\frac{C_A(0) - C_M(0) + C_M(w)}{C_A(w)} < 1 - \int_0^w e^{-\rho x} f(x) \, dx \tag{20.23}$$

20.8.3 Warranties Beneficial to Both Consumer and Producer

For the long-run-time-average cost and profit, a renewing FRW is acceptable to both consumer and producer if

$$\frac{C_A(0) - C_M(0) + C_M(w)}{C_A(w)} < 1 - F(w) < \frac{C_A(0)}{C_A(w)} \tag{20.24}$$

This condition combines formulas (20.17) and (20.21). Note that formulas

Table 20.1 Definition of Lifetime Distributions and Formulas Needed for Long-Run-Time-Average Cost and Profit

Distribution	$f(x)$	$1 - F(x)$
Exponential	$\lambda e^{-\lambda x}$	$e^{-\lambda x}$
Gamma	$\dfrac{\lambda^\alpha x^{\alpha-1} e^{-\lambda x}}{\Gamma(\alpha)}$	$1 - \dfrac{\lambda^\alpha}{\Gamma(\alpha)} \displaystyle\int_0^x y^{\alpha-1} e^{-\lambda y}\, dy$
Weibull	$\dfrac{\beta x^{\beta-1}}{\theta^\beta} \exp\left[-\left(\dfrac{x}{\theta}\right)^\beta\right]$	$\exp\left[-\left(\dfrac{x}{\theta}\right)^\beta\right]$

for $1 - F(w)$ are given in column 3 of Table 20.1 for three commonly used lifetime distributions: the exponential, the gamma, and the Weibull.

For the expected discounted cost and profit, a renewing FRW is acceptable to both consumer and producer if

$$\frac{C_A(0) - C_M(0) + C_M(w)}{C_A(w)} < 1 - \int_0^w e^{-\rho x} f(x)\, dx < \frac{C_A(0)}{C_A(w)}$$

$$(20.25)$$

This condition combines formulas (20.19) and (20.23). Note that formulas for $1 - \int_0^w e^{-\rho x} f(x)\, dx$ are given in column 2 of Table 20.2 for three commonly used lifetime distributions: the exponential, the gamma, and the Weibull.

Table 20.2 Formulas Needed for Discounted Cost and Profit

Distribution	$1 - \displaystyle\int_0^w e^{-\rho x} f(x)\, dx$	$\dfrac{1}{w}\displaystyle\int_0^w x e^{-\rho x} f(x)\, dx$
Exponential	$\dfrac{\rho}{\lambda + \rho} + \dfrac{\lambda}{\lambda + \rho} e^{-(\lambda+\rho)w}$	$\dfrac{\rho}{w(\lambda + \rho)^2}\left(1 - e^{-(\lambda+\rho)w} - (\lambda + \rho)we^{-(\lambda+\rho)w}\right)$
Gamma	$1 - \dfrac{\lambda^\alpha}{\Gamma(\alpha)} \displaystyle\int_0^w x^{\alpha-1} e^{-(\lambda+\rho)x}\, dx$	$\dfrac{\lambda^\alpha}{w\Gamma(\alpha)} \displaystyle\int_0^w x^\alpha e^{-(\lambda+\rho)x}\, dx$
Weibull	$1 - \dfrac{\beta}{\theta^\beta} \displaystyle\int_0^w x^{\beta-1} \exp\left[-\left(\dfrac{x}{\theta}\right)^\beta - \rho x\right] dx$	$\dfrac{\beta}{w\theta^\beta} \displaystyle\int_0^w x^\beta \exp\left[-\left(\dfrac{x}{\theta}\right)^\beta - \rho x\right] dx$

Example 20.3 (Choosing a Price Given a Warranty Length)

Suppose that the producer wishes to offer a warranty of length 5, with the pricing scheme $C_A(5) = BC_A(0)$ for an acceptable B. Suppose that 0.9 is a good estimate of $\gamma(5)$. Also, suppose that $C_M(0) = \$70$ and that $C_A(0) = \$100$.

To use the long-run-time-average cost and profit as performance measure, these values should be substituted into formula (20.24). This results in the following condition for a renewing FRW acceptable to both consumer and producer:

$$\frac{100 - (70)(1.0 - 0.9)}{100B} < 1 - F(5) < \frac{1}{B}$$

In this condition, the consumer's and producer's curves appear as functions of B. The determining curve is constant with respect to B. If the consumer's curve is graphed as dots, the producer's curve is graphed as dashes, and the determining curve is graphed as a horizontal solid line, then acceptable values of B are those values where the solid line lies between the other two curves. In Figure 20.5, this is done simultaneously for three separate hypotheses on the distribution of the product lifetime. The three distributions are the exponential, the gamma, and the Weibull. The appropriate formulas can be found in the third column of Table 20.1. The parameters of the distributions have been deliberately chosen so that the expected value is equal to 4 in each case. For the exponential distribution, λ is set to 1/4. For the gamma distribution, λ is set to 1 and α is set to 4. For the Weibull distribution, β is set to 1/2 and θ is set to 2. Figure 20.5 shows that expected value analysis is not sufficient for determination of an acceptable pricing policy.

Example 20.4 (Choosing a Warranty Length Given a Pricing Policy)

Suppose that $C_M(0) = \$85$, that $C_A(0) = \$100$, and that the pricing scheme $C_A(w) = C_A(0) + Aw$ is used, with $A = 35$. Suppose that management has decided to approximate $\gamma(w)$ by the functional form $\gamma(w) = 0.9 + 0.1 \exp(-w^2/2)$. This function satisfies $\gamma(0) = 1.0$ and decreases toward 0.9 as w increases. Thus, the maximal reduction in the producer's per item cost is 10%.

In addition, suppose the producer believes that the product's lifetime has an exponential distribution with parameter λ. Suppose the MARR, ρ, is 0.10.

To use the expected discounted cost and profit as a performance measure, these values should be substituted into formula (20.25). This results in the following condition for a renewing FRW acceptable to both consumer and producer:

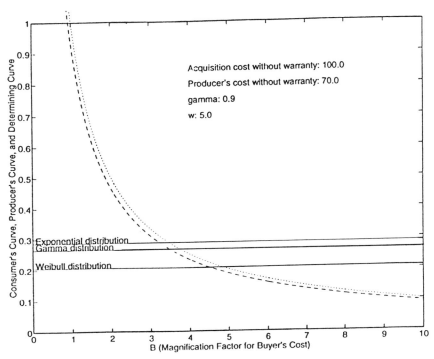

Figure 20.5 Choice of warranty price given warranty period. (Renewing FRW, long-run-time-average cost and profit.)

$$\frac{100 - 70[1 - \gamma(w)]}{100 + 25w} < \frac{0.1}{\lambda + 0.1} + \frac{\lambda}{\lambda + 0.1}e^{-(\lambda + 0.1)w} < \frac{100}{100 + 25w}$$

In this condition, the consumer's and producer's curves appear as functions of w and the determining curve as a function of w and λ. If all three curves are graphed for a particular λ, with the determining curve as a solid curve, the consumer's curve as dots, and the producer's curve as dashes, then acceptable warranty periods are easily identified as the values of w, where the solid curve lies between the other two curves. This is shown in Figure 20.6 for $\lambda = 1/3$. The figure shows both short and long acceptable warranties. The long warranties may be seen as acceptable extended warranties (H. Liggett, personal communication).

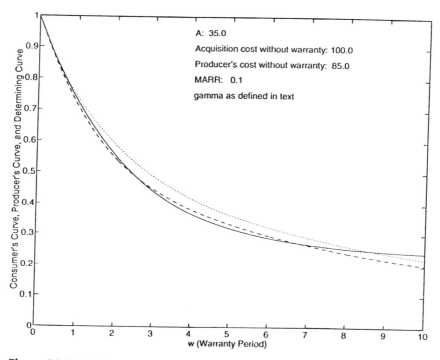

Figure 20.6 Choice of warranty period given cost parameters. (Renewing FRW, expected discounted cost and profit.)

20.9 RENEWING PRO-RATA WARRANTY

20.9.1 Consumer's Expected Cost

The *consumer's long-run-time-average cost for renewing pro-rata warranties* is

$$\frac{[C_A(w)/w] \int_0^w [1 - F(x)]\, dx}{E[X]} \tag{20.26}$$

For warranties beneficial to the consumer, this should be less than formula (20.1). A few algebraic steps lead to the following condition for the long-run-time-average cost to be less than the corresponding cost for no warranties:

$$\frac{1}{w} \int_0^w [1 - F(x)] \, dx < \frac{C_A(0)}{C_A(w)} \qquad (20.27)$$

The *consumer's expected discounted cost for renewing pro-rata warranties* is

$$C_A(w) \left(\frac{1 - \int_0^w e^{-\rho x} f(x) \, dx + 1/w \int_0^w x e^{-\rho x} f(x) \, dx}{1 - \hat{f}(\rho)} \right) \qquad (20.28)$$

For warranties beneficial to the consumer, this should be less than formula (20.2). A few algebraic steps lead to the following condition for the expected discounted cost to be less than the corresponding cost for no warranties:

$$1 - \int_0^w e^{-\rho x} f(x) \, dx + \frac{1}{w} \int_0^w x e^{-\rho x} f(x) \, dx < \frac{C_A(0)}{C_A(w)} \qquad (20.29)$$

20.9.2 Producer's Expected Profit

The *producer's long-run-time-average profit for renewing pro-rata warranties* is

$$\frac{[C_A(w)/w] \int_0^w [1 - F(x)] \, dx - C_M(w)}{E[X]} \qquad (20.30)$$

For warranties beneficial to the producer, this should be greater than formula (20.3). A few algebraic steps lead to the following condition for the long-run-time-average profit to be greater than the corresponding profit for no warranties:

$$\frac{C_A(0) - C_M(0) + C_M(w)}{C_A(w)} < \frac{1}{w} \int_0^w [1 - F(x)] \, dx \qquad (20.31)$$

The *producer's expected discounted profit for renewing pro-rata warranties* is

$$\frac{C_A(w) \left[1 - \int_0^w e^{-\rho x} f(x) \, dx + 1/w \int_0^w x e^{-\rho x} f(x) \, dx \right] - C_M(w)}{1 - \hat{f}(\rho)}$$

$$(20.32)$$

For warranties beneficial to the producer, this should be greater than formula (20.4). A few algebraic steps lead to the following condition for the expected discounted profit to be greater than the corresponding profit for no warranties:

$$\frac{C_A(0) - C_M(0) + C_M(w)}{C_A(w)} < 1 - \int_0^w e^{-\rho x} f(x)\, dx + \frac{1}{w} \int_0^w xe^{-\rho x} f(x)\, dx$$

$$(20.33)$$

20.9.3 Warranties Beneficial to Both Consumer and Producer

For the long-run-time-average cost and profit, a renewing PRW is acceptable to both consumer and producer if

$$\frac{C_A(0) - C_M(0) + C_M(w)}{C_A(w)} < \frac{1}{w} \int_0^w [1 - F(x)]\, dx < \frac{C_A(0)}{C_A(w)}$$

$$(20.34)$$

This condition combines formulas (20.27) and (20.31). Note that for the exponential, gamma, and Weibull distributions, the value of $(1/w) \int_0^w [1 - F(x)]\, dx$ can be obtained by first integrating column 3 of Table 20.1 and then multiplying by $1/w$.

For the expected discounted cost and profit, a renewing PRW is acceptable to both consumer and producer if

$$\frac{C_A(0) - C_M(0) + C_M(w)}{C_A(w)} < 1 - \int_0^w e^{-\rho x} f(x)\, dx$$

$$+ \frac{1}{w} \int_0^w xe^{-\rho x} f(x)\, dx < \frac{C_A(0)}{C_A(w)} \quad (20.35)$$

This condition combines formulas (20.29) and (20.33). Note that for the exponential, gamma, and Weibull distributions, the value of $1 - \int_0^w e^{-\rho x} f(x)\, dx + (1/w) \int_0^w xe^{-\rho x} f(x)\, dx$ can be obtained by adding together columns 2 and 3 of Table 20.2.

Example 20.5 (Choosing a Price Given a Warranty Length)

As in Example 20.2, suppose that the producer wishes to offer a warranty of length 5 but with the pricing scheme $C_A(5) = C_A(0) + 5A$. Suppose that 0.85 is a good estimate of $\gamma(5)$. Also, suppose that $C_M(0) = \$35$ and that $C_A(0) = \$50$.

To use the long-run-time-average cost and profit as a performance measure, these values should be substituted into formula (20.34). This results in the following condition for a renewing PRW to be acceptable to both consumer and producer:

$$\frac{100 - (70)(1.0 - 0.9)}{100 + 5A} < \frac{1}{5}\int_0^5 [1 - F(x)]\,dx < \frac{100}{100 + 5A}$$

In this condition, the consumer's and producer's curves appear as functions of A. The determining curve is constant with respect to A. If the consumer's curve is graphed as dots, the producer's curve is graphed as dashes, and the determining curve is graphed as a horizontal solid line, then acceptable values of A are those values where the solid line lies between the other two curves. In Figure 20.7, this is done simultaneously for three separate hypotheses on the distribution of the product lifetime. The three distributions are the exponential, the gamma, and the Weibull. The formulas needed for the determining curve can be found in the second column of Table 20.2. The parameters of the distributions have been deliberately chosen so that the expected value is equal to 4 in each case. For the exponential distribution, λ is set to 1/4. For the gamma distribution,

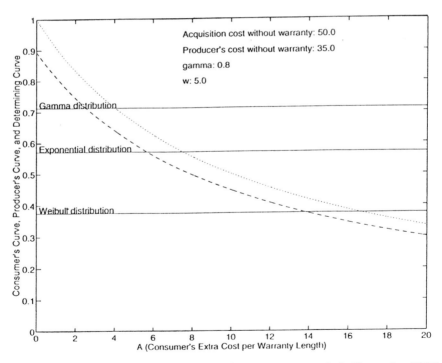

Figure 20.7 Choice of warranty price given warranty period. (Renewing PRW, long-run-time-average cost and profit.)

λ is set to 1 and α is set to 4. For the Weibull distribution, β is set to 1/2 and θ is set to 2. The figure shows that expected value analysis is not sufficient for determination of an acceptable pricing policy.

Example 20.6 (Choosing a Warranty Length Given a Pricing Policy)
Suppose that $C_M(0) = \$70$, that $C_A(0) = \$100$, and that the pricing scheme $C_A(w) = C_A(0) + Aw$ is used, with $A = 25$. Suppose that management, as in Example 20.4, has decided to approximate $\gamma(w)$ by the functional form $\gamma(w) = 0.9 + 0.1 \exp(-w^2/2)$. This function satisfies $\gamma(0) = 1.0$ and decreases toward 0.9 as w increases. Thus, the maximal reduction in the producer's per-item cost is 10%.

Suppose the producer believes that the product's lifetime has a gamma distribution with parameters λ and α. Suppose the MARR, ρ, is 0.10.

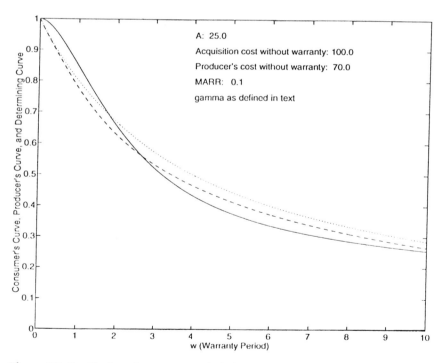

Figure 20.8 Choice of warranty period given cost parameters. (Renewing PRW, expected discounted cost and profit.)

To use the expected discounted cost and profit as a performance measure, these values should be substituted into formula (20.35). This results in the following condition for a renewing PRW to be acceptable to both consumer and producer:

$$\frac{100 - 70[1 - \gamma(w)]}{100 + 25w} < 1 - \frac{\lambda^\alpha}{\Gamma(\alpha)} \int_0^w x^{\alpha - 1} e^{-(\lambda + 0.1)x} \, dx$$

$$+ \frac{\lambda^\alpha}{w\Gamma(\alpha)} \int_0^w x^\alpha e^{-(\lambda + 0.1)x} \, dx < \frac{100}{100 + 25w}$$

In this condition, the consumer's and producer's curves appear as functions of w. The determining curve is a function of w, λ, and α. If all three curves are graphed for a particular λ and α pair, with the determining curve as a solid curve, the consumer's curve as dots, and the producer's curve as dashes, then acceptable warranty periods are easily identified as the values of w where the solid curve lies between the other two curves. This is shown in Figure 20.8 for $\lambda = 4/3$ and $\alpha = 2$. The determining curve for the gamma distribution is qualitatively different from that for the exponential distribution shown in Figure 20.6. The determining curve for the gamma distribution is concave near zero, which rules out the acceptability of extremely short warranties.

20.10 CONCLUSION

This chapter presents, for three of the most commonly used warranty policies, a relatively complete quantification of the consumer's and producer's costs and benefits.

The producer can use these results to examine the possibility of offering a warranty attractive to the consumer that at the same time decreases the producer's amortized cost per item. Given a pricing policy, attractive warranty lengths can be found; given a warranty length, economic pricing policies can be found.

The consumer who has the option of buying the product with or without warranty can use the results to determine whether or not the asking price for the warranty is reasonable. The results can also be used to determine whether it is reasonable to purchase an extended warranty at increased cost. Of course, in both cases, the consumer must know enough about the product to estimate its lifetime distribution.

The results depend on a simple and basic model of the relationship between consumer and producer. It is possible to extend these basic results to more complicated models (see Chapter 6), at the expense of increased difficulty in application. For example:

1. The life-cycle costs and profits can be calculated for more varieties of warranty policy. The idea of finding warranty policies acceptable to both consumer and producer is explored for some more complicated warranties by Pirojboot [19].

2. The life-cycle costs and profits can be calculated for finite life cycles.

3. The individual consumer–producer transactions can be more carefully modeled by including such things as repair, service, investigation, and administrative costs for the producer, and inconvenience costs for the consumer.

4. The individual consumer–producer transactions can be assumed to occur over a non-negligible amount of time, or there may be a time delay between a product failure and the submission of a warranty claim.

5. There can be a nonzero probability that the consumer will abandon the producer after a certain number of transactions, as in Ref. 13. This probability can be a function of the warranty lengths, decreasing with the warranty length.

6. The lifetime distribution of the initially purchased product may differ from the lifetime distributions of the replacements.

7. Not all failures under warranty may result in actual claims.

8. There may be invalid warranty claims.

9. If the producer or consumer is not risk-neutral, then the quantity K_i should be replaced by its utility (if it is a profit) or its disutility (if it is a cost). Brown [9], Ritchken [14], and Ritchken and Tapiero [15] quantify benefit and risk in terms of utility and disutility functions.

REFERENCES

1. Lowerre, J. M. (1968). On warranties, *Journal of Industrial Engineering*, 19(7), 359–360.

2. Menke, W. W. (1969). Determination of warranty reserves, *Management Science*, 15(10), B542–B549.

3. Blischke, W., and Murthy, D. N. P. (1994). *Warranty Cost Analysis*, Marcel Dekker, Inc. New York.

4. Thomas, M. U. (1983). Optimum warranty policies for nonreparable items, *IEEE Transactions on Reliability*, R-32(3), 282–288.

5. Glickman, T. S., and Berger, P. D. (1976). Optimal price and protection period decisions for a product under warranty, *Management Science*, 22(12), 1381–1390.

6. Anderson, E. E. (1977). Product price and warranty terms: An optimization model, *Operational Research Quarterly*, 28(3), 739–741.

7. Murthy, D. N. P. (1989). Optimal product choice in product design, *Engineering Optimization*, **15**, 281–294.

8. Mitra, A., and Patankar, J. G. (1993). An integrated multicriteria model for warranty cost estimation and production, *IEEE Transactions on Engineering Management*, **EM-40**(3), 300–311.

9. Brown, J. P. (1974). Product liability: The case of an asset with random life, *The American Economic Review*, **64**(1), 149–161.

10. Blischke, W. R., and Scheuer, E. M. (1975). Calculation of the cost of warranty policies as a function of estimated life distributions, *Naval Research Logistics Quarterly*, **22**, 681–696.

11. Blischke, W. R., and Scheuer, E. M. (1981). Applications of renewal theory in analysis of the free-replacement warranty, *Naval Research Logistics Quarterly*, **28**, 193–205.

12. Mamer, J. W. (1982). Cost analysis of pro rata and free-replacement warranties, *Naval Research Logistics Quarterly*, **29**(2), 345–356.

13. Mamer, J. W. (1987). Discounted and per unit costs of product warranty, *Management Science*, **33**(7), 916–930.

14. Ritchken, P. H. (1985). Warranty policies for non-repairable items under risk aversion, *IEEE Transactions on Reliability*, **R-34**(2), 147–150.

15. Ritchken, P. H., and Tapiero, C. S. (1986). Warranty design under buyer and seller risk aversion, *Naval Research Logistics Quarterly*, **33**(4), 657–671.

16. Biedenweg, F. M. (1981). *Warranty Policies: Consumer Value vs. Manufacturer Risk*, Technical Report 198, Department of Operations Research and Department of Statistics, Stanford University, Stanford, California.

17. Udell, J. G., and Anderson, E. E. (1968). The product warranty as an element of competetive strategy, *Journal of Marketing*, **32**, 1–8.

18. The MathWorks, Inc. (1992). *MATLAB: User's Guide*, The MathWorks, Inc., South Natick, MA.

19. Pirojboot, B. (1991). *Warranty Policies Beneficial to Both Consumer and Producer*, Master's thesis. Northern Illinois University.

Part F

Warranty and Engineering

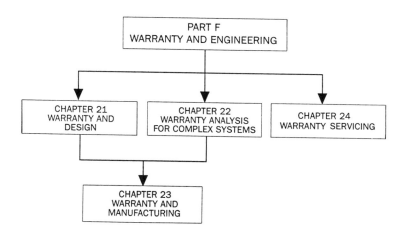

Introduction

During the warranty period, claims occur due to product failures. Failures depend on the reliability of the product and this, in turn, is influenced by engineering decisions at the design and manufacturing stages. In addition, warranty costs are influenced by servicing strategies. Part F contains four chapters that examine these topics.

Chapter 21, by D. N. P. Murthy, discusses the design process and its impact on product reliability. It deals with a variety of models for determining optimal design strategies. These take into account not only the cost of designing the product, but also the subsequent warranty costs. This approach integrates design with marketing and servicing issues and hence assists the manufacturer in the overall optimization. The design strategies include options such as reliability improvement through redundancy and product development.

Most products can be viewed as complex systems, comprising many interconnected components. As a result, the reliability of the system is dependent on the reliability of its components. In Chapter 22, S. Chukova and B. Dimitrov examine warranty analysis for such complex systems, where different components are covered by different warranty terms. This is the first attempt at analysis of such systems in the context of warranty costing. Many new problems are encountered and much research in this area remains to be done.

Due to quality variations in manufacturing, not all items produced conform to design specifications. Those failing to meet the design specification are termed nonconforming items, and releasing such items increases the warranty cost to the manufacturer. Chapter 23, by Murthy, looks at alternative quality control strategies to reduce nonconforming items being produced and/or reaching the customer.

Finally, Chapter 24, also by Murthy, examines the impact of warranty terms on warranty servicing. This includes issues such as warranty reserves (for nonrenewing pro-rata warranties), spares (for nonrepairable items sold with free-replacement warranties), and repair facilities (for repairable items sold with free-replacement policies). In the case of repairable items, alternative strategies involving different types of repair as well as replacement of the failed item are considered. In addition, the author briefly touches on other issues, such as use of loaners and product recall.

21

Warranty and Design

D. N. Prabhakar Murthy

The University of Queensland
Brisbane, Queensland, Australia

21.1 INTRODUCTION

Before a product (consumer durable or industrial) is released on the market, many stages of decision making are involved. Some of these deal with economic and marketing aspects (e.g., the sale price, amount spent on advertising, and so on) and others deal with more technical or engineering aspects. Decisions with regard to engineering aspects are made at three levels,

1. Design
2. Manufacture
3. Presale testing

and are sequentially linked, as shown schematically in Figure 21.1.

When the product is sold with warranty, additional costs in the form of warranty servicing costs are incurred. The expected value of this warranty cost depends on the failure distribution for the product. This, in

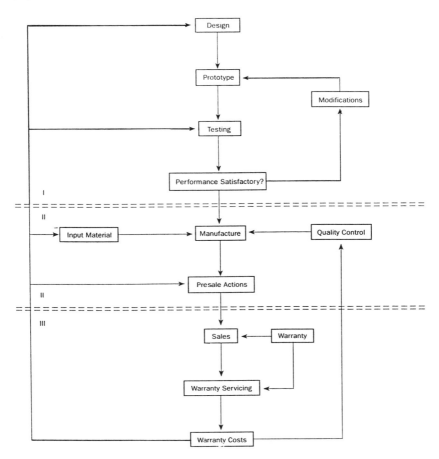

Figure 21.1   Design (I)–manufacture (II)–sales/service (III) sequence.

turn, depends on the engineering of the product. As a result, the expected warranty cost can be influenced by engineering decisions made with regard to product design, development, and manufacture. Higher reliability, or a lower failure rate, can be achieved through improvements in design and/or manufacture. Such improvements reduce the expected warranty cost, but at the expense of increased development and manufacturing cost. Thus, optimal engineering decisions must be based on a sensible trade-off between development and manufacturing cost on the one hand and expected warranty cost on the other. In this chapter, we focus our attention on warranty and design and the next chapter deals with warranty and manufacturing.

The outline of the chapter is as follows. Section 21.2 deals with a brief introduction to the design process. We first consider the problem of optimal reliability choice and allocation. For multicomponent products, the product reliability depends on component reliability. This is discussed in Section 21.3. The reliability of the product can be improved through improvement of reliability at the component level. We discuss two approaches. The first approach uses redundancy at the component level. We consider both active and passive redundancy and these are discussed in Section 21.4. The second approach involves research and development effort to improve reliability at component and product levels. A variety of models has been developed for reliability growth and improvement over time, incorporating factors such as development effort. We discuss some of these models and the optimal developmental plans in the context of product warranty in Section 21.5.

21.2 DESIGN PROCESS

As indicated earlier, the design of the product is the first step in the engineering of a product. Traditionally, product design dealt mainly with the functional aspect of the product, without much consideration given to the manufacturing and postsale maintenance or servicing issues. The design decisions are mainly technical in nature and done at two levels: the component level and the product level. At the component level, the designer deals with issues such as the maximum acceptable stress to which a component of a mechanical product can be subjected. At the product level, issues such as choice of components and subassemblies to ensure some specified product performance are handled. In using this traditional approach, problems often arise at the manufacturing level which required design modifications.

This type of design iteration can be very costly. As a result, the design philosophy has been changing to incorporate manufacturing aspects into the design phase. With the advent of computer-integrated manufacturing, this integration has become more common and fewer design changes are needed as a result. However, the incorporation of postsales economic issues, resulting from the consequences of the design, into the decision-making process at the design level is a relatively new concept, the importance of which is more and more becoming recognized. This approach requires a strong interaction among design, manufacturing, marketing, and sales. Thus, many groups or departments of a manufacturing organization must be involved cooperatively at very early stages of product development. As a result, optimal decision making at the design stage must take into account the postsale warranty costs.

There are many books which deal with reliability and engineering design (see, e.g., Refs. 1–3). Marshall [4] discusses design trade-offs in the context of a special class of warranties. Many articles deal with reliability requirements and specification in the context of design; Refs. 5 and 6 are an illustrative sample.

21.3 RELIABILITY CHOICE AND ALLOCATION

Let J denote the number of components in each item of the product. We first consider the single component case ($J = 1$) to illustrate the problem of reliability choice and later consider the multicomponent case ($J > 1$).

21.3.1 Single-Component Product

Let $F(x; \theta)$ be the product-failure distribution. We first consider the case where the manufacturer has the option to manufacture the component based on one of K different design choices. For the kth design choice, the parameter θ has a value θ_k, $1 \le k \le K$. Let $C_m(\theta_k)$ denote the manufacturing cost per unit for the kth design choice. We assume that the K design choices are ordered in terms of decreasing reliability; that is, design choice j implies a less reliable product than design choice i if $j > i$, $1 \le i$, and $j \le K$. As a result, $C_m(\theta_k)$ is a decreasing sequence in k, $1 \le k \le K$, implying that a less reliable product is also less expensive.

Let the expected warranty cost per unit sale with design choice k and a warranty period W be given by $\omega(W; \theta_k)$. This cost depends on the type of the warranty policy and the servicing strategy used by the manufacturer. Expressions for $\omega(W; \theta_k)$ for different types of one-dimensional warranty policies are given in Chapters 10–12. Then the manufacturer's total expected cost per unit sale, $E[C_s(W; \theta_k)]$, is given by

$$E[C_s(W; \theta_k)] = \omega(W; \theta_k) + C_m(\theta_k) \tag{21.1}$$

The optimal reliability design choice is given by k^*, the optimal k ($1 \le k \le K$), which minimizes $E[C_s(W; \theta_k)]$. Derivation of k^* requires an enumerative method in which $E[C_s(W; \theta_k)]$ is first computed for all values of k ($1 \le k \le K$) and then k^* obtained by comparison.

When K is large, one can model the design option as choosing θ from a continuum over a specified interval. This allows one to use powerful optimization techniques to decide on the optimal design. We propose a model based on the following assumptions:

1. θ is a scalar, with smaller values of θ corresponding to more reliable product. (The extension to a vector θ is straightforward.) θ can take on any value in the interval $\theta^- \leq \theta \leq \theta^+$. θ^- and θ^+ reflect the limits for the allowable and achievable reliability.
2. $C_m(\theta)$, the manufacturing cost per unit, is a continuous function of θ with $dC_m(\theta)/d\theta < 0$, so that θ increasing corresponds to a less reliable, and hence less expensive, product.

As an example, for the exponential distribution, $1/\theta$ would represent the mean time to failure, so that θ decreasing would correspond to a more reliable product, which costs more to manufacture.

As before, $\omega(W; \theta)$, the expected warranty service cost per unit sale, depends on the design parameter θ, the type of warranty, and the warranty period W. The manufacturer's total expected cost per unit sale, $E[C_s(\theta; W)]$, is given by

$$E[C_s(W; \theta)] = \omega(W; \theta) + C_m(\theta) \tag{21.2}$$

The optimal reliability choice, θ^*, is the value of θ which minimizes $E[C_s(W; \theta)]$ subject to the constraint

$$\theta^- \leq \theta \leq \theta^+ \tag{21.3}$$

This is a standard nonlinear programming problem that can be solved by the usual Lagrangian multiplier method, in which the constraint (21.3) is adjoined to $E[C_s(W; \theta)]$ given by (21.2). Depending on the shape of $E[C_s(W; \theta)]$, θ^* can be either inside the constraint interval (i.e., $\theta^- \leq \theta^* \leq \theta^+$) or equal to either θ^- or θ^+.

21.3.2 Multicomponent Product

In this case, the reliability of the product depends on the reliability of each component and the manner in which they are interconnected. For a general network configuration, the reliability of the system can be expressed as a function of the reliability of the components (see, e.g., Ref. 7). The usual reliability allocation problem deals with allocation of reliability at the component level to ensure that the product meets some specified performance and cost limits (see Ref. 3). Here, we deal with reliability allocation to minimize the manufacturer's total expected cost per unit sale. We confine our attention to the case in which the reliability choice for each component is from a continuum over a specified interval.

Let the number of components in the product be J and let $F_j(t, \theta_j)$ denote the failure distribution of the jth component, with θ_j constrained by the relationship $\theta_j^- \leq \theta_j \leq \theta_j^+$ to indicate the limits. As in the earlier case, we assume that θ_j is a scalar variable, with a smaller value implying more reliable component. Let Θ denote the set $\{\theta_j, 1 \le j \le J\}$. The distribution for the product failure time can be obtained, in principle, as function of Θ. Let $F(t, \Theta)$ denote this distribution. The manufacturing cost per unit is a function of Θ. This can be obtained in terms of component costs, which, in turn, are functions of component reliability. Let $C_{mj}(\theta_j)$ denote the manufacturing cost per unit of component j, with reliability parameter θ_j. The production cost per item is given by

$$C_m(\Theta) = \sum_{j=1}^{J} C_{mj}(\theta_j) \qquad (21.4)$$

As mentioned earlier, the expected warranty cost depends on whether a failed item can be made operational by rectification of the failed component or not, whether a failed component can be repaired or not, the reliabilities of the different components, and the type duration of the warranty. As a result, the expected warranty cost per item sold is a function of Θ; let $\omega(W; \Theta)$ denote this cost. Expressions for $\omega(W; \Theta)$ for different types of one-dimensional warranty policies are given in Chapters 10–12. The manufacturer's total expected cost per unit sale, $E[C_s(W; \Theta)]$, is given by

$$E[C_s(W; \Theta)] = \omega(W; \Theta) + C_m(\Theta) \qquad (21.5)$$

Θ^*, the optimal Θ, is the value which minimizes $E[C_s(W; \Theta)]$ subject to the constraints on the θ_j's.

As can be seen, the problem is similar to the problem of the single-component case, except that minimization involves J parameters and the form of $\omega(W; \Theta)$ is more complicated. A detailed study of this model formulation can be found in Ref. 8, in which conditions for the existence of Θ^* are given. The following example is also from Ref. 8.

Example 21.1

The product consists of J components connected in series. This implies that an item fails whenever a component fails. We assume (1) that component failures are statistically independent, (2) that the probability of two or more components failing at the same time is zero, and (3) that the failure distributions for all components are exponential, i.e.,

$$F_j(x) = 1 - \exp(-\theta_j x)$$

The mean time to failure for component j is $1/\theta_j$. We assume that θ_j is constrained to satisfy

$$\theta_j^- \le \theta_j \le \theta_j^+$$

Note that the failure rate of the item is a constant given by

$$r = \sum_{j=1}^{J} \theta_j$$

As part of design specification, we want to ensure that the failure rate r is not greater than a prespecified value γ. This will ensure a certain minimum reliability for the product.

Suppose the item is sold with a free-replacement warranty (FRW) with warranty period W. We examine two cases.

Case 1 (repairable items). Suppose that the components are not repairable but that failed or defective components are replaceable. In this case, the manufacturer can repair failed items by replacing only the failed component. If the failure is due to component j failing, the failed item is made operational by replacing the failed component. The cost of this is $C_{mj}(\theta_j) + C_h$, where $C_{mj}(\theta_j)$ is the cost of the new component and C_h is the average extra cost incurred in handling the warranty claim. Because components fail independently and component-failure distributions are exponential, the expected number of component j failures, $1 \leq j \leq J$, over the warranty period is given by $W\theta_j$. As a result, the expected cost of servicing the warranty per unit sale is given by

$$\omega(W; \Theta) = \sum_{j=1}^{J} \{(W\theta_j)[C_{mj}(\theta_j) + C_h]\}$$

From (21.5), we have

$$E[C_s(W; \Theta)] = \left\{ \sum_{j=1}^{J} W\theta_j[C_{mj}(\theta_j) + C_h] \right\} + C_m(\Theta)$$

where $C_m(\Theta)$ is given by (21.4).

Case 2 (Nonrepairable items). Suppose that failed items cannot be repaired by replacing failed components (or any other way) or that such repairs would cost more than a new item. In this case, whenever an item fails, the entire item must be replaced by a new one. This is done at a cost $C_m(\Theta)$. The item failures are assumed to occur as in Case 1. As a result, we have

$$E[C_s(W; \Theta)] = \left\{ \sum_{j=1}^{J} W\theta_j \right\} [C_m(\Theta) + C_h] + C_m(\Theta)$$

where $C_m(\Theta)$ is given by (21.4).

For both cases, the optimal Θ^* is obtained by minimizing $E[C_s(W; \Theta)]$ subject to the constraints.

We illustrate these results by means of a numerical example with $J = 4$ and the following form for $C_{mj}(\theta_j)$:

$$C_{mj}(\theta_j) = b_j + a_j(\theta_j)^{-d^j}$$

with a_j and $b_j > 0$ and $d_j > 1$. This implies that as θ_j decreases (so that the component becomes more reliable), the unit cost increases. The cost parameters and the limits on θ_j (per year), $1 \leq j \leq 4$, are shown in Table 21.1. The parameters were selected so that (1) the cost per unit with minimum reliability for component j (i.e., $\theta_j = \theta_j^+ = 0.5$ per year), $1 \leq j \leq 4$, are the same and equal \$10 per unit and (2) the cost per unit for component j with reliability $\theta_j^+ - \delta\theta$, for a given $\delta\theta$, is decreasing in j. This implies that for the same reliability, cost per unit for component 4 > cost per unit for component 3 > cost per unit for component 2 > cost per unit for component 1. The warranty period is $W = 1$ year and c_h is taken to be \$10. We consider two values of γ (prespecified maximum value for the failure rate), namely, $\gamma = 1$ per year and $\gamma = 2$ per year. The former corresponds to a product with greater reliability. Table 21.2 shows the optimal reliability allocation (obtained using a standard computational procedure). The optimal allocation of reliability to different components is shown in Figures 21.2 and 21.3 for the repairable and the nonrepairable cases.

We make the following observations regarding these results:

1. The optimal cost per unit for $\gamma = 1$ is greater than that for $\gamma = 2$. This is to be expected because the design requirements specify a more reliable product when $\gamma = 1$.
2. For both cases, the constraint is tight when $\gamma = 1$ and W small. Furthermore, whenever the constraint is not tight, θ_j^* decreases with W. This is to be expected because W increasing corresponds

Table 21.1 Model Parameter Values for Example 21.1

	1	2	3	4
a_j	0.20	0.24	0.30	0.40
d_j	3.00	2.50	2.00	1.50
b_j	8.40	8.64	8.80	8.87
θ_j^-	0.00	0.00	0.00	0.00
θ_j^+	0.50	0.50	0.50	0.50

Table 21.2 Optimal Reliability Allocation [Example 21.1]

γ	θ_1^*	θ_2^*	θ_3^*	θ_4^*	$\sum \theta_j^*$	$C_m(\Theta^*)$	$\omega(W; \Theta^*)$	$E[C_s(W; \Theta^*)]$
				Case 1: Repairable items				
2	0.260	0.334	0.398	0.454	1.446	44.9	30.6	75.5
1	0.162	0.224	0.281	0.333	1.000	60.0	22.4	82.4
				Case 2: Nonrepairable items				
2	0.219	0.282	0.338	0.387	1.226	49.4	72.9	122.3
1	0.168	0.226	0.279	0.327	1.000	60.0	65.9	125.9

Note: $\Theta = [\theta_1, \theta_2, \theta_3, \theta_4]$.

to a longer warranty period and, to reduce the expected warranty cost, the reliability must improve or, equivalently, θ_j^* must decrease.

3. For case 1, individual θ_j^* may increase or decrease with W when the constraint is tight. This is again to be expected because the

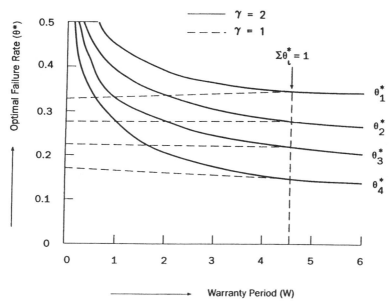

Figure 21.2 Optimal failure rate allocation (repairable products).

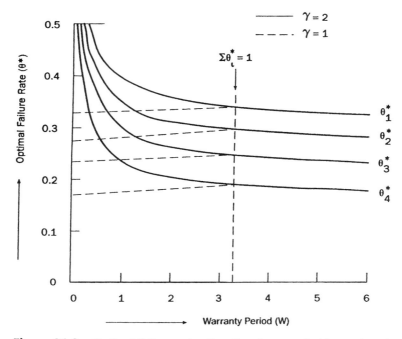

Figure 21.3   Optimal failure rate allocation (nonrepairable products).

warranty cost depends on each θ_j. In contrast, for case 2, the individual θ_j^*'s do not change with W when the constraint is tight because the warranty cost depends on $\Sigma \, \theta_j$ as opposed to individual θ_j.

21.3.3   Integration of Design and Marketing Choices

Our discussion so far has looked at design choice in isolation of marketing choice. The total product sales depend on the marketing variables such as price and advertising. According to the signaling theory of warranty (see Chapter 15 and Refs. 9 and 10), warranty terms play an important role in the promotion of the product—better warranty terms implying a better (or more reliable) product. Hence, it is more sensible to jointly decide on the optimal design and marketing choices.

A variety of models have been proposed by Glickman and Berger [11], Anderson [12] Ritchken and Tapiero [13], Menezes [14], and Murthy [15]. We discuss a model from Murthy [15] in which the design choice [reliability parameter θ determining the product failure distribution $F(x, \theta)$], sale price (P), and the warranty duration (W) are selected optimally by the manufacturer to maximize the total expected profits over the product life cycle (L). The salient features of the model are as follows:

1. The total first-purchase sales is given by

$$Q(P, W) = KP^{-\alpha}W^{\beta} \tag{21.6}$$

where $\alpha > 1$ and $0 < \beta < 1$. This implies that as P increases and/or W decreases, the total sales decrease. α and β are the price and warranty period elasticities.

2. The items are nonrepairable and are sold with a linear rebate warranty policy. If the age at failure is X, then the fraction refunded is given by $S(X) = 1 - aX/W$ if $X < W$.

3. Some of the first-purchasers are not happy with the product and cease to buy any replacements. The rest are satisfied customers and continue to buy the product over the life cycle (L) of the product. Let γ denote the fraction of satisfied first-purchasers. As a result, the repeat purchases of satisfied first-purchasers occur according to a renewal process associated with the distribution function $F(x, \theta)$. (See Chapter 7.)

4. As in earlier models, a smaller θ corresponds to a more reliable product. As a result, the manufacturing cost per unit, $C_m(\theta)$, is a decreasing function of θ over a specified interval representing the range of achievable reliability.

The manufacturer's total expected profit is given by

$$\Pi(P, W, \theta) = Q(P, W)[P - C_m(\theta) - \omega(W, \theta)][1 + \gamma M_F(x, \theta)] \tag{21.7}$$

where $\omega(W, \theta)$ is the expected warranty cost per item sold and is given by

$$\omega(W, \theta) = \frac{(1 - a)WF(W, \theta) + a \int_0^W F(t, \theta)dt}{W} \tag{21.8}$$

and $M_F(x, \theta)$ is the renewal function associated with the distribution function $F(x, \theta)$.

The optimal design choice (θ^*) and market choice $(P^*$ and $W^*)$ are given by the values of θ, P, and W which maximize $\Pi(P, W, \theta)$. The following example from Ref. 15 illustrates this.

Table 21.3 Optimal Design and Marketing Choice

L	P^*	W^*	θ^*	$\Pi(P^*, W^*, \theta^*)$
0	6.90	7.60	0.125	237
1	6.06	6.38	0.149	269
2	5.23	4.97	0.191	311
3	4.18	2.85	0.333	357
4	3.98	2.38	0.400	422

Example 21.2

Let the unit manufacturing cost be given by $C_m(\theta) = \theta^{-0.1}$ with $0 \leq \theta \leq 0.4$, indicating the limits of achievable reliability. Let $K = 1000$, $\alpha = 2$, $\beta = 0.8$, $a = 1$, and $\gamma = 1$. The optimal design choice (θ^*) and market choice (P^* and W^*) are shown in Table 21.3 for five values of L.

As seen, when $L = 0$ (i.e., no repeat purchases), P^* and W^* are high and θ^* low. This is to be expected because the manufacturer aims to maximize the expected profits based solely on first-purchase sales. As a result, W^* has to be high to attract more customers. This, in turn, requires θ^* to be low (implying greater reliability) to reduce warranty costs and forces P^* to be high. This indicates that W^* is an important variable, acting as a signal to attract more customers. As L increases, the repeat purchases become more important. Both P^* and W^* decrease with L and θ^* increases. In this case, the joint effort of P^* and W^* is important. As a result, θ^* must increase (implying a less reliable product) to maximize the expected profits. Finally, as γ decreases, for a given L, P^*, and W^* increase, whereas θ^* decreases as is to be expected.

21.4 REDUNDANCY

As mentioned earlier, the product reliability is a function of component reliabilities. If a critical component has low reliability, then it can result in a large number of failures (and claims) under warranty. This leads to high warranty costs. One way of reducing the warranty cost is to build in redundancy for the critical component so that it improves the overall reliability of the product. Typically, redundancy involves replication of the critical components. This is possible only for certain components for which incorporation of such replication is permissible by the functional design of the item. Building in redundancy results in greater manufacturing cost per item and this is justified only if the reduction in warranty costs exceed this increase.

Two types of redundancy have been used for improving product reliability:

1. Active (also called hot standby) redundancy
2. Passive (also called cold standby) redundancy

We shall consider both of these in the context of a critical component for which failure of component results in item failure. We confine our discussion to the case where the redundancy involves the component and a duplicate. The extension to two or more replicates is relatively straightforward conceptually although the analysis can be fairly complex.

The literature on redundancy is vast and deals with various aspects of analysis and optimization (see, e.g., Refs. 16–22). Murthy and Hussain [23] deal with redundancy in the context of warranties and the results of this section are based on that work.

21.4.1 Active Redundancy

In the case of active redundancy, the two components are connected as shown in Figure 21.4. The component as well as its duplicate are in use at the same time and the item functions as long as one of them is working. Thus, if the individual failure times of the two components are X_1 and X_2 respectively, then the failure time, as a pair, is given by

$$X_{12} = \max\{X_1, X_2\}$$

Let $F(x)$ denote the failure distribution of the component. Then both X_1 and X_2 are distributed according to $F(x; \theta)$. We assume the following: (1) the failures are statistically independent, (2) the components are nonrepairable so that failed ones need to be replaced by new ones, (3) the time to replace is negligible so that it can be treated as being zero, and (4) the product is sold with a FRW policy with a warranty period W. (*Comment*: We confine our attention to items sold with FRW policy. The analysis of

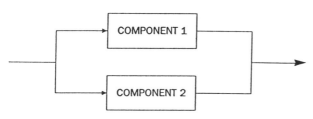

Figure 21.4 Configuration for active redundancy.

redundancy for other types of policies, for example, PRW or a combination, can be carried out in a similar manner.)

With no redundancy, the expected number of component failures over the warranty period is given by $M_F(W)$, where $M_F(x)$ is the renewal function associated with the distribution function $F(x)$. If C_m is the component manufacturing cost per unit, then the expected cost to the manufacturer without redundancy is given by

$$E[C_s(W; \text{no redundancy})] = C_m + (C_m + C_h)M_F(W) \qquad (21.9)$$

where C_h is the handling cost associated with each warranty claim.

With redundancy, the time between warranty claims due to the critical component failure is given by X_{12} as indicated earlier. Let $G(x)$ denote the distribution function for X_{12}. It is related to $F(x)$ by

$$G(x) = [F(x)]^2 \qquad (21.10)$$

As a result, the expected number of warranty claims (due to the critical component failures) over the warranty period is given by $M_G(W)$, where $M_G(x)$ is the renewal function associated with the distribution function $G(x)$. Each warranty claim costs the manufacturer $2C_m$ as opposed to only C_m in the case with no redundancy. The total expected cost to the manufacturer with redundancy is given by

$$E[C_s(W; \text{redundancy})] = 2C_m + (2C_m + C_h)M_G(W) \qquad (21.11)$$

The use of redundancy is justified only if

$$E[C_s(W; \text{redundancy})] < E[C_s(W; \text{no redundancy})] \qquad (21.12)$$

where $E[C_s(W; \text{redundancy})]$ is given by (21.9) and $E[C_s(W; \text{no redundancy})]$ is given by (21.11). Whether the inequality in (21.12) is satisfied or not depends on the failure distribution $F(x)$ and the various cost parameters. The following proposition characterizes when building in redundancy is the optimal strategy.

Proposition 21.1

Building in redundancy is the optimal strategy if

$$\frac{C_h}{C_m} > \frac{1 + 2M_G(W) - M_F(W)}{M_F(W) - M_G(W)}$$

∎

The implication of this is that redundancy is worthwhile if C_h/C_m is high.

Example 21.3

Let $F(x)$ be the exponential distribution, that is,

$$F(x) = 1 - e^{-\theta x}$$

with mean time to failure $1/\theta = 4$ years. With no redundancy, the expected number of claims over the warranty period is given by

$$M_F(W) = W\theta$$

With redundancy, the time between claims is distributed according to $G(x)$ with the density function $g(x)$ $[= dG(x)/dx]$ given by

$$g(x) = 2\theta(e^{-x\theta} - e^{-2x\theta})$$

As a result, the expected number of claims over the warranty period is given by

$$M_G(W) = \frac{2}{3}(W\theta) - \frac{2}{9}(1 - e^{-3W\theta})$$

The item is relatively inexpensive to manufacture, but the handling cost (relative to the manufacturing cost) is high so that the ratio $C_h/C_m > 1$. Table 21.4 shows $E[C_s(W; \text{redundancy})]/C_m$ for two different ratios of C_h/C_m and a range of warranty periods W (years). We first consider the case $C_h/C_m = 2$, that is, handling cost is twice the cost per unit. For $W = 1$ year, the optimal strategy is no redundancy. The same is true for $W = 2$ years. For $W = 3$ and 4 years, building in redundancy is the optimal strategy. When $C_h/C_m = 5$ (i.e., handling cost is five times the unit cost of component), the results are different. For all four values of W, the optimal strategy is to build in redundancy. For these W values, C_h/C_m is sufficiently large so that the inequality of Proposition 21.1 is satisfied and, hence, the result.

Table 21.4 $E[C_s(W)]/C_m$ with Active and Without Redundancy

	No redundancy	Redundancy
	(a) $C_h/C_m = 2.0$	
1	1.75	2.1977
2	2.50	2.6428
3	3.25	3.2048*
4	4.00	3.8220*
	(b) $C_h/C_m = 5.0$	
1	2.50	2.3459*
2	4.00	3.1249*
3	5.50	4.1084*
4	7.00	5.1886*

Asterisks indicate that redundancy is optional.

21.4.2 Passive Redundancy

In the case of passive redundancy, the two components are connected as shown in Figure 21.5. Note that, in contrast to the previous case, here we have a switch. At the start, only one of the two components is in use. When it fails, whether the duplicate (or spare) is switched on or not depends on the switch. In case of a perfect switch, there is no uncertainty in the operation of the switch and the duplicate is always connected when needed. In the case of an imperfect switch, the operation is uncertain in the sense that the spare is connected with probability q ($0 \le q \le 1$) and not connected with probability $1 - q$. Because the perfect switching is a special case of the imperfect switching, we confine our analysis to the imperfect switching case.

The literature on passive redundancy with imperfect switch is also vast (see, e.g., Refs. 24–27), but none deals with warranty. Because the switch does not function properly, the time between item failures (or equivalently, between warranty claims) is given by

$$X_{12} = \begin{cases} X_1 & \text{with probability } 1 - q \\ X_1 + X_2 & \text{with probability } q \end{cases}$$

The distribution for X_{12}, $H(x; q)$, is given by

$$H(x; q) = (1 - q)F(x) + qF(x)*F(x) \tag{21.13}$$

where the asterisk is the convolution operator (see Chapter 7). Note that $q = 1$ corresponds to perfect switch; that is, the spare is always switched on when needed.

The expected number of claims due to critical component failures over the warranty period is given by $M_H(W; q)$, where $M_H(x; q)$ is the renewal function associated with the distribution function $H(x; q)$. The cost associated with each warranty claim is the cost of each module (com-

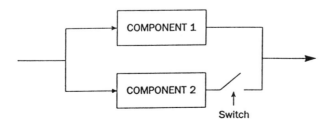

Figure 21.5   Configuration passive redundancy.

prising of two components, each costing C_m, and the switching mechanism, costing δC_m) and is equal to $(2 + \delta)C_m$. As a result, the manufacturer's expected cost per unit sale is given by

$$E[C_s(W; \text{redundancy})] = (2 + \delta)C_m + [(2 + \delta)C_m + C_h]M_H(W; q)$$

$$(21.14)$$

As before, building in redundancy is the optimal strategy if the inequality in (21.12) is satisfied with the left-hand side given by (21.14) and the right-hand side given by (21.9). Note that $M_H(W; q)$ decreases as q decreases. The following proposition characterizes when building in redundancy is the optimal strategy.

Proposition 21.2

Building in redundancy is the optimal strategy if

$$\frac{C_h}{C_m} > \frac{(2 + \delta)[1 + M_H(W; q)] - [1 + M_F(W)]}{M_F(W) - M_H(W; q)}$$

∎

The implication of this is that redundancy is worthwhile if C_h/C_m is high and δ small and q large.

Example 21.4

As in Example 21.3, let $F(t)$ be the exponential distribution with mean time to failure $1/\theta = 4$ years. With redundancy, the time between claims is distributed according to $H(x)$ with the density function $h(x; q)$ $[= dH(x)/dx]$ given by

$$h(x; q) = \theta[(1 - q)(1 - x\theta) + x\theta]e^{-x\theta}$$

Note that if $q = 1$ (perfect switch), then $H(x)$ is distributed according to a Gamma distribution with mean time to failure $= 2/\theta$ (or 8 years).

The expected number of claims over the warranty period is given by

$$M_H(W; q) = \left(\frac{1}{1 + q}\right)(W\theta) - \left(\frac{q}{q + 1}\right)^2 (1 - e^{(1 + q)W\theta})$$

Table 21.5 shows $E[C_s(W; \text{redundancy})]/C_m$, with perfect switching ($q = 1$), for two different ratios of C_h/C_m and a range of warranty periods W (years) and δ. We first consider the case $C_h/C_m = 2$, that is, handling cost is twice the cost per unit. For $W = 1$ year, the optimal strategy, for all three values of δ, is no redundancy. For $W = 2$ and 3 years, we see that building in redundancy is the optimal strategy for $\delta = 0$ and 0.1 due to the advantage resulting from fewer warranty claims. When $\delta = 0.5$, this

Table 21.5 $E[C_s(W)]/C_m$ with Passive and Without Redundancy
$(q = 1.0)$

	No	Redundancy		
W	Redundancy	$\delta = 0.0$	$\delta = 0.10$	$\delta = 0.50$
		(a) $C_h/C_m = 2.0$		
1	1.75	2.0799	2.1826	2.6199
2	2.50	2.3679*	2.4771*	2.9139
3	3.25	2.7231*	2.8412*	3.3135
4	4.00	3.1353*	3.2638*	3.7773*
		(b) $C_h/C_m = 5.0$		
1	2.50	2.1864*	2.2890*	2.6997
2	4.00	2.6438*	2.7530*	3.1898*
3	5.50	3.2655*	3.3836*	3.8559*
4	7.00	3.9868*	4.1152*	4.6288*

Asterisks indicate redundancy is optimal.

advantage is negated by the increase in the module price due to the high cost of the switch. For $W = 4$, the warranty duration is sufficient to make building in redundancy the optimal strategy for all the values of δ considered. For the case $C_h/C_m = 5$ (i.e., handling cost is five times the unit cost of component), the results are different. For $W = 1$ year, building in redundancy is optimal when δ is small (0 or 0.1). When $\delta = 0.5$, the cost of replacement negates the reduction in the number of claims and, hence, the optimal strategy is no redundancy. For the remaining values of W and δ considered, building in redundancy is the optimal strategy.

Tables 21.6 gives the results for the case $q = 0.8$ (imperfect switch). The results for $C_h/C_m = 2$ indicates that for $W = 1$ and 2, no redundancy is the optimal strategy for the three values of δ considered. For $W = 3$ and 4, building in redundancy is optimal strategy for $\delta = 0$ and 0.1 and no redundancy is optimal for $\delta = 0.5$. The reason for this is the same as discussed earlier; that is, the reductions in the expected number of failures is negated by the high cost of each replacement module. For the case $C_h/C_m = 5$, the optimal strategy is to have no redundancy when $W = 1$ year and $\delta \geq 0.1$ and to build in redundancy for all other combinations of W and δ.

On comparing the results of Tables 21.5 ($q = 1$) with Table 21.6 ($q = 0.8$), the interesting thing to note is that the expected warranty costs are larger for the latter case. This is to be expected for obvious reasons. Also note that for $C_h/C_m = 2$, $W = 2$ years, and $\delta = 0$ or 0.1, building

Table 21.6 $E[C_s(W)]/C_m$ with Passive and Without Redundancy
($q = 0.8$)

W	No redundancy	Redundancy		
		$\delta = 0.0$	$\delta = 0.10$	$\delta = 0.50$
		(a) $C_h/C_m = 2.0$		
1	1.75	2.2692	2.3760	2.8029
2	2.50	2.6422	2.7583	3.2225
3	3.25	3.0814*	3.2084*	3.7166
4	4.00	3.5627*	3.7018*	4.2580
		(b) $C_h/C_m = 5.0$		
1	2.50	2.4712*	2.5799	3.0048
2	4.00	3.1239*	3.2399*	3.7042*
3	5.50	3.8924*	4.0194*	4.5276*
4	7.00	4.7347*	4.8738*	5.4300*

Asterisks indicate redundancy is optimal.

in redundancy is optimal when $q = 1$ (perfect switch) and no redundancy is optimal when $q = 0.8$ (imperfect switch). In other words, the advantage of redundancy is lost due to uncertainty in the switching operation.

21.4.3 Relative Comparison

We now compare the case of no redundancy with active redundancy and passive redundancy with perfect switch. Note that the times between warranty claims are as follows:

X_1: with no redundancy
$\max\{X_1, X_2\}$: with active redundancy
$X_1 + X_2$: with passive redundancy and perfect switch ($q = 1$)

Because

$$X_1 \le \max\{X_1, X_2\} \le (X_1 + X_2)$$

we have the following relationship (see Ref. 28):

$$M_F(W) > M_G(W) > M_H(W)$$

This implies that the expected number of claims with no redundancy is greater than that with active redundancy, which, in turn, is greater than that with passive redundancy with perfect switch. However, the cost for each warranty claim is smallest with no redundancy, and largest for passive redundancy with perfect switch. This is to be expected as increased reliability is achieved at a cost. The total expected warranty cost over the warranty period is the warranty cost per claim times the expected number of claims over the warranty period.

21.5 PRODUCT DEVELOPMENT

One way of reducing the expected warranty cost is to improve product reliability. This involves research and development—subjecting the product to a process of test–fix–test–fix iterations. During this process, the product is tested until a failure mode appears. Design and/or engineering modifications are then made as attempts to eliminate the failure mode and the product is tested again. As this continues, the product reliability improves.

The development program costs money. To decide on an optimal development plan, one needs to model the product improvement process. A variety of deterministic and stochastic models have been proposed for reliability growth (see, e.g., Refs. 29–32). Using these models, one can decide on optimal development plans which minimize the total expected cost of development plus the servicing cost for warranty. We consider two such models.

21.5.1 Deterministic Model for Product Reliability Improvement

We first consider the case where the failure distribution is exponential and characterized by a single parameter λ which represents the failure rate. Let the failure rate at the start of the development program be λ_0 and let $\lambda(t)$ denote the failure rate after development for a period t. Duane [31] proposed the following model for product reliability improvement:

$$\lambda(t) = \lambda_0(1 + t)^{-\alpha} \tag{21.15}$$

with $0 \leq \alpha \leq 1$, α represents the improvement rate. As a result, if the development program is stopped after T_d units of time, the failure rate is given by (21.15) with $t = T_d$. Note that with no development, the failure rate remains at λ_0.

One can extend this approach to model reliability improvement where the failure distribution is not exponential but is given by a more general distribution function $F(t, \theta)$. We assume that θ is scalar and that smaller values of θ correspond to a more reliable product. In this case, reliability improvements can be modeled by

$$\theta(t) = \theta_0(1 + t)^{-\alpha} \tag{21.16}$$

where θ_0 represents the value before product development commences and $\theta(t)$ is the value after a development period of t.

21.5.2 Stochastic Model for Product Reliability Improvement

We consider a model where, during the development period, failures (and subsequent modifications to eliminate them through design) are assumed to occur according to a nonhomogeneous Poisson process with intensity $v(t)$ (see Chapter 7). $v(t)$ is called the modification rate. $v(0)$ is the failure rate before the start of the development program and $v(t)$ decreases with development time t.

The number of modifications made over $[0, t)$, $N(t)$, is a random variable with

$$E[N(t)] = \Lambda(t) = \int_0^t v(x)\, dx \tag{21.17}$$

If the development program is stopped at time T_d, then the failure rate for the product is given by

$$\theta = v(T_d) \tag{21.18}$$

As $v(t)$ is a nonincreasing function of t, larger values of T_d imply smaller failure rates. In other words, the longer the duration of product development, the smaller the failure rate for the product at the end of the development program.

Various forms for $v(t)$ have been proposed. The one proposed by Crowe [32] is given by

$$v(t) = \lambda\beta t^{\beta-1} \tag{21.19}$$

where $\lambda > 0$ and $0 < \beta < 1$. In the literature on reliability growth, this model is called the nonhomogeneous, Poisson-process, reliability growth model. (Note this model can be viewed as the stochastic version of the deterministic model proposed by Duane [31].)

21.5.3 Optimal Product Development

One can build a variety of models to determine the optimal product development. We outline a model proposed by Murthy and Nguyen [33].

The reliability improvement is given by Crowe's model, where at the end of development period T_d, the failure rate is given by (21.18). Because development costs money, we assume the following costs for development:

1. C_d: the cost per unit time of running the development program
2. C_a: the expected cost of each design modification to fix a failure mode

Let $N(T_d)$ denote the number of design modifications made during the development period. Then the cost of the development program is given by

$$C_d T_d + C_a N(T_d) \tag{21.20}$$

Note that $N(T_d)$ is a random variable. The failure rate at the end of the development program is given by (21.18).

We assume that (1) the product is sold with a one-dimensional free-replacement (FRW) policy with warranty period W and (2) the item is repairable and the manufacturer always chooses to minimally repair items which fail under warranty. Let C_r denote the expected cost of each repair. (*Comment*: One can easily modify the model to handle other types of warranties such as PRW or a combination.)

Finally, let Q denote the total number of items sold. Then the expected warranty cost $\omega(\theta; W)$ is given by

$$\omega(\theta; W) = C_r Q \theta W \tag{21.21}$$

Let $E[C_s(\tau; W)]$ be this total expected cost to the manufacturer. Then from (21.20) and (21.21), we have

$$
\begin{aligned}
E[C_s(T_d; W)] &= C_d T_d + C_a E[N(T_d)] + C_r Q W \theta \\
&= C_d T_d + C_a \Lambda(T_d) + C_r Q W \nu(T_d)
\end{aligned}
\tag{21.22}
$$

T_d^*, the optimal T_d, is the value of T_d which minimizes $E[C_s(T_d; W)]$. If T_d^* exists, it can be obtained by the usual first-order condition. The following example is from Ref. 33.

Example 21.5

Suppose that the product in question is an expensive industrial product sold with a long warranty period. Let the modification rate $\nu(t)$ be given by (21.19) with $\lambda = 1$ per week so that time during development is measured in weeks. Let $C_a = \$1000$, $C_d = \$100$ per day, $C_r = \$100$, and $Q = 1000$.

Table 21.7 gives T_d^*, the optimal development period (in weeks), and the expected cost to the manufacturer, $E[C_s(T_d^*; W)]$, for different combinations of W (in years) and β. Note that for a given β, as W increases, the optimal development period also increases. This is to be expected, as one needs a more reliable product to reduce the expected warranty cost. For a given W, as β increases, again the development period increases.

Murthy and Nguyen [33] propose two other models for optimal product development. In the first model, the development is stopped after n modifications and in this case, the duration of the development testing is a random variable. The optimal value of n is obtained by minimizing the

Table 21.7 Optimal Product Development: T_d^* versus W and β
(Example 21.4)

β	W (years)	T_d^* (weeks)	$\theta(T_d^*)$ (per year)	$E[C_s(T_d^*; W)]$ ($\times$ 100)
0.4	3	50	0.038	213
	5	71	0.031	281
	7	89	0.027	339
0.5	3	59	0.065	331
	5	87	0.054	448
	7	112	0.047	549

total expected cost to the manufacturer. In the second model, the failure rate changes after each modification and changes in an uncertain manner in contrast to the first model. As a result, the analysis of this model is more complex.

21.6 OTHER DESIGN ISSUES

One way of reducing warranty cost is to build diagnostic features into the item so that, for most common type failures, the user is guided to rectifying the problem without any external help. Failures which cannot be fixed by the user need to be returned to the manufacturer for rectification. A typical example of such diagnostic features being built into an item is a modern photocopier. Building in such features increases the cost of the item and is worthwhile only if the reduction in the expected warranty cost is more than the additional costs incurred. Diagnostic design is a relatively new concept and very little work has been done in this area. See Refs. 34 and 35 for some interesting results.

In some items, the design includes either a warning signal which requires action on the part of the consumer to avoid a failure (e.g., engine oil being low: unless action is taken immediately, there is the possibility of engine failure) or a mechanism which automatically initiates action to minimize item failure (e.g., a fuse acting as a protective device against electrical damage).

Finally, fault location is a challenging problem in multicomponent systems. Fault detectors help in reducing the time needed to locate the fault. Takami et al. [36] deal with optimal location of fault detectors. The use of modular design helps in reducing the time needed to rectify a defective item. However, this is achieved at a cost. The use of the modular approach involves replacement of a module, although only a component of the module has failed. This increase in replacement cost is justified only if the savings in the labor cost is greater. The optimal design of modularization in the context of products sold with warranty is still an open and challenging problem.

21.7 IMPLEMENTATION ASPECTS

A scheme to implement the design procedures discussed in this chapter is as follows.

21.7.1 New Product Design

Step 1: Decide on the minimum acceptable reliability for the product.

Step 2: Design the product using standard components. Evaluate the product reliability and the unit manufacturing and expected warranty costs.

Step 3: If the product reliability exceeds the minimum specification of Step 1 and the total (manufacturing + expected warranty) unit cost is acceptable, then go to Step 8.

Step 4: Can redundancy be used to improve product reliability and/ or reduce the total unit cost? If not, go to Step 6.

Step 5: Design optimal redundancy which minimizes the total unit cost for the following two cases: (1) no constraint on product reliability and (2) product reliability to meet the minimum specification of Step 1.

Step 6: Determine optimal reliability improvement plan which minimizes the total unit cost. If the product reliability is below the minimum specified in Step 1, evaluate the unit cost based on a design which meets the specified reliability.

Step 7: Compare the results of Steps 5 and 6 with that of Step 1. Decide on the best strategy. This might involve trade-offs between total unit cost and reliability of product, and possibly going back to Step 1 to alter the minimum acceptable reliability for the product.

Step 8: Stop.

27.7.2 Existing Product

In this case, the design modification is to reduce the total unit cost. The modified design might also require that the product reliability exceed some specified minimum value.

Step 1: Can redundancy be used to improve product reliability and/ or reduce the total unit cost? if not, go to Step 3.

Step 2: Design optimal redundancy which minimizes the total unit cost for the following two cases: (1) no constraint on product reliability and (2) product reliability exceeds the minimum specification.

Step 3: Determine optimal reliability improvement plan which minimizes the total unit cost. If the product reliability is below the minimum specified, evaluate the unit cost based on a design which meets the specified reliability.

Step 4: Compare the results of Steps 2 and 3 with total unit cost for the existing product. Decide on the best strategy. This might involve trade-offs between total unit cost and reliability of product.

Step 5: Stop.

21.8   SUMMARY AND CONCLUSIONS

Product reliability and unit manufacturing cost are determined by product design. The expected warranty cost is a function of product reliability. As a result, the total unit cost is critically dependent on product design. In this chapter, we have discussed alternate design approaches to improve product reliability and/or reduce total unit cost.

There are still many unresolved problems that need to be studied. We discuss one such problem in which the product failures depend on the usage intensity. For many products, the usage intensity varies across the consuming population, for example, washing machines in light-duty use in a domestic situation versus heavy-duty use in a hospital. In this case, the manufacturer can either design one machine for both usages or design two different machines. The optimal choice between the two would depend not only on warranty and manufacturing unit cost but also on the market size for the two different usages and the extra cost involved in the design and manufacturing of two different machines as opposed to a single machine.

REFERENCES

1. Kapur, K. C., and Lamberson, L. R. (1977). *Reliability in Engineering Design*, John Wiley & Sons, New York.
2. Kececioglu, D. (1991). *Reliability Engineering Handbook*, Vol. 1, Prentice-Hall, Englewood Cliffs, NJ.
3. Kececioglu, D. (1991). *Reliability Engineering Handbook*, Vol. 2, Prentice-Hall, Englewood Cliffs, NJ.
4. Marshall C. W. (1981). Design trade-offs in availability warranties, *Proceedings of the Annual Reliability and Maintenance Symposium*, pp. 95–100.

5. Court, E. T. (1981). To specify a reliability requirement does not ensure its achievement, *Reliability Engineering*, **2**, 243–258.

6. Hutchinson, G. M. (1983). The specification and achievement of reliability in military equipment, *Reliability Engineering*, **4**, 158–168.

7. Barlow, R. E., and Proschan, F. (1981). *Statistical Theory of Reliability and Life Testing*, To Begin With, Silver Spring, MD.

8. Nguyen, D. G., and Murthy, D. N. P. (1988). Optimal reliability allocation for products sold under warranty, *Engineering Optimization*, **13**, 35–45.

9. Spence, A. M. (1977). Consumer misperception, product failure and producer liability, *Review of Economic Studies*, **44**, 561–572.

10. Grossman, S. (1981). The informational role of warranties and private disclosure about product quality, *Journal of Law and Economics*, **24**, 261–283.

11. Glickman, J. S., and Berger, P. D. (1976). Optimal price and protection period for a product under warranty, *Management Science*, **22**, 1381–1396.

12. Anderson, E. E. (1977). Product price and warranty terms: An optimization model, *Operations Research Quarterly*, **28**, 739–741.

13. Ritchken, P. H., and Tapiero, C. S. (1986). Warranty design under buyer and sellers risk aversion, *Naval Research Logistics Quarterly*, **33**, 657–671.

14. Menezes, M. A. J. (1985). *Product Warranty as an Element of Marketing Mix*, Ph.D. thesis, University of California, Los Angeles.

15. Murthy, D. N. P. (1990). Optimal reliability choice in product design, *Engineering Optimization*, **15**, 281–294.

16. Brooks, A. C. (1983). Realistic cold redundant systems, *Reliability Engineering*, **6**, 127–131.

17. Sharma, G. N. (1981). Hot redundant versus cold redundant systems, *Reliability Engineering*, **2**, 193–197.

18. Rao, S. S., and Natrajan, R. (1970). Reliability with standbys, *OPSEARCH*, **7**, 23–35.

19. Iyer, S. (1989). The increase in reliability and the mean time to failure of a system due to component redundancy, *OPSEARCH*, **26**, 143–150.

20. Osaki, S., and Nakagawa, T. (1971). On a two unit standby redundant system with standby failure, *Operations Research*, **19**, 510–523.

21. Misra, K. B. (1971). A method of solving redundancy optimization problems, *IEEE Transactions on Reliability*, **R-20**, 117–120.

22. Misra, K. B. (1972). A simple approach for redundancy optimization problem, *IEEE Transactions on Reliability*, **R-21**, 30–34.

23. Murthy, D. N. P., and Hussain, O. (1993). Warranty and optimal redundancy design, Engineering Optimization, **23**, 301–314.

24. Osaki, S. (1971). A note on two-unit standby redundant system with imperfect switch over, *RAIRO*, **5**, 103–109.

25. Nielson, D. S., and Runge, B. (1974). Unreliability of a standby system with repair and imperfect switching; *IEEE Transactions on Reliability*, **R-23**, 17–24.

26. Nakagawa, T., and Osaki, S. (1975). Stochastic behavior of a 2-unit standby redundant systems with imperfect switchover, *IEEE Transactions on Reliability*, **R-24**, 143–145.

27. Chow, D. K. (1971). Reliability of two items in sequence with sensing and switching, *IEEE Transactions on Reliability*, **R-20**, 254–256.
28. Stoyan, D. (1983). *Comparison Methods for Queues and Other Stochastic Models*, John Wiley & Sons, New York.
29. Dhillon, B. S. (1980). *Reliability growth*: A survey, *Microelectronics and Reliability*, **20**, 743–751.
30. Robinson, D., and Dietrich, D. (1989). A nonparametric-Bayes reliability growth model, *IEEE Transactions on Reliability*, **R-38**, 591–598.
31. Duane, J. T. (1964). Learning curve approach to reliability monitoring; *IEEE Transactions on Aerospace*, **A-2**, 563–56.
32. Crowe, L. H. (1974). Reliability analysis for complex repairable systems, in *Reliability & Biometry* (F. Proschan & R. J. Serfling, eds.), SIAM, Philadelphia, pp. 397–410.
33. Murthy, D. N. P., and Nguyen, D. G. (1987). Optimal development testing policies for products sold with warranty; *Reliability Engineering*, **19**, 113–123.
34. Hegde, G. G., and Kubat, P. (1989). Diagnostic design: A product support strategy; *European Journal of Operations Research*, **38**, 35–43.
35. Malcolm, J. G., and Foreman, G. L. (1984). The Need: Improved diagnostic—Rather than improved R, *Proceedings of the Annual Reliability and Maintenance Symposium*, pp. 315–322.
36. Takami, I., Inagaki, T., Sakiuno, E., and Inoue, K. (1978). Optimal allocation of fault detectors, *IEEE Transactions on Reliability*, **R-27**, 360–362.

22

Warranty Analysis for Complex Systems

Stefanka Chukova and Boyan Dimitrov

GMI Engineering and Management Institute
Flint, Michigan

22.1 INTRODUCTION

Complex items and complex systems are products that exhibit a multicomponent structure (e.g., involve two or more components). This results in the necessity of considering more than one warranty parameter or cost component in the analysis. The main feature of these types of policies, models, and problems is that warranties can be different for different components.

Examples of complex items are as follows:

Car: battery, tires, drive train, and the rest, having warranties separate for each.
Television: picture tube and the rest. Recently, warranty on labor has also been separate.
Refrigerator: compressor and the rest.

Airplane: engines, avionics, windshield, other significant components, and the rest.
Computer: processor, monitor, hard-drivers, and the rest.

Typically, the components are warranted by their producers rather
than by the system's manufacturer. However, the manufacturer is responsible for determining the warranty policy related to the entire item.

In studying warranty problems relative to complex items, various
approaches are possible. Among them we list

Construction of a single failure distribution to characterize warranty
claims for the entire item
The view of the system as a multicomponent structure, with joint
multivariate probability distributions for warranted components
Modeling the warranty based on considering failure interactions
among the components
Modeling warranty of complex item as a network structure

However, warranty models for complex systems is a new topic in
warranty analysis. There is very limited literature on the subject. But the
questions of most importance are still the same: types of warranty, how
to model warranty claims, and cost analysis. At this point, we note the
following peculiarities of the warranty cost analysis for complex items:

Dependent on the failures of components
Dependent on some known and already established probability parameters, and also on some new parameters that must be poste-
riorly determined
Straightforward, if component failures are independent and if there
are no competing warranty restrictions among the components
Complex when there are failure dependencies or stochastic interactions among the components

This chapter is an attempt to formulate and discuss some models
and problems relevant to warranties of complex items. Four cases will be
considered: independent or dependent failures, and, for each, warranties
that are the same for all components or that are different for different
components. Under independence and identical warranty terms, the cost
analysis is a simple extension of previous results. (See, e.g., Chapters
10 and 11.) A principal difference in what follows occurs if a warranted
component fails. This may lead to a claim for the entire item, or a claim
for just the failed component. Dependent failures lead to a more complex
characterization of claims, and different warranties further complicate the
analysis.

The outline of the chapter is as follows: In Section 22.2, several warranty policies are introduced. In Section 22.3, various setups of modeling failures (warranty claims) are discussed. Corresponding warranty cost evaluation for the cases of independent and dependent component failures is conducted in Sections 22.4 and 22.5, respectively. Numerical examples illustrate all the involved theoretical results. Calculations are performed by the use of advanced program techniques from Ref. 1. Conclusions and topics for further research are formulated in the final section.

22.2 WARRANTY POLICIES

We confine our consideration to systems with two components. The extension to more than two components is conceptually straightforward. The diversity of warranty policies for complex systems is based on the concepts and terminology introduced in Chapter 1, and on the variety of warranty policies and models for the components as shown in Chapters 6, 10, and 11. The combinations of warranties assigned to each component lead to consideration of the following nonrenewing warranty policies for two-component systems:

Policy 1

Component 1 is covered by a free-replacement warranty (FRW) for a period W_1, and Component 2 is covered by a FRW for a period W_2.

Example

Consider a refrigerator. The compressor may have a FRW for a period of 5 years, whereas the rest may have a FRW for a period of 3 years.

Policy 2

Component 1 is covered by a pro-rata warranty (PRW) for a period W_1, and Component 2 is covered by a FRW for a period W_2.

Example

In a car, the battery has a 4-year PRW and the new battery will be purchased for a price with a discount proportional to the residual warranty time, whereas the rest has a FRW for 3 years or 80,000 miles, whichever comes first.

Policy 3

Component 1 is covered by a PRW for a period W_1, and Component 2 is covered by a PRW for a period W_2.

Example

In a new car, the battery has a 4-year PRW as shown in the previous example, whereas the tires have a PRW for 60,000 miles each.

Policy 4

Component 1 is covered by a reliability improvement warranty (RIW) for a period W_1, and Component 2 is covered by a FRW for a period W_2.

Example

For an airplane, the engine is covered by RIW for the contract period, whereas the remaining parts are covered by a FRW for, say, 6 years.

We will refer to Policies $1'$, $2'$, $3'$, and $4'$ whenever both components are under renewing warranty. Policies $1''$, $2''$, and $3''$ will denote that one component is under renewing warranty and the other component is under nonrenewing warranty.

Note that the variety of warranty policies increases with the increase in the number of components whose individual warranties have to be viewed within the item.

22.3   MODELING FAILURES (WARRANTY CLAIMS)

The modeling of warranty claims for complex items depends on the adopted failure model. The claims are caused by the component's and/or system's failures. The claimed amount depends on respective costs and policy. When a single component fails, its repair or replacement cost differs from the cost of repair or replacement of a set of components. Sometimes it will be preferable to replace the entire item instead of repairing or replacing just failed components. These possibilities lead to the consideration of a variety of models. Here we point out some models with an emphases on the model structure rather than on their explicit analysis. The formulation is done in terms of failure distribution functions. Again, we will consider two-component systems. Assume that the marginal operating lifetimes of Components 1 and 2 are X_1 and X_2, with marginal cumulative distribution functions (CDFs) $F_1(x_1)$ and $F_2(x_2)$, respectively.

22.3.1   Independent Component Failures

If instantaneous repairs or replacements are assumed for each component, then the total amount of claims is a superposition of the two separate independent flows of claims caused by each component.

If it is assumed that the item is replaced by a new one after any component failure, the time between failures is then modeled by the univariate distribution function of $\min\{X_1, X_2\}$, that is,

$$F(x) = 1 - [1 - F_1(x)][1 - F_2(x)] \tag{22.1}$$

The claim process will be generated by a renewal process with CDF $F(x)$ given by (22.1). (See Chapter 7.)

Under the rectification actions of some policies, subsequent failures can also be involved. Assume that the nth repair of Component i changes the next lifetime distribution to $F_{i,n}(x)$. Then the renewal function $M_i(t)$ associated with the failures of the ith component will be

$$M_i(t) = \sum_{n=1}^{\infty} F_{i1} F_{i2} \cdots F_{in}(t) \tag{22.2}$$

In particular, the CDFs $F_{in}(t) = F_i(\beta^n t)$, $n = 1, 2$, with $\beta > 1$ and $F_i(t)$ the same as the original lifetime distribution of Component i, could represent the wearing of this component due to failures, expressed in terms of accelerated time to failure. The renewal function $M_i(t)$ is then the solution of the following integral equation:

$$M_i(t) = F_i(t) + \int_0^t M_i(\beta(t - x)) \, dF_i(x) \tag{22.3}$$

A renewal function for an RIW policy is given by exactly the same equation, for suitably chosen $\beta < 1$. Such a model could correspond to a wearing process slowed down after any rectification. This may be true when new materials or better technologies invested into the operating component lead to proportional prolongations of its life.

This analysis can be extended by introducing item replacements after each component failure when the cost of the failed component is high. Then the claim process will follow a renewal function associated with $F(x)$ given by (22.1), with the claim size depending on just the failed component. These cases and related questions are discussed in detail in Chapters 6 and 7.

22.3.2 Component Failures Not Independent

In most cases of complex items, there will be dependence among the components' lifetimes. They are functioning in common environmental conditions, sharing the same energy sources and operation loads, and have undergone the same sets of shocks and stresses. Therefore, the component failures of a complex item will act similarly. Failure of a component will degrade the performance of all subsystems that contain it. For example, failure of a fuse leads to operating failures of all electrical circuits loaded through it. A high stress in a circuit will affect all components involved. A switch between function and failure of a component contributes to a decrease or increase in the load shared with the remaining components and may increase or decrease their failure rates afterward. Thus, we come to the concept of the associated random lifetimes used in the description of the reliability of coherent systems (see Ref. 2). Moreover, when considering problems related to warranty analysis, the effect of possible maintenance and/or rectification actions also have to be taken into account. Then the complexity of corresponding analyses becomes significant. Therefore, inclusion of certain specifications and simplifying assumptions are unavoidable.

Bivariate (Joint) Life Distribution

To model dependent failures for two-component systems, the joint CDF

$$F(x_1, x_2; \theta) = P\{X_1 \leq x_1, X_2 \leq x_2\}$$

is used. Here, x_1 and x_2 are resource variables measuring, in appropriate units, the marginal life times X_1 and X_2 (resources) of the two components, and θ is the parameter of the joint distribution. Other equivalent representations can be given by the joint survival function

$$\overline{F}(x_1, x_2, \theta) = 1 - F_1(x_1) - F_2(x_2) + F(x_1, x_2; \theta)$$

where $F_1(x_1)$ and $F_2(x_2)$ are the marginal life distributions for the components. In addition, the joint probability density function (p.d.f.)

$$f(x_1, x_2; \theta) = \frac{\partial^2}{\partial x_1 \, \partial x_2} F(x_1, x_2; \theta)$$

can be used, provided that it exists.

In estimation of the probability of having warranty claims during given warranty periods W_1 and W_2 for Components 1 and 2, respectively, the following Fréchet's inequalities

$$\max\{0, F_1(W_1) + F_2(W_2) - 1\}$$
$$\leq F(W_1, W_2) \leq \min\{F_1(W), F_2(W_2)\} \quad (22.4)$$

can be used successfully to approximate the c.d.f. The joint (bivariate) life distribution $F(x_1, x_2)$ contains all the information one needs to answer any question in modeling failures in the item with nonrepairable components. For instance, the p.d.f. $f(x_1, x_2; \theta) \, dx_1 \, dx_2$ gives the probability of failure of both components when their resources are in a neighborhood of the point (x_1, x_2). The expression

$$P\{a_1 < X_1 \leq b_1, a_2 < X_2 \leq b_2\}$$
$$= F(a_1, a_2) - F(a_1, b_2) - F(a_2, b_1) + F(b_1, b_2)$$

gives the probability for a failure in the region $\{a_1 < X_1 \leq b_1\} \cap \{a_2 < X_2 \leq b_2\}$. This can be used to calculate probabilities of claims.

Here we have the first problem: When either of the two components fails, is the claim for the entire item or for the failed component only? Another case is when individual component failures are not detectable, and a claim for the entire item will be made only when both components fail. These cases are important in modeling warranty claims and refunds, especially under PRW policies. Possible repairs and their influence on further component interactions make the picture much more complicated, and failure modeling is difficult without other specific assumptions.

As an example, consider a two-component system with nonrepairable components and a FRW policy (W_1, W_2). Assume that failures of the components are detectable and lead to claims. Then the probability for no claim during the assigned warranty times is $P\{X_1 > W_1, X_2 > W_2\}$ = $\bar{F}(W_1, W_2)$. The sources of claims can be separated as follows: The probability of a claim due to failure of Component 1 is

$$p_1 = P\{X_2 \geq \min(X_1, W_2), X_1 \leq W_1\}$$

$$= \int_0^{W_1} \left(\int_{\min(x_1, W_2)}^{\infty} f(x_1, x_2) \, dx_2 \right) dx_1$$

The corresponding probability for Component 2 is obtained from this by changing the subscripts 1 to 2, and 2 to 1. The probability of a claim due to failure of both components simultaneously is $p_{12} = P\{X_1 = X_2, X_1 \leq W_1, X_2 \leq W_2\}$ which equals 0 when (X_1, X_2) has a joint continuous probability distribution. Shock models (see the next section) show that there are cases when this probability can be positive.

Now let us assume that the system fails only when both components fail. We again consider two cases:

1. System failure leads to a warranty claim only if both components fail within their warranty times. This occurs with probability $P\{X_1 \leq W_1, X_2 \leq W_2\} = F(W_1, W_2)$. From (4), one can get useful bounds for this probability.

2. System failure leads to a claim if at least one of its component fails within the assigned warranty. Then the probability of having no claim is $P\{X_1 > W_1, X_2 > W_2\} = \bar{F}(W_1, W_2)$, and the probability of having a claim is $1 - \bar{F}(W_1, W_2)$.

The discussion of these cases continues in Section 22.5.1.

Example 22.1

Suppose that the freezing and cooling sections of a refrigerator have a bivariate normal distribution of lifetimes X_1 and X_2, given by the p.d.f.

$$f(t_1, t_2) = \frac{1}{2\pi\sigma_1\sigma_2 \sqrt{1 - \rho^2}} \exp\left\{ - \frac{1}{2(1 - \rho^2)} \left(\frac{(t_1 - \mu_1)^2}{\sigma_1^2} \right.\right.$$

$$\left.\left. - 2\rho \frac{(t_1 - \mu_1)(t_2 - \mu_2)}{\sigma_1\sigma_2} + \frac{(t_2 - \mu_2)^2}{\sigma_2^2} \right)\right\}$$

with $\mu_1 = 5$, $\mu_2 = 6$, $\sigma_1 = \sigma_2 = 1$, and $\rho = 0.6$ (ρ is the correlation coefficient between the two lifetimes). The two sections also have a common 4-year FRW (Policy 1). The marginal lifetimes have univariate normal distributions with parameters ($\mu_1 = 5$, $\sigma_1 = 1$) and ($\mu_2 = 6$, $\sigma_2 = 1$).

Suppose that failures of each section can be recognized and that these are claimed immediately. Then the probability of a claim due to failure of the freezing section only during the warranty period is

$$p_1 = \int_{-\infty}^{4} \left(\frac{1}{\sqrt{2\pi}} \exp\left(- \frac{(t_1 - 5)^2}{2} \right) - \int_{-\infty}^{4} f(t_1, t_2)\, dt_2 \right)$$

$$\times\, dt_1 = 0.15863$$

The probability of a claim due to failure of the cooling section only is

$$p_2 = \int_{-\infty}^{4} \left(\frac{1}{\sqrt{2\pi}} \exp\left(- \frac{(t_2 - 6)^2}{2} \right) - \int_{-\infty}^{4} f(t_1, t_2)\, dt_1 \right)$$

$$\times\, dt_2 = 0.006927$$

The probability of simultaneous failure of both sections is 0 because the bivariate normal distribution is continuous.

Suppose that a claim is made only when both sections fail and that the warranty period is 4 years. Then the probability of claim is found by numerical calculation of the integral to be

$$F(4, 4) = \int_{-\infty}^{4} \int_{-\infty}^{4} f(t_1, t_2)\, dt_1\, dt_2 = 0.015828$$

Its Fréchet upper bound is $F_2(4) = \Phi(-2) = 0.0228$ (the Fréchet lower bound is 0).

Suppose that claims are made if at least one of the refrigerator's sections has failed before the 4-year warranty period expires. Then the probability of no claim is

$$\overline{F}(4, 4) = \int_4^\infty \int_4^\infty f(t_1, t_2) \, dt_1 \, dt_2 = 0.8343$$

The probability of having a claim due to at least one component failure during the warranty is 0.1656. Note that this is very different from the value of $F(4, 4) = 0.0158$, which is the probability of claim only if both components fail in 4 years.

22.3.3 Shock Models: Marshal–Olkin's Bivariate Exponential Distribution

Failure dependence associated with the influence of environmental conditions may be modeled by Marshal–Olkin's bivariate exponential distribution (here it will be abbreviated MO'BVE) [2]. Suppose that the environment may be represented by three independent sources of shocks. A shock from source 1 destroys Component 1, and it occurs at exponentially distributed random time Y_1 with parameter λ_1. A shock from source 2 destroys Component 2 at an exponentially distributed random time Y_2 with parameter λ_2. Finally, a shock from source 3 destroys both components at another exponentially distributed random lifetime Y_{12} which has parameter λ_{12}. All three random variables Y_1, Y_2, and Y_{12} are independent. Then the random life X_1 of Component 1 is represented by $\min(Y_1, Y_{12})$ and has exponential marginal distribution

$$F_1(x_1) = P\{X_1 \le x_1\} = 1 - \exp[-(\lambda_1 + \lambda_{12})x_1] \quad \text{for } x_1 \ge 0$$

(22.5)

Component 2 has lifetime $X_2 = \min(Y_2, Y_{12})$ and marginal c.d.f.

$$F_2(x_2) = P\{X_2 \le x_2\} = 1 - \exp[-(\lambda_2 + \lambda_{12})x_2] \quad \text{for } x_2 \ge 0$$

(22.6)

The joint survival function of the lifetimes of the two components in this item is then given by the expression

$$\overline{F}(x_1, x_2) = \exp[-\lambda_1 x_1 + \lambda_2 x_2 + \lambda_{12} \max(x_1, x_2)]$$
$$\text{for } x_1 \ge 0, x_2 \ge 0$$

(22.7)

This is the MO'BVE distribution. It can be used whenever the item's component behaves as described above. Moreover, the following nonfatal shock model also gives a MO'BVE distribution: In this shock model, a shock from source 1 with probability p_1 may cause a failure of Component 1; a shock from source 2 causes failure of Component 2 with probability p_2, and finally, a shock from source 3 may cause failure of both components with probability p_{11}. Only Component 1 will fail with probability p_{10}; only Component 2 will fail with probability p_{01}; and neither component fails with probability p_{00} ($p_{11} + p_{01} + p_{10} + p_{00} = 1$). Then the joint distribution of the life lengths (X_1, X_2) of components 1 and 2 is again a MO'BVE with parameters

$$\lambda_1^* = \lambda_1 p_1 + \lambda_{12} p_{10}, \qquad \lambda_2^* = \lambda_2 p_2 + \lambda_{12} p_{01}, \qquad \lambda_{12}^* = \lambda_{12} p_{11}$$

(22.8)

When $p_1 = p_2 = p_{11} = 1$, this shock model reduces to the original one (with fatal shocks). Suppose the failed components are immediately replaced by new ones with the same distribution and under the same environment. Let $N_1(t)$ and $N_2(t)$ be the number of replacements of Components 1 and 2 during an interval of operation $[0, t]$. Then [2, p. 137]

$$P\{N_1(t) = n_1, N_2(t) = n_2\} = \exp[-\lambda t(q_{11} + q_{10} + q_{01})]$$

$$\times \sum_{m=0}^{\min(n_1,\, n_2)} \frac{(\lambda t q_{11})^m (\lambda t q_{10})^{n_1 - m} (\lambda t q_{01})^{n_2 - m}}{m!(n_1 - m)!(n_2 - m)!}$$

(22.9)

Here, $q_{11} = p_{11}\lambda_{12}/\lambda$, with $\lambda = \lambda_1 + \lambda_2 + \lambda_{12}$, is the probability that both components must be replaced; $q_{10} = \lambda_1 p_1/\lambda + \lambda_{12} p_{10}/\lambda$ is the probability that only Component 1 will be replaced; $q_{01} = \lambda_2 p_2/\lambda + \lambda_{12} p_{01}/\lambda$ is the probability that just Component 2 will be replaced; and $q_{00} = 1 - q_{11} - q_{10} - q_{01}$ is the probability that neither component will be replaced. The process $\{N_1(t), N_2(t), t \geq 0\}$ is called the bivariate Poisson process. Note that the individual components $N_1(t)$ and $N_2(t)$ on any time interval have one-dimensional Poisson distributions with parameters $\lambda_1^* + \lambda_{12}^*$ and $\lambda_2^* + \lambda_{12}^*$, respectively. The time to first failure is modeled by the survival function $P\{N_1(t) = 0, N_2(t) = 0\} = P\{X_1 \geq t, X_2 \geq t\}$.

Induced Failures

This type of failure interaction for nonexponential life distributions is frequently observed in practice. (See Refs. 3 and 4.) Whenever Component 1 (2) fails, it can induce a failure of Component 2 (1) (with respective probabilities p_1 and p_2). In the case of induced component failures, the whole system also fails. Note that if $p_i = 0$, $i = 1, 2$, there is no failure interaction and the failures are independent. If $p_i = 1$ for $i = 1, 2$, then failure of each component causes item failure, that is, a series reliability scheme describes the item's structure. For the general case $0 \leq p_1, p_2 \leq 1$, the time to first item's failure, say Y, has a survival function given by

$$\overline{G}(t) = P\{Y > t\} = \overline{G}_1(t)\overline{G}_2(t)$$

(22.10)

where

$$\overline{G}_i(t) = \sum_{n=0}^{\infty} (1 - p_i)^n \{F_i^{(n)}(t) - F_i^{(n+1)}(t)\} \qquad (22.11)$$

and $F_i^{(0)}(t) = 1$ and $F_i^{(n)}(t)$ is the n-fold Stieltjes convolution of $F_i(t)$. Each of the survival functions $\overline{G}_i(t)$ represents the time to first item failure under the absence of failures induced by the other component. The probabilities of induced failures can also be functions of the elapsed time (usually the age of the system in use), that is, $p_i = p_i(t)$, $i = 1, 2$, can be assumed.

Shock-Damage Models

When assuming that a failure of Component 1 (2) results in some damage to Component 2 (1), without inducing an instantaneous failure but affecting its failure rate, we obtain a shock-damage model. Then the failure rate of each component at time t is a random variable depending on its natural failure rate as well as on the number of other component failures in [0, t]. To keep the item in operation during warranty time (W_1, W_2) by covering the costs of replacing the failed components in each of corresponding intervals will form the warranty expenses.

22.3.4 Modular Structures

Complex systems comprise a modular structure of subsystems. A precise description of modules of coherent systems can be found in Ref. 2. For a two-component system, we consider the following model of dependent failures. When Component 1 fails, it is replaced; when component 2 fails, both are replaced. A model for this can be also obtained from the above model of induced failures as a particular case, namely, when $p_1 = 0$ and $p_2 = 1$. For systems with more than two components, this structure allows other possibilities that make this class of situations more complex.

22.3.5 Network Structures

Network structures for complex items can be recognized as one of a variety of discrete reliability structures considered in Chapter 21. Here, we recall the parallel structure, where, in principal, the components can be considered to be independent (e.g., redundancy, discussed in Chapter 21), the series system (e.g., in a stereo system the sound subsystem is considered to be the tuner and amplifier connected in series), or more general structures such as bridge or star structures, discussed in standard reliability texts. In each case, one has to consider the time to failure for the entire system as a function of the components' failures and of the corresponding engineering structure, where possible interactions are involved.

Suppose that a system has r components. Then the system can be in one of 2^r states, varying from that with all components working, to that with one or more not working, to that with all failed. As a result, modeling can be done by a stochastic process $\{\xi_t, t \geq 0\}$ (ξ_t denotes the system state at time t) with $n = 2^r$ states, suitably numbered $i = 1, 2, \ldots, n$. Note that the number of states could be significantly reduced if some states have properties equivalent to those of other states. This is characterized by a matrix of transition probabilities $\mathbf{P} = \{p_{ij}\}$ which incorporate the possible dependencies of component failures. p_{ij} is the probability that the system is next in state j given that at the present time it is in state i. (The matrix $\mathbf{P}$ itself defines an imaginary discrete time process known as an *imbedded Markov chain*.) Let $F_{ij}(t)$ be the c.d.f. of the time ξ_t spent in state i, having started in state j (the "sojourn time" in state i). Two important approaches in modeling network structures are developed under special assumptions about the transition probabilities p_{ij} and the functions $F_{ij}(t)$.

22.3.6 Markov Formulation

The Markovian approach is based on the assumption that the p_{ij} depend neither on time nor on the number of changes of states already made. Another important assumption is that $F_{ij}(t)$ is exponential for all pairs (i, j). This approach has been used to model dependent failures for purposes of warranty cost analysis in Refs. 5–7.

We illustrate the elements of the Markovian approach by means of a simple two-component system ($r = 2$). Suppose that a failed component can be repaired while the other operates, and that the item also operates. Both components can simultaneously fail, as well as each component failing separately. Two failed components require replacement of the entire item. The possible transitions between the four states are shown on Figure 22.1. The states are as follows:

State $\{0\}$: Components 1 and 2 both failed
State $\{1\}$: Component 1 functioning and Component 2 failed
State $\{2\}$: Component 1 failed and Component 2 functioning
State $\{3\}$: Components 1 and 2 functioning

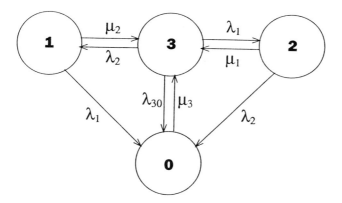

Figure 22.1   Model of state transitions.

The corresponding Markovian transition matrix

$$\mathbf{P} = \begin{bmatrix} 0 & 0 & 0 & 1 \\ p_{10} & 0 & 0 & p_{13} \\ p_{20} & 0 & 0 & p_{23} \\ p_{30} & p_{31} & p_{32} & 0 \end{bmatrix}$$
(22.12)

is used to calculate the expected number of claims of each kind during periods of warranty. The sojourn times at each state also are involved in construction of corresponding renewal processes.

For a continuous Markovian model of the same system, another approach, based on the so-called birth and death process, can be applied. It uses the following *infinitesimal* (or *intensity*) *matrix*

$$\mathbf{A} = \begin{bmatrix} -\mu_3 & 0 & 0 & \mu_3 \\ \lambda_1 & -(\lambda_1 + \mu_2) & 0 & \mu_2 \\ \lambda_2 & 0 & -(\lambda_2 + \mu_1) & \mu_1 \\ \lambda_{30} & \lambda_2 & \lambda_1 & -(\lambda_1 + \lambda_2 + \lambda_{30}) \end{bmatrix}$$
(22.13)

Here λ_i and μ_i are the intensities of failure (death) and repair (birth) of the ith component, respectively, and λ_{30} and μ_3 are the intensities of item failures and repairs when both components are functioning (failed). The products $a_{ii}p_{ij}$ of diagonal entries of matrix $\mathbf{A}$ and transition probabilities are the parameters of the exponential c.d.f.'s $F_{ij}(t)$, representing times spent at state i when next state is j. The exponential assumptions make admissible the use of powerful mathematical analytical techniques in studies. For instance, differential equations can be written for the conditional probabilities $P\{\xi_t = j | \xi_0 = i\}$ which describe the process evolution during the assigned warranty times and policies (see Ref. 5). These are also involved in analysis of warranty costs related to the item. Numerical methods are needed to complete the necessary calculations.

Explicit illustration of the Markovian approach to the analysis of claims generated by a three-component system can be found in Ref. 5. Some r-component models with rectifications are considered in Ref. 6. A review of related results is given in Ref. 7. The technique of application of Markovian modeling is discussed in detail in Ref. 8. More about calculation of characteristics of warranties is given in Section 22.5.4.

22.3.7 Semi-Markovian Formulation

If in the Markovian approach the $F_{ij}(t)$ are not exponential, then the result is a semi-Markovian process $\{\xi_t; t \geq 0\}$, briefly SMP. A SMP is defined by the matrices $\mathbf{P}$, $\mathbf{F}(t) = \{F_{ij}(t)\}$, and by a vector $\mathbf{a} = (a_0, a_1, \ldots, a_n)$ of initial probabilities where the process starts. The family of random variables

$$\{N_0(t), N_1(t), \ldots, N_n(t), t \geq 0\} \tag{22.14}$$

where $N_i(t)$ denote the number of visits to state i on the interval $[0, t]$ made by the SMP, is called a *Markov renewal process* (see Refs. 9 and 10). The function

$$F_i(t) = \sum_{j=0}^{n} p_{ij} F_{ij}(t) \tag{22.15}$$

is the c.d.f. of the unconditional sojourn time at state i. Let μ_i and μ_{ij} be the respective unconditional and conditional (given that the next state is j) expected sojourn times for state i. Then the mean time σ_{ij} to go from state i to state j for the first time is the solution of the system of linear equations [2, 10]

$$\sigma_{ij} = p_{ij}\mu_{ij} + \sum_{k \neq i}^{n} p_{ik}(\mu_{ik} + \sigma_{kj}) \tag{22.16}$$

When the embedded Markov chain is irreducible (i.e., from each state i, ξ_t can pass to any other state j in a finite number of steps), then there exists a stationary distribution $\boldsymbol{\pi} = (\pi_0, \pi_1, \ldots, \pi_n)$ of that chain. It is given as the only solution of the following system of linear equations:

$$\pi = \pi P, \quad \pi_0 + \pi_1 + \cdots + \pi_n = 1 \tag{22.17}$$

The expected time to first return to state i is determined by the equation

$$\sigma_{ii} = \frac{1}{\pi_i} \sum_{k=0}^{n} \pi_k \mu_k \tag{22.18}$$

These times characterize claims related to the visits to the corresponding states.

Example 22.2

Consider the above two-component system. Assume that any total system failure is immediately removed (i.e., $\mu_0 = 0$). This corresponds to $p_{03} = 1$ for the embedded Markov chain. Also, let the transition matrix (22.12) be

$$P = \begin{bmatrix} 0 & 0 & 0 & 1 \\ 0.5 & 0 & 0 & 0.5 \\ 0.4 & 0 & 0 & 0.6 \\ 0.2 & 0.3 & 0.5 & 0 \end{bmatrix} \tag{22.19}$$

Suppose that the conditional sojourn time at state i, given that the next state is j, is a random variable with a Gamma distribution with scale parameter $\lambda_{ij} = 1$ and shape parameter α_{ij}. Let the corresponding numerical values be as follows:

$$\alpha_{10} = 2; \quad \alpha_{13} = 3; \quad \alpha_{20} = 1; \quad \alpha_{23} = 4; \quad \alpha_{30} = 1; \quad \alpha_{31} = 2;$$

$$\alpha_{32} = 4; \quad \text{and all others } \alpha_{ij} = 0.$$

Then, from (22.15)–(22.18), we have the following:

The expected time μ_3, during which both components simultaneously function, is

$$\mu_3 = (0.20)(0.1) + (0.30)(0.3) + (0.50)(0.2) = 2.1$$

The expected time while only Component 1 (2) works is

$$\mu_1 = (0.50)(0.2) + (0.50)(0.3) = 2.5$$
$$\mu_2 = (0.40)(0.1) + (0.60)(0.4) = 2.8$$

The expected times to first passage from state i to state j are solutions to the system (22.16)–(22.18). The stationary probabilities for the embedded Markov chain are

$$\pi = \left(\frac{55}{235}, \frac{30}{235}, \frac{50}{235}, \frac{100}{235} \right)$$

The expected times until next return to the same state for each of the four states are

$$\sigma_{00} = \frac{1}{55\{30(2.5) + 50(2.8) + 100(2.1)\}} = \frac{435}{55}$$

$$\sigma_{11} = \frac{435}{30}; \qquad \sigma_{22} = \frac{435}{50}; \qquad \sigma_{33} = \frac{435}{100}$$

The expected first passage times σ_{ij} between the states are entries of the matrix

$$\Sigma = \begin{bmatrix} \dfrac{435}{55} & \dfrac{35}{3} & \dfrac{57}{10} & 0 \\[2mm] \dfrac{75}{13} & \dfrac{435}{30} & \dfrac{82}{10} & \dfrac{25}{10} \\[2mm] \dfrac{548}{130} & \dfrac{434}{30} & \dfrac{435}{50} & \dfrac{28}{10} \\[2mm] \dfrac{85}{13} & \dfrac{35}{3} & \dfrac{57}{10} & \dfrac{435}{100} \end{bmatrix}$$

Semi-Markovian Characteristics During the Time to First Failure

Assume that failure of a complex item leads to a warranty claim only in the case of total item failure. In cases of component failures, claims are only for components that fail. Then the claim process between total failures can be modeled by an absorbing semi-Markovian process. The embedded Markov chain now is absorbing, that is, there are states from which, once entered, the process cannot leave. For Example 22.2, this case is obtained when we let state $\{0\}$ be absorbing, and states $\{1\}$, $\{2\}$, and $\{3\}$ be transient. Then $p_{00} = 1$ and $p_{01} = p_{02} = p_{03} = 0$ will replace the first row in the matrix (22.19). An absorbing Markov chain with k absorbing states is defined by a transition matrix of the form

$$\mathbf{P} = \left[\begin{array}{c|c} \mathbf{I} & \mathbf{O} \\ \hline \mathbf{R} & \mathbf{Q} \end{array}\right] \tag{22.20}$$

where $\mathbf{I}$ is the identity matrix, $\mathbf{O}$ is a zero matrix, and $\mathbf{R}$ and $\mathbf{Q}$ are $(n - k) \times k$ and $(n - k) \times (n - k)$ submatrices of $\mathbf{P}$. Then the (i, j)th entry of the inverse matrix

$$\mathbf{N} = (\mathbf{I} - \mathbf{Q})^{-1} \tag{22.21}$$

represents the expected number of visits to state j given that the system starts at state i before the process stops in the absorbing set.

Example 22.3 (Continuation of Example 22.2)

Let us assume that the system fails if either both components simultaneously fail (transition from {3} to {0} or if one component fails during the recovery of the other (transition from {1} or {2} to {0}). Then the entire item must be replaced. With the data from Example 22.2 and absorbing state {0}, we have the following auxiliary values:

$$I = [1]; \qquad O = [0, 0, 0]; \qquad R = \begin{bmatrix} 0.50 \\ 0.40 \\ 0.20 \end{bmatrix};$$

$$Q = \begin{bmatrix} 0 & 0 & 0.50 \\ 0 & 0 & 0.60 \\ 0.30 & 0.50 & 1 \end{bmatrix}$$

Therefore

$$N = \begin{bmatrix} 1 & 0 & -0.5 \\ 0 & 1 & -0.6 \\ -0.3 & -0.5 & 1 \end{bmatrix}^{-1} = \begin{bmatrix} \dfrac{130}{55} & \dfrac{15}{55} & \dfrac{50}{55} \\ \dfrac{18}{55} & \dfrac{85}{55} & \dfrac{60}{55} \\ \dfrac{30}{55} & \dfrac{50}{55} & \dfrac{100}{55} \end{bmatrix}$$

This means that for a new item with repairable components, prior to its failure, it will be completely restored on the average $N_{33} = 100/55$ times, only Component 2 will fail $N_{31} = 30/35$ times, and only Component 1 will fail $N_{32} = 50/55$ times. These numbers are the entries of the last row of N. Each visit to state {1} ({2}) is a failure of Component 2 (1). In an analogous way, one can interpret the other entries of matrix N. For instance, assume that the item starts operating just after a failure of Component 1, that is, from state {2}. From the second row, we find that the expected number of failures of Component 2 only (visits to state {1}) is $N_{21} = 18/55$, the expected number of failures of Component 1 only (visits in state {2}) is $N_{22} = 85/55$ (the initial failure of Component 1 is included), and finally, the system is expected to have both components working (visits into state {3}) $N_{23} = 60/55$ times.

We end consideration of the semi-Markovian approach with some results about the expected number $M_{ij}(t)$ of visits to state j on $[0, t]$, given that the system starts from state i at $t_0 = 0$. These could be of relevance in modeling expected warranty claims and/or possible approximations thereto on long time intervals. The functions $M_{ij}(t)$ are solutions of a complex system of integral equations (see Refs. 9 and 10.) Solutions can be obtained by suitable numerical methods (see, e.g., Ref. 11).

For purposes of warranty analysis, the following representation is useful:

$$M_{ij}(t) = \frac{t}{\sigma_{ii}} - \frac{\sigma_{ij}}{\sigma_{jj}} + \frac{1}{\sigma_{jj}} \sum_{k=0}^{n} P_{ik}(t)\sigma_{kj} \tag{22.22}$$

where $P_{ij}(t)$ is the conditional probability that the system is in state j at epoch t, given that it starts at $t_0 = 0$ from state i. These functions also are solutions of a system of integral equations. The σ_{ij} are determined by (22.16). Asymptotic results for the behavior of $M_{ij}(t)$ for large values of t may also be used in many applications. These are given by

$$M_{ij}(t) \sim \frac{t}{\sigma_{ii}} - \frac{\sigma_{ij}}{\sigma_{jj}} + \frac{1}{\sigma_{jj}} \sum_{k=0}^{n} P_k^* \sigma_{kj}, \quad t \to \infty \tag{22.23}$$

Example 22.4 (Continuation)

We again consider the two-component item of the previous examples. With the values of μ_j and σ_{ij} found earlier, we obtain the following steady-state probabilities:

$$P_0^* = 0 \qquad P_1^* = \frac{(2.5)(30)}{435} = \frac{75}{435}; \qquad P_2^* = \frac{140}{435}; \qquad P_3^* = \frac{210}{435}$$

Thus, if the new item starts operating at time 0, then the expected number of failures of Component 2 only (counted as visits to $\{1\}$), the expected number of failures of Component 1 only (visits to state $\{2\}$), and the expected number of failures of the entire item (visits to state $\{0\}$) are asymptotically expressed by (22.23). The results are

$$M_{31}(t) \approx \frac{30}{435}t - \frac{57/10}{435/50} + \frac{30}{435}\left(\frac{76}{435}\frac{435}{30} + \frac{140}{435}\frac{434}{30} + \frac{210}{435}\frac{35}{5}\right)$$

$$= 0.06897t + 0.7734$$

$$M_{32}(t) \approx 0.11494t + 0.14546; \qquad M_{30}(t) \approx 0.12644t - 0.13031$$

The expected number of visits to state $\{3\}$, due either to replacements of the item or to successful component renewals, may also be approximated by (22.23). The result is $M_{33}(t) \approx 0.22989t - 0.02110$.

Remark. Results for Markov chains in continuous time can be obtained from those for semi-Markovian processes when all the underlying distributions are exponential. An important new approach, based on a concept of repeated Markovian models, is developed in Ref. 12.

22.4 WARRANTY COST ANALYSIS: INDEPENDENT FAILURES

22.4.1 Nonrenewing Warranties

Independent failures of nonrenewing warranties of the various policy types as defined in Section 22.2 are now discussed.

Policy 1 Warranties with Nonrepairable Components

We assume that claims are made after each component failure of a component, that failures occur independently, and that the ith failed component in replaced instantaneously at a cost of C_i, $i = 1, 2$. Let $M_i(t)$ be the renewal function associated with $F_i(t)$. Then the expected warranty cost (EC) under policy 1 is

$$EC = C_1M_1(W_1) + C_2M_2(W_2) \tag{22.24}$$

Example 22.5

Two compressors are built to transmit air conditioning in a building. The larger and more powerful one (Component 1) is warranted for $W_1 = 3$ years under a FRW and its replacement in case of failure costs $C_1 = \$2700$. The smaller (Component 2) has a FRW for $W_1 = 2.5$ years and its emergency replacement costs $C_2 = \$1650$. The lifetimes of the compressors have Weibull distributions

$$F_1(t) = 1 - e^{-(0.05t)^2}; \qquad F_2(t) = 1 - e^{-(0.08t)^{3/2}}$$

where time is measured in years.

The expected component lifetimes, given by $E(T) = \Gamma(1/\alpha + 1)/\lambda$, are $\mu_1 = 17.725$ years and $\mu_2 = 11.284$ years. For this system,

$$EC = 2700M_1(3) + 1650M_2(2.5)$$

The renewal function $M(t)$ associated with a Weibull distribution $F(t) = 1 - e^{-(\lambda t)^\alpha}$ is (see Ref. 13)

$$M(t) = \sum_{k=1}^{\infty} \frac{(-1)^{k-1}A_kt^{k\alpha}}{\lambda^{k\alpha}\Gamma(k\alpha + 1)} \tag{22.25}$$

where the coefficients A_k are recursively determined by the equations

$$A_1 = \Gamma(\alpha + 1); \qquad A_k = \frac{\Gamma(\alpha k + 1)}{k!} - \sum_{v=1}^{k-1} \frac{\Gamma(v\alpha + 1)}{v!} A_{k-v} \tag{22.26}$$

and $\Gamma(x)$ is the Gamma function. For this example, we obtain

$$M_1(3) = 0.0223335; \qquad M_2(2.5) = 0.0878403$$

Thus,

$$EC = (2700)(0.223325) + (1650)(0.0878403) = 60.30 + 144.94$$
$$= \$205.24$$

Component 1 Nonrepairable and Component 2 Repairable (Minimal Repair)

Assume that each component is replaced or repaired without any effect on the other functioning component and that repair is minimal [14], that is, the component is returned to its condition immediately prior to failure, and the failure rate is as it was. Then we have

$$EC = C_1 M_1(W_1) + C_{2r} \int_0^{W_2} r_2(t)\, dt \tag{22.27}$$

where $r_2(t) = f_2(t)/[1 - F_2(t)]$ is the failure rate associated with $F_2(t)$, and C_{2r} is the average cost of a minimal repair of Component 2.

Example 22.6

Assume for the data of Example 22.5 that Compressor 2 is maintained under minimal repairs for an average cost of $C_{2r} = \$480$ each time it fails. For the Weibull distribution with parameters (λ, α), $r(t) = \lambda^\alpha \alpha t^{\alpha-1}$, and we have

$$EC_2 = C_{2r} \int_0^{W_2} \lambda^\alpha \alpha t^{\alpha-1}\, dt = C_{2r}(\lambda W_2)^\alpha \tag{22.28}$$

From (22.27) and (22.28), with $M_1(3)$ from Example 22.5, we obtain

$$EC = 2700[M_1(3)] + 480[(0.08)(2.5)]^{1.5} = 60.30 + 42.93 = \$103.23$$

Both Components Repairable Under Minimal Repair

Then

$$EC = C_{1r} \int_0^{W_1} r_1(t)\, dt + C_{2r} \int_0^{W_2} r_2(t)\, dt \tag{22.29}$$

where $r_i(t)$ is the failure rate associated with $F_i(t)$, and C_{ir}, $i = 1, 2$, are costs of minimal repair for component i.

Example 22.7

For the data of Example 22.5, let both compressors be subject to minimal repair under FRW. Let the average repair costs be $C_{1r} = \$840$, and $C_{2r} = \$480$. From (28) and (29), we obtain

$$EC = 840[(0.05)(3.5)]^2 + 42.93 = 25.73 + 42.93 = \$68.66$$

Repairs That Modify Subsequential Failure Rates

Consider a system with repairable components in a situation where repairs lead to worse statistical properties of the functioning component. Specifically, assume that both components experience accelerated wear after each failure, so that their renewal functions are given by Eq. (22.3) for some $\beta > 1$. Then the computational algorithm and formula (22.25) hold with the recursive expression (22.26) replaced by the new relation [13]

$$A_1 = \Gamma(\alpha + 1); \qquad A_k = \frac{\Gamma(\alpha k + 1)}{k!} - \sum_{\nu=1}^{k-1} \frac{\Gamma(\nu \alpha + 1)}{\nu!} A_{k-\nu} \beta^{\alpha(k-\nu)}$$

(22.30)

Example 22.8

For the data from Example 22.5, assume that any failure of Compressor 1 accelerates its current time to next failure by a coefficient $\beta_1 = 1.05$, whereas failures of Compressor 2, after being removed, accelerate the time to next failure by a coefficient $\beta_2 = 1.1$. Under Policy 1, with $W_1 = 3$ and $W_2 = 2.5$, we find

$$M_{1,\beta} = 0.0223411; \qquad M_{2,\beta} = 0.0881935$$

hence, EC $= 2700 M_{1,\beta}(3) + 1650 M_{2,\beta}(2.5) = \$205.84$. Comparing this result to EC $= \$205.24$ in Example 22.5, we see that the increase is not significant. The reason is that the expected times to first failure of the components are much longer than the given warranty times. For instance, if $W_1 = 12$ and $W_2 = 10$, then the result would be EC $= 868.12 + 408.63 = \$1276.75$ for Example 22.5, and EC $= 873.32 + 421.18 = \$1294.50$ for this example. Thus, the difference between the warranty costs when repairs are "good as new" and when repairs decrease the expected life increase with an increase in warranty times and with an increase in the coefficient β, which represents the accelerated wearing time.

Policy 2 Warranties

Under Policy 2 the expected warranty costs are given by

$$EC = \int_0^{W_1} q_1(t) \, dF_1(t) + EC_2(W_2)$$

(22.31)

where $q_1(t)$ is the rebate function for Component 1 when its actual service time at the time of failure equals t, and $EC_2(W_2)$ is the expected warranty cost coming from the second component. Any particular choice of $q_1(t)$ and $F_1(t)$ determines the contribution of the first component to the overall EC. The contribution of the other component to EC is specified analogously.

Note. Special attention must be paid if Component 1 is covered under a PRW. The buyer would like to buy (with the discount) a replacement for a failed Component 1 in order to keep the system operational. The replacement would come with a new warranty which in no way is related to the original warranty.

Example 22.9

Suppose that a car battery is sold under a PRW with $W_1 = 4$ years and that the lifetime of the battery has a Gamma distribution with parameters $\lambda = 0.5$ and $\alpha = 2.5$. Suppose the battery costs $C_{b1} = \$90$ to the buyer and that the rebate function is proportional linear prorated, given by

$$q_1(t) = \max\left\{(0.8)\left(1 - \frac{t}{4}\right)(90), 0\right\}$$

The amount $q_1(t)$ is the discount to the buyer on the purchase of a new battery to replace the failed one. The rest of the car is sold under a FRW, that is, all failures will be removed at the seller's expense, and the car is maintained under minimal repair (with an estimated cost of $C_{m,2} = \$280$ per claim) for $W_2 = 3$ years. The overall claim intensity associated with the rest of this car model is a U-shaped function

$$r_2(t) = (0.4)\left(\frac{t}{10}\right)^{-0.03}\left(1 - \frac{t}{10}\right)^{-0.6}$$

which reflects more intensive claims for new cars and more intensive breakdowns toward the end of the car's lifetime (here assumed to be 10 years). Then the calculations of EC, excluding the seller's cost, give

$$\text{EC} = \int_0^4 (0.8)\left(1 - \frac{t}{4}\right)(90)\frac{(0.5)^{2.5}t^{3/2}}{\Gamma(2.5)}\,dt$$

$$+ 280\int_0^3 (0.4)\left(\frac{t}{10}\right)^{-0.03}\left(1 - \frac{t}{10}\right)^{-0.6}\,dt$$

$$= 36.02 + 411.48 = \$447.50$$

Policy 3 Warranties

For this case, we again have $\text{EC} = \text{EC}_1(W_1) + \text{EC}_2(W_2)$, with $\text{EC}_i(W_i)$ given by the first term of (22.31), where $q_i(t)$ is the rebate function for Component i, $i = 1, 2$. The sum will have more terms if more independent components are involved.

Example 22.10

A battery and four new tires are mounted on a car. The battery is sold under a FRW with the parameters given in Example 22.9. The tires are sold with a PRW of $W_2 = 60,000$ miles each. Wearing "time" is expressed in terms of the mileage that a tire can serve in various environmental conditions. Suppose that this is a normally distributed random variable X with mean $\mu = 60,000$ and standard deviation $\sigma = 5000$, that the rebate function for tires is

$$q(t) = 50 \left\{ 1 - \frac{t}{(60,000)} \right\}$$

and that the seller's cost per tire is $C_{s,2} = \$32$. Then the expected warranty cost associated with this purchase is (using the first cost component of EC for Example 22.9)

$$EC = 36.02 + 4 \left\{ 32 + \int_0^{60,000} 50 \left(1 - \frac{t}{60,000} \right) \frac{1}{5000 \sqrt{2\pi}} \right.$$

$$\left. \times \exp\left(- \frac{(t - 60,000)^2}{2(5000)^2} \right) dt \right\}$$

$$= 36.02 + 4[32 + (1.6623)] = 36.02 + 134.65 = \$170.67$$

22.4.2 Renewing Warranties

The cases considered above are extended in a natural way to renewing warranties if these are offered for some components. The only difference is in the expressions for calculating the contribution $EC_i(W_i)$ of the ith component to the overall warranty cost. We obtain the following:

For the FRW

$$EC_i(W_i) = C_{s,i} \left(1 + \frac{F_i(W_i)}{1 - F_i(W_i)} \right) \tag{22.32}$$

where $C_{s,i}$ is the seller's repair/replacement cost per component of type i.

For the PRW

$$EC_i(W_i) = \left(C_{s,i} + \int_0^{W^i} q_i(t) \, dF_i(t) \right) \left(1 + \frac{F_i(W_i)}{1 - F_i(W_i)} \right) \tag{22.33}$$

Example 22.11

Suppose that a new battery and a new engine have been mounted on a used car under renewing PRW and renewing FRW, respectively (Policy 2'). For the engine, the warranty period is $W_2 = 2$ years. An engine costs the service station $S_{s,2} = \$950$, and it is known that for this kind of engine, the lifetime is normally distributed with a mean of 4 years and a standard deviation of 2 years. The battery lifetime and warranty parameters are as in Example 22.9. Assume that a battery costs the seller $C_{s,1} = \$40$. Then, using the results of Example 22.9 and (22.32) and (22.33), we have

$$EC = (40 + 36.02)\left(1 + \frac{G(4; 2.5, 0.5)}{1 - G(4; 2.5, 0.5)}\right)$$

$$+ 950\left(1 + \frac{\Phi(2 - 4)/2}{1 - \Phi(2 - 4)/2}\right)$$

$$= 76.02(2.494) + 950(1.1886) = 190.02 + 1149.18 = \$1339.19.$$

Here $G(x; \alpha, \lambda)$ is the incomplete Gamma function, and $\Phi(\cdot)$ is the standard normal c.d.f.

22.5 WARRANTY COST ANALYSIS: DEPENDENT FAILURES

22.5.1 Joint Lifetime Distribution Known

Policy 1 Warranties

Suppose that an item consist of two nonrepairable components and that each component is replaced by a new one without changing the joint life distribution. Then the bivariate renewal process $\{N_1(t), N_2(t)\}$ must be known in order to analyze the system. Under Policy 1, the overall EC will again be represented by Eq. (22.24). The only difference is that now $M_i(t) = E\{N_i(t)\}$ must be calculated for dependent renewal processes.

Example 22.12

Suppose the two components have a MO'BVE distribution with parameters λ_1, λ_2, and λ_{12}. Then $\{N_1(t), N_2(t)\}$ is the bivariate Poisson process, given in (22.9), with the specific values of the probabilities q_{ij} as given there. More specifically, we now have

$$M_1(t) = (\lambda_1 + \lambda_{12})t; \qquad M_2(t) = (\lambda_2 + \lambda_{12})t$$

Thus,

$$EC = C_{1s}W_1(\lambda_1 + \lambda_{12}) + C_{2s}W_2(\lambda_2 + \lambda_{12}) \qquad (22.34)$$

where C_{is} is the seller's cost per component i, $i = 1, 2$.

For this example, consider the following situation. A computer (Component 1) and a monitor (Component 2) are sold under a FRW for $W_1 = 2.5$ years and $W_2 = 3$ years respectively. Their joint distribution is MO'BVE with parameters $\lambda_1 = 0.2$, $\lambda_2 = 0.15$, and $\lambda_{12} = 0.1$. The seller's cost per computer is $C_{1s} = \$750$ and per monitor is $C_{2s} = \$350$. The total seller's cost is $C_{ss} = \$1100$ per set. The expected warranty cost per set sold, from (22.34) is

$$EC = 750(2.5)(0.2 + 0.1) + 350(3)(0.15 + 0.1)$$
$$= 562.5 + 262.5 = \$825.00$$

Policy 3 Warranties

Consider a two-component, nonrepairable system where the ith component has a PRW for a period W_i. The rebate function $q(t_1, t_2)$ gives the buyer's discount if Component 1 fails at time t_1 and Component 2 fails at time t_2. Construction of $q(t_1, t_2)$ reflects a possible interpretation of the bivariate warranty: (W_1, W_2) is understood as the amount of service guaranteed to be provided by each component. Then $\{X_1 = t_1, X_2 = t_2\}$ is equivalent to the amount of service provided by time (t_1, t_2). This interpretation justifies the following forms of the rebate function:

$$q(t_1, t_2) = C_{b1}\left(1 - \frac{\min(t_1, W_1)}{W_1}\right) + C_{b2}\left(1 - \frac{\min(t_2, W_2)}{W_2}\right),$$
$$0 \le t_i, i = 1, 2 \qquad (22.35)$$

or

$$q(t_1, t_2) = C_{bs}\left(1 - \frac{\min(t_1, W_1)}{W_1} \frac{\min(t_2, W_2)}{W_2}\right),$$
$$0 \le t_i, i = 1, 2 \qquad (22.36)$$

or some derivatives of these forms.

Other forms of $EC(W_1, W_2)$ depend on the claim pattern. There are two cases: component failures are "visible" (i.e., detectable) or "invisible"; see Section 22.3.2. Moreover, there are two claim possibilities:

First, the claim may be for refund of the component that has just failed.
[In this case, $q_i(t_i)$ could be the ith component of the sum on the right-
hand side of (22.35), where C_{bi} is the buyer's cost for that component].
Note that here the user has to buy a replacement to keep the system
running, but this is of no relevance for computing the expected warranty
cost. Second, the claim is for refund of the entire item. This is determined
by (22.36), where the amount of work (t_1, t_2) must be established at the
time of claim. C_{bs} is the buyer's cost for the system. In this way, the EC
can be classified as one of the following forms:

(i) The claim is only for the component that fails first. The other
component will be claimed only if it fails later within its warranty. Then

$$EC_1 = EC(W_1, W_2)$$
$$= \int_0^{W_1} \left(\int_{\min(t_1, W_2)}^{\infty} [q_1(t_1) + q_2(t_2)]f(t_1, t_2)\, dt_2 \right) dt_1$$
$$+ \int_0^{W_2} \left(\int_{\min(W_1, t_2)}^{\infty} [q_2(t_2) + q_1(t_1)]f(t_1, t_2)\, dt_1 \right) dt_2$$

$$(22.37)$$

(ii) At least one failure must be in warranty limits. At the time of
first failure, if it occurs within warranty limits, the entire item
will be claimed. The warranty of the nonfailed component will
be no longer be valid. Then

$$EC_2 = EC(W_1, W_2) = \int_0^{W_1} \left(\int_{\min(t_1, W_2)}^{\infty} q(t_1, t_1)f(t_1, t_2)\, dt_2 \right) dt_1$$
$$+ \int_0^{W_2} \left(\int_{\min(W_1, t_2)}^{\infty} q(t_2, t_2)f(t_1, t_2)\, dt_1 \right) dt_2$$

where $q(t_1, t_2)$ can be either (22.35) or (22.36).

(iii) Invisible failures. A claim arises only when both components
fail within the assigned warranty. We assume that the service
time of each component can be precisely evaluated. Then

$$EC_3 = EC(W_1, W_2) = \int_0^{W_1} \int_0^{W_2} q(t_1, t_2)f(t_1, t_2)\, dt_1\, dt_2$$

(iv) Invisible failures. Both components must fail as in (iii). A refund is claimed for the entire item. Then

$$EC_4 = EC(W_1, W_2) = \int_0^{W_1} \left(\int_0^{\min(t_1, W_2)} q(t_1, t_1) f(t_1, t_2) \, dt_2 \right) dt_1$$

$$+ \int_0^{W_2} \left(\int_0^{\min(W_1, t_2)} q(t_2, t_2) f(t_1, t_2) \, dt_1 \right) dt_2$$

Other forms may arise in practice. Policies 3' and 3" require considerably more complicated analyses.

Example 22.13

Consider the two-component refrigerating system from Example 22.1. In addition, let us assume that Component 1 (the freezing) and Component 2 (the cooling section) cost $C_{b1} = \$640$ and $C_{b2} = \$380$, respectively, and that the whole system costs $C_{bs} = \$900$. Thus, for a PRW with $W_1 = 4$ and $W_2 = 4$, the following rebate functions could be proposed:

$$q_1(t_1) = 640 \left(1 - \frac{\min(t_1, 4)}{4} \right); \quad q_2(t_2) = 380 \left(1 - \frac{\min(4, t_2)}{4} \right);$$

$$q_{12}(t_1, t_2) = q_1(t_1) + q_2(t_2)$$

$$q(t_1, t_2) = 900 \left[1 - \left(\frac{\min(t_1, 4)}{4} \right) \left(\frac{\min(4, t_2)}{4} \right) \right]$$

Warranty claims under the forms discussed above will then be as follows:

(i) If each component is claimed separately, then by (22.37), we obtain

$$EC_1 = \int_0^4 \left(\int_x^\infty [q_1(x) + q_2(y)] f(x, y) \, dy \right) dx$$

$$+ \int_0^4 \left(\int_y^\infty [q_2(y) + q_1(x)] f(x, y) \, dx \right) dy$$

$$= 13.49 + 0.65 = \$14.14$$

(ii) If a claim is made for the entire system at the time of first failure of either of its components, then

$$EC_2 = \int_0^4 \left(\int_x^\infty q(x, x) f(x, y) \, dy \right) dx$$

$$+ \int_0^4 \left(\int_y^\infty q(y, y) f(x, y) \, dx \right) dy$$

$$= 19.06 + 1.26 = \$20.32$$

(iii)–(iv)   If a claim is made for the system only when both components fail within the assigned warranty, then (for this particular example)

$$EC_3 = EC_4 = \int_0^4 \left(\int_0^x q(x, x) f(x, y) \, dy \right) dx$$

$$+ \int_0^4 \left(\int_0^y q(y, y) f(x, y) \, dx \right) dy$$

$$= 0.81 + 3.29 = \$4.10$$

Example 22.14

A computer (Component 1) and a monitor (Component 2) are sold under a PRW for $W_1 = 2.5$ years and $W_2 = 3$ years, respectively. Their joint distribution is MO'BVE with parameters $\lambda_1 = 0.2$, $\lambda_2 = 0.15$, and $\lambda_{12} = 0.1$, as in Example 22.12. If the components are bought separately, then $C_{b1} = \$1200$ and $C_{b2} = \$450$. If the set is bought as an item, then $C_{bs} = \$1500$. The rebate function is given by (22.35) and (22.36). Then the expected warranty cost per set sold is as follows:

(i)   If claims are made separately for each failed component, then from (22.37)

$$EC_1 = \int_0^{2.5} \int_{\min(x,\,3)}^{\infty} \left[1200 \left(1 - \frac{\min(x,\,2.5)}{2.5} \right) \right.$$

$$+ 450 \left(1 - \frac{\min(y,\,3)}{3} \right) \Bigg] d_{x,y}$$

$$\times \exp\{-[0.2x + 0.15y + 0.1 \min(x,\,y)]\}$$

$$+ \int_0^3 \int_{\min(2.5,\,y)}^{\infty} \left[450 \left(1 - \frac{\min(y,\,3)}{3} \right) \right.$$

$$+ 1200 \left(1 - \frac{\min(x,\,2.5)}{2.5} \right) \Bigg] d_{x,y}$$

$$\times \exp\{-[0.2x + 0.15y + 0.1 \min(x,\,y)]\}$$

Direct calculation of this expression offers some analytical complications due to the absence of a p.d.f. for this distribution. By using the equivalent shock model of Section 22.2.3, we can derive a decomposition of EC_1 into three components that correspond to the events causing the claims. A claim for Component 1 will come about only if a shock from source 1 (whose intensity is λ_1) destroys it before the shock from source 3. Thus, the overall warranty cost will have the component

$$
\begin{aligned}
EC_1^{(1)} &= \int_0^{W_1} q_1(t)\lambda_1 e^{-\lambda_1 t} e^{-\lambda_{12} t}\, dt \\
&= \frac{C_{b1}\lambda_1}{\lambda_1 + \lambda_{12}}\left(1 - \frac{1 - e^{-(\lambda_1 + \lambda_{12})W_1}}{(\lambda_1 + \lambda_{12})W_1}\right)
\end{aligned}
\tag{22.38}
$$

Similarly, $EC_1^{(2)}$, the warranty cost for Component 2, is given by (22.38) with λ_1, W_1, and C_{b1} replaced by λ_2, W_2, and C_{b2}, respectively. Finally, a third cost component $EC_1^{(12)}$ arises from a simultaneous claim for both components. This occurs if a shock from source 3 with intensity λ_{12} comes first and destroys both. Therefore, if $W_1 < W_2$ and $\lambda = \lambda_1 + \lambda_2 + \lambda_{12}$, we have

$$
\begin{aligned}
EC_1^{(12)} &= \int_0^{W_1} q(t,\, t)\lambda_{12} e^{-\lambda_{12} t} e^{-\lambda_1 t} e^{-\lambda_2 t}\, dt \\
&\quad + \int_{W_1}^{W_2} q_2(t)\lambda_{12} e^{-\lambda_{12} t} e^{-\lambda_1 t} e^{-\lambda_2 t}\, dt \\
&= (C_{b1} + C_{b2})\frac{\lambda_{12}}{\lambda}\left(1 - \frac{1 - e^{-\lambda W_1}}{\lambda W_1}\right) \\
&\quad + C_{b2}\frac{\lambda_{12}}{\lambda}\left(1 - \frac{1 - e^{-\lambda(W_2 - W_1)}}{\lambda(W_2 - W_1)}\right)
\end{aligned}
\tag{22.39}
$$

With the numerical data for the case considered, from (22.38) and (22.39), we obtain

$$
EC_1^{(1)} = \frac{1200\,(0.2)}{0.2 + 0.1}\left(1 - \frac{1 - e^{-(0.2 + 0.1)(2.5)}}{(0.2 + 0.1)(2.5)}\right) = \$237.19
$$

$$
EC_1^{(2)} = \frac{450\,(0.15)}{0.15 + 0.1}\left(1 - \frac{1 - e^{-(0.15 + 0.1)(3)}}{(0.15 + 0.1)(3)}\right) = \$80.05
$$

$$
\begin{aligned}
EC_1^{(12)} &= (1200 + 450)\frac{0.1}{0.2 + 0.15 + 0.1} \\
&\quad \times \left(1 - \frac{1 - e^{-(0.45)(2.5)}}{(0.45)(2.5)}\right) + \frac{(450)(0.1)}{0.45} \\
&\quad \times \left(1 - \frac{1 - e^{-(0.45)(3 - 2.5)}}{(0.45)(3 - 2.5)}\right) = \$188.88
\end{aligned}
$$

Therefore, $EC_1 = EC_1^{(1)} + EC_1^{(2)} + EC_1^{(12)} = 237.19 + 80.05 + 188.88 = \$516.12$.

(ii) Now let us assume that each failure within the assigned warranty causes a claim for the entire item. Then, after similar considerations of the shock model as in (i), with the rebate function

$$q_1(x, y) = 1200 \left(1 - \frac{\max(x, 2.5)}{2.5} \right) + 450 \left(1 - \frac{\max(y, 3)}{3} \right)$$

we obtain

$$EC_2^{(1)} = \int_0^{2.5} (0.45)q_1(x, x)e^{-0.45x}\, dx$$

$$+ \int_{2.5}^3 (0.25)q_1(x, x)e^{-0.45x}\, dx$$

$$= 680.17 + 1.41 = \$681.58$$

Suppose that the entire set is bought at the price of $C_{bs} = \$1500$. Then we may consider the case when either a failure of the monitor only (the component with larger warranty) or a simultaneous failure of monitor and computer leads to a warranty claim for the entire set. A restriction of the claim to only the failed component is made when the rebate function

$$q_2(x, y) = 1500 \left(1 - \frac{\max(x, 2.5)}{2.5} \frac{\max(y, 3)}{3} \right)$$

is specified. Analogous considerations as above give the following results:

$$EC_2^{(2)} = \int_0^{2.5} (0.25)q_2(x, x)e^{-0.45x}\, dx$$

$$+ \int_0^3 (0.15)q_2(x, x)e^{-0.45x}\, dx$$

$$+ \int_0^3 (0.1)q_2(x, x)e^{-0.45x}\, dx$$

$$= 358.34 + 271.58 + 181.06 = \$810.98$$

This discussion shows that claim pattern must be a part of rebate function and can affect the EC.

22.5.2 Failure Interaction

In models assuming failure interaction, even if the marginal (component) life distributions are known, additional information is needed to construct the failure interaction process. To conduct warranty cost analysis, cost coefficients corresponding to all significant actions must also be known.

Induced Failures

Consider a two-component system in which failure of the ith component induces a simultaneous failure of the other component with probability p_i. The item fails only if both components fail. Then the entire item is replaced by a new one (at a cost of C_r). Otherwise, just the failed component is replaced (each costs C_{ri}, $i = 1$, 2). We assume that the system is sold under a FRW with warranty period W.

The sequence of system failures forms a renewal process associated with the random variable Y whose survival function is defined by (22.10) and (22.11). The expected total cost of maintaining the item on $[0, W]$ is given by the expression [3]

$$EC(W) = C_r M_Y(W) + \int_0^W (1 + M_Y(W - t))C_0(t)\, dt \qquad (22.40)$$

Here

$$C_0(t) = C_{r1} \frac{1 - p_1}{p_1} g_1(t)\, \overline{G}_2(t) + C_{r2} \frac{1 - p_2}{p_2} g_2(t)\, \overline{G}_1(t) \qquad (22.41)$$

gives the expected operating cost for component replacements up to the moment t when the first system failure occurs; $g_i(t)$ is the p.d.f. of $G_i(t)$.

Calculations using (22.10), (22.11), (22.40), and (22.41) are difficult. However, the following particular cases offer some simplifications. First, we note that if $p_i = 0$, then (22.41) still holds with the two factors [(one $- p_i)/p_i]g_i(t)$ replaced by the factor $m_i(t)$, which is the renewal intensity for component i. Then it follows that

(i) If $p_1 = p_2 = 0$ (no failure interactions), then EC is given by (22.23).

(ii) If $p_1 = p_2 = 1$ (each component failure causes also a system failure), then $C_0(t) = 0$ and $EC(W) = C_r M(W)$.

(iii) If $p_1 = 0$ and $p_2 = 1$, then $G(t) = G_2(t)$ and $C_0(t) = C_{r1}[1 - G_2(t)]m_1(t)$. Here $m_1(t)$ is the renewal intensity of Component 1. Also,

$$EC(W) = C_r M_2(W) + C_{r1} \int_0^W [1 + M_2(W - t)]\overline{G}_2(t)\, dM_1(t)$$

$$(22.42)$$

(iv)   If $p_1 = 0$ and $0 < p_2 < 1$, then there will be an influence of Component 2 on Component 1 but not conversely. Also, $M(t) = p_2 M_2(t)$,

$$EC(W) = (C_r p_2 + C_{r2})M_2(W)$$
$$+ C_{r1} \int_0^W [1 + p_2 M_2(W - t)]\overline{G}_2(t)\, dM_1(t) \qquad (22.43)$$

and

$$C_0(t) = C_{r1} m_1(t)\overline{G}_2(t) + C_{r2} \frac{1 - p_2}{p_2} g_2(t) \qquad (22.44)$$

(v)   If the two component have exponential lifetimes with parameters λ_i, $i = 1, 2$, then for any values of the probabilities of induced failure, it is true that

$$EC(W) = C_r(p_1\lambda_1 + p_2\lambda_2)W$$
$$+ C_{r1}(1 - p_1)\lambda_1 W + C_{r2}(1 - p_2)\lambda_2 W \qquad (22.45)$$

where the distribution of Y is determined from the survival function as indicated in (22.10) and (22.11).

When an item is in use for long time, several life times or more (see Chapters 10 and 12), then the long-run average maintenance cost per unit of time can be of interest. This cost is given by the expression [3]

$$\overline{EC} = \left(C_r + \int_0^\infty C_0(t)\, dt \right) \frac{1}{\mu_Y} \qquad (22.46)$$

Example 22.15
An infrared light and a ventilation system are the two major components of a kitchen air purifier. Suppose that the components have exponentially distributed lifetimes with parameters $\lambda_1 = 0.45$ and $\lambda_2 = 0.6$. Suppose also that failure of the light causes a simultaneous failure of the fan with probability $p_1 = 0.05$, whereas failure of the fan induces a failure of the light with probability $p_2 = 0.15$, and that marginal costs are $C_{r1} = \$8$, $C_{r2} = \$15$, and $C_r = \$30$. The system is sold under warranty for 2 years. The expected costs are determined as follows.

By (22.45), we immediately calculate

EC = 6.75 + 6.84 + 15.30 = $28.89

For the exponential distributions chosen, we have

$$\mu_Y = \frac{1}{p_1\lambda_1 + p_2\lambda_2} = 8.89 \text{ years}$$

and

$$\int_0^\infty C_0(t)\,dt = C_{r1}(1 - p_1\lambda_1)\mu_Y + C_{r2}(1 - p_2\lambda_2)\mu_Y$$

$$= \{8[1 - (0.45)(0.05)] + 15[1 - (0.6)(0.15)]\} = \$190.87$$

From (22.46), the EC per unit time (1 year) is $14.45.

Shock-Damage Model

Here, each failure of Component 2 induces failure of Component 1 with probability p_2, whereas failures of Component 1 act as shocks to Component 2, without inducing a simultaneous failure but affecting its failure rate. The system fails when a failure of Component 2 causes a simultaneous failure of Component 1. Let $F_2^*(t)$ denote the modified c.d.f. of the time to failure of Component 2 under above conditions. Then the expected cost of keeping the system in operation during the warranty period can be obtained from the induced failure model (iii) with $p_1 = 0$, and p_2 as specified here.

In general, it is difficult to obtain $F_2^*(t)$ explicitly. Let us assume as in Ref. 3 that Component 1 has an exponential lifetime with parameter λ. Component 2, if it has not failed after k shocks, has failure rate $k\alpha$ until the next shock or until its own failure occurs. Then we have

$$F_2^*(t) = 1 - \exp\left[-\lambda\left(t - \frac{1 - e^{-\lambda t}}{\alpha}\right)\right] \tag{22.47}$$

Thus, from (22.43), we get, as in Ref. 3,

$$EC(W) = C_r p_2 M_2^*(W) + C_{r2} M_2^*(W) + C_{r1}\lambda W \tag{22.48}$$

and

$$\overline{EC} = \frac{C_r + C_{r1}p_2 + C_{r1}\lambda\mu_2^*}{\mu_2^*} \tag{22.49}$$

Here $M_2^*(t)$, $F_2^*(t)$, and μ_2^* are the renewal function, c.d.f., and expected value associated with Y^*, the time to failure of Component 2.

Example 22.16

Suppose costs are as in Example 22.15. Assume that failures of the light system cause only an increase in the failure rate of the ventilation system, with a constant rise in the failure rate of $\alpha = 0.1$, without causing any simultaneous failures, and that breakdowns of the fan induce a simultaneous breakdown of the whole system with probability $p_2 = 0.25$. Suppose that lifetimes of the light system are exponentially distributed with $\lambda = 0.75$ and that the system has a FRW for $W = 2.5$ years. Warranty costs are calculated as follows.

The time to failure for the fan has a c.d.f. determined by (22.47), namely,

$$F_2^*(t) = 1 - \exp\left[-0.75\left(t - \frac{1 - e^{-0.1t}}{0.1}\right)\right]$$

Its expected life is $\mu^* = 5.0772$. Expected renewals for the ventilation system during the warranty time are found by use of an algorithm for the discretized (with respect to the time parameter) distribution of Y^*. The result is

$$M_2^*(2.5) = 0.201412$$

Thus, the expected warranty costs, found from (22.48) and (22.49), are

$$EC(2.5) = 30(.25)(.201412) + 15(.201412) + 8(.75)(2.5) = \$26.33;$$
$$\overline{EC} = [30 + 8(.25) + 8(.75)(5.0772)]/5.0772 = \$12.30.$$

22.5.3 Modular Structure

Nonrepairable Components or Renewal "Good as New"

We consider only Policy 1. Suppose the modular structure as explained in Section 22.3.2 describes the item's behavior. If Component 2 fails, the whole system must be replaced. Failures of Component 1 can be rectified only by replacement or repair. This is equivalent to model (iii) of induced failures. Therefore, the EC for a FRW of length W for such an item is given by (22.42). The long-run EC per unit time is given by (22.46).

Example 22.17

For modeling purposes, a TV set is partitioned into two parts: a servicing subsystem (Component 1) and the picture tube (Component 2). Component 1 is assumed to have an exponentially distributed lifetime X_1 with parameter $\lambda_1 = 0.3$, whereas Component 2 has a Weibull distribution with c.d.f. $F_2(t) = 1 - e^{-(0.2t)^{2.5}}$. Failures of Component 1 are repaired at an average cost of $C_{r1} = \$60$, and this completely restores it as new.

It is assumed that failures of the picture tube require complete renewal of the whole set, and this costs the seller $C_r = \$210$. Suppose that the set is sold with a FRW of $W_1 = 2$ years for the service system (and also for the whole system) and $W_2 = 3$ years for the picture tube alone.

An obvious modification of (22.42) determines the EC in this case:

$$EC(W_1, W_2) = C_r M_2(W_2) + C_{r1} \int_0^{W_1}$$

$$\times [1 + M_2(W_1 - t)][1 - F_2(t)] \, dM_1(t) = 210 M_2(3)$$

$$+ 60 \int_0^2 [1 + M_2(2 - t)] \, e^{-(0.2t)^{2.5}}(0.3) \, dt$$

In the derivation of this expression, it is taken into account that $M_1(t) = \lambda_1 t = 0.3t$ is the renewal function corresponding to Component 1. $M_2(t)$ is the renewal function for the Weibull distribution. Using the numerical algorithm of [11, 13], we compute

$$EC(2, 3) = 210(0.249362) + 119.997 = \$179.36$$

Also numerically, with (22.46) used in case (iii), we find

$$\mu_Y = \int_0^\infty e^{-(0.2t)^{2.5}} \, dt = 4.436 \text{ years}$$

$$\overline{EC} = \frac{210 + \int_0^\infty 60 e^{-(0.2t)^{2.5}}(0.3) \, dt}{\mu_Y} = \$65.34$$

Induced Failures Under Minimal Repair

Consider the model of induced failures, where, in the case of a single repair, Component i is immediately repaired (at cost C_{mi}) without changing its current failure rate $r_i(t)$, $i = 1, 2$. In the case of simultaneous repair of both components, the item is replaced by a new one. Then [3, 4] the life distribution of the system is specified by the survival function

$$\overline{G}(t) = P\{Y > t\} = \exp\left(-\int_0^t [p_1 r_1(x) + p_2 r_2(x)] \, dx\right) \quad (22.50)$$

The expected cost is still given by (22.40), but with expected operating costs until the first system failure, now given by the expression

$$C_0(t) = [C_{m1}(1 - p_1)r_1(t) + C_{m2}(1 - p_2)r_2(t)]$$

$$\times \exp\left(-\int_0^t [p_1 r_1(x) + p_2 r_2(x)] \, dx\right) \quad (22.51)$$

The long-run expected maintenance costs per unit time can be calculated by use of the expression

$$\overline{EC} = \left\{ C_r + \int_0^\infty [C_{m1}(1 - p_1)r_1(t) + C_{m2}(1 - p_2)r_2(t)] \right.$$

$$\left. \times \exp\left(-\int_0^t [p_1 r_1(x) + p_2 r_2(x)] \, dx \right) dt \right\} \frac{1}{\mu_Y} \qquad (22.52)$$

In some cases, the two components will have separate warranties. Under Policy 1, with $W_1 < W_2$, EC(W_1, W_2) will be obtained by combining (22.40) and (22.51). This yields

$$EC(W_1, W_2) = \left(C_{r2} + C_{m2} \int_0^\infty (1 - p_2) \, r_2(t)\overline{G}(t) \, dt \right)$$

$$\times [M_Y(W_2) - M_Y(W_1)] + C_r M_Y(W_1)$$

$$+ \int_0^{W_1} [1 + M_Y(W_1 - t)]C_0(t) \, dt \qquad (22.53)$$

Here (22.50), (22.51), and (22.53) have to be explicitly specified for any particular probability distribution.

Example 22.18

Consider the system of Example 22.16 under the assumptions that failures of the picture tube do not lead to failure of the entire system but that they may cause simultaneous failure of the service system with probability $p_2 = 0.4$. Failures in the service system can cause failure of the picture tube with probability $p_1 = 0.1$. The cost coefficients, warranty periods, probability distributions, and numerical values of their parameters are as in Example 22.16. Suppose that repair of the picture tube costs an average of $C_{m2} = \$120$ and replacement by a new tube costs $C_{r2} = \$150$.

From the exponential and Weibull assumptions, the failure rates of the two components are found to be, for $t \geq 0$,

$$r_1(t) = \lambda_1 = 0.3; \qquad r_2(t) = \alpha\lambda^\alpha t^{\alpha - 1} = (2.5)(0.2)^{2.5}t^{1.5}, \quad t > 0. \tag{22.54}$$

The time to failure Y is determined by the survival function

$$\overline{G}(t) = e^{-0.03t}e^{-0.4(0.2t)^{2.5}}$$

From this, we obtain $\mu_Y = 2.672$ years. From (22.53), we find EC(W_1, W_2) to be

$$EC(2.5, 3) = \left(150 + 120 \int_0^\infty (0.6)(2.5)(0.2t)^{2.5} e^{-0.03t} e^{-0.4(0.2t)^{2.5}} dt\right)$$

$$\times [M_Y(3) - M_Y(2.5)] + 210M_Y(2.5) + 60 \int_0^{2.5} [1 + M_Y(2.5 - t)]$$

$$\times [60(0.9)(0.3) + 120(0.6)(2.5)(0.2)^{2.5} t^{1.5}] \times e^{-0.03t - 0.4(0.2t)^{2.5}} dt$$

$$= \$318.34$$

Results for some more complicated models of multicomponent systems with a variety of failure interactions can be found in Refs. 3 and 4.

22.5.4 Network Structures

Standard Structures

Parallel Systems

Consider complex systems consisting of two nonrepairable components connected in parallel, with lifetimes X_1 and X_2. We need to consider two cases. First, the system failure is observed only when both components fail. In this case, the time to failure is given by $Y = \max(X_1, X_2)$. Warranty claims occur according to Y, whose c.d.f. is $F_Y(x) = F_1(x)F_2(x)$. Then each warranty claim requires replacement of both components. Second, failure of a component is detected immediately. This results in an immediate warranty claim for the failed component. Then the warranty costs are the sum of warranty costs of the two components, each failing independently. In either case the warranty costs are derived from the results of Section 22.4.

Example 22.19

Two speakers are components of a stereo sound system. Both have exponentially distributed lifetimes with the same parameters $\lambda_1 = \lambda_2 = 0.55$. In case of complete failure, both speakers are replaced by a new item for a cost of $C_r = \$260$. The system is installed under a FRW for $W_1 = W_2 = 1.5$ years.

The time to failure of this system is determined by the c.d.f.

$$F_Y(t) = F_1(t)F_2(t) = (1 - e^{-0.55t})^2 = 1 - 2e^{-0.55t} + e^{-1.1t}$$

Its expected lifetime is $\mu_Y = 2.73$ years, and $M_Y(1.5) = 0.344466$. The expected cost during the given warranty period is

$$EC(1.5) = (260)M_Y(1.5) = \$89.56$$

Series Systems

Consider a complex item with a series structure and independent components. Define $Y = \min(X_1, X_2)$. Then claims occur according to Y, whose c.d.f. is $F_Y(x) = 1 - [1 - F_1(x)][1 - F_2(x)]$. However, the warranty cost of each failure is uncertain, as it depends on which component has failed. If we let $\alpha_1 = \int_0^\infty [1 - F_2(x)] \, dF_1(x)$, $\alpha_2 = 1 - \alpha_1$, then the expected warranty cost for each claim is given by the sum $\alpha_1 C_{r1} + \alpha_2 C_{r2}$. Suppose that only the failed component has to be replaced, and the failure of one component does not affect the distribution of nonfailed components. Then the results are the same as for independent components, as given in Section 22.4. If the item is replaced by a new one, but just for the cost of the failed component only, then for a FRW with warranty period (W_1, W_2), the EC is given by

$$EC(W) = M_Y(W)(\alpha_1 C_{r1} + \alpha_2 C_{r2}) \tag{22.55}$$

Example 22.20

Suppose that the control and mechanical subsystems are the two components of a machine, having exponentially distributed lifetimes with parameters $\lambda_1 = 0.2$ and $\lambda_2 = 0.3$, respectively. Replacement of the control subsystem costs $C_{r1} = \$400$, whereas replacement of the mechanical component costs $C_{r2} = \$2300$. The machine is sold under a FRW with $W_1 = 3$ years for the control and $W_2 = 2.5$ years for the mechanical system. Each component is replaced separately in case of failure.

Because the series reliability model describes this system, we have the survival function, the renewal function, and probabilities for first failure of the item given by

$$G_Y(t) = e^{-(0.2 + 0.3)t}; \qquad M_Y(t) = 0.5t; \qquad \alpha_i = \frac{\lambda_i}{\lambda_1 + \lambda_2};$$

$$\alpha_1 = 0.4; \qquad \alpha_2 = 0.6$$

From this and (22.55), we obtain

$$EC(3) = (0.5)(3)(0.4)(400) + (0.5)(2.5)(0.6)(2300)$$
$$= 240 + 1725 = \$1965$$

Markovian Approach

In the Markovian approach, the two-component system is represented by a Markov model determined by an intensity matrix of the form (22.13).

The time that the system stays in state i is an exponentially distributed random time X_i with parameter equal to the ith diagonal entry of (22.13). For the continuous case, we then need the nonstationary time-dependent state probabilities, specified by the corresponding system of differential equations (see, e.g., Ref. 8), and derivation of various measures relevant to analysis of the warranty policy, such as the expected number of claims due to particular components, the expected number of repairs, and so forth (see Examples 22.2–22.4). To conduct numerical calculations when working with a continuous-time finite Markov model, discretization with respect to the time parameter is often a recommended technique.

For both the discrete and continuous Markov models, we can then use the following result to determine the expected number of visits to any state:

For a discrete-time Markov chain with transition matrix $\mathbf{P} = \{P_{ij}\}$, the expected number of visits $M_{ij}(n)$ to state j in a total of n steps, given that the process starts at state i, is determined by the equation [15, Chap. 9]

$$M_{ij}(n) = \sum_{k=1}^{n} P_{ij}^{(k)}, \quad i, j \in S, \text{ the set of states} \tag{22.56}$$

where $P_{ij}^{(k)}$ is the (i, j)th entry of the matrix $\mathbf{P}^k$. Expected costs can then be calculated as functions of the number of returns and the costs of each type of failure.

More complicated examples of the use of Markov chains in warranty cost analysis can be seen in Refs. 5 and 6. It is assumed in Ref. 5 that the item consists of three components, A, B, and C, with independent exponential lifetimes. The service policy under warranty is as follows: the first two failed components will be repaired; each subsequent failed item will be removed and replaced by an identical new component. A three-dimensional Markov chain is used with 27 states $\epsilon = (\epsilon_1, \epsilon_2, \epsilon_3)$, where $\epsilon_i = 0, 1$ and 2 indicate the number of failures of the ith component. Time is measured in years. Weekly discretization transforms the continuous-time process to a discrete Markov chain. Transition probabilities between states for a week are determined, costs C_{ij} for passages between the states are given, and EC is determined for an arbitrary future time period (in weeks) given that the present state of the item is known.

In Ref. 6, the optimal warranty period W for the entire multicomponent item under some overall cost restrictions is obtained.

22.6 CONCLUSIONS AND RECOMMENDATIONS

22.6.1 Scope for New Research

A number of new and interesting questions arise in the study of warranties for complex systems. These relate mostly to the reliability model of the system and of the underlying random processes. It is still difficult to match the functional use of a product with its engineering and the mathematical models of multicomponent systems. In addition, different warranty policies for distinct components and the entire item must reflect possible maintenance. Available information for implementation of warranty cost analysis is of crucial importance. This may significantly restrict the use of a model in many applications.

The use of an appropriate large variety of stochastic processes for description of the operational behavior and failures of the system is a new topic in investigation of warranties. Even for independent components, specific kinds of interactions arise because of possible common restrictions (e.g., a common limit for the overall warranty costs, common restrictions in the overall number of failures, etc.). The claim process may also reflect user's behavior patterns. This creates new classes of models for warranty policies relevant to this issue. Lack of information about probability distributions used in the models (e.g., we may know only the first several moments) creates new unsolved problems.

Exploration of multivariate distributions in warranty analysis for complex systems is an open field. Assigning warranties for multicomponent systems involves a series of optimization problems. The relationship between particular warranty policies and the optimal warranty for the whole system is of specific importance and analytical interest. Studies of age-specific warranty policies, component and block assigned warranties (analogous to component and block replacement policies in reliability maintenance), statistical issues, financial analysis of complex warranties, data reporting, and warranty forecasting are all areas that need development of specific new analytical methods and models. Appropriate computational algorithms and numerical methods that will make theoretical developments useful in practice must also be developed.

22.6.2 Some Topics for Study

The questions considered in this chapter and the examples given indicate a number of specific topics for study. Among them are the following:

> Development and collection of various warranty policies for multicomponent items used in practice, with special attention to claim patterns

Classification (taxonomy) of warranty models and warranty problems for complex items

Development of general and particular models when joint multivariate distributions are known, for FRW and PRW policies

Extension of these results to other warranty policies, such as combination warranties

Models with failure dependence and failure interactions in a random environment

Computational probability models for Markovian multicomponent systems in continuous time (e.g., use of phase-type distributions and Markovian arrival process (MAP) models [12] in warranties)

Use of marked point processes [16] in warranty analysis, claim description, and warranty cost forecasting

Development of software for computing warranty characteristics of complex items

Statistical studies

Warranty servicing problems (use of opportunistic replacement, i.e., overhaul or replacement of other components while removing a failed one)

ACKNOWLEDGMENTS

We are very thankful to Professor D. N. P. Murthy for his help, suggested constructions, and advice throughout preparation of this text. We warmly thank Professor W. Blischke for his excellent job in editing our several versions of this text, suggesting changes and fixing numerous errors without affecting our style of presentation.

REFERENCES

1. Wolfram, S. (1992). *MATHEMATICA. A System for Doing Mathematics by Computer*, 2nd ed., Addison-Wesley, Reading, MA.
2. Barlow, R. E., and Proschan, F. (1983). *Statistical Theory of Reliability and Life Testing. Probability Models*, Holt, Rinehart and Winston, New York.
3. Murthy, D. N. P., and Nguyen, D. G. (1985). Study of two-component system with failure interaction, *Naval Research Logistics Quarterly*, **32**, 239–247.
4. Nguyen, D. G., and Murthy, D. N. P. (1989). Optimal replace-repair strategy for servicing products sold with warranty, *European Journal of Operations Research*, **39**(2), 206–212.
5. Balachandran, K. R., Maschmeyer, R. A., and Livingstone, J. L. (1981). Product warranty period: a Markovian approach to estimation and analysis of repair and replacement costs. *The Accounting Review*, **LVI**(1), 115–124.

6. Chukova, S. (1988). *Mathematical Models and Methods for Determination of Optimal Warranty Periods*, Ph.D. thesis, Sofia University. [In Bulgarian].

7. Chukova, S., Dimitrov B., and Rykov V. (1993). Warranty analysis. A survey, *Journal of Soviet Mathematics*, **67**(6), 3486–3508.

8. Kemeny, J., and Snell, J. (1960). *Finite Markov Chains*, D. Van Nostrand Co., Princeton, NJ.

9. Pyke, R. (1961). Markov Renewal processes: definition and preliminary properties, *Annals of Mathematical Statistics*, **32**(4), 1231–1242.

10. Pyke, R. (1961). Markov renewal processes with finitely many states, *Annals of Mathematical Statistics*, **32**(4), 1243–1259.

11. Cleroux, R., and McConalogue, D. J. (1976). A numerical algorithm for recursively-defined convolution integrals involving distribution functions, *Management Science*, **22**(10), 1138–1146.

12. Neuts, M. F. (1981). *Matrix-Geometric Solutions in Stochastic Models. An Algorithmic Approach*, The Johns Hopkins University Press, Baltimore, MD.

13. Smith, W. L., and Leadbetter, M. R. (1963). On the renewal function for the Weibull distribution, *Technometrics*, **5**(3), 393–396.

14. Block, H. W., Borges, W. S., and Savits, Th. (1985). Age-dependent minimal repair, *Journal of Applied Probability*, **22**, 370–385.

15. Feller, W. (1957). *An Introduction to Probability Theory and Its Applications*, 2n ed., Vol. 1, John Wiley & Sons, New York.

16. Shnyder, D. L. (1975). *Random Point Processes*, John Wiley & Sons, New York.

23

Warranty and Manufacturing

D. N. Prabhakar Murthy

The University of Queensland
Brisbane, Queensland, Australia

23.1 INTRODUCTION

Once the design of a product is finalized, the next step is the manufacturing of the product. Manufacturing can be viewed as a process which transforms inputs (raw material and/or components), through a sequence of operational stages (involving operations such as machining, assembly, and testing), to produce the product as the output of the final stage.

Due to variability in the manufacturing process, the quality of items produced (also called output quality) varies. In the simplest characterization, items can be classified as being either conforming or nonconforming. An item is termed conforming if it meets the design specification in terms of its performance, and nonconforming if it does not. Nonconforming items are produced in an uncertain manner and influenced by factors such as process deterioration, quality variations in input, operator, and so forth.

One needs to differentiate two types of nonconformance: Type A and Type B. A Type-A nonconforming item is not operational when put into use and is due to defective assembly and/or defective components that make the item inoperable. A Type-B nonconforming item is operational when put into use but has performance characteristics (e.g., mean time to failure) that are considerably inferior to those of a conforming item. Type-A nonconforming items can be detected either by inspection or testing for a very short (or nearly zero) time duration. In contrast, Type-B nonconforming items can only be detected, in a statistical sense, by usage over time.

The expected warranty servicing cost for a nonconforming item is considerably greater than that for a conforming item. The total expected warranty cost can be reduced by either (1) reducing the number of nonconforming items produced and/or (2) reducing the number of nonconforming items reaching consumers. The former is achieved by controlling the deterioration in the manufacturing process and the latter by inspection or life-testing at the final stage (and possibly at one or more intermediate stages) to weed out nonconforming items. Both these forms of quality control actions involve additional costs and are justified (all other things being equal) only when the reduction in the expected warranty cost exceeds the additional cost involved in controlling the quality.

The outline of the chapter is as follows. Section 23.2 deals with the modeling of product quality variations. Section 23.3 examines quality control schemes for Type-A nonconforming items and Section 23.4 deals with schemes for Type-B nonconforming items. Even when all items meet the design specification, if the product has a bathtub failure rate (i.e., high failure rate during the initial phase of an item's life), it is often more economical to subject the items to life-testing (also called burn-in) before they are released for use. This is examined in Section 23.5. The last section gives recommendations for the practitioners.

23.2 MODELING PRODUCT QUALITY VARIATIONS AND CONTROL

Variations in product quality depend on many factors, as mentioned earlier. Modeling of variations is influenced by the type of manufacturing process. In this section, a brief discussion of the different types of manufacturing processes is presented first before proceeding to the characterization of quality variations.

23.2.1 Manufacturing Processes

The three main types of manufacturing systems are as follows:

1. Continuous production
2. Batch (or lot) production
3. Flexible manufacturing

Continuous production is used for large-volume production and limited variety in the range of products produced. In contrast, flexible manufacturing is used for small-volume production and large variety. Batch production is used for intermediate-volume production, and as the name suggests, here the items are produced in batches or lots. In the remainder of the chapter, the words "batch" and "lot" will be used interchangeably.

A variety of models have been proposed for modeling quality variations for each of the three types of manufacturing. However, in the remainder of the chapter, the attention is confined to items produced in batches.

23.2.2 Product Quality Variations

The output quality of a manufacturing process depends on the input quality and the operations at different stages of the manufacturing process.

In a "black-box" approach, one models the output quality by an appropriate mathematical formulation based on intuition or empirical data. In the simplest characterization, the quality variations are modeled in terms of the probability of an item conforming or not. Three characterizations that will be used later are as follows:

Case 1. The probability that an item is nonconforming is the same for all items in a batch and this does not vary from batch to batch. This situation represents the case in which the manufacturing process has been stabilized and is operating under a steady-state condition.

Case 2. The probability that an item is nonconforming is the same for all items within a batch, but this probability varies from batch to batch. This characterizes the case in which the variations in the input and/or the operations at different stages differ from batch to batch.

Case 3. The probability that an item is nonconforming changes from item to item in a batch. This captures the deterioration in the manufacturing process over time and its effect on product quality. Hence, it models the dynamic nature of the manufacturing process. In the simplest characterization, the process can be viewed as being in one of two states: "in-control" or "out-of-control." The process state starts in-control and changes to out-of-control after a certain length of time (or after a certain number of items are produced) and stays there until some action is initiated to bring it back to in-control.

In the "white-box" approach, one models the output quality in terms of the input quality and the operations at different stages of the manufacturing process. A variety of models based on this approach have been developed to model the output quality of a general multistage production system (see Ref. 1 for a brief review of such models). These models are more complex than the models based on the black-box approach. As such, models linking quality and warranty are based mainly on the black-box approach.

23.2.3 Type A and Type B Nonconformance

The items are produced in lots of size L and this subsection deals with the modeling of quality variations based on the black-box approach.

Type A Nonconformance

In this case, the differentiation between conforming and nonconforming items is done through discrete binary random variables. Let X_{ji} (a binary random variable) denote the quality of item i in batch j, $1 \leq i \leq L$ and $j \geq 1$. It is assigned a value 1 if the item is conforming and a value 0 if it is not. Let p_{ji} denote the probability that the item is conforming, that is,

$$p_{ji} = \text{Prob}\{X_{ji} = 1\}$$

Then $1 - p_{ji}$ is the probability that the item is nonconforming. As a result, the modeling of the three cases discussed in the previous subsection is as follows:

> *Case 1*: $p_{ji} = p$ for all i and j.
> *Case 2*: $p_{ji} = p_j$ for $1 \leq i \leq L$ and p_j varies (randomly) from lot to lot; for example, p_j being distributed according to beta distribution.
> *Case 3*: p_{ji} changes with i and j; for example, $p_{ji} = p_0$ for $1 \leq i < i_0$ (a random variable) and $p_{ji} = p_1$ ($< p_0$) for $i \geq i_0$ models the situation where the process state changes suddenly from in-control to out-of-control after i_0 items are produced. Note that the probability that an item is nonconforming is higher when the process is out-of-control. In contrast, if for each j, p_{ji} is a decreasing sequence in i, then it models the situation where the process state is deteriorating gradually.

Type B Nonconformance

Here the differentiation between conforming and nonconforming items is done through a continuous-valued variable or function; for example, the mean time between failures or the failure rate. The modeling is as follows.

Let $F_1(t)$ $[F_2(t)]$ denote the failure distribution function for a conforming (nonconforming) item. Let $f_i(t)$ $[= dF_i(t)/dt]$, $\overline{F}_i(t)$ $[= 1 - F_i(t)]$, and $r_i(t)$ $[= f_i(t)/\overline{F}_i(t)]$ denote the density function, survivor function, and the failure rate, respectively, associated with $F_i(t)$, $i = 1, 2$. Because the performance of a nonconforming item is inferior to that of a conforming item, one can characterize it as

$$r_1(t) < r_2(t), \quad 0 \le t < \infty \tag{23.1}$$

that is, a nonconforming item has a higher failure rate in relation to a conforming item. This implies that

$$\mu_1 = \int_0^\infty \overline{F}_1(t)\, dt > \int_0^\infty \overline{F}_2(t)\, dt = \mu_2 \tag{23.2}$$

that is, the mean time to failure for a nonconforming item is much smaller than that for a conforming item.

Let p_{ji} $(1 - p_{ji})$ denote the probability that item i in batch j is conforming (nonconforming). The modeling of the three cases (discussed in the previous subsection) is as follows:

Case 1: $p_{ji} = p$ for all i and j.

Although items are produced in batches, an item for sale, or use in replacement under warranty, is picked randomly from the batch. As a result, the failure distribution, $F(t)$, of an item picked randomly for use is given by

$$F(t) = pF_1(t) + (1 - p)F_2(t) \tag{23.3}$$

where p is as previously defined.

Case 2: $p_{ji} = p_j$, for $1 \le i \le L$, and p_j being a random variable varying from batch to batch.

For Case 2, an alternate approach is to let θ, the parameter of the failure distribution function, vary from batch to batch. For the jth batch, let the parameter value (a random variable) be denoted θ_j. The expected warranty cost for batch j depends on θ_j. Mann and Saunders [2,3] use such a model formulation to decide on the best warranty period for each batch based on limited life-testing of items from each batch. Mamer [4] also uses a similar model formulation in the context of selecting optimal warranty terms.

Case 3: p_{ji} varies with i and j as in Type A nonconformance.

23.2.4 Quality Control Schemes

As mentioned earlier, the two different approaches to reduce the number of nonconforming items reaching consumers are as follows:

1. Testing to weed out nonconforming items (a reactive approach)
2. Preventing nonconforming items from being produced (a proactive approach)

Often, the word "inspection" is used instead of testing. One can view inspection as a form of testing, and henceforth we will use "testing" to include "inspection" in the remainder of the chapter.

Testing

The case of Type-A nonconforming items is considered first. As no life-testing is involved, testing does not affect the failure distribution of items. One can employ either complete (100%) testing or limited (less than 100%) testing based on some sampling plan. The testing can be either perfect or imperfect. If the testing is perfect, the outcome of the test reveals the true status (conforming or not) of the item being tested. If the testing is imperfect, a conforming item can sometimes be wrongly classified as being nonconforming and vice versa.

A variety of sampling plans have been studied and these can be found in any standard text on quality control, for example, Refs. 5 and 6. Two such plans are as follows.

Plan 1 (Acceptance Sampling Plan)

The sampling plan is characterized by two integers n ($0 < n \leq L$) and c ($0 \leq c \leq n$). A sample of n items is selected randomly from each batch and each item in the sample tested to see if it is conforming or not. Let $\bar{n}$ denote the number of nonconforming items in the sample. If $\bar{n} \leq c$, then the items in the batch are released with no further testing. However, if $\bar{n} > c$, then all the remaining items in the batch are subjected to testing. In other words, 100% testing is used only when the sample contains more than a specified number of nonconforming items. The rationale for this is that a batch which contains few nonconforming items will, on the average, have a small ($\leq c$) number of nonconforming items in the sample, and in this case, there is no need for further testing of the batch. In contrast, if a batch contains a large number of nonconforming items, then, on the average, the number of nonconforming items in the sample is more likely to be greater than c, and in this case, it is worthwhile testing the whole batch to weed out all of the nonconforming items.

Plan 2 (Curtailed Sampling Plan)

The sampling plan is characterized by an integer n. For each batch, items are tested one at a time until a nonconforming item is found or until n items have been tested. If a nonconforming item is found, then all items in the batch are tested. If no nonconforming item is found, then the batch is released with no further testing.

In all sampling schemes, the fate of nonconforming items depends on whether they are repairable or not. In the nonrepairable case, they are scrapped. In the repairable case, the choice between repair or scrap depends on the relative costs of manufacturing and repair. The optimal design of a sampling plan involves a trade-off between the expected cost of testing and the expected savings in the warranty cost.

In the case of Type B nonconformance, weeding nonconforming items requires testing items (on a test bed) for a duration T (also called as "burn-in" period). Because a nonconforming item has a higher failure rate, it is more likely to fail during the testing period than a conforming item. Hence, the number of items failing in the sample tested can be viewed as an indicator of the quality of the batch—a large number implying a poor quality, that is, a high fraction of nonconforming items in the batch. As with Type-A nonconformance, the testing can be either complete (100%) or limited (less than 100%) when a sampling plan is used. A variety of sampling plans have been studied and one of them is as follows:

Plan 3 (Acceptance Sampling Plan (n, c) Involving Life-Testing)

The sampling plan is characterized by two integers n $(0 < n \leq L)$ and c $(0 \leq c \leq n)$ and a real variable T (>0). A sample of n $(0 < n \leq L)$ items is selected randomly from each batch and life-tested for a period T. Let $\tilde{n}$ denote the number of items from the sample which fail during the testing. If $\tilde{n} \leq c$, then the remaining items in the batch are not tested. However, if $\tilde{n} > c$, then the remaining items in the batch are subjected to life-testing for a period T. All items which fail during testing are either scrapped (in the case of nonrepairable product) or reworked (in the case of repairable product) and released along with the nonfailed and nontested items. Note that the larger the value of T, the greater is the likelihood of a nonconforming item failing during testing and, hence, getting weeded out. However, this is achieved at the expense of reducing the life of items released for sale.

Prevention

As mentioned earlier, the state of the process affects the probability that an item is nonconforming. This probability is low (high) when the state is in-control (out-of-control). As a result, the number of nonconforming items produced in a batch can be reduced (in a statistical sense) by actions which either reduce the likelihood of the state changing from in-control to out-of-control or minimize the duration for which the state is out-of-control once a change occurs. In batch manufacturing, lot sizing is used to achieve both of these.

A model which characterizes the impact of lot sizing on the output quality is due to Porteus [7]. In this model, the process is always in-control at the start of each batch production. If the state is in-control at the start of an item production, it can switch to out-of-control with probability (1 − q) or stay in-control with probability q. Once the switch occurs, the state remains unaltered until the end of the batch production. This is appropriate when it is not desirable to interrupt the process during the production of a batch. All items produced are conforming (nonconforming) when the process is in-control (out-of-control). Note that both the probability that the process state changes before a lot is processed or while it is being processed and the expected number of nonconforming items produced (should the process state change) decreases (increases) as L decreases (increases).

This model has been extended in Ref. 8 to solutions in which not all items produced are conforming (nonconforming) when the process is in-control (out-of-control).

Rosenblatt and Lee [9] model the change in process state and its effect on output quality slightly differently. In their model, the number of items produced before the state changes is modeled by a discrete distribution function. However, the final effect of lot size on output quality is similar to that in the Porteus model.

23.3 QUALITY CONTROL MODELS FOR TYPE A NONCONFORMANCE

This section deals with the case in which the nonconforming items are of Type A and presents several models for quality control. The models assume that nonconforming items can be made conforming by appropriate rectification actions. Rectification of nonconforming items before they are released will be termed as "rework" and these are carried out at the manufacturing plant. Rectification of nonconforming items returned under warranty will be termed as "servicing" and are carried out either at a service facility or at the plant. The per-unit cost of rectification under servicing is larger than that under rework. This is because of the additional costs associated with the setting up of a service center and/or the handing of warranty claims. (See Chapter 24.)

In the models, the focus will be on the expected warranty cost resulting from nonconforming items being returned for rectification. The expected warranty costs associated with failures over the warranty period is the same for all items (conforming or nonconforming) once the nonconforming items have been rectified. As such, this cost is of no relevance in the context of quality control schemes of this section.

Let C_r (C_s) denote the rework (service) cost to rectify a nonconforming unit before (after) sale. Note that $C_s > C_r$ for reasons mentioned above.

23.3.1 Models Involving Testing

Let C_t denote the cost of testing each item.

Model 1

The process is assumed to be in a steady state. Hence, the items produced are statistically similar, that is, each item is conforming with probability p and not conforming with probability $\bar{p} = 1 - p$. If no testing is used, then the expected number of nonconforming items released per batch is $\bar{p}L$.

For each batch, n ($0 \leq n \leq L$) items are tested and the nonconforming ones are rectified through rework. The remaining items, if any, are released with no testing. As a result, the expected number of nonconforming items per batch reaching consumers is given by $\bar{p}(L - n)$. The average outgoing quality (AOQ), the expected fraction of conforming items in a batch after testing, is given by

$$\text{AOQ} = 1 - \frac{\bar{p}(L - n)}{L} \tag{23.4}$$

This increases from p (when $n = 0$) to 1 (when $n = L$). The savings in the expected warranty cost per batch is $n\bar{p}(C_s - C_r)$ and the cost of testing is nC_t. As a result, testing is preferred to no testing if and only if $\bar{p}(C_s - C_r) > C_t$. When n ($0 \leq n \leq L$) is a decision variable, n^*, the optimal n which minimizes the expected total (warranty + testing) cost is characterized as follows.

If there are no capacity constraints (see Model 2), then $n^* = 0$ when $\bar{p}(C_s - C_r) < C_t$ and $n^* = L$ when $\bar{p}(C_s - C_r) > C_t$. With capacity constraints (see Model 2), n^* can be an interior point, that is, $0 < n^* < L$.

Model 2

This is an extension of Model 1 and was proposed by Tapiero and Lee [10, Section 3]. The quality control program involves testing n items in each batch. All nonconforming items detected are rectified through rework. The remaining $L - n$ items are released with no testing. Nonconforming items sold are returned to the service center for rectification. As a result, the expected total (rework + servicing + testing) cost per lot depends the servicing center capacity (s) and the number tested per lot (n).

The cost of a service center with capacity s is given by $C_0(s) = c_0 + sc_3$ for $s > 0$, where c_0 is the fixed cost and c_3 is the variable cost. The number of items per lot returned under warranty for servicing is a random variable distributed binomially. Let u denote the number of items per lot returned under warranty, then the cost of servicing (CS) a lot is given by

$$CS = \begin{cases} C_0(s) + c_1 u & \text{if } u \le s \\ C_0(s) + c_1 s + c_2(u - s) & \text{if } u > s \end{cases} \tag{23.5}$$

where c_1 (c_2) is the variable cost per unit serviced below (above) capacity for items returned. $c_2 > c_1$ reflects the cost of overtime or obtaining service elsewhere as the capacity is exceeded. Using a normal approximation, Tapiero and Lee derive an expression for the expected total (test + rework + service) cost per lot. It is given by

$$J = J(n, N, s) = n(C_i + C_r \bar{p}) + C_0(s) + c_2(N - n)\bar{p} - (c_2 - c_1)$$

$$\times \left\{ \int_0^s u g(u, N - n)\, du + s[1 - G(s, N - n)] \right\} \tag{23.6}$$

where

$$G(u, N - n) = \frac{1}{\sqrt{2\pi}} \int_0^v \exp(-x^2)\, dx$$

$$v = \frac{u - (N - n)\bar{p}}{\sqrt{(N - n)p\bar{p}}}$$

and

$$g(u, N - n) = \frac{dG(u, N - n)}{du}$$

Let s^* and n^* denote the optimal values of s and n which minimize J. Tapiero and Lee [10] obtain the following result.

Proposition 23.1

Suppose $c_3 < c_2 - c_1$.

1. If $\bar{p} \ge C_i/(c_1 - C_r)$, then $n^* = L$ (100% testing) and $s^* = 0$.
2. If $\bar{p} \le C_i/(c_2 - C_r)$, then $n^* = 0$ (no testing) and s^* is given by the solution to $c_3 = (c_2 - c_1)[1 - G(s, N)]$.
3. If $C_i/(c_2 - C_r) < \bar{p} < C_i/(c_1 - C_r)$, then $0 < n^* < L$ and $s^* > 0$.

(Note that n^* and s^* in case 3 above need to be obtained by numerical methods.)

If $c_2 = c_1$ (implying no service capacity constraint as servicing can be outsourced at no extra cost), then situation case 3 above does not arise and in this case $n^* = 0$ or L and s^* is zero in the former case and is L in the latter case.

Model 3 [10]

This is an extension of Model 2 and was proposed by Tapiero and Lee [10, Section 4]. Here the quality varies from lot to lot and as a result, $\bar{p}_j$ $(= 1 - p_j)$ is modeled by a density function $f(\bar{p})$. The quality control process involves testing a sample of n items. All nonconforming items detected during testing are rectified through rework. Nonconforming items (from the items not tested and released for sale) are rectified under warranty at a service center as in Model 2. Tapiero and Lee derive an expression for the expected total (rework + servicing + testing) cost per lot. It is given by

$$J = n(C_i + C_r\bar{p}) + y(N - n)(c_2\bar{p} - C_i - C_r\bar{p}) + C_0(s)$$
$$+ y(c_1 - c_2)\left[\int_0^s (u - s)g(u, N - n)\, du\right] \quad (23.7)$$

where

$$y = \sum_{i=0}^{r} \binom{n}{i} \bar{p}^i p^{n-i} \quad (23.8)$$

and $g(u, N - n)$ is the same as in Model 2. Tapiero and Lee obtain results similar to Proposition 23.1, but they are more complex.

Model 4 [11]

In this model, the item is a noncritical component of a complex system. The system is functional even when the item is nonconforming. The quality varies from batch to batch. For batch j, the probability that an item is nonconforming is given by $\bar{p}_j (= 1 - p_j)$, a random variable with density function $f(\bar{p})$. Should the system fail and it is established that the item (noncritical component) was defective, then the manufacturer is held liable and incurs a cost. In other words, the claim is under implied warranty. The quality control is based on an acceptance sampling plan of (n, r). Thus, when testing is less than 100%, some systems sold can have nonconforming items which can result in such claims.

The sampling plan is chosen to ensure that the average outgoing quality (AOQ) is above a guaranteed value. This can be viewed as a self-insurance (by the manufacturer) to reduce the liability risk. Maruchek [11] then discusses the choice between quality control (involving acceptance sampling scheme) and no quality control based on the expected profit (revenue − total cost) for two scenarios: (1) all rejected lots (lots where the number of failures in the sample tested is greater than r) are 100% tested and nonconforming items are either rectified or scrapped and (2) all rejected lots are scrapped. Note the total cost is the sum of the manufacturing cost, testing cost, and liability cost.

The model involves the following variables:

r: the revenue generated per unit sale
b: manufacturing cost per unit
d: probability that a consumer receiving a nonconforming item will demand settlement
C_L: average cost per claim

The other variables (for example, C_r and C_i) are as defined earlier. Then the choice between acceptance sampling and no sampling is given by the following proposition.

Proposition 23.2 [11]

Acceptance sampling is preferred to no sampling if

$$C_i \le \frac{N(1 - y)\overline{p}(C_L d - C_r)}{N - y(N - n)}$$

for Scenario 1, and

$$C_i \le \frac{(C_L d\overline{p} - r)(1 - y)N}{n}$$

for Scenario 2, where y is the probability that a batch is released for sale without 100% testing and is given by (23.8).

Maruchek carries out a detailed study of the cost trade-offs for quality improvement through (1) reduction in process variability (measured by the standard deviation of p_j) and (2) improving the process mean (measured by the mean of p_j). She provides a numerical example for the case where $f(\overline{p})$ is a beta distribution.

Model 5 [12]

In this model, the lot size corresponds to the number of items produced in each time period and is modeled as varying randomly. The control scheme involves a sampling plan and is characterized by a parameter ϕ. The expected service cost to rectify nonconforming items released for sale is obtained using a formulation similar to Model 4 [10] with service capacity s. Finally, the output quality is modeled by a parameter θ, which represents the probability that an item is conforming. Better manufacturing results in a higher value for θ, but this is achieved at a higher manufacturing cost. Hence, the decision variables in the model are ϕ, θ, and s. The model examines the optimal choice of these to maximize the expected profit per unit time for two different sampling schemes: (1) acceptance sampling plan of (n, r) and (2) curtailed sampling plan.

23.3.2 Models Based on Prevention

The output quality goes down as the process state deteriorates. As mentioned earlier, the deterioration can be either gradual or sudden. In this subsection, a model due to Porteus [7] is outlined. Under this model, the process change is modeled as occurring suddenly, and lot sizing is used to reduce the likelihood of the change occurring as well as the duration for which the system is in out-of-control state once a change in state occurs.

Model 6 [7]

Items are produced in lots of size L to meet a constant demand of m items per unit time. The time to produce a lot is relatively small so that it is treated as being zero. This implies that the time between the production of two successive lots is L/m.

At the start of each lot production, the state is checked to make sure that it is in-control. The change in process state occurs in an uncertain manner and is characterized by the parameter q as discussed in Section 23.2.4. Should the process change from in-control to out-of-control, then a certain number of items are nonconforming. As a result, the number of conforming items in a lot ($\tilde{N}$) is a random variable with an expected value given by

$$E[\tilde{N}] = \frac{q(1 - q^L)}{1 - q} \tag{23.9}$$

The manufacturing cost per lot is $C_f + C_m L$, where C_f is the fixed cost and C_m is the variable cost (material + labor) to produce a unit. Let h denote the inventory holding cost for an item per unit time. The total inventory holding cost per lot is $hL^2/2m$. Finally, all nonconforming items are reworked at a cost of C_r for each nonconforming item.

Porteus derives an expression for the expected total (manufacture + inventory + servicing) cost per unit time. It is given by

$$J(L) = \frac{(C_f + C_m L) + (hL^2/2m) + C_r[L - q(1 - q^L)]/(1 - q)}{L/m}$$

(23.10)

Note that as L increases, the manufacturing cost per unit time decreases and the inventory holding cost and the expected rework cost per unit time increase. This implies that L^*, the optimal L which minimizes $J(L)$, achieves a sensible trade-off between manufacturing and the rework costs.

Porteus uses the following approximation (appropriate when q is close to 1)

$$q^L \simeq 1 + \ln(q)L + [\ln(q)L]^2$$

(23.11)

which yields an approximation to L^* given by

$$L^* \simeq \sqrt{\frac{2mC_f}{h + mC_r(1 - q)}}$$

(23.12)

Note that this model is in the spirit of inventory models as it incorporates inventory holding cost. Porteus also examines the cost trade-offs of investing in quality improvement. Let q_0 denote the original process quality. This can be increased to a value q ($q_0 \leq q \leq 1$) by an investment effort $a(q)$. Porteus suggests the optimal choice of q and L to minimize $J(L) + ba(q)$, where b is a constant. For the case in which

$$a(q) = a_0 - a_1 \ln(1 - q)$$

(23.13)

q^*, the optimal q, is given by

$$q^* = \max\left[q_0, 1 - \frac{(ba_1)^2 + (ba_1)[(ba_1)^2 + 2mC_fh]^{1/2}}{C_r m^2 C_f}\right]$$

(23.14)

Example 23.1   [7]

Let $m = 1000$ units/unit time and the cost parameters be $C_f = \$100$, $C_m = \$50$, $h = \$0.5$/unit time, $C_r = \$25$, and $q_0 = 0.9996$. Using only lot sizing to improve the output quality, L^* [the optimal L obtained from (23.12)] is 105 and the expected cost per unit time [obtained from (23.10)] is \$1895. The percentage of nonconforming items in a lot is 2.1.

With investment, q can be increased. Let $a_0 = -1485$, $a_1 = 190$, and $b = 0.15$. This implies that each 10% improvement in q costs \$20. In this case, q^* [the optimal q obtained from (23.14)] is 0.999985 and L^* (the optimal lot size) is 155. The expected cost per unit time goes down from \$1895 to \$1388 and the percentage of nonconforming items in a lot goes down from 2.1 to 0.11. The implication of this is that by investing in quality improvement, the optimal lot size is larger, the expected cost per unit time smaller, and output quality higher.

Finally, Porteus [7] looks at investments to reduce C_f, the setup cost, and its effect on lot sizing, the expected cost per unit time, and the quality of the output.

23.4 QUALITY CONTROL MODELS FOR TYPE B NONCONFORMANCE

As in the previous section, models involving testing are considered first. This is followed by models based on prevention, and, finally, models which involve both prevention and testing are discussed at the end of the section.

23.4.1 Models Involving Testing

Because nonconforming items are of Type B, the quality control schemes to weed out such items involve life-testing which affect the failure distribution of the items tested.

Model 7 [13]

In this model, the process is assumed to be in steady state so that the probability that an item is conforming (nonconforming) is p ($\bar{p} = 1 - p$) for all items. Testing involves an acceptance sampling plan of (n, r). However, the model assumes that testing involves negligible time and does not affect the failure distribution of the item. This is a rather restrictive assumption which is not easily justifiable. Lee and Chun [13] study the effect of n ($r \le n \le L$) and r ($0 \le r \le n$) on the expected total (manufacturing + testing + rework + warranty servicing) cost when items are sold with FRW and PRW policies and derive conditions which yield n^*, the optimal n, which minimizes this total expected cost for a given r. They show that n^* is either r or L and give conditions for n^* to equal r or L.

Model 8 [14]

The model assumes that the process is in steady state. Each item produced is conforming (nonconforming) with probability p ($\bar{p} = 1 - p$). Let $F_1(t)$ [$F_2(t)$] denote the failure distribution function for a conforming (nonconforming) item.

The scheme to weed out defective items involves testing items (on a test bed) for T units of time. As all items are statistically similar, either one tests all items or none. The choice between the two depends on the savings in expected warranty cost versus the increased manufacturing cost per item due to testing and scrapping (or reworking) of items which fail during the testing period. Murthy et al. [14] consider the case in which the number of items manufactured is large and derive results for the asymptotic case (i.e., the number of items produced approaching infinity).

If no item is tested, then the asymptotic total (manufacturing + warranty) cost per item sold is given by

$$J(0) = C_m + C_w(0) \tag{23.15}$$

where C_m is the manufacturing cost per unit and $C_w(0)$ is the asymptotic warranty cost per unit. $C_w(0)$ depends on the type of warranty, the warranty duration (W), failure distributions F_1 and F_2, and p, the probability that an item is conforming.

If all items are tested for a period T, then the manufacturing cost per lot is $(C_m + C_i)L$, where $C_i (= \phi_0 + \phi_1 T)$ is the cost of testing an item for period T. ϕ_0 represents the fixed cost and ϕ_1 the variable cost. The expected number of items per lot which fail and get discarded is given by $[1 - \alpha(T)]L$, where

$$\alpha(T) = p\overline{F}_1(T) + \overline{p}\overline{F}_2(t) \tag{23.16}$$

is the probability that an item picked randomly will survive the test. Let C_d denote the cost of disposing each failed item. As a result, the asymptotic total (manufacturing + testing + discarding) cost per item released is given by

$$\gamma(T) = \frac{C_m + C_i + \alpha(T)C_d}{1 - \alpha} \tag{23.17}$$

Note that the failure distribution of conforming and nonconforming items which survive the test are given by

$$\hat{F}_i(t) = \frac{F_i(t + T) - F_i(T)}{\overline{F}_i(T)} \tag{23.18}$$

with $i = 1$ [2] for conforming (nonconforming) items. Finally, the asymptotic probability that an item released is conforming (nonconforming) is given by $\hat{p}$ $[1 - \hat{p}]$, where

$$\hat{p} = \frac{p\overline{F}(T)}{\alpha(T)} \tag{23.19}$$

with $\alpha(T)$ given by (23.16). Note that $\hat{p} > p$ and that $\hat{p}$ increases with T [since $\overline{F}_2(T)/\overline{F}_1(T)$ decreases as T increases]. The asymptotic (manufacturing + testing + discarding + warranty) cost per item is given by

$$J(T) = \gamma(T) + C_w(T) \tag{23.20}$$

where $C_w(T)$ is the asymptotic warranty cost per item. $C_w(T)$ depends on the type of warranty, the warranty duration (W), failure distributions $\hat{F}_1$ and $\hat{F}_2$, and $\hat{p}$, the asymptotic probability that an item is conforming.

Testing is the optimal strategy if $J(T) > J(0)$ for some $T > 0$ and no testing is the optimal strategy if $J(T) < J(0)$ for all $T > 0$. In the former case, the optimal testing period, T^*, is the value of T which minimizes $J(T)$.

FRW Policy (with minimal repair)

The free-replacement warranty (FRW) is discussed in Chapters 1 and 10. Let C_r denote the cost of each repair. Then

$$C_w(0) = C_r \left[p \int_0^W r_1(t)\, dt + (1 - p) \int_0^W r_2(t)\, dt \right] \tag{23.21}$$

and

$$C_w(T) = C_r \left[\hat{p} \int_0^W \hat{r}_2(t)\, dt + (1 - \hat{p}) \int_0^W \hat{r}_2(t)\, dt \right] \tag{23.22}$$

Consider the special case where $F_1(t)$ and $F_2(t)$ are exponential distributions with parameters λ_1 and λ_2, respectively, with $\lambda_2 > \lambda_1 > 0$. This characterization is often used for modeling failure of electronic systems (with either the whole system or a subsystem being the item under consideration) and here the failure of an item is not dependent on its age. In this case, the choice between testing and no testing is given by the following propositions.

Proposition 23.2 [14]

Testing is the optimal strategy if

$$p\bar{p}(\lambda_2 - \lambda_1)[\bar{F}_1(T) - \bar{F}_2(T)] > \frac{(1 - \alpha)(C_m + C_d) + C_i}{WC_r} \tag{23.23}$$

for some $T > 0$.

Comments:

1. The inequality is more likely to be satisfied if (1) C_m, C_d, and C_i are small relative to WC_r, (2) $\lambda_2 - \lambda_1$ is large, and (3) p is close to 0.5 (i.e., 50% defectives!).
2. If the inequality is satisfied for $p = p_c$, then it is also satisfied for $p_c \le p \le 1 - p_c$.

Proposition 23.3 [14]

No testing is the optimal strategy if

$$C_m + C_d + C_i > WC_r p\bar{p}(\lambda_2 - \lambda_1) \tag{23.24}$$

Comments:

1. The inequality is more likely to be satisfied if (1) C_m, C_d, and C_i are large relative to WC_r, (2) $\lambda_2 - \lambda_1$ is small, and (3) p is close to 0. This is just the reverse of Proposition 23.1.
2. The inequality is always satisfied when $\bar{p}$ is very close to zero. In this case, because the majority of the items are defective, there is no advantage in doing 100% testing to weed out the defectives. The more sensible option is to scrap the whole batch.
3. If the inequality is satisfied for $p = p_c$, then it is also satisfied for all $p \leq p_c$.

Example 23.2

Let $\lambda_1 = 0.2$ per year and $\lambda_2 = 4.0$ per year. This implies that the mean time to failure is 5 years for conforming items and 0.25 year for nonconforming items. Let $L = 100$ and the remaining values of the model parameters be as follows: $p = 0.75$, $C_m = \$300$, $\phi_1 = 0.0$, $\phi_2 = \$45$ per year, $C_d = \$5$, and $C_r = \$100$. This implies that the cost of testing a unit for a month is approximately \$3.80. $p = 0.75$ implies that 25% of the items are defective. This is not an unusual figure for certain electronic systems using integrated chips where the fraction of defectives can be unusually high.

The above situation corresponds to the case where the item is an expensive electronic system composed of three modules, each costing roughly the same. Item failures occur due to one of the modules failing and the item being repaired by replacing the failed module by a new one. Hence, the cost of each replacement is roughly one-third the cost of a new item.

Table 23.1 shows T^* and $J(T^*)$ for four different warranty periods ranging from 1 to 4 years. Note that $T^* = 0$ implies that no testing is the optimal strategy, and $T^* > 0$ implies that testing is the optimal strategy. For $W = 1$, no testing is the optimal strategy. For $W \geq 2$, testing is the optimal strategy. Figure 23.1 shows T^* versus W. Note that T^* is zero for $W \leq 1.25$ and increases with W for $W > 1.25$. The reason for this is as follows. Because all items released for sale are rectified through minimal repair, the warranty servicing cost increases rapidly with W for any nonconforming item released. As a result, as the warranty period in-

Table 23.1 Optimal Strategy, T^* and $J(T^*)$ Under FRW

W	Optimal strategy	T^*	$J(T^*)$
1	No testing	0.0	415.00
2	Testing	0.2259	514.30
3	Testing	0.4231	567.13
4	Testing	0.5343	606.47

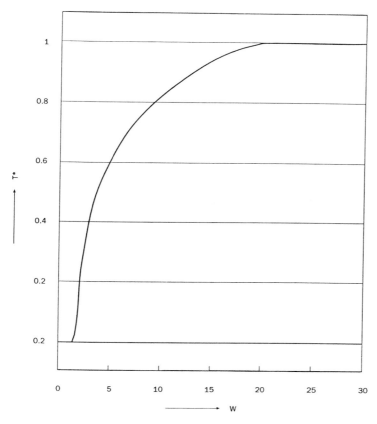

Figure 23.1 T^* versus W (FRW policy with minimal repair).

creases, longer testing is needed to reduce the number of nonconforming items being released. This is seen more clearly in Table 23.2 where as W increases, T^* increases, and the expected number of nonconforming items released decreases.

The optimal testing period varies from 0.2259 year for $W = 2$ years to 0.5343 year for $W = 4$ years. Obviously, it is not possible to test items for such long periods. However, life-testing is usually carried out in accelerated manner. Under accelerated life-testing, the item is subjected to a harsher environment which hastens the aging process. As a result, testing for one unit of time in the accelerated mode corresponds to β (> 1) units of testing under normal conditions. This implies that testing for T/β units of time in the accelerated mode corresponds to testing for T units under normal conditions. If the testing in accelerated mode is done with $\beta = 50$, testing for 0.5343 year in normal mode would require accelerated testing for only approximately 4 days. This practical solution corresponding to the optimal period may be obtainable.

Figure 23.2 shows the influence of p on T^* with the remaining parameters held at their nominal values. Note that $T^* = 0$ for small p (close to 0) and large p (close to 1) as to be expected from Proposition 23.3. For $W = 1$, T^* is zero for all values of p. For $W = 2$, $T^* > 0$ for $0.48 < p < 0.92$. Over this range, T^* first increases and then decreases as shown in Figure 23.2. The reason for this is as follows. When $p = 1$, no item is nonconforming and, hence, there is no need for testing. As p decreases (for $0.92 \leq p \leq 1$), the fraction of nonconforming is sufficiently small so that testing to weed out nonconforming items is not justified and, as a result, $T^* = 0$. As p decreases still further (for $0.48 \leq p < 0.92$), the fraction of nonconforming items increases and testing is worthwhile. Note that testing for greater time reduces the warranty cost per item but in-

Table 23.2 Expected Number of Nondefective and Defective Items per Batch

	Expected number of items per batch of 100 items					
	Weeded out during testing			Released for use		
W	Nondefect.	Defect.	Total	Nondefect.	Defect.	Total
1	0.00	0.00	0.00	75.000	25.000	100.00
2	3.317	14.872	18.185	71.687	10.128	81.815
3	6.085	21.397	26.282	68.915	4.603	73.718
4	7.601	22.050	29.651	67.399	2.950	70.349

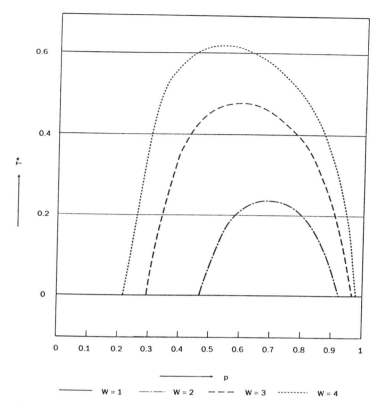

Figure 23.2 T^* versus p (FRW policy with minimal repair).

creases the manufacturing cost per item as a greater fraction of items fail
during testing and are discarded. The reason that T^* increases as p decreases in the interval $0.69 \leq p < 0.92$ is that the increase in the manufacturing cost per item is less than the decrease in the warranty cost as p
decreases. In the interval $0.48 \leq p < 0.69$, T^* decreases with p decreasing
as the increase in the manufacturing cost per item is more than the reduction in the warranty cost per item. Finally, for $0 < p \leq 0.48$, the fraction
of nonconforming items is very high. The cost of any testing is not worth
the reduction in the warranty cost and, hence, $T^* = 0$. As W increases,
the interval over which $T^* > 0$ also increases and T^* has a shape similar
to that for $W = 2$ except that the values for any given p are increasing
with W. This is to be expected for reasons discussed earlier.

Pro-Rata Warranty (Linear Proration)

Under this policy, the manufacturer agrees to refund a fraction of the original sales price S to the consumer if the item sold fails within the warranty period $[0, W]$. (See Chapter 11.) With linear proration, the asymptotic warranty cost per item with no testing is given by

$$J(0) = p \left[kS \int_0^W \left(1 - \frac{\beta t}{W} \right) f_1(t) \, dt + C_h F_1(W) \right]$$

$$+ \bar{p} \left[kS \int_0^W \left(1 - \frac{\beta t}{W} \right) f_2(t) \, dt + C_h F_2(W) \right] \qquad (23.25)$$

and with testing, it is given by

$$J(T) = \hat{p} \left[kS \int_0^W \left(1 - \frac{\beta t}{W} \right) \hat{f}_1(t) \, dt + C_h \hat{F}_1(W) \right]$$

$$+ (1 - \hat{p}) \left[kS \int_0^W \left(1 - \frac{\beta t}{W} \right) \hat{f}_2(t) \, dt + C_h \hat{F}_2(W) \right] \qquad (23.26)$$

where $\hat{F}_i(t)$, $1 \le i \le 2$, and $\hat{p}$ are given by (23.18) and (23.19), respectively.

For the special case where $F_1(t)$ and $F_2(t)$ are exponential distributions with parameters λ_1 and λ_2, respectively, with $\lambda_2 > \lambda_1 > 0$, sufficiency conditions for optimality, similar to Propositions 23.3 and 23.4, are given in Ref. 14.

Example 23.3

As in Example 23.2, let F_1 and F_2 be exponential distributions with parameters λ_1 and λ_2. Let $\lambda_1 = 0.2$ per year, $\lambda_2 = 4.0$ per year, $C_h = \$20$, $k = 1.0$, $\beta = 1.0$, and $S = \$1000$. The remaining parameter values are the same as in Example 23.2.

Table 23.3 shows T^* and $J(T^*)$ for the four different warranty periods. For $W = 1$ to 4, $T^* > 0$, implying that testing is the optimal strategy in all four cases. Figure 23.3 shows T^* as a function of W. For $W < 0.5$,

Table 23.3 Optimal Strategy, T^* and $J(T^*)$ Under PRW

W	Optimal strategy	T^*	$J(T^*)$
1	Testing	0.1539	560.03
2	Testing	0.1871	650.45
3	Testing	0.1565	720.22
4	Testing	0.1090	778.35

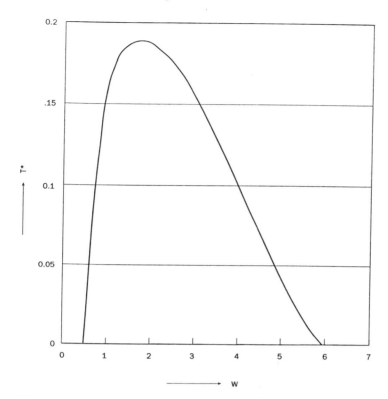

Figure 23.3 T^* versus W (PRW policy).

$T^* = 0$. This is to be expected because small W implies a smaller warranty service cost even with all nonconforming items released and the cost of testing not justified. For $0.5 < W < 6.0$, $T^* > 0$. In contrast to the FRW policy with minimal repair (Example 23.2), here T^* increases with W for $0.5 < W < 1.7$ and decreases for $1.7 < W < 6$. For $0.5 < W \le 1.70$, the testing period increases with W because the warranty cost for released nonconforming items increasingly dominates the cost per item released. Thus, by reducing the number of released nonconforming items through increased testing, the warranty cost can be decreased by more than the increase in the manufacturing cost per item released. For $1.70 < W < 6.0$, the influence of the warranty cost for released nonconforming items decreases while the influence of the other costs increases, resulting in the testing period decreasing as W increases. This continues until, for $W > 6.0$, testing is again not justified, so that $T^* = 0$. Table 23.4 shows the

Table 23.4 Expected Number of Nondefective and Defective Items per Batch

| | Expected number of items per batch of 100 items | | | | | |
| | Weeded out during testing | | | Released for use | | |
W	Nondefect.	Defect.	Total	Nondefect.	Defect.	Total
1	2.273	11.490	13.763	72.727	13.510	86.237
2	2.755	13.172	15.927	72.245	11.828	84.073
3	2.311	11.633	13.944	72.689	13.367	86.056
4	1.617	8.834	10.451	73.383	16.166	89.549

expected number of nonconforming items released and weeded out in each batch.

The effect of p on T^*, with the remaining parameters held at their nominal values, is similar to Figure 23.2 for reasons discussed earlier.

Finally, the case in which the items are sold with FRW policy and the items are nonrepairable has been studied in Refs. 8 and 14. The results are similar to the two policies discussed in this subsection.

23.4.2 Models Based on Improvement

Djamaludin et al. [15] deals with lot sizing as a decision variable to reduce the number of nonconforming items produced. The process deterioration is as in Ref. 7 and the model is as follows.

Model 8 [15]

Before a lot production starts, the process is checked to make sure that it is in-control. This costs C_f if the process is in-control and $C_f + \eta$ ($\eta \geq 0$) if it is out-of-control. η is the cost of bringing an out-of-control process in-control. This setup cost formulation is slightly different from that in Section 23.3.2.

The asymptotic manufacturing (fixed + variable) cost per item is given by

$$\gamma(L) = C_m + \frac{C_f + (1 - q^L)\eta}{L} \tag{23.27}$$

where C_m is the variable cost (material and labor) to produce one unit and q is the probability that the process remains in control.

The expected number of conforming items in a batch of size L is given by (23.9). As a result, the asymptotic probability that an item is conforming is given by

$$p = \frac{E[\tilde{N}]}{L} = \frac{q(1 - q^L)/(1 - q)}{L} \tag{23.28}$$

The asymptotic warranty cost per item, $C_w(L)$, depends on the type of warranty, the warranty duration (W), failure distribution of conforming items (F_1) and nonconforming items (F_2), and the probability that an item is conforming (p). The asymptotic total (manufacturing + warranty) cost per item is given by

$$J(L) = \gamma(L) + C_w(L) \tag{23.29}$$

L^*, the optimal L, is the value of L which minimizes $J(L)$.

FRW Policy (Minimal Repair)

The items are sold with FRW warranty. Whenever an item fails under warranty, it is minimally repaired. In this case, the asymptotic warranty cost per item is given by

$$C_w(L) = C_r \left[p \int_0^W r_1(x)\, dx + (1 - p) \int_0^W r_2(x)\, dx \right] \tag{23.30}$$

where p is given by (23.30) and C_r is the cost of each repair.

L^* is obtained by minimizing $J(L)$ given by (23.29) with $\gamma(L)$ given by (23.27) and $C_w(L)$ given by (23.28). In general, one would need to use a computational scheme to obtain L^*. However, if q is very close to 1, then q^L can be approximated by (23.11), and in this case, L^{**} [an approximation to L^* obtained by solving $dJ(L)/dL = 0$] is given by

$$L^{**} = q \left(\frac{2C_f}{(1 - q)\left(C_r \left[q \int_0^W r_2(x)\, dx - \int_0^W r_1(x)\, dx \right] - \eta \right)} \right)^{1/2} \tag{23.31}$$

Example 23.4

Suppose the failure distribution of both conforming and nonconforming items are exponential distributions with parameter λ_1 and λ_2, respectively. Let $\lambda_1 = 0.1$ and $\lambda_2 = 1.0$. This implies that the mean time to failure is 10 years for a conforming item and 1 year for a nonconforming item.

Let L_m denote the upper limit on L. Take $L_m = 100$. Let the nominal values for the remaining parameters be $q = 0.99$, $C_m = \$5.0$, $C_f = \$50.0$, $\eta = \$10.0$, $C_r = \$5.0$, $\theta_1 = 0.95$, and $\theta_2 = 0.04$. Four different values for the warranty period W, ranging from 1 to 4 years, are considered. Also included is the case $W = 0$, which corresponds to the product being sold with no warranty.

Table 23.5 shows L^* and $J(L^*)$ [obtained by evaluating $J(L)$ for $L = 1, \ldots, L_m$], L^{**} [obtained from (23.31)], $J(L^{**})$, and $J(L_m)$, the cost per unit if the lots are of size L_m ($= 100$). The percentage reduction in cost, RC, given by

$$\text{RC} = 100 \left(\frac{J(L_m) - J(L^*)}{J(L_m)} \right)$$

is also shown in Table 23.5.

For $W = 0$, $L^* = L_m$. As W increases, L^* decreases, as a longer warranty period implies increased warranty costs for nonconforming items released. Hence, smaller lot sizes are required to ensure that the expected fraction of nonconforming items released is smaller. The percentage reduction, RC, obtained by using L^*, increases with W, indicating that lot sizing becomes more critical as the warranty period increases. Note that the error between L^* and L^{**} decreases as W increases and that L^{**} is always less than L^*, indicating that the true optimal lot size is somewhat larger than that given by the approximation. Also note, however, that the increase in cost using the suboptimal result is less than 1%.

The effect of q on L^*, holding the remaining parameters at their nominal values, is shown in Figure 23.4. Note that there is a critical value

Table 23.5 Exact and Approximate Optimal Lot Sizes and Related Costs (FRW Policy)

	W				
	0	1	2	3	4
L^*	100	58	38	30	26
$J(L^*)$	5.5634	7.8439	9.5526	11.0515	12.4429
L^{**}	100	47	33	27	23
$J(L^{**})$	5.5634	7.8739	9.5758	11.0701	12.4653
$J(L_m)$	5.5634	8.0383	10.5131	12.9879	15.4628
RC (%)	0.0	2.42	9.14	14.91	19.53

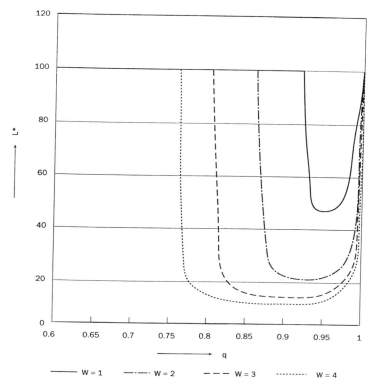

Figure 23.4 L^* versus q (FRW policy with minimal repair).

$\hat{q}$ (dependent on W) such that, for $q < \hat{q}$, $L^* = L_m$, and for $\hat{q} \le q < 1$, $L^* < L_m$. First, if $q = 1$, then the process is always in-control. Reducing the lot size increases the manufacturing cost per item with no effect on the warranty cost per item and so $L^* = L_m$. Now, as q decreases with L kept at L_m, the expected fraction of nonconforming items in a lot increases, resulting in a higher warranty cost per item. By decreasing L as well, the warranty cost can be reduced, but the manufacturing cost per item will then increase. Decreasing L is justified if the reduction in warranty cost compensates for the increase in manufacturing cost. This occurs, so that L^* decreases as q decreases until some value (depending on W) is reached. At this point, the increase in manufacturing cost is greater than the decrease in the warranty cost as q decreases, resulting in higher values for L^*. Consequently, L^* increases to the maximum L_m as q decreases from this point and then, for $q < \hat{q}$, L^* is constrained to be the maximum L_m. Note that $\hat{q}$ decreases as W increases; this is to be expected, as the warranty costs increase as W increases.

PRW Policy (Linear Rebate)

Here, whenever an item fails under warranty, a fraction of the sale price S is refunded. Under linear rebate, the asymptotic warranty cost per unit is given by

$$C_w(L) = p \int_0^W \left[S \left(1 - \frac{x}{W} \right) + C_h \right] dF_1(x) + (1 - p)$$

$$\int_0^W \left[S \left(1 - \frac{x}{W} \right) + C_h \right] dF_2(x) \tag{23.32}$$

where C_h is the handling cost for each warranty claim.

L^* is obtained by minimizing $J(L)$ given by (23.29) with $\gamma(L)$ given by (23.27) and $C_w(L)$ given by (23.32). If q is very close to 1, then using the approximation given in (23.11), L^{**}, an approximation to the optimal lot size, is given by

$$L^{**} = q \left(\frac{2C_f}{(1 - q)[q(A_2 - A_1) - \eta(1 - q)]} \right)^{1/2} \tag{23.33}$$

where

$$A_1 = (S + C_h)F_1(W) - S \int_0^W \left(\frac{\beta t}{W} \right) f_1(t) \, dt \tag{23.34}$$

and

$$A_2 = (S + C_h)F_2(W) - S \int_0^W \left(\frac{\beta t}{W} \right) f_2(t) \, dt \tag{23.35}$$

Example 23.5

Suppose the failure distributions of both conforming and noncon- forming items are exponential distributions with parameter λ_1 and λ_2, respectively. Let the parameter values be the same as in Example 23.4 and let $S = \$30$.

Table 23.6 shows L^*, $J(L^*)$ [obtained by evaluating $J(L)$ for $L = 1, \ldots, L_m$], L^{**} [obtained by using (23.33)], $J(L^{**})$, $J(L_m)$ [the cost per unit if the lots are of size L_m (= 100)], and RC, the percentage reduction in cost, defined in Example 23.4. The results are similar to those for the

Table 23.6 Exact and Approximate Optimal Lot Sizes and Related Costs (PRW Policy)

			W		
	0	1	2	3	4
L^*	100	38	31	28	27
$J(L^*)$	5.5634	9.9418	12.2504	13.9478	15.3331
L^{**}	100	33	27	25	24
$J(L^{**})$	5.5634	9.9658	12.2722	13.9714	15.3620
$J(L_m)$	5.5634	10.8939	14.1262	16.28927	17.8809
RC (%)	0.0	8.74	13.28	14.39	14.25

FRW policy. Note that for the nominal values used, for $W = 1, 2$, and 3, the optimal lot sizes are smaller than those for the FRW policy and larger for $W = 4$. Also, the percentage reduction, RC, is bigger than that for the FRW policy for $W = 1, 2$, and 3 and smaller for $W = 4$. Again, the approximate optimal lot sizes are close to the optimal lot sizes.

The influence of q on L^* is similar to that for FRW policy.

The case in which items are sold with FRW policy and the items are nonrepairable is studied in Ref. 8 and the results are similar to the two policies considered in this subsection.

23.4.3 Models Based on Improvement and Testing

In these models both lot sizing and testing are used to improve output quality and, hence, reduce the warranty cost. Note that the process is in-control at the start of each batch production and can change to out-of-control before the batch is processed. Also, the probability that an item is nonconforming is relatively high when the process is out-of-control. Hence, testing to weed out nonconforming items is appropriate only for items produced with the process out-of-control. This subsection outlines two models. The details of the model and numerical examples can be found in Refs. 8 and 16.

Model 9 [16]

In this model, the process state is checked at the end of each lot production run. If the state has changed from in-control to out-of-control, the last K

items are life-tested for a period T and items which fail are discarded (or reworked) and the rest released with no testing. If there is no change in the state, then the batch is released with no testing. Expressions for $J(L, K, T)$ are derived in Ref. 16 (and also in Ref. 8) along with the optimal L, K, and T which minimize $J(L, K, T)$.

Model 10 [8]

In this model, it is assumed that the state of the process cannot be observed. Hence, the scheme to weed out nonconforming items (due to possible change in state) is as follows. The L items are sequentially numbered with the last item being item L. Item L is life-tested for a period T. If it does not fail, then the batch is released with no further testing. However, if it fails, then item $L - K$ is life-tested for a period T. If it fails, then items $L - K + 1$ to $L - 1$ are life-tested. Those which fail are either scrapped (or reworked). If item $L - K$ does not fail, then the batch is released with no further life-testing. Expressions for $J(L, K, T)$ are derived in Ref. 8 along with the optimal L, K, and T which minimize $J(L, K, T)$.

23.5   PRESALE TESTING

When the failure rate of a product has a bathtub shape, the number of failures in the early period of usage (the infant mortality region) can be high. As a result, the fraction of items returned under warranty in a short period subsequent to the sale can be high. This not only implies high warranty cost but results in a low product reputation. Burn-in is a testing program (similar to that used in Section 23.4) which aims to eliminate the frequently occurring early failures before an item is sold. The rationale for the burn-in is as follows. For nonrepairable products, items with very short life should not be sold. For repairable products, failures occurring during the initial phase, corresponding to the infant mortality region, are rectified cheaply by testing before sale, as opposed to fixing under warranty, which entails not only additional costs but affects the goodwill of the consumer.

Although the testing scheme is similar to that discussed in Section 23.4, the motivation is different. There, the testing was done to weed out nonconforming items. If the proportion of nonconforming items is very low, then there is no need for testing. Here, in contrast, all items are similar and the testing is to eliminate early failures before an item is sold. For more on burn-in, see Ref. 17.

With burn-in, the expected warranty cost is reduced. However, this is at the expense of the costs involved in the burn-in. Nguyen and Murthy [18] proposed a model to obtain the optimal burn-in period which achieves a trade-off between these two costs. The model is as follows:

Model 11 [18]

Let $F(t)$ denote the failure distribution for the item. It is assumed that the failure rate, $r(t)$, has a bathtub shape with $r(t)$ strictly decreasing over $0 \leq t \leq \tau_m$ (with $\tau_m > 0$). If an item is subjected to a burn-in for a period τ, the failure distribution of the item after burn-in, $F_\tau(t)$, is given by

$$F_\tau(t) = \frac{F(\tau + t) - F(\tau)}{\overline{F}(\tau)} \qquad (23.36)$$

Let $f_\tau(t)$ and $r_\tau(t)$ denote the failure density function and the failure rate associated with the distribution function $F_\tau(t)$.

For a repairable product, with failures during burn-in fixed minimally, the expected total cost (manufacturing + testing + rework + warranty) per item is given by

$$J(\tau) = (C_m + c_1 + c_2\tau) + C_r \int_0^\tau r(t) \, dt + C_w(\tau) \qquad (23.37)$$

where C_m is the manufacturing cost per item, C_r is the expected cost of each repair, c_1 is the fixed setup cost of burn-in per unit, c_2 is the cost per unit time of burn-in per unit, and $C_w(\tau)$ is the expected warranty cost per unit. $C_w(\tau)$ depends on the type of warranty, the warranty period (W), and the duration of testing (τ).

For a repairable product, the burn-in cost per unit is

1. $c_1 + c_2t, 0 \leq t < \tau$, if the item fails during burn-in
2. $c_1 + c_2\tau$ if the item survives the burn-in

As a result, the expected total cost (manufacturing + testing + rework + warranty) per item is given by

$$J(\tau) = \frac{C_m + c_1 + c_2 \int_0^\tau \overline{F}(t) \, dt}{\overline{F}(\tau)} + C_w(\tau) \qquad (23.38)$$

τ^*, the optimal τ, minimizes $J(\tau)$.

The expressions for $J(\tau)$ for a variety of warranty policies are as follows.

FRW (Minimal Repair)

The expected total cost per unit is given by

$$J(\tau) = (C_m + c_1 + c_2 \tau) + C_r \int_0^\tau r(t)\, dt$$

$$+ (C_r + C_h) \int_0^W r_\tau(t)\, dt \tag{23.39}$$

where C_h is the handling cost for each warranty claim.

FRW (Replace by New)

The expected total cost per unit is given by

$$J(\tau) = \left[(C_m + c_1 + c_2 \tau) + C_r \int_0^\tau r(t)\, dt + C_h \right] M_\tau(W) \tag{23.40}$$

where $M_\tau(W)$ is the renewal function associated with the distribution function $F_\tau(t)$. (See Chapter 6.)

PRW (Linear and Lump-Sum Rebates)

In this case, the expected total cost per unit is given by (23.38) with

$$C_w(\tau) = \int_0^W R(t) f_\tau(t)\, dt \tag{23.41}$$

where S is the sale price per unit and $R(t) = kS(1 - \alpha t)/W$ with $\alpha = 1$ and $k = 1$ for linear rebate and $\alpha = 0$ for lump-sum rebate.

Example 23.6 [18]

Let $F(t)$, a mixture of two Weibull distributions, be given by

$$F(t) = p[1 - \exp(-\lambda_1 t^{\beta_1})] + (1 - p)[1 - \exp(-\lambda_2 t^{\beta_2})]$$

with $\lambda_1, \lambda_2 > 0$; $0 < \beta_1 < 1$, $\beta_2 > 1$, and $0 \le p \le 1$. When $p = 0$, $F(t)$ has an increasing failure rate, and when $p = 1$, it has a decreasing failure rate. When $0 < p < 1$, it has a bathtub failure rate.

Let $p = 0.1$, $\lambda_1 = 4$ per year, $\lambda_2 = 0.08$ per year, $\beta_1 = 0.5$, and $\beta_2 = 3$, with cost parameters $C_m = \$5$, $C_r = \$2$, $c_1 = 0.2$, $c_2 = 5$, $C_h = \$10$, and $S = \$20$. Figure 23.5 shows τ^* as a function of W for the four warranty policies mentioned above. The interesting thing to note is that τ^* is zero when $W = 0$. As W increases, τ^* increases to a critical value W_c (different for each policy), beyond which it starts decreasing and becomes zero for large W.

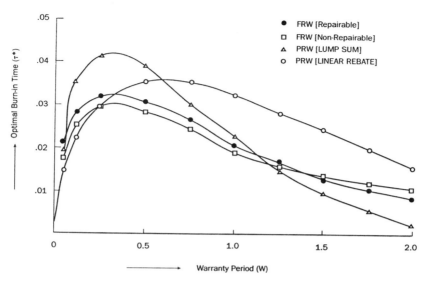

Figure 23.5 τ^* versus W for four warranty policies.

To study the variations in the magnitude of savings in the expected total cost, define η as

$$\eta = \frac{J(0) - J(\tau^*)}{J(0)}$$

where $J(0)$ is the expected total cost per unit if no burn-in is employed. Thus, η is the relative savings using a burn-in as opposed to no burn-in. Figure 23.6 shows a plot of η as a function of W for the four different cases. Note that the savings are positive for $W_1 < W < W_2$ (and these vary with each policy) and negative for values outside this. The reason for this is the fixed burn-in cost c_1; if c_1 is zero, then η is always positive and approaches 0 as W approaches 0 or ∞.

Comment: Often it is not possible (or economical) for the manufacturer to carry out life-testing to weed out nonconforming items. In this case, items are released without any testing. The consequence of this is that nonconforming items, which tend to fail early, can lead to loss of goodwill and subsequent sales. One way of countering this is to compensate consumers who experience such failures and offer some incentives to retain them. In a sense, this approach can be viewed as consumers

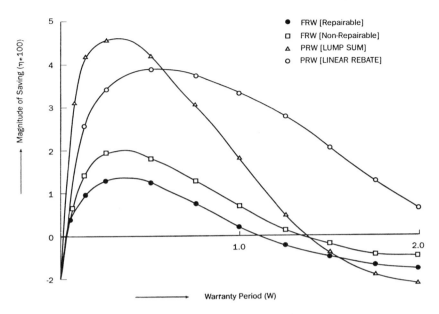

Figure 23.6 η^* versus W for four warranty policies.

carrying out the testing and being compensated for being supplied with nonconforming items. A model which deals with this is given in Ref. 19.

23.6   IMPLEMENTATION ASPECTS

The implementation of the results of this chapter involves the following steps.

Step 1: Decide on the model for the production of nonconforming items. Select life distribution for conforming and nonconforming items. (See Chapters 6–9.) Estimate the model parameters using statistical methods of Chapter 8.

For example, if the process has been stabilized, then the probability that an item is conforming (nonconforming) is p $(1 - p)$. If items are picked randomly and life-tested to failure, then the time to failure is distributed according to

$$F(t) = pF_1(t; \theta_1) + (1 - p)F_2(t; \theta_2)$$

where F_1 (F_2) is the distribution of conforming (nonconforming) items. The parameters to be estimated are p, θ_1, and θ_2.

Step 2: Compute the expected warranty cost per item sold. If this is acceptable, go to Step 12. If not, can life-testing be used to improve output quality? If yes, go to Step 3. If not, go to Step 7.

Step 3: Choose a testing scheme to weed out nonconforming items.

Step 4: Compute the expected net gain, that is, the savings in the expected warranty cost minus the testing cost per item sold.

Step 5: Can the net gain be increased by modifying the testing scheme? If yes, modify the testing scheme and go to Step 4.

Step 6: Is the reduction in warranty cost acceptable? If yes, go to Step 11.

Step 7: Can lot size be modified? If yes, go to Step 8. If not, go to Step 11.

Step 8: Determine the optimal lot size.

Step 9: Compute the expected net gain, that is, the savings in the expected warranty cost minus the testing cost per item sold.

Step 10: Compare the expected net savings of Step 8 with that of Step 4 to decide on the optimal strategy for quality control.

Step 11: Implement the optimal quality control strategy.

Step 12: Stop.

REFERENCES

1. Murthy, D. N. P., and Djamaludin, I. (1991). Quality control in a single stage production system: Open and closed loop policies, *International Journal of Product Research*, **28**, 2219–2242.

2. Mann, N. R., and Saunders, S. C. (1969). On evaluation of warranty assurance when life has a Weibull distribution, *Biometrika*, **56**, 615–625.

3. Mann, N. R. (1970). Warranty periods based on three ordered sample observations from a Weibull populations, *IEEE Transactions on Reliability*, **R-19**, 167–171.

4. Mamer, J. W. (1987). Discounted and per unit costs of product warranty, *Management Sciences*, **33**, 916–930.

5. Duncan, A. J. (1975). *Quality Control and Industrial Statistics*, 4th ed., Irwin, Homewood, IL.

6. Montgomery, D. C. (1985). *Statistical Quality Control*, John Wiley & Sons, New York.

7. Porteus, E. L. (1986). Optimal lot sizing, process quality improvement and setup cost reduction, *Operations Research*, **34**, 137–144.

8. Djamaludin, I. (1993). *Warranty and Quality Control*, Ph.D. thesis, The University of Queensland, Brisbane, Australia.

9. Rosenblatt, M. J., and Lee, H. (1986). Economic production cycle with imperfect production processes, *IIE Transactions*, **18**, 48–55.

10. Tapiero, C. S., and Lee, H. L. (1989). Quality control and product servicing: A decision framework, *European Journal of Operations Research*, **39**, 61–73.

11. Marucheck, A. S. (1987). On product liability and quality control, *IIE Transactions*, **19**, 355–360.
12. Ritchken, P. H., Chandramohan, J., and Tapiero, C. S. (1989). Servicing, quality design and control, *IIE Transactions*, **21**, 213–220.
13. Lee, C. H., and Chun, Y. H. (1987). Optimum single-sample inspection plans for products sold under free and rebate warranty, *IEEE Transactions on Reliability*, **R-36**, 634–637.
14. Murthy, D. N. P., Djamaludin, I., and Wilson, R. J. (1993). Product warranty and quality control, *Quality and Reliability Engineering*, **9**, 431–443.
15. Djamaludin, I, Wilson, R. J., and Murthy, D. N. P. (1994). Quality control through lot sizing for items sold with warranty, *International Journal of Production Research*, **33**, 97–107.
16. Djamaludin, I, Murthy, D. N. P., and Wilson, R. J. (1994). Lot sizing and inspection for items sold with warranty, Journal of Mathematical and Computer Modelling, in press.
17. Leemis, L. M., and Benke, M. (1990). Burn-in models and methods: A review, *IIE Transactions*, **22**, 172–180.
18. Nguyen, D. G., and Murthy, D. N. P. (1982). Optimal burn-in time to minimize cost for products sold under warranty, *IIE Transactions*, **14**, 167–174.
19. Murthy, D. N. P., Djamaludin, I., and Wilson, R. J. (1994). A warranty policy for products with uncertain quality, Quality and Reliability Engineering, in press.

24

Warranty Servicing

D. N. Prabhakar Murthy

*The University of Queensland
Brisbane, Queensland, Australia*

24.1 INTRODUCTION

For products sold with warranty, the manufacturer has to service all claims made under warranty. The actions on the part of the manufacturer depend on the type and the terms of the warranty. Warranty servicing deals with the study of such actions and the related planning issues.

Warranty servicing provided by the manufacturer is viewed by consumers as service support for the product. Over the last few years, effective service support for products has assumed an important role in the strategic marketing of products (see, e.g., Ref. 1). Planning to ensure effective product support service is a topic which is receiving considerable attention in the management literature. Bleuel and Bender [2] discuss some of these issues.

Under a nonrenewing PRW (pro-rata warranty) policy, the manufacturer has to refund a fraction of the sale price. In order to carry this out, the manufacturer has to set aside a fraction of the sale price. This is called warranty reserving. For nonrepairable items sold with FRW (free-replacement warranty) or renewing PRW policies, the manufacturer has to supply a replacement item for failures under warranty. In this case, the number of spares needed is of interest and this is of importance in the context of production and inventory control. For repairable products sold with a FRW policy, the planning of a repair facility requires evaluation of demand for repairs over the warranty period. This depends on the type of repair action.

For single-item sale, the results of Chapters 10–14 yield the answers. In this chapter, we discuss warranty servicing with multiple sales, either as a single lot or distributed over time and each item covered with a separate warranty.

In the case of repairable products, Chapters 10 and 13 dealt with the case in which the failed items are always repaired. Often this is not the best strategy. Some simple strategies involving choice between repair and replacement to minimize the expected warranty servicing cost per item are discussed in this chapter.

The outline of the chapter is as follows. Sections 24.2 and 24.3 deal with warranty servicing for nonrepairable items. The former deals with the warranty reserving issue and the latter deals with spares needed for replacements under warranty. Sections 24.4–24.6 deal with warranty servicing for repairable items. Section 24.4 deals with the case in which all failed items are repaired and the next two sections deal with repair replacement strategies. Finally, Section 24.7 discusses a variety of issues of relevance to warranty servicing and Section 24.8 deals with implementation aspects.

24.2   WARRANTY RESERVES (NONRENEWING PRW POLICIES)

This section deals with nonrepairable items sold with a nonrenewable pro-rata warranty policy under which the manufacturer refunds a fraction of the purchase price should the item fail within the warranty period. The single-lot sale is considered first and this is followed by sales occurring continuously over the product life cycle.

In a single-lot sale, the manufacturer sells a lot of N items which are put into use immediately. Examples of this range from fleet purchases of automobiles, buses, and so forth, to the sale of bulbs to replace all the lights in a factory at regular intervals. In the case of a continuous sale, items are purchased individually and are put into use immediately after purchase. Note that in both cases, each item is covered by an individual warranty. As a result, the analysis in this section makes extensive use of the results of Chapters 10 and 11. The failure distribution of items is given by $F(x)$.

24.2.1 Single-Lot Sale

This topic was first studied by Menke [3] and he dealt with a special case in which the failure distribution is exponential. Amato and Anderson [4] extended the model of Menke to include the effect of discounting. Thomas [5] deals with the general case in which the failure distribution is arbitrary.

Let Y_i denote the refund for the ith item in the lot. With a linear refund, Y_i is given by

$$Y_i = \begin{cases} c_b \left(1 - \dfrac{X_i}{W}\right) & \text{if } X_i < W \\ 0 & \text{if } X_i \geq W \end{cases} \tag{24.1}$$

where X_i is the age of item i at failure and c_b is the sale price. The mean, $\mu(W)$, and the variance, $\sigma^2(W)$, of Y_i are given by

$$\mu(W) = c_b \int_0^W \left(1 - \frac{x}{W}\right) f(x)\, dx = c_b \left(F(W) - \frac{\mu_W}{W}\right) \tag{24.2}$$

and

$$\sigma^2(W) = c_b^2 \int_0^W \left(1 - \frac{x}{W}\right)^2 f(x)\, dx - [\mu(W)]^2 \tag{24.3}$$

with μ_W given by

$$\mu_W = \int_0^W x f(x)\, dx$$

Because items fail independently, the total warranty refund, TR(W), is a random variable given by

$$\text{TR}(W) = \sum_{i=1}^{N} Y_i \tag{24.4}$$

To meet these payouts, the manufacturer must set aside a sum $R(W)$, as warranty reserves, from the revenue generated from the sale of the lot. If $R(W)$ is selected to equal the expected refunds, then

$$R(W) = E[\text{TR}(W)] = N\mu(W) \tag{24.5}$$

This implies that the manufacturer must set aside a fraction γ of the sale price of each item to create the warranty reserve; γ is given by

$$\gamma = \frac{R(W)}{Nc_b} = \frac{\mu(W)}{c_b} \qquad (24.6)$$

Note that the reserves might not cover the total payouts if TR(W) exceeds the reserves $R(W)$. In order to evaluate the probability of this happening, one needs the distribution function for TR(W). This can be obtained in terms of the convolutions of the distribution functions of Y_i, which, in turn, is related to the distribution functions of the X_i. Although, in principle, this is straightforward, it is analytically intractable for even the simplest form of $F(x)$. However, if N is large and under the assumption that items fail independently, then by the Central Limit Theorem (see Ref. 6), TR(W) is approximately normally distributed with mean and variance given by $N\mu(W)$ and $N\sigma^2(W)$, respectively. In this case, selecting reserves $R(W) = N\mu(W)$ implies that the probability of the total payouts exceeding the reserves is approximately 0.5. By increasing $R(W)$, this probability can be reduced. The argument is as follows.

The normal approximation can be used to obtain a $100(1 - \alpha)\%$ prediction interval for TR(W). This is given by

$$N\mu(W) \pm z_{\alpha/2} \sqrt{N}\, \sigma(W) \qquad (24.7)$$

where $z_{\alpha/2}$ is the $\alpha/2$ fractile of the standard normal distribution. (See Ref. 6). It follows that if the warranty reserves $R(W)$ is chosen according to the relation

$$R(W) = N\mu(W) + \epsilon\sqrt{N}\, \sigma(W) \qquad (24.8)$$

then by proper choice of ϵ (> 0), the probability of the total payouts exceeding the reserves can be made as small as desired. However, this implies setting aside a larger fraction of the revenue generated by sales to service the warranty.

Example 24.1 (Exponential Distribution)
Let the failure distribution be exponential with parameter λ, that is,

$$F(x) = 1 - \exp(-\lambda x)$$

The unit for time is years, so that λ has dimension (year)$^{-1}$. From (24.2) and (24.3), we have

$$\mu(W) = c_b \left[1 - \left(\frac{1 - e^{-\lambda W}}{\lambda W} \right) \right]$$

and

$$\sigma(W) = c_b \left(\frac{1 - e^{-\lambda W}}{\lambda W} \right)$$

Note that in both $\mu(W)$ and $\sigma(W)$, λ and W appear as a product. Hence, the warranty reserve is a function of this product. Let $N = 100$ and $c_b = \$10$. Values of $R(W)$ (in \$) obtained from (24.8) with $\epsilon = 0$ for two sets of W (in years) and a range of λ values are shown in Table 24.1. Note that the mean time to failure is 2 years for $\lambda = 0.5$ and 10 years for $\lambda = 0.1$. The ratio of the expected warranty service costs between a high-reliability product sold with shorter warranty ($\lambda = 0.1$ and $W = 1$) and a low-reliability product sold with longer warranty ($\lambda = 0.5$ and $W = 2$) is roughly 8.

Table 24.2 shows $R(W)$ and $P(\epsilon)$, the probability of the payout $TR(W)$ exceeding the warranty reserve $R(W)$, as a function of ϵ for the case $\lambda = 0.5$ and $W = 1$ year. Note that as ϵ increases $P(\epsilon)$, the probability that the warranty reserve is not adequate to meet the payout decreases as to be expected.

Formulation Involving Investment of Warranty Reserves

In the previous subsection, the warranty reserves $R(W)$ are set aside at the time of the sale although payouts occur over a period W subsequent to the sale. The reserves can be invested (e.g., in short-term money market

Table 24.1   Warranty Reserve $R(W)$ (in \$) (Lot Size $= 100$; Exponential Failure Distribution)

	W (years)	
λ (year^{-1})	1	2
0.10	48	213
0.50	213	367
1.00	367	567
2.00	567	754

Table 24.2   Warranty Reserve $R(W)$ and Probability of Payouts Exceeding Warranty Reserves $P(\epsilon)$ versus ϵ

	ϵ							
	0.0	0.1	0.5	1.0	1.5	2.0	2.5	3.0
$R(W)$	213	221	252	292	331	370	409	449
$P(\epsilon)$	0.500	0.460	0.301	0.159	0.067	0.023	0.005	0.001

or as interest-bearing deposits) to generate income. This implies that the total amount of funds to be set aside as reserves is smaller than it would be if the reserves were not invested. In order to compute the reserves needed, a model for returns on investments is required. Two model formulations for this purpose are as follows.

Let $I(t)$ denote the value at time t of an investment of value I_0 at time $t = 0$. For Model I, $I(t)$ is given by

$$I(t) = I_0 \exp(\phi t) \tag{24.9}$$

and for Model II, it is given by

$$I(t) = I_0(1 + \phi)^t \tag{24.10}$$

where ϕ (> 0) is the rate of return on investment. Let $R_I(W; \phi)$, the reserves set aside (and invested), be selected to equal the expected total payout over the warranty period. For Model I, the payout $(1 - x/W)$ for an item failing at time x ($0 < x < W$) can be met by an amount $(1 - x/W) \exp(-\phi x)$ of the initial reserves invested for a period x. As a result, it is easily seen that

$$R_I(W; \phi) = N c_b \int_0^W \left(1 - \frac{x}{W}\right) \exp(-\phi x) f(x) \, dx \tag{24.11}$$

for Model I and

$$R_I(W; \phi) = N c_b \int_0^W \left(1 - \frac{x}{W}\right) (1 + \phi)^{-x} f(x) \, dx \tag{24.12}$$

for Model II. Note that $R_I(W; \phi)$ is a function of ϕ and decreases as ϕ increases for both models. In this case, the fraction of the selling price per item that must be set aside to service the warranty is given by

$$\gamma(\phi) = \frac{R_I(W; \phi)}{N c_b} \tag{24.13}$$

with $R_I(W; \phi)$ as given by (24.11) or (24.12).

Example 24.2 (Exponential Distribution)
    Suppose that the failure distribution is exponential with parameter λ. Then from (24.11), $R_I(W; \phi)$ for Model I is given by

$$R_I(W; \phi) = \left(\frac{\lambda}{\lambda + \phi}\right) R(W)$$

where $R(W)$ is the warranty reserve requirement obtained from (24.5) with an exponential failure distribution with parameter $\lambda + \phi$.

Suppose that $\lambda = 1$ and $W = 1$ year. Values of $\gamma(\phi)$ for different values of ϕ are shown in Table 24.3(a). $\gamma(\phi)$ decreases as ϕ increases as is to be expected. Table 24.3(b) shows $\gamma(\phi)$ for $\lambda = 2$ and $W = 2$ years, and the results are similar the previous case.

An Illustrative Comparison

Thomas [5] compares $\gamma(\phi)$ for four different failure distributions with the same mean. His results are for the following parameter values: $W = 1.5$ years, $\phi = 0.100$, and the mean time to failure is 0.9 year. The results are as given below.

Distribution	$\gamma(\phi)$
Exponential ($\lambda = 1.11$)	0.335
Uniform ([0,1.8])	0.398
Gamma ($\alpha = 2.8$, $\beta = 0.32$)	0.439
Weibull ($\alpha = 1$, $\beta = 1.5$)	0.428

It is seen that warranty reserves for the exponential distribution are smaller than those for either the Gamma or Weibull distributions for the parameters of the distribution indicated. This will not be true in general.

24.2.2 Continuous Sales

Let L denote the product life cycle, that is, sales occur over the interval $[0, L]$. Let $s(t)$, $0 \le t \le L$, denote the sales rate (i.e., sales per unit time) over the life cycle. This includes first and repeat purchases for the total consuming population. The total sales over the life cycle, S, is given by

$$S = \int_0^L s(t)\, dt$$

Table 24.3 Fraction of Selling Price Set Aside as Warranty Reserve $\gamma(\phi)$ Versus Rate of Return on Investment ϕ

	(a) $\lambda = 1$ and $W = 1$			
ϕ	0.00	0.075	0.100	0.125
$\gamma(\phi)$	0.367	0.3605	0.3582	0.3560
	(b) $\lambda = 2$ and $W = 2$			
ϕ	0.00	0.075	0.100	0.125
$\gamma(\phi)$	0.754	0.7359	0.7302	0.7246

It is assumed that the life cycle L exceeds W, the warranty period, and that items are put into use immediately after they are purchased. Because the manufacturer must provide a refund for items which fail before reaching age W and because the last sale occurs at or before time L, the manufacturer has an obligation to service warranty claims over the interval $[0, L + W]$. (Note that the manufacturer has to provide a refund only if the failed item is of age less than W.)

The refund rate (i.e., amount refunded per unit time) is a random variable because item failures and the resulting claims occur randomly. Let $v(t)$ denote the expected value of the refund rate at time t. It can be shown (for details, see Chapter 9 of Ref. 7) that $v(t)$, $0 \leq t \leq W + L$, is given by

$$v(t) = c_b \left[\int_\psi^t s(\tau) f(t - \tau) \left(\frac{t - \tau}{W} \right) d\tau \right] \tag{24.14}$$

where

$$\psi = \max(0, t - W). \tag{24.15}$$

Let γ denote the fraction of the sale price set aside to form the warranty reserves and selected so that the input to warranty reserves over the product life cycle equals the total expected payout. This yields

$$\gamma = \frac{\displaystyle\int_0^{L+W} v(t)\, dt}{c_b S} = \frac{S\mu(W)}{c_b S} = \frac{\mu(W)}{c_b} \tag{24.16}$$

which is identical to (24.6).

Example 24.3   (Exponential Distribution)

Let $L = 4$ years and suppose that the sales rate over the product life cycle is given by $s(t) = kt \exp(-t)$, $0 \leq t \leq 4$, with $k = 1100$. This value was chosen to yield total sales of $S = 1000$.

Let $F(x)$ be an exponential distribution with parameter $\lambda = 1$. Thus, the mean age of items at failure is 1 year. Let $W = 1$ year and $c_b = \$10$. From (24.16), we have $\gamma = 0.367$. The expected refund rate $v(t)$, $0 \leq t \leq 5$, obtained from (24.14) is given by

$$v(t) = \frac{ke^{-t} t^3}{6}$$

for $0 \leq t \leq 1$ and by

$$v(t) = \frac{ke^{-t}(3t - 2)}{6}$$

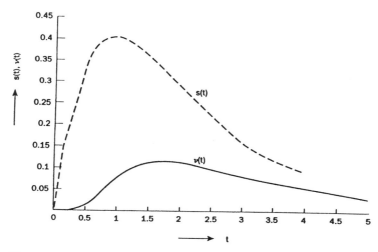

Figure 24.1 Plot of $v(t)$ and $s(t)$ versus t.

for $1 \leq t \leq 5$. A plot of $s(t)$, $0 \leq t \leq 4$, and $v(t)$, $0 \leq t \leq 5$, is shown in Figure 24.1. Note that $v(t)$ lags behind $s(t)$ as failures occur on the average 1 year after purchase, as the mean time to failure is 1 year.

24.3 DEMAND FOR SPARES (FRW POLICY)

When a nonrepairable product is sold with a nonrenewing free-replacement policy, then the manufacturer has to replace all items which fail within warranty period W. This section deals with the expected number of spares needed to service the warranty.

24.3.1 Single-Lot Sales

For each item in the lot, failures (and hence replacements) over the warranty period occur according to a renewal process because each failed item is replaced by a new one. Let $\tilde{N}(t)$ denote the (random) number of replacements needed over the interval $[0, t)$. Because item failures are statistically independent, the expected number of replacements over $[0, t)$ is given by

$$E[\tilde{N}(t)] = NM(t) \tag{24.17}$$

where $M(t)$ is the renewal function (see Chapter 7) associated with the failure distribution $F(t)$. Let $\rho(t)$ denote the expected demand rate for spares at time t, $0 \leq t \leq W$. It is given by

$$\rho(t) = \frac{dE[\tilde{N}(t)]}{dt} = Nm(t) \tag{24.18}$$

where $m(t)$ $[= dM(t)/dt]$ is the renewal density function (see Chapter 7) associated with the failure distribution $F(t)$.

The total expected number of spares needed over the warranty period is given by

$$E[\tilde{N}(W)] = NM(W) \tag{24.19}$$

The variance of the total number of spares needed to service the warranty is given by

$$\text{Var}[\tilde{N}(W)] = NV(W), \tag{24.20}$$

where $V(W)$ is given by

$$V(W) = \sum_{n=1}^{\infty} F^{(n)}(W) - M(W) - M^2(W) \tag{24.21}$$

Renewal tables (e.g., Ref. 8) give values for $M(W)$ and $V(W)$ for many different failure distributions.

From this, using the normal approximation for large N, one can obtain the prediction interval for the number of spares needed to service the warranty in a manner indicated in Section 24.2.1.

Example 24.4 (Erlang Distribution)

Let F be an Erlangian distribution with two stages, that is,

$$F(x) = 1 - (1 + \lambda x) \exp(-\lambda x)$$

with parameter λ. The lot size N is assumed to be 1000. The total expected number of spares, $E[\tilde{N}(W)]$, needed to service the warranty [obtained from (24.19) and using renewal tables] for different values of W (years) and λ are as shown in Table 24.4. Note that the expected number of spares needed increases as the warranty period and/or the failure rate increase.

Table 24.4 Expected Number of Spares Needed to Service Warranty (Lot Size = 1000; Erlangian Failure Distribution)

$\lambda \backslash W$	1.0	2.0	3.0
0.1	4.7	17.6	37.3
0.2	17.6	62.4	125.4

Example 24.5 (Weibull Distribution)

Let F be a Weibull distribution function given by

$$F(x) = 1 - \lambda \exp(-\lambda x)^\beta$$

with parameters λ and β. The lot size N is assumed to be 1000. The total expected number of spares, $E[\tilde{N}(W)]$, needed to service the warranty [obtained from (24.19) and using renewal tables] with $\beta = 2$ for different values of W (years) and λ are as shown in Table 24.5. The results are similar to that in the previous example.

24.3.2 Continuous Sales

The expected demand for spares in the interval $[t, t + \delta t)$ is due to failure of items sold in the period $[\psi, t)$ where ψ is given by (24.15). It can be shown (details of derivation can be found in Chapter 9 of Ref. 7) that the expected demand rate for spares at time t, $\rho(t)$, given by

$$\rho(t) = \int_\psi^t s(\tau)m(t - \tau)\, d\tau \tag{24.22}$$

where $m(t)$ is the renewal density function (see Chapter 7) associated with the distribution function $F(t)$. The expected total number of spares needed to service the warranty is given by

$$E[\tilde{N}(W)] = \int_0^{L+W} \rho(t)\, dt \tag{24.23}$$

Example 24.6 (Erlangian Distribution)

Let F be as in Example 24.4. Let $L = 4$ years and suppose that the sales rate over the product life cycle is given, for $0 \le t \le 4$, by

$$s(t) = kt^{\beta-1} \exp(-\lambda t^\beta)$$

Suppose that $k = 722$, $\beta = 2.0$, and $\lambda = 0.36$. These values were chosen to yield total sales of 1000. Figure 24.2 shows $s(t)$, $0 \le t \le L (= 4)$ and

Table 24.5 Expected Number of Spares Needed to Service Warranty (Lot Size = 1000; Weibull Failure Distribution; $\beta = 2.0$)

$\lambda \backslash W$	1.0	2.0	3.0
0.1	10.0	39.5	87.4
0.2	39.5	151.9	321.6

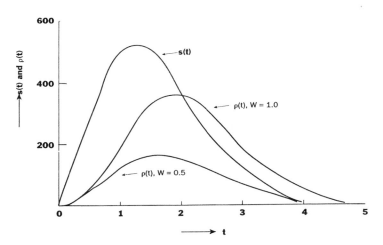

Figure 24.2 Plot of $\rho(t)$ and $s(t)$ versus t.

$\rho(t)$, $0 \leq t \leq L + W$, for two different values of W. For both values, the peak of $\rho(t)$ lags the peak of $s(t)$, as is to be expected because failures occur after the item has been in use for some time. Table 24.6 shows the expected demand for spares on an annual basis for years 1 through 5 and $E[\tilde{N}(W)]$, the total expected demand over the product life cycle, for $W = 0.5$ and 1.0. Both $s(t)$ and $\rho(t)$ initially increase then decrease with time. $\rho(t)$ lags $s(t)$ because the mean time to failure is equal to 1.0 years. The expected number of replacements increases with the warranty period W, as to be expected.

24.4 DEMAND FOR REPAIRS

This section deals with the case in which failed items are repairable and the items are sold with a free-replacement warranty policy. It is assumed that whenever a failed item is returned under warranty, the manufacturer always repairs the failed item and returns it to the owner. Nguyen and Murthy [9] deals with the case in which failed items are minimally repaired. Under minimal repair, the failure rate of the item after repair is the same as that just before failure. Biedenweg [10] deals with the case in which all repaired items are identically distributed with a failure distribution $G(x)$ which is different from $F(x)$, the failure distribution for new items. Gerner and Bryant [11] develop a regression model for the demand for repairs.

Table 24.6 Annual Sales and Expected Demand for Spares to Service
Warranty (Erlangian Failure Distribution)

		Years					
	W	0–1	1–2	2–3	3–4	4–5	$E[\tilde{N}(W, L)]$
$\int s(t)\,dt$		303	463	198	36	0	
$\int p(t)\,dt$	0.5	46	137	81	19	1	284
	1.0	59	325	273	87	11	755

24.4.1 Single-Lot Sales

The expected demand for repairs depends on the type of repair.

Minimal Repair

For each item sold, failures over the warranty period occur according to
a nonstationary Poisson process with an intensity function $\lambda(t)$ given by

$$\lambda(t) = r(t) \tag{24.24}$$

where $r(t)$ is the failure rate (see Chapter 7) associated with the failure
distribution function $F(t)$.

Because item failures are independent, the number of items returned
for repair under warranty in the interval $[0, t)$, $\tilde{N}(t)$, is distributed according
to a nonstationary Poisson process with intensity $p_r(t)$ given by

$$p_r(t) = Nr(t) \tag{24.25}$$

$0 \le t \le W$. Hence, the total expected number of times items are returned
for repair over an interval $[0, t)$, $t \le W$, is given by

$$E[\tilde{N}(t)] = N \int_0^t r(x)\,dx \tag{24.26}$$

As a result, the expected return rate at time t is given by

$$\frac{dE[\tilde{N}(t)]}{dt} = p_r(t) \tag{24.27}$$

and the total expected number of times items are returned for repair over
the warranty period is given by

$$E[\tilde{N}(W)] = N \int_0^W r(t)\,dt \tag{24.28}$$

Example 24.7   (Erlangian Distribution)

Let F be as in Example 24.4 and the lot size N is assumed to be 1000. The failure rate is given by

$$r(t) = \frac{\lambda^2 t}{1 + \lambda t}$$

and from (24.28), we have

$$E[\tilde{N}(W)] = N[\lambda W - \ln(1 + \lambda W)]$$

The total expected number of repairs, $E[\tilde{N}(W)]$, carried out over the warranty period for different values of W (years) and λ are shown in Table 24.7. Note that this increases with λ and W as to be expected.

Example 24.8   (Weibull Distribution)

Suppose the failure distribution is a Weibull distribution with parameters β and λ. In this case, $r(t)$ is given by

$$r(t) = \beta\lambda(\lambda t)^{\beta - 1}$$

From (24.28), we have

$$E[\tilde{N}(W)] = N(\lambda W)^\beta$$

For the special case $\beta = 1$ (exponential failure distribution)

$$E[\tilde{N}(W)] = N\lambda W$$

Table 24.8 gives $E[\tilde{N}(W)]$ for a range of λ and W values for $\beta = 2$ and $\beta = 1$. The interesting thing to note is that $E[\tilde{N}(W)]$ for $\beta = 1$ is considerably larger than that for $\beta = 2$. This follows as the probability of item failure under warranty for the exponential case is much higher than that for the Weibull case.

Table 24.7   Expected Number of Items Returned for Repair over the Warranty Period (Lot Size = 1000; Erlangian Failure Distribution)

$\lambda\backslash W$	1	2	3
0.1	4.7	17.7	37.6
0.2	17.7	63.5	130.0

Table 24.8   Expected Number of Items Returned for Repair over the Warranty Period (Lot Size = 1000; Weibull and Exponential Failure Distributions)

β	λ\W	1	2	3
2 (Weibull)	0.1	10	40	90
	0.2	40	160	360
1 (Exponential)	0.1	100	200	300
	0.2	200	400	600

Repaired Items with Identically Distributed Lifetimes

In this case, for each item sold, the failures under warranty occur according to a modified (or "delayed") renewal process with the time to first failure given by $F(x)$ and subsequent interfailure times distributed according to $G(x)$. As a result, the expected return rate, $\rho_r(t)$, is given by

$$\rho_r(t) = N m_d(t) \tag{24.29}$$

where $m_d(t)$ is the renewal density function for the delayed renewal process (see Chapter 7) and is given by

$$m_d(t) = f(t) + \int_0^t m_g(t - x) f(x) \, dx \tag{24.30}$$

with $m_g(t)$ the renewal density function associated with $G(t)$.

The total expected number of repairs over the warranty period is given by

$$E[\tilde{N}(W)] = N \int_0^W m_d(t) \, dt \tag{24.31}$$

Example 24.9   (Exponential Distributions)

Let $F(x)$ and $G(x)$ be exponentially distributed with parameters λ_0 and λ_1 with $\lambda_1 > \lambda_0$. This implies that repaired items have a shorter mean life to new items. Then it is easily seen that

$$m_g(t) = \lambda_1$$

and

$$m_d(t) = \lambda_1 + (\lambda_0 - \lambda_1) \exp(-\lambda_0 t)$$

It follows from (24.31) that

$$E[\tilde{N}(W)] = N\left[\lambda_1 W + (1 - e^{-\lambda_0 W})\left(\frac{\lambda_0 - \lambda_1}{\lambda_0}\right)\right]$$

Let $N = 1000$ and $\lambda_0 = 0.1$. Table 24.9 gives $E[\tilde{N}(W)]$ for two values of λ_1 (0.3 and 0.5) and W varying from 1 to 3 years. When $W = 1$, the difference between $\lambda_1 = 0.3$ and 0.5 is small and this gets bigger as W increases as to be expected.

24.4.2 Continuous Sales

The approach for continuous sales is similar to that given in Section 24.3.2. The final results are as follows.

Minimal Repair

The expected return rate for repair is given by

$$\rho_r(t) = \int_\psi^t s(\tau)r(t - \tau) \, d\tau \tag{24.32}$$

$0 \le t \le L + W$, and ψ is given by (24.15). The total expected demand for repair over the warranty period is given by

$$E[\tilde{N}(W)] = \int_0^{L+W} \rho_r(t) \, dt \tag{24.33}$$

Example 24.10 (Erlangian Distribution)
     Let $F(x)$ and $s(t)$ be as in Example 24.6. Figure 24.3 shows $s(t)$, $0 \le t \le L \, (= 4)$ and $\rho_r(t)$, $0 \le t \le L + W$, for two different values of W. For both values, the peak of $\rho_r(t)$ lags the peak of $s(t)$, which is to be expected because failures occur after the item has been in use for some time. Table 24.10 shows the expected demand for spares on an annual basis for years 1 through 5 and the total expected demand for repair $E[\tilde{N}(W)]$.
     On comparing with the results for Example 24.6, it is seen that the

Table 24.9   Expected Number of Repairs over the Warranty Period (Lot Size = 1000, $\lambda_0 = 0.1$)

$\lambda_1\backslash W$	1	2	3
0.3	109.67	237.46	381.64
0.5	119.35	274.92	463.27

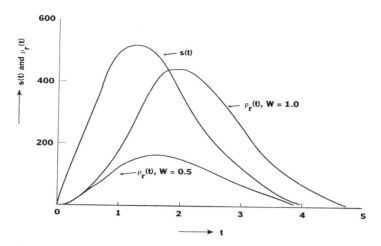

Figure 24.3 Plot of $\rho_r(t)$ and $s(t)$ versus t.

number of returns per 1000 items in each period is larger for the present model. This is to be expected because repaired items have a higher failure rate than that of a brand new item. Also, Table 24.10 shows that the expected demand on the repair facility changes from period to period. This aspect is important in the context of repair facility planning and will be discussed further in a later section.

Repaired Items with Identically Distributed Lifetimes

The expected return rate for repair, at time t, for $0 \leq t \leq L + W$, is given by

$$\rho_r(t) = \int_{\psi}^{t} s(\tau) m_d(t - \tau) \, d\tau \tag{24.34}$$

where ψ is given by (24.15) and the total expected demand for repair under warranty is given by (24.33), using $\rho_r(t)$ given in (24.34).

Example 24.11

Let $F(x)$ and $G(x)$ be exponential distributions with parameters λ_0 and λ_1 with $\lambda_0 < \lambda_1$. It is easily shown that $m_g(t) = \lambda_1$ and

$$m_d((t) = \lambda_1 + (\lambda_0 - \lambda_1) \exp(-\lambda_0 t)$$

Let $s(t)$ as in Example 24.3. Then the expected return rate for repair, at time t, for $0 \leq t \leq L + W$, is given by

$$\rho_r(t) = \int_\psi^t k\tau \exp(-\tau)\{\lambda_1 + (\lambda_0 - \lambda_1) \exp[-\lambda_0(t - \tau)]\}\, d\tau$$

with ψ is given by (24.15). From (24.33), the total expected demand for repair over the warranty period is given by

$$E[\tilde{N}(W)] = \int_0^{L+W} \left\{ k\lambda_1 - [(k\lambda_1 - \tilde{k})t \exp(-t)] \right.$$
$$\left. - \left[\left(k\lambda_1 + \frac{\tilde{k}}{k_1}\right) \exp(-t)\right] + \left[\left(\frac{\tilde{k}}{k_1}\right) \exp(-\lambda_0 t)\right] \right\} dt$$

for $0 \le t \le W$, where $k_1 = (\lambda_0 - 1)$ and $\tilde{k} = k(\lambda_0 - \lambda_1)/k_1$. This can rewritten as

$$E[\tilde{N}(W)] = (\mathcal{Q} + \mathcal{S})\{1 - \exp[-(L + W)]\}$$
$$- \mathcal{Q}\{(L + W) \exp[-(L + W)]\}$$

where

$$\mathcal{Q} = k\lambda_1[\exp(W) - 1] + \tilde{k}[1 - \exp(-k_1 W)]$$

and

$$\mathcal{S} = k\lambda_1[(1 - W) \exp(W) - 1]$$
$$+ \tilde{k}\left\{ \left(-\frac{1}{k_1}\right) + \left[W + \left(\frac{1}{k_1}\right)\right] \exp(-k_1 W) \right\}$$

Let $L = 4$ years, $\lambda_0 = 0.1$, and $k = 1100$. This implies that the total sales over the 4-year period is 1000. Table 24.11(a) gives the expected demand for repairs for each of the 4 years for two different warranty periods ($W = 0.5$ and 1 year) when $\lambda_1 = 0.3$. Similar results for the case when $\lambda_1 = 0.5$ are given in Table 24.11(b). The values are roughly the same when $W = 0.5$ and differ by a small amount (roughly less than 10%)

Table 24.10 Annual Sales and Expected Demand for Repairs Under Warranty

| | W | \multicolumn{5}{c}{Years} | $E[\tilde{N}(W, L)]$ |
		0–1	1–2	2–3	3–4	4–5	
$\int s(t)\, dt$		303	463	198	36	0	
$\int \rho_r(t)\, dt$	0.5	49	148	88	21	1	307
	1.0	65	384	331	107	14	901

Table 24.11 Annual Sales and Expected Demand for Repair Under Warranty (Exponential Distributions; $k = 1100$, $\lambda_0 = 0.1$)

	W	0–1	1–2	2–3	3–4	4–5	$E[\tilde{N}(W, L)]$
				Years			
			(a) $\lambda_1 = 0.3$				
$\int s(t)\, dt$		291	363	228	118	0	
$\int \rho(t)\, dt$	0.5	10	20	14	7	4	55
	1.0	12	40	33	19	9	113
			(b) $\lambda_1 = 0.5$				
$\int s(t)\, dt$		291	363	228	118	0	
$\int \rho(t)\, dt$	0.5	10	21	14	8	4	57
	1.0	13	44	36	20	10	123

when $W = 1$. Figures 24.4 and 24.5 show plots of $s(t)$ and $\rho_r(t)$ for the two cases. The shapes are similar to that in Figure 24.2 for reasons similar to that in Example 24.6.

24.5 REPAIR VERSUS REPLACE

When a repairable item is returned to the manufacturer for repair under free-replacement warranty, the manufacturer has the option of either repairing it or replacing it by a new one. The optimal strategy is one which minimizes the expected cost of servicing the warranty over the warranty period. This section examines several simple strategies which are characterized by a single parameter chosen optimally to minimize the expected cost of servicing the warranty. The case in which the choice is between minimal repair and replacement is considered first. Later on, the case in which all repaired items have identical life distributions different from that for new items is considered.

24.5.1 Minimal Repair

Nguyen [12] considers the following two simple strategies:

Strategy 1. An item is replaced by a new one if it fails in (0, $W - \alpha$] and subjected to minimal repair if it fails in ($W - \alpha$, W].

Strategy 2. An item is subjected to minimal repair if it fails in (0, α] and is replaced by a new one if it fails in (α, W].

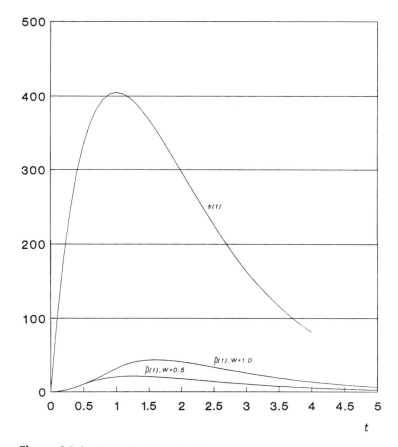

Figure 24.4   Plot of $\rho_r(t)$ and $s(t)$ versus t.

In both cases, the parameter α $(0 \le \alpha \le W)$ is selected to minimize the expected cost of servicing the warranty.

Comment. Strategy 1 is more appropriate when the initial failure rate is high due to a small fraction of the items being of an inferior quality (not conforming to design specification and having a very high failure rate; see Chapter 23). As a result, replacing items which fail early by new ones can be viewed as an effective way of weeding out inferior quality items. Strategy 2 is more appropriate when items have a decreasing failure rate in the early stages of their life and failed items can be repaired relatively cheaply. As the item ages, the failure rate increases and, hence, it is more sensible to replace failed items by new ones when they are old.

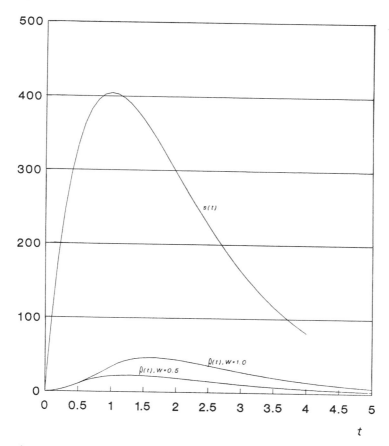

Figure 24.5 Plot of $\rho_r(t)$ and $s(t)$ versus t.

Let c_s and c_r be the manufacturing cost and the repair cost per unit. From Ref. 12, the expected cost of servicing the warranty per-item sale, $\omega(\alpha; W)$, for Strategy 1 is given by

$$\omega(\alpha; W) = c_s M(W - \alpha) + c_r \ln[\overline{F}_\gamma(\alpha)] \qquad (24.35)$$

where

$$F_\gamma(x) = F(W - \alpha + x) - \int_0^{W-\alpha} \overline{F}(W - \alpha + x - y) \, dM(y),$$

$$x \geq 0 \qquad (24.36)$$

and $M(x)$ is the renewal function associated with $F(x)$.

Similarly, for Strategy 2, the total expected warranty service cost per item sale, $\omega(\alpha, W)$, is given by

$$\omega(\alpha; W) = c_s M_d(W - \alpha) - c_r \ln \overline{F}(\alpha) \tag{24.37}$$

where $M_d(x)$ is the expected value of a modified renewal process with the distribution for the first failure, $F_\gamma(x)$, given by

$$F_\gamma(x) = \frac{F(x + \alpha) - F(\alpha)}{\overline{F}(\alpha)} \tag{24.38}$$

and the distribution for subsequent failures given by $F(x)$.

Optimal α

α^*, the optimal α which minimizes $\omega(\alpha, W)$, can be obtained from

$$\frac{d\omega(\alpha; W)}{d\alpha} = 0 \tag{24.39}$$

provided α^* exists. From this, it follows that for both strategies, α^* is the solution of

$$-c_s + \frac{c_r \overline{F}(\alpha)}{\overline{F}_\gamma(\alpha)} = 0 \tag{24.40}$$

In general, one needs to use a computational scheme to obtain α^*.

Example 24.12   (Erlang Distribution)
Let $F(x)$ be an Erlangian distribution with two stages and parameter λ, that is,

$$F(x) = 1 - (1 + \lambda x) \exp(-\lambda x)$$

Let $\lambda = 2.0$ per year. This implies that $F(x)$ is IFR (increasing failure rate) with mean $\mu = 1$ year. Let $c_s = \$100$ and $W = 1$ year.
The existence of α^* depends on the ratio c_r/c_s (i.e., expected repair cost/manufacturing cost). A numerical approach is needed to obtain α^*. Nguyen [12] uses a search methd, in which the left-hand side of (24.40) is evaluated for values of c_r/c_s ranging from 0.00 to 1.00 in steps of 0.01 and for α ranging from 0.00 to 1.00 in steps of 0.01, with a finer grid for search when the quantity evaluated is close to zero. The results are as follows.
For Strategy 1, for $c_r/c_s < 0.78$, there is no α, with $0 \le \alpha \le 1$, satisfying (24.40); $\omega(\alpha; W)$ is minimum for $\alpha = 1$. Because repairs are carried out over the interval $(W - \alpha, W]$, this implies that the optimal strategy for $c_r/c_s < 0.78$ is that failed items are always repaired. For $0.78 \le c_r/c_s < 1$, there are two solutions which we denote by α_1 and α_2, respectively. α^* is the one which gives the minimum value for $\omega(\alpha; W)$.

Table 24.12 gives results for a range of c_r values. Also included are $\omega(0, W)$ and $\omega(W, W)$. The former corresponds to always replace, and the latter to always repair. For $c_r = 80$, the total expected cost for each of the two solutions is larger than that for $\alpha = W$. Hence, in this case $\alpha^* = 1$. Note that as c_r increases (from \$80 to \$95), α^*, the optimal period over which minimal repair is to be carried out, decreases.

The corresponding results for Strategy 2 are given in Table 24.13. We see that the values of α^* are the same as in Table 24.12. However, the resulting optimal expected warranty service cost, $\omega(\alpha^*, W)$, is larger.

A comparison between Strategies 1 and 2 for two values of c_r (= \$80 and \$90) is as follows: For $c_r = 80$, the optimal decision in both cases is to repair all failures over the warranty period. For $c_r = 90$, the optimal decision under Strategy 1 is to replace all failures in (0, 0.87] and repair all failures in (0.87, 1.0]. In contrast, the optimal decision under Strategy 2 is to repair all failures in (0, 0.13] and replace all failures in (0.13, 1.0].

24.5.2 Repaired Items with Identically Distributed Lifetimes

Nguyen and Murthy [13] examine the optimal replace versus repair strategy for the case where all repaired items are identical in the sense that their failure distribution is given by $G(x)$ which is different from $F(x)$, the failure distribution for new items. Biedenweg [10] and Nguyen [12] discuss a variety of repair–replace strategies. See also Ref. 14, in which replace-

Table 24.12 Optimal Repair Versus Replace Decisions for Strategy 1 (Repair Failures in [0, $W - \alpha$) and Replace Failures in [$W - \alpha$, W)]

c_r	α_1	$\omega(\alpha_1, W)$	α_2	$\omega(\alpha_2, W)$	$\omega(0, W)$	$\omega(W, W)$	α^*	$\omega(\alpha^*, W)$
80	0.39	72.3	0.72	72.7	75.5	72.1	1.00	72.1
85	0.23	73.9	0.84	76.8	75.5	76.6	0.23	73.9
90	0.13	74.9	0.91	81.2	75.5	81.1	0.13	74.9
95	0.06	75.3	0.96	85.6	75.5	85.6	0.06	75.9

Table 24.13 Optimal Repair Versus Replace Decisions for Strategy 2 (Replace Failures in [0, $W - \alpha$) and Repair Failures in [$W - \alpha$, W)]

c_r	α_1	$\omega(\alpha_1, W)$	α_2	$\omega(\alpha_2, W)$	$\omega(0, W)$	$\omega(W, W)$	α^*	$\omega(\alpha^*, W)$
80	0.39	74.5	0.72	74.9	75.5	72.1	1.00	72.1
85	0.23	75.1	0.84	78.0	75.5	76.6	0.23	75.1
90	0.13	75.4	0.91	81.7	75.5	81.1	0.13	75.4
95	0.06	75.5	0.96	85.8	75.5	85.6	0.06	75.5

ment involves both new and used items. In general, expressions to obtain the optimal decision are complex.

24.6 COST REPAIR LIMIT STRATEGY

This section deals with the case in which the cost to repair a failed unit, C_r, is a random variable characterized by a distribution function $H(z)$. Analogous to the notion of a failure rate, one can define a repair cost rate given by $h(z)/\overline{H}(z)$, where $h(z)$ is the derivative of $H(z)$. Depending on the form of $H(z)$, the repair cost rate can increase, decrease, or remain constant with z. A decreasing repair cost rate is usually an appropriate characterization for the repair cost distribution (see, e.g., Ref. 15). Murthy and Nguyen [16] consider a model in which items are sold with a free-replacement warranty (FRW) and is as follows: When an item is returned under warranty, the failed item is inspected and an estimate of the repair cost determined. If the estimate is less than a specified limit ϑ, then the failed item is repaired and returned to the owner. If not, the failed item is junked and the customer is supplied with a new item at no cost. The repair is minimal repair. $\omega(\vartheta; W)$, the expected warranty service cost per-unit sale, is given by

$$\omega(\vartheta; W) = M_g(W; \vartheta) \left[c_s + \left(\frac{\int_0^\vartheta z \, dH(z)}{\overline{H}(\vartheta)} \right) \right] \tag{24.41}$$

where $M_g(W; \vartheta)$ is the renewal function associated with the distribution function $G(u; \vartheta)$ given by

$$G(u; \vartheta) = 1 - [\overline{F}(u)]^{\overline{H}(\vartheta)} \tag{24.42}$$

and ζ, the expected cost of each repair carried out, given by

$$\zeta = E[C_r | C_r < \vartheta] = \frac{\int_0^\vartheta z \, dH(z)}{H(\vartheta)} \tag{24.43}$$

(For details of derivation, see Ref. 16). The optimal ϑ, if it exists, is the value of ϑ which minimizes $\omega(\vartheta; W)$. This can be obtained from the usual first-order condition given by

$$\frac{d\omega(\vartheta; W)}{d\vartheta} = 0 \tag{24.44}$$

In general, it is difficult to obtain ϑ^* analytically. One needs to use a computational scheme to obtain ϑ^* using (24.44). The following proposition gives bounds for ϑ^*:

Proposition 24.1 [16]

1. If $F(t)$ is IFR, then $0 \leq \vartheta^* \leq c_s$.
2. If $F(t)$ is DFR, then $\vartheta^* \geq c_s$.

(See Chapter 6 for IFR and DFR.) The proposition states that if the failure rate is decreasing, then it is worth spending more than the replacement cost for repair, as the repaired item has a smaller failure rate and, hence, is more reliable than a new item. However, this situation happens seldom in the real world. For increasing failure rate, repaired items are less reliable than new items and, hence, the optimal repair limit must be less than the price of a new unit.

Example 24.13   (Weibull Distribution)
Let $F(t)$ be a Weibull distribution with parameter λ and β, that is,

$$F(t) = 1 - \exp(-\lambda t)^\beta$$

with $\beta = 2$ and $\lambda = 0.886$. This results in $F(t)$ having an increasing failure rate with mean time to failure of 1.0. The costs are normalized so that $c_s = 1.0$.

Suppose that the repair cost distribution, $H(z)$, is also a Weibull distribution with parameters $\bar{\beta}$ and $\bar{\lambda}$. Consider the following three sets of parameter values:

1. $\bar{\beta} = 0.5$ and $\bar{\lambda} = 2.0$; this implies a decreasing repair cost rate.
2. $\bar{\beta} = 1.0$ and $\bar{\lambda} = 1.0$; this implies a constant repair cost rate.
3. $\bar{\beta} = 3.0$ and $\bar{\lambda} = 0.883$; this implies an increasing repair cost rate.

The above values of $\bar{\lambda}$ were chosen so that the expected repair cost, c_r, is equal to 1. The corresponding values of ζ will depend on ϑ but are always less than c_r.

Figure 24.6 shows the optimal ϑ (ϑ^*) as a function of W for these three sets of parameters values obtained using a computational scheme to find the solution to (24.44) using an iterative gradient method (see, e.g., Ref. 17). Note that ϑ^* is always less than c_s ($= 1$) as is to be expected. Also, note that ϑ^* decreases with W increasing. The results imply that the longer the warranty period, the smaller the repair limit, which, in turn, implies that the failed unit is more often replaced by a new item, as is expected.

Two limiting cases are as follows:

1. $\zeta = \infty$: This corresponds to always repair and the expected service cost be given by $\omega(\infty; W)$.
2. $\zeta = 0$: This corresponds to always repair and the expected service cost is given by $\omega(0; W)$.

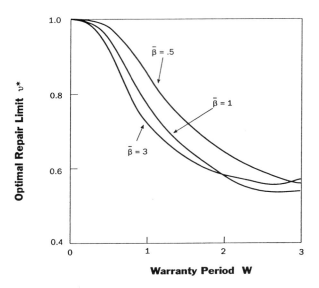

Figure 24.6 ϑ^* versus W.

If F is IFR, then it is easily seen that $\omega(0; W) < \omega(\infty; W)$ for all W. To compare $\omega(\vartheta^*; W)$ with $\omega(0; W)$, define the percentage savings, $\eta(W)$, as

$$\eta(W) = \left(\frac{\omega(0; W) - \omega(\vartheta^*; W)}{\omega(0; W)} \right) \times 100$$

Figure 24.7 shows $\eta(W)$ for different $\overline{\beta}$'s as a function of W. For a given $\overline{\beta}$, $\eta(W)$ decreases with W as is to be expected. For a given W, $\eta(W)$ decreases as $\overline{\beta}$ increases, as is to be expected for the following reason. When $\overline{\beta}$ is less than 1, the repair cost rate is decreasing. The smaller the value of $\overline{\beta}$, the greater the probability of the repair being carried out at a small cost. When $\overline{\beta}$ is greater than 1, the repair cost rate is increasing. This implies that the probability of the repair cost being small decreases with increasing $\overline{\beta}$ and, hence, the advantage of repair over replacement is reduced. Thus, $\eta(W)$ decreases as $\overline{\beta}$ increases.

24.7 SOME ADDITIONAL TOPICS

This section discusses a variety of topics of relevance to warranty servicing.

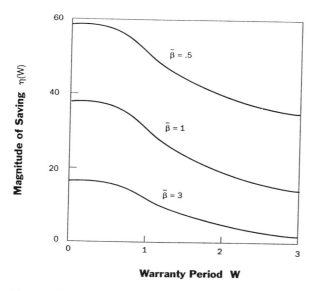

Figure 24.7 $\eta(W)$ versus W.

24.7.1 Repair Facility Planning

Repair facility planning involves issues such as the number of repairmen needed to service the warranty, inventory of components for repair replacements, and strategies to reduce the mean repair service time.

Repairman Problem

The repair facility can be viewed as a service station with failed items arriving randomly over time. The repair time can be either deterministic or random. In either case, the service facility can be viewed as a queuing system with single input and the number of servers equal to the number of repairmen. The statistical characterization of the arrival rate of items depends on the number of items in use and the individual ages of these items. Because these change randomly, the arrival rate changes stochastically and is not easily characterized. This makes it difficult to use standard queuing models because they deal with simpler characterizations of the arrival rates.

The service repair time depends on the arrival rate and the number of repairmen. If the number is small, then the service repair time increases. Increasing the number of repairmen decreases this at the expense of additional labor cost. Determining a sensible trade-off between the service repair time and the cost of labor can only be done through simulation studies.

Another interesting feature of warranty service is that the demand on the service facility changes over the product life cycle, as illustrated in Examples 24.10. This implies that to optimize the repair facility operation, one should have either a variable size labor force or use an overtime option.

Also, for many heavy and nonportable items, the repairman has to travel to different locations to provide warranty service. This adds an extra dimension to operations management. The literature on the Travelling Repairman problem (see Ref. 18) is highly relevant to the analysis of this situation.

Inventory of Spares

As seen from Example 24.10, the expected demand for spares varies over the product life cycle. The inventory level of spares depends on the frequency of replenishment and the safety stock level selected. This frequency, in turn, can vary over the life cycle. A simple inventory policy is the following: The inventory is updated every Δ units of time. The quantity added to the inventory is selected so that the level of inventory after receipt of the order equals the expected demand at the next ordering point plus the safety stock at some selected level. The safety stock must be sufficient to ensure that the probability of the inventory being reduced to zero between ordering points is small. The extensive literature on inventory control (e.g., see Refs. 19 and 20) is relevant to this problem.

If the item is not repairable, the manufacturer has to carry new items for replacements. The need for replacements occurs randomly and the expected demand changes over the product life cycle. Determining the level of replacement items to be held in stock is similar to the stocking of spares discussed earlier.

Often components are repairable. In this case, when a failed component is replaced by a new one, the manufacturer can repair the failed item at a later time and add it to the inventory of repaired spares. As a result, there is an inventory of new spares and an inventory of repaired spares. The optimal levels for each is critical for minimizing the expected cost of operating the repair service facility. There are many articles under the heading of "Logistics" which deal with the problem of spares (new and used) for the effective maintenance of systems and are relevant to the analysis of the operation of the repair service facility.

Use of Loaners

For certain expensive items, the downtime cannot exceed some specified value. Often, the warranty contract specifies a penalty should this happen. One way for the manufacturer to reduce the probability of this happening is to have a stock of loaners which are issued to the owners of failed items when they are undergoing repair. This implies additional servicing costs and the manufacturer must optimally decide on the number of loaners to be held in stock. Again, because of the complexity of the model needed to study this problem, the optimal number can only be decided by simulation studies. A simple model dealing with loaners can be found in Ref. 21.

24.7.2 Warranty Claims Not Exercised

The models discussed so far implicitly assume that all claims under warranty are exercised. If a fraction of warranty claims are not exercised, then the analysis needs to be modified. Patankar [22] has studied this case in depth and suggests the following model. Let $e(x)$ denote the probability that the warranty claim is not exercised if the age of the item at failure is x ($0 < x < W$). One can choose many different forms for the function $h(x)$. One simple form is

$$e(x) = \begin{cases} 0 & \text{for } 0 < x < l \\ \dfrac{x - l}{W - l} & \text{for } l \le x < W \end{cases}$$

This implies that all items which fail before age l are returned for rectification under warranty; for items which are of age greater than l at failure, only a fraction of the failed items are returned. This is a reasonable model because it accounts for the fact that for items failing very close to the end of the warranty period, there is very little (great) incentive for the owner to exercise the warranty claim under pro-rata (free-replacement) warranty. This aspect is discussed in greater detail in Chapter 17.

24.7.3 Product Recall

So far, the discussion has focused on warranty servicing in which a failed item is returned by the owner for rectification under warranty. Occasionally, the manufacturer has to recall either a fraction or all of the items sold, for some rectification action. The recall of only a fraction arises when items are produced in batches and some of the batches are defective due inferior component(s) being used and this is not detected under quality control. The manufacturer is held responsible for damages resulting from such defective components, either under express or implied warranty. (See Chapters 4 and 5 for further discussions.) A hypothetical example is the following. Due to poor quality of insulation, certain batches of a domestic appliance (e.g., an electric frying pan) are prone to result in an electrical shock under normal use. In this case, under the terms of *implied warranty*, the manufacturer can be held liable for damage caused by such defective items and it is more economical for the manufacturer to recall items from such batches. A total-recall situation usually arises because of poor design specifications which can lead to malfunction and serious damage under certain conditions and is discovered only after the items have been produced and sold. A hypothetical example of this is where the brakes of an automobile malfunction under certain conditions of driving. In such cases, the manufacturer can be held responsible for damages caused under the terms of warranty for fitness.

Under the conditions discussed above, the manufacturer has the option of either recalling the items, either for replacement of defective components or for replacement of one or more old components by newly designed ones. The optimal decision depends on many factors. Failure to act can result in huge payouts. On the other hand, any recall is not only costly but can do serious damage to the reputation of the manufacturer and may affect sales for a long time.

Chandran and Lancioni [23] and Fisk and Chandran [24]) discuss the topic of product recall. Min [25] deals with strategies for recall based on a mathematical programming formulation. The model proposed in Tapiero and Posner [26] is appropriate for describing the building up of reserves where large claims can occur, as is the case in product recall situation. Dardis and Zent [27] carry out a simple cost–benefit analysis of the "Pinto Recall." In their model, they incorporate various costs, such as tracing of items, rectification, and so forth. Some analytical research has been carried out on the logistics of product recall and optimal strategies for recall, such as items being returned to the manufacturer or to some retail-level outlets. There is considerable scope and need for further research into this topic.

24.7.4 Servicing by Third Party

Often, it is not economical for a manufacturer to own and operate a service facility to repair items which fail under warranty. In such a case, the manufacturer might delegate the warranty servicing to either the retailer or some independent servicing operator to service all item failures under warranty. This raises many interesting problems, for example:

1. If payment to the service agent is based on the number of claims serviced, there is the possibility of the agent acting in a fraudulent manner, for example, overservicing, repairing items failing outside warranty and reporting these as failures under warranty, to name a few.
2. On the other hand, if the agent is paid a fixed amount, then the agent can do an unsatisfactory job and minimize his expenses. This action can affect the reputation of the product and, in the long run, the profits of the manufacturer.

The literature on this topic is again very limited.

24.8 IMPLEMENTATION ASPECTS

24.8.1 Nonrenewing PRW Policy

Step 1. Compute the mean and variance of the refund per item sold.
Step 2. Decide on ϵ (the probability that the pay out exceeds the warranty reserve).
Step 3. Create warranty reserve using (24.6). The fraction of sale price set aside as warranty reserve, γ, is computed using (24.6).

Note: If the reserves are invested, then Step 3 gets modified. The warranty reserves are given by either (24.11) or (24.12).

24.8.2 Nonrenewing FRW Policy (Nonrepairable Items)

Step 1. Determine the sales rate over the life cycle of the product.
Step 2. Compute expected demand rate, $\rho(t)$, using (24.22). Create an inventory with input rate equal to the expected demand rate.

Items from the inventory are used as needed. The inventory at time t is negative if the total number of replacements exceeds the total input. Two ways of reducing the chances of this happening are as follows:

1. Increase the input rate so that it is given by $(1 + \epsilon)\rho(t)$, with $\epsilon > 0$. The higher the value of ϵ, the smaller the probability that the inventory is zero over an interval of time.
2. Establish a minimum safe level for the inventory. Adjust the input rate to the inventory depending on the inventory level. If the level is above the minimum safe level, the input is given by $\rho(t)$, and if it is below, then it is given by $(1 + \epsilon)\rho(t)$ as in method 1 above. In this case, by a proper choice of the minimum safe

level and ϵ, the probability that the inventory is negative can be made sufficiently small.

24.8.3 Nonrenewing FRW Policy (Repairable Items)

Step 1. Determine the optimal repair versus replacement strategy for each claim.

Step 2. Determine the expected demand rate for repairs and plan repair facility to cope with this demand. If the capacity of repair facility is based on $1 + \epsilon$ times the demand rate, with $\epsilon > 0$, then the probability that the actual demand rate (a random variable) exceeds the planned capacity becomes smaller. In this case, the waiting time is reduced if the demand exceeds capacity.

Step 3. Determine the expected demand for replacement items. Plan an inventory policy as in Section 24.8.2.

If the strategy is always to repair, then only Step 2 is needed; if it is always to replace, then only Step 3 is needed.

REFERENCES

1. Lele, M. M., and Karmarkar, U.S. (1983). Good product support is smart marketing; *Harvard Business Review*, **66**(6), 124–132.
2. Bleuel, W. H., and Bender, H. E. (1980). *Service–Marketing–Engineering Interactions*, AMACOM, New York.
3. Menke, W. W. (1969). Determination of warranty reserves, *Management Sciences*, **15**, B542–B549.
4. Amato, H. N., and Anderson, E. E. (1976). Determination of warranty reserves: An extension, *Management Sciences*, **22**, 1391–1394.
5. Thomas, M. U. (1989). A prediction model for manufacturer warranty reserves, *Management Sciences*, **B–35**, 1515–1519.
6. Hines, W. M., and Montgomery, D. C. (1990). *Probability and Statistics in Engineering and Management Science*, John Wiley & Sons, New York.
7. Blischke, W. R., and Murthy, D. N. P. (1993). *Warranty Cost Analysis*, Marcel Dekker, Inc., New York.
8. Baxter, L. A., Scheuer, E. M., McConalogue, D. J., and Blischke, W. R. (1982). On the tabulation of the renewal function, *Technometrics*, **24**, 151–156.
9. Nguyen, D. G., and Murthy, D. N. P. (1984). A general model for estimating warranty costs for repairable products, *IIE Transactions*, **16**, 379–386.
10. Biedenweg, F. M. (1981). *Warranty Analysis: Consumer Value vs Manufacturers Cost*, Ph.D. Thesis, Stanford University.
11. Gerner, J. L., and Bryant, W. K. (1980). The demand for repair service during warranty, *Journal of Business*, **53**, 397–414.

12. Nguyen, D. G. (1984). *Studies in Warranty Policies and Product Reliability*, Ph.D. thesis, The University of Queensland, Australia.
13. Nguyen, D. G., and Murthy, D. N. P. (1989). Optimal replace–repair strategy for servicing warranty sold under warranty, *European Journal of Operations Research*, **39**, 206–212.
14. Nguyen, D. G., and Murthy, D. N. P. (1986). An optimal policy for servicing warranty, *Journal of the Operations Research Society*, **37**, 1081–1088.
15. Mahon, B. H., and Bailey, R. J. M. (1975). A proposed improved replacement policy for army vehicles, *Operations Research Quarterly*, **26**, 477–494.
16. Murthy, D. N. P., and Nguyen, D. G. (1988). An optimal repair cost limit policy for servicing warranty, *Mathematical and Computer Modelling*, **11**, 595–599.
17. Issacson, E., and Keller, H. B. (1968). *Analysis of Numerical Methods*, John Wiley & Sons, New York.
18. Agnihotri, S. R. (1988). A mean value analysis of the travelling repairman problem, *IIE Transactions*, **20**, 223–229.
19. Brown, R. G. (1967). *Decision Rules for Inventory Management*, Drysden Press, Hinsdale, IL.
20. Hadley, G., and Whitin, T. M. (1963). *Analysis of Inventory Systems*, Prentice-Hall, Englewood Cliffs, NJ.
21. Karmarkar, U.S., and Kubat, P. (1983). Value of loaners in product support, *IIE Transactions*, **15**, 5–11.
22. Patankar, J. G. (1978). *Estimation of Reserves and Cash Flows Associated with Different Warranty Policies*, Ph.D. Thesis, Clemson University, South Carolina.
23. Chandran, R., and Lancioni, R. A. (1981). Product recall: A challenge for the 1980's; *International Journal of Physical Distribution Materials Management*, **11**, 46–55.
24. Fisk, G., and Chandran, R. (1975). How to trace and recall products, *Harvard Business Reviews*, **53**(6), 90–96.
25. Min, H. (1989). A bicriterion reverse distribution model for product recall, *Omega*, **17**, 483–490.
26. Tapiero, C. S., and Posner, M. J. (1988). Warranty reserving, *Naval Research Logistics Quarterly*, **35**, 473–479.
27. Dardis, R., and Zent, C. (1982). The economics of the Pinto recall, *Journal of Consumer Affairs*, **16**, 261–277.

Part G

Warranty and Society

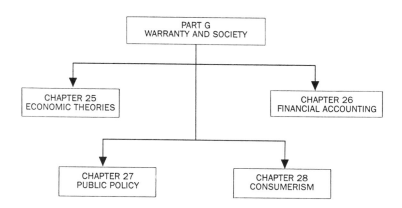

Introduction

Part G contains four chapters, starting with a treatment of economic and accounting issues of warranties. An understanding of these two and the relationship between warranty and consumerism sets the scene for understanding warranty from a public policy perspective.

In Chapter 25, N. A. Lutz looks at warranty from a microeconomic viewpoint. The author builds on some of the concepts from Chapter 15 to characterize the demand for warranties and the actions of the suppliers of warranties to meet this demand. The resulting outcome determines the market structure. A variety of fairly stylized models are discussed, taking into account issues such as the effect of consumer attitude toward risk, moral hazard, and so on.

Chapter 26, by R. A. Maschmeyer and K. R. Balachandran, looks at financial accounting and reporting for warranties in the United States, as specified by the Financial Accounting Standards Board (FASB) and the Internal Revenue Service (IRS). Alternative accounting methods for warranty accounting and some problem areas are discussed.

In Chapter 27, G. C. Mosier and J. L. Wiener discuss public policy efforts that address warranty issues. They start with a brief historical overview of the provisions of the Uniform Commercial Code (UCC) and the Magnuson–Moss Act. Following this, they carry out a study of warranty performance before and after the Act in the context of automobile and consumer durable warranties, addressing topics such as warranty content, warranties as signals, warranty coverage, and quality.

Chapter 28, by J. Burton, looks at warranties from a consumerist perspective and describes three different consumer movements and their impact on the legislative actions that led to the formulation of the public policies related to warranties discussed in Chapter 27. The author then critically evaluates the Magnuson–Moss Act from a consumerist perspective and suggests improvements to the Act and possible future legislation.

25

The Economic Theory of Warranties

Nancy A. Lutz

Virginia Polytechnic Institute and State University
Blacksburg, Virginia

25.1 INTRODUCTION

The economic theory of warranties is one of many applications of microeconomics, the branch of economics that studies the choices made by individual firms and consumers and how these choices interact within a market. Microeconomics emphasizes rational decision making, assuming first that consumers seek to maximize their well-being, or utility, given the information available to them at the time of decision; second, that firms act to maximize profits, again given the available information. Using these assumptions, a wide variety of conclusions can be reached about the behavior of consumers and firms.

The great strength of microeconomics when applied to warranties is that it calls for the careful consideration of all the factors affecting warranty provision.* For example, if we ask what warranty will maximize the profit associated with the sale of some product, we must consider not only the costs of providing the warranty but also the revenue associated with the warranty. The costs depend on the kind of compensation promised in the event of a failure and the likelihood that the compensation is claimed. The warranty increases revenues if it allows the firm to sell more units or to charge a higher price while selling the same number of units. Similarly, if we consider when a consumer will, in fact, be willing to pay more for a product sold with a warranty, we must consider not only the cost of the warranty to the consumer but also the benefits conferred by the warranty.

Microeconomics, then, answers two questions about warranties. First, we can analyze why consumers demand warranties (in the sense that they are willing to pay more for a product covered by some sort of warranty). Second, given this demand on the part of consumers, we can understand when warranties maximize a producer's profit and the kind of warranty that does so. This chapter explains the basic economic answers to these questions. Consumers demand warranties because they provide insurance against the risk of product failure and because a warranty may mean that a product is more durable. Firms offer warranties because the cost of insurance is low and because warranties are a tool to convince consumers about a product's durability.

25.2 ATTRIBUTES OF ANY ECONOMIC MODEL OF WARRANTIES

Consider the basic elements of any economic model of warranty provision. Such a model will begin with the producer of the product being warranted. The first thing to be determined is market structure: Is the producer a monopolist, or do competing firms also produce a more or less equivalent product? Market structure defines the options available to consumers; if there are many producers, consumers have a wide range of options and the producer will face stiff competition for consumers. The next element of the model is technology. This determines the product's features and the cost of production. Technology also determines how likely it is that the product will break down in use. Durability is an innate attribute of

---

* For the purposes of this chapter, I will define a warranty as a contract obligating the manufacturer of a product to make some specified compensation to the consumer in the event that the product fails to function. A warranty is thus distinct from a money-back guarantee, under which the consumer can return a product for any reason and receive a refund of the purchase price. Unlike a warranty, a guarantee compensates the consumer who simply does not like the product.

any product.* The probability of a failure depends not just on a product's durability but also on how it is used and maintained by the consumer. Certain kinds of use may increase the probability of a failure; for example, an automobile is more likely to break down if it is used for stop and go driving in extreme weather conditions. An automobile is also more likely to break down if it is not regularly maintained. Finally, the repair technology must be considered. Assumptions must be made about whether or not the product can be repaired, the cost of the repair, and probability that a repaired unit will fail again. Assumptions must also be made about the distribution system for the product. Here there are several options. The manufacturer may sell directly to consumers or through a dealer. Warranty service may be provided by the manufacturer or by some independent repair service.

Any economic model must also include information about the consumers who may purchase a product. Several issues are crucial. First, are the consumers risk-averse? Risk-averse consumers prefer the expected value of any gamble to the gamble itself; in the context of a warranty, a risk-averse consumer will be willing to pay more for a warranty than the expected (actuarial) cost of providing the warranty. Risk-neutral consumers, on the other hand, are indifferent between a gamble and its expected value; they will not be willing to pay a premium above cost for warranty protection. The model must also specify whether or not consumers take any actions that affect the probability of failure; in other words, whether consumers make decisions about maintaining the product. The model will include assumptions about the entire group of consumers. The consumers in the market may be identical. Alternatively, they may vary along one or more dimensions. The distribution of the varying characteristics in the population as a whole will then be an important determinant of the demand for warranty protection. Finally, assumptions must be made about the kind of information available to the consumer at the time of purchase. The simplest case is when consumers know the product's durability, through direct observation or word of mouth. If consumers do not know the product's durability, we will be interested in how they formulate expectations about durability and whether the warranty affects these expectations.

Assumptions are also made (often implicitly) about various other aspects of the market for the good being modeled. These assumptions concern the social and legal institutions that constrain the producer and

---

* It is common in the economics literature to refer to "quality" in terms of the failure probability; high-quality products are less likely to fail. We will use durability, the more precise term. There are many other dimensions of quality beyond durability.

consumers; for example, there may be minimum-warranty legislation, or common-law warranties. Contracts are generally considered to be legally enforceable, assuming that an outside observer (the court) can determine that a breach occurred.* If the product does fail, repairs may be available only from the manufacturer; alternatively, there may be independent repair providers.

Let us summarize the model of warranty provision most frequently used by economists. Most of the work to be surveyed in the rest of this chapter uses this framework, with small variations. There is a single producer, that manufactures a product for sale to a large number of consumers who will use the product for a single period. Although the outcome of the product's use is a random variable, it has only two possible values: the product fails or it works. The product is sold at a price p, with a warranty w that obligates the producer to pay w dollars to the consumer in the event of a failure. The probability that the product works is π. This probability is, in turn, a function of q, the durability or quality of the unit, and e, the effort spent by the consumer in maintenance. The cost of producing a unit is constant at $c(q)$. The cost to the consumer of maintenance is $g(e)$. The consumer has a utility function U, with a valuation V for a working product and valuation 0 if the product does not work. The consumer's utility if the product works is then $U(V - p)$, and the utility if it fails is $U(w - p)$, so that the consumer's expected utility from the purchase is

$$\pi(q, e)U(V - p) + [1 - \pi(q, e)]U(w - p) - g(e) \qquad (25.1)$$

If the consumer does not purchase the product, he has utility $\overline{U}$. The consumer decides whether or not to purchase, and what level of effort to take after purchase, with the goal of maximizing expected utility. The producer chooses p, w, and q to maximize profits; profits per unit sold are

$$p - [1 - \pi(q, e)]w - c(q) \qquad (25.2)$$

price minus the expected cost of warranty claims and the cost of production.

It is useful to think carefully about the role played by the warranty. Because the product's performance is a random variable—the product may work or it may fail—we can consider two different costs associated with the product. The first is the cost of producing the product, $c(q)$ in

---

* Warranty contracts are binding as long as the firm is in business: However, if the firm declares bankruptcy, its warranties may not be honored. See Ref. 1 for a more extensive discussion.

our model; this cost is obviously borne by the manufacturer. The second "cost" is the (expected) cost of product failure. In the model above, a failure costs V, as this is the difference between the value to the consumer of a working and a broken unit. The expected failure cost is then $(1 - \pi)V$. This cost is borne by the consumer. However, if the manufacturer offers a warranty with the product, the failure cost is divided between the manufacturer and consumer, with the manufacturer paying $(1 - \pi)w$ of the cost and the consumer paying the remaining $(1 - \pi)(V - w)$. If the manufacturer offers what we will term a full warranty, with $w = V$, it bears the full cost of any breakdown. Warranties are therefore most usefully viewed as a way for the manufacturer to bear at least some of the cost associated with unsatisfactory product performance.

A warranty thus promises the consumer some compensation in the event of certain product outcomes. The model we have just considered is, of course, quite stylized; there are only two possible outcomes (the product works or it fails), and the compensation takes the form of a monetary payment by the producer to the consumer. However, the insights we will develop are not dependent on the stylized nature of the model. For example, we could extend the model to consider a wide range of outcomes, with more than one kind of failure and a different warranty policy for each. Instead of making a payment to the consumer, the manufacturer could repair or replace the product in the event of a breakdown. The compensation provided by the warranty may not be complete; the warranty may not fully compensate the consumer for all the costs incurred in a breakdown, and the warranty may not cover any possible failure. For example, manufacturers commonly offer warranties that cover breakdowns occurring within a limited span of time after purchase; the warranty is also often limited to breakdowns of certain components. Even when a warranty commits the firm to cover all failures, the warranty may not include full coverage. For example, the cost of parts but not of labor necessary for repair may be covered, or the consumer may have to mail the product at his own expense to a service center. There is also the cost to the consumer of forgoing use of the product while it is being repaired. A full warranty would compensate the consumer for all costs associated with any failure during the product's promised lifetime. Anything less than this level of coverage would be a partial warranty. Of course, the longer the warranty's duration, the more kinds of failures covered; the larger the fraction of the costs associated with a given failure that are borne by the manufacturer, the more extensive the warranty will be. In what follows, we will for simplicity continue to think of the warranty as being one dimensional.

25.3 ECONOMIC REASONS FOR WARRANTIES

In the absence of government regulation, if a producer offers a warranty with a product, it must earn higher profits by doing so.* Economists have identified several main reasons why offering a warranty can increase a producer's profit. First, warranties provide insurance against the risk that the product will fail. Second, a warranty may convince consumers that the product is of high quality or durability. Third, when consumers vary in some dimension that affects their willingness to pay for a warranty, the producer may use different warranties to segment the market.† We will consider each of these motivations in turn, beginning with the intuition for the motive and then surveying work that analyzes the motive in more detail.

25.3.1 Warranties as Insurance

A warranty insures a consumer against the risk that a product will break down, by promising the consumer some compensation in the event of certain failures. Whether or not the insurance is complete, it will be valued by a risk-averse consumer. As we have already mentioned, a risk-averse consumer is one who is willing to pay more than an actuarially fair expected cost premium for insurance. If consumers are risk-averse, warranty insurance is profitable; the higher price that consumers are willing to pay for a product with a warranty exceeds the cost to the risk-neutral firm of providing the warranty. Note that this explanation for warranties rests on the assumption that consumers are, in fact, risk-averse with respect to the kind of risks imposed by product failure. This seems a fair assumption when the loss to the consumer in the event of a breakdown is large, when the product is expensive, and the cost of repair is high: Most consumers are probably risk-averse when it comes to their automobiles. However, this may be a relatively poor assumption when a product is inexpensive. The insurance motive for warranty provision was first discussed by Heal [2].

---

* There are two major regulations on warranties in the United States. The Magnuson–Moss Warranty and Federal Trade Commission Improvement Act requires that the terms of the warranty be available to consumers before they purchase a product but does not require a minimum warranty. (See Chapters 4 and 28.) The Uniform Commercial Code includes implicit warranties that are part of every transaction. The "express" warranties we discuss here, which offer compensation in the event of a breakdown, generally go beyond the UCC implicit warranties.

† A fourth important motive has not yet been addressed in the economics literature. If there is more than one producer, they may compete with each other on warranty as well as on price.

A risk-averse consumer is willing to pay more for full insurance than for incomplete insurance; thus, the insurance motive for warranty provision suggests that all warranties should fully cover the losses associated with product failure. This is also the socially efficient division of the expected costs of product failure; because the producer is risk-neutral, it is cheaper for the manufacturer to bear the costs of product failure.

However, this is only true when the consumer cannot affect the probability that the product will fail.* Suppose, instead, that the failure probability is jointly determined by the manufacturer's choice of durability and the consumer's choice of maintenance. If the consumer is risk-averse, he would prefer to be fully insured. However, full insurance gives the consumer no incentive at all to maintain the product, as a failure would not impose a loss, and maintenance is expensive. More generally, any insurance reduces the consumer's incentive to make a maintenance effort, with higher insurance producing weaker incentives. This phenomenon is known as consumer moral hazard.† Less effort on the part of the consumer increases the probability that the product will fail, in turn increasing the cost to the manufacturer of providing the warranty. In general, incomplete insurance will be optimal when consumers are risk-averse but there is consumer moral hazard. At a very low level of warranty coverage, the increase in revenues from an increase in warranty coverage outweighs the increase in the manufacturer's costs; because consumers value the product more highly when the warranty is larger, the firm can charge a higher price while continuing to sell its product. Consumers value the first dollar of insurance coverage more than they value subsequent dollars. Thus, although consumers continue to demand increased coverage, as the warranty increases, the marginal effect on revenues of further increases decreases. At the same time, the marginal cost of providing the warranty increases as the warranty increases, because consumers are taking less effort: The first dollar of warranty coverage has a smaller impact on con-

---

* The effect of warranties on consumer's incentive to perform maintenance was first recognized by Oi [3]. See Chapter 3 for additional discussion.

† The manufacturer may be able to make the warranty contingent on the consumer performing certain kinds of maintenance. For example, automobile warranties are contingent on regular oil changes and so on. The warranty contract can only include such contingencies when maintenance is observable and verifiable. This is true when the manufacturer can observe the consumer's effort, and this effort can be verified in court, in the event that the consumer disputes the firm's claim that the required effort was not taken. Warranties can be made contingent of regular oil changes because the consumer can be required to produce receipts and other evidence that the maintenance was performed. The assumption in a situation of consumer moral hazard is that at least some of the actions taken by a consumer are either unobservable or unverifiable.

sumer effort than any subsequent dollar. The result is that neither zero nor full coverage is optimal. With or without the effects of consumer moral hazard, the insurance motive for warranty provision is often also a part of richer, more complicated models of warranty provision.

25.3.2 Warranties to Convince Consumers About Durability

Besides insuring against breakdown, warranties may also increase a manufacturer's profits by convincing consumers that the product is quite durable or of high quality. Of course, a warranty can only play this role when consumers do not already know the durability; the assumption is that durability is neither directly observable at the time of purchase nor can it be inferred from past experience with the product. This might be especially true in the case of a new product. Even if consumers have access to good information about a product's past performance, such information may not be a good predictor of the performance of a new unit.

Models of warranties as a method to convince consumers of product durability must include an explicit treatment of how a consumer deduces the unobservable durability from the observable warranty. It is commonplace to suggest that if more durable products also come with longer or more comprehensive warranties, a rational consumer will realize this fact and draw the appropriate conclusions. However, such a story does not in itself explain why the manufacturers of more durable products might offer longer warranties. After all, if consumers use the warranty to deduce durability, it would perhaps be optimal for the producer of a low-durability product to fool consumers by offering an extensive warranty. This story also does not address exactly how consumers draw conclusions from warranties.

Recent advances in the theory of consumer and firm behavior under conditions of asymmetric information (in this context, situations in which the manufacturer knows the product durability but the consumer does not) have been applied to the study of warranties. These models fall into two general categories. In the first, the manufacturer can choose the durability of the product, and an extensive warranty provides the manufacturer with an incentive to choose a high level of durability in order to reduce the expected cost of the warranty. Models of this sort can be grouped together as "incentive" or "producer moral hazard" models of warranty provision. The second category includes models in which durability is an exogenous parameter known to the manufacturer but not the consumer. A warranty can be used to validate the firm's claim of high durability. Models of this sort are "signaling" models of warranty provision.

When the firm can choose durability, it chooses a profit-maximizing level. If consumers cannot observe durability, the firm cannot earn higher revenues with a more durable product. Therefore, the durability will be chosen to minimize the manufacturer's costs. If the manufacturer offers no warranty with the product, the manufacturer minimizes costs by choosing the minimum level of durability, as this will result in low production costs. However, if the manufacturer does offer a warranty, the situation changes. Suppose that the firm offers a warranty committing the firm to pay the consumer w if the product fails. Now the firm's cost has two components: the production cost (which is increasing in durability) and the warranty cost. The latter, of course, is decreasing in durability, because more durable products will incur fewer warranty claims. The manufacturer will still choose durability to minimize his costs. The chosen level is larger than the minimum durability level. The manufacturer's chosen durability will be an increasing function of the warranty payment w, because higher warranties mean higher expected warranty costs.

Given that a warranty implies a higher level of durability, we must then ask when the manufacturer will find it profitable to offer a warranty. Although consumers cannot directly observe durability, a rational consumer, understanding the manufacturer's incentives, will be able to correctly deduce durability from the (observable) warranty. This argument works whether or not consumers are risk-averse: We have a noninsurance motive for warranty provision.

The profit-maximizing warranty for a monopoly manufacturer is again a full warranty, with the manufacturer compensating the consumer for the entire loss associated with a breakdown. The manufacturer, by choosing a higher level of w, increases the price that it can charge for the product. However, this price is still less than the maximum that the consumer is willing to pay, for two reasons. First, the warranty coverage is incomplete, and second the durability is at less than the optimal level. Under a full warranty, the consumer bears none of the costs of a breakdown. The manufacturer, being solely responsible for the cost of a breakdown, will choose the efficient level of durability, the level that minimizes the total cost associated with the product.

The predictions of the incentive theory change if consumers as well as the manufacturer affect the probability of breakdown. A warranty gives the producer an incentive to choose a high level of durability, with a full warranty yielding the most durable product. A warranty has the opposite effect on incentives when a consumer decides how much effort to take to prevent breakdown. The more extensive the warranty, the smaller the share of the breakdown cost that is borne by the consumer. Because effort is costly, the consumer will take less effort under a more extensive warranty. Consumers work hard to prevent breakdown when there is no warranty, and they take little or no effort under a full warranty. The optimal warranty must then trade off the two opposing incentive problems.

The profit-maximizing warranty will provide partial coverage to the consumer.

The incentive theory of warranties was first analyzed by Priest [4]. Later work, with formal mathematical models incorporating the key features we have discussed, includes Cooper and Ross [5,6], Dybvig and Lutz [7], and Emons [8]. Cooper and Ross [5] analyze a basic one-period model, finding that it is optimal for the manufacturer to offer a partial warranty. Emons [8] looks at double moral hazard in a situation in which consumers are risk-averse, and choose whether or not to perform some specified action to maintain the product. Cooper and Ross [6] extend their earlier work to a two-period model and show that the optimal warranty does not offer a higher level of coverage in the second period than was offered in the first. Dybvig and Lutz [7], in a continuous-time model with more general conditions on the strategies available to the manufacturer and consumer, find that the optimal warranty has a constant level of coverage for some time immediately following purchase, with no coverage of later breakdowns.

An alternative to the incentive model of warranty provision is the signaling model of warranty provision.* The signaling model applies when the product's durability is exogenous, rather than being chosen by the manufacturer. The consumer cannot observe the level of durability, but this level is known by the firm. Obviously, the manufacturer would like to convince the consumer that the product is highly durable. The consumer is willing to pay more for the product when it is less likely to break down. The manufacturer cannot expect that the consumer will believe a simple claim that the product is highly durable. After all, the manufacturer of a low-durability product would be able to increase profits by making such a statement. In this context, a warranty makes the manufacturer's claims about durability credible.

A simple way to make a claim of high durability credible is to offer the product with a very high warranty at a low price. The expected warranty costs depend on the product's durability: The less durable the product, the higher the cost. Therefore, the manufacturer cannot afford to offer a high warranty while charging a low price if its product is of low durability. Any claim of high durability will be credible if it is accompanied by such this sort of price and warranty.

---

* I am using the term signaling in its technical economic sense. More broadly, the incentive and signaling theories discussed here have been lumped together by some authors as signaling theories; in both, the consumer uses the warranty to derive information about the product's durability.

Let us be more specific. A consumer, examining a product with warranty w selling at a price p, can consider whether the manufacturer could afford to offer such a warranty at this price if its product were of low rather than high durability. If the product is less durable, the manufacturer could announce the fact and choose a price and warranty to maximize profits, given that the consumers believe the product to be of low durability. Let p^L and w^L be this price and warranty, and let P^L be this level of profit. Consider a price and warranty combination (p, w) such that the manufacturer's profit at this (p, w) when the product is of low durability is no more than P^L:

$$p - [1 - \pi(L)]w \le P^L \qquad (25.3)$$

where $\pi(L)$ is the probability that a low-durability product will fail. Price–warranty combinations satisfying this requirement would include those with a more extensive warranty than w^L with an increase in price that does not compensate for the increase in warranty costs, and those with a less extensive price than w^L with a decrease in price that outweighs the decrease in warranty costs. The manufacturer has no reason to offer any such (p, w) if its product is of low durability, because it can earn P^L by announcing its low durability. For this reason, consumers can conclude upon observing such a (p, w) that the product is highly durable.* The manufacturer of a high-durability product then has a range of price–warranty combinations that will convince consumers that the product is highly durable. This set of price–warranty combinations can be thought of as the "signaling constraint": The manufacturer of a highly durable product must choose a price and warranty from this set to signal its durability to consumers. The more durable the product, the lower the warranty costs; a high-durability manufacturer can afford to charge a lower price for a given warranty than a low-durability manufacturer can charge. When the product is highly durable, the manufacturer chooses from this signaling constraint the price–warranty combination that maximizes profits.

There are two formal models developing the implications of this signaling theory. Spence [10] considers a competition among manufacturers and finds that the scope of the warranty can, in fact, signal durability, with more durable products having longer warranties. Grossman [11] assumes that there is no consumer moral hazard, that all consumers are risk-averse, and that the product is sold by a single manufacturer. He finds that all products, whether they are of high or low durability, are sold

---

* Using the formal language of game theory, we are assuming that the way consumers reach conclusions about product quality satisfies the Intuitive Criterion. See Ref. 9 for more detail.

with a full warranty. If the manufacturer's product is of high durability, full insurance maximizes profit subject to the signaling constraint. If the product is of low durability, full insurance again maximizes profit, for the insurance reasons we have already explored. Under full insurance, of course, the price the consumer is willing to pay for the product does not depend on durability, and so price and warranty are identical across all products.

Lutz [12] introduces consumer moral hazard to Grossman's model. With moral hazard, a full warranty is not optimal; this also means that the profit-maximizing warranty will depend on the level of durability. Lutz first shows that when consumers can observe durability, highly durable products may or may not be sold with more extensive warranties than less durable products. This seemingly paradoxical result is actually easy to explain. It is less expensive to offer a warranty on a more durable product. But consumers are willing to pay less for the insurance, as they face a lower risk of breakdown. Thus, the manufacturer of a more durable product might offer a more extensive warranty (because it is cheap to do so) or a less extensive warranty (because consumers are not willing to pay much for the extra insurance). Lutz then goes on to consider the situation when quality is not observable. As described earlier, if the product is of low durability, the manufacturer simply offers the price and warranty that maximize its profit. If the product is of high durability, the manufacturer communicates this to the consumers by choosing a price and warranty that satisfy the signaling constraint. The manufacturer in this case would choose the price and warranty that maximize profit subject to this constraint. Lutz again finds that the manufacturer of a more durable product may offer a higher or a lower warranty than that offered with less durable products.

25.3.3 Market Segmentation Through Warranties

Up until now, we have ignored possible differences across the consumers in a market. Of course, in reality, consumers are not identical, and they may vary in ways that affect their demand for the product as well as their demand for warranty coverage.

For example, some consumers may place a very high value on a working product, whereas others place a lower value.* We would expect that the high valuation consumers would be willing to pay more for the

---

* An example here would be the case of two automobile owners. The first commutes daily to work by car and, therefore, places a high value on her automobile. The second uses his car only for pleasure driving and places a lower value on the car.

product. If the manufacturer sells to a market made up of consumers with varying valuations, it faces a difficult problem. If it charges a price that reflects the high value some consumers place on its product, many consumers will find the product too expensive. If it charges a lower price, it is losing potential revenue from sales to high valuation customers. The firm would like to charge a high price to high valuation consumers and a low price to low valuation consumers. However, such a scheme will not succeed, because all the high valuation consumers will refuse to pay the high price and chose instead to buy the product at the lower price. The manufacturer can increase profits by segmenting the market: offering the product with different warranties designed to appeal to different consumers. High valuation consumers, who face a large loss in the event of a breakdown, are willing to pay more for warranty insurance. The manufacturer can then bundle the identical product with two different warranties: a full warranty for high valuation consumers at a high price and an incomplete warranty intended for low valuation consumers at a lower price. Low valuation consumers will not purchase the expensive bundle intended for high valuation consumers; high valuation consumers will prefer to pay a high price for full insurance than pay a lower price but bear the risk of a large loss in the event of breakdown.

Models of warranties as a screening device in markets in which consumers vary in their valuation of a working product include Mann and Wissink [13], Matthews and Moore [14], and Holmes [15]. In Ref. 13, consumers are risk-neutral and take maintenance effort. Firms choose the quality of the products they sell to minimize warranty costs; the market for the product is competitive. They find that high valuation consumers pay a high price for an extensive warranty: This is also a highly durable product. Matthews and Moore consider a monopoly producer's screening problem when consumers are risk-averse and take no maintenance effort. The monopolist chooses both the warranty and the quality of the product.* They find that high valuation consumers always pay a higher price but may purchase a product with lower quality or a lower warranty. Holmes considers the Matthews–Moore model with a single change: Consumers cannot observe durability directly, so they depend on the warranty to draw inferences about durability. The results are broadly similar to those obtained by Mann and Wissink.

Other sources of variation across consumers can also lead the producer to use a warranty as a tool for segmenting markets. Kubo [16]

---

* In the Matthews–Moore formulation, durability is observable to consumers. Therefore, consumers do not draw inferences about the product's durability from its warranty.

considers consumers who vary in income, are risk-averse, and take no actions. High-income consumers are willing to pay more for the product: By offering a more extensive warranty to this group, the producer can charge them a higher price, selling a product with a less extensive warranty to low income consumers at a lower price. Padmanabhan and Rao [17] consider a market of consumers who vary in their degree of risk aversion. Each takes effort. Less risk-averse consumers are willing to pay more for the product: however, they are willing to pay less for an increase in the warranty. The manufacturer segments this market by offering the less-risk-averse consumers the product with a small warranty. The price charged is less than these consumers would be willing to pay but is more than more-risk-averse consumers are willing to pay. More-risk-averse consumers then are offered a more extensive warranty.

Padmanabhan [18] considers a market made up of consumers who vary in their usage of the product. Usage affects the probability of failure. High-usage consumers therefore face a higher risk of failure than low-usage consumers. The product is again sold with two different warranties, an extensive warranty designed for the high-usage customers and a less extensive warranty intended for low-usage customers.

25.4 CONCLUSION

The economic theory of warranties makes it clear that a producer can maximize profits by offering its product with a warranty when consumers are risk-averse, when consumers do not know the product's durability, or when the market is made up of groups of consumers, each group placing a different value on a warranty. In deciding the scope of a profit-maximizing warranty, a producer must consider how consumers can affect the probability that the product will break. The more extensive the warranty, the less the consumer will do to keep the product working; because of this consumer moral hazard, in many situations the producer will maximizes profits by offering a partial rather than a full warranty.

ACKNOWLEDGMENTS

The author wishes to thank Sheryl Ball and Pamela Peele for helpful comments.

REFERENCES

1. Bigelow, J., Cooper, R., and Ross, T. (1993). Warranties without commitment to market participation, *International Economic Review*, **34**, 85–100.

2. Heal, G. K. (1977). Guarantees and risk sharing, *Review of Economic Studies* **44**, 549–560.
3. Oi, W. (1973). The economics of product safety, *Bell Journal of Economics and Management Science*, **4**, 3–28.
4. Priest, G. (1981). A theory of consumer product warranty, *Yale Law Journal* **90**, 1297–1352.
5. Cooper, R., and Ross, T. (1985). Product warranties and double moral hazard, *RAND Journal of Economics*, **16**, 103–113.
6. Cooper, R., and Ross T. (1988). An intertemporal model of warranties, *Canadian Journal of Economics*, **21**, 72–86.
7. Dybvig, P. H., and Lutz, N. A. (1993). Warranties, durability and maintenance: two-sided moral hazard in a continuous-time model, *Review of Economic Studies*, **60**, 575–599.
8. Emons, W. (1989). On the limitation of warranty duration, *Journal of Industrial Economics*, **37**, 287–301.
9. Cho, I. K., and Kreps, D. M. (1987). Signaling games and stable equilibria, *Quarterly Journal of Economics*, **102**, 179–222.
10. Spence, A. M. (1977). Consumer misperceptions, product failure, and producer liability, *Review of Economic Studies*, **87**, 561–572.
11. Grossman, S. J. (1981). The informational role of warranties and private disclosure about product quality, *Journal of Law and Economics*, **24**, 461–483.
12. Lutz, N. A. (1989). Warranties as signals under consumer moral hazard, *RAND Journal of Economics*, **20**, 239–255.
13. Mann, D. P., and Wissink, J. P. (1989). Hidden actions and hidden characteristics in warranty markets, *International Journal of Industrial Organization*, **8**, 53–71.
14. Matthews, S., and Moore, J. (1987). Monopoly provision of product quality and warranties: an exploration in the theory of multidimensional screening, *Econometrica*, **55**, 441–467.
15. Holmes, T. J. (1984). *Monopoly Bundling of Warranty and Quality When Quality is Unobservable*, Northwestern University Center for Mathematical Studies in Economics and Management Science Discussion Paper 612s.
16. Kubo, Y. (1986). Quality uncertainty and guarantee: a case of strategic market segmentation by a monopolist, *European Economic Review*, **30**, 1063–1080.
17. Padmanabhan, V., and Rao, R. (1993). Warranty policy and extended service contracts: theory and an application to automobiles, *Marketing Science*, **12**, 230–247.
18. Padmanabhan, V. (1992). *User Heterogeneity and Extended Service Contracts*, Stanford University Graduate School of Business Research Paper 1197.

26

Financial Accounting and Reporting for Warranties

Richard A. Maschmeyer

University of Alaska Anchorage
Anchorage, Alaska

Kashi Ramamurthi Balachandran

New York University
New York, New York

26.1 INTRODUCTION

This chapter discusses the financial accounting and reporting requirements for product warranties. In Chapter 29, "Cost Management Planning and Control for Product Quality and Warranties," several issues related to the provision of requisite information for cost management decisions are presented. Here we are concerned with the accumulation and presentation of information in periodic financial statements for decision-making purposes of investors, creditors, and others external to the reporting entity. The accounting for product warranty costs includes the fulfillment of financial reporting requirements as established by various rule-making bodies. This chapter's discussion is limited to the recording and reporting requirements in the United States, as specified by the Financial Accounting Standards Board (FASB) and the Internal Revenue Service (IRS).

26.2   FINANCIAL ACCOUNTING AND REPORTING REQUIREMENTS

26.2.1   The Framework of Financial Accounting

Accounting is an activity designed to identify, measure, and communicate information about economic entities for the purpose of making economic decisions. Users of accounting information are both internal and external of the accounting–reporting entity. To meet the needs of external users, primarily investors and creditors, financial report preparers follow a common set of standards and procedures called "generally accepted accounting principles" (GAAP). The authoritative sources of GAAP are the promulgations by the FASB and its predecessors, the Accounting Principles Board (APB), and the Committee on Accounting Procedure (CAP). Where specific transactions are not covered by these primary sources, second- and third-level sources are considered authoritative. Figure 26.1 designates the sources of authority at each level. As the APB was a board set up by the American Institute of Certified Accountants (AICPA) in 1959 to establish accounting standards, controversy over the independence of the APB led to the creation of the FASB in 1973. The FASB, unaffiliated with other professional organizations and independently financed, is now the primary organization responsible for the establishment of financial accounting standards and for the research program required to support the standard-setting process. Financial statement user groups that influence the formulation of accounting standards include CPA accounting firms, the AICPA and other professional organizations, academicians, organizations in the financial community (banks, financial analysts, etc.), government agencies (IRS, Securities and Exchange Commission, etc.), public investors, and industry associations (e.g., the Machinery and Allied Products Institute).

Accounting standards prolmugated by the FASB must be in accordance with the conceptual framework for financial accounting as specified in the FASB's Statements of Accounting Concepts. Definitive explanations of accounting within the context of a conceptual framework is thought to be useful for enhancing the consistency between accounting standards, improving the consistency and comparability of financial accounting among reporting entities, providing a frame of reference for emerging accounting issues, and instilling in the user a greater confidence and understanding of financial statements. Figure 26.2 provides a graphic illustration of the current conceptual framework of accounting articulated by the FASB's Statements of Financial Accounting Concepts (SFAC).

The objectives of financial accounting are to provide information useful in investment, credit, and similar decisions; to provide information useful in assessing cash-flow prospects; and to provide information about enterprise resources, claims to those resources, and changes in them.

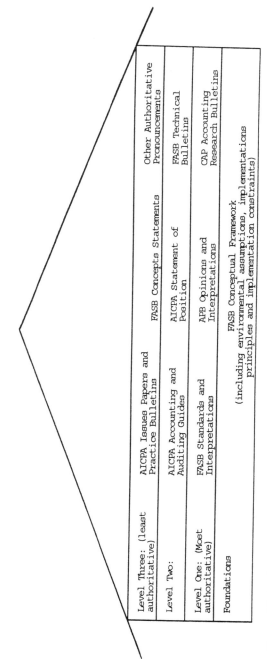

Figure 26.1 House of GAAP. [Adapted from Rubin, S. (1984), the house of GAAP, *Journal of Accounting*, June, 123.]

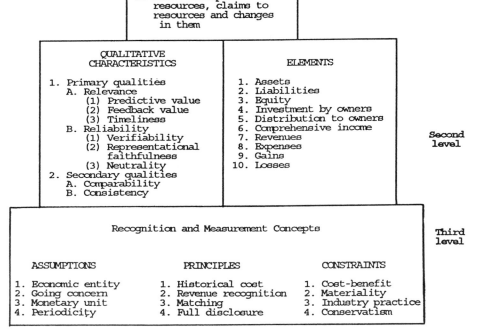

Figure 26.2 Conceptual framework of accounting. (Adapted from Kieso, D. E. and Weygandt, J. J. *Intermediate Accounting*, 7th ed., Copyright © 1992. Reprinted by permission of John Wiley and Sons, New York.)

Financial reports are the principal medium by which accounting information is provided about an entity's economic resources and claims to those resources and about the changes in those resources and claims. A full set of financial statements, including a balance sheet, an income statement, a statement of cash flows, and a statement of retained earnings, is necessary to satisfy the FASB's stated objectives of financial reporting. As the reporting of assets, liabilities, or other events in the financial statements often does not provide all the information necessary for investors, creditor, and others, disclosure of other information is necessary and is considered an integral part of financial reporting. This additional information may be in the form of narrative disclosures or supplemental schedules. Additional disclosure may be necessary for items already recognized in the financial statements or for information related to nonrecognized events (possibly due to the level of uncertainly associated with the event's financial consequences).

A major issue in financial accounting is distinguishing between financial events that require formal recognition and those that require some level of disclosure in the financial statement notes or no disclosure at all. Accounting standards of the FASB may include recognition and disclosure requirements or focus on one or the other. FASB Concepts Statement No. 5., "Recognition and Measurement in the Financial Statements of Business Enterprises" [1], defines recognition as

> . . . the process of formally recording or incorporating an item into the financial statements of an entity as an asset, liability, expense, or the like. Recognition includes depiction of an item in both words and numbers, with the amount included in the totals of the financial statements.

Guidance on the nature of required disclosures was included in their discussion of SFAS No. 105 [2]. Four major purposes of disclosure were advanced:

1. To describe both recognized and unrecognized items
2. To provide a useful measure of unrecognized items and relevant measures of recognized items other than the measure recognized in the statement of financial position
3. To provide information to help investors and creditors assess risks and potentials of both recognized and unrecognized items
4. To provide important information in the interim while other accounting issues are being studied in more depth

The extent of financial accounting and reporting of product warranties should conform to the stated accounting objectives as well as the stated intent of the FASB with respect to recognition, measurement, and full disclosure.

26.2.2 Warranties as a Contingent Liability

An explicit product warranty is a promise made by the seller to cover all or a pro-rata share of the repair/replacement cost associated with deficient products. The accounting requirements resulting from the seller's obligations associated with product warranties and product defects are specified by Statement of Financial Accounting Standard (SFAS) No. 5, "Accounting for Contingencies" [3]. A contingency is "an existing condition, situation, or set of circumstances involving uncertainty as to possible gain (hereinafter a gain contingency) or loss (hereinafter a loss contingency) to an enterprise that will ultimately be resolved when one or more future events occur or fail to occur." Product warranties create a loss contingency for the seller at the time the product is sold, but the specific accounting treatment depends on the likelihood that a future event will confirm the loss or impairment of an asset. The FASB has defined three terms that span the range of likelihood:

1. Probable. The future event or events are likely to occur.
2. Reasonably possible. The chance of the future event or events occurring is more than remote but less than likely.
3. Remote. The chance of the future event or events occurring is slight.

The judged likelihood of the future event or events occurring dictates whether the cash or accrual method is to be followed and also signals the level of disclosure to be included in the financial statements. The accrual method is used in accounting to match revenues and expenses in an accounting period. For example, when a product is sold, the selling price is considered revenue whether or not cash is received and where warranty expenditures are probable and estimable, a product warranty expense is incurred in the same financial reporting period.

According to SFAS No. 5, an estimated loss from a loss contingency shall be accrued by a charge to income if both of the following conditions are met:

1. Information available prior to issuance of the financial statements indicates that it is probable that an asset had been impaired or liability had been incurred at the date of the financial statements. It is implicit in this condition that it must be probable that one or more future events will occur confirming the fact of the loss.
2. The amount of loss can be reasonably estimated.

Thus, loss contingencies associated with current period product sales should be recognized as an expense with a corresponding increase in liabilities for current financial reporting purposes. Where a judgment is made that both requirements do not conform to FASB guidelines, loss contingencies are recognized under the cash basis. This method calls for the recognition of the loss contingency only after the future event or events occur and the impairment of an asset or the incurrence of the liability becomes certain.

The major difference between the accrual and cash method of accounting for a loss contingency is the timing of recognition of expenses and related liabilities for periodic financial reporting. The accrual method attempts to match the products' loss contingencies with the corresponding revenues where the loss is probable and estimable. This methodology conforms to the matching principle, which specifies that costs should be recognized as expenses when the goods and services represented by those costs contribute to revenue. SFAC No. 3 [4] states, "The goal of accrual accounting for a business enterprise is to account in the periods in which they occur for the effects of transactions and other events and circumstances, to the extent that those financial effects are recognizable and measurable." Contrary to the accrual method, warranty expenses and associated liabilities under the cash method may relate to efforts and activities that occurred in prior financial accounting periods. Therefore, the judgment on the likelihood of a loss contingency occurring has the effect of shifting income between accounting periods.

Even when the two criteria for accrual are met, the cash method may be justified on the basis of materiality or when the warranty period is usually confined to the year of sale. Regardless of GAAP guidelines, the Internal Revenue Service requires the cash method to be used for all product-warranty-related costs. As a result, the possibility exists that net income for financial statement purposes can differ from net income determined for tax purposes.

26.2.3 Disclosure of Warranty Loss Contingencies

The accounting profession in the United States has adopted the full-disclosure concept for financial reporting; that is, all financially related information that is considered significant enough to influence an informed report reader is to be included in the financial reports of the reporting entity. Therefore, information useful in investment and credit decisions extends beyond what might appear in the line items of financial statements. Figure 26.3 illustrates the types of information that may be included in financial reports under the concept of full disclosure.

Notes are an integral part of the financial statements, as they provide the text required for the reader's full understanding of information presented on each of the main statements. Disclosures of loss contingencies may be found in the notes to the financial statements. Whether warranty disclosures are present depends on the characteristics of the loss contin-

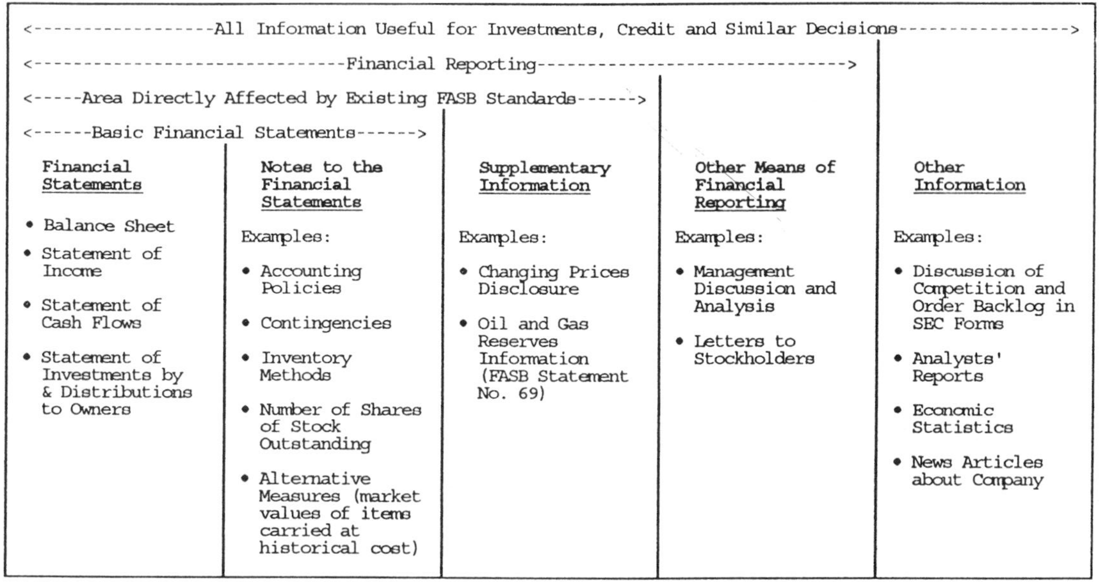

Figure 26.3 Types of financial information. (From Ref. 1, p. 5.)

gency. If the nature of the loss contingency requires an accrual, disclosure may still be necessary so that financial statements are not misleading. If the nature of the loss contingency does not require accruing the liability, disclosure of the contingency is to be made where there is at least a reasonable possibility that a loss may have been incurred. Disclosures should include the nature of the contingency and an estimate of the possible loss. If an estimate cannot be determined, then a statement to that effect is required.

26.2.4 Accounting for Warranty Litigations

The accounting treatment of pending or threatened warranty litigation is also dependent on the probability of occurrence and the ability to estimate the loss contingency. If an unfavorable outcome is probable, the loss is reasonably estimable, and the action occurred on or before the financial statement date, then an expense with a corresponding liability should be accrued in the amount of the loss contingency during the current financial period. If the loss contingency is either not probable or estimable, yet there is a reasonable possibility that a loss will occur, then only disclosure in the notes to the financial statements is required. The disclosure should include the nature of the litigation and an estimate of the possible loss.

26.2.5 Extended Warranties and Product Maintenance Contracts

In addition to the usual warranty protection received on consumer durables, buyers may be offered the opportunity to purchase extended (multiyear) warranty protection. These warranties either provide additional coverage to the standard warranty or simply extend the warranty period. GAAP accounting treatments for transactions related to extended warranty contracts is provided in Financial Accounting Series, FASB Technical Bulletin No. 90-1, "Accounting for Separately Priced Extended Warranty and Product Maintenance Contracts" [5]. According to the bulletin, revenue from separately priced extended warranty and product maintenance contracts should be deferred as a liability at the time of sale and subsequently recognized as income over the contract period. Unless evidence exists to the contrary, earned revenue should be recognized evenly throughout the term of the contract. Where exceptions exist, revenue should be recognized over the contract period in proportion to the costs expected to be incurred in the performance of the contract [6]. All cost associated with the performance of the contract should be charged to expense as incurred. In this manner, conformance to the matching principle is observed when revenues are recognized during the period in which they are earned and are matched with expenses incurred in the same financial reporting period.

Tax reporting of extended warranties differs from GAAP requirements in that the full price received by an accrual method seller of a multiyear warranty contract is taxable in the year in which the customer pays [6]. Where multiyear contracts are between automobile dealers and customers, dealer obligations under the contract may be insured by third parties, including the manufacturer. In this case, the IRS requires the dealer to capitalize the amount of insurance premiums paid to the manufacturer and amortise the amount over the contract's life. Alternatively, IRS Letter Ruling 9218004 permits a portion of the dealers advanced payment to be recognized as taxable income over the life of a multiyear contract rather than taxing lump-sum cash amounts in the year received. The alternative method is available only to those accrual taxpayers who [6]

1. Have never received advanced payments for warranty contracts before their first tax year subject to the alternative method, or
2. Use the proper method of accounting for advanced payments received under warranty contracts (i.e., advanced payments for services to be rendered in the future are income in the year received by an accrual tax payer).

The amount of income that may be deferred under the IRS alternative method is called the "qualified advanced payment amount" (QAPA). This amount is the proportion of the advanced payment to be paid within 60 days of receipt to a third party for the insurance covering the taxpayer's risk under the warranty contract. The balance of the advanced payment for extended warranty contracts is fully taxable to the dealer in the year payment is received. A similar accounting treatment would likely be accepted by the IRS if the facts were similar to the automobile dealer situation.

In summary, the IRS treatment of extended warranty transactions is primarily based on the cash method, whereas GAAP reporting primarily requires the accrual method. This difference is consistent with what we observed earlier in the discussion on the accounting for loss contingencies of product warranties.

26.3 CURRENT REPORTING AND DISCLOSURE PRACTICES

The financial reporting of product warranties may be found within the main body of the financial statements or disclosed in the notes to the statements. Disclosures may include the recognition of expenses and liabilities related to loss contingencies, the method of recognition, the recognition of revenues, expenses and liabilities associated with extended warranty contracts, reconciliations between tax expenses and tax liabilities due to the different accounting treatments between GAAP and the IRS, and disclosures on litigations resulting from product warranties.

The financial statement preparer has an obligation to provide sufficient disclosure so as not to make the statements misleading to the reader. In this context, the location and specificity of financial statement disclosures is a result of the preparer's judgment, within the guidelines prescribed by GAAP. This discretion results in a wide variety of disclosures that the financial statement reader is assumed to interpret correctly for the purpose intended.

```
Fruehauf Trailer Corporation and Subsidiaries
As of December 31st

                                               1991
CURRENT LIABILITIES
Trade accounts payable                 $    42,937
Accrued compensation and benefits           14,454
Accrued warranties                           5,114
Accrued workers compensation                 8,537
Accrued cost of facility realignment         7,485
Other current liabilities                   24,442
Current portion of long-term debt
  to third parties                          11,900
Total Current Liabilities              $   114,869

Partial Notes to the Financial Statements:
    Accrued Warranties:  The Company's financial statements
    reflect accruals for potential product liability and
    warranty claims based on the Company's claim experience.
    Such cost are accrued at the time revenue is recognized.


Data General Corporation
As of September 26
                                               1992
CURRENT LIABILITIES:
Notes payable                          $     3,940
Accounts payable                            85,078
Other current liabilities                  239,142
Total Current Liabilities              $   328,160

Partial Notes to the Financial Statements:
    Revenue Recognition.  Product revenues are recognized at
    the time of shipment.  Services are recognized ratably
    over applicable periods or as services are performed.
    Warranty Costs.  Estimated warranty cost are accrued.
```

Figure 26.4 Disclosure variations—current liability section of the balance sheet (in thousands). (Data provided by Disclosure Inc.)

To illustrate the variation of disclosure, we will look at some examples. First, let us look at two companies offering warranties on their products and both judged their loss contingencies are probable and estimable. Both companies appropriately recognize the estimated loss as a liability, but they differ in terms of how the liability is shown on the balance sheet. Figure 26.4 demonstrates the varying specificity of information provided within the liability section of each company's balance sheet. As indicated by their respective notes, each company is accruing their warranty expense at the time of sale.

Fruehauf Trailer Corporation has elected to show accrued warranties as a separate line item on the balance sheet, whereas Data General Corporation has elected to lump their accrued warranty amount into the general category of other current liabilities. Other companies follow an intermediate scenario where warranties accruals are lumped with other current liabilities and a special schedule is presented in the notes to the financial statements that details the specific costs contained in the general category. Figure 26.5 provides an illustration of this method of disclosure. In terms of understanding the warranty liability, a worst-case scenario is the situation where a company accrues estimated warranty costs, lumps the cost

```
EMC Corporation
                                    December 28,    December 29,
                                        1991            1990

CURRENT LIABILITIES
Notes payable and current portion
  of long-term obligations          $   6,055       $     706
Accounts Payable                       10,703           9,950
Accrued Expenses                       17,658          11,392
Income taxes payable                    7,016           4,295
Deferred income taxes                     419             709
Deferred revenue                        2,277           1,184
Total current liabilities           $  44,128       $  28,236

Partial Notes to the Financial Statements

Note 4.   Accrued Expenses
    Accrued expenses consist of:
                                    December 28,    December 29,
                                        1991            1990
    Salaries and benefits           $   5,979       $   2,629
    Warranty and service                2,379           2,077
    Other                               9,300           6,686
    Total accrued expenses          $  17,658       $  11,392
```

Figure 26.5   Disclosure variations—current liability section of the balance sheet (in thousands). (Data provided by Disclosure Inc.)

into the "other current liability" account, and does not refer to warranties at all in the notes to the financial statements. In this case, the financial statement reader may not be aware that liabilities and corresponding expenses related to warranties even exist. Companies that follow this procedure argue that they are still in compliance with GAAP.

Where financial reporting results under GAAP differ from those under IRS requirements, deferred taxes relative to the temporary differences may be disclosed so the financial statements are not misleading. The diversity of disclosures methods on deferred tax effects ranges from no disclosure at all, a minimal narrative indicating the presence of deferred taxes resulting from product warranties to the disclosure of a schedule

---

A O Smith Corporation, December 31, 1991

Income taxes
The Company accounts for income taxes using the liability method prescribed by FASB Statement No. 96. Certain income and expense items (principally depreciation, pension, model change, product liability and warranty, group health insurance, and reserves for discontinued and restructured operations) are recorded on different bases for financial statement and income tax purposes.

Harley Davidson Inc., December 31, 1990

Provision for Income Taxes
Deferred income taxes result from timing differences in the recognition of revenues and expenses for financial statements and tax returns. The principal sources of these differences and the related effect of each on the Company's provision for income taxes are as follows:

	1991	1990	1989
		(in thousands)	
Excess of tax over depreciation	$ 1,654	$ 943	$ 1,319
Warranty accrual	**(1,356)**	**(845)**	**402**
Product liability	(2,190)	205	(323)
Vacation accrual	(1,034)	(640)	(176)
Employee benefits	(710)	(1,190)	(1,408)
Hedge Contracts	-	(1,024)	1,024
Lawsuit judgement	-	(2,748)	-
Foreign tax credits	(1,500)		
Other	2,155	(1,091)	(17)
Total Deferred Income Taxes	(2,981)	$(6,390)	$ 821

Figure 26.6 Disclosures of deferred taxes from notes to the financial statements. (Data provided by Disclosure Inc.)

showing the specific amounts due to tax deferrals. Figure 26.6 presents one example of minimal disclosure and a second example with more specific information on deferred taxes related to warranties.

The difference in disclosure specificity between A O Smith Corporation and Harley Davidson again highlights the difference in interpretation as to what is a sufficient level of disclosure, relative to satisfying the stated accounting objectives. Related to this issue, SFAS No. 5 [2] stresses, ". . . additional information beyond that on the face of the financial statements may need to be disclosed in the notes or parenthetically for more complex items and for heterogeneous categories." In this case, deferred income taxes is a heterogeneous item in that the total balance is determined from a number of sources.

Companies may also disclose their method of revenue recognition for extended warranty contracts. An example is provided in Figure 26.7.

Disclosure of information related to pending litigation resulting from product warranties is illustrated in Figure 26.8. Once again, there appears to be significant divergence on the level of detail provided on litigation issues. Without sufficient detail on the legal issue involved, how can the statement reader assess the potential risk of either the recognized or unrecognized contingent legal liability?

```
Intercity Products, Inc., December 31, 1990

Deferred Income
The Company offers and sells extended warranty contracts for
its products through certain distributors.  The revenue for
such contracts is deferred and recognized over the life of the
contract on a straight-line basis.  The costs related to
extended warranties are offset against income recognized from
those contracts.


Note 13 WARRANTY
Estimated product warranty is accrued and charged against
earnings in the year related products are sold.  Warranty and
extended protection plan accruals of $31.7 million and $40.2
million as of September 30, 1992 and 1991, respectively, are
reflected in Accrued Liabilities-Other and Other Non-Current
Liabilities in the Statement of Consolidated Financial Position
```

Figure 26.7   Disclosure of extended warranty contracts in notes to the financial statements. (Data provided by Disclosure Inc.)

Avondale Industries, Inc.
Financial Statements - 1990

Partial Notes-Litigation
The Company is a principal co-defendant in a lawsuit arising out of the
sinking of a vessel built by the Company. During 1989, the United States
Court of Appeals for the Fifth Circuit rendered a decision affirming a
lower court's decision that the Company had breached its contractual
warranty to deliver a seaworthy vessel built by it, but concluded that the
construction agreement limited the Company's liability to repair costs
rather than the value of the vessel. The case has now returned to the
lower court for resolution of the damage portion of the lawsuit.

Emerson Radio Corporation
Financial Statements - 1991

Note K - Litigation

In March 1992, an action was commenced in the Superior Court of the State
of California for the County of Los Angeles by the California State
Electronics Association (CSEA) and 19 other plaintiffs against the Company
and 33 other defendants. CSEA, a non-profit corporation and trade
association of individuals and entities performing warranty and nonwarranty service on customer electronics products of a multitude of
manufacturers, and 19 individual plaintiffs (collectively, "plaintiffs")
purporting to bring the action on behalf of a class of all persons who
performed service warranty work for various manufactures and distributors
of consumer electronics products ("defendants") during the past 4 years and
prior thereto, commenced the action alleging that the defendants violated
California's Song-Beverly Consumer Warranty Act ("Song-Beverly Act") based
upon, among other things, a claimed failure to pay negotiated rates for
warranty repairs individually with the service providers, failure to
calculate rates with relation to actual and reasonable costs of the service
providers, failure to pay full costs for parts and transportation thereof,
and failure to provide good faith discounts. The complaint also alleges
that the defendants violated the California Unfair Practices Act ("Unfair
Practices Act") based upon, among other things, contracting with plaintiffs
to perform warranty repairs at rates below the plaintiffs cost for such
repairs, as defined in the Unfair Practices Act, and requiring plaintiffs
to repay for parts. The complaint seeks damages from the defendants in an
undetermined amount and preliminary and permanent injunctions requiring
defendants to pay plaintiffs for warranty work at the rate the service
provider charges for out-of-warranty work. Defendants' time to answer has
not expired.

Figure 26.8 Disclosure on litigation related to product warranties. (Data provided by Disclosure Inc.)

26.4 THE NEED FOR DISCLOSURE REFORM ON WARRANTIES

The success of financial accounting and reporting practices is determined
by the usefulness of the information provided to the users. Questions have
been raised over the diverse interpretations of SFAS No. 5, ''Accounting
For Contingencies.'' Only 3 years after the standard was issued, the Commission on Auditors' Responsibilities [7] recommended further development of the disclosure requirements.

26.4.1 Concerns over the Interpretation of "Remote, Reasonably Possible and Probable"

A major issue in the call for additional disclosure concerns the use of the expression "remote, reasonably possible and probable" in SFAS No. 5. Two areas of previous investigations that have focused on the interpretation of these terms are (a) auditors' interpretations of the three terms and the probability threshold at which the decision shifts from one range to another and (b) the relative agreement between auditors and attorneys on the interpretation of the terms. A third area of study has been the determination of adequate disclosure requirements for litigation purposes.

The studies of Schultz and Reckers [8], Jiambalvo and Wilner [9], and Harrison and Tomassini [10] exhibited some degree of disagreement among the participating auditors with respect to the probability thresholds that differentiate the three expressions. This was particularly true when moving from "remote" to "reasonably possible." Reimers [11] found general agreement between auditors' and managers' interpretations of these terms, but he did conclude that "the three expressions from SFAS No. 5 appear inadequate to communicate the entire range of probabilities." Although not specifically addressing the interpretation of the terminology in SFAS No. 5, Chesley [12] warned that "common words such as probable or likely will cause wide variations in interpretation and should not be used for specific communication of uncertainty in general situations sensitive to inaccurate interpretations." For example, the decision to accrue a loss contingency from warranties or simply disclose the warranty information in the notes to the financial statements is dependent on the statement preparer's interpretation of probable and reasonably possible, respectively. This decision may have a significant consequence on stated liabilities and expenses recognized in the current financial statements. However, rather than assign probability percentages to these categories, the FASB relies on the professional judgment of each account to make the assessment.

Harrison and Pearson [13] studied the problems that might result from communications between lawyers and auditors on contingent liability issues. As auditors are required to have corroborating evidence for representations made within the clients' financial statements, lawyers are a primary source to corroborate the client's assertions. However, auditors and lawyers have different definitions for the terms "remote" and "probable." In a study to determine the sufficiency of disclosing pending litigation, Fesler and Hagler [14] emphasized the "complexity of litigation contingencies and the apparent lack of consensus among policy makers are reflected in the considerable diversity in interpretation and application of

SFAS No. 5." In fact, their data indicate that nondisclosure of litigation contingencies is common practice, "even at relatively high materiality levels."

26.4.2 The Potential Effects on Financial Statement Analysis

Stockholders and other investors are major users of financial statements, interested in making purchasing, selling, or retention decisions on corporate shares of stock. Where the decision has an investment focus, a portfolio's selection is a function of the investor's preference for liquidity, return on investment, dividend yield, and risk [15]. As a result, a wide array of information is necessary to choose a portfolio.

There are two general approaches to evaluate investment opportunities in the securities market: technical analysis and fundamental analysis. Although the technical analysis approach does not rely on financial statement information, financial statements play a major role in the fundamental analysis approach. An important component of this approach is "predicting the timing, amounts, and other uncertainties of the future cash flows of the firm" [15]. The analysis may be undertaken by individuals with limited resources, large firms with extensive resources, or by intermediaries such as securities analysts.

Financial statement analysis is an evaluation of the relationships among financial items reported in the financial statements. This analysis may be based on a period of time or study the trends in relationships over a period of time. Whether cross-sectional or time series techniques are used, financial ratio analysis plays a major role in the decision maker's study of security investments. Of the numerous individual ratios that exist, many could be affected by a firm's chosen accounting treatment of their contingent liabilities on products covered by warranties. A few important examples are as follows:

$$\text{Current ratio} = \frac{\text{Current assets}}{\text{Current liabilities}} \tag{26.1}$$

$$\text{Capital structure ratio} = \frac{\text{Current liabilities} + \text{Long-term liabilities}}{\text{Shareholders' equity}} \tag{26.2}$$

The current ratio provides an indication of the firms' liquidity, that is, their ability to meet short-term obligations in a timely manner. The capital structure ratio provides insight into the extent to which nonequity funds are used to finance the enterprise's assets. The decision to formally recognize a warranty contingency as a liability (either short or long term) or simply disclose the presence of the warranty in the financial statement notes will impact the results of these ratios and many others. Dependent on the financial magnitude associated with a firm's product warranties and the influence product warranties play in the decision maker's preference criteria, the statement preparer's decision over formal recognition or only disclosure may significantly affect the investment decision.

26.4.3 The Warranty Expense/Liability Estimation Problem

Estimating the accrued warranty expenses for financial accounting purposes is compounded by the following considerations:

1. The specific cost to repair and replace components of a product
2. The specific failure rates and corresponding failure rate distributions associated with a product's components
3. The interdependence of product components relative to wearout and failure
4. Multiple products/components with different ages within the same arbitrary accounting period

Preliminary evidence indicates that most accrued expense/liability estimates for financial reporting are determined on the basis of past experience; that is, estimates of current accrued warranty expense are usually calculated as a percentage of current period sales. For example, in the notes to their 1991 financial statements, Fruehauf Trailer Corporation states, "the Company's financial statements reflect accruals for potential product liability and warranty claims based on the Company's claim experience."

Several authors, including Menke [16], Blischke and Scheuer [17], Amato et al. [18], Balachandran et al. [19], Mamer [20], Nguyen and Murthy [21], Frees [22,23], Mitra and Patankar [24], Frees and Nam [25], Thomas [26], Patankar and Mitra [27], and Balachandran and Maschmeyer [28], have proposed various techniques for estimating warranty costs. However, there is little evidence to indicate that models more sophisticated than the simple percentage of sales method are used for financial accounting reporting purposes.

26.5 FUTURE RESEARCH ON THE ACCOUNTING AND DISCLOSURE OF WARRANTIES

Controversy continues on the proper accounting treatment and disclosure of contingencies related to warranties. The major issue of distinguishing between recognition and disclosure raises several questions for future study. In particular, the three probability levels for a contingency, "probable, reasonably possible and remote," need further study to assess the sufficiency of definitions and the range of uncertainty defined by each. There are many questions on disclosures that relate to the level specificity necessary to satisfy generally accepted accounting principles. Johnson [29] advocated that the FASB would be interested in many areas of research on disclosure and suggested several avenues that would provide a better understanding of the required disclosure for warranty contingencies. Some of his suggestions are as follows:

1. When, if ever, is disclosure an adequate substitute for recognition in financial statements?
2. Is less cost incurred in providing information by means of disclosure versus recognition? What are the costs incurred to disclose information on a first time and ongoing basis?
3. Are the purposes of disclosure stated in SFAS 105 (paragraphs 71–86) complete? Are they appropriate? What criteria should be used to determine their appropriateness? Can other purposes of disclosure be imputed from standards or practices?
4. Do the purposes of disclosure lead the FASB to focus too much on each separate topic and not enough on the total package of disclosures?
5. To what extent is disclosure actually used?
6. Are U.S. disclosure requirements significantly different from those of other countries? What are the implications of those differences for global harmonization?
7. It has been asserted that there is a trend toward greater use of databases in financial analysis. How would that affect disclosure requirements? Does the database suggest the need for a more structured form of information in the financial statements and required disclosures?
8. Should there be one or more categories of required disclosure beyond the notes to the financial statements? Concepts Statement 5 suggests the need for "supplementary information" beyond the notes. In that case, how should the location or "geography" of the individual disclosures be determined?

Research on these questions could assist the FASB in either validating their current pronouncements regarding the recognition and disclosure of warranty contingencies or provide information that will be useful in the evaluation of new disclosure requirements.

REFERENCES

1. FASB (1984). *Recognition and Measurement in Financial Statements of Business Enterprises*, Statement of Financial Accounting Concepts No. 5, Financial Accounting Standards Board, Stamford, CT.
2. FASB (1990). *Disclosure of Information about Financial Instruments with Off-Balance-Sheet Risk and Financial Instruments with Concentrations of Credit Risk*, Statement of Financial Accounting Standards No. 105, Financial Accounting Standards Board, Stamford, CT.
3. FASB (1975). *Accounting for Contingencies*, Statement of Financial Accounting Standards No. 5, Financial Accounting Standards Board, Stamford, CT.
4. FASB (1980). *Elements of Financial Statements of Business Enterprises*, Statement of Financial Accounting Concepts No. 3, Financial Accounting Standards Board, Stamford, CT.
5. FASB (1990). *Accounting for Separately Priced Extended Warranty and Product Maintenance Contracts*, Financial Accounting Series, FASB Technical Bulletin No. 90-1, Financial Accounting Standards Board, Stamford, CT.
6. Green, G. L. (1993). More on Extended Warranties, *The National Public Accountant*, **38**, 36–37.
7. AICPA (1978). *Commission on Auditors' Responsibilities*: *Report, Conclusions, and Recommendations*, American Institute of Certified Public Accountants, New York.
8. Schultz, J., and Reckers, P. (1981). The impact of group processing on selected audit disclosure decisions, *Journal of Accounting Research*, **19**, 482–501.
9. Jiambalvo, J., and Wilner, N. (1985). Auditor evaluation of contingent claims, *Auditing*: *A Journal of Practice and Theory*, **5**, 1–11.
10. Harrison, K., and Tomassini, L. (1989). Judging the probability of a contingent loss: An empirical study, *Contemporary Accounting Research*, **5**, 642–648.
11. Reimers, J. (1992). Additional evidence on the need for disclosure reform, *Accounting Horizons*, **6**, 36–41.
12. Chesley, G. (1986). Interpretation of uncertainty expressions, *Contemporary Accounting Research*, **2**, 179–199.
13. Harrison, K., and Pearson, T. (1989). Communications between auditors and lawyers for the identification and evaluation of litigation, claims, and assessments, *Accounting Horizons*, **3**, 76–84.
14. Fesler, R., and Hagler, L. (1989). Litigation disclosures under SFAS No. 5: A study of actual cases, *Accounting Horizons*, **5**, 10–20.
15. Foster, G. (1986). *Financial Statement Analysis*, 2nd ed., Prentice-Hall, Englewood Cliffs, NJ.
16. Menke, W. (1969). Determination of warranty reserves, *Management Science*, **15**, 542–549.

17. Blischke, W., and Scheuer E. (1975). Calculations of the cost of warranty policies as a function of estimated life distribution, *Naval Research Logistics Quarterly*, **22**, 681–696.
18. Amato, H., Anderson, E., and Harvey, D. (1976). A general model of future period warranty costs, *The Accounting Review*, **51**, 854–862.
19. Balachandran, K., Maschmeyer, R., and Livingstone, J. (1981). Product warranty period: A Markovian approach to estimation and analysis of repair and replacement costs, *The Accounting Review*, **56**, 115–124.
20. Mamer, J. (1982). Cost analysis of pro rata and free-replacement warranties, *Naval Research Logistics Quarterly*, **29**, 345–356.
21. Nguyen, D., and Murthy, D. (1984). Cost analysis of warranty policies, *Naval Research Logistics Quarterly*, **31**, 525–543.
22. Frees, E., (1986). Warranty analysis and renewal function estimation, *Naval Research Logistics Quarterly*, **33**, 361–372.
23. Frees, E., (1988). On estimating the cost of a warranty, *Journal of Business and Economic Statistics*, **6**, 79–86.
24. Mitra, A., and Patankar, J. (1988). Warranty cost estimation: A goal programming approach, *Decision Sciences*, **19**, 409–423.
25. Frees, E., and Nam, S. (1988). Approximating expected warranty costs, *Management Science*, **34**, 1441–1449.
26. Thomas, M. (1989). A prediction model for manufacturer warranty reserves, *Management Science*, **35**, 1515–1519.
27. Patankar, J., and Mitra, A. (1989). A multiple-objective model for warranty cost estimation using multiple products, *Computers and Operations Research*, **16**, 341–351.
28. Balachandran, K., and Maschmeyer, R, (1992) Accounting for product wear-out cost, *The Journal of Accounting, Auditing and Finance*, **7**, 49–66.
29. Johnson, L. (1992). Research on disclosure, *Accounting Horizons*, March, 101–103.

27

Warranties and Public Policy

Gregory C. Mosier and Joshua Lyle Wiener

Oklahoma State University
Stillwater, Oklahoma

27.1 INTRODUCTION

There have been three major public policy efforts which have focused upon warranty issues: the provisions of the Uniform Commercial Code (UCC), the Magnuson–Moss Warranty and Federal Trade Commission Improvement Act of 1976, and the recent series of lemon laws passed at the state level. In this chapter, the UCC and Magnuson–Moss Act initiatives are evaluated. Lemon laws are discussed in the context of evaluating the extent to which the Magnuson–Moss Act has improved the extent to which consumers receive high-quality warranty performance. The evaluation of the Magnuson–Moss Act is based on earlier research by Wiener [1].

27.1.1 Uniform Commercial Code Goals

In terms of the Uniform Commercial Code, the indirect goal of warranties is to serve public policy aimed at protecting consumers and imposing strict quality standards on manufacturers. The protection afforded consumers is avoidance of the burden of loss when goods fail to meet the expectations of the buyer. Warranties allow this burden to be transferred to the manufacturers, distributor, or retailer because they are the parties in the best position to bear the loss and spread it among many individuals in the form of higher costs.

The public policy that underlies warranty theory is the same as that underlying negligence and strict liability, the other principle sources of product liability law. This policy favors shifting the risks of loss that are a consequence of defective products to those persons responsible for placing the goods in the marketplace. The underlying assumption is that as a consequence of this burden, persons in the chain of distribution and, ultimately, the manufacturer will accurately price products to reflect their true cost. This true cost will include the risk of injury and or property damage that may result from defects in the product. In the case of certain products, the true cost may outweigh the utility of the product and force a manufacturer to discontinue the product and/or find new and innovative ways to reduce costs associated with personal injury and damages from use of the product.

27.1.2 UCC Implied Warranties

In order to accomplish its goal of risk allocation, the Uniform Commercial Code uses certain implied warranties that, absent a valid disclaimer, define a minimum standard of consumer expectation and allocate responsibility for meeting those expectations to those in the chain of distribution. It rejects the doctrine of *caveat emptor*, which regards the seller and buyer as on equal footing. Instead it recognizes that, in a modern economic environment, buyers rarely possess the level of knowledge or expertise about a product of the seller/manufacturer.

The Uniform Commercial Code warranties have been interpreted to provide the only means for allocating purely economic loss when only the product is damaged. These damages are generally interpreted to include qualitative defects, including reduced value, reimbursement, repair, replacement, or lost profits. As a consequence, in cases where there are no personal injuries resulting from a defective product, UCC warranties provide the only means of recovery. In cases in which personal injuries do result from use of a defective product, alternative theories of negligence and strict liability in tort may be pursued.

As noted, the UCC focuses on very basic inferences. The direct goals of the UCC warranties are to allow parties to a transaction to determine what it is that the seller has in essence agreed to sell. As an example, one inference is that a product being offered for sale will meet minimum performance standards, for example, an automobile will provide transportation services. A second is that verbal claims made by a seller will be honored. Whether the Code has actually accomplished these goals is the subject of intense debate that focuses on the many issues relating to product liability.

Regardless of whether or not the UCC met its initial goals, the perception grew that state law was not providing an adequate means of consumer protection. Consumer advocates and policy-makers began to reach the conclusion that warranties themselves were the source of significant consumer problems. For example, the Uniform Commercial Code allowed a disclaimer of implied warranties under certain circumstances, following certain procedures with language dictated by the Code. This meant that a product with a written warranty might actually have less protection than a product without such a warranty. In response to these problem, hearings which led to the passage of the Magnuson–Moss Warranty Act and Federal Trade Commission Improvement Act were initiated.

27.2 MAGNUSON–MOSS: AN OVERVIEW

The Magnuson–Moss Act was passed after many years of investigations, consumer complaints, and Congressional hearings.* The key problem with warranties, according to those who testified, was that a product's warranty was not a source of useful information. Most of the consumer representatives testifying before the House and Senate Committees focused on the extent to which consumers drew false inferences from product warranties [3]. For example, Federal Trade Commission Chairman Lewis Engman [4] testified that many warranties led a buyer to "believe that he is getting broader warranty coverage than he actually is" and that many consumers perceived a product's warranty as an "assurance of product reliability" when all too often it was not. In addition, both a Federal Trade Commission staff [5] and consumer representatives [6] claimed that manufacturers used warranty promises as a substitute for (rather than signal of) product reliability. In the words of Robert Klein, economics editor of *Consumer Reports*:

> Quality controls on automobile assembly lines and in dealer make-ready began a long decline in 1955 when Detroit revved up production for its first 7-million-car sales year. It has been the observa-

---

* See Refs. 2 and 3 for reviews of the act's legislative history.

tion of Consumers Union that quality control has remained poor
ever since. The industry, conscious of a crisis of confidence in
its products, began extending the warranty period. [7]

After reviewing the testimony, the U.S. Senate Committee on Commerce summarized the evidence by first enumerating four fundamental
consumer needs: (1) the need for consumer understanding; (2) the need
to ensure that written warranties provided a minimum level of protection;
(3) the need to assure adequate warranty performance; and (4) the need for
better product reliability [7]. After discussing these needs, the committee
argued that an underlying cause of these problems was that consumers
could not use a product's warranty as an accurate signal of its reliability.

Many possible solutions to these problems were proposed and debated. The adopted approach explicitly attacked the warranty performance, availability, and content problems. The warranty performance
problems were dealt with by clarifying the nature of adequate warranty
performance. This clarification allowed warranty problems to be handled
under Section 5 of the original Federal Trade Commission Act.* The availability problem was addressed by requiring retailers to provide customers
with the opportunity to read a product's warranty prior to purchasing
the product.† Warranty content problems were attacked by a number of
specific rules. The disclaiming of implied warranties during the warranty
period was prohibited. Every warranty had to be written in "easy and
understandable" language. All exclusions and limitations had to be clearly
displayed. Finally, a "full warranty" designation was created. A full warranty provides complete (elusion-free) coverage for the specified warranty
period.

It was argued that consumers would benefit from these rules because
they would no longer be deceived by misleading product warranties, they
would be able to use a product's warranty as a signal of product reliability,
and they would receive greater warranty protection.

Supporters of the Magnuson–Moss Act argued that the act would
end deception because it would ensure that consumers would receive the
warranty protection they expected to receive. The warranty performance

---

* See Ref. 8 for an excellent discussion of the act's requirements.

† The alternatives a retailer could follow to make a warranty available as explained by Stern
and Eovaldi [8] are "(1) displaying the text of the warranty in close proximity to the displayed
product; (2) providing "ready access" to an indexed binder prominently entitled "Warranties" and containing copies of warranties for all products sold in the department (these
binders must be prominently displayed, or signs must be prominently dispayed that inform
the consumer of the availability of the binder); (3) displaying the package on which the text
of the warranty is clearly visible; or (4) posting near the product a notice that contains the
text of the warranty."

rules would ensure that manufacturers kept their literal warranty promises, whereas the warranty availability and content rules would ensure that consumers would base their expectations about a warranty's coverage on a reading of the warranty. Moreover, the warranty content rules would ensure that the consumers who read a product's warranty would expect to receive only the level of protection literally promised by the warranty.

The argument that the rules would transform a product's warranty into an accurate signal of product reliability was based on straightforward application of the economic theory of signaling to the issue of product warranties. The theory postulates that the total cost to the firm of providing a better warranty is inversely related to the likelihood that the product will fail [9]. In other words, if a firm agrees to provide owners with compensation each time a product fails, then it has a private incentive to reduce the product's failure rate. The theory presumes that consumers are aware of this relationship and, consequently, use a product's warranty as a signal of its reliability. Like most economic theories, its derivation uses a number of behavioral assumptions that may or may not be valid (see Ref. 10 for a detailed critique of these assumptions). These assumptions include the following: warranty service is provided; consumers understand what is being promised; consumers who have products that fail exercise their warranty rights; and consumers know the cost of production. The act explicitly sought to ensure that the first two assumptions would be met. Data collected for the Federal Trade Commission [11] support the third assumption that consumers will exercise their warranty rights. However, the last assumption does not appear to be reasonable, a priori. If consumers do not know the producer's cost structure, a product's warranty may not be an accurate signal of product reliability.

Finally, the act's sponsors' argument that the rules would lead to an overall increase in warranty protection was based on two lines of reasoning. The first was simply that if consumers received better warranty performance then, *ceterus paribus*, they would receive better protection for any given level of promised protection. The second was that the act would lead to warranty competition. The sponsors argued that due to the content, availability, and performance problems, the manufacturers of more reliable products were discouraged from offering better warranties. They were discouraged because they could not differentiate their warranties from the misleading warranties offered by their competitors. Because the warranty rules would put an end to misleading warranties, the rules would encourage the manufacturers of very reliable products to gain a competitive advantage over other manufacturers by offering more warranty protection.

The Magnuson–Moss Act sought to remedy the warranty performance, availability, and content problems so that consumers would not be deceived by warranties, could use warranties to judge product reliability, and would receive greater warranty protection. Research, both published and unpublished, has focused on the issues of warranty performance, warranty availability, use of warranties as prepurchase signals, and warranty content. The findings and implications of these studies are discussed below.

27.2.1 Warranty Performance

Three major empirical studies of warranty performance have been conducted. Two were conducted for the Federal Trade Commission and the third was conducted by Consumers Union. The first Federal Trade Commission study [12] collected information about purchases made during 1976. The sample included 4300 households drawn from an established national consumer mail panel; the response rate was 71.2%. A second Federal Trade Commission study [13] collected information about purchases made in 1982. The sample included 8691 households drawn from the same national consumer mail panel; the response rate was 73.8%. The Consumers Union study [14] collected information about automobile warranty experiences during 1969. The sample consisted of readers of *Consumer Reports*; both the sample size and the response rate are unknown.

All three studies can be used to gain insight into how pre-act automobile warranty performance compares to post-act performance. However, one major problem is that the pre-act data are based on a different sample (*Consumer Reports* readers) than that used by the two post-act samples. It is well established that readers of *Consumer Reports* are very atypical consumers. However, there is no empirical basis for ascertaining whether this difference introduces a bias and, if so, in which direction the bias might be. A second problem is that the questions used in the three surveys differed.*

---

* The Consumers Union survey asked whether the individual agreed that the repair work done on their automobile under factory warranty was satisfactory. The only othe available response was unsatisfactory. The initial FTC-sponsored study asked, "Overall, how satisfied were you with the way the problem was handled?" A balaned five-point scale anchored by very satisfied and very dissatisfied was used. The follow-up Federal Trade Commission report asked, "Overall, how would you describe your whole experience in getting or trying to get the problem taken care of under warranty?" An unbalanced scale (excellent, very good, good, fair, and poor) was used. The discussion assumes that receiving a satisfactory repair is a necessary but not sufficient cause for considering one's warranty experience to be "good."

Despite the obvious problems caused by the use of differing samples and differing questions, a comparison of their respective findings is of interest. In 1969, only 63% of responding *Consumer Reports* readers agreed that the warranty work performed on their automobiles had been satisfactory. In 1976, the year the performance rules were implemented, 64% of responding consumer panel members reported that they were at least somewhat satisfied with the way their problem was handled, and another 5.6% reported that they were not dissatisfied. By 1980, over 71% of responding consumer panel members reported that their overall experience with their warranty problem was "good," and an additional 13.7% described their experience as fair. Because it is unlikely that consumers who received unsatisfactory warranty work would describe their overall experience as good, or even fair, the above data suggest a modest trend toward improved automobile warranty performance.

There is additional evidence that warranty performance after the act's passage was better than it was prior to the act's passage. In 1968, the Major Appliance Consumer Action Panel (MACAP) began to receive, classify, and resolve consumer complaints about consumer durables. During the 8 years preceding the implementation of the Magnuson–Moss Act (1968–1975), a total of 1358 warranty-related complaints were received; these comprised 2.7% of all complaints received. However, in the 10 years between 1976 and 1985, only 510 warranty-related complaints were received: These comprise only 1.7% of all complaints [15–16].

A final piece of evidence is provided by two studies (using a small local sample) presented in Ref. 17. They investigated consumer attitudes toward warranties by conducting one survey prior to the act's passage and a second after the act's passage. They found that consumer attitudes toward warranties were more favorable after the act's passage; however, their evaluation of how well their products were repaired was unchanged.

Both the results of the FTC-sponsored surveys and the MACAP data suggest that warranty performance was better after the act's passage. Unfortunately, it cannot be proven that this difference is attributable to the act's passage. A potential alternative explanation is that the improvement was due to changes in product design and product repair practices. Prior to 1975, poor warranty performance was often provided because there was an insufficient number of qualified personnel [2]. During the past decade, both automobiles and consumer durables have been increasingly designed so that repairs do not require highly trained personnel. They are repaired by simply removing and replacing the malfunctioning part; for example, a service man does not need to know how to repair a dishwasher's motor—only how to snapout the bad one and put in the new one. Diagnosis can be performed by operating a machine or by interchanging existing parts with new parts until the product works properly. Whether these technological changes were a response to the act's passage is not known.

Better may not be good enough. During the past decade, many states have passed lemon laws. These laws were passed in response to the complaints of the purchasers of new automobiles. In essence, they argued that they were not receiving adequate service when they brought in a new automobile for warranty service (see Refs. 18 and 19). Although there are examples of new automobile warranty performance problems, there is no evidence which substantiates the claim that these problems are widespread. The advocates of lemon laws [18] argue that the lack of evidence stems from consumer unfamiliarity with their rights, whereas the critics (e.g., Ref. 19) argue that the lack of evidence reflects the actual degree of the problem.

The public policy significance of the lemons laws (needed or not) should not be discounted. They reveal how consumer expectations regarding their legitimate "rights" regarding product performance continue to expand. In particular, they illustrate how the issue of "loss of product use" is becoming a "right." It also illustrates how the political power of the injured group, in the case of new automobiles, middle-class voters, can drive legislation.

27.2.2   Warranty Availability and Use

A comparison of the two Federal Trade Commission sponsored surveys provides some insight into the issue of availability. In 1976, 18.4% of all respondents reported warranties were sometimes not available. In 1982, only 0.5% (of those who said they could remember looking) reported that warranties were sometimes not available. Since the availability rules were implemented in 1977, this contrast (despite the change in question format) suggests that warranties were more available in post-act retail establishments.

A major reason for requiring availability was to promote the prepurchase reading of warranties. In 1982, somewhere between 4.8% and 14.8% of all purchases were preceded by a careful reading of the product's warranty, and the incidence of brief or careful reading was in the 7.4% to 26.6% range [12]. The author of the 1985 FTC report argues that the true incidence of prepurchase reading is probably near the bottom of the respective ranges. Although there are no comparable pre-act data, one conclusion can be drawn from the 1982 survey results: In the post-act marketplace, most purchases are not preceded by the reading of a product's warranty.

A second important dimension of warranty use is the extent to which warranties influence the purchase decision. Although the unavailability of pre-act data makes it impossible to estimate whether warranty usage has increased, there is evidence that many post-act purchases are influenced by a product's warranty. For example, 75% of the respondents to an FTC [12] survey reported that they would pay a premium for a product that has a superior warranty. This empirical finding corresponds to descriptions of the pre-act market [3].

27.2.3 Warranty Content

Prior to the act's passage, there was general agreement that many warranties were not easy to understand. Studies using the Flesch readability index as a measure have concluded that post-act warranties are more comprehensible than pre-act warranties [2,13,20]. However, both these studies and those conducted by Shuptrine and Moore [21] have found that post-act warranties are not written in "simple and readily understood" language. In addition, only 35% of a consumer mail panel agreed that "warranties are usually written in plain English" [12].

Prior to the act's passage, many warranties contained unfair exclusions. Schmitt et al. [2] investigated the act's impact on the use of unfair exclusions by comparing 40 pre-act and post-act warranties. They concluded that post-act warranties were more likely to provide full coverage during the warranty period (58% versus 85%) and less likely to disclaim implied warranties during the warranty period (55% versus 5%). In addition, the warranties were generally longer (83% increased in length between 1974 and 1977).

The findings about warranty comprehensibility are frequently cited as evidence that the act has failed to reach its goal of transforming a product's warranty into an accurate signal of a product's reliability (e.g., Refs. 21 and 22). However, whether consumers who fail to fully comprehend a written warranty are deceived by the warranty is unknown. As the most common form of deception is that the warranty contains unanticipated exclusions, the results of Schmitt et al.'s 1980 study [2] suggests that uninformed consumers may be deceived less in post-act markets than they were in pre-act markets. However, their (Schmitt et al.) sample is small, and there is evidence that some post-act markets contain unfair exclusions.

27.3 WARRANTIES AS SIGNALS OF RELIABILITY

Existing research suggests that warranty performance may be better in post-act markets than it was in pre-act markets. Although warranty availability and comprehensibility are somewhat better in the post-act period, there is no basis for concluding that there has been a significant increase in the number of consumers who read a product's warranty, fully comprehend the warranty, and then use the warranty to evaluate the product. The next two sections will empirically investigate the consequences of increasing the literal accuracy of a message that is read by only a minority of consumers.

A major goal of the Magnuson–Moss Act was to transform a product's warranty into an accurate signal of a product's reliability. Wiener [23] has empirically investigated this issue. This work is briefly reviewed below.

The data used by Wiener consisted of paired product reliability and product warranty measures. The data sets include all products with available pre-act and post-act measures of both warranty quality and product reliability. The reliability data are drawn from *Consumer Reports*. Wiener constructed and analyzed two data sets. One was composed of consumer durables, the other of automobiles. See Ref. 1 for a discussion of the methodology used to construct the data sets.

27.3.1 Consumer Durables

The consumer durables data are presented in Table 27.1. The data in each product category are summarized by comparing the reliability of products with better-than-standard warranties to the reliability of all other products and/or by comparing the reliability of products with standard warranties to the reliability of products with worse-than-standard warranties. In 53% of the pre-act cases and 88% of the post-act cases, products that offer more warranty coverage are associated with greater reliability. In other words, a greater proportion of these products is associated with above-average reliability and a lesser proportion of these products is associated with below-average reliability. Conversely, in 35% of the pre-act markets and in 11% of the post-act markets, products that offer more warranty coverage are associated with lower reliability. However, because the number of products in each category is so small, none of these associations are significant at the .05 level. For this reason, the data presented in Table 27.1 are aggregated so that an analysis can be performed.

The data are aggregated by classifying warranty coverage as being either "better than" or "worse than" the warranties offered by the other models in the product category. When only standard or below standard coverage is offered, the standard coverage is classified as "better" and the below standard coverage is classified as "worse." Similarly, when

Table 27.1 The Warranty as a Signal of Consumer Durable Reliability

| Year/product | Better-than-standard coverage warranty | | | | Standard coverage warranty | | | | Worse-than-standard coverage warranty | | | | Better-than-standard vs. other | Other vs. worse-than-standard |
| | N | Reliability[a] | | | N | Reliability[a] | | | N | Reliability[a] | | | Smirnov D[b] | Smirnov D[b] |
		H	A	L		H	A	L		H	A	L		
1967 color console TV	2	0%	50%	100%	10	40%	70%	100%	0				−0.4	—
1968 color console TV	3	67	100	100	9	44	78	100	0				+0.29	—
1972 color console TV	2	50	50	100	10	20	60	100	2	0	50	100	X[c]	+0.25
1980 color console TV	2	100	100	100	9	11	56	100	0				+0.89	—
1973 color 19″ TV	6	33	83	100	5	20	60	100	0				+0.13	—
1975 color 19″ TV	1	100	100	100	10	10	70	100	2	50	100	100	+0.83	−0.32
1978 color 19″ TV	5	80	100	100	10	30	50	100	0				+0.5	—
1972 washing machine	10	50	70	100	9	44	44	100	0				+0.26	—
1975 washing machine	7	43	100	100	6	17	100	100	2				+0.30	+0.32
1978 washing machine	1	100	100	100	8	25	75	100					1	—
1974 6000 BTU air conditioner	3	0	100	100	12	17	100	100	1	0	100	100	−0.15	+0.13
1979 5700–6300 BTU air conditioner	0				4	75	100	100	1	0	100	100	—	+0.75
1975 7700–8500 BTU air conditioner	2	0	50	100	8	38	63	100	1	0	0	100	−0.33	+0.70
1979 7700–8500 BTU air conditioner	0				4	75	100	100	6	0	67	100	—	+0.75
1974 range	0				10	20	80	100	3	0.33	100	100	—	−0.23
1979 range	0				d	2			d				—	d
1974 dishwasher	1	0	100	100	12	58	75	100	0				X[c]	—
1980 dishwasher	2	100	100	100	13	62	69	100	0				+0.38	—
1974 top-freezer refrigerator	1	0	0	100	14	36	71	100	0				−0.71	—
1978 top-freezer refrigerator	0				12	73	91	100	0				X[c]	—
1975 side-by-side refrigerator	1	0	100	100	9	22	78	100	0				X[c]	—
1978 side-by-side refrigerator	1	0	100	100	8	25	75	100	0				X[c]	—

[a] All data is in cumulative percentages; an H denotes better than average reliability; an A denotes average reliability; an L denotes below average reliability.
[b] The Smirnov D statistic is the maximum cumulative percentage at a particular level of product reliability over two coverage categories.
[c] The Smirnov D cannot be calculated because there is no ordering.
[d] *Consumer Reports* groups "average reliability" and "unknown reliability" together.
Note: A positive (+) D statistic means that higher warranty quality is associated with greater reliability, a negative (−) D indicates that it is associated with lower reliability, and an X indicates that there is no relationship. Due to the small sample sizes none of the D statistics are significant at the .05 level.

only above standard and standard coverage is offered, the above standard coverage is classified as "better" and the standard coverage is classified as "worse." If three coverage levels are offered, then the products offering standard coverage are combined with the coverage level that has the smaller sample size. The aggregated data are presented in Table 27.2.

A one-tail two-sample Kolmolgorov–Smirnov test procedure is used to test the hypothesis that a product's warranty is an accurate signal of

Table 27.2 Consumer Durables: A Comparison of the Pre- and Post-Act Totals

Sample	Better warranty coverage				Worse warranty coverage				Smirnov D	Hildebrand's Y[b]
		Reliability score[a]				Reliability score[a]				
	N	H	A	L	N	H	A	L		
Pre-act	54	31	73	100	118	31	77	100	0.04	—
Post-act	19	79	100	100	53	30	66	100	0.49[c]	0.43

[a] All data are in cumulative percentages; an H denotes better than average reliability; an L denotes below average reliability.
[b] Hildebrand's Y range is −20 to 1; it is a measure of association in which ties count as errors.
[c] Significant at .01.

its reliability.* The Smirnov D statistic is the maximum cumulative percentage at a particular level of product reliability over two coverage categories. The results of the analysis are shown in Table 27.2. The results of this analysis reveals that a warranty is a signal of product reliability in post-act markets but not in pre-act markets. Further investigation reveals that in all product categories except dishwashers and ranges the signaling relationship is more accurate in post-act markets because manufacturers adjusted their warranty to reflect their product's reliability. In the dishwasher case, a brand that offered better-than-standard warranty coverage improved its reliability. Only in the case of ranges is the pre-act versus post-act difference potentially attributable to the loss of brands from the data set.

27.3.2 Automobiles

Both the automobile data analyzed in this study (1974–1976) and Wiener's [24] data are presented in Table 27.3. These data suggest that prior to the implementation of the Magnuson–Moss Act, an automobile's warranty was not an accurate signal of its reliability. According to this analysis, there was no association between an automobile's warranty and its reliability in 1974. Although there was an association between warranty and

---

* The two-sample Kolmorgov–Smirnov procedure is a nonparametric test that can be used when the dependent variable is ordinal and the sample size is very small. The Student's *t*-test is the parametric analog.

Table 27.3 Automobile Reliability

Product	Better-than-standard warranty coverage						Standard warranty coverage						Smirnov D[b]	Hildebrand Y
		Reliability[a]						Reliability[a]						
	N	5	4	3	2	1	N	5	4	3	2	1		
1974 All	19	6	38	75	100	100	93	11	27	61	79	100	X	—
1974 Foreign	19	6	38	75	100	100	16	47	63	79	84	100	X	—
1975 All	38	24	39	61	89	100	70	6	27	60	81	100	0.18	—
1975 Domestic	13	8	8	23	77	100	59	2	22	61	80	100	X	—
1975 Foreign	25	32	56	80	96	100	11	27	55	55	82	100	0.25	—
1976 All	48	35	50	73	75	100	108	6	32	65	81	100	X	—
1976 Domestic	17	9	18	45	55	100	97	2	15	66	81	100	X	—
1976 Foreign	30	53	73	100	100	100	11	45	54	54	73	100	0.46[c]	0.58
1977–79 All	42	43	74	98	100	100	119	4	22	66	87	100	0.52[c]	0.47
1977–79 Foreign	42	43	74	98	100	100	12	33	42	50	75	100	0.48[d]	0.56

[a] All data are cumulative percentages; a score of 5 indicates highest reliability and a score of 1 lowest reliability.
[b] X indicates no relationship.
[c] Significant at $p = .01$.
[d] Significant at $p = .05$.
Note: In 1974 and 1977–1979, all domestic automobiles provided standard warranty coverage.

reliability in 1975, this association was not statistically significant. The year of the act's implementation, 1976, was a year of transition. In this year, a foreign automobile's warranty was an accurate signal of its reliability, but there was no association between a domestic automobile's warranty and its reliability. Finally, in the post-act period (1977–1979), an automobile's warranty was an accurate signal of its reliability [25].

A detailed analysis of the automobile market reveals why the above changes transpired. In 1975, Toyota, Datsun, Subaru, and Chrysler increased their warranty coverage. Because Toyota, Datsun, and Subaru are all foreign manufacturers of highly reliable automobiles, this warranty change had the effect of bringing about an association between a foreign automobile's warranty and its reliability. In 1976, BMW improved its warranty and the automobiles produced by Volkswagen (which offered a better-than-standard warranty) recorded significant gains in reliability. The consequence of these changes was that in the foreign market a warranty was an accurate signal of a product's reliability, at the .05 level. However, because of the dismal reliability record of Chrysler products (11 of 17 models received the lowest possible rating), there was no warranty–reliability relationship for the total market. In 1977, Chrysler returned to the standard warranty and, as Wiener [24] reported, a warranty became an accurate signal of an automobile's reliability in the total market as well as in the foreign market.

It should be noted that the conclusion that the Magnuson–Moss Act has improved the signaling quality of warranties is not the same as the conclusion that a warranty is a perfect signal. Critics of the act have argued that warranties are poor signals because (1) product reliability is a continuum and (2) in most markets there are only two or three distinct levels of warranty coverage. In other words, the very lack of variability in warranty coverage is taken as proof of the failure of the warranty to be a high-quality signal of product reliability.

A related criticism is that the intellectual basis for the signaling remedy (the work of Spence) is flawed. It is flawed in that it does not take into account factors such as moral hazard (see Chapter 25). As is pointed out in preceding chapters, when a firm is a monopoly, there may not be even an ordering relationship between the warranty it offers and the reliability of its products.

Both of these criticisms serve to emphasize the limited signaling goals advanced by the sponsor's of the act. Specifically, there was the hope of turning a warranty into a gross, relative signal. In other words, it was hoped that by comparing product warranties a consumer would be able to gauge the relative reliability of competing products. Nowhere is it suggested that a consumer should be able to develop a precise prediction of the product's actual failure rate. Moreover, it was understood that a consumer's estimate of a product's reliability would be based on a variety of sources of information. In other words, it was not argued that a product's warranty would become the sole gauge. Finally, it should be noted that given the limitations of the reliability data (i.e., how gross it is), there is no way to tell whether more refined warranty signals would provide more useful information.

27.4 WARRANTY COVERAGE AND QUALITY

Both the act's supporters and opponents claimed that the act would lead to a change in the general level of warranty coverage. The supporters emphasized that the act would increase the consumer segment that would pay a premium for a better warranty and would, therefore, encourage manufacturers to provide better warranties. The critics argued that the act would increase the cost of offering a high level of protection and would lead to a general decline in warranty protection.

Research by Mazis et al. [26] suggests that both sides could be right. If a signal is created, then manufacturers will compete by trying to provide a more favorable level of the signal. However, they will seek to gain this competitive advantage at as little cost as possible. Applied to warranties, Mazis et al.'s [26] framework suggests that some manufacturers will seek to gain a competitive advantage over their competition by offering a relatively better warranty. The framework also suggests that if some firms respond to the act by reducing their coverage, then other manufacturers will be able to gain their advantage by leaving their warranties unchanged. In other words, when the competitive firms decrease their coverage, it is possible to forge ahead by standing still. Mazis et al. [26] emphasize that, all too often, information remedies have produced unintended costs because public policy-makers have ignored the tendency of manufacturers to follow the least cost route to competitive advantage.

27.4.1 Methodology

The question of whether the act produced the unintended consequence of diminishing warranty protection is investigated by comparing the pre-act and post-act warranties offered by manufacturers. A brand's warranty is evaluated in terms of its main warranty's duration, exclusions, and additions. A product's main warranty covers the entire product. It provides parts and/or labor coverage. Exclusions exempt specific expensive components, such as an air conditioner's compressor from warranty coverage. An addition provides parts and/or labor coverage for a specific component, such as an air conditioner's housing.

The data set was constructed by selecting all products for which either *Consumer Reports* or the FTC [13] furnished both pre-act and post-act warranty information. Only one model per brand was counted. When multiple pre-act product category descriptions were available, the latest pre-act data were used. When multiple post-act data were available, the earliest data were always used. In some cases, the post-act data set is based on more than one issue of *Consumer Reports*. A second issue of *Consumer Reports* was used to provide information on brands not described by the earliest issue of *Consumer Reports*. This method of overcoming missing data was only used when there was no difference in the reported product warranties.

27.4.2 Results

Table 27.4 presents the results of the warranty coverage study. In five product categories (color console televisions, 19″ color televisions, washing machines, 5000 BTU air conditioners, and 7700–8500 BTU air conditioners), there were significant declines in warranty coverage after the act's passage. In three product categories (12″ color televisions, dishwashers, and automobiles), there were significant increases. In other words, the data suggest that there was no uniform manufacturer response. Neither the sponsors' hopes nor the critics' claims were fully met.

Table 27.4 Brand-Specific Changes in Warranty Coverage

Product	N	Main duration		Exclusions	Additions[a]	Years
		Increase	Decrease			
Consoles	8	0	6	0	0	1974–1977
19″ color TV	13	0	9	0	0	1975–1978, 1981
12″ color TV	12	4	0	0	0	1972–1977, 1980
Table radios	8	0	0	0	0	1973–1976
Portable radios	7	1	2	0	0	1975–1978
Washing machines	13	0	5	0	− 1	1972–1982
Dryers	9	1	0	0	0	1975–1978
Refrigerators	12	0	1	0	+ 1	1974–1979
Ranges	9	1	0	0	0	1974–1980
Dishwashers	8	0	0	0	+ 5	1971–1979
Vacuum cleaners	10	1	0	0	0	1973–1986
Toasters	11	0	0	0	0	1971–1983
Steam irons	7	0	0	0	0	1973–1977, 1980
5000 BTU air cond.	10	0	0	7	1	1974–1979
6000 BTU air cond.	5	0	1	0	0	1974–1979
7700–8500 BTU air cond.	8	1[b]	1	2	+ 1	1975–1977, 1979
Automobiles	25	7	0	0	+ 4	1974–1976
Automobiles	25	1	0	0	+ 1	1975–1976

[a] A plus (+) means added coverage. A minus (−) means additional parts coverage eliminated.
[b] Despite the increase, the brand still provides less than standard coverage.

A detailed review of the products that changed their warranty coverage reveals an interesting pattern. Significant declines in warranty coverage occurred for products that were not highly reliable. In other words, prior to the act's passage, some manufacturers of products that were not more reliable than the competition offered warranties that were better than the competition. After the act's passage, only one brand continued to offer a 2-year warranty. This brand was also the only brand that had an above-average reliability record. A parallel trend is found for the products that increased their coverage. In the case of automobiles, most of the increases in warranty coverage came from the actions of those foreign manufacturers who produced highly reliable automobiles. In the case of dishwashers, most of the additions reflected specific changes in dishwasher construction. Because all of these changes reflect adjustments of warranty quality to reflect product reliability, they coincide with the expected consequences of the act.

Finally, as Mazis et al. [26] predicted, when some manufacturers decreased their coverage, the others were content to remain at their pre-act level. However, if competitors did not decrease their coverage, then the manufacturers of highly reliable products increased their warranty coverage. The phenomenon of some manufacturers increasing and others decreasing their category (relative to the old standard) is observed only in the portable radio category.

Although there have been no formal studies since Wiener [1], many reports in the popular and trade press suggest that the downturn in warranty coverage he reports was transitory. This issue of increasing warranty coverage is discussed in more detail below.

27.5 WARRANTY COMPETITION

In many ways, the key goal of the Magnuson–Moss act was to create warranty competition. As noted above, because a consumer could judge neither product reliability nor warranty performance from a product's warranty, an honest manufacturer could not effectively compete along the warranty dimension. Behavior research which has investigated the impact of warranty claims on consumer purchase decisions suggests that this problem was most severe for companies that enjoyed modest reputations. In other words, a company with a very positive reputation did not need to use a warranty to bolster their claims, and an unknown company would be viewed as making a "too-good-to-be-true" claim [27].

Menezes and Quelch [28] conclude that the use of a product's warranty as a competitive weapon has increased significantly over the years. This view is consistent with both popular and business press reports about warranty wars, and the use of product warranties in product classes where they were once unknown (e.g., Refs. 29–30). In particular, warranties are now being used by service providers (e.g., the real estate company, ERA) and are being promoted by manufacturers (e.g., Reliance Electric), who are selling to industrial customers (e.g., electric motors for commercial vehicles). Consequently, there does appear to be some basis for concluding that the Magnuson–Moss act met its goal of increasing warranty competition.

27.6 SUMMARY AND IMPLICATIONS

The preceding analyses suggest that between 1975 and 1980 the Magnuson–Moss Warranty and Federal Trade Improvements Act had an impact on the markets investigated. The major goal of transforming a warranty into an accurate signal of product reliability was achieved. In addition, there is some evidence that the act has increased the level of warranty competition. However, it was at the cost of temporally reducing promised warranty coverage in a number of markets.

The finding that the accuracy of the warranty signal is greater in post-act markets than it was in pre-act markets illustrates how an information remedy can lead to a substantive change in the marketplace without changing how information is processed and used. The Magnuson–Moss Act required manufacturers to provide the protection promised. Moreover, it sought to ensure that a reader of a product's warranty would correctly infer the level of promised projection. As Spence [31] and Grossman [32] have shown, under this set of conditions the only manufacturers that can afford to offer better warranties than their competitors are those manufacturers that produce products that are more reliable than their competitors. These firms have an incentive to offer products accompanied by better warranties because there is a small segment of consumers who both read warranties and are willing to pay a premium for products accompanied by better warranties [11].* This segment of warranty-reading, premium-paying consumers may be identical to the segment of warranty-reading, premium-paying consumers that existed prior to the act's passage. The critical difference is that since the act's passage, they receive the protection which they expect to receive. In other words, the act may have transformed a misinformed segment of consumers into an informed segment of consumers without altering how these people processed or used the information. The finding that warranty coverage either rose or fell in individual markets suggests that sellers took advantage of the opportunity to signal their product's superior reliability by following a least cost strategy, as predicted by Mazis et al. [26]. Unlike past studies which have found that firms follow this approach, this study cannot conclude that the strategy was due to a conscious effort to exploit an improperly designed information remedy (see Ref. 34, and Ref. 26 for examples of exploitation). Rather, the least cost approach was simply dictated by the fact that consumers judge products relative to one another.

---

* The market reacted to this segment by using a product's warranty as a signal of product reliability. This finding illustrates the significance of prior research by Thorelli and Engeldow [34] and Salop [35]. Thorelli and Engeldow [34] have established that there is an informed segment of consumers who seek out and process information. Salop [35] has shown how the existence of a small informed minority of consumers can influence the market as a whole. The belief in the existence and power of an informed minority of consumers underlies many consumer information remedies (see Ref. 35 for a more detailed discussion of this issue).

This research has implications for public policy-makers, marketing managers, and researchers. For public policy-makers, the findings strongly support Mazis et al.'s [26] argument that incentive-compatible remedies are potentially effective remedies. However, the findings also suggest that unintended costs can be generated no matter how well the information program's testing and communication system is designed. Second, the findings emphasize the role which can be played by the informed segment. If the goal of the remedy is to induce a market response, then communication strategies designed to inform a small segment of consumers might be more cost-effective than strategies designed to educate all consumers. Finally, the findings suggest that an information remedy can have distributional consequences, at least in the short run. Informed consumers may benefit from the addition of accurate information, whereas uninformed consumers may be harmed by an overall reduction in the level of the regulated attribute provided by the market. Recent events, such as the automobile industry's warranty competition, suggest that this distributional consequence may be temporary.

The key managerial implication is that if an attribute is regulated, it acquires the potential to be a tool for product differentiation. The initial consequence of the regulation is that the manufacturers of products that have higher endowments of the attribute are provided with a means of communicating their relative superiority. However, the long-run consequences may be far different. The existence of an accurate signal supplies manufacturers lacking a current reputation for providing a superior level of a product attribute with an effective means of claiming superior attribute levels. In the television market, firms entering the market after 1976 have used their products' warranties as a means of substantiating their reliability claims. Currently, domestic automobile manufacturers are using warranty claims to bolster their improved reliability claims. Similar patterns of response have been observed with regard to regulations governing tar and nicotine, nutrients, automobile gas mileage, and automobile tire longevity [25]. The pattern suggests that the regulation of an attribute may provide a long-run threat to the manufacturers who, when the regulation is passed, enjoy the best reputation for the regulated dimension.

REFERENCES

1. Wiener, J. (1988). An evaluation of the Magnuson–Moss Warranty and Federal Trade Commission Improvement Act of 1975, *Journal of Public Policy and Marketing*, **7**, 65–82.
2. Schmitt, J., Kanter, L., and Miller, R. (1980). Impact report on the Magnuson–Moss Warranty Act: A comparison of 40 major consumer product war-

ranties from before and after the act, Bureau of Consumer Protection, Washington, DC.

3. Feldman, L. (1976). New legislation and the prospect for real warranty reform, *Journal of Marketing*, **40** (July), 41–47.

4. U.S. Senate Committee on Commerce (1970). *Hearings on Consumer Products Guaranty Act*, 91st Cong., 2d sess.

5. Federal Trade Commission (1968). *Staff Report on Automobile Warranties*, Federal Trade Commission, Washington, DC.

6. Federal Trade Commission (1973), *Magnuson–Moss Warranty–Federal Trade Commission Improvement Act*, 93d Cong., S. Rept. 93-151, Serial Set 13017-2.

7. U.S. Senate (1973). Magnuson–Moss Warranty–Federal Trade Commission Improvement Act, 93d Cong., Report of the Committee on Commerce.

8. Stern, L., and Eovaldi, T. (1984). *Legal Aspects of Marketing Strategy*, Prentice-Hall, Englewood Cliffs, NJ.

9. Grossman, S. (1981). The information role of warranties and private disclosures about product quality, *The Journal of Law and Economics*, **24**, 461–484; Spence, M. (1976). Consumer misperceptions, product failure and producer liability, *Review of Economic Studies*, **44**(3), 561–572.

10. Gerner, J., and Bryant, K. (1981). Appliance warranties as a market signal, *The Journal of Consumer Affairs*, **15** (Summer), 75–86.

11. Federal Trade Commission (1981). *Comparative Performance Information Remedies*, Federal Trade Commission, Washington, DC; Federal Trade Commission (1985). *An Evaluation of the Warranty Rules of the Magnuson–Moss Act*, Federal Trade Commission, Washington, DC.

12. Federal Trade Commission (1985). *An Evaluation of the Warranty Rules of the Magnuson–Moss Act*, Federal Trade Commission, Washington, DC.

13. Federal Trade Commission (1979). *Consumer Information Remedies, Warranties Rules Consumer Baseline Study, and Warranties Rules Warranty Content Analysis*, Federal Trade Commission, Washington, DC.

14. U.S. House Committee on Interstate and Foreign Commerce (1970). *Hearings on Warranties and Guarantees*, 91st Cong., 2d sess.

15. U.S. House Committee on Interstate and Foreign Commerce (1972). *Hearings on Consumer Warranty Protection*, 92d Cong., 1st sess.

16. Major Appliance Consumer Action Panel (1983), Statistical Report, unpublished report, Chicago, IL.

17. McDaniel, S. W., and Rao, C. P. (1982). Consumer attitudes toward and satisfaction with warranties and warranty performance—before and after Magnuson–Moss, *Baylor Business Studies*, 47–62.

18. Dahringer, L. D., and Johnson, D. R. (1988). Lemon laws: Intent, experience, and a pro-consumer model, *The Journal of Consumer Affairs*, **22**(1), 158–170.

19. Smithson, C. W., and Thomas, C. R. (1988). Measuring the cost to consumers of product defects: The value of "lemon insurance," *Journal of Law & Economics*, **31**, 485–502.

20. Lehman, J. C., Manzer, L. L., Gentry, J. W., and Ellis H. W. (1983). The

readability of warranties: Did they improve after the Magnuson–Moss Act and are they more complex than other related communications? *Proceedings, Southwest Marketing Association Conference*, pp. 19–22.

21. Shuptrine, F. K., and Moore, E. M. (1980). Even after the Magnuson–Moss Act of 1975, warranties are not easy to understand, *Journal of Consumer Affairs*, **14**(Winter), 394–404.

22. Kelly, C. (1986). Consumer product warranties under the Magnuson–Moss Warranty Act: A review of the literature, in *AMA Educators' Conference Proceedings* (T. Shimp et al., eds), pp. 369–373.

23. Wiener, J. (1985), Are warranties accurate signals of product reliability, *Journal of Consumer Research*, **12** (Sept.), 245–250.

24. Wiener, J. (1985), The inferences consumers draw from a warranty: Did they become more accurate after the Magnuson–Moss Act? in *AMA Educators' Conference Proceedings* (R. Lusch et al., ed.), pp. 309–312.

25. Wiener, J., and Murphy, P. (1985), A case for comparative performance information, *Journal of Macromarketing*, **5**(Fall), 32–42.

26. Mazis, M., Staelin, R., Beales, H., and Salop, S. (1981). A framework for evaluating consumer information regulations, *Journal of Marketing*, **45**(Winter), 11–22.

27. Goldberg, M. E., and Hartwick, J. (1990). The effects of advertiser reputation and extremity of advertising claim on advertising effectiveness, *Journal of Consumer Research*, **17**(Sept.), 172–179.

28. Menezes, M., and Quelch, J. A. (1990). Leverage your warranty program, *Sloan Management Review, SMR Forum* (Summer), 69–80.

29. Avery, S. (1989). Still no slowdown in sight, *Purchasing*, **107** (Oct), 84–85.

30. Cullen, D. (1990). Warranty wars: How goes the battle? *Fleet Owner*, **85**, 97–100.

31. Spence, M. (1977). Consumer misperceptions, product failure and producer liability, *Review of Economic Studies*, **44** (3), 561–572.

32. Grossman, S. (1981). The informational role of warranties and private disclosures about product quality, *The Journal of Law and Economics*, **24** (Dec.), 461–484.

33. Beales, H., Craswell, R., and Salop, S. (1981). The efficient regulation of consumer information, *Journal of Law and Economics*, **24** (Dec), 491–539.

34. Beales, H., Mazis, M. Salop, S., and Staelin, R. (1981). Consumer search and public policy, *Journal of Consumer Research*, **8** (June), 11–22.

28

Warranty Protection: A Consumerist* Perspective

John R. Burton

University of Utah
Salt Lake City, Utah

28.1 INTRODUCTION

In 1960, *Business Week* [1] reported that Ford was implementing a new policy to gain a competitive advantage by showing the public that Ford

---

* As the definition of *consumerist* may lie in the eyes of the beholder, it is important for the author of this chapter to define whom he regards as a consumerist. A brief definition to which I might subscribe is, "One who advocates for the rights of the consumer in the marketplace." However, a person associated with such conservative organizations as the Hoover Institute or the American Enterprise Institute might also make this claim because he/she would argue that minimum government regulation is always in the best interest of the consumer. Therefore, a more descriptive definition is in order. I will describe myself as a consumerist who lies a little to the left of center on a continuum of economic philosophy ranging from Ralph Nader to Milton Friedman. This means that I do not believe that market forces alone will solve problems of the consumer in the marketplace. When market failure adversely affects the consumer, government intervention where the benefits exceed the cost is justified.

was working hard to build good cars. Starting in September 1960, Ford replaced their 90-day, 4000-mile warranty, the industry standard for decades, with a 12-month, 12,000-mile warranty. For 30 years prior to this new warranty policy, Ford, in unison with other manufacturers, had given their dealers a 12-month, 4000-mile warranty on their cars and the dealer would, in turn, give the lesser warranty to the buyers. Although the manufacturer seldom balked at making repairs on major parts after the 90-day, 4000-mile period (but before 12 months), the public was not aware of this policy. It appears that "secret warranties" were common over the previous 30 years ago.

Ford, in breaking ranks with the other American manufacturers, brought consternation to its competitors. General Motors, for example, thought that everything was going fine under the old warranty system. The shorter warranty period was a gate to "shut out the cranks" from demanding frivolous warranty service. Under the old system, customers would never bring their cars into the dealership to claim an adjustment for relatively small claims. "But then, nobody told them they could" [1, p. 56]. This same *Business Week* article contains a very prophetic statement: "What happens now that the consumers are being let in on the 12 month, 12,000 mile secret is a big question" [1, p. 56].

This chapter describes what eventually did happen when Ford opened the warranty wars in 1960. It will tie growing consumer dissatisfaction with warranties, in particular automobile warranties, to a social movement—the consumer movement—and to the eventual passage of a major piece of consumer legislation, the Magnuson–Moss Warranty and Federal Trade Improvement Act. (The shorter terms for this act, Magnuson–Moss and the act, will be used throughout this chapter.) It will be argued in this chapter that although consumers have long been frustrated with the lack of warranty coverage on their durable goods, protection did not come until the political and social environment was conducive to the passage of a substantive piece of consumer legislation. The chapter concludes with a postpassage evaluation of Magnuson–Moss from a consumerist perspective and with speculation on possible new legislation to address what consumerists might perceive as deficiencies of this act. Additional information concerning the evolution of warranty protection in the twentieth century is contained in Chapter 2 of this Handbook.

Although Magnuson–Moss eventually passed during a consumer movement that gained momentum during the era of the Great Society, two other consumer movements preceded this last one. The first two movements will be described, accompanied by speculation as to why no federal warranty protection was legislated prior to the 1970s. The third consumer movement will be discussed from the perspective of how this movement provided the impetus for the passage of Magnuson–Moss.

28.2 THE FIRST CONSUMER MOVEMENT

28.2.1 History of the First Consumer Movement

The first consumer movement began around the turn of the century [2,3]. This movement was in reaction to marketplace excesses that were rooted in the industrial revolution of the nineteenth century. The production of goods was rapidly moving from the home to factories where the user no longer oversaw the quality of the process. In addition, whereas most goods had been produced and purchased locally, many goods were now being manufactured in distant cities and purchased through mail order. This growing impersonality of the marketplace made the seller feel less responsible for buyer satisfaction.

The major consumer problems during the first consumer movement, as exposed by writers of the period, were the quality and safety of foods and drugs, and the growing marketplace power of the business trusts [2,3]. These problems were address around the turn of the century by a new group of political reformers who emerged in the urban middle and professional classes, forming the Progressive Movement. These reformers were concerned with many of the social conditions of the day and sought to use the power of government to bring about the and economic changes, including marketplace reforms [2]. Their influence resulted in the first federal consumer laws ever enacted, the 1906 Pure Food and Drug Act and the Meat Inspection Act. Although these laws were somewhat ineffective in getting at the root problems of unhealthy and unsafe foods and drugs and had to be amended several times in the succeeding decades, they set a precedent for federal involvement in protecting the consumer. Another milestone federal action took place in 1914 with the creation of the Federal Trade Commission (FTC), which was intended to bring the power of trusts under control. However, this action only indirectly benefited the average citizen, as the mission of the FTC was not to give direct protection to the consumer but to proscribe unfair competition in the marketplace. However, by the end of the decade, unfair competition was interpreted by the FTC as, "to cheat the consumer is to cheat the competitor." Using this legal theory, the FTC embarked upon a program to regulate deceptive advertising.

The first consumer movement came to an end with the onset of World War I when national attention was refocused from domestic consumption to winning the war [2]. Although consumer protection legislation during this period was somewhat limited, the first consumer movement set the stage for consumer action in later decades. "This appearance of consumer consciousness was perhaps the most important result of the first consumer movement" [2, p. 8].

28.2.2   Warranties and the First Consumer Movement

Why were warranty problems not on the agenda of the early consumerists? Presumably warranties were a significant consumer problem at that time. Preston [4] documents a gross lack of warranty protection, from 1534 when the legal concept of *caveat emptor* entered common law, through the passage of the Uniform Commercial Code in most states by 1969.

This first consumer movement was not characterized by any significant effort to directly protect the consumer from fraud and deception either in the sale or in the warranting of goods and services (other than in the sale of food and drugs). The nation was still in an era dominated by the principle of *caveat emptor*; if the marketplace was to be regulated, it was the responsibility of the individual states. In addition, the moral principles of the time stressed the attitude of self-reliance. The individual, not the state, had the responsibility for self-protection in the marketplace [4].

Caveat emptor and self-reliance could be more readily justified in the early part of this century. Most products were relatively simple. The electric home was mostly just a dream, and the automated home was beyond the realm of fantasy. The lack of technologically advanced products, and the household resources to purchase them, meant that advanced durable goods were not a part of the household stock. One could compare different brands of ice boxes, stoves, or person-powered clothes washers with manual wringers and easily discern which brand was more sturdily made. Most products did not rely on a source of nonhuman power, and they did not require automatic controls that are often the cause of problems in more modern household products. Other than a wood or coal stove, the primitive hand-powered washing machine, the spring-powered phonograph, the electric lamp, and an occasional powered vacuum cleaner, the typical household had few durable goods that needed repair by a commercial specialist. The exception might have been the automobile. Although the automobile did become a part of the household stock during the first consumer movement, the automobile owned by the average citizen was a relatively simple machine that could often be repaired in the backyard with common tools.

When one takes into consideration that the early consumerist's resources needed to be directed toward safe foods and drugs, the prevalent belief in self-reliance, and the fact that most households owned a few very simple durable goods, one can see why the literature might not document a public outcry for warranty protection legislation.

28.3 THE SECOND CONSUMER MOVEMENT

28.3.1 History of the Second Consumer Movement

In his book on consumer movements, Mayer [3, p. 19] states:

> Whereas the first wave of consumerism in the United States was an outgrowth of the massive changes wrought by the early stages of the Industrial Revolution, the second wave may be conceived as a response to the revolution's broadening impact—beyond the factories and transportation systems to the home, the domain of consumption itself.

After World War I, the United States went into an era of "normalcy" and relative affluence. Consumers had greater real income than ever before, and a result of this affluence was a heightened need for consumer information due to improvements in technology and widening choices [3]. The marketplace was chiefly characterized by an increase in both the volume and sophistication of advertising. Many new durable goods were coming into the market and were diffused widely among an increasingly affluent public. These durable goods included such electric-powered devices as radios, washing machines, toasters, kitchen ranges, irons, refrigerators, and some personal care appliances. Most of these were a quantum leap in technological development compared with earlier manual models. In addition, the automobile was becoming much more complex.

Added to technological advances and affluence were rising consumer expectations. These were often a function of advertising claims for the good life that could result from owning all the new products on the market. Yet, much advertising was either deceptive or noninformative. It was in this environment that the second consumer movement began.

Books began appearing that chastised marketers for wasting the consumer's dollars and endangering the public's health. One of the more influential books of the time was *Your Money's Worth*, by Stuart Chase and Frederick Schlink [5]. The authors summarize the consumer situation of the 1920s: "We are Alices in a Wonderland of conflicting claims, bright promises, fancy packages, soaring words, and almost impenetrable ignorance'' [5, p. 2]. This book proposed the formation of a consumer-sponsored organization to test products. The proposal resulted in the formation of Consumers' Research, from which Consumers Union, the publisher of *Consumer Reports*, arose as a splinter group in 1936.

Although the consumer may have had enough discretionary income to waste some of it in the 1920s, by the 1930s the Great Depression had forced most Americans to shop more efficiently. ''The depression made the problems of the consumer far more immediate and compelling than they had been in the palmy Twenties'' [2, p. 10]. In addition, because business took credit for the affluence of the 1920s, it was only logical for citizens to blame business for the economic woes of the thirties. The second consumer movement accomplished many things for the consumer. Roosevelt's New Deal recognized the rights of the consumer to some degree. More importantly, consumers were given additional health protection in 1938 through the Food and Drug amendments, and the passage of the Wheeler–Lea Act strengthened the FTC. But like the first consumer movement, the second movement came to an end with a war, this time World War II [2,3].

28.3.2 Warranties and the Second Consumer Movement

During the 1920s and 1930s one might have expected some legislative attention to consumer warranties. Herrmann notes, ''The problems of consumers were aggravated by a flood of shoddy merchandise at bargain prices'' [2, p. 11]. In addition, the courts still gave consumers little warranty protection; implied warranties for consumers were mostly unknown. Furthermore, the advertising of this era was confusing and heightened expectations of greater utility from consumer products.

Given the growth of unmet expectations, the increased complexity and diffusion of durable goods, the newness in product design and yet-to-be-ironed-out defects, and the need to conserve one's economic resources, we would expect a public outcry for legislation for warranty protection in the 1920s and 1930s. Yet the literature from this period does not indicate any such demand. The person most often recognized as the father of consumer education, Henry Harap, in his 1923 book, *The Education of the Consumer* [6], thoroughly discusses such topics as the consumption of food, fuel, housing, clothing, and furnishings but does not even mention guarantees. Even Chase and Schlink in their 267-page book devote scarcely 3 pages to guarantees and this is summarized in one sentence: ''It [a warranty] is usually just another selling word; sound and fury signifying nothing at all'' [5, p. 248].

Other prominent consumer economists of the time devoted little space to the discussion of warranties. In 1939, Leland Gordon [7], the father of college consumer economics texts, devoted only 3 of 631 pages to guarantees. He argues that guarantees were used to give the reputable seller a competitive advantage and used by less scrupulous businessmen to their own advantage. Gordon summarizes the value of guarantees to the consumer as totally a function of the reliability of the guarantor. Babson and Stone [8], in their 1938 text on consumer protection, devoted only one paragraph to guarantees; they summarize by saying, "On this topic [warranties] the schools can probably do the most good by teaching the dangers of relying blindly upon warranties, to the neglect of other measures of self protection" [8, p. 161].

Consumer activists of the 1930s also seemed to give low priority to consumer warranty problems. In his recollections of his consumer activities, Colstron Warne [9], the founder of Consumers Union in 1936, never mentions any efforts to get greater warranty protection for consumers. Neither *Consumer Reports* nor *Consumers Research*, the two primary consumer magazines of that era, editorialized for warranty legislation.

If problems with consumer goods were increasing both absolutely and in relation to expectations during the second consumer era, and if there were no improvements in warranty protection, then why was there no apparent consumer demand for legislation? We can only speculate. First of all, as with the first consumer era, a major consumer problem continued to be with drugs. Also, economic problems such as warranty protection were lower in priority when competing against health and safety issues. Weak and deceptive warranties cannot compare with the specter of children dying as the result of dangerous products.

As stated earlier, the second consumer movement came to an end with World War II. Ironically, the nation's productivity, honed during that war, would eventually lead to the passage of warranty legislation on the national level.

28.4 THE THIRD CONSUMER MOVEMENT

28.4.1 Precursors to the Third Consumer Movement

Mayer [3, p. 25] summarizes the reason why consumerism was not a social force immediately following the war:

The decade following World War II was not hospitable to any form of social movement, including consumerism. After the shortages of the war years, consumers were eager to start spending their wartime savings and make up for past deprivation. It wasn't necessary to have the best refrigerator for the money as long as you had a new one.

The post-World War II era was a time of massive consumption of durable goods. Those on the market before World War II were now refined with added bells and whistles. For example, automobiles had more cylinders, frills, air conditioners, and automatic transmissions. Clothes washers were now automatic and phonographs were now hi-fi and stereo. New and technologically advanced products were entering the market. Consumers began to equip their homes with clothes dryers and dishwashers. Alongside the car were power lawn mowers and snow blowers. Electric personal care products became commonplace. Advertising that raised expectations with overstatements and deception during the 1920s was still commonplace according to *The Hidden Persuaders*, the 1957 best seller by Vance Packard [10]. By the end of the 1950s, society had many but not all of the ingredients for a third consumer movement.

Some of the critical ingredients for a consumer movement that were missing in the 1950s emerged in the 1960s. According Mayer [3], the consumer movement needed such things as exposé books, widely publicized tragic events, and leadership to become a social movement that catches the public's interest. The sixties began with symbolic consumerist leadership at the presidential level. In 1962, President Kennedy gave a message to Congress in which he outlined a Consumer Bill of Rights which confirmed the rights to safety, information, choice among a variety of products and services at competitive prices, and the right to a fair hearing in the formation of consumer policy. There was no mention of warranties.

28.4.2   The Third Consumer Movement Gains Momentum

In the 1960s, consumerism gained momentum quickly on two fronts. First, there was the thalidomide incident in which approval of a drug that caused deformed babies was narrowly averted. This event created pressure for further strengthening of drug laws. The second catalytic event was the congressional automobile hearings wherein Ralph Nader became synonymous with consumerism through his testimony about the auto industry's lack of concern for the motorist's safety. These hearings severely damaged the image of the auto industry. The early sixties also witnessed the passage of the first major pieces of federal consumer legislation, the Truth in Packaging Act. This act, as weak as it was, helped pave the way for additional consumer legislation in this era of consumerism.

After embarrassing the auto industry, Ralph Nader became the highly visible symbol of America's fight against the corrupt practices of big business. Nader gave consumerism the leadership it needed. His influence was important in getting consumer legislation enacted.

With President Johnson's Great Society, legislation to protect the consumer focused on laws intended to help the economic well-being of the poor. Beginning with Truth-in-Lending legislation in 1969, Congress passed a series of credit legislation that included the Fair Credit Reporting Act (1970), Equal Credit Opportunity Act (1974), and the Fair Debt Collection Practices Act (1977). Not all legislation was focused on helping the poor, however. In the area of safety, the National Traffic and Motor Vehicle Safety Act and the Consumer Product Safety Act (1966) were enacted. When it came to legislation in the period between the mid-1960s and the mid-1970s, consumer activists were on a roll.

According to Americans for Democratic Action ratings, the Senate Consumer Subcommittee, which was the source of much consumer legislation, including Magnuson–Moss Warranty, was controlled by much more liberal senators than Congress as a whole during the early 1970s [11]. An additional reason for the rise of consumerism during the late 1960s and early 1970s may be attributed to the loss of public confidence in big business from 55% in 1966 to 27% in 1971 [12].

Even though consumerism was in full swing during the late 1960s and early 1970s, consumerists did not get everything they wanted. In his in-depth study of consumer warranties and puffery in advertising, Preston notes [4, p. 31]:

> . . . the vast movement of sellerism which began so long ago is not yet over. Consumerism predominates, but sellerism is not yet dead. What we have in the 1970s might be called an incomplete consumerism flavored heavily by some stubborn features of now-incongruous sellerism which refuses to be pushed off the scene. For the time being, these features are immovable objects, while consumerism is an irresistible force.

Even this era of "incomplete consumerism" was too much for the sellerist forces. Coupled with the growing conservatism in the Senate was the stagflation of the late 1970s and the regulatory learning curve of industry. Industry belatedly became adept at organizing and fighting legislation that was not in its best interest [13]. In addition, whereas public confidence in business had eroded from 1966 to 1971, this was reversed between 1975 and 1977 [14]. These events culminated in the regulatory backlash of the late 1970s. Finding government regulation a good scapegoat for the nation's economic problems, Congress began to weaken the authority of regulatory agencies, particularly the FTC. When Ronald Reagan took office, one of his primary objectives was "to get government off the backs of people," which really meant eliminate or not enforce what Republicans considered antibusiness regulation. From many perspectives, the third consumer movement was now history.

28.4.3 Warranties and the Third Consumer Movement

Warranties as a Nonissue

In 1971, Michael West wrote in the *Journal of Consumer Affairs* [15, p. 155]:

> Effective warranties are most important to consumers, especially in this age of product complexity and sophistication, and it is precisely at this age that warranties put on a disappearing act. Manufacturers have succeeded in so disclaiming the warranties that are provided by law that the expectations of the consumer are invariably frustrated.

There may have been a lot of consumer frustration with warranties, yet consumer protection concerning warranties was not the highest priority for consumer advocates during the heyday of the third consumer movement. Interest in warranties did not emerge until almost the end of the 1960s. In an analysis of the first proposed federal warranty law, Schoenfeld noted in 1971 [16, p. 891], "Consumer groups have not yet become active on warranty and FTC regulation."

In 1963 the federal Consumer Advisory Council identified 10 fields of pressing interest to the consumers. Warranty protection was not on the list. A 14-page report to the president from the Consumer Advisory Council in 1966 made only a passing comment on warranties [17]. The only concern was that the new extended warranties (service contracts) being offered might induce the consumer to feel obligated to have all maintenance performed by the manufacturer's authorized dealer, thus limiting competition. It is interesting that this report ignored warranty problems even though the council included several prominent consumer advocates in its membership. In his 1968 consumer message, President Johnson indicated that he would not press for passage of warranty legislation until he had time to explore with businessmen possible ways to improve quality control and coverage [18].

A search of the periodicals from 1945 to 1975 reveals few articles that addressed consumer problems with warranties. Table 28.1 shows that the articles that did appear were mostly published in *Consumer Reports* and *Consumers Research*. The only news magazine that consistently reported on warranty issues was *Business Week*. Even in these publications,

Table 28.1 Warranty Articles in Popular Magazines 1945–1975[a]

Years	Consumer reports	Consumers research[b]	Changing times	Business week[c]	Other	Totals
1945–1952	0	0	0	0	0	0
1952–1960	1	3	1	4	2	11
1961–1965	4	6	2	3	1	16
1966–1970	3	1	1	3	5	13
1971–1975	1	2	1	2	8	14

[a] Information for this table taken from *Readers Guide to Periodic Literature*, 1945 to 1975. Only articles that address warranties as an *issue* are counted. "How-to" articles (e.g., how to read a warranty) were omitted from the tabulation.
[b] *Consumers Research* went under the title of *Consumers Research Bulletin* and *Consumer Bulletin* from 1957 to 1973.
[c] *Business Week* was the only popular magazine during the period of time studied that had more than two articles on warranties.

the emphasis focused on specific warranty problems while an explicit need for passage of federal warranty protection was not addressed.

David Caplovitz [19,20] a well-known author of books on the problems of low-income consumers, never mentioned consumer warranties in his writings. Alan Andreasen [21], in his book on the consumer problems of the disadvantaged, does not cite warranties as an underlying cause of financial problems of the poor and minorities. The well-known consumer writer Sidney Margolius [22], listed six overriding problems of consumers. Although the list included quality problems that have led to high repair costs, especially for automobiles, he pointed to quality standards as the problem and not to warranties. Nader appears not to have addressed the need for warranty legislation. In a *Ladies Home Journal* [23] article, Nader only goes as far as to state that the implied warranty exemption that often occurs in express warranties should be eliminated and that consumers should not be required to pay the cost of returning goods covered by warranty. A *Business Week* [24, p. 31] article quotes Nader as saying that the 1969 FTC report critical of automobile warranties was ". . . a personal letter to the automobile companies." However, in a 1968 piece, Nader [25] listed 10 major forces that need to be strengthened for there to be a decent consumer society, and warranty protection was not one of them. Most surprisingly, even Senator Magnuson never mentions warranty problems in his 1967 book with Jean Carper, *The Dark Side of the Marketplace* [26]. College consumer texts by the two major authors of the time,

Troelstrup [27a] and Gordon [27b], did not emphasize warranty problems in their respective texts. In a review of high school level consumer education texts from 1938 to 1978, Herrmann [28] does not list warranties as a major topic addressed by these books.

In addition to many consumer writers and advocates not pressing for warranty protection for the consumers, at least one study concerning major appliance warranties conducted prior to the passage of Magnuson–Moss found that consumers held favorable attitudes toward existing industry trade practices [29].

Automobiles Create Warranty Issue

The warranty problems that precipitated the eventual passage of the Magnuson–Moss are attributed by most writers to problems with automobiles. Americans were buying cars in record numbers in the 1950s and 1960s; as stated in the beginning of this chapter, the warranty wars started in 1960 when Ford expanded its warranty from 90 days/4000 miles to 1 year/ 12,000 miles. The automobile industry was quick to see warranties as a marketing tool, and by 1962 Chrysler decided it could reverse its sagging market share by offering a 5-year/50,000-mile warranty. Then the warranty wars among the big three auto manufacturers, plus American Motors, began in earnest. By the late sixties, the battle was abating as the manufacturers began to realize the substantial expense of offering these expanded warranties. In 1969, Ford paid out $300 million in warranty protection, General Motors $580 million, Chrysler $200 million, and American Motors $35 million. "Now, many harried Detroit executives rue the day in August 1962, when Chrysler first expanded its warranties" [30, p. 44]. Warranties were not only costing the manufacturers, consumers were paying about $75 each year for the extra maintenance to keep the warranty in effect [18]. Presumably because of their own costs, manufacturers began decreasing warranty coverage. The reduction of warranty coverage in the late sixties after manufacturers had expanded it in the early sixties caused ill feelings among consumers [24].

One might expect that the reduction in warranty coverage would give rise to consumer action on the warranty front. Consumers were losing protection to which they might have thought they had rights. Yet in recent years, auto manufacturers have alternately increased and decreased warranty coverage and this seems to have stirred no great consumer reaction.

Not only were manufacturers doing a yo-yo trick with the length of coverage, they did not bundle them with a higher-quality product nor were they willing to give the warranty service to which consumers thought they were entitled. A July 1970 *Business Week* quote is quite telling [30, p. 44]:

Action [expanded warranties] raised hopes of millions of car buying Americans and seemed to promise a whole new era of problem-free autos and inexpensive service—two giddy expectations that Detroit not only failed to fulfill, but which the auto capital in a gigantic U-turn has lately been trying to discourage through decreased warranty coverage.

During this period, *Consumer Reports* consistently reported poor workmanship in the automobiles they tested. In April 1967, *Consumer Reports* [31] stated that if quality and workmanship were criteria for their category of "Not Acceptable" (this category is usually reserved for substantial and incurable safety defects), most of the cars tested for that year would be rated as such. In the same article, *Consumer Reports* stated [31, p. 194]:

> Indeed, the new car warranty is sometimes an inverse indicator of quality control. When a manufacturer states it is "liberalizing" that warranty, watch out. Faced with a choice of tightening inspection procedures or extending terms of warranty, Detroit tends to choose the latter course as the one yielding more immediate profit.

Consumer Reports' annual questionnaires revealed that of those respondents whose cars had defects, 25% failed to get proper attention and that the manufacturer was blamed more than the dealer [32].

If consumers thought that greater warranties were an indicator of greater quality, they were greatly disappointed. In addition to poorer quality of manufacture was the problem of poorer quality service. As durable goods became more complex and therefore more prone to breakdown, they also became more difficult to repair. Due to the advanced skills and not-so-common tools needed, backyard auto repairs were becoming obsolete. The quality of auto repair at service centers also appeared to deteriorate. Auto manufacturers consistently complained that the dealers were doing a poor job of making repairs [30]. A Conference Board [33] report also cites that a major obstacle to adequate repair service is lack of sufficient numbers of competent repairmen. This left the consumers in a double bind. They possessed defect-prone goods that needed repair and then found that it was difficult to get the repairs done. This disappointment also exhibited itself in warranty studies conducted in the late sixties and early seventies.

In 1969 the Federal Trade Commission issued a report on automobile warranties. A field investigation on auto warranties was begun in 1965 after the FTC received numerous complaints about auto warranties. "This [warranty complaints] is one of the largest totals of letters received regarding one topic of consumer complaints since the Commission was established in 1914'' [34, p. 66]. In citing *Consumer Reports* articles, *Newsweek* surveys, numerous complaints received by the FTC, and other sources, the report revealed that consumers have expectations of warranties that are not part of the warranty terms. Furthermore, a substantial portion of vehicles are delivered with defects; a significant amount of warranty work is not performed to customer satisfaction (partly because of the way the manufacturer compensates the dealer for warranty work); manufacturers' costs of underwriting warranties are extremely low considering the purchase price of new cars; and performance of manufacturers and dealers under the warranty has not achieved the levels implied by the warranty. In addition, the report states: "Despite all disclaimers, *sales* are still foremost and *service* retains the status of 'necessary evil' in much of the automobile business'' [34, p. 194]. The report concludes that it ". . . is inescapable that general performance under warranty often is unsatisfactory to a great many—actually millions—of new car buyers'' [34, p. 83]. A 1974 staff report to the House Subcommittee on Commerce and Finance found after reviewing 200 warranties from 51 manufacturers that warranties were replete with limitations and that actions taken by manufacturers and trade associations to clean up their warranties had minimal results [35].

In a study of dispute settlement procedures under automobile warranties, Whitford [36] interviewed dealers, service personnel, manufacturers, and new-car purchasers. His most substantive finding was that consumers were at a disadvantage in dispute resolution because there was a large disparity of power in favor of the manufacturer who set the terms for the procedures on the signing of the purchase contract. Whitford concluded that there was room for some regulation in this area.

Warranty Problems for Appliances

Consumer warranty complaints were not confined to just automobiles. There was also dissatisfaction with appliance warranties. According to *Business Week* [37], warranty coverage on appliances also started to expand in the early 1960s. The Federal Trade Commission [38] came out with a report on appliance warranties and service. Using complaint letters to the Special Assistant to the President for Consumer Affairs and the Federal Trade Commission, the report concluded that the basic cause of consumer dissatisfaction with service provided under a warranty is the failure to fulfill the obligations set forth in the warranty to the extent and in the manner expected by the consumer. The basic consumer problems as outlined in the report were defective products either in design or assembly, delays in making repairs, problems in finding facilities for authorized repairs, excessive labor charges when labor is not covered, failure to honor guarantees, and excessive disclaimers in guarantees.

In 1971 the Major Appliance Consumer Action Panel (MACAP), a mediation organization financed by appliance manufacturers, studied industry's compliance with 10 warranty guidelines that MACAP had written. For some of the guidelines (e.g., name and address of warrantor, geographical limitations, and specific time limitations), there was high compliance. However, for items such as accuracy of headings and titles and to whom the warranty is extended, the warranties complied in little more than half of the cases. In addition, for one important item, "In case of a claim: a) exactly what the warrantor will do and at whose expense. b) exactly what the consumer must do and at whose expense" there was only 31% compliance [39]. In a 1973 update, MACAP [40] studied how 106 warranties complied with nine of MACAP's guidelines. They found a 75% compliance and that 17 warranties fulfilled all guidelines compared to only 9 in the 1971 study. So a comparison of the two MACAP studies indicated warranties were improving.

28.4.4 Summary of Pre-1975 Warranty Issues

In summarizing four consumer studies (the two FTC studies mentioned here, the Whitford study, and the Staff of Subcommittee on Commerce and Finance Report on Consumer Product Warranties), Rothschild [41, p. 353] concludes:

> First, warranties do not provide consumers with notice of their rights, duties, or remedies under contract. Second, disclaimers are couched in legal jargon that is unfair to the consumer, and they do not protect the warrantor from unjustified claims. Third, warranty administration in marketplaces is confusing and ineffective due to the complex distributive structures of consumer products. Fourth, consumer remedies for defective products are impractical. Fifth, consumers are not aware of the scope of their warranties or their obligations of care at the time of purchase. Finally, consumer frustration and hostility has resulted in excessive and unjustified claims.

These studies often pointed to a classic example of a gulf between consumer expectations and seller delivery and this typically results in consumer disillusionment, frustration, anger, and a cry for action. Consumers' expectations of product quality and reliability had been steadily rising due to the introduction of new products and the manufacturers' advertised emphasis on improvements. However, according to Aaker and Day [14], many improvements make products more complex with new possibilities for malfunction. As consumers became more aware of warranties, manufacturers were using warranties much more extensively as a sales gimmick. In April 1963, *Consumer Reports* [42, p. 158] declared: "The automobile warranty has come a long way—from the seldom-read pages of the owner's manual to the nation's billboards."

Yet a cry for action did not occur to the extent that one might expect. As noted earlier, there were relatively few magazine articles calling for action and there was no indication that consumer leaders had warranty protection high on their agendas. Ralph Nader apparently did not put warranty protection in the same league as automobile crash protection or protection from unsafe foods and drugs. Also, as indicated earlier, the writings of consumer advocates did not show concern with protecting consumers from warranty problems.

We might assume that the lack of attention by the media and activists was a reflection of public disinterest. Why was there not a great public demand for action? Bishop and Hubbard [43, p. 174] put the blame on consumer resignation: ". . . consumers are so inculcated with the doctrine of *caveat emptor* that they accept occasional bad luck on a new purchase as part of the game, as though it were a lottery." In addition, consumers may have seen warranty coverage offered by the manufacturer as a marketing tool within the rights of the manufacturer to manipulate and not as a specific consumer right. Another possible explanation that is implicit in Bishop and Hubbard's book and the works of others is that many of the warranties are written in such a way that it puts most of the responsibility on the consumer to adhere to the conditions of the warranty. For example, an automobile warranty may require oil changes at specific intervals and if the consumers do not comply, they may feel that they have not kept up their part of the bargain; or, it might have been that consumers had been so used to not getting warranty protection, they did not expect it when it was supposed to be available.

28.5 CONSUMER ENTREPRENEURS AND THE PASSAGE OF MAGNUSON–MOSS

28.5.1 Role of Consumer Entrepreneurs

In his book, Michael Pertschuk [13], who was Warren Magnuson's legislative director and who later became FTC chair, discusses the principle of *consumer entrepreneurial politics.* According to Pertschuk, and Wilson

[44], entrepreneurial politics is where a proposed policy will confer general benefits (though perhaps small) at a cost to be borne chiefly by a small segment of society. "The entrepreneurs serve as the vicarious representatives of groups not directly part of the process" [44, p. 310].

Pertschuk wrote that consumer issues by nature do not assume priority among citizens' competing economic concerns and may dominate the political agenda only under special circumstances, as in rent strikes or problems of the elderly. Pertschuk [13] believed that there was not public ambivalence about consumer initiatives, but that consumers had a limited stake in each issue, which when taken together are viewed as consumer legislation.

Although there were several congressional consumer entrepreneurs, the primary entrepreneur was Senator Warren Magnuson, a Democrat from Washington. Consumer advocacy seemed a natural for politicians. "Consumer issues were homey, usually simple in conception, and of broad general interest" [13, p. 20]. According to Pertschuk, Magnuson had not always been a consumer advocate; he assumed this role after nearly losing an election in 1962. He appointed Michael Pertschuk as his legislative director with the assignment ". . . to help build a consumer record for Magnuson, to identify opportunity, develop strategies, shepherd bills, and make certain that Magnuson received appropriate acknowledgement for his achievements" [13, p. 25]. A consumer law that would fit Magnuson's political agenda would be one that addressed warranties as a consumer concern, as such a law had a "motherhood and apple pie" appeal. It evoked the image of the poor consumer being ripped off by the large corporation mainly for an item that devoured a large part of the family budget: the automobile. Also, it was an opportune time to legislate on an automobile issue because it was not long after Nader had diminished the credibility of the auto industry.

According to Hasin [45], Pertschuk was a primary political entrepreneur behind Magnuson–Moss. Tolchin and Tolchin [46, p. 148] stated that his power was "so extensive that he was sometimes called the 101st Senator."

Members of Congress and their staffs were not the only consumer entrepreneurs on the warranty bandwagon. Consumers Union actively pushed for passage of Magnuson–Moss through articles in *Consumer Reports* as well as through the Consumer Federation of America, US Public Interest Research Group (a group founded by Nader) and the National Consumer Law Center and organized labor.

There seemed to be no particular advocacy or investigative journalists crusading on the issue of legislating consumer protection in warranties. Referring to Table 28.1, the only nonconsumer periodical that had more than one or two articles about warranties was *Business Week*. There were no books such as *Unwarranted at Any Speed* or *Warrantors Pay Less* to catch the public interest.

28.5.2 Joining of Warranty Protection and FTC Improvement

Warranty protection and political advancement was not the only motivation behind the passage of the Magnuson–Moss. Recall that the full title of the warranty act is "Magnuson–Moss Warranty and Federal Trade Improvement Act." Consumerists had long been critical of the FTC for not being active enough on behalf of consumers. Two studies, one by a Nader Group [47] and another by the American Bar Association [48], had been highly critical of the FTC in its failure to fulfill its mission. Consumerists wanted to expand the powers of the FTC by allowing it to issue rules that covered practices of an entire industry rather than proceeding against abuses on a case-by-case basis. The FTC improvement part of Magnuson–Moss was just as important, if not more so, than the warranty provisions [13,45]. Yet, changing the internal working of the FTC as it bore on protecting consumers was a more abstract concept to voters than the much more tangible issue of warranty protection. Because the warranty portion of the act had greater appeal to the public, these two issues were bundled in order to assure passage of a more comprehensive consumer protection package [45]. Pertschuk [13] states that consumer entrepreneurs took advantage of the energy behind the warranty bill to get FTC reform legislation passed.

28.5.3 Impotence of Business Community Resistance

Although the political entrepreneurs were the primary force behind the passage of Magnuson–Moss, their job would have been much more difficult had there been solid opposition from the business community and its political allies. The Nixon administration wanted weaker legislation for both the warranty provisions and for the FTC improvements (see testimony of Virginia Knauer, Special Assistant to the President for Consumer Affairs, before the Subcommittee on Commerce and Finance, 1973 [49]); thus, business had a potential ally in the White House. The Commerce Department testified in Congress that voluntary action by industry was a better solution to consumer problems than legislation [50]. However, during the early part of the third consumer movement, business and its trade group representative had relatively low credibility and influence and were generally on the defensive [13].

28.5.4 Magnuson–Moss Warranty Act Passed

The Magnuson–Moss Warranty Act—Federal Trade Commission Improvement Act finally passed in 1974 and became law in 1975. The evidence indicates that passage was facilitated by the impact of the third consumer movement in general and consumer entrepreneurs in particular. There was general unrest about automobile warranties, but this issue did not have the dramatic appeal that had inspired health and safety laws. Warranty problems did not arouse the images of deformed children with no arms and legs that inspired the drug amendments of 1962, of mangled bodies in auto wrecks that were a force behind automobile safety legislation, or of maimed children and elderly that were the basis for the Consumer Product Safety Act. Consumer entrepreneurs operating in the political and social environment of the third consumer movement were the necessary catalyst to galvanize the consumer unrest with warranties into passage of Magnuson–Moss.

Did the passage of Magnuson–Moss achieve the aims of consumerists in giving consumer protection from warranty problems? Or did business and business-oriented members of Congress let this bill slip by because it really only gave cosmetic relief to consumers? The next section evaluates the benefits of Magnuson–Moss to the consumers.

28.6 EVALUATION OF MAGNUSON–MOSS FROM A CONSUMERIST PERSPECTIVE

28.6.1 Basic Goals of Magnuson–Moss

The three basic goals of Magnuson–Moss were to provide consumers with information through mandated disclosures and labeling (this supposedly would also increase competition among companies to offer better warranties), to improve the quality of warranties by providing for full warranties, and to create procedures for consumer remedies through informal dispute resolution programs and through private action, including class action. Have these goals been met and do they give *real* warranty protection to the consumer?

Readability of Warranties

An early study by the FTC found three possible changes in the content of automobile warranties that may have come about from the passage of the act. One change was the increased readability of warranties [51]. Schmitt et al. [52] found that warranties became slightly more readable when measured by a readability index; however, most still were far short of the act's standard of "simply and readily understood." Using a sample of 1988 warranties, Moore and Shuptrine [53] analyzed the readability of 121 warranties using standard readability tests. They found that the readability levels in warranties are beyond the reading abilities of most Americans.

Disclosures and Disclaimers in Warranties

Although disclosure is one of the primary sections of Magnuson–Moss, the act has many limitations that impede this goal. According to Brickey [54, p. 85], "The restrictive approach of the Act raises fundamental questions regarding the wisdom of selecting disclosure as the primary mechanism for achieving the stated legislative goals." Brickey goes on to say that the mechanics of compliance with disclosure rules may prove to be a disincentive to making disclosures and that the legislation does nothing to prevent consumer deception nor does it provide a remedy in the event that the product characteristics are misrepresented. Brickey also makes the interesting argument that Magnuson–Moss errs in not requiring information that would enable a consumer to evaluate a risk associated with the benefits.

Warranty disclaimers, a critical issue for consumerists during the congressional hearings of 1970, 1971, and 1973, still haunt the vast majority of warranties. The act may give the consumer an illusion of protection, but the fine print of the warranty may still take it away.

One of the primary disclaimer problems with warranties, and one that is addressed often in the literature and in the subcommittee hearing, is the disclaimer of responsibility by manufacturers. *Consumer Reports* [55] found that such disclaimers often negate other stated terms of the warranty, terms which a consumer might typically expect from a warranty and from implied warranties. This implied warranty disclaimer is most troublesome. It has been claimed by consumer advocates that the primary purpose of the express warranty is to eliminate consumer rights under an implied warranty of merchantability. Although consumerists wanted a total elimination of disclaimers, especially those disclaimers of implied warranties, congressional hearings on the bill show clearly that business was totally opposed to it. Business and the United States Department of Commerce thought that the elements of a warranty should remain as a point of negotiation between the buyer and the seller. A representative of the Commerce Department stated, "We think that is a basic commercial right—both buyer and seller—to freely and knowingly allocate the economic risks relating to the quality of goods" [56, p. 35].

Although a few states have prohibited the disclaimer of implied warranties, most states continue to permit them. Therefore, consumers still may have fewer rights under an express warranty while being under the illusion that the warrantee has given them special rights. For example, the typical VCR has a warranty that covers labor for 3 months and parts for 1 year [57]. Yet, a study done by a national service contract company found that only 7% of the 45 million VCRs need servicing in the first year and 43% need service within the first 5 years [58]. Would it not be reasonable for a consumer, or for that matter a court of law, to expect a $300 VCR to be trouble-free for more than 3 months? Yet, in all but 43 states, the limited warranty offered with these VCRs gives considerably less coverage than an implied warranty of merchantability. An appliance longevity study reported in *Changing Times* [58] found the life span of color TVs to be 8 years; dishwashers 10 years; stereo systems, air conditioners, and microwaves 11–14 years; clothes washers 13 years; and refrigerators 17 years. Again, most of the basic warranties on these products only cover a very small portion of the time that a consumer owns it—a much shorter time than one might reasonably expect. It appears that manufacturers of these products have minimized their warranty risks.

In addition, there are all sorts of other disclaimers that negate assumed benefits of the warranty. For example, who determines if a warranted product failure is due to a factory defect or consumer abuse? Presently, it is the burden of the consumer to prove he or she is not at fault. Shipping, handling, deductions, copayments, and sometimes "diagnostic" charges can make warranty repairs almost as costly as nonwarranty repairs. Even with Magnuson–Moss protection, consumers may still find themselves with a warranty that may be no more than a pretty piece of paper contributing to the bulk of home files filled with other pretty pieces of paper.

If a consumer has a full warranty, he or she does not have many of these disclaimer problems that are cited above because most are prohibited under the act. Although several products do have full warranties, albeit full warranties on only part of the good and limited warranties on other parts (e.g, full warranty on picture tube but limited warranty on television chassis), the vast number of warranties fall in the limited category.

28.6.2 Two-Tier Warranties

Congressional testimony regarding passage of warranty protection often reflected the fear that by having two categories of warranties (full and limited), most manufacturers would opt for limited warranties. An industry group survey conducted soon after the passage of the act concluded that consumers were getting less coverage, as two-thirds of the manufacturers surveyed switched from what in effect a full warranty to a limited warranty when the act took effect [59]. McDaniel and Rao [60] came to similar conclusions. However, an early FTC study [52] that compared warranties before and after passage of the act found more warranties would fit the definition of a full warranty before than after passage. In a more recent study, Moore and Shuptrine [53] concluded that act may have increased the relative number of limited warranties and that limited warranties are more difficult to read than full warranties.

28.6.3   Warranties and Competition

A primary rationale for the passage of Magnuson–Moss was that it would stimulate competition among manufacturers to offer better warranty protection. As mentioned above, Wisdom [61] investigated the change in warranty protection, pre- and post-Magnuson–Moss and found that increased disclosure was not improved by the act. Wiener [62] also studied changes in coverage in warranties pre- and post-Magnuson–Moss. He found significant declines in coverage for five products and significant increases in coverage for three sets of goods. Although an in-depth analysis of post-act warranty coverage is not attempted here, a review of coverage offered on 1992 automobiles and on recent models of other household durable goods was attempted to determine if there is substantial variation among warranties offered by different manufacturers. Lack of variation might indicate that disclosure provisions of Magnuson–Moss have not prevented market failure, which often results from the actions of insufficiently informed consumers.

A review of automobile warranties summarized in *Consumers Digest* [63] shows little variation among the warranties offered by foreign and domestic automobile manufacturers. With minor variations within warranties and with few exceptions among manufacturers, all the major manufactures offered 3-year/36,000-mile warranties. Luxury vehicles had better warranties and the Yugo had only a 1-year/12,000-mile warranty.

In order to gauge the variability of warranties for durable goods other than automobiles, warranty coverage of 16 common household products tested in *Consumer Reports* were examined. As shown in Table 28.2, with the exceptions of televisions, which had a bimodal distribution, there was little variation among home entertainment equipment. For the kitchen and laundry appliances, where there were more brands that differed in warranty coverage from the modal warranty, the substance of the divergence was often minimal. For example, for the refrigerators, there were different warranties on the cabinet liner and on the crisper drawers—items that are unlikely to fail.

One might speculate that the reason for greater diversity in warranties among kitchen and laundry appliances in relation to entertainment equipment is that the former group represents more mature goods, whereas the latter group represents many goods that are competing on new technological features. Therefore, one might conclude that when competitors have only minor differences to offer in regard to product features, they may compete in the quality of warranties offered.

The review of the automobile warranty data and the warranty data for other household products appears to indicate that competition among manufacturers regarding warranty coverage is minimal. One explanation might be that manufacturers may just be responding to consumer demand in not using warranty coverage as a competitive tool. A *Wall Street Journal* article [64] states that although warranties have been used as an alternative to price competition, only 7% of major appliances were purchased primarily because of the warranty protection offered. In evaluating war-

Table 28.2 Warranty Coverage on Selected Household Durable Goods[a]

Durable good and typical warranty[b]	Date reported[c]	Number tested	Better warranties[d]	Worse warranties[e]	Percentage deviation[f] (%)
Entertainment Goods					
Stereo receivers					
2 yr, parts/labor	March '94	9	0	2	22.2
CD players					
1 yr, parts/labor	March '94	14	0	2	14.2
Speakers					
5 yr, parts/labor	March '94	23	1	2	13.0
Camcorders					
1 yr, parts/3 mo, labor	March '94	14	1	1	14.2
Cassette decks					
1 yr, parts/labor	March '94	12	0	0	0.0
VCRs—stereo					
1 yr, parts/3 mo, labor/					
1 yr, heads	March '94	14	1	1	14.2
TV—color, 27 in					
1 yr, parts/3 mo, labor/					
2 yr, picture tube	March '94	21[g]	8	1	42.8
Rack entertainment systems					
1 yr, parts/labor	December '93	16	2	0	12.5

Table 28.2   Continued

Durable good and typical warranty[b]	Date reported[c]	Number tested	Better warranties[d]	Worse warranties[e]	Percentage deviation[f] (%)
Kitchen and Laundry Appliances					
Breadmaker					
1 yr, parts/labor	December '93	11	1	1	0.9
Dishwashers					
1 yr, parts/labor					
10 yr, tub	October '93[h]	N/A	N/A	N/A	N/A
Slow cookers					
1 yr, repair/replace	January '93	12	1	0	0.8
Microwaves					
1 yr, parts/labor					
5 yr, megatron	December '92	16	7	0	43.7
Refrigerators					
1 yr, parts/labor					
5 yr, sealed unit	July '92	15	5	1	40.0
Clothes Dryers—Elec.					
1 yr, parts/labor	July '93[i]	N/A	N/A	N/A	N/A
Miscellaneous Goods					
Room AC—small					
1 yr, parts/labor					
5 yr, on sealed system	July '93	10	0	0	0.0
Power circular saws					
1 yr, parts/labor	November '92	15	1	1	13.3

[a] All brands selected were recently tested in *Consumer Reports*.
[b] "Typical warranty" is the modal warranty for that class of goods.
[c] "Date reported" is the issue of *Consumer Reports* in which the product tests were reported.
[d] Number of brands with a warranty better than the typical warranty.
[e] Number of brands with a warranty worse than the typical warranty.
[f] Percentage of brands warranties with that were better or worse than the modal warranty.
[g] This category was bimodal with 11 brands limiting labor to 3 months and 8 brands limiting labor to 12 months. Also, the top-rated Sony model was not included in this summary because it was an outlier in the category.
[h] There were 10 different warranties for the 18 models tested, making it difficult to establish a "typical" warranty for this class of goods.
[i] There were five different warranties for the seven models tested making it difficult to establish a "typical" warranty for this class of goods.
N/A, not applicable.

ranties as a market signal, Gerner and Bryant [65] assert that although consumers may spend some time learning about product warranties for a general category of products, they are reluctant to spend much time and energy in evaluating actual warranty provisions on a specific appliance. Economists may offer additional explanations.

Looking at the same data, one economist may be optimistic and conclude that the lack of variation may indicate near-perfect competition or that the market is in equilibrium; another economist might pessimistically conclude there is collusion to fix prices among the sellers. If the same differential in price paid for warranties and the service received for warranties exists from 1992 to 1994 that existed when MIT did its extensive study of warranties (the price was about double the cost) [66], then the market is not in equilibrium, as there would be substantial differences between price and service. In addition, because express warranty periods seem shorter than one would reasonably expect on most durable goods and because warranty coverage is not typically a primary dimension for product selection, we might conclude that there is minimal competition on the basis of warranty coverage. The only exceptions might be a few periods since 1960, when automobile manufacturers used warranties as a significant form of product differentiation.

28.6.4 Warranties as Indicators of Quality

One of the objectives of Magnuson–Moss was that the warranty should be an indicator of the quality of the product; that is, the consumer would be able to use the quality of the warranty as a cue to the quality of the product, thus leading to more efficient search and purchases and to more competition among sellers. Wiener [67] empirically tested whether warranties serve as an accurate market signal. For the four household durables and for automobiles, he found that warranties were an accurate signal of product quality. In 1988, Wiener [62] reported a second study in which he compared the accuracy of warranties as a market signal before and after the passage of Magnuson–Moss. Using *Consumers Union* data, he found that a warranty is a better signal of product reliability for automobiles and some other consumer durables in the post-act era than in the pre-act era. Wiener does caution his readers that other variables not included in his study may have influenced his findings.

If Wiener's results were to hold when the other variable are factored, this would give credence to a goal of Magnuson–Moss—to create a market where a product's warranties are an accurate signal of its reliability. However, Kelley [68] found that warranties were an accurate market signal both before and after the passage of Magnuson–Moss. Both Kelley's and Wiener's studies may be more difficult to undertake in the 1990s because there appears to be less variance in the terms of the warranties for many products, with the exception of certain kitchen and laundry appliances. This is addressed later in this chapter in a discussion concerning variability in warranties.

28.6.5 Informal Dispute Resolution

Although it was stated earlier that the results of the informal dispute resolution program under the act are mixed, consumerists might give it an even lower rating. Magnuson–Moss permits manufacturer-run informal dispute resolution programs and most states recognize them. However, 13 states have established their own arbitration programs. A California study [69] of over 3000 consumers reported that only a little over half of consumers who *received a decision in their favor* in arbitration were satisfied with the process. *Consumer Reports* [70] found that 14% of automobile arbitration cases handled by the Better Business Bureau, 9.6% handled by Chrysler, and 30% handled by Ford resulted in consumers receiving a car replacement or refund, whereas 55% of the cases presented before state-run arbitration panels were awarded a refund or replacement car. After reviewing hundreds of lemon law cases (consumers usually must go through arbitration before qualifying for lemon law action), *Consumer Reports* [70] major finding was that, "Consumers are much more likely to get satisfaction from a state-run arbitration program than from a manufacturer or a private arbitrator" [70, p. 40]. For a more comprehensive discussion of informal dispute resolution, see Ref. 71.

28.6.6   Consumer Protection

Although most elements of Magnuson–Moss may be considered to be protecting the consumer, especially in the areas of warranty disclaimers, there are two substantial consumer problems with warranties that seem to fit into the areas of more flagrant deception—service contract and so-called "secret" warranties. Of these two, the greatest problem pertains to service contracts, sometimes called extended warranties. Although the act does address service contracts, the FTC has not promulgated regulations on disclosures of the terms and conditions of service contracts [72].

Service Contracts

Service contracts are purchased by nearly half of auto buyers who pay from about $300 to $900 to enhance the profits of the sellers [73]. Sales have increased by about 10% a year during the late 1980s and into the 1990s to the point where service contracts are now estimated to be a $5 billion a year business [74]. More than 90% of all major appliance dealers offer service contracts with sales of over $1 billion a year and with a sales ratio of 14% on all major appliances [75]. In an exploratory study, Kelley and Conant [76] found that 47% of their sample purchased extended warranties. Of those who insured using service contracts, they purchased 51% on new cars, 29% on major appliances, and 20% on miscellaneous products.

Service contracts are extremely profitable for the seller with typical profits ranging from 100% to 600% [73,75,77–79]. Often more profit is made on service contracts than on the sale of the product itself [78,80]. The loss ratios are from 4% to 15% on appliances [81], with gross profit margins reaching as high as 77% [78]. The ratios may be higher for automobiles. The New Jersey State Division of Consumer Affairs reports an average profit on automobile service contracts as 600% [73]. An MIT warranty study [66] found the ratio of service contract cost to the amount spent on repairs of appliances to be 10 to 1. However, when considering loss ratios for service contracts, one must weigh the fact that the service contract sellers probably pay wholesale rates for repairs. Because the consumer would typically pay much higher retail rates, this tends to overstate the loss ratio from the perspective of the individual consumer.

Usually when an item has a high profit margin for the seller, this is an indication of market failure, either because of insufficient information on the part of the consumer, deception on the part of the seller, or both. Consumers must be ignorant of the negative cost–benefit ratio, as only an extremely risk-averse person would buy a service contract. Deception appears to be rampant. Because service contracts are a high-profit good, there are great incentives to overstate both the need for the coverage and the extent of the coverage itself. In addition, because the consumer knows little about how service contracts work and of their cost–benefit ratio, it is easy to misrepresent the contract's power [80].

Not only do sellers exaggerate the protection of service contracts, they contain "exclusions galore" [73,80] and consumers often find it difficult to even get the covered repairs paid [73,80,82]. One of the greatest problems with service contracts is the high rate of failure of non-manufacturer-sponsored contracts [78]. These insurers have folded in substantial numbers, leaving both consumers and the selling dealers high and dry [66,74,82]. In addition, service contracts often overlap with the express and implied warranty for the product and, therefore, do not give as much protection as a consumer might assume [66,81].

There have been complaints that those who insure their products with service contract are often dissatisfied with the service rendered when a problem is encountered. Kelley and Conant [76] found that the insured who had exercised their right to request service were only moderately satisfied.

Some service contract arrangements include extra disincentives for performance of covered repairs. According to Deenan [83], some automobile dealers use administrators to conduct their service contract business. The administrators are paid a specific amount. A portion is used to cover administrative fees and excess losses. The balance is placed in a reserve in the dealer's name. If few or no repairs are reimbursed, the dealer then receives a kickback amounting to the balance of the reserve.

One may speculate that because service contracts are so profitable, this might increase the incentives for the manufacturers to minimize regular warranties. There is less motivation for a consumer to purchase a service contract if there is a generous warranty. Because of this, retailers may be less likely to carry products that have longer-than-average regular warranties because this, in turn, might cut their chances of selling highly profitable service contracts [80].

Kelley and Conant [76] found that 39% bought service contracts because they believed it was cheaper than paying for repairs when they occur. However, are consumer fears of product failure and the expense of repair justified? *Consumer Reports* [78] estimates that fewer than 20% of the products covered by a service contract are ever brought in for repair.

Service contracts are not a recent problem; as long ago as the 1960s they were a concern. The Task Force on Appliance Warranties [38] found service contracts to be overpriced and designed only to increase profit margins.

Some states such as Florida and California strictly regulate service contracts as insurance [78]. Other states indirectly regulate them under consumer protection law, and in still others, like Utah, regulations fall between the cracks of consumer protection law and insurance law. Because the purchase of service contracts is so economically detrimental to the consumer, both *Consumer Reports* [73,77,78,81] and *Changing Times* [75,79] strongly advise against their purchase. According to testimony of Congressman Chandler of Washington, attorneys general of several states view service contracts are a "serious consumer problem" [84, p. H8019]. The Consumer Federation of America and United States Public Interest Research Group (a Nader group) have agreed in the past that congressional action is needed on service contracts [84]. Kelley and Conant [76] state that the Federal Trade Commission may want to consider promulgating rules concerning service contracts.

Secret Warranties

Secret warranties, sometimes called "warranty extensions" or "goodwill adjustments," give protection "only to car owners who complain long and loudly to the right person to get reimbursed for repairs" [85, p. 214]; that is, only automobile owners who have the perseverance and sophistication to aggressively push their case for an automobile repair that might reasonably be covered under an implied warranty, if not an express warranty, will get satisfaction from a manufacturer. The manufacturer will take care of such matters only on a case-by-case basis.

Consumer groups say that the regular long-term warranty on an automobile often does not cover the items that a secret warranty usually covers [86]. The FTC has not taken any action on secret warranties since a Ford case in 1988 [87]. The FTC's position is that, "There is a real question as to whether 'secret' warranties help or hurt consumers. If we make adjustment policies more expensive [for the manufacturer], then fewer companies will offer them" (William McLeod, director of the FTC's Bureau of Consumer Protection, quoted in Ref. 85, p. 215). Clarence Ditlow, director of the Center for Automobile Safety disagrees and calls secret warranties "one of the greatest economic abuses perpetrated by automobile manufacturers" [88, p. 7]. The consumerist Attorney General of New York, Robert Abrams [89], claims that secret warranties are among the nation's greatest consumer abuses. Both Consumers Union and the Center for Auto Safety believe that the service given under secret warranties should be available for all automobile owners with similar problems. Secret warranties exist not only in the automobile domain but also in other areas of consumer durables (e.g., telephones) [90].

Although several states (Florida, Indiana, Maryland, Minnesota, Missouri, New York, Pennsylvania, and Washington) have considered legislation to extend the protection of secret warranties to all vehicle owners [87], only Virginia and Connecticut require manufacturers to notify consumers of postwarranty adjustment programs to handle nonsafety defects [87,90].

28.7 SUMMARY OF THE EVALUATION OF WARRANTY POLICY

Brickey [54] concludes her detailed analysis of Magnuson–Moss (several additional shortcomings that are not addressed here are also mentioned in her law review piece) with the following statement:

> As a tool for achieving consumer protection, the Magnuson–Moss Warranty Act is flawed. It is unfortunate that the legislation is not drafted with the clarity demanded of warranty documents. Much of the Act is obscure in purpose and effect, and this factor has hindered the FTC in its efforts to develop a cohesive policy. [54, p. 117]

Not all legal scholars have such negative views of Magnuson–Moss. Clark and Davis [91] concluded that although the act was flawed in that it permits disclaimers of consequential damages and fails to deal with salesperson's oral statements, it was a giant step forward in consumer protection. Rothschild [41], who has written extensively on warranties, concludes in a 1975 article that Magnuson–Moss will have a considerable beneficial effect on the marketplace. According to Pridgen [72, p. 14], "The Magnuson–Moss Act does not do much in the area of *substantial regulation of warranties*, but what it does has proved very valuable to consumers."

Most consumerists will probably agree more with Brickey than with the scholars just mentioned. Although Magnuson–Moss does a credible job of providing its information and, according to Pridgen [72], has met with mixed success in the area of informal dispute resolution goals, it still falls short of its goal of providing quality warranty protection for the consumer.

In summary, from a consumerist perspective, the benefits of Magnuson–Moss are mixed at best. Ralph Nader has been known to comment that a weak regulatory agency is worse than no agency at all because it gives the consumer an illusion of protection when none exists. In a corollary, the author of this chapter often tells his students that a weak law is worse than no law at all. Weak laws often have a lulling effect. Consumers who know of the law's existence but not the specifics of the law may feel they are protected when they really are not. Also, there may be less motivation for a legislature or for Congress to pass a strong law on an issue because they can point to the protection already offered by the existing law.

If Magnuson–Moss is weak, as has been argued here, should it be amended to give greater consumer protection? Is the political environment conducive to more warranty regulation? These questions will be discussed in the next section.

28.8 A CONSUMERIST'S SUGGESTIONS FOR IMPROVEMENTS IN MAGNUSON–MOSS

During the hearings on the passage of Magnuson–Moss, many consumerists testified that this legislation should go much further in protecting the consumer than the final act actually did. Most of the desired additional protection is summarized in a statement by Robert J. Klein, economics editor for *Consumers Union*, before the congressional hearings on Magnuson–Moss [49]:

1. Prohibit disclaimers of implied warranties.
2. Exclusions in warranties should be limited to items specifically named, eliminating the term "such as" or its equivalent.
3. Prohibit disclaimers for consequential damages.
4. Create minimum standards for warranty terms (e.g., include a requirement that new-car owners must be supplied with a

"loaner" during warranty repairs). The MIT warranty study also made minimum standard recommendations [66].

5. Establish a national standard for minimum warranty periods under which most types of parts failure would automatically fall.

Although some of these suggestions currently apply to full warranties, Klein meant that they should apply to all warranties. In fact, several consumerists testified in the congressional hearings that the two-tier system of warranties in Magnuson–Moss was not in the consumers' best interest.

With the dramatic increase in the sale of service contracts and public awareness of secret warranties, legislation also might be considered in these areas. The MIT warranty study [66] recommended that the FTC should write comprehensive disclosure statements for service contracts under the authority granted in Magnuson–Moss. They also recommended that bankruptcy laws be altered to protect those who hold service contracts from defunct firms and that the service contract should be automatically extended when delays in service are rendered.

Service contract legislation on the federal level was attempted in 1989 when Congressman Chandler of Washington introduced a bill that would:

Subject extended service contracts to a 60 percent excise tax unless they comply with certain reasonable requirements; allow contract sellers to mark up the price of the contract by only 100 percent; make sure that the seller of the contract is liable for any necessary repairs; exempt those contracts which reveal the expected cost of performance on the face of the document; exempt those contracts which are subject to full State insurance commission regulation; and require that contract sellers maintain adequate reserves to pay for expected payouts. [84, p. 8019]

Even this relatively weak bill failed in Congress.

A comprehensive service contract policy that is directed at most of the abuses of service contracts was outlined by Burton [92] in a presentation to the American Council on Consumer Interests. This suggested policy is based on principles taken from Magnuson–Moss, state service contract law, insurance law, truth-in-lending, implied warranty law, consumer protection law, and information theory.

Legislation might address the secret warranty problem by considering legislation similar to legislation in the two states that have laws on this problem. According to Gillis [87], this legislation would require:

Direct notice to consumer within a specific time after the adoption of warranty adjustment policy.

Notice of the disclosure law to new car buyers.

Reimbursement, within a number of years after payment, of owners who paid for covered repairs before they learned of the extended warranty service.

Dealers to inform consumers who complain about a covered defect that is eligible for repair under warranty.

New York has an additional requirement that provides for the establishment of a toll-free number for consumer questions on secret warranties [87]. Consumerists might also advocate that automobile service bulletins and indexes to these bulletins be made available to the consumers at little or no cost.

Blame for automobile warranty problems has been partially attributed to the fact that dealers receive less compensation for warranty work. A federal law might be considered similar to that of Tennessee's in which a manufacturer is required to pay the same rates for warranty work as for nonwarranty work. Although such legislation is also desired by dealers (in addition to some consumerists), manufacturers have fought it and commissioned a study that found that equal pay for warranty repairs would raise automobile prices [91].

Lemon laws, first adopted in 1982 and currently in effect in 48 states, are in some respects an extension of warranty law and have come under criticism. Coffinberger and Samuels [93] found many inconsistencies in lemon laws among the states. They believe that lemon laws are beneficial to consumers; however, they suggested a federal lemon law. *Consumer Reports* is very critical of many state lemon laws.

> But the road to satisfaction for lemon owners has turned out to be badly potholed. Weak laws, including many that allow auto manufacturers to run the arbitration process; poor oversight by the states; and miles of red tape have stymied consumers at every turn. [70, p. 40]

In 1983, Congressman Lantos introduced the Automobile Consumer Protection Act in an attempt to create a federal lemon law, but it was not given serious consideration by Congress [93]. The automobile industry attempted to have the FTC establish uniform lemon rules under the FTC 703 rule-making powers that would preempt state laws; however, the industry proposal was opposed by the National Association of Attorneys General, National Conference of State Legislators, the Center for Auto Safety, and other consumer groups [94]. The FTC turned down the request based on a cost–benefit analysis [94]. Government officials and consumerists are saying that if there is to be a uniform federal lemon law, it should be a strong one.

An MIT report [66] recommended warranty policy that would unbundle warranties from the product. This proposal has not been addressed by consumer advocates. Unbundling would not require legislation and is permitted in Magnuson–Moss. With an unbundled warranty, a person who was not risk-averse, who was a light user, or who was proficient at his or her own repairs could reject warranty coverage in exchange for a lower price for the good. A major problem with unbundling might be adverse selection [66] and another might be moral hazard. Unbundling would have appeal to consumerists who advocate choice, fairness, and efficiency. However, consumerists who see unbundling as nothing more that expanding the scope of service contracts and who feel that cross-subsidies are justified for warranties as they are for some other consumer commodities (e.g., local telephone service) would strenuously object to unbundling.

The MIT study [66] also recommended that appliance warranties be based on usage and/or time. Under this policy, the warranty would expire after a certain number of uses or a certain time, whichever came first; this is similar to an automobile warranty. This action would incorporate fairness because lighter users would gain longer warranty periods and heavy users would be subject to shorter periods. With advances in microchip technology, the imbedding of timing or counting devices in appliances is more feasible now than it was in 1978.

These are some warranty policies that might be considered by consumerists if they wish to strengthen consumer protection under Magnuson–Moss. Some of these proposals might not need legislation because they could be initiated by manufacturers or could be mandated by FTC Trade Regulation Rules. However, because the FTC has enacted few trade rules since the 1970s, and even those were weak, (e.g., Funeral Rule, Used Car Rule), and has enacted no rules under Rule 703, it is unlikely it will act now without a dramatic change in the philosophy and leadership of the FTC.

28.9 POSSIBLE FUTURE LEGISLATION

Is warranty legislation currently contemplated? Other than the Chandler bill on service contracts and the Lantos bill on secret warranties, there has been little or no substantive warranty legislation proposed on the federal level since 1975. During the 12 years of the Reagan and Bush administration, consumer protection legislation was considered excessive government regulation and most consumer legislation was left to the states. As noted above, several states have enacted substantive warranty protection laws (e.g., California's Song–Beverly Consumer Warranty Act and its amendments); however, consumers in many other states have been left vulnerable by the weaknesses of Magnuson–Moss.

Other than legislation concerning service contracts and secret warranties, there do not seem to be any political or organizational consumer entrepreneurs pushing for dramatic changes in Magnuson–Moss. Telephone calls in January 1993 to the major players in consumer advocacy at the federal level—Consumer Federation of America, United States Public Interest Research Group, Center for Auto Safety, and Consumers Union—found that these organizations do not have federal warranty reform on their legislative agenda. In an open letter to President Clinton, Consumers Union lists seven major consumer concerns that should be addressed by the Clinton administration. Problems with warranties were not on the list [95]. With consumer concerns regarding health care costs, the banking system, and nutrition absorbing the energies of consumer entrepreneurs, warranty reform is probably a low priority. Another consideration is that consumers may be experiencing less dissatisfaction with warranties. Because a consumer uses a warranty only when there is a product defect and with the increase in the quality of American automobiles, consumers may be experiencing fewer warranty problems. Even *Consumer Reports* [78] finds that products are more reliable today.

Will warranties be on the agenda of consumerists in the future? There will need to be a mix of consumer dissatisfaction with the performance and reliability of warranted goods, discontent with warranty promises that have been expanded beyond what the warrantor is either willing or able to service, creation of a political environment conducive to consumerism, and an increase of consumer entrepreneurs to carry the banner for reform.

REFERENCES

1. *Business Week* (1960). Trouble with new car warranties, November 6, 55–56.
2. Herrmann, R. O. (1979). *The Consumer Movement in Historical Perspective*, Pennsylvania State University, University Park,
3. Mayer, R. N. (1989). *The Consumer Movement: Guardians of the Marketplace*, Twayne Publishers, Boston, MA.
4. Preston, I. L. (1975). *The Great American Blow-up: Puffery in Advertising and Selling*, The University of Wisconsin Press, Madison.
5. Chase, S., and Schlink F. J. (1927). *Your Money's Worth*, Macmillan Company, New York.
6. Harap, H. (1924). *The Education of the Consumer*, The Macmillan Company, New York.
7. Gordon, L. J. (1939). *Economics for Consumers*, American Book Company, New York.

8. Babson, R. W., and Stone, C. N. (1938). *Consumer Protection*, Harper and Brothers Publishers, New York.
9. Warne, C. E. (1982). Consumer Union's contribution to the consumer movement, in *Consumer Activists: They Made a Difference: A History of the Consumer Movement* (E. Angevine, ed.), Consumer Union Foundation, New York, pp. 85–1111.
10. Packard, V. (1957). *The Hidden Persuaders*, David McKay Company, Inc., New York.
11. Congressional Quarterly, Inc. (1982). *Regulations: Process and Politics*, Congressional Quarterly, Inc., Washington, DC.
12. *Business Week* (1972). America's growing antibusiness mood, June 17, 100–103.
13. Pertschuk, M. (1982). *Revolt Against Regulation: The Rise and Pause of the Consumer Movement*, University of California Press, Berkley.
14. Aaker, D. A., and Day, G. S. (1978). Introduction: A guide to consumerism, in 3rd ed. (D. A. Aaker and G. S. Day, eds.), *Consumerism: Search for the Consumer Interest*, The Free Press, New York, pp. 1–19.
15. West, M. G. (1971). Disclaimer of warranties–its curse and possible cure, *Journal of Consumer Affairs*, **5**, 155–173.
16. Schoenfeld, A. F. (1971). Consumer report/administration, FTC are at odds over proposals to strengthen the agency, *National Journal*, **3**(17) (April 24), 887–894.
17. President's Committee on Consumer Interests (1966). *Consumer Issues '66: A Report to the President from the Consumer Advisory Council*, Consumer Advisory Council, Washington, DC.
18. *Consumer Reports* (1968). Warranties: This year's retreat, April, 176–179.
19. Caplovitz, D. (1974). *Consumers in Trouble: A Study of Debtors in Default*, The Free Press, New York.
20. Caplovitz, D. (1967). *The Poor Pay More: Consumer Practices of Low Income Families*, The Free Press, New York.
21. Andreasen, A. R. (1975). *The Disadvantaged Consumer*, The Free Press, New York.
22. Margolius, S. (1975). The consumer's real needs, *Journal of Consumer Affairs*, **9**, 129–138.
23. Nader, R. (1972). Ralph Nader reports, *Ladies Home Journal*, October, 58.
24. *Business Week* (1970). Can FTC put a governor on Detroit? proposed automobile quality act, February 28, 31–33.
25. Nader, R. (1968). The great American gyp, *New York Review of Books*, **11**, 27–34.
26. Magnuson, W. G., and Caper, J. (1968). *The Dark Side of the Marketplace: The Plight of the American Consumer*, Prentice-Hall, Englewood Cliffs, NJ.
27a. Troelstrup, A. W. (1957). *Consumer Problems and Personal Finance*, McGraw-Hill Book Company, Inc., New York.
27b. Gordon, L. J. (1939). *Economics for Consumers*, American Book Company, New York.
28. Herrmann, R. O. (1979). *The Historical Development of the Content of High*

School Level Consumer Education, 1938–1978, A report to the Office of Consumer Education, U. S. Department of Health, Education and Welfare.

29. Rao, C. P., and Weinrauch, J. D. (1976). Consumer perceptions and satisfaction with major appliance warranties, *Baylor Business Studies* 7(109) (August–October), 17–27.

30. *Business Week* (1970). Detroit tries a U-turn on warranties, July 25, 44–48.

31. *Consumer Reports* (1967). Warranties: The promises and the reality, April, 194–197.

32. *Consumer Reports* (1970). Some variable costs of ownership: Repairs, insurance, warranties, April, 201–204.

33. Conference Board (1971). *The Challenge of Consumerism,* Conference Board, New York.

34. Federal Trade Commission (1968). *Federal Trade Commission Staff Report on Automobile Warranties,* Federal Trade Commission, Washington DC.

35. House Interstate and Commerce Committee, Subcommittee on Finance (1974) *Staff Report on Consumer Product Warranties,* September 17.

36. Whitford, W. C. (1968). Law and the consumer transaction: A case study of the automobile warranty, *Wisconsin Law Review,* **1968**(4), 1006–1098.

37. *Business Week* (1963). Guaranteeing more, enjoying it less, January 26, 1963, 46.

38. Federal Trade Commission (1969). *Report of the Task Force on Appliance Warranties and Service,* Superintendent of Documents, Washington, DC.

39. MACAP (1971). *MACAP Study of Appliance Industry Warranties,* unpublished report.

40. MACAP (1973). *MACAP Analysis of Major Appliance Warranties,* unpublished report.

41. Rothschild, D. P. (1976). The Magnuson–Moss Warranty Act: Does it balance warrantor and consumer interests?, *George Washington Law Review,* **44**, 335–380.

42. *Consumer Reports* (1963). Warranties: The consumer's rights on paper and in the law, April, 158–159.

43. Bishop, J., and Hubbard, H. W. (1969). *Let the Seller Beware,* The National Press, Washington, DC.

44. Wilson, J. Q. (ed.) (1980). *The Politics of Regulation,* Basic Books, New York. Cited in Pertschuk, M. (1982). *Revolt Against Regulation: The Rise and Pause of the Consumer Movement,* University of Calif Press, Berkeley.

45. Hasin, B. R. (1986). *Consumers, Commissions, and Congress: Law, Theory and the Federal Trade Commission,* Transaction Books, New Brunswick, NJ.

46. Tolchin, S. J., and Tolchin, M. (1983). *Dismantling America* Houghton Mifflin, Boston. Cited in Hasin, B. R. (1986). *Consumers, Commissions, and Congress: Law, Theory, and the Federal Trade Commission,* Transaction Books, New Brunswick, NJ.

47. Cox, E. F., Fellmeth, R. C., and Shultz, J. E. (1969). *The Nader Report on the Federal Trade Commission,* Baron Press, New York.

48. American Bar Association (1969). *Report to the ABA Commission of the Study of the Federal Trade Commission*, American Bar Association
49. 93rd Congress, House (1973). Hearings before the House Subcommittee on Commerce and Finance of the Committee on Interstate and Foreign Commerce, House of Representatives. *Consumer Warranty Protection*. Serial No. 93-17. U.S. Government Printing Office, Washington, DC.
50. 92nd Congress, House (1971). Hearings before the House Subcommittee on Commerce and Finance of the Committee on Interstate and Foreign Commerce, House of Representatives. *Consumer Product Protection*. Serial No. 92-50. U.S. Government Printing Office, Washington, DC.
51. Federal Trade Commission (1979). *Warranty Rules: Warranty Content Analysis*, Federal Trade Commission, Washington, DC.
52. Schmitt, J., Kanter, L., and Miller, R. (1980). *Impact Report on the Magnuson–Moss Warranty Act*, Federal Trade Commission, Washington, DC.
53. Moore, E. M., and Shuptrine, F. K. (1993). Warranties: Continued readability problems after the 1975 Magnuson–Moss Warranty Act, *Journal of Consumer Affairs*, **27**(1), (Summer), 23–36.
54. Brickey, K. F. (1978). The Magnuson–Moss Act: An analysis of the efficacy of the federal warranty regulation as a consumer protection tool, *Santa Clara Law Review*, **18**, 73–118.
55. *Consumer Reports* (1984). Is that warranty any good? July, 408–410.
56. Hearings before the consumer subcommittee of the committee on commerce. *Consumer Products Guarantee Act*, U.S. Government Printing Office, Washington, DC, p. 35.
57. *Consumer Reports* (1994). Guide to the gear. March, 154–191.
58. Moreau, D. (1989). Why the dealer wins: Appliance service contracts, *Changing Times*, January, 83–84, 86.
59. *Marketing News* (1978). Consumers are getting less—not more—warranty protection (February 10), 2. Cited in McDaniel, S. W., and Rao, C. P. (1980). A post-evaluation of the Magnuson–Moss Warranty Act, *Akron Business and Economic Review*, **7**(2) (Summer), 38–41.
60. McDaniel, S. W., and Rao, C. P. (1980). A post-evaluation of the Magnuson–Moss Warranty Act, *Akron Business and Economic Review*, **7**(2) (Summer), 38–41.
61. Wisdom, M. J. (1979). An empirical study of the Magnuson–Moss Warranty Act, *Stanford Law Review*, **31**, 1117–1146.
62. Wiener, J. L. (1988). An evaluation of the Magnuson–Moss Warranty and Federal Trade Commission Improvement Act of 1975, *Journal of Public Policy and Marketing*, **7**, 65–82.
63. Krebs, M. (1992). Complete guide to warranties, *Consumers Digest*, January/February, 54, 56.
64. Koten, J. (1984). Aggressive use of warranties is benefiting many concerns, *Wall Street Journal*, April 5, 33.
65. Gerner, J. L., and Bryant, W. K. (1981). Appliance warranties as a market signal? *Journal of Consumer Affairs*, **15**(1) (Summer), 75–86.
66. Center for Policy Alternatives (1978). *Consumer Durables: Warranties Ser-*

vices Contracts and Alternatives, Vol. 1, Institute of Technology, Cambridge, MA.

67. Wiener, J. L. (1985). Are warranties accurate signals of product reliability? *Journal of Consumer Research*, **12**(2), (September), 245–250.

68. Kelley, C. A. (1988). An investigation of consumer product warranties as signals of product reliability, *Marketing Science*, **16**, 72–78.

69. California Department of Consumer Affairs, *Arbitration Review Program*, *Consumer Satisfaction Survey*, (1991). Unpublished report.

70. *Consumer Reports* (1993). The sour truth about lemon laws, January, 40–42.

71. Selnick, D., Burton, J., Heinzenling, B., and Janson, J. (1994). Informal dispute resolution: Past, present, and future. *Consumer Issues Annual*, 40, 357–358.

72. Pridgen, D. (1991). *Consumer Protection and the Law*, Clark, Boardman and Callaghan, New York.

73. *Consumer Reports* (1990). Insure against future repairs?, April, 226–227.

74. Del Prete, D. (1991). Looks like a hot summer for extended car warranties. *Marketing News*, July 8, 9.

75. Moreau, D. (1989). Why the dealers wins: Appliance service contracts. *Changing Times*, January, 83–84.

76. Kelley, C. A., and Jeffrey S. Conant (1991). Extended warranties: Consumer and manufacturer perceptions. *Journal of Consumer Affairs*, **15**(1) (Summer), 68–83.

77. *Consumer Reports* (1986). Auto service contracts, October, 663–667, 685.

78. *Consumer Reports* (1994). Should you pay for piece of mind? January, 42.

79. Henry, E. (1988). Extended service contracts: Still a bad idea, *Changing Times*, July, 76–77.

80. Snider, J., and Ziporyn, T. (1992). *Future Shop*, St. Martins Press, New York.

81. *Consumer Reports* (1991), Who needs an extended warranty? January, 21–22.

82. Thomas, C. M. (1990). Service contracts under siege. *Automotive News*, May 28, 1.

83. Deneen, D. G. (1993). Automobile finance, warranty, and insurance extras. *Loyola Consumer Law Reporter*. **6**(1) (Fall), 5–10.

84. *Congressional Record* (1989). November 7, H8019–H8020.

85. *Consumer Reports* (1989). Is your car's warranty a secret? April, 214–215.

86. White, J. (1987). Detroit's secret warranties, critics say, cause big difference in auto repair bills, *Wall Street Journal*, December 8, 31.

87. Gillis, J. (1992). *The Car Book 1993*, Harper Perennial, New York.

88. *Common Cause* (1991). Car trouble, January/February, 7.

89. *Modern Maturity* (1991). Dirty secrets: What you don't know can cost you, April/May, 10.

90. Brown, W. (1991). Letting everyone in on the secret, *The Washington Post National Weekly Edition*, September 30–October 6, 22.

91. Clark, B., and Davis, M. J. (1975). *Beefing Up Product Warranties: A New Dimension in Consumer Protection*, The Macmillan Company, New York.
92. Burton, J. R. (1994). Warranty law: Appropriate for the 1990's? *Consumer Interest Annual*, 40, 223–230.
93. Coffinberger, R. L., and Samuels, L. B. (1985). Legislative responses to the plight of new car purchasers, *Uniform Commercial Code Law Journal*, **18**, 168–181.
94. Kahn, H. (1991). Equal pay for warranty repairs will cost consumers, study concludes, *Automotive News*, May 13, 4.
95. *Consumer Reports* (1993). Dear Bill Clinton, January, 6.

Part H

Warranty and Management

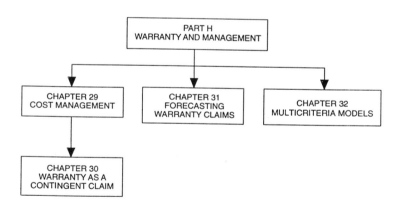

Introduction

Part H contains four chapters that address some issues related to administration and management of warranties.

Chapter 29, by R. A. Maschmeyer and K. R. Balachandran, building on their earlier chapter (Chapter 26), looks at cost management planning and control for warranties. After a brief discussion on internal versus external accounting, the authors cover several issues of importance in internal accounting. In this context, the link between warranty and quality (developed in Chapter 23) is viewed in a cost context, and the tradeoffs between the two are highlighted from a strategic cost management view.

In Chapter 30, E. W. Frees looks at warranty as a contingent claim, that is, monetary events based on contingent random events. This viewpoint allows for developing models for the study and management of warranties based on concepts and models from both the insurance and financial risk literatures. The chapter develops the relationship between warranty policy and a variety of actuarial models from the insurance arena.

In Chapter 24, the forecasting of warranty claims was based on the failure distribution of the product and its sales over time. This involves modeling failures for each item by a point process formulation (discussed in detail in Part D) and modeling sales rates separately using a deterministic model. In Chapter 31, J. Chen, N. J. Lynn and N. D. Singpurwalla develop an alternate approach to modeling claims using a time series formulation and Kalman filter models. This approach can be viewed as empirically driven, where the model parameters are determined using the claims data, and hence avoids the need for more detailed information such as the failure distribution.

In the Part's final chapter (Chapter 32), A. Mitra and J. G. Patankar deal with the optimal choice of warranty policy parameters. Chapter 20 addressed this problem based on consumer and manufacturer risks. In contrast, this chapter develops a more comprehensive framework to decide on the policy parameters by taking into account the business goals and objectives defined by top management. A multicriteria model involving manufacturing, marketing, and economic goals is discussed in detail. The authors indicate how the parameters can be optimally selected to meet these goals and identify the trade-offs needed when the goals are not met. Such a model allows for better management of warranty from a business viewpoint.

29

Cost Management Planning and Control for Product Quality and Warranties

Richard A. Maschmeyer

University of Alaska Anchorage
Anchorage, Alaska

Kashi Ramamurthi Balachandran

New York University
New York, New York

29.1 INTRODUCTION

The accounting for product warranty costs includes the fulfillment of financial reporting requirements and the provision of requisite information for cost management decisions. The purpose of this chapter is to discuss the role warranty costs have on the managerial planning, control, and internal evaluation processes over a product's life cycle. Our discussion will focus on the long-term financial relationships of product warranty costs to the overall product quality costs.

Section 29.2 addresses the differences between cost management accounting and the accounting required for external financial reporting. This discussion will provide the rationale for appropriate accounting treatments adopted for internal management purposes as opposed to financial accounting and reporting for investors and creditors. Section 29.3 deals with the role of warranty issues during product life cycles. Section 29.4 considers the trade-offs among quality enhancement costs of which warranty costs is one component. Section 29.5 provides a discussion of strategic cost management methods that considers the quality enhancement and warranty costs trade-offs. This section includes a target analysis approach to planning and control and the requisite concepts of activity-based accounting. Cost estimation models for product warranties and maintenance agreements are discussed in Section 29.6.

29.2   EXTERNAL VERSUS INTERNAL ACCOUNTING

29.2.1   Financial Reporting for Investors and Creditors

Accounting information is useful to users that are external to the reporting entity, such as present and potential shareholders and creditors. Overall, the objectives of financial reporting are to provide (1) information that is useful in making rational investment and credit decisions, (2) information that is useful in assessing the amounts, timing, and uncertainty of cash flows, and (3) information about the enterprise's economic resources, claims to those resources, and the effects of transactions (and/or events), and the circumstances that change its resources and claims to those resources [1].

Information provided to external users is presented in quarterly, semiannual, or annual financial reports that are prepared in accordance with generally accepted accounting principles (GAAP). These reports will include a balance sheet, an income statement, a cash-flow statement, and a statement of changes in retained earnings. GAAP requires and achieves uniformity and consistency in reporting by different companies to afford comparability. External financial reports are prepared on an overall reporting entity; thus, historical cost information is presented on an aggregate, firmwide basis. Although the objectives of financial accounting emphasize "prospective cash receipts," the accounting profession believes that information based on accrual accounting provides the best indication of a firm's present and future ability to generate positive cash flows. The essence of accrual accounting is embodied within the revenue realization principle and the matching principle. Revenues are generally recognized in the accounting period earned, regardless of when the corresponding cash is collected, and expenses are generally recognized in the period incurred, regardless of when cash is disbursed. The matching principle requires that recognized revenues of a given accounting period should be matched on the same income statement with all expenses that were incurred to generate those revenues. Conceptualizing this principle, Williams et al. [2] clarify:

> . . . the matching principle implies that the accountant should
> determine the extent to which goods and services represented by

historical costs have contributed to revenues during the accounting period. Costs that have contributed to revenues should be reported as expenses, and costs that are expected to contribute in the future should be reported as assets.

Where costs appear to have uncertain associations with future revenues, costs are reported as expenses in the accounting period incurred. In summary, the objective of accrual accounting is to reflect the economic consequences of an enterprise's financial events in the accounting period in which the event occurred.

29.2.2 Cost Management Accounting: A Different Vantage Point

Whereas managers of the reporting enterprise have many specific uses of accounting information, they usually require accounting reports that are configured quite differently from external financial reports. Cost management accounting, as defined by Horngren et al. [3], is "the actions by managers to satisfy customers while continuously reducing and controlling costs." Cost management accounting differs from financial accounting in the following ways:

1. Reports are prepared in a timely manner to meet management's ongoing decision tasks. Reports may capture very short time frames or may be designed to provide information across a product's full life cycle.
2. Reports are nonuniform in that they are designed to meet the specific informational needs of internal management at any point in time.
3. Reports are not based on generally accepted accounting principles. Managers are free to reshape data to a form that is more useful for the specific decision task; that is, statements or reports may be based on cash flows rather than accrual accounting.
4. Reports are usually designed to capture the operating results of specific segments of the firm, that is, product lines.
5. Emphasis is placed on the relevance of information for future events. A manager's planning framework may rely on forecasts, budgets, statistical estimates, and/or appraisals.

There is a daily demand for managerial decision making on a variety of operational matters and these decisions have both short- and long-term consequences on the firm's profitability and competitiveness. Managers relying on the information contained within external financial reports, prepared on the basis of generally accepted accounting principles, may be including distorted information into their decision process. For example, the actual amount of cash spent to repair products under warranty during a specific period is not provided in financial statements prepared for external users. Instead, an estimate of the ultimate warranty cost associated with the products sold in the current period is accrued as an expense. Following this method, revenues and expenses related to the same products are matched and recognized on the income statement in the period the products are sold. For internal planning and cost control purposes, managers may desire an accumulation of all warranty expenditures on a per-product-line basis in a single financial reporting period. Typically, products requiring current period repairs have been sold in both current and previous accounting reporting periods. Further, the products requiring current period repairs may be in different stages of their respective warranty life.

Other cost distortions may occur when the internal user perspective is ignored by management in favor of the externally prepared financial statements. For example, the cost of lost sales due to customer dissatisfaction is an opportunity cost which is not reported under GAAP accounting, but this cost component is a necessary consideration for management in setting the warranty policy with respect to warranty durations and the extent of repair/replacement coverage. Prevention costs that reduce product failure rates are incurred, generally, during the design phase of the product rather than during the manufacturing and sales periods. As such, external financial statements do not include these costs as part of the product's unit cost that is completed and inventoried prior to sale. Rather, these prevention costs are expensed in the period in which the design costs were incurred and, thus, the full product cost from management's perspective is understated on the financial statements. Management's use of an understated full product cost for choosing optimal product quality levels will lead to distortion and suboptimal solutions. These examples illustrate the importance of developing accounting reports that provide information that is useful in optimizing the internal management decision process. There is a definite need to study the information requirements for management's planning, control, and evaluation processes before deciding on the accounting information model to utilize. Whereas all firms must comply with GAAP for external accounting, the best internal information system for management varies and is dependent on the time frame of the decision, the product, the human aspects of the decision-maker, and technical characteristics of each situation.

29.3 THE ROLE OF WARRANTIES IN PRODUCT LIFE CYCLES

Warranties are usually provided when a product or service is sold in order to protect the purchaser against quality deficiencies for specified periods of time. For a number of products, extended warranty contracts may be purchased beyond the initial warranty period. Consumers are generally interested in these extended agreements when they are apprehensive about the product's quality during downstream phases of a product's life cycle. Thus, when evaluating the full cost implications of specific warranty policies or extended maintenance agreements, management should give consideration to a product's life-cycle cost analysis.

There are several forms of life cycles that are relevant to product characteristics and the determination of the products' corresponding warranty policies. Product life cycles may be defined and discussed from the marketing and production perspectives. The marketing perspective characterizes the life-cycle stages of a product as start-up, growth, maturity, and harvest or decline. The production perspective may be presented in terms of the various business functions required to deliver a completed unit to the customer and reflects the product's life-cycle costs to be borne by the manufacturer. The production cycle includes the following business functions: selection of the product and target pricing, research and development, product design, manufacturing, marketing, distribution, and customer service.

Each stage of the marketing life-cycle approach incorporates features of the production life-cycle perspective. However, the relative importance given to specific business functions differs across the marketing life-cycle stages. For example, the start-up stage emphasizes research and development more than the customer service function. Further, the functional emphasis during a product's maturity stage will usually be on manufacturing, marketing, and distribution as compared to other business functions. Figure 29.1 graphically portrays the six business functions as a series of activities that results in a product's buildup of cost. Although essentially similar, Berliner and Brimson [4] describe a product life cycle from the production perspective in the following terms: product planning, preliminary design, detailed design, production, and logistical support. An understanding of product life-cycle stages, across each perspective, is

Product Selection and Target Pricing	Research and Development	Product Design 1) Pilot Testing 2) Financial Planning 3) Target Costing 4) Quality Setting Analysis	Manufacturing	Marketing	Distribution	Customer Service

Figure 29.1 Product cost accumulations across business functions.

important for management to consider the costs of quality enhancements during the product's design and production phases, the product's potential conformity to the performance expectations of management, and the resulting cost associated with the repair and/or replacement process under the warranty contract.

Product development occurs at the start-up stage in which the desired quality of the product is determined in relation to its countervailing parameters that include warranty costs. Choices among various materials to be used, alternative production processes, manufacturing design complexities, frequencies of production setups, and so forth, affect quality and warranty cost specifications. During the product and process development stage, there are many decisions that will have long-term implications on the structure of a product's whole-life costs. For example, management must decide on product performance requirements, the physical properties of material inputs, and the performance capabilities of the production facility. Decisions on these factors have a direct influence on the product's performance during the reliability testing phase and the product's useful life stage. Warranties represent a company's promise to guarantee a product's performance beyond the time of sale. The magnitude of warranty costs resulting from the external repair and replacement process is dependent on management's commitment toward product quality and reliability during the earlier stages of the product's life cycle. Management has the discretion during the product and process development stages and the manufacturing stage to require a particular level of product quality prior to the time of sale or to accept a level of warranty cost after the sale. For competitive strategy, management promises good product performance by investing in elements of product quality and/or promises good performance by implementing a repair and replacement policy after the time of sale. Costs associated with investments in a product's quality and reliability prior to the time of sale may be viewed as an implicit warranty for good product performance.

Good reliability is indicated by high mean time to failure and low mean time to repair. High mean time to failure can be achieved by spending more on functions upstream from production, where a careful analysis has been made on product designs, material specifications, and manufacturing processes. (See Chapter 21.) Appropriate investments in upstream functions will likely be more than offset by lower costs incurred downstream of the production, that is, warranty costs. A low mean time to repair is achieved with high quality and low complexity of product design. Reliability and quality aspects are further discussed in Chapter 23.

Sales during the start-up stage will be scrutinized carefully by consumers. Whether the product reaches the growth and maturity stages may depend on the consumer's perception of the product's quality and warranty acceptance. From a marketing perspective, high product quality coupled with a generous product warranty may be desirable at this stage, irrespective of the cost trade-offs.

During the growth stage, the product's design, development, and manufacturing processes are refined and continually improved with respect to meeting management's targeted cost and targeted quality. As a product's target costs are progressively reduced and target quality is increased, increased demands are placed on improving the effectiveness and efficiency within each business function. Meeting these improved targets contributes to lowering the warranty duration and the warranty policy's related cost.

A traditional approach to minimize the cost subject to reaching a specified level of product quality is to use a constrained optimization program. According to Balachandran and Srinidhi [5], no trade-off is necessary between cost and quality. To demonstrate their assertion, they provide an example of the effect of prespecifying a quality criterion at the body welding stage of automobile manufacturing. Any prespecified quality other than zero defects would allow some defective parts to enter the subsequent painting and assembly processes. The defective part would be worked on in all the subsequent stages and would finally be rejected in quality testing or by the customer. Thus, relaxing the quality a little in each stage increases the cost at subsequent processes and finally impacts warranty costs and the product's reputation as well.

In the maturity stage, standards may be used to specify quality, product cost, warranty duration, and estimated warranty costs. In this stage, production is generally steady, warranties are in place, and warranty costs are estimable. Effective control of warranty costs is performed by comparing a product's estimated warranty cost per unit and estimating total warranty cost in an arbitrary accounting period with their actual cost counterparts. This analysis will be discussed in more detail later in the chapter.

The harvest stage is characterized by a decline in product sales. The firm's strategy is generally to wind up the production and sale of this product and move on to new products with expanded growth potential. High quality and proper warranty maintenance are still extremely important at this stage because the reputation of the firm continues to be at stake.

29.4 THE RELATIONSHIP OF PRODUCT WARRANTY COST TO OTHER QUALITY ENHANCEMENT COSTS

Consumer expectation for higher-quality products has largely increased over the past few years. Consumers believe that paying a higher cost for a higher-quality product results in their paying a lower total cost over the life of the product [6]. Consequently, management should reevaluate their product strategies to understand the consequences associated with quality costs. However, executive awareness of the significance of product quality cost is still low. Although some management consultants have shown quality costs to be as much as 60% of sales, a study revealed that 50% of the managers surveyed thought their cost of quality was less than 5% of sales [7]

Although quality costs are due to several factors, the costs associated with nonconformity to product specifications and lost sales opportunities predominate. Products are classified as defective or nondefective depending on whether they conform to product specifications. However, not all products that perform within specifications are equally desirable. Market share can be lost due to customer dissatisfaction if products are not totally defect-free and do not perform well to their expectations. Warranties cover not only total breakdown of products but customer dissatisfaction with its performance as well. In this regard, robust quality is preferred to acceptable quality levels. Robust quality requires product parts to be made exactly on the target quality, whereas accepted quality levels allow for the part to be within specified tolerances. A zero-defect requirement for all parts may result in every part lying on the boundary of the acceptance level with a consequent complete product failure to meet customer expectations due to the cumulative errors. This will not happen if parts are required to meet target quality. For example, Ford Motor Company used parts with zero defects based on a specified tolerance level and experienced significant product failure and customer dissatisfaction. On the other hand, Mazda (then known as Toyo Koygo), which required robust quality, experienced low product failure and high customer satisfaction [8].

29.4.1 Components of Quality Costs

Quality costs are divided broadly into four components. Figure 29.2 gives an example list of costs under the categories.

Prevention costs are incurred to enhance the quality of production and include costs incurred for new machinery, automation with new technology, and education programs. Appraisal costs are typically internal inspection costs incurred to detect defective units. In advanced manufacturing environments that desire zero defects, inspections are carried out at several stages of the production process so as to reduce the possibility of utilizing defective subcomponents in a later stage. This increases the total appraisal cost but reduces defective outputs and, hence, the ultimate warranty expense.

Prevention Costs
- Product research and design
- Quality engineering
- Quality circles
- Quality education and training
- Supervision of prevention activities
- Pilot studies
- Systems development and implementation
- Process controls
- Technical support provided to vendors
- Auditing the effectiveness of the quality system

Appraisal Costs
- Supplies used in testing and inspection
- Test and inspection of incoming materials
- Component inspection and testing
- Review of sales orders for accuracy
- In-process inspection
- Final product inspection and testing
- Final inspection at customer site prior to final release of product
- Reliability testing
- Supervision of appraisal activities
- Plant utilities in inspection area
- Depreciation of test equipment
- Internal audits of inventory

Internal Failure Costs
- Net cost of scrap
- Net cost of spoilage
- Disposal of defective product
- Rework labor and overhead
- Reinspection of reworked product
- Retest of reworked product
- Downtime due to quality problems
- Net opportunity cost of products classified as "seconds"
- Data reentered due to input errors
- Defect cause analysis and investigation
- Revision of in-house computer programs due to software errors
- Adjusting entries necessitated by quality problems

External Failure Costs
- Cost of responding to customer complaints
- Investigation of customer claims on warranty
- Warranty repairs and replacements
- Out-of-warranty repairs and replacements
- Product recalls
- Product liability
- Returns and allowances due to quality problems
- Opportunity cost of lost sales due to bad quality reputation

Figure 29.2 Components of quality costs. [Adapted from Morse, W. J., Roth, H. P., and Poston, K. M. (1987). *Measuring, Planning and Controlling Quality Costs*, National Association of Accountants, Montvale, NJ.]

Internal failure costs are due to defective components discovered in the production process. These costs should be detected quite early in the production process in order to minimize future internal costs and reduce subsequent costs that the consumer may suffer. If defective units are sold at a discount, the includable internal failure cost is the discount offered. Any machine breakdown, labor downtime and/or time taken to readjust a malfunctioning machine should also be considered internal failure costs.

The main components of external failure costs are warranty costs and cost of lost sales due to customer dissatisfaction. Although there are numerous models to estimate warranty costs, the cost of lost sales is more difficult to estimate. Heagy [9] illustrates an approach to estimate loss sales using market research findings.

Among the four categories of quality costs, only prevention costs are value-added costs. Proper expenditure in this category can lead to considerable savings in the other three non-value-added quality costs areas. As warranty cost represents a major component of external failure costs, management should analyze the product warranty's role within the entire spectrum of quality costs.

29.4.2 Costs of Quality Optimality Considerations

A fundamental model describing the relationships between the components of quality costs and the optimal cost of quality was proposed by Juran and Gryna [10]. Figure 29.3 is a graphic display of their now traditional model of optimum quality cost. Quality cost is a convex function with a well-defined minimum cost and optimal quality assurance level.

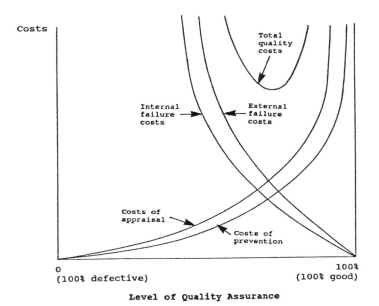

Figure 29.3 Quality cost relationships.

The traditional cost model of Juran did not consider several aspects of optimizing quality costs. For example, the quality costs components of prevention, appraisal, internal, and external are sequentially interrelated; that is, insufficient expenditures in prevention will lead to greater expenditures in appraisal, and insufficient expenditures in both prevention and appraisal will lead to increased expenditures on failure costs. Given the timing difference in quality cost incurrences, management should consider the time value of money associated with the relevant cost flows in their planning of an optimal quality cost level.

Although warranty costs were included as part of the external quality cost component in Juran's model, the cost associated with lost sales due to customer dissatisfaction was omitted. In an experiment to build a more robust model, Heagy [9] added lost sales to the external failure component of the traditional model. The cost of lost sales was determined as a linear function of the number of defective units, which was gleaned from warranty records. The basic shape of the curves in Figure 29.3 did not change even though the optimal point moved significantly to the right, demanding higher quality.

Models that consider only one or two of the quality cost components may lead to suboptimal or poor solutions. An example would be models that consider only the trade-off cost between appraisal costs and warranty costs. The magnitude of appraisal costs is partially influenced by the sample size chosen for testing. A larger sample size likely results in larger appraisal costs and a better estimate of quality. Although an appropriate determination of warranty costs can be made, testing does not improve quality per se and, hence, does not reduce warranty costs in any way. The size of the test sample should be determined as an integral part of the entire spectrum of appraisal function costs which, in turn, should be analyzed considering all other quality cost components.

29.4.3 Reporting Warranty Costs as a Component of Quality Costs for Control

A useful and timely reporting system for quality cost will provide management with a frame of reference to make resource allocation decisions. A report showing summary data, highlighting cost trends and the trade-offs among various quality cost activities, is helpful to management in their task to reduce overall quality cost while maintaining or increasing product quality. The reporting format may facilitate a comparison with normal practice by other manufacturers and with the company's own experience in controlling and reducing quality costs. A useful accounting format for the quality report is provided in Figure 29.4. The report provides each

	Quarter	Year to Date	Current Budget	Long Run Average Target
Prevention Costs				
•Quality control administration	1.2	1.0	1.0	--
•Reliability Engineering	2.7	2.6	2.4	1.0
•Quality Training	0.9	0.7	0.7	0.6
•Quality improvement programs	0.7	0.6	0.6	0.3
Total prevention costs	5.5	4.9	4.7	1.9
Appraisal Costs				
•Material Inspection	0.6	0.5	0.5	0.2
•Machine Calibration	0.3	0.2	0.2	--
•Product Acceptance	0.7	0.6	0.4	--
•Process Acceptance	0.7	0.7	0.5	0.4
•Outside Testing	0.6	0.6	0.4	--
•Quality Audits	0.3	0.3	0.2	--
Total appraisal costs	3.2	2.9	2.2	0.6
Internal Failure Costs				
•Scrap	1.1	1.1	1.0	--
•Rework	2.1	2.1	2.0	--
•Retesting	0.1	0.1	0.1	--
•Discarded	0.2	0.3	0.2	--
Total internal failure costs	3.5	3.6	3.3	--
External Failure Costs				
•Customer Handling	0.1	0.09	0.2	--
•Warranty Rework	0.2	0.1	0.2	--
•Recalls	0.2	0.1	0.1	--
•Product Liability	0.6	0.3	0.2	--
•Adjustments	1.0	0.9	0.9	--
Total external failure costs	2.1	1.49	1.6	--
Total quality costs	14.3	12.89	11.8	2.5

Figure 29.4 Quality cost report (quality costs as a percentage of sales dollars).

cost category under the four broad sources of quality cost with warranty costs listed under the external failure costs section. Each quality cost is stated as a percentage of sales by quarter, year to date, current budget, and long-run average target. With the introduction of just-in-time (JIT) inventory and production systems, a focus on manufacturing cells and reliable suppliers, the target column does not show cost percentages for quality control, machine calibrating, project acceptance, outside testing, quality audits, internal failure costs, and external failure costs.

The report should not be used indiscriminately to reduce quality costs, either across the board or individually by category. As Ostrenga [7] suggests, the cost of prevention is value added, whereas appraisal costs, internal failure costs and external failure costs are all non-value-added. Thus, by isolating the specific costs in each quality cost category, management has additional information to formulate their quality enhancement and cost reduction decisions. For example, an appropriate trade-off that increases a prevention activity and reduces an internal failure activity can conceivably reduce overall quality costs and increase the product's quality. Companies with well-run quality management programs can reduce quality costs to less than 2.5% of sales and, at the same time, maintain their excellence in quality products [11]. For example, Tennant Company decreased their total cost of quality from 17% of sales in 1980 to 7.9% of sales in 1986, with a further reduction to 2.5% by 1988 [12]. However, many companies consistently incur quality cost that are 12–15% of sales.

Although sales is an easily understood basis for management to use in their analysis of quality costs, there may be more appropriate bases. For instance, the cost of production is a better basis for internal failure costs, appraisal costs, prevention costs, as well as the total cost of quality. Sales is recommended as a base of comparison for external failure costs only. Reports on product quality may also provide nonfinancial quantitative information. For example, it is also important to keep track of the number of defectives from internal failures, the appraisal process, and those reported by external sources. A graph highlighting the trends is useful to management to see how fast the company is approaching its target.

29.5 STRATEGIC MANAGEMENT OF WARRANTY COSTS

As we have seen from the previous sections, warranty costs should be planned and controlled in relation to other considerations, such as product quality. This requires management to assess the interdependencies of functional costs across a product's life cycle. This section will address the cost accumulation process necessary to derive a product's functional cost per unit and overall life-cycle cost per unit for cost trade-off analysis.

A complete accumulation of all business function costs on a product-by-product basis is important for a company to effectively plan and control costs over a product's life cycle. Traditional accounting, prepared for external accounting purposes, does not accumulate product cost beyond the manufacturing business function. Rather, traditional external accounting recognizes many product and process development costs as current period cost for income statement purposes, and as a result, total product cost from a life cycle perspective are dramatically understated. Figure 29.5 shows the difference between cost accumulation under the traditional cost accounting approach and product life-cycle costing. Because more than 90% of a product's costs are usually determined by management upstream of the manufacturing phase, early planning should include the cost consequences of all possible alternatives to effect appropriate product warranty planning and control measures. Berliner and Brimson [4] sug-

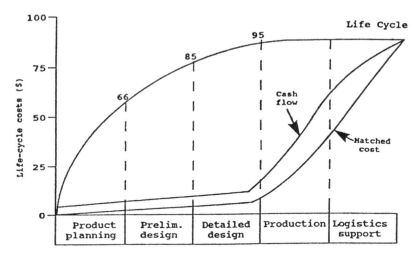

Figure 29.5 Conflicts between current accounting practices and the life-cycle concept. (From Ref. 14.)

gests that both product and process development costs should be charged to the products that benefit from these resources within each of the business functions. Only by accumulating specific product costs on the basis of business functions can management assess a product's long-term cost structure and profitability. The most accurate product cost per unit, regardless of the business function being considered, is achieved by increasing the percentage of direct costs to total product costs and allocating the remaining indirect cost by the most appropriate cost driver. A cost driver may be defined as a measure of activity, such as labor hours, that is considered a causal factor in the resulting indirect cost.

29.5.1 Identifying Direct and Indirect Product Costs

Costs incurred in each business function may be defined as direct or indirect with respect to the specific products benefiting from the respective business function's activities. Direct product costs are those that can be economically and physically traced to the particular product under consideration. For example, the cost of replacement components used in the repair process of a product under warranty is a direct product cost incurred within the customer service function. Indirect costs do not have a traceable relationship with a particular product. Rather, these costs are incurred as a consequence of operating the overall function or activity and all products serviced by the business function under consideration are assumed to generally benefit from these indirect costs. For example, the supervisor's salary in the repair and replacement center is an indirect product cost. Indirect costs must be allocated to the products benefiting from the indirect costs in order to determine the product's full function cost per unit.

29.5.2 Activity-Based Cost Accounting

Indirect costs of business functions may be allocated to products from a single cost pool by a single cost driver. For example, if the indirect cost of the customer service department was estimated to be $100,000 for the year and the selected cost driver was estimated to be 5000 direct labor hours, the application rate for indirect costs would be $20 of overhead applied to a product for every direct labor hour charged to a product being serviced. The basis of allocation ought to reflect the cause of the indirect cost incurrence. Therefore, accurate product costing requires a reasonably high correlation between the allocation base and the indirect costs being allocated. Where the indirect costs of a business function are heterogeneous and the overall correlation between the cost driver and the indirect costs is low, the accuracy of the product's resulting unit cost suffers. Further, if the allocation of indirect cost relies on a volume indicator (such as direct labor hours) where products differ regarding their complexity of servicing, then unit product cost may again be distorted because the volume measure may not capture the relative benefits received by the different products. To resolve these problems, indirect business function costs may be partitioned into more homogeneous cost pools on the basis of specific activities within the department. Greater accuracy in cost allocation is then achieved when a highly correlated cost driver is selected for each homogeneous cost pool. The practice of allocating indirect cost on the basis of specific activities is referred to as activity-based cost accounting (ABC).

Let us suppose we are interested in determining the amount of indirect costs allocated to products that receive attention in the customer service department. We will assume the department handles customer claims on a variety of defective products that are covered under a warranty or maintenance agreement. The department's resources are used in three broad ways: solving customer problems, processing returns, and testing returned products. The traditional and potentially less accurate approach would apply all indirect costs of the service center to the products serviced using a single cost driver, such as direct labor hours. Figure 29.6 illustrates the potential product cost distortions that may result from a single cost

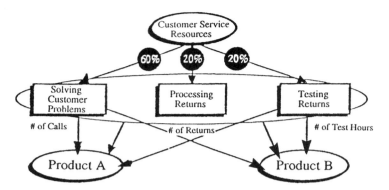

Figure 29.6   Activity-based costing in a customer service department. [Adapted from Blanchard, B. S. (1978). *Design and Manage to Life-Cycle Costs*, M/A Press, Portland, OR.]

driver. For example, if product A is a high-volume, low-variety type with simple design features, many of the problems arising in the product may be similar to each other and are easily detectable and correctable. On the other hand, if product B is a low-volume, high-variety type with complex designs, it may take many test runs for each claim to ascertain the defects. Essentially, product B may require more attention, both in terms of personnel time and the utilization of other resources, as compared to product A. To apply all the service costs using a single rate based on direct labor hours could overcost the high-volume, low-variety, simple-design product A and undercost the low-volume, high-variety, complex-design product B.

To remedy this problem, the ABC method may be used to allocated indirect cost to products on a more equitable basis. Figure 29.6 shows that the overall costs of the customer service department are first broken down into more homogeneous cost pools on the basis of service activities, that is, solving customer problems, processing returns, and testing returns. In our example, the relative indirect resources consumed by these activities were 60% : 20% : 20%, respectively, of the total indirect cost incurred within the customer service department. Next, a cost driver is identified for each activity. As the cost driver is assumed to be linearly related to the level of the activity's cost, the best choice may be ascertained by using linear regression analysis. The cost drivers chosen in the illustration are number of calls for the activity solving customer problems, number of returns for processing returns, and number of test hours for testing returns. Activity costs are subsequently allocated to products according to the product's consumption of the activity's resources as measured by the relevant cost driver. Given that the allocation of indirect cost is achieved on the basis of specific product causal factors, the ABC analysis is considered a more accurate method for unit cost buildup. As a comparison, Figure 29.7 illustrates the costs allocated to products using ABC as a percentage of conventional cost allocated using traditional methods.

Another cost that needs to be considered from a cost management perspective is the opportunity cost due to the delay caused to customer claims at the service center. If a product claim is very complex, it may require special attention. This will tie up the service center and delay the processing of other claims that may arrive. Such a delay will result in opportunity costs to the firm due to customer impatience and dissatisfaction. These costs should be attributed to the claim that is causing the delay. For analytical work in this area, see Ref. 13.

29.5.3 Target Analysis for Warranties and Maintenance Agreements

Product cost management has traditionally taken an internal perspective; that is, after the design stage is completed and manufacturing facilities

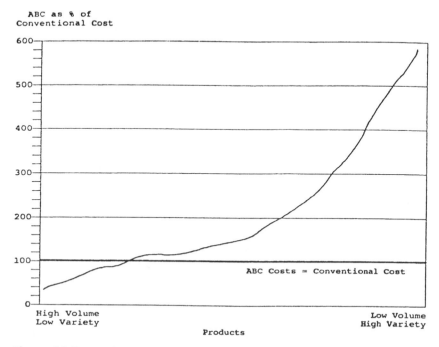

Figure 29.7 ABC as a percentage of conventional costs. (From Ref. 14.)

become operational, steps are then taken to minimize input resource costs and other operating costs to reduce the product's unit cost. On the other hand, target analysis takes an external perspective and emphasizes the management of product unit cost prior to the manufacturing stage. A product's initial target cost is determined by subtracting management's desired level of profit (over the product's life cycle) and anticipated distribution costs from the expected selling price required to retain a specified market share. With the target cost identified, a cost analysis of alternative product designs and manufacturing processes is performed to determine the product design that achieves the targeted cost as well as product quality and functionality. This process links engineering decisions to the marketing requirements [14], as each alternative includes varying product complexities, production technologies, and material specifications.

Target analysis strives to achieve an efficient point solution in a systematic manner in which an increase in quality is not possible without increasing the cost or a decrease in cost is not possible without decreasing quality. Target analysis is a powerful tool in the implementation of the continuous improvement philosophy. On a continuous basis, opportunities are sought to either increase quality at the current cost level or decrease the cost level while maintaining the current target quality. Once changed, the new levels become the targets.

The target analysis begins with a determination of customer requirements as to the product's functions, quality, and warranty policy. These identified attributes will not be changed during the target analysis process unless a change in the market takes place. The market or desired selling price, as well as the desired markup for the product, is also determined in the first step.

In the second step, the aggregate unit target cost and target quality of the product are operationally determined. The unit target cost is obtained by subtracting the desired markup from the selling price. The target quality is derived from an analysis of customer expectations and competition that impact on the market share.

In the third step, a preliminary design of the product with a consideration of the requisite production facilities, material and parts, human resource skills, and quality control functions is developed. The product's aggregate quality is estimated from this preliminary design phase. If the quality is not greater than the target quality, the process goes back to the design phase for further work in developing new or improved parts, manufacturing techniques, and/or quality control methods. If the quality is greater than the target, the analysis moves to the fourth step.

The fourth step involves estimation of the product's unit cost. The unit cost is determined by identifying all direct and indirect costs of the product across all business functions. This includes all costs associated with the cost of quality, that is, prevention, appraisal, internal failure, and external failure costs. As noted earlier, the most accurate cost per unit will result when the proportion of direct cost to total cost per unit is maximized and activity-based cost accounting is used to assign the remaining indirect costs of each business function to the products. If the estimated total unit cost of the product is less than the target cost, the process may go back to the second step to study the desirability of either lowering the target cost or increasing the target quality or both. If the total estimated unit cost exceeds the target cost, alternative product designs, manufacturing processes, product quality testing, marketing plans, distribution channels, and warranty policies are considered. The activity-based accounting system provides management with explicit information on the means to reduce the cost per unit, yet maintain a product that is comparable to the competition. If the unit cost cannot be reduced to the target cost level, the analysis is taken back to step one to consider a change in the markup. If a sufficient markup does not seem feasible given the cost, management can terminate this venture prior to major financial investments. Note that the selling price or desired quality are not changed in this analysis. Rather, the emphasis is on alternatives, assuming price and product quality are givens. This is in contrast to the traditional practice where costs are first determined and the selling price is subsequently determined by adding a markup.

If the total estimated unit cost equals the target cost, the analysis moves to step five which sets the target cost for each product on the basis of direct costs and cost per activity. In this step, the availability of the identified resources necessary to achieve the targeted cost and quality is considered. If this process does not conclude satisfactorily, either in achieving the target cost or the target quality, a cost analysis is performed on a sufficient alternative or the analysis is taken back to step one to adjust the markup. If the requisite resources are available, suppliers are notified and production plans can be started. This process will eventually reduce all four types of quality costs.

29.6 ESTIMATION MODELS FOR PLANNING AND CONTROL

In this section, we shall present selected models to demonstrate the estimation of costs associated with product warranties and maintenance agreements. With reliable estimates of repair and replacement cost that are associated with particular warranty or maintenance policies, management has a valuable tool to aid their planning and control processes. As an introduction to the models presented, we will first provide a brief discussion on the concepts of reliability.

29.6.1 The General Concepts of Product Reliability

The reliability of a product is related to the product's maintainability, the product's life cycle, and the probability of a failure-free operation. The maintainability of a product provides important insight into the product's overall reliability. For example, product components that cannot be restored are basically nonrepairable. Repairable product components, on the other hand, are susceptible to various renewal operations. Maintenance, whether preventive or corrective, involves time and money invested in repairing or replacing components to achieve continued satisfactory product performance.

Product failures may occur in any one of the three product life-cycle periods; the burn-in period, the useful life period, and the wearout period. (See Chapter 6 for a discussion of the "bathtub curve," which describes this reliability process.) The burn-in or debugging period is characterized by "early failures." To estimate repair and replacement cost during a product's life cycle, the probability of a failure-free operation must be known. This probability is characterized by a probability distribution function that is a function of the failure rate and the stochastic nature of the process. The debugging period generally reflects a decreasing failure rate. These failures typically result from component design and/or production deficiencies and are usually detected either during the appraisal function or the internal failures stage occurring in the factory. Quality testing groups, whose related costs are part of the appraisal cost component, may detect further deficiencies before the product is sold. Components found to be faulty are replaced with fault-free components and the product is then retested for other early failures. With additional care and increased investment in the prevention mode, early failures can be virtually eliminated and, as a result, products sold under warranty may be free of any manufacturing deficiencies. With increased attention paid to the prevention of defects in both the design and manufacturing, a consideration of these failures in the estimation of warranty costs less likely. Chapter 23 gives a rigorous treatment of the burn-in phenomenon.

The useful life period, which is usually the period product warranties are in effect, is generally characterized by "chance failures." If extreme care is taken at prevention, chance failures should not include failures due to manufacturing deficiencies. The term "chance" implies relatively unexpected failures. Accordingly, the incidence of chance failures during the duration of the useful life period and particularly during the shorter warranty period should be low as compared to the product's wear-out period. Chance failures occur at random and during sufficiently long periods of time, the failures tend toward a constant rate which may be suitably approximated by the Poisson process. (See Chapters 6 and 7).

The most inevitable type of failure occurs during the wear-out period. These failures are caused by deterioration of the design strength. The wear-out phase experiences an increasing failure rate. When the failure rate associated with repairing the component becomes unacceptably high, component replacement is required.

29.6.2 Estimating Warranty Costs

The hierarchical issues relating to the estimation of warranty cost for control purposes are estimating the product's repair and replacement costs during its warranty period followed by the estimation of the aggregate repair and replacement costs for all the product units under warranty during an arbitrary accounting period. The second phase of the estimation process allows management to monitor whether the actual warranty costs during an arbitrary period correspond to the estimate. The cost estimates can also be integrated into a variance analysis scheme to evaluate various responsibility centers associated with the warranty repair.

Many models have been proposed to estimate the warranty cost per unit of the product. For a comprehensive coverage of these models see Parts C and D of this book. Here we shall refer to Ref. 15 that models this estimation and the estimation for the arbitrary accounting period as well. Their models assume that warranties are in effect only during the useful life portion of the product. Consequently, failures occur only due to chance at random and exhibit a constant failure rate phenomenon. They assume an exponential distribution for the time to failure. The distribution in addition to constant failure rate has a memoryless property as well; that is, the probability of a failure does not depend on how long the unit has been operating without failure. The only probability distribution satisfying the memoryless property is the exponential distribution.

Given a configuration of components for a product, Balachandran et al. [15] describe the failure process of the product as a Markov chain. The state of the Markov chain is specified by a vector of dimension equal to the number of components. The elements of the vector, which is the state variable, denote the number of failures in each component up to the present time. For example, a state variable (0, 2, 1) indicates that the product comprises three components. The first component has not failed because it was replaced with a new one, the second component has failed twice and has been repaired both times, and the third component has failed once and was repaired. With the assumption of the Markovian property, given the present state of the process, the probability of reaching another state in the future is independent of past states. The failure processes of the components are mutually independent and the repair or replacement costs of these components are also mutually independent. Any component that fails will be repaired or replaced immediately with no downtime. This means that the state can move from (0, 2, 1) to (1, 2, 3) in the next time period (which may be 1 week) when there is one failure–repair in component 1, no failures in component 2, and two failure–repairs in component 3. Given the independence assumption, the individual probabilities of failure in each component are multiplied to calculate the probability of movement from one state to another. To make the state space tractable, they assume a specific replacement policy dependent on the number of accumulated failures of a component. A specific policy could be that components are replaced with new ones after four failures. The first three failures in any component, according to this policy, will result in repairs. Attaching a cost per repair and per replacement for each component, the expected value of the cost per week due to warranty can be calculated by multiplying the probabilities of the possible movements to a state from a starting state by the cost incurred to repair/replace consequent to such movement.

The authors use a recursive relationship to estimate the repair/replacement cost of a new product unit during a warranty period that may last several weeks. To estimate the warranty costs during an arbitrary accounting period is a little more complicated. There are many units that have been sold under warranty with different unexpired warranty durations outstanding. It is necessary to estimate the cost for a product that is not new and does not have the entire original warranty period left. The model developed by Balachandran et al. [15] handles these complications to provide a total estimate for the repair/replacement cost. This approach is useful for the control function which requires a comparison of actual warranty compliance cost with expected cost during the current accounting period. Products under warranty during the current accounting period could have been sold during the current or the prior accounting periods.

Balachandran and Maschmeyer [16] use the Markovian model of their earlier paper to illustrate the accounting variance calculations from a product's repair and replacement process. An accounting variance is defined as the deviation of the actual cost from the expected. They partition the sources for these deviations and provide an approach to identifying the causes for the failures. An analysis of product cost deviations or accounting variances requires two decision tasks: (1) the reporting decision and (2) the investigating decision. On the reporting of deviations, consideration should be given to nonaggregation or fine partitioning of sources prior to observing and investigating deviations; that is, a planning decision should be made on the number of individual sources to be monitored and the deviations to be reported. In the absence of cost-effective partitioning of sources prior to observations, investigation costs subsequent to the reporting of deviations from aggregated sources could conceivably be more and the control process itself is more difficult.

There are three types (i.e., sources) of deviations of actual from expected costs. They include nonrandom deviations resulting from permanent changes, controllable nonrandom deviations that are due to temporary causes, and random fluctuations in the process that deserve no attention under the accounting profession's concept of materiality. Identification of these sources of deviations form the basis for the investigation decision.

Balachandran and Maschmeyer [16] provide an approach to analyzing the deviation of actual warranty cost from its estimate during an accounting period. They illustrate the process of separating the effects of repair and replacement activities into spending and reliability measures. The analysis provides input into the performance evaluation and control process by allowing the investigator to identify performance deviations from products sold in both current and prior accounting periods. By selecting particular variance combinations, several interdepartmental or functional dependencies can be analyzed. The analysis provides an effective cost control to improve product performance by investigating the specific sources of product warranty cost deviations. With a better understanding of warranty cost sources, management can assess cost–benefit relationships in terms of improved product designs, material specifications, quality control procedures, reliability testing, and repair and replacement policies. An improved understanding of the period's component replacement requirement will contribute to a more effective inventory control policy for product components.

To implement the estimation and control models, it is necessary to determine the failure rates of electronic and mechanical subcomponents and components and to estimate the average costs due to component breakdowns, dependent on the number of prior breakdowns. The failure rates for electronic components are generally available, whereas the rates for mechanical components are determined from reliability testing. Where companies manufacture components for their own products and a reliability testing program for the quality control operation is operable, failure rate data should be obtainable without additional costs. If the components are purchased, the manufacturer may be able to provide the failure rates. A less desirable source of failure rates may be the data gathered from past experience with the component. However, precise determination from past data may be considerably more difficult to obtain because the component may be used in different products and undergo changes over time. As a last resort, it may be necessary to initiate reliability testing to obtain the failure rate data for which the cost may be high and needs to be compared with the benefits of the estimation and control processes.

29.6.3 Estimating Maintenance Agreement Costs

After an explicit warranty on a product has elapsed, many manufacturers offer maintenance agreements that cover failure costs during the product's remaining operational life. Some maintenance agreements also provide preventive maintenance that is to be performed periodically. Typical manufacturers providing maintenance agreements are those that make mechanical components or machines such as major household appliances, office products, power-driven vehicles, and other production equipments. Such products that are beyond their useful life period exhibit wear-out phenomena of increasing failure rates.

The concern here, as in the previous section, is on two issues: estimating a product's repair costs during the maintenance agreement period and estimating the aggregate repair costs due to maintenance agreements during an arbitrary accounting period. A valid determination of expected repair costs for a product unit can facilitate management's pricing decisions relative to the maintenance agreements. If prices are more influenced by the market, these estimates help the provider of the agreements to check whether they fall within the target costs. The second focus can be used to integrate the cost estimates into an accounting variance analysis scheme to evaluate various responsibility centers associated with the repair of products under a maintenance agreement. For an analysis of the effect of extended warranties on the marketplace, see Chapter 18.

Balachandran and Maschmeyer [17] provide a model to estimate repair costs of products under maintenance agreements. They use a Markov chain with state vectors defined as the number of accumulated failures of each component. As maintenance agreements are likely to be in effect during the wear-out period, the Weibull probability distribution was assumed, as an increasing failure rate is appropriate during this stage of a product's life cycle. The Weibull distribution is one of the most widely accepted and used lifetime probability distributions across many types of manufactured items. (See Chapters 6 and 8 and Part D.) Although the Balachandran and Maschmeyer model does not explicitly consider costs associated with preventive maintenance, these costs are easy to estimate and can be added to the total maintenance costs. Of course, the failure rates will decrease with the introduction of preventive maintenance.

The offering of maintenance-type agreements is expanding across several industries. Office products retailers and appliance dealers have offered such agreements for years. More recently, automobile dealers have offered "extended warranties" beyond the normal warranty period for an annual cost to the purchaser.

This chapter has presented issues on warranties and quality that are relevant to management accounting. Warranties should not be studied in isolation. Warranty costs should be controlled in relation to the other quality costs. After all, warranties guarantee quality of the product's performance.

REFERENCES

1. FASB (1978). *Objectives of Financial Reporting by Business Enterprises*, Statement of Financial Accounting Concepts No. 1, Financial Accounting Standards Board, Stamford, CT.
2. Williams, J., Stanga, K., and Holder, W. (1992). *Intermediate Accounting*, 4th ed., Dryden Press, Harcourt Brace Jovanovich College Publishers, San Diego, CA.
3. Horngren, C., Foster, G., and Datar, S. (1994). *Cost Accounting, A Managerial Emphasis*, 8th ed., Prentice-Hall, Englewood Cliffs, NJ.
4. Berliner, C., and Brimson, J. (eds.) (1988), *Cost Management for Today's Advanced Manufacturing, The CAM-I Conceptual Design*, Harvard Business School Press, Boston, MA.
5. Balachandran, K., and Srinidhi, B. (1991). Target analysis: Cost, quality or both? *Vikalpa: The Journal for Decision Makers* (April–June), 19–25.
6. Simpson, J., and Muthler, D. (1987). Quality costs: Facilitating the quality initiative, *Cost Management*, **61**, 25–34.
7. Ostrenga, M. (1991). Return on investment through the cost of quality, *Cost Management*, **65**, 37–44.
8. Taguchi, G., and Clausing, D. (1990). Robust Quality, *Harvard Business Review*, **68**, 65–75.
9. Heagy, C. (1991). Determining optimal quality costs by considering cost of lost sales, *Cost Management*, **65**, 64–72.
10. Juran, J., and Gryna, F. (1980). *Quality Planning and Analysis*, McGraw-Hill, New York.
11. Crosby, P. (1979) *Quality is Free: The Art of Making Quality Certain*, McGraw-Hill, New York, p. 18.
12. Carr, L., and Tyson, T. (1992). Planning quality cost expenditures, *Management Accounting*, **74**, 53–56.
13. Balachandran, K., and Srinidhi, B. (1987). A rationale for fixed charge application, *Journal of Accounting, Auditing and Finance*, **2**, 151–169.
14. Turney, P. (1991). *Common Cents, The ABC Performance Breakthrough*, Cost Technology, Hillsboro, OR.

15. Balachandran, K., Maschmeyer, R., and Livingstone, J. (1981). Product warranty period: A Markovian approach to estimation and analysis of repair and replacement costs, *The Accounting Review*, **56**, 115–124.
16. Balachandran, K., and Maschmeyer, R. (1992). Cost analysis for product warranties, *Pacific Accounting Review*, **4**, 59–76.
17. Balachandran, K. and Maschmeyer, R. (1992) Accounting for product wear out costs, *Journal of Accounting, Auditing and Finance*, **7**, 49–66.

30

Warranty as a Contingent Claim

Edward W. Frees

University of Wisconsin
Madison, Wisconsin

30.1 WARRANTIES AND FINANCIAL RISKS

With increasing emphasis on the quality of products, consumers are looking to warranties to serve as a financial guarantee regarding the reliability
of a product. A warranty is an agreement provided to the consumer by
the manufacturer to repair or replace a purchased product upon failure.
A warranty is a cost associated with a product, and this cost is implicitly,
or explicitly, a part of the pricing of the product. Many warranties are
long term in the sense that their guarantees exceed the length of the current
financial reporting period. Hence, costs that occur in the future, with
respect to the current reporting period, are liabilities rather than current
expenses. These future costs should be estimated and recognized in the
financial statements. Financial obligations are realized only upon failure
of the product, which can be viewed as a random event. The random
nature of these obligations can be easily interpreted using analogies to an
insurance policy, a common contract having characteristics similar to the
warranty agreement. These analogies, and related financial characteris-

tics, are developed more fully from a broad managerial perspective in Section 30.2.

Randomness, and consequential financial risks, is inevitable in most monetary systems. Gambling is a classic example of a monetary system endowed with randomness and, indeed, games of chance have been used to suggest many stochastic models. Stochastic models are also regularly used to model insurance and financial risk systems. A key distinction is that gambling systems create risk, although insurance and financial systems are devoted to the transfer and reduction of risk. Insurance risk systems feature methods of pooling to reduce risk in the system. In financial systems, pooling, or diversification of assets, is also useful. However, in financial risk systems, pooling is limited as a method for risk reduction due to a common economic environment shared by many of the risks. Contingent claims, that is, monetary events based on contingent, or random, events, and models thereof can be found in both the insurance and financial risks literatures. A good overview of the distinctions of these two literatures can be found in Ref. 1.

This chapter shows the relationship between the financial consequences of a warranty policy and the well-developed stochastic models in the insurance arena. These latter models are often labeled as "actuarial" models. Sections 30.3 and 30.4 discuss the quantification of warranty costs from an actuarial perspective. The idea here is that the time of failure of a product, at the inception of the warranty, is unknown and can be modeled stochastically, that is, as a random event. Briefly, the analyst makes various assumptions about the shape of the distribution of failures, for example, Weibull, new better than used, and so forth, or about the moments of the distribution. (See Chapters 6 and 8.) From these assumptions, costs associated with warranty can be quantified as summary measures of the random cost. Chapter 6 provides a detailed treatment of these stochastic models. By establishing the relationship between warranty policy and actuarial models, for many warranty contracts we will immediately be able to specify important financial quantities, including prices and reserves. For other products, certain complications arise in the renewal guarantees of the warranty or assumptions concerning the economic lifetime of the product. These complications are the source of a number of discussions in the literature. Sections 30.3 and 30.4 provide an introduction to some of the complexities that arise for certain warranty contracts.

30.2 MANAGEMENT PERSPECTIVE

In the insurance industry, it has long been understood that the timing of recognition of book profits and losses has a tremendous impact on the financial statements of the company. The potential variability in the timing of recognition is due to the long-run nature of the insurance contract. As an extreme example, a married couple purchasing a life insurance policy in their early twenties may enjoy annuity payments on conversion of the policy until age 100, a contract span of 80 years. Because of the long-run nature of the contract, a liability, or reserve, for the contract must be established in order to ensure a smooth recognition of profits from financial period to financial period.

Due to marketing and other pressures, in several manufacturing industries the warranty is being used to a greater and greater extent to ensure the consumer of the reliability of the manufacturer's product. A warranty is an agreement provided to the consumer by the manufacturer to repair or replace a purchased component upon failure. This agreement is a contractual liability and, if considered material, must be itemized in the firm's financial statements as required in the Statement of Financial Accounting Standards No. 5; *Accounting for Contingencies* [2]. Thus, as in insurance accounting, it is important to establish a reserve (liability) for the warranty contract in order to establish an orderly flow of profits through the business. Too large a reserve will deflate surplus funds which could be used for other purposes. Too small a reserve may conceal poor operating results. Due to the minor financial impact that warranties historically have had, it is not surprising that this item has traditionally been ignored in the manufacturer's balance sheet.

A warranty is similar to an insurance contract in that a warranty is an *aleatory* contract, that is, a contract depending on the outcome of uncertain events. Similar to insurance or gambling agreements, the uncertain events under warranties can be modeled stochastically. Unlike insurance agreements, typically the purchase of the manufacturer's product is the main agreement and the warranty is an ancillary agreement. Thus, the warranty may be viewed more akin to a rider, or attachment, in insurance contracts. Because the financial importance of each contract is relatively small compared to the main agreement, the manufacturer usually agrees to accept the risk of component failure. In insurance jargon, this practice is known as "self-insurance." There have, however, been some attempts to model insurance systems associated with warranty policies; see, for example, Refs. 3–5. For any particular contract, both the amount and timing of the warranty cost is random. This warranty cost is, thus, only a potential liability of the manufacturer and is all too easily ignored.

Insurance models provide ways to organize data and forecast costs. For example, consider a particular carline, assembly plant, and type of warranty coverage. It is often useful to organize aggregate warranty costs into the following triangular array:

$$\begin{bmatrix} \text{Cost}_{1990,1} & \text{Cost}_{1990,2} & \text{Cost}_{1990,3} & \text{Cost}_{1990,4} \\ \text{Cost}_{1991,1} & \text{Cost}_{1991,2} & \text{Cost}_{1991,3} & \cdot \\ \text{Cost}_{1992,1} & \text{Cost}_{1992,2} & \cdot & \\ \text{Cost}_{1993,1} & \cdot & \cdot & \cdot \end{bmatrix}$$

Here, $\text{Cost}_{i,j}$ represents the incremental cost arising from model year i in year $i + j - 1$. For example, $\text{Cost}_{1990,1}$ and $\text{Cost}_{1990,3}$ are the incremental costs from 1990 models in 1990 and 1992, respectively. The triangle is incomplete because costs from 1990 models in 1994 have not yet been realized. Similar triangles can be constructed for number of vehicles or some other exposure measure method. Other refinements, including breakdown by month or component, may be handled in a similar fashion.

Forecasting for this type of data is discussed in detail in the actuarial literature. Without other covariate information, the {Cost} variable is a longitudinal series with two dimensions of time, one for the incurral (i) and one for the development (j). Autoregressive models for forecasting the {Cost} series have appeared in the literature, see, for example, Ref. 6. However, most forecasting approaches use regression methodologies by treating i and j as explanatory variables. Various types of nonlinear regression and smoothing techniques can be found in the literature. Taylor [7] provides an overview.

Despite the similarities, it should also be emphasized that forecasting warranties is not the same as insurance costs. There are two fundamental differences. One, experience under many lines of insurance is stable, or nearly so, even though the reliability of many products under warranty is constantly improving. Two, one of the key issues in forecasting warranty costs is to provide a reliable estimate of exposure to potential liabilities from warranty agreements. However, in the insurance industry, a direct charge is made for the contract and, thus, an "earned premium" figure serves as a useful proxy for the level of exposure.

The traditional actuarial approach to valuation in North American may be termed a "micro" model, that is, a representation of an individual contract. By adding up the contracts, one gets a model for a block of business or a firm. The other approach is to develop a model directly of a block of business or a firm, called a "macro" or stochastic processes approach, adopted by many European actuaries; see Ref. 8. In the classical actuarial literature, micro and macro models are called individual and collective models, respectively. In economics, they are referred to as disaggregate and aggregate models, respectively. Historically, computations associated with macro models have been much more tractable than those for micro models, although the computing facilities available now make this advantage less important. Another argument for macro models is that the basic collection of claims can be thought of as arising as the sum of independent claims. This motivation is unattractive when trying to analyze the behavior of a block of claims when the independence assumption no longer holds.

One advantage of a model at a contract level is that when basic situations change, such as when the reliability of the underlying product is improved, we have some inkling as to how to change the structure of the model. Of course, when basic conditions change, any model needs to be validated. The disadvantage of working with smaller components is that we must have some idea of how to put them together.

30.3 ACTUARIAL MODELS OF INDIVIDUAL CONTRACTS

Assume that the random cost of each warranty agreement depends on some underlying random variables in some complex fashion. From the consumer's point of view, the value of the warranty is dependent on the realized lifetime of the product available and, thus, is a random variable. For example, typically the product's lifetime exceeds the warranty duration and, thus, the value of the warranty is zero. The manufacturer often makes many such contracts, however, and therefore is interested in various summary measures of this random variable. The goal here is to understand these summary measures based on knowledge of some aspects of the distribution of the underlying random variables and certain model parameters.

In the warranty literature, random times to product failure drive cost considerations. If the cost amount at failure is also random, this is assumed to be modeled by exogenous information. For example, for a warranty on the tire of an automobile, only the time to failure is important because one could assume that the replacement cost of the tire is reasonably well known. Conversely, costs associated with the warranty on a microcomputer may vary depending on the extent of failure and, hence, the repair amount could also be modeled stochastically. The repair amount model is not generally explicitly addressed in the warranty literature. The most important model parameter typically considered is the duration of the warranty, W (see Part D). In principle, this could be a two-dimensional vector of parameters. For example, one could use one dimension for calendar time and one for operational time, for example, automobile mileage (see Chapters 13 and 14). An important application of these types of models is comparing expected costs associated with a warranty duration with the costs of a longer duration. This difference is interpreted to be the incremental cost of extending the warranty guarantee period. Other types of parameters include the amount of warranty reimbursement as a function of failure (see Chapter 11).

In this section, some financial aspects of a basic contract are provided. The financial aspects studied are the basic value (or liability) of a warranty contract at contract inception date and at an arbitrarily chosen later date. Value at contract inception may be used for pricing a product so that the warranty cost can be built into the price of the underlying product. Even if warranty costs are not explicitly recognized in the product pricing, for example due to marketing pressures, at least the values of the warranty obligations will be available to the manufacturer's financial managers. Values of the warranty at later contract dates are useful in determining an approximation to the manufacturer's liability that arises from the obligations in the warranty contract. We first examine contracts without renewal options in detail and then examine more briefly contracts with renewal options. For pedagogic purposes, all values presented are based on exact, or continuous, time. In actuarial practice, discrete-time approximations are used. Details of these approximations can be found in Ref. 9.

30.3.1 Models of Contracts Without Renewal Options

Consider a warranty with random time to failure X at contract inception, labeled time $= 0$ for convenience. The random present value of the cost of a basic contract without renewal options to the manufacturer, or holder of this liability, is denoted by

$$L(X) = D(X)\, C(X)$$

Here, $D(x)$ is a nonrandom discount factor. In practice, typically we have $D(x) \equiv 1$ for no discounting or $D(x) = (1 + i)^{-x}$ for a continuously compounded interest factor, where i is the effective yield. In this chapter, assume that $D(x)$ is a known function. For some implications of a random discounting factors, one reference is the work of this author [10]. The term $C(x)$ is the cost factor in constant dollars terms. Typically, we might use $C(x) = 100$ if $x \le W$ or 0 otherwise. For this example, the warranty pays \$100 only if the product fails by time W, the warranty duration. As another example, consider $C(x) = 100\,(W - x)/W$ if $x \le W$ and $= 0$ otherwise, a standard pro-rata warranty contract. Chapter 1 introduces a wide variety of warranty contracts that are encountered in practice.

For any given warranty contract, there is a winner and a loser. For example, consider the simple case of paying \$100 only if the product fails before the warranty duration W. If the product is good, the manufacturer pays nothing. If it is a lemon, the manufacturer pays $100 at the time of failure. The closer the failure is to time of contract inception, the larger the present value of the liability. The average worth, or liability, of the contract can be expressed through expectations as

$$E[L(X)] = \int_0^\infty D(x)C(x)f(x)\,dx$$

However, the expected value is not the only way to establish the liability of a contract. Consider the following example.

Suppose that, at time 0, a manufacturer issues 100 identical warranties, each of which pays $100 only if the product fails before the warranty duration W. Assume that the failure times underlying each contract are identical and independent of the others. Suppose that we wish to determine (a) the maximum obligation of the manufacturer, (b) the minimum obligation of the manufacturer, and (c) an approximate fund amount so that the manufacturer is 95% certain of meeting the obligations undertaken by the warranties issued.

To see the solution, let us begin by assuming, for convenience, that $X_1, X_2, \ldots, X_{100}$ represent the random failure times for the 100 contracts issued. For part (a), if the manufacturer is unlucky in the extreme, each of the X's will be less than W, and the manufacturer's obligation is 100×100, or $10,000. On the other hand, for part (b) each of the failure times may exceed W, and the manufacturer's obligation turns out to be $0. More realistically, in part (c), because we are dealing with the sum of 100 independent policies, we can use the Central Limit Theorem to obtain reliable solutions. Thus, we would like to determine a FUND so that

$$0.95 = \text{Prob}(\text{Sum of obligations} \leq \text{FUND}) = \text{Prob}\left(\sum_{i=1}^{100} L(X_i) \leq \text{FUND}\right)$$

$$= \text{Prob}\left(\frac{\sum_{i=1}^{100} L(X_i) - E[L(X)]}{\sqrt{100\,\text{Var}[L(X)]}} \leq \frac{\text{FUND} - E[L(X)]}{\sqrt{100\,\text{Var}[L(X)]}}\right)$$

$$\approx \text{Prob}\left(N(0,1) \leq \frac{\text{FUND} - E[L(X)]}{\sqrt{100\,\text{Var}[L(X)]}}\right)$$

Here, $N(0, 1)$ means a standard normal random variable and $\text{Var}[L(X)]$ is the variance of the random cost $L(X)$. This variance would be calculated in a similar fashion to $E[L(X)]$. Thus, using standard probability tables, we calculate the desired fund to be approximately FUND $\approx E[L(X)]$ + $1.96(100 \text{ Var}[L(X)])^{1/2}$.

In part (c) of the above example, the fund necessary to comfortably meet the obligations is calculated as the expected value of the contract plus a factor times the standard deviation of $L(X)$. Interpret this second term as a loading for the uncertainty of the failure times, or risk. Clearly, other loadings may also be important; for example, administrative expenses associated with the contract, the amount of capital available to the manufacturer, and so on. See Ref. 11 for a nice introduction to a variety of methods of "premium principles," as this issue is referred to in the actuarial literature. Bowers et al. [9] provide an introduction to expense allocation.

Now suppose that we would like to determine the value of contract at some time t. For example, we may be at the end of the financial period and have issued a warranty contract t time periods before. In the actuarial literature, the value at time t is denoted by $_tV$. If the warranty has expired or the product has failed, there is no contingent obligation and $_tV = 0$. Otherwise, using expectations, a basic definition of the value is $_tV = E_t[L(X)]$. Here, the notation $E_t[\cdot] = E[\cdot|X > t]$ means an expectation conditional on the fact that X exceeds t. Note that if $X \leq t$, then at time t, we have observed the failure of X and there is no randomness. With this premium principle we have

$$_tV = E_t[L(X)] = \int_t^\infty D(x - t)C(x)\frac{f(x)}{\overline{F}(t)}\,dx$$

$$= \int_0^\infty D(s)C(s + t)\frac{f(s + t)}{\overline{F}(t)}\,ds$$

This calculation is simply due to the fact that $\text{Prob}(X > x|X > t) = \overline{F}(x)/\overline{F}(t)$ for $x > t$. Note that at any time point, we may interpret $C(t) - _tV$ as the "amount at risk" for the manufacturer; that is, the liability of the manufacturer is $_tV$, and presumably the manufacturer has available an equal amount of assets to cover this liability. If a warranty claim occurs at time t, the amount of the claim is $C(t)$. Thus, the difference, $C(t) - _tV$ is the amount of potential claims that the manufacturer does not have ready assets for and is "at risk" for this amount.

To illustrate the reserve calculation, consider a warranty that pays \$100 if the product fails within 3 years, and 0 otherwise. Suppose that we wish to calculate the reserve $_tV$, for $t = 0, 0.25, 0.50, \ldots, 2.75, 3$, corresponding to quarterly financial statements for the first 3 years of the product. For convenience, we assume that the product survival is governed by the Gamma distribution with mean 5 years and variance 5 years. Thus, we assume that the product survival follows the density

$$f(x) = \frac{\lambda^r \, x^{r-1} \, e^{-\lambda x}}{\Gamma(r)}$$

with $r = 5$ and $\lambda = 1$. With this notation, we denote the distribution function as

$$F(r, \lambda, t) = \int_0^t f(x) \, dx = \int_0^t \frac{\lambda^r \, x^{r-1} \, e^{-\lambda x}}{\Gamma(r)} \, dx$$

Consider the continuously compounded interest case, so that $D(x) = e^{-\delta t}$, where $\delta = \ln(1 + i)$ is the force of interest and i is the effective interest rate. Thus, the reserve at time t is

$$
\begin{aligned}
{}_t V &= \int_t^\infty D(x - t) C(x) \frac{f(x)}{F(t)} \, dx = \frac{100}{1 - F(5,1,t)} \int_t^3 e^{-\delta(x-t)} \frac{x^{5-1} e^{-x}}{\Gamma(5)} \, dx \\
&= \frac{100 e^{\delta t}}{[1 - F(5,1,t)](1 + \delta)^5} \int_t^3 \frac{(1 + \delta)^5 x^{5-1} e^{-(1+\delta)x}}{\Gamma(5)} \, dx \\
&= \frac{100 e^{\delta t}}{[1 - F(5,1,t)](1 + \delta)^5} [F(5, 1 + \delta, 3) - F(5, 1 + \delta, t)]
\end{aligned}
$$

Table 30.1 and Figure 30.1 summarize the calculation for $i = 0\%$ and $i = 10\%$ rate of interest. It is interesting that the reserve is relatively flat from $t = 0$ through 1.75. This is due to the relatively small number of failures during this time period. From these summaries, we see that the greatest effect of the interest rate is at the warranty initiation at $t = 0$. Note that from about $t = 2$ on, the reserves at different interest rates become much closer to one another. This is because there is little discounting effect in the remaining time period. Finally, because the warranty duration expires at 3 years, all reserves are 0 at $t = 3$.

At the end of the financial period, ${}_t V$ is an obligation of the manufacturer that should be recognized in the financial statements. Changes in financial statements represent sources of revenues, or expenses, to the manufacturer and thus are important. For convenience, consider a continuously compounded interest case with $D(x) = (1 + i)^{-x}$ and define $\delta = \ln(1 + i)$ to be the force of interest. Under appropriate smoothness conditions, straightforward calculus shows that

$$\frac{d}{dt} {}_t V = (\delta + \mu_{x+t}) \, {}_t V - C(t) \mu_{x+t}$$

Table 30.1 Reserve Sample Calculation

t	$F(5, 1, t)$	$_tV$ (Calculated using 0% interest)	$F(5, 1 + \delta, t)$	$_tV$ (Calculated using 10% interest)
0.00	0.000000	18.4737	0.000000	14.7413
0.25	0.000007	18.4731	0.000010	14.7407
0.50	0.000172	18.4596	0.000266	14.7273
0.75	0.001065	18.3868	0.001613	14.6567
1.00	0.003660	18.1742	0.005435	14.4567
1.25	0.009124	17.7230	0.013289	14.0443
1.50	0.018576	16.9306	0.026543	13.3410
1.75	0.032902	15.7001	0.046140	12.2804
2.00	0.052653	13.9425	0.072496	10.8089
2.25	0.078014	11.5753	0.105507	8.8831
2.50	0.108822	8.5185	0.144621	6.4649
2.75	0.144621	4.6898	0.188955	3.5173
3.00	0.184737	0.0000	0.237410	0.0000

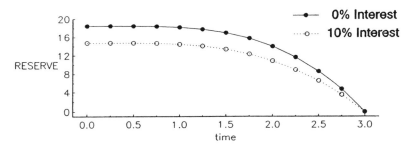

Figure 30.1 Plot of reserves versus time. Reserves denoted by the closed circle (●) were calculated at 0% interest. Reserves denoted by the open circle (○) were calculated at 10% interest.

Here, $\mu_x = F(x)/\overline{F}(x)$ is the hazard rate, also called the force of mortality in the actuarial literature. With this expression, we interpret the left-hand side as the change in value over a small period of time. The first term on the right-hand side is the increase in the fund due to interest plus survivorship. The second term on the right-hand side is the decrease due to payment of realized obligations. A rearrangement of terms provides another useful expression:

$$\frac{d}{dt}\,_tV = \delta\,_tV - \mu_{x+t}[C(t) - \,_tV]$$

Here, the change in value can be interpreted as the increase due to interest earnings minus the decrease due to realized failures. Recall that the financial risk of a failure is expressed in the "amount at risk," $C(t) - \,_tV$. In actuarial practice, the discrete-time approximations provide useful financial statement information concerning changes in value from one financial statement to the next.

30.3.2 Models of Contracts with Renewal Options

We now briefly consider warranty contracts with options to renew; that is, at product failure prior to the end of a warranty period, the consumer may purchase, usually at reduced or no cost, a new product with a warranty that extends to the original time, say, W. This renewal right is an option of the original contract, and, hence, the liability should be recognized at the contract initiation date. To this end, we use $\{X_i\}$ to be a sequence of independently and identically distributed random variables that represent the lifetime of current and potential replacement products. As earlier, use $C(x)$ to represent the reimbursement amount for failure of a product that is x time units old, $D(\cdot)$ is the known discounting function, and $L_r(W)$ is the potential liability with the renewal option. Assume that replacement time is negligible (or if not, the manufacturer will compensate the consumer by extending the warranty duration). Thus, if X_k is the lifetime of the current unit, then the partial sum $S_k = X_1 + X_2 + \cdots + X_k$ is the time of failure of the current unit (since contract initiation). The potential, random, liability can be expressed as

$$L_r(W) = \sum_{k=1}^{\infty} D(S_{k-1})C(X_k)I(S_k \le W)$$

where $I(\cdot)$ is the indicator of a set, that is, $I(S_k \le W) = 1$ if $S_k \le W$ and $= 0$ otherwise. For example, in the case of no discounting with constant replacement cost of \$100, we have $D(x) \equiv 1$ and $C(x) \equiv 100$, so $L_r(W) = 100 \sum_{k=1}^{\infty} I(S_k \le W)$, which is 100 times the number of failures by time W. Using expectations, a value of this contract is

$$_0V_r = E[L_r(W)] = \sum_{k=1}^{\infty} \left(\int_0^W \int_0^{W-u} C(u)D(v)f^{(k-1)}(v)f(u)\,dv\,du \right)$$

Here, $f^{(k)}$ is the density associated with the k-fold convolution of F, $F^{(k)}$; that is, $f^{(k)}(x) = (d/dx)F^{(k)}(x)$, where $F^{(k)}(x) \equiv \text{Prob}(X_1 + \cdots + X_k \le$

x). In the case of no discounting with constant replacement cost of $100, we have $E[L_r(W)] = 100M(W)$, where M is the renewal function at time W. See Chapters 7 and 8 for further discussions of the renewal function.

As in the case of no renewal options, it is straightforward to calculate $\text{Var}[L_r(W)]$ to arrive at prices adjusted for risk loadings or to calculate $E_t[L_r(W)]$ to arrive at values at time t. Several authors discuss estimation of model parameters when the data are from a simple random sample without censoring or covariates; that is, under the basic sampling scheme, we can think of $X_1, X_2, \ldots, X_n$ as a simple random sample. In a series of articles [12–16], this author discussed estimation of warranties whose cost is a related to the renewal function under this sampling scheme. This work was recently extended to include situations where covariate information is available (Frees [17]). Under this type of sampling scheme, associated with each failure time X_i is a set of independent variables that may help explain the failure time. For example, for an automobile warranty, this set of independent variables may include the type of driver, geographic area where the car is primarily operated, primary purpose that the car is being used for, and so on. By including these explanatory variables, more reliable estimates of the renewal function may be obtained.

30.4   ACTUARIAL MODELS OF GROUPS OF CONTRACTS

Models of individual contracts can be tailored to special features of each contract. For example, in life insurance, special features include age and sex of the insured, health status, size of the policy, and so on. In warranty analysis, special features include the type of product underlying the warranty and concomitant survival characteristics, the warranty duration, renewal options, and so on. There is little conceptual difficulty in accommodating the many special features of each contract.

In contrast, under models of groups, it is assumed that several contracts are similar and can be categorized into homogeneous blocks. In the actuarial literature, these models are called *collective* models of risk. The underlying potential obligation for a block of contracts can be represented as

$$S = \sum_{i=1}^{N} C_i$$

Here, S is the sum of claims and N is the number of claims in a given financial period. The random variables C_i represent the random claim amounts. For the claim amounts, there is a key distinction between the individual and collective models. For the collective models, C_i represents the amount of the ith claim, conditional on its occurrence. In contract, the random variables L in Section 30.3 represented the unconditional loss. Hence, in applications, many of the random losses would be zero although the claims C are typically modeled as positive random variables.

To consider claims over several periods, we next index S and N by time, resulting in

$$S(t) = \sum_{i=1}^{N(t)} C_i$$

The basic stochastic model used in the literature is to assume that $N(t)$ is a Poisson process. Thus, if the claims number process $\{N(t)\}$ is independent of the claims amount process $\{C_i\}$, then $S(t)$ is a compound Poisson process. It is interesting to note that the basic value of the group of claims can be represented as $E[S(t)] = M(t)E[C]$, that is, the expected number of claims times the expected claim amount. Again, the renewal function M plays an important role in the analysis of claims. See Ref. 11 for an introduction to this field of study. Panjer and Willmot [18] provide an alternative introduction, as well as a discussion of several quantitative implementation issues. Beard et al. [8] also provide a classical overview, as well as some of the practical implementation issues.

30.5 SUMMARY AND CONCLUDING REMARKS

Warranty analysis can roughly be described as the study of financial implications associated with the reliability, or lack thereof, of a manufacturer's product. Because a warranty is a contractual obligation of the manufacturer, accounting procedures and responsible management require that current costs and future obligations be estimated. In this chapter, analogies to insurance, another type of contract where the outcome is stochastic, are made. There is a rich literature on quantitative models of insurance. In particular, a certain subset of these models, called actuarial models, provide well-developed responses to the study of financial obligations of insurance contracts. The purpose of this chapter is to provide an overview of these models with an emphasis on applications to warranty contracts.

REFERENCES

1. Boyle, P. P. (1992). *Options and the Management of Financial Risk*, Society of Actuaries, Schaumburg, IL.

2. FASB (1979). *Accounting for Contingencies*, Statement of Financial Accounting Standards No. 5, Financial Accounting Standards Board, Stamford, CT.

3. Cheng, J. S., and Bruce, S. J. (1993). A pricing model for new vehicle extended warranties, *Casualty Actuarial Society Forum*, Special Edition, 1–24.

4. Noonan, S. J. (1993). The use of simulation techniques in addressing auto warranty pricing and reserving issues, *Casualty Actuarial Society Forum*, Special Edition, 25–52.

5. Taylor, J. M. (1986). Extended warranty insurance, *Journal of the Institute of Actuaries Student Society*, **29**, 1–24.

6. Taylor, G. C. (1986). *Claim Reserving in Non-Life Insurance*, Elsevier Science, Amsterdam.

7. LeMaire, J. (1982). Claims provisions in liability insurance, *Journal of Forecasting*, **1**, 303–318.

8. Beard, R. E., Pentikäinen, T., and Pesonen, E. (1984). *Risk Theory: The Stochastic Basis of Insurance*, Chapman & Hall, New York.

9. Bowers, N. L., Gerber, H. U., Hickman, J. C., Jones, D. A., and Nesbitt, C. J. (1986). *Actuarial Mathematics*, Society of Actuaries, Schaumburg, IL.

10. Frees, E. W. (1990). Stochastic life contingencies with solvency considerations, *Transactions of the Society of Actuaries*, **42**, 91–148.

11. Gerber, H. U. (1979). *An Introduction to Mathematical Risk Theory*, S. S. Huebner Foundation, University of Pennsylvania, Philadelphia.

12. Frees, E. W. (1986). Warranty analysis and renewal function estimation, *Naval Research Logistics Quarterly*, **33**, 361–372.

13. Frees, E. W. (1986b). Nonparametric renewal function estimation, *Annals of Statistics*, **14**, 1366–1378.

14. Frees, E. W. (1988). On estimating the cost of a warranty, *Journal of Business and Economic Statistics*, **6**, 79–86.

15. Frees, E. W., and Nam, S. H. (1988). Approximating expected warranty costs, *Management Science*, **34**, 1441–1449.

16. Frees, E. W. (1989). Infinite order U-statistics, *Scandinavian Journal of Statistics*, **16**, 29–45.

17. Frees, E. W. (1995). Semiparametric Estimation of Warranty Costs, *Journal of Nonparametric Statistics*.

18. Panjer, H. H., and Willmot, G. (1992). *Insurance Risk Models*, Society of Actuaries, Schaumburg, IL.

31

Forecasting Warranty Claims

Jingxian Chen, Nicholas J. Lynn, and Nozer D. Singpurwalla

The George Washington University
Washington, D.C.

31.1 INTRODUCTION

31.1.1 Overview

Suppose a company offers a warranty on its products. For a certain year of manufacture, it holds data on warranty claims, indexed by months in service. Suppose the company wants an idea of warranty claims over the whole warranty period, say 5 years. Then, based on warranty claims data collected over a fraction of these 5 years, it wishes to forecast future claims.

Current methods use standard regression techniques, implementing log–log plots. Such methods suffer from numerous drawbacks; these are outlined in Section 31.1.3. Techniques have also recently been developed that approach the problem from a Bayesian perspective, through the use of dynamic linear models, also known as Kalman filtering. Although an improvement over the primitive log–log plots, for forecasting warranty claims these methods still ignore important information because they do not consider data for products from other years of manufacture, which typically incorporate a similar trend. What is required is a method for incorporating data from previous model years. The solution we describe here is Kalman filtering with leading indicators, where the leading indicators represent previous model year data; this approach was developed Chen and Singpurwalla [1]. This chapter extends the work of that report, introducing an alternative approach that is shown to result in much improved forecasts.

31.1.2 Claims Data

For a particular commodity, warranty claims data may be recorded in a variety of forms. For the purposes of this chapter, we shall restrict our attention to two of the more common types of claims data. The first form of data pertains to the cost (in dollars) of repairing a unit; the second form pertains to the number of repairs per unit. Both these measures are reported as a function of the amount of usage, usually months in service.

The field data described in this chapter comes from both of these sources. The first set of data, which will be referred to as Repairs/1000, concerns repairs per thousand units of a specific commodity of the unit, say the motherboard of a personal computer. The second set of data concerns the repair cost per unit of this same commodity; this data will be referred to as Cost/Unit. The data covers three model years of manufacture, which we will refer to as Year 1, Year 2, and Year 3, representing three consecutive years, and is expressed in terms of the number of months in service (MIS). The data are cumulative, meaning that Cost/Unit refers to the cumulative cost of repairing a unit that has perhaps failed many times, rather than the cost of a single repair.

In Figures 31.1 and 31.2, we display plots of the data, as a function of the number of months in service. These plots display differing characteristics. The former, Repairs/1000, is concave, whereas the latter, Cost/Unit, is convex, reflecting a diminishing repair frequency and an increasing repair cost. Also, the plot of Repairs/1000 displays an obvious change of slope at 12 on the MIS axis (12-MIS) corresponding to 1 year in service. This can be attributed to the termination of a first year warranty that typically has a higher level of coverage than the warranty over subsequent years.

The differing features of these plots suggest that a model that works for one set of data is unlikely to work for the other. Also, our model for Repairs/1000 will have to take into account the nature of warranties and make adjustments for the fact that coverage in the first year of use is usually higher than that of subsequent years.

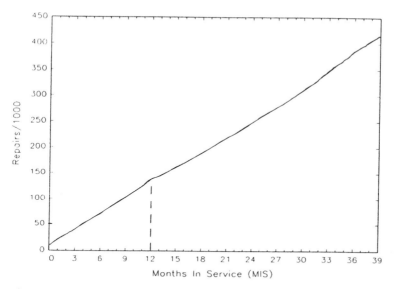

Figure 31.1 Repairs/1000 data for a given commodity.

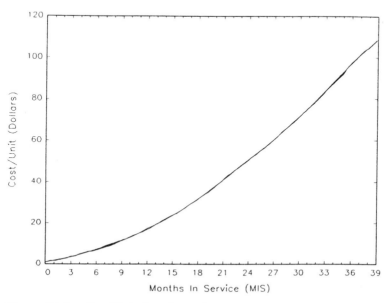

Figure 31.2 Cost/Unit data for a given commodity.

We make one final comment regarding the nature of claims data. Because units are typically produced throughout the year, there is variation in the age of units from one model year of manufacture. This means there will be more units that have survived 1-MIS than, say, 10-MIS. Hence, data associated with low values of MIS are more credible than that for high values, as there is a larger sample. For this reason, one may wish to weight data differently, depending on the value of MIS.

31.1.3 Current Methods

In the forecasting of warranty claims, there are two distinct approaches that are in common use. One approach is based on sales over time and the failure distribution for the product. The method assumes that units share a common, known failure distribution. By collecting data on item failures, one may estimate the parameters of the failure distribution, and hence produce forecasts of future failures, and hence warranty claims. This is the approach used in Chapter 24.

The alternative approach is the time-series approach, in which warranty claims are plotted as a function of time in use. This technique offers numerous advantages: There is no requirement to know the form of a possibly complex failure distribution, and effects like the slope change at $t = 12$ can easily be incorporated. Hence, this chapter concerns itself with the time-series approach to forecasting claims.

The usual time-series approach is motivated by plotting warranty claims data against months in service on a log–log scale; a plot of Repairs/1000 data for the three model years is displayed in Figure 31.3. The near linearity of the Year 1 data suggests a model of the form $Y_t = At^B$, where t represents the MIS, A and B are unknown constants, and Y_t represents the Repairs/1000 after t-MIS. A similar relationship holds for the Year 2 and Year 3 data. Whenever the relationship $Y_t = At^B$ holds, for unknown constants A and B, where A and B are static, the log–log plots provide an easy to use procedure for prediction, calling for few judgments from the user. Unfortunately, the simplicity of log–log plots is their downfall. They confine the user to a regimented structure and a slow rate of adaptivity to changes in A and B. In essence, the log–log plot is designed to function for static models, whereas what is more realistic for warranties is a dynamic model, which allows A and B to change with time. Other disadvantages of the log–log plot stem from the fact that they do not possess a formal mechanism for incorporating the effects of anticipated changes in the forecasts; that is, the plots are entirely data driven, responding solely to observations rather than to engineering or judgmental inputs. When the plots display unusual characteristics, such as the change

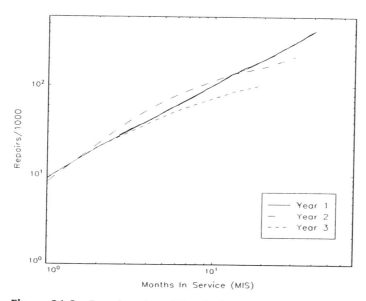

Figure 31.3 Log–log plot of Repairs/1000 data for a given commodity.

in slope at 12-MIS, subjective judgments are required, and forecasts are then inevitably colored by the nature of these judgments. Log–log plots suffer from further problems: They neither provide prediction intervals nor coverage probabilities for the forecasts, and although it may be appropriate to give more weight to data indexed by low values of MIS, log–log plots are restricted to an equal weighting of all observations.

Methods implementing dynamic linear models solve many of these problems. These models do not require the constants A and B to be static but let them dynamically change over time as more data are collected. Hence, changes in slope, such as the one observed at 12-MIS, can easily be incorporated. A further advantage of these models is that they provide prediction intervals and coverage probabilities.

31.1.4 Aims and Needs

With knowledge as to the type of data we are considering, the aim is to forecast warranty claims for specified time horizons, particularly the 60-MIS horizon, corresponding to 5 years, and often the limit of warranty coverage. These forecasts should provide us with prediction intervals for specified probabilities of coverage, typically 90% or 95%. In other words, we require a predictive distribution. We also require a mechanism allowing us to incorporate anticipated changes in the forecast due to modified design, different warranty coverage, and other factors.

As noted above, methods based on the commonly used log–log plot will not achieve these aims. We require an alternative approach and, to this aim, we turn to dynamic linear models (DLMs). In Section 31.2, we give an overview of DLMs, and in Section 31.3 we propose the *DLM with leading indicators* that will be used to incorporate data from previous model years. Section 31.4 applies these new methods to the two data sets described and compares the results with those given by current techniques. We offer some concluding comments in Section 31.5, with ideas as to future research and extensions to this work.

31.2   DYNAMIC LINEAR MODELS

31.2.1   Overview

A simple statistical model for Y_t, the warranty claim at t months in service, assumes that observations can be described through the following equation:

$$Y_t = \mu_t + u_t, \quad u_t \sim N[0, U_t] \tag{31.1}$$

where μ_t represents the state, or level, of the claims process at t-MIS, and u_t is the observational error at time t. Hence, any observation Y_t is characterized in terms of the actual state of the process, μ_t, and the error in observing this state, u_t. The error is assumed to be symmetrically distributed around zero and it is, hence, common practice to assume that u_t has a normal distribution with zero mean and variance U_t. For obvious reasons, Eq. (31.1) is known as the observation equation, because it relates observations to the actual state of the process.

A further equation, called the system equation, describes the actual level of the claims process, in terms of previous levels. For example, one simple model is the random walk, wherein

$$\mu_t = \mu_{t-1} + v_t, \quad v_t \sim N[0, V_t] \tag{31.2}$$

v_t describes the random changes in level between time $(t - 1)$ and time t. If V_t is small in relation to U_t, the evolution is smooth.

The error sequences, $\{u_t\}$ and $\{v_t\}$, are assumed to be independent. In practice, a simplifying assumption is made; namely, that $U_t = U$ and $V_t = V = bU$ for all values of t, where U is an unknown constant. Then, for mathematical convenience, we describe our uncertainty about U through an inverse Gamma distribution with scale parameter α and shape parameter β. Under these assumptions, Eqs. (31.1) and (31.2) constitute what is known as a First-Order Constant Dynamic Model. The ratio V/U is called the signal-to-noise ratio and measures the relative variation of the state to the observation. High values indicate that observations are dominated by the system state, whereas low values indicate that observations are dominated by observational error.

To make use of this model, we require a starting value for the level of the process, μ_0. If, at time t, we let D_t represent all the available information, including past observations, then we incorporate our prior beliefs about μ_0 by defining

$$(\mu_0|D_0) \sim N[m_0, s_0] \tag{31.3}$$

m_0 and s_0 are specified by the user based on his knowledge of the process being modeled.

Thus, given data ($Y_1, \ldots, Y_t$), our aim is to make forecasts of Y_{t+k} ($k = 1, 2, \ldots$), the future observables, and prediction intervals for these forecasts. This can be achieved through a sequence of updating equations, that are best derived via a Bayesian argument. Under the assumptions outlined, we arrive at a predictive distribution of the k-step ahead forecasts, Y_{t+k}, ($k = 1, 2, \ldots$), and updated distributions for the system level, μ_t. The distribution theory which is operative centers around the normal, Gamma, and Student's t-distributions. (For a review of Kalman filtering techniques, refer to Ref. 2.)

31.2.2 System Equations for Data with Trends

Careful consideration should be given to the choice of model. Clearly, the simple random walk as described in Eq. (31.2) is not suitable for describing data with underlying trends, such as the Repairs/1000 and Cost/Unit data that we presented earlier. For data with underlying trends, we require a more complex model, and, to this aim, we introduce the following modified system equations:

$$\begin{aligned}
\mu_t &= \mu_{t-1} + \theta_{t-1} + v_t, \quad v_t \sim N[0, bU] \\
\theta_t &= K_t\theta_{t-1} + w_t, \qquad w_t \sim N[0, cU] \\
(\mu_0|D_0) &\sim N[m_0, s_0]. \\
(\theta_0|D_0) &\sim N[b_0, t_0]
\end{aligned} \tag{31.4}$$

The series $\{\theta_t\}$ describes the changes in system level over time; we represent our initial beliefs by specifying b_0 and t_0. If b_0 is positive, we believe $\{\mu_t\}$ to be increasing; if it is negative, we expect the reverse. Observations will modify our beliefs. The constants, $\{K_t\}$, allow one to incorporate changes in slope at given times. If the default value of these constants is 1, then values less than 1 will translate to downward bends, whereas values greater than 1 will translate to upward bends. Here, we have specified the variance of the random system changes in terms of the variance of the observational error. b and c are constants that are specified by the user based on his experience of the data and his views as to the signal-to-noise ratio.

In choosing a model, we take into account both our prior beliefs and the general shape of the data plot. If no previous data were available to guide model selection, we may base our model on data for an analogous product. For the Cost/Unit data introduced earlier, the data seems to follow an upward trend; hence, we would adopt a *linear growth model*, achieved by setting $K_t = 1$ for all t. For the Repairs/1000 data, we note the change of slope after 1 year and, hence, adopt a *growth model with a bend*. This is achieved by setting $K_{12} = 0.9$ and $K_t = 1$ for all other t. If our data followed an S-shaped pattern, we could describe this with two bends; other contortions can be incorporated in a similar manner.

31.3 DYNAMIC LINEAR MODELS WITH LEADING INDICATORS

31.3.1 Overview

When analyzing warranty claims data from multiple model years, such as that presented in Figure 31.3, or when generally looking at time series with leading indicators, the system equations of the previous section need to be extended. To describe such extended models, we introduce a more elaborate notation.

Suppose that we have data from K model years, where Model Year 1 is the oldest. For model year ℓ, denote the warranty claims (Repairs/ 1000 or Cost/Unit) by $Y_{\ell t}$ ($\ell = 1, 2, \ldots, K\ t = 1, 2, \ldots$). Corresponding to $Y_{\ell t}$, let $\mu_{\ell t}$ and $u_{\ell t}$ denote the level and the observation error of the claims process. Then

$$Y_{\ell t} = \mu_{\ell t} + u_{\ell t}, \quad u_{\ell t} \sim N[0, U_\ell] \tag{31.5}$$

is the observation equation, and, as before, we assume that U_ℓ has an inverted Gamma distribution, with scale and shape parameters given by α_ℓ and β_ℓ.

31.3.2 Model Year 1

Consider first the system equation for Model Year 1. We recall the general model of Section 31.2, which was given by

$$\mu_{1t} = \mu_{1(t-1)} + \theta_{1(t-1)} + v_{1t}, \quad v_{1t} \sim N[0, bU_1]$$
$$\theta_{1t} = K_t\theta_{1(t-1)} + w_{1t}, \qquad\qquad w_{1t} \sim N[0, cU_1] \tag{31.6}$$

If Y_{1t} pertains to Cost/Unit data, we adopt the linear growth model where $K_t = 1$ for all t. If Y_{1t} pertains to Repairs/1000, we use the growth model with a bend at 12-MIS, specifying K_{12} as 0.9.

The parameters of the inverted Gamma distribution, α_1 and β_1, are specified by the user to reflect his prior beliefs about the variances. The default values of m_{10} and b_{10}, the means of the normal distributions of μ_{10} and θ_{10}, respectively, are based on Y_{10}, an initial guess, and Y_{11}, the observed claim at 1-MIS. Specifically, we set $m_{10} = Y_{10}$ and $b_{10} = (Y_{11} - Y_{10})$.

Now suppose that Model Year 1 consists of T_1 observations $Y_{11}, \ldots, Y_{1T_1}$, denoted $\mathbf{Y}_1$. Then under the foregoing modeling assumptions and inputs, these observations enable us (via the updating equations) to obtain filtered and smoothed estimates of $\mu_{10}, \ldots, \mu_{1T_1}$ and $\theta_{10}, \ldots, \theta_{1T_1}$, plus forecasts of the future claims, $Y_{1(T_1+j)}$ ($j = 1, 2, \ldots$). They also provide us with updated values of α_1 and β_1, the shape and scale parameters of the inverted Gamma distribution of U_1. We note that for Model Year 1, the absence of a leading indicator sequence dictates that our updating procedure is exactly the same as for Section 31.2.

31.3.3 Other Model Years

For Model Year 2, we have as the observation equation

$$Y_{2t} = \mu_{2t} + u_{2t}, \quad u_{2t} \sim N[0, U_2] \tag{31.7}$$

U_2 is assumed to have an inverted Gamma distribution, with parameters α_2 and β_2, where these are the new values of α_1 and β_1, after updating by the data $\mathbf{Y}_1$. This choice of α_2 and β_2 is one mechanism by which information from Year 1 is brought into the analysis of data for Year 2.

For the system equation of Model Year 2, we propose the following model:

$$\mu_{2t} = \mu_{2(t-1)} + \theta_{2(t-1)} + v_{2t}, \qquad v_{2t} \sim N[0, bU_2]$$
$$\theta_{2t} = \gamma K_t\theta_{2(t-1)} + (1 - \gamma)\theta_{1t} + w_{2t} \quad w_{2t} \sim N[0, cU_2] \tag{31.8}$$

This model is a special case of the two-dimensional Kalman filter and is described as a *DLM with leading indicators*. In this case, we assume that μ_{2t}, the Year 2 system level, can be described solely in terms of $\mu_{2(t-1)}$, the previous level, $\theta_{2(t-1)}$, the previous slope, and $\theta_{1(t-1)}$, the equivalent slope for the data from Model Year 1. We note that the equivalent system level from Year 1, μ_{1t}, has no effect on the data for Model Year 2, due to the nature of the data and is, hence, excluded from the model. We further note that in Eq. (31.8), θ_{1t} can be replaced in by its posterior distribution given Y_1.

The quantity $(1 - \gamma)$ is a weighting constant that determines how much weight is given to the trend from Model Year 1. Clearly, low values of γ indicate that the previous year's trend is significant, whereas high values indicate that the trend is governed by the trend at $t - 1$. Note that when $\gamma = 1$, Eq. (31.8) reduces to Eq. (31.6), the one-dimensional DLM. The use of the weight γ is another mechanism by which we bring information from the previous year into our analysis of the current year's data.

Regarding a specification of m_{20}, b_{20}, s_{20}, and t_{20}, the parameters needed for describing the starting distributions for μ_{20} and θ_{20}, our strategy is to set s_{20} and t_{20} equal to the variances of the posterior distributions of μ_{10} and θ_{10} given $Y_{11}, \ldots, Y_{1T_1}$, respectively. Regarding the means of μ_{20} and θ_{20}, we proceed as for Year 1, setting $m_{20} = Y_{20}$, an initial guess as to the warranty claim for Model Year 2, and $b_{20} = (Y_{21} - Y_{20})$.

Thus, to summarize, our approach to dealing with data involving two model years is to analyze the data of Model Year 1 using the methodology of Section 31.2, and then use the results of this analysis for analyzing the data of Model Year 2 via the system equations given in this section. This procedure is understood as being able to _correlate_ with previous years claims history.

If the data involves three model years, then the above exercise is repeated by treating the Y_{2t} series as the Y_{1t} series, and the Y_{3t} series as the Y_{2t} series. In this way, we can analyze any number of model years by working with any pair of model years, one at a time. Note, however, that if the data for Model Year 2 consist of observations Y_2, then α_3 and β_3 will be derived from α_2 and β_2 through updating with this new data.

The updating equations become quite complex and are given in a report by Chen and Singpurwalla [1]. These equations form the basis of our computational approach that has been implemented for use on a personal computer (PC).

31.4 RESULTS OF DATA ANALYSIS

To evaluate the effectiveness of the dynamic linear model with leading indicators, we applied the methods of the previous section to the previously described data, namely, Repairs/1000 data and Cost/Unit data. To see whether these techniques provide any improvement over current methods, the forecasts obtained using these new methods were compared to those resulting from one-dimensional dynamic linear model analysis. As a measure of performance, we considered the cumulative square error.

For model year ℓ, if we have observations $y_{\ell 1}, \ldots, y_{\ell t}$ and n forecasts based on these observations, then the cumulative square error (CSE) is defined to be

$$\text{CSE}(n, t) = \sum_{k=1}^{n} (y_{\ell(t+k)} - E[Y_{\ell(t+k)}|D_{\ell t}])^2 \tag{31.9}$$

where $(Y_{\ell(t+k)}|D_{\ell t})$ is the k-step ahead predictive distribution based on $D_{\ell t}$, our knowledge, including past observations, at time t. The smaller the CSE, the better the forecasts are because CSE reflects the total loss of the n forecasts when the loss function is quadratic.

The results of applying the two methods to Cost/Unit data from Years 2 and 3 are displayed in Table 31.1. Here the techniques have been used to make 10 forecasts, 5 based on 5 months of data, and 5 based on 10 months of data; the previous year's data are used as a leading indicator series. The values of the constants b and c were set to 0.01 and 0.2, respectively, and our uncertainty about the variance, U_1, was reflected through an inverted Gamma distribution with α_1 and β_1 set to 1 and 2.5, respectively. The weighting, γ, was set at 0.3, reflecting our belief that the leading indicator series was a significant indicator of trend.

For the forecasts based on 5 months of data, the techniques developed in this chapter clearly result in much improved forecasts, reflected

Table 31.1 Forecast Values (Cost/Unit Data Years 2 and 3)

		Prediction $\bar{y}_t$			
	Observed	Based on first 5		Based on first 10	
t	y_t	2-d	1-d	2-d	1-d
6	0.693	0.6875	0.7294		
7	0.752	0.7597	0.8191		
8	0.814	0.8286	0.9088		
9	0.845	0.8949	0.9984		
10	0.915	0.9589	1.0881		
11	0.974			0.9874	0.9677
12	1.098			1.0460	1.0210
13	1.162			1.1007	1.0742
14	1.196			1.1524	1.2724
15	1.196			1.2025	1.1807
Cumulative square error { Total		0.0047	0.0680	0.0086	0.0186
Average		0.0009	0.0137	0.0017	0.0037

in a much reduced cumulative square error. The forecasts based on 10 months of data also show a significant improvement, although not as marked, due to the fact that there is more data available from which to forecast, thus reducing the significance of the leading indicator series.

A similar analysis was performed with Repairs/1000 data, on Year 2 and Year 3 data. Referring back to Figure 31.3, one can see that these plots have markedly different slopes. The results are displayed in Table 31.2. Here, 5 forecasts have been made, based on 5, 10, and 15 observations.

For these data, these new methods give worse results than the one-dimensional DLM. This is probably due to the choice of γ, which was again set at 0.3, although the dissimilarity between the various years plots would suggest a higher value.

We conclude that these methods, when used correctly, will result in much improved forecasts. However, prior to using them, careful consid-

Table 31.2   Forecast Values (Repairs/1000 data Years 2 and 3)

		Prediction $\bar{y}_t$					
		Based on first 5		Based on first 10		Based on first 15	
t	Observed y_t	2-d	1-d	2-d	1-d	2-d	1-d
6	68.09	61.61	61.86				
7	63.50	69.57	68.90				
8	68.35	77.54	76.63				
9	72.95	85.51	84.35				
10	77.34	93.46	92.07				
11	81.7			90.05	82.49		
12	86.5			97.18	87.15		
13	89.01			102.74	91.34		
14	91.87			107.45	95.53		
15	94.27			111.54	99.72		
16	96.71					100.87	97.68
17	99.45					104.81	100.56
18	101.8					108.77	103.46
19	104.2					112.81	106.33
Cumulative square error { Total		551.8	460.65	913.3	49.57	168.7	21.9
Average		110.36	92.13	182.66	9.915	42.18	5.48

eration should be given to specifying the weighting of the leading indicator series, as incorrect specification may lead to worse performance than current techniques. For similar data sets, these new methods perform well; dissimilar data sets should be handled with care.

31.5 CONCLUDING COMMENTS

The key contributions in this chapter on developing methods for forecasting warranty claims has been the use of dynamic linear models with various forms of trends and the further development of this notion to allow correlation with data from previous model years. As an attempt toward facilitating an easy implementation of this approach, interactive, user-friendly computer software for use on PCs has been developed. Incorporated into the software are default values for several input parameters that are needed for executing the approach. Whereas some knowledge of the underlying models is desirable for getting the most out of the procedure, a potential user who does not feel inclined toward investing time in grasping the underlying details need not be discouraged. The software, its built in default values, and the accompanying on-line notes and comments will guide one through the necessary steps after an initial (but limited) exposure to it.

Despite the encouraging results and success that we have had so far, improvements on the proposed themes and the exploration of alternate approaches promise to bear further fruit. These are listed below:

- A procedure to make adjustments for the "maturing data" issue, that is prevalent with warranty claims data, needs to be formally developed. The maturing data issue is brought about by a changing population of models under warranty; see, for example, Ref. 3 or Ref. 4.
- A procedure to give differing weights to the observed claims because of averaging over different months of production needs to be developed.

As weighting of data can be achieved by changing the variance of the error terms, we believe that these two procedures can be achieved through some judicious choices of U_t.

- A procedure for the automatic estimation of the values of K_t, the multipliers of the trend coefficient, needs to be developed. Currently, default values of the K_t are specified; see, for example, Ref. 3 or Ref. 4.

- A procedure for the automatic estimation of the weighting constant γ when considering previous years data needs to be developed. Also needed is a procedure which allows γ to change with t. Currently, γ is specified by a default value.

Implementing these two procedures requires complex mathematical integration. However, we believe that Gibbs sampling strategies may facilitate the implementation. These techniques involve repeated sampling from the full conditional distributions; fortunately, for the model of Section 31.3, these conditionals are readily obtained. (For a review of Gibbs sampling, refer to Ref. 5.)

- An alternate approach for incorporating the effects of previous years data. The current approach has a structure similar to that of Box–Jenkins-type ARIMA models. Modern methods involving Markov random fields (used in image processing) should prove to be superior.
- Forecasting based on count data via nonhomogeneous Poisson processes with covariates incorporating other variables such as design changes, manufacturing changes, month of manufacture, user profiles, and so on. The current approach does not include the effect of covariates.
- A two-dimensional point process procedure that, in addition to the MIS, accounts for the amount of use as an index needs to be developed. We need to forecast warranty claims in a two-dimensional plane involving both time and usage.

We believe that the theory of point processes and planar processes offers some possibilities here.

ACKNOWLEDGMENTS

This work was supported by the U.S. National Science Foundation, Grant SES-9122494.

REFERENCES

1. Chen, J., and Singpurwalla, N. D. (1994). *Dynamic Linear Models with Leading Indicators*, Technical Report GWU/IRRA/Serial TR-94/1, The Institute for Reliability and Risk Analysis, The George Washington University, Washington, DC.
2. Meinhold, R., and Singpurwalla, N. D. (1983). Understanding the Kalman filter, *The American Statistician* **37**, 123–127.

3. Robinson, J. A., and McDonald, G. C. (1991). Issues related to field reliability and warranty data, *Data Quality Control: Theory and Pragmatics*, Marcel Dekker, Inc., New York.

4. Singpurwalla, N. D., and Wilson, S. P. (1992). Warranties, *Proceedings of the Fourth Valencia Conference* (J. Berger, J. Bernado, P. Dawid, D. Lindley, and A. F. M. Smith, eds.), Clarendon Press, Oxford.

5. Smith, A. F. M., and Roberts, G. O. (1993). Bayesian computation via the Gibbs sampler and related Markov chain Monte Carlo methods, *Journal of the Royal Statistical Society*, Series B, **55**, 3–23.

32

Multicriteria Models for Determining Warranty Parameters

Amitava Mitra

Auburn University
Auburn, Alabama

Jayprakash G. Patankar

The University of Akron
Akron, Ohio

32.1 INTRODUCTION

Manufacturers are usually faced with the burden of determining parameters associated with a warranty policy. Typically such parameters involve the selling price and warranty period of the product. This chapter considers some possible approaches for determining the values of these parameters under certain criteria or objectives that are chosen by the manufacturer. The chosen criteria may vary from manufacturer to manufacturer. Whereas minimization of the total cost of warranty reserves might be a desirable criteria for one, maximization of market share might be that for another. The goals and objectives set by the top management of a company will dictate the operational objectives in determining the parameters of a warranty policy. Feedback from the customer on product satisfaction will provide the manufacturer with some direction in selecting suitable criteria.

32.1.1 Single Versus Multiple Criteria

A majority of the formulated models in the literature have focused on a single criteria or objective. This objective is either maximized or minimized, for example, warranty reserve cost minimization or market share maximization, and suitable warranty parameters are determined. One of the earliest articles by Menke [1] estimated warranty costs for a nonrepairable product using a linear pro-rata and lump-sum rebate plans. As an extension of Menke's article, Amato and Anderson [2], and Amato et al. [3] incorporated the time value of money to estimate warranty costs in situations in which the warranty period is long.

Several other single-objective models that deal with renewable warranty policies abound in the literature. Blischke and Scheuer [4] used renewal theory to estimate warranty costs for renewable policies. They considered a free-replacement policy and a pro-rata rebate policy under a variety of product failure distributions. For the type of warranty policies studied by Blischke and Scheuer, Mamer [5] estimated short-run total costs and long-run average costs of products under warranty. Mamer [6] subsequently extended his previous research using present value analysis using the time value of money and analyzed the trade-off with warranty and quality control. Articles by Blacer and Sahin [7], Frees and Nam [8], and Frees [9] dealt with the estimation of expected warranty costs.

From a practical point of view, modern-day manufacturers operate in a complex decision-making environment. A variety of constraints influence the manager in making a decision on the parameters of a warranty policy. It is not just simply the minimization of a single objective, such as warranty reserve costs, that is solely responsible for influencing a warranty policy. There are other factors such as the adhering to environmental constraints, maximizing market share, or minimizing overtime. Some of these objectives may be conflicting. A decision-maker must arrive at a "satisficing" solution that may meet several of the stipulated objectives. Whereas solutions for single-objective problems could be labeled as optimal, based on meeting the specified criteria specified in the objective function, those for multiobjective problems might not simultaneously optimize all of the selected criteria. Thus, it may not be possible to minimize total warranty reserve costs and maximize market share at the same time.

Although single-objective problems are more clearly defined and are conducive to established optimization procedures, they are nevertheless not as realistic as multicriteria models. Decisions made by managers impact various constituencies, each with their separate objectives. Thus, the framework in which decisions are made is more adequately represented by multicriteria models. It is, therefore, the purpose of this chapter to introduce some criteria that may impact the selection of warranty parameters. The terms *criteria* and *objectives* are used interchangeably in this chapter. Although the exact selected criteria will vary from company to company, this chapter presents an example of some of these criteria.

32.1.2 Solution Procedures

In problems with multiple objectives, it is possible for some objectives to be conflicting with others. Consequently, it might not be feasible to obtain a solution that will maximize or minimize all objectives as defined in the problem. One approach to solving multiobjective problems is through the use of goal programming. Here, it is assumed that the user has an idea of the desirable levels of achievement of each of the objectives. The objectives are now stated as goals based on these desirable levels. For example, a goal may be that the expected warranty reserve costs should not exceed a desirable level (say $\alpha\%$) of the total sales. Another goal may be that the market share of a product should be at least a chosen level ($\beta\%$). The solution of the problem is then based on achieving the specified desirable levels of each goal. Note that, given the above-mentioned goals, as the market share of a product increases, the expected warranty reserve costs will also increase because the more products there are in the market, the more the opportunities for failure within the warranty period, necessitating more warranty reserves. Thus, a feasible solution might not be found that would satisfy any combination of the desirable goal levels.

A process that facilitates the solution procedure is the preemptive or lexicographic goal programming formulation. Here, it is assumed that the decision-maker is able to prioritize the selected goals and thereby distinguish which goals are more important than others. The priorities assigned to the goals, however, may vary from one decision-maker to another. The approach seems rational because if users are presented with multiple goals, all of which may not necessarily be achieved due to limited resources, they are likely to specify a priority ranking of the various goals. In this chapter we will consider the priorities as given by the goals, that is, Goal No. 1 must be satisfied before Goal No. 2, and so on.

In preemptive goal programming, the goals are grouped according to priorities. The goals at the highest priority level are considered to be much more important than the goals at the second priority level, and so on. Thus, the solution procedure solves a series of problems in a lexicographic format. The objective function is stated in terms of a deviational variable, which represents the deviation from the stipulated goal value. If the goal is satisfied, the value of this deviational variable in the objective function will be zero. As all variables are assumed to be non-negative, a positive value of a deviational variable in the objective function indicates the magnitude by which the goal was not met.

In the solution procedure using the preemptive approach, the goal with the highest priority is considered first in the objective function, subject to the other system constraints in the problem. If this goal is achievable, the solution set is then constrained so that the achieved goal level is maintained, and a second problem is solved while attempting to optimize the second goal. This procedure continues while a series of problems is solved. At each stage, optimality information from the previous stage is used, and a sacrifice is not made in the attainment of a goal level at the expense of a goal at a higher priority level. This is also sometimes stated as one of the drawbacks of the preemptive approach, as higher-priority goals are not compromised with lower-priority goals.

It should be noted that other forms of goal programming formulations exist [10]. The "goodness" of a solution depends entirely on one's philosophy as to its measurement. One alternative way of measuring the goodness of a solution is to minimize the sum of weighted goal deviations. This is known as the minisum approach of Archimedean goal programming. In this case, there is a single objective function and conventional mathematical programming methods can be used to determine optimal solutions. The concept here assumes that the goals are commensurable, which might not always be true in real problems. One other drawback of this method is the determination of the weights to use in the objective function. In the problem under investigation, it is felt that the decision-maker is in a better position to indicate a preferential ranking of which goals are important to others, rather than to specify valid weights. Another alternative way of measuring the goodness of a solution is to minimize the maximum goal deviation, which is known as the fuzzy goal programming approach. The formulation there is also in the form of a single objective model. However, some disadvantages of the approach [11] are as follows. First, commensurability of goals is assumed and the underachievement of one goal can have a major impact on the solution. Second, in the formulation in fuzzy goal programming, associated with each goal is a fuzzy membership function. The form of this membership function is open to question. Third, the formulation involves aspiration levels as well as allowable degradations for each objective. A precise choice of these values may be difficult. Because of these reasons, we chose to use the lexicographic goal programming approach.

32.2 MULTIPLE CRITERIA IN MODELING

This chapter presents a model that attempts to link various subsystems in the organization when dealing with warranty issues. Much of the previous research has considered estimation of warranty reserve costs. Some have considered the marketing aspect by modeling the market share component as influenced by the price and warranty time. The integration into the model of production quantities, that meet capacity constraints on resources, is an important aspect. This model considers production quantity (on a unit time, say a year), lot size, product unit price, and warranty time as decision variables in the context of a multiproduct framework. The approach demonstrates an example that integrates the production, marketing, and financial concepts in the setting of a multiobjective model.

In the formulated multicriteria model, since the preemptive goal programming procedure is used based on selected priority levels associated with the goals, the focus will be the attainment of the desirable goal levels to the extent possible. Thus, unlike the traditional optimization formulation such as in linear or nonlinear programming where production capacity is set up as a rigid constraint, in this case we would prefer to meet the capacity constraint. If, however, the attainment of objectives at a higher priority level requires more capacity or resources than is normally available, the solution will not be treated as infeasible. On the contrary, we may attempt to determine ways in which higher-priority goals may be satisfied while remaining as close as possible within the bounds of resource constraints. An explanation for this is that in addition to the normal capacity or resource that is available, extra capacity or resource could be obtained through means such as overtime, subcontracting, improvement in efficiency, or procurement of additional resources.

One of the advantages of using the goal programming method for the multicriteria model is that the stated levels of the goals are formulated using goal constraints. These constraints are not necessarily "rigid"; for example, they do not have to satisfied exactly. They represent only desirable levels. Thus, management can determine the extent to which these stipulated goal levels can be met (achieved, overachieved, or underachieved) given the available resources. Such analysis could assist top management in setting realistic goals and determining warranty parameters that would help in achieving such goals. Trade-offs can take place between goals to the extent that if a higher priority goal is overachieved, the level of achievement can be sacrificed or reduced in order to better achieve a lower-priority goal. However, if a higher-priority goal is not achieved, that is, underachieved, then its level of achievement will not be sacrificed for a lower-priority goal.

Several applications of multicriteria models in the warranty area may be found in the literature. Mitra and Patankar [12] considered a formulation in the single-product situation. In addition to the rigid or system constraints on the product price and warranty time, three other goals are considered. These were the minimization of total warranty reserve costs per unit price, offering a warranty time that is greater than or equal to a specified value that is based on the allowable proportion of failures during that time and the achievement of a certain level of market share of the product. The decision variables were the product price and warranty time. Patankar and Mitra [13] formulated a multiobjective model in the context of a firm with multiple products, where products are assumed to be substitutes. The decision variables were the product price and warranty time. Along with the system constraints on product price and warranty time, the goals included the achievement of a specified market share, limitation of the total warranty cost as a given proportion of the total sales, limitation of the warranty reserve for a given product as a proportion of the total warranty reserve for all products, and attainment of a minimum level of warranty reserve for a given product as a proportion of the total warranty reserve costs. A similar model was considered by Mitra and Patankar [14] in the multiproduct case, in which the products are assumed to be complements of each other. Mitra and Patankar [15] also developed a model for the multiproduct situation in which the decision variables included the price, warranty time, production quantity, and lot size. System constraints on the decision variables are based on absolute minimum and maximum values between which those variables should lie. The goals considered include operation within a limited resource capacity, the achievement of a specified market share, limitation on the total warranty cost as a given proportion of total sales, limitation of the warranty reserve for a given product as a proportion of the total warranty reserve for all products, and attainment of a minimum level of warranty reserve for a given product as a proportion of the total warranty reserve cost.

The goals considered in this model are as follows. Some absolute or system constraints (rigid goals) are that the product price, warranty time, production quantity, and lot size must be within certain bounds. Another goal is that the total time requirements for setup and production, including that for replacement units, should not exceed the available capacity. The achievement of a certain market share for the products is yet another goal. A fourth goal is that the total warranty reserve costs for all the products should not exceed a given proportion (α) of total sales. The priorities associated with each of these goals, except for the first one involving a rigid goal, will be dependent on the decision-maker. In the application section, we demonstrate some results based on selected choices of the priorities associated with each of the goals.

The following notation is employed in formulating the model:

m = number of products
w_i = duration of warranty period of product i

c_i = unit product price including warranty cost of product i

N_i = production quantity of product i per unit time period (typically annual)

Q_i = lot size for product i

w_{i1} = minimum warranty time of product i

w_{i2} = maximum warranty time of product i

c_{i1} = minimum price of product i

c_{i2} = maximum price of product i

N_{i1} = minimum production quantity of product i

N_{i2} = maximum production quantity of product i

Q_{i1} = minimum lot size of product i

Q_{i2} = maximum lot size of product i

p_i = rate of production of product i

s_i = setup time for product i

A_j = available capacity (typically on an annual basis) of resource j

h = number of shared resources

T_j = set of indices of the products which use resource j

$G(w_i, c_i)$ = market share function for product i

32.2.1 Bounds on Warranty Parameters

We first consider the top priority goal (at level P_0) which are the system constraints associated with the decision variables on product price, warranty time, production quantity, and lot size. A manufacturer having knowledge of the unit cost of production and desirable profit margin can typically specify a minimum price (c_{i1}), below which it would not be feasible to sell the product. Similarly, knowing the market and the competition, he has a notion of the maximum price (c_{i2}) at which the product should be priced. Along the same lines, the manufacturer might be able to specify a minimum (w_{i1}) and maximum (w_{i2}) bound on the warranty time to be offered with the product. Management may have as a minimum bound on a resource, the normal capacity that is available to manufacture the products. For example, such available time includes setup and processing times. Furthermore, if the products share the use of some common production facilities, then the availability of the resource would provide an upper bound on the production quantities of the several products given their individual unit requirements of the resource. An upper bound on the production quantity per unit time (N_{i2}) may be set based on such capacity constraints. Additionally, if management has a certain desirable level of resource utilization in mind, a minimum production quantity (N_{i1}) may be stipulated so as to meet this guideline.

Along the same lines, a bound may be developed for the lot size. The lot size should not be below a certain level (Q_{i1}) which may otherwise cause too many setup changes. Production leveling, lack of availability of labor skilled in multifunctions, and procurement of raw material in certain bulk orders are other factors which may influence the level of this lower bound. Further, the lot size should not be more than an upper bound (Q_{i2}), which may otherwise increase carrying costs. Storage limitations based on available warehouse capacity may also influence the choice of this bound.

The system constraints on product price, warranty time, production quantity, and lot sizes are formulated based on the preceding discussion and are expressed as follows:

$$c_{i1} \leq c_i \leq c_{i2}, \quad i = 1, 2, \ldots, m$$
$$w_{i1} \leq w_i \leq w_{i2}, \quad i = 1, 2, \ldots, m \qquad (32.1)$$
$$N_{i1} \leq N_i \leq N_{i2}, \quad i = 1, 2, \ldots, m$$
$$Q_{i1} \leq Q_i \leq Q_{i2}, \quad i = 1, 2, \ldots, m$$

In the goal programming approach, we would add goal deviation variables to these constraints and in the objective function, at the highest priority level, we would minimize the appropriate deviational variable. The constraints may be expressed as follows:

$$c_i - c_{i1} + d_{i1}^- - d_{i1}^+ = 0, i = 1, 2, \ldots, m$$
$$c_i - c_{i2} + d_{i2}^- - d_{i2}^+ = 0, i = 1, 2, \ldots, m$$
$$w_i - w_{i1} + d_{i3}^- - d_{i3}^+ = 0, i = 1, 2, \ldots, m$$
$$w_i - w_{i2} + d_{i4}^- - d_{i4}^+ = 0, i = 1, 2, \ldots, m \qquad (32.2)$$
$$N_i - N_{i1} + d_{i5}^- - d_{i5}^+ = 0, i = 1, 2, \ldots, m$$
$$N_i - N_{i2} + d_{i6}^- - d_{i6}^+ = 0, i = 1, 2, \ldots, m$$
$$Q_i - Q_{i1} + d_{i7}^- - d_{i7}^+ = 0, i = 1, 2, \ldots, m$$
$$Q_i - Q_{i2} + d_{i8}^- - d_{i8}^+ = 0, i = 1, 2, \ldots, m$$

In the objective function, we would minimize the following terms:

$$\text{minimize } P_0 \left[\sum_{i=1}^{m} (d_{i1}^- + d_{i2}^+ + d_{i3}^- + d_{i4}^+ + d_{i5}^- + d_{i6}^+ + d_{i7}^- + d_{i8}^+) \right]$$

where P_0 represents the highest priority level.

32.2.2 Production Capacity or Resource Goal

Here, it is assumed that the goal is for the total time requirements for production should be within the available capacity. The availability of limited resources influences the capacity. The setup and processing times for products that share a common resource influence the amount of production given an available capacity of the resource. We assume that the goal of the manufacturer is to not exceed the normal capacity. Even though capacity could be increased through measures such as overtime and subcontracting, their effect would be to cause an increase in the cost. For product i, the total setup time is given by

$$\left(\frac{N_i}{Q_i}\right) s_i, \quad i = 1, 2, \ldots, m$$

The processing time for product i is

$$\frac{N_i}{p_i}, \quad i = 1, 2, \ldots, m$$

The warranty policy assumed is that of a cash refund to the customer, based on a linear pro-rata basis if failure occurs before the expiration of warranty. Furthermore, it is also assumed that there is no replacement purchase expressed in the warranty. However, for the purpose of creating an integrative model, we assume that production processes must have not only the capacity for producing original demand units but also the capacity to repair defective units and/or produce new replacement units. This is because of the desire of the manufacturer to have, on hand, sufficient items to satisfy demand. This will create an integrative model.

The product falure distribution can be general, such as those modeled by Gamma or Weibull distributions. For the purpose of analysis, it is assumed to be exponential, which is a special case of both distributions. Assuming an exponential failure density function, the probability of an item failing by time w_i, for product i, is given by

$$
\begin{aligned}
F(w_i) &= \int_0^{w_i} \lambda_i e^{-\lambda_i t}\, dt \\
&= (1 - e^{-\lambda_i w_i}), \quad i = 1, 2, \ldots, m
\end{aligned}
$$

Therefore, if N_i units are sold, the expected number of failures by time w_i is $N_i(1 - e^{-\lambda_i w_i})$, $i = 1, 2, \ldots, m$. These units, if they are to be produced, will require additional setup time and processing, which may be calculated as earlier. The total time requirements for all the products should not exceed the available capacity. The capacity constraint for resource j becomes

$$\sum_{i \in T_j} \left[\left(\frac{N_i}{Q_i}\right) s_i \{2 - e^{-\lambda_i w_i}\} + \frac{N_i}{p_i} \{2 - e^{-\lambda_i w_i}\} \right]$$

$$\leq A_j, \quad j = 1, 2, \ldots, h \quad (32.3)$$

where T_j represents the set of indices of the products that use resource j. If goal deviational variables are included, the constraint becomes

$$\sum_{i \in T_j} \left[\left(\frac{N_i}{Q_i} \right) s_i \{2 - e^{-\lambda_i w_i}\} + \frac{N_i}{p_i} \{2 - e^{-\lambda_i w_i}\} \right]$$
$$- A_j + d_{j9}^- - d_{j9}^+ = 0, j = 1, 2, ..., h \quad (32.4)$$

For the objective function, we would minimize

$$P_1 \left(\sum_{j=1}^{h} d_{j9}^+ \right)$$

where P_1 represents the priority associated at this level.

32.2.3 Market Share Goal

The goal at this priority involves obtaining at least some specified minimum market share for each product. Glickman and Berger [16] have presented a demand function which has the properties of decreasing exponentially with respect to price and increasing exponentially with warranty time. The market share function should be bounded between 0 and 1. The market share function proposed for product i is as follows:

$$G(w_i, c_i) = D_i c_i^{-a_i}(w_i + k_{2i})^{b_i} \quad \text{for } w_i \le w_{i2} \quad (32.5)$$

where

$$D_i = \frac{1}{c_{i1}^{-a_i}(w_{i2} + k_{2i})^{b_i}}, \quad i = 1, 2, ..., m$$

In Eq. (32.5), k_{2i} is a constant of warranty time displacement, allowing for the possibility of nonzero market share when warranty time is 0, a_i and b_i are constants such that $a_i > 1$, $0 < b_i < 1$, representing the price elasticity and displaced warranty period elasticity, and c_{i1} and w_{i2} represent the minimum price and maximum warranty time, respectively, for product i as defined before. The quantity D_i represents a normalizing constant to ensure that the market share is between 0 and 1. Note that when the price of the product is at its minimum and the warranty time at its maximum, the market share function takes on a value of unity. The market share function, $G(w_i, c_i)$, is bounded between 0 and 1.

Now, if our goal is to achieve at least a certain proportion (g_i) of the market share, the corresponding goal constraint becomes

$$D_i c_i^{-a_i}(w_i + k_{2i})^{b_i} \ge g_i, \quad i = 1, 2, ..., m \quad (32.6)$$

The selection of the value of g_i would be influenced by the goals and objectives of the company, that is, the aspiration levels of the proportion of the market share that they desire to achieve for that product based on their relative standing with respect to the competitors. Additionally, consideration should be given to the production capacities and the cost feasibility when selecting a value of g_i. Under the assumption that the unit price of a product is proportional to a unit fixed cost and a unit variable cost which is influenced by the production quantity, we have

$$c_i = f_i + u_i = f_i + \frac{v_i}{N_i}, \quad i = 1, 2, \ldots, m \tag{32.7}$$

where f_i and v_i are constants as explained below.

Suppose that the unit product price (c_i) includes the manufacturing cost and the cost of warranty per unit, and allowances for administrative overhead, shipping charges, and profit margin. Also, assume that the inclusion of the warranty cost is achieved by increasing the manufacturing unit cost by a certain percentage. Because unit manufacturing costs can be explained to consist of two components, a direct cost and an indirect cost, so also the unit price can be thought of as the sum of these two components. Typical components of direct costs are direct material and direct labor, whereas those for indirect costs are overheads for insurance, taxes, maintenance, and depreciation. The quantity u_i, which is equal to v_i/N_i, represents the unit direct cost component with the added increase for the direct cost factors mentioned previously. It implies that as the production quantity increases, the unit direct cost decreases, which could happen due to purchase quantity discounts and learning curve/performance improvement efficiencies. The quantity f_i represents the unit indirect cost portion with the added increase due to the other indirect cost factors mentioned above.

Assuming that the minimum unit price of product i is achieved when the production quantity (N_i) is at its maximum and, similarly, the maximum unit price of product i is realized when the production quantity is at its minimum, we have the following equations:

$$c_{i1} = f_i + \frac{v_i}{N_{i2}}, \quad i = 1, 2, \ldots, m \tag{32.8}$$

$$c_{i2} = f_i + \frac{v_i}{N_{i1}}, \quad i = 1, 2, \ldots, m \tag{32.9}$$

The solution of these equations yields the values of the constants f_i and v_i as

$$f_i = \frac{c_{i1}N_{i2} - c_{i2}N_{i1}}{N_{i2} - N_{i1}}, \quad i = 1, 2, \ldots, m \tag{32.10}$$

$$v_i = \frac{(c_{i2} - c_{i1})N_{i1}N_{i2}}{N_{i2} - N_{i1}}, \quad i = 1, 2, \ldots, m \tag{32.11}$$

Substituting the values of f_i and v_i into Eq. (32.7), we have

$$c_i = \frac{c_{i1}N_{i2} - c_{i2}N_{i1}}{N_{i2} - N_{i1}} + \frac{(c_{i2} - c_{i1})N_{i1}N_{i2}}{(N_{i2} - N_{i1})N_i}, \quad i = 1, 2, \ldots, m \tag{32.12}$$

This value of c_i could be used to reexpress the market share goal constraint given by Eq. (32.6) as follows:

$$D_i \left(\frac{c_{i1}N_{i2} - c_{i2}N_{i1}}{N_{i2} - N_{i1}} + \frac{(c_{i2} - c_{i1})N_{i1}N_{i2}}{(N_{i2} - N_{i1})N_i} \right)^{-a_i} (w_i + k_{2i})^{b_i} \geq g_i,$$

$$i = 1, 2, \ldots, m \tag{32.13}$$

Adding the goal deviation variables, the constraint becomes

$$D_i \left(\frac{c_{i1}N_{i2} - c_{i2}N_{i1}}{N_{i2} - N_{i1}} + \frac{(c_{i2} - c_{i1})N_{i1}N_{i2}}{(N_{i2} - N_{i1})N_i} \right)^{-a_i} (w_i + k_{2i})^{b_i} - g_i$$

$$+ d_{i,10}^- - d_{i,10}^+ = 0, \quad i = 1, 2, \ldots, m \tag{32.14}$$

The corresponding terms to minimize in the objective function would be

$$\text{minimize} \quad P_2 \left(\sum_{i=1}^{m} d_{i,10}^- \right) \tag{32.15}$$

32.2.4 Expected Warranty Reserve Costs Goal

The goal is for the total expected warranty reserve costs not to exceed a given proportion (α) of total sales. The following notation is now defined to formulate the goal associated at this level:

R_i = total warranty reserve cost for lot size N_i associated with product i, $i = 1, 2, \ldots, m$

λ_i = failure rate of product i

t = time of product failure

The pro-rata customer rebate for a product (i) that fails at time t, assuming a linear pro-rata rebate plan (see Chapter 11), is given by

$$C_i(t) = c_i \left(1 - \frac{t}{w_i} \right), \quad 0 < t < w_i, \, i = 1, 2, \ldots, m \tag{32.16}$$

Assuming exponential distribution of failures, the total number of failures of product i in the interval between t and $t + dt$ is given by

$$\Delta F_i = N_i \lambda_i e^{-\lambda_i t} \Delta t \tag{32.17}$$

The cost of these failures in the time interval t to $t + \Delta t$ is

$$\Delta R_i = C_i(t) N_i \lambda_i e^{-\lambda_i t} \Delta t \tag{32.18}$$

$$= c_i \left(1 - \frac{t}{w_i}\right) N_i \lambda_i e^{-\lambda_i t} \Delta t, \quad i = 1, 2, \dots, m$$

Therefore, the total cost of all failures of product i occurring within the duration of the warranty time w is obtained from

$$R_i = \int_0^{w_i} N_i c_i \lambda_i \left(1 - \frac{t}{w_i}\right) e^{-\lambda_i t}\, dt \tag{32.19}$$

$$= N_i c_i \left[1 - \left(\frac{1}{\lambda_i w_i}\right)(1 - e^{-\lambda_i w_i})\right], \quad i = 1, 2, \dots, m$$

The total sales from all the products is given by $\sum_{i=1}^{m} N_i c_i$. Sometimes, based on this amount of total sales, the manufacturer may place an upper bound on the total warranty reserve cost. Obviously, the higher the total sales, the greater the flexibility needed to increase the total amount of warranty reserve. Let α denote a desirable proportion of total sales that the manufacturer would be willing to spend on total warranty reserve costs. The goal may be expressed as

$$\sum_{i=1}^{m} R_i \le \alpha \left(\sum_{i=1}^{m} N_i c_i\right)$$

Using Eq. (32.19) and adding the goal deviational variables, we have

$$\sum_{i=1}^{m} N_i c_i \left[(1 - \alpha) - \left(\frac{1}{\lambda_i w_i}\right)(1 - e^{-\lambda_i w_i})\right] + d_{11}^- - d_{11}^+ = 0 \tag{32.20}$$

The corresponding term to minimize in the objective function at this priority level would be

$$\text{minimize} \quad P_3(d_{11}^+)$$

32.3   MODEL FORMULATION AND APPLICATION

Using the derivation of the previous section, the decision-maker now se-
lects the priorities associated with the various goals based on his/her preferences and the goals and objectives of the company.

32.3.1   Goal Priorities

The set of bounds on warranty parameters is treated as a system constraint
at the highest priority level, indicated by P_0. System constraints are rigid;
they have to be satisfied. If the warranty parameters do not lie within the
specified bounds, the solution is infeasible. The goal constraints at the
other levels are, however, treated as flexible with those at a higher priority
level taking preference over those below it. One possible combination
could be the production capacity or resource goal at the highest priority
level, P_1, the market share goal a the next priority level, P_2, followed by
the warranty reserve costs goal at the lowest priority level, P_3. If the
relative preferences of the decision-maker change, corresponding changes
would be made in priority rankings of the formulated goals.

32.3.2   Multiobjective Model

If the relative priority ranking is as specified in the previous subsection,
the model is expressed as follows. The objective function is expressed as
follows:

Lexicograpically minimize:

$$P_0\left(\sum_{i=1}^{m}(d_{i1}^- + d_{i2}^+ + d_{i3}^- + d_{i4}^+ + d_{i5}^- + d_{i6}^+ + d_{i7}^- + d_{i8}^+)\right),$$

$$P_1\left(\sum_{j=1}^{h}d_{j9}^+\right), \qquad P_2\left(\sum_{i=1}^{m}d_{i,10}^-\right), \qquad P_3(d_{11}^+) \quad (32.21)$$

The goal constraints would be the collection of the equations given by
Eqs. (32.2), (32.4), (32.14), and (32.20). Sensitivity analyses on the relative
priorities associated with the goals could be conducted to determine their
impact on the solution. Such analyses would provide valuable information
to top management as they formulate policies for the company.

32.3.3   Application

A sample problem involving two products, sharing a common resource,
is solved to illustrate the model formulated in the previous section. Table
32.1 shows the values of the problem parameters chosen. The units of the

Table 32.1 Problem Parameters for Sample Problems

Production price bounds and elasticities	Warranty period bounds and elasticities	Production quantity bounds
$c_{11} = 2$, $c_{12} = 8$, $a_1 = 2$	$w_{11} = 4$, $w_{12} = 6$, $b_1 = 0.05$	$N_{11} = 200$, $N_{12} = 600$
$c_{21} = 4$, $c_{22} = 6$, $a_2 = 2$	$w_{21} = 6$, $w_{22} = 8$, $b_1 = 0.10$	$N_{21} = 300$, $N_{22} = 500$
	$k_{21} = 2$, $k_{22} = 2$	

Lot size bounds	Setup time	Production rate
$Q_{11} = 100$, $Q_{12} = 200$	$s_1 = 20$ hours	$p_1 = 40$ per hour
$Q_{21} = 100$, $Q_{22} = 200$	$s_2 = 40$ hours	$p_2 = 80$ per hour

Failure rate	Resource capacity	Total desirable market share
$\lambda_1 = 0.10$	$A_1 = 200$ hours	$g_1 + g_2 = 0.3$
$\lambda_1 = 0.05$		

failure rate (λ_i) and of the warranty period (w_i) should be consistent. Thus, if λ_1 is 0.10 units/month, then the warranty time (w_1) should be considered in time units of months. The available resource capacity is 200 hours, with the goal being to not exceed this value. The next goal, in order of priority, is to achieve a total market share of at least 0.3. The value of α (proportion of the total revenue to be assigned, at most, to warranty costs) is 0.10. This is designated as Problem 1.

A solution is obtained by using a nonlinear programming package entitled GAMS [17]. GAMS is designed to find solutions that are locally optimal. The nonlinear functions must be smooth (i.e., their first derivatives must exist). A certain region is defined by the bounds on the variables (in our case, the system constraints) and linear constraints in the problem, if any. If the nonlinear objective and constraint functions are convex within this region, any optimal solution obtained will also be a global optimum.

GAMS can handle up to 32,000 constraints, with virtually no limit on the number of variables. It, thus, has the ability to solve practical problems. For the problems investigated, a personal computer (486/33C) with a math coprocessor was used. Execution times are less than 6 seconds. The model may be built with any word processing package. The solutions are written to a list file which can be read using the word processor.

The solution approach chosen is the sequential goal programming procedure using a preemptive approach, that is, the objective function value of the top priority goal (namely, production capacity requirements) is entered as a constraint at the second priority level where the objective is to achieve at least a desirable level of total market share. Using the system constraints and the production capacity goal, the optimal solution is $N_1 = 200$, $N_2 = 300$, $Q_1 = 111$, $Q_2 = 108$, $w_1 = 4$ time units, and $w_2 = 6$ time units. The production capacity or resource goal (at priority level P_1) is exactly satisfied with the objective function value being 0. This means that the resource usage is 200 hours and is not exceeded. Entering the achieved resource usage as a constraint, optimization is conducted for the attainment of the combined market share at priority level P_2. The optimal solution is $c_1 = 3.52$, $c_2 = 4.0$, $w_1 = 4.0$, $w_2 = 6.0$, $Q_1 = 200$, $Q_2 = 200$, $N_1 = 398$, and $N_2 = 500$. The objective function value is -0.056. This indicates that the market share goal was not achieved. In fact, it is underachieved by 5.6%. In other words, instead of achieving the goal of a market share of 30%, the achieved market share is 24.4%. Observe that in order to meet the market share goal, while not exceeding the level of resource usage in Goal 1, the lot sizes and production quantities have changed. Also, the product unit prices have been adjusted to meet the market share goal without sacrificing the resource usage goal.

Now, the achieved levels of resource usage (Goal 1) and combined market share (Goal 2) are entered as constraints, and the warranty reserve costs goal at level P_3 (using $\alpha = 0.1$) is optimized. The goal is exactly achieved, with the value of the objective function being 0. This means that the expected warranty reserve costs equal 10% of total sales. The decision variable values for this solution are $c_1 = 2.92$, $c_2 = 4.24$, $w_1 = 4.0$, $w_2 = 6.0$, $Q_1 = 200$, $Q_2 = 200$, $N_1 = 459$, and $N_2 = 462$. The proportion of the market shares for products 1 and 2 are $g_1 = 0.069$ and $g_2 = 0.175$, respectively. The decision variable values of the unit prices are observed to have changed from the previous level in order to accommodate the goal at this level. Thus, the first and third goals are met, with the second goal being underachieved. The results are shown in Table 32.2.

Let us discuss some of the managerial implications of the results. The concern to the manufacturing manager would be issues such as required capacity, production quantity, and lot size. For this sample problem, no overtime is necessary. From the perspective of the marketing manager, market share could be a concern. Given the available resources, it is not feasible to achieve the goal value of a combined market share of 30%. The market shares for the individual products may provide some insights.

Table 32.2 Solutions and Goal Achievements

Goal	Priority level	Objective function value	c_1	c_2	w_1	w_2	Q_1	Q_2	N_1	N_2	g_1	g_2
Problem 1												
Production capacity or resource	P_1	0.0	—	—	4.0	6.0	111	108	200	300	—	—
Market share	P_2	−0.056	3.52	4.0	4.0	6.0	200	200	398	500	0.048	0.196
Warranty reserve costs	P_3	0.0	2.92	4.24	4.0	6.0	200	200	459	462	0.069	0.175
Problem 2												
Market share	P_1	0.0	2.00	4.7	6.0	6.0	—	—	—	—	0.150	0.150
Production capacity or resource	P_2	−244.63	2.45	4.0	6.0	8.0	100	100	522	500	0.100	0.200
Warranty reserve costs	P_3	0.0	2.45	4.0	5.25	8.0	122	104	522	500	0.100	0.200
Problem 3												
Warranty reserve costs	P_1	0.0	2.0	4.0	4.0	6.0	—	—	600	500	—	—
Market share	P_2	0.0	2.39	4.02	4.53	6.88	—	—	531	497	0.104	0.196
Production capacity or resource	P_3	−244.63	2.5	4.0	6.0	8.0	100	100	521	500	0.100	0.200

Since product 2 achieves a significant portion (19.6%) of the goal value, this may motivate the manager to develop and or modify promotional and marketing strategies for the products. To the financial manager, the amount of funds to be set aside for warranty reserve costs is of importance. In the present problem, based on profitability objectives and the corresponding values of the warranty parameters, it is possible for the financial manager to allocate 10% of sales toward warranty reserves, which was the goal of management. Therefore, the manager would not have to look for additional funds for the warranty reserve pool. Another implication of the results could be that the selected goal values are not very stringent under the present circumstances. It may help management set benchmarks for achievement levels.

A sensitivity analysis is conducted to determine the impact of changing the relative priorities of the goals. Two additional problems are demonstrated in Table 32.2. In Problem 2, after the system constraints, the market share goal is at the highest level (P_1), followed by the production capacity or resource goal at priority level P_2, after which follows the warranty reserve costs at priority level P_3. We find that the market share goal, being at the highest priority level, is now exactly achieved. A market share of 30% is achievable, with 15% being the share for each product, with the warranty parameter values of $c_1 = 2.0$, $c_2 = 4.7$, $w_1 = 6.0$, and $w_2 = 6.0$, for the unit product prices and warranty times of products 1 and 2, respectively. However, at the next priority level (P_3), we find that the production capacity or resource goal is underachieved, in order to maintain the higher-priority goal of achieving a market share of 30%. This implies that an additional 244.63 hours of capacity or resource is needed to maintain the desired market share. Thus, the production manager might have to subcontract part of the production or seek other means of increasing capacity, such as through overtime. The next goal of warranty reserve costs at level P_3 is exactly achieved. So, 10% of total sales is allocated to warranty reserve costs.

Problem 3 considers a priority ranking where the goal associated with warranty reserve costs is at level P_1, followed by the market share goal at level P_2 and production capacity or resource goal at level P_3. We observe that the first two goals are achievable. The production capacity goal is underachieved. An additional 244.63 hours of capacity is required to maintain the desired level of warranty reserves and market share.

These results demonstrate to management the impact of a change in the relative priorities of formulated goals. Such analysis will aid top management in selecting preferential rankings of goals. They indicate to management that a trade-off may take place among the various goals depending on the importance that they assign to the goals.

32.4 CONCLUSIONS

A multiproduct warranty cost estimation problem along with production and market share constraints has been described using a preemptive goal programming solution procedure. Based on certain stipulated goal values, warranty decision parameter values are found and the attainment of goals are discussed.

Modern-day managers tackling realistic problems are faced with a variety of objectives in the process of decision making. They operate under fiscal and environmental restraints and have to achieve customer satisfaction under these conditions. In this context, warranty related problems have to be solved in the context of multiple objectives. This chapter has presented a methodology for modeling such problems.

REFERENCES

1. Menke, W. W. (1969). Determination of warranty reserves, *Management Science,* **15**(10), 542–549.
2. Amato, H. N., and Anderson, E. E. (1976). Determination of warranty reserves: An extension, *Management Science,* **22**(12), 1391–1394.
3. Amato, H. N., Anderson, E. E., and Harvey, D. W. (1976). A general model of future period warranty costs, *Accounting Review,* **51**, 854–862.
4. Blischke, W. R., and Scheuer, E. M. (1981). Applications of renewal theory in analysis of the free-replacement warranty, *Naval Research Logistics Quarterly,* **28**, 193–205.
5. Mamer, J. W. (1982). Cost analysis of pro-rata and free-replacement warranties, *Naval Research Logistics Quarterly,* **29**(2), 345–356.
6. Mamer, J. W. (1987). Discounted and per unit costs of product warranty, *Management Science,* **33**(7), 916–930.
7. Blacer, U., and Sahin, I. (1986). Replacement costs under warranty: Cost moments and time variability, *Operations Research,* **34**(4), 554–559.
8. Frees, E. W., and Nam, S. H. (1988). Approximating expected warranty cost, *Management Science,* **43**(12), 1441–1449.
9. Frees, E. W. (1988). Estimating the cost of a warranty, *Journal of Business and Economic Statistics,* **4**(1), 79–86.
10. Ignizio, J. P. (1983). Generalized goal programming, *Computers and Operations Research,* **10**, 277–289.
11. Ignizio, J. P. (1982). *Linear Programming in Single and Multiple Objective Systems,* Prentice-Hall, Englewood Cliffs, NJ.
12. Mitra, A., and Patankar, J. G. (1988). Warranty cost estimation: A goal programming approach, *Decision Sciences,* **19**(2), 409–423.
13. Patankar, J. G., and Mitra, A. (1989). A multiobjective model for warranty cost estimation using multiple products, *Computers and Operations Research,* **16**(4), 341–351.
14. Mitra, A., and Patankar, J. G. (1990). A multi-objective model for warranty estimation, *European Journal of Operational Research,* **45**, 347–355.
15. Mitra, A., and Patankar, J. G. (1993). An integrated multicriteria model for warranty cost estimation and production, *IEEE Transactions on Engineering Management,* **EM-40**(3), 300–311.
16. Glickman, T. S., and Berger, P. D. (1976). Optimal price and protection period decisions for a product under warranty, *Management Science,* **22**(12), 1381–1390.
17. Brook, A., Kendrick, D., and Meeraus, A. (1988). *GAMS, A Users Guide,* The Scientific Press, Redwood City, CA.

Bibliography

33

Bibliography on Warranties

Istiana Djamaludin and D. N. Prabhakar Murthy

The University of Queensland
Brisbane, Queensland, Australia

Wallace R. Blischke

University of Southern California
Los Angeles, California

33.1 INTRODUCTION

This chapter contains a list of articles, reports, books, and theses that deal with one or more issues of relevance to product warranties. The references included are mainly technical and applications oriented. Thus, we have made no effort to cover the vast legal literature on the subject, and some other areas may not be completely covered.

In using this bibliography, please note the following:

1. The citations are listed alphabetically based on the author (or first author in the case of multiple authors) surname and numbered sequentially within each set. For example, R3 is the third entry under R.
2. The list deals primarily with material written in English, but contains a few references to non-English material.
3. Although the authors have diligently searched many sources in compiling this list, it is inevitable that some articles may have

been inadvertently missed. The authors would appreciate having such omissions brought to their attention.
4. Each reference is classified into one or more categories (see Section 33.2). This classification is not comprehensive. A good number of the references can be classified into many different categories; the authors have restricted the classification of each reference to only the most significant and relevant categories. The user seeking papers on a particular topic should look under related topics as well.
5. The classification for a large number of references was based solely on abstracts as the authors could not access the complete paper, report, or thesis in a timely manner.
6. Finally, the authors would like to thank the authors of the chapters of the *Handbook,* who supplied lists to help in the compilation of this comprehensive list of references.

33.2 CATEGORIES FOR CLASSIFICATION

Classification is based on the following categories:

ACC	Accounting for Warranties
BOC	Breach of Contract
BOW	Breach of Warranty
CBH	Consumer Behavior
CCO	Consumer Complaints
CEI	Consumer Education/Information
CMD	Consumer Manufacturer Dispute
COM	Consumer Movement
COR	Consumer Risk
CPT	Consumer Protection
CSD	Consumer Satisfaction/Dissatisfaction
CTR	Contract Risk
CUS	Consumer Usage
CUW	Cumulative Warranty
CWP	Combination Warranty Policy
DCV	Decision Variable
EAW	Economic Assessment
ESC	Extended Service Contract
EST	Estimation
ETH	Economic Theory of Warranty
EXT	Extended Warranty
FCT	Forecasting

FLW	Fleet Warranty
FRA	Fraud
FRW	Free Replacement Warranty
LAW	Law
LCC	Life Cycle Costing
LIT	Litigation
MAI	Maintenance
MAM	Mathematical Modelling
MCM	Multicriteria Models
MDW	Multidimensional Warranty
MOR	Moral Hazard
OPT	Optimization
PDE	Product Design
PDS	Product Distribution
PDV	Product Development
PLI	Product Liability
PQU	Product Quality
PRC	Product Recall
PRM	Probability Model
PRS	Producer Risk
PRW	Pro-Rata Warranty Policy
PSU	Product Support
PUB	Public Policy
QAS	Quality Assurance
QCO	Quality Control
REG	Regulation
REL	Reliability
REP	Renewal Process
RFT	Renewal Functions
RIW	Reliability Improvement Warranty
RLW	Review of Literature on Warranty
SCT	Service Contracts
SIM	Simulation
SST	Servicing Strategy
STA	Statistical Model
STO	Stochastic Model
TDW	Two-Dimensional Warranty
TOR	Tort
TQM	Total Quality Management
TYW	Type of Warranty
WAD	Warranty Administration
WAP	Warranty Application

WAM   Warranty and Marketing
WCA   Warranty Cost Analysis
WCT   Warranty Contracts
WDA   Warranty Data Analysis
WLE   Warranty Legislation
WMA   Warranty Management
WMS   Warranty and Market Structure
WPR   Warranty and Price
WRE   Warranty Reserves
WSE   Warranty Servicing
WSL   Warranty Selection
WTE   Warranty Term
WVE   Warranty Verification

33.3 CLASSIFICATION OF REFERENCES

The following is a compilation of references with significant content for each of the categories listed in the previous section:

ACC   Accounting for Warranties
A13, B16, B17, B18, B19, B58, C83, C103, D43, F1, F33, G33, H5, H43, J16, K7, L55, M82, M96, M99, P33, T7, T14.

BOC   Breach of Contract
A2, A22, A38, B37, C6, C12, C45, D28, E2, G11, H23, H59, H72, K30, M118, P8, R32, S1, S34, S36, S59, V1, W23, W24.

BOW   Breach of Warranty
H81, K30, K44, L10, S65, S67, S78, S81, W6, W18, Y6.

CBH   Consumer Behavior
A8, A26, B28, B95, B96, B114, C85, C94, D7, D14, D19, E11, G15, K10, K22, K26, K27, K35, L49, M14, M100, P4, P54, R19, R37, S41, T15, W30, W36, W37.

CCO   Consumer Complaints
A24, A34, B33, B34, B54, B55, B114, B127, C76, D14, D15, D16, D17, D18, D19, D20, F19, F27, F28, F29, G21, G22, H7, I1, K31, K34, L6, L7, L25, M59, N8, R19, R30, S27, S47, S82, S84, S85, S91, W27, Z2.

CEI   Consumer Education/Information
A30, A31, B49, C60, C67, C70, C72, C73, C82, C99, D5, D27, F14, F19, F27, F38, H76, K29, L18, L40, M25, M47, M100, N9, P39, S28, S45, S46, S70, S71, S100, T19, W22, W32, W33, W40.

CMD **Consumer Manufacturer Dispute**
A3, B36, C81, M3, N29, O8, P5, R10, S7, S9, S10, S50, S82, S84, S85, U10.

COM **Consumer Movement**
A25, C50, C53, C54, C55, C56, C57, C61, M1, M2, N3, P29.

COR **Consumer Risk**
A18, A27, B35, B51, B66, E4, F3, G39, H35, I12, K8, K24, L51, L52, L53, M65, M140, P24, P28, P31, R27, R37, S22, S41, U1, Z3.

CPT **Consumer Protection**
A19, A36, B2, B24, B102, B115, C20, C23, C32, C46, C58, C61, C75, C84, C86, C87, C102, D1, D41, D42, D63, E10, F6, F7, F8, F10, F13, F20, G6, G31, H41, K9, K25, K43, L10, M15, M38, M56, M57, M58, M61, M145, N6, N8, P40, P41, P45, R6, R20, R38, S40, S68, S77, S80, T17, W31, W34, W43.

CSD **Consumer Satisfaction/Dissatisfaction**
A23, A24, A33, A34, A35, A42, B34, B48, B52, B53, B55, B114, C3, C82, D6, D12, D14, D15, D16, D17, D18, D51, D64, E15, E16, F18, F19, F28, G21, G22, H7, H20, H26, H27, H53, H66, H76, K51, L6, L41, M45, M59, M94, M128, O6, R14, R18, R19, R30, R42, S47, S52, S87, S91, S95, S104, U10, W1, W8, W19, W20, W21, W22, W41, W45.

CTR **Contract Risk**
G5.

CUS **Consumer Usage**
A35, M126, M127, N25, P1, P6, T22.

CUW **Cumulative Warranty**
B85, B86, D52, G43, K52.

CWP **Combination Warranty Policy**
B59, B76, B77, B86, C79, E4, G43, H51, I11, M124, N15, N17, N22, T11.

DCV **Decision Variable**
A4, C38, D44, J14, J15, K38, L38, M39, M54, M98, N11, S18, S69, T1.

EAW **Economic Assessment**
B81, B83, C81, C92, D26, D62, G27, G28, H6, I4, K14, K56, L2, M20, M117, O4, O5, P47, R24, S28, S29, S44, T2, W39.

ESC   Extended Service Contract
B135, F33, H48, L34, M64, O1, P1, P4.

EST   Estimation
A13, B5, B19, B20, B80, C30, D38, E14, F34, F35, F36, F37, G8,
G17, H54, H55, K3, K4, K19, K39, L1, L43, M72, M73, M84, M89,
M90, M92, M127, N17, N18, P14, P15, P16, P17, P19, P20, P38, R1,
R7, S38, S55, S89, S98, S103, T13, Y5.

ETH   Economic Theory of Warranty
A27, B63, B106, C79, D8, D65, E6, G1, G19, H28, H61, M13, M14,
M15, W16.

EXT   Extended Warranty
A36, B22, B44, B121, C51, C71, C83, C84, C85, C86, C90, C100,
C101, D11, D22, D34, D35, D43, F2, F30, F32, G17, G33, G35, H6,
H80, K26, K27, K28, K48, L24, L51, L54, M21, M76, M91, M102,
N1, P33, P37, R8, R18, S6, S8, S20, S49, S57, T4, T7, W3, W10,
W28, W39.

FCT   Forcasting
B136, S73, W10.

FLW   Fleet Warranty
B52, B65, B66, C100, C101, D33, D34, D35, F22, F23, Z1.

FRA   Fraud
B87, D5, F40, S84.

FRW   Free Replacement Warranty
A12, A13, B19, B75, B76, B77, B79, B80, B82, B84, B86, D56, D57,
D58, D59, E13, L32, M10, M11, M70, M124, M135, M136, M141,
M142, M144, N2, N15, N18, N19, N20, P10, P12, P13, R23, R26,
U16.

LAW   Law
A1, A5, A15, A38, B42, B43, B67, B68, B132, C77, C91, D1, D42,
E1, E2, E9, F5, F17, F39, F40, F41, F42, H13, H14, H15, H25, H31,
H47, H65, H72, H78, K18, M6, M24, M78, M85, M97, M119, N23,
P45, P53, R20, R35, R38, S4, S26, S31, S51, S60, S61, S77, U3, U5,
V1, W12, W26, W34, W46, Y3.

LCC   Life Cycle Costing
A44, B13, B58, B86, C8, D45, D48, R39, S22.

LIT   Litigation
A1, A2, A38, B23, B31, B36, B38, B50, B88, B99, C5, C20, C22,

C23, C40, C63, D27, D28, D42, F20, F39, F40, F41, F42, G2, G40, G41, H19, H23, H52, H63, H72, J4, J6, J7, J9, J10, J11, K30, K44, L25, L28, M7, M77, M78, M108, M118, M119, M120, M121, M122, M123, O10, P8, P46, R10, S1, S7, S9, S10, S11, S15, S50, S60, W2, W7, W23, W24.

MAI Maintenance

B105, C98, D4, D11, D39, D69, G3, H64, I12, J1, M75, M139, P7, P10, R33, S92, W44, Y1.

MAM Mathematical Modelling

A6, A17, B47, B75, C18, C24, C78, C79, C80, C92, D7, D8, D56, D69, E6, E13, E14, F29, F35, F37, G1, G15, G26, G38, H22, H54, H56, I8, I9, J12, K10, K17, K19, K55, L4, L16, L43, L44, L45, L49, M9, M11, M32, M33, M67, M73, M88, M89, M103, M112, M125, M126, M127, M130, M131, M134, M140, M141, M143, N18, N19, N20, N22, P4, P5, P16, P52, P54, R23, R24, S97, S98, W16, W17, W30, Y2.

MCM Multicriteria Models

M90, M92, P17, P19.

MDW Multidimensional Warranty

W38.

MOR Moral Hazard

B105, C12, C78, D69, E4, F3, K8, L47, L48, L49, L51, L52, M13.

OPT Optimization

A20, B59, C16, C25, C26, C29, C79, D26, G26, H22, K8, K46, K57, M125, M133, M137, M138, M140, N2, N16, N19, N21, N22, P2, P13, P18, R26, R34, S55, S101, T12, W15.

PDE Product Design

B120, D39, H6, H45, H52, H66, J15, K41, K54, K57, L33, M5, M46, M125, P41, R11, S16, S56, S62, S67, V3.

PDS Product Distribution

B62, G3, M103.

PDV Product Development

C24, D68, H24, H44, H46, H63, L39, M109, M137.

PLI Product Liability

A11, A16, A22, A28, A32, A37, A38, A40, A41, B27, B37, B38, B41, B42, B43, B49, B56, B57, B61, B64, B71, B90, B92, B94, B99, B106, B107, B108, B110, B117, B131, B132, B133, B134, C1, C5, C7, C10,

C11, C14, C20, C22, C31, C33, C34, C35, C36, C37, C43, C45, C77, C91, C96, D23, D24, D49, D61, D62, D67, E2, E8, F4, F5, F17, F38, F43, F44, G9, G10, G24, G28, G29, G31, G34, G38, H3, H8, H9, H10, H16, H23, H25, H30, H47, H52, H58, H60, H63, H72, H73, H74, H75, H77, H78, H81, I2, I3, I7, J2, J4, J5, J7, J8, J10, J11, J13, J17, K18, K20, K21, K32, K36, K43, K44, K47, L2, L11, L12, L13, L14, L15, L19, L22, L23, L28, L35, L42, M5, M6, M7, M12, M24, M35, M36, M39, M52, M53, M55, M60, M74, M77, M80, M86, M87, M99, M101, M104, M105, M106, M107, M108, M109, M113, M114, M116, M117, M119, M121, N5, N13, N14, N25, O4, O5, O7, O12, O13, P22, P24, P25, P27, P39, P41, P42, P43, P46, P48, R2, R9, R11, R28, R29, R31, R35, R43, S3, S11, S13, S16, S24, S25, S35, S39, S51, S60, S62, S63, S67, S71, S76, S78, S80, S81, S90, S99, T5, T10, T16, T21, T22, U15, V5, W2, W5, W6, W7, W12, W13, W46, W48, Y3, Y4, Y6.

PQU   Product Quality

A9, B95, B96, B101, B104, B107, B118, B119, B122, B128, B136, C19, C35, C39, C41, C78, C80, C82, C97, C102, D63, D65, D69, E5, E8, E15, E16, F16, F19, F32, G1, G6, G12, G28, G39, H12, H28, H36, H67, H70, J4, J6, J7, J20, K8, K53, K55, L16, L20, L29, L30, L37, L49, M8, M15, M18, M34, M42, M69, M81, M83, P5, P23, P55, R25, S5, S50, S75, T2, T15, T16, T24, W16, W17.

PRC   Product Recall

A39, B104, B130, C15, C34, D8, F21, H1, H11, H28, H52, H61, H62, J2, J5, K35, K37, L15, M7, M22, M23, M40, M41, M48, M50, M88, M109, M117, N26, O9, P54, R22, R40, S90, T3, T17.

PRM   Probability Model

A12, A13, B20, B96, F36, G15, H5, K17, L43, L44, M10, M70, N17, P16, P20, S103, T13.

PRS   Producer Risk

A27, E4, G36, H35, M61, M111, O8, R27.

PRW   Pro-Rata Warranty Policy

A7, A12, A13, B20, B59, B60, B75, B76, B77, B79, B80, B86, D56, D57, D58, D59, H56, I10, L32, M10, M11, M70, M91, M124, M142, N2, N15, N17, P10, P20, R23, T12.

PSU   Product Support

D9, H44, H45, H46, H57, K13, K14, L21, M147, N3, S19.

PUB   Public Policy
   A14, A30, K28, S29.

QAS   Quality Assurance
   B64, B72, B90, B110, B136, C49, C87, C95, K50, M131, O2, R18,
   S19, S44, V6, W39.

QCO   Quality Control
   B94, B120, C17, C21, C25, C87, D2, D51, D56, D57, D58, D59, D60,
   F4, H24, H66, I4, K38, L32, M35, M46, M62, M84, M113, M132,
   M142, Q1, R4, R12, R24, S58, S66, S95, S102, V7, W29, Y2.

REG   Regulation
   A29, B102, C48, C91, D29, M145.

REL   Reliability
   A28, B7, B30, B62, B97, C2, C22, C25, C28, C49, C90, C93, C98,
   D26, D38, D40, D46, D70, F4, F12, F24, H4, H11, H21, H49, H68,
   J3, J9, K3, K11, K25, K39, K41, K46, K50, L3, L4, L17, L33, L35,
   M116, M125, M147, N10, N15, N16, N21, P7, P10, P26, P38, R3,
   R41, R44, S20, S32, S38, S41, S54, S94, T8, T19, V2, W32, W34,
   W42, W44, W47, Y1.

REP   Renewal Process
   B29, E13, S93.

RFT   Renewal Functions
   A7, B4, B82, B84, F34, F35, F37, H54, R7, Z1.

RIW   Reliability Improvement Warranty
   A21, A29, B5, B6, B7, B8, B9, B11, B12, B13, B14, B15, B25, B32,
   B69, B74, B75, B76, B86, B91, C18, C99, D13, D45, D48, D55, D66,
   F24, F26, G4, G7, G8, G20, G25, G37, G42, H17, H29, H33, H57,
   K42, K49, K52, L46, M8, M26, M27, M28, M29, M30, M31, M32,
   M75, M93, N10, N12, P49, R15, R16, R17, R39, S21, S22, S32, S37,
   S42, S43, S72, S74, S93, T18, T20, U16.

RLW   Review of Literature on Warranty
   K23.

SCT   Service Contracts
   B44, B114, B116, C16, C51, C68, C84, C85, D11, D37, E17, F45,
   H22, H42, K12, K56, L31, L45, L50, L52, L53, L55, M21, M37,
   M102, M145, N7, O1, P11, P28, P51, R21, S14, S31, S77, T4, T9,
   U2, Y7.

SIM   Simulation
   B4, D39, H2, H55, I8, I9, K19, K41, M136, P15, R34, S98.

SST   Servicing Strategy
   A6, B62, B122, C42, G15, H44, H46, L21, L39, M132, P13, S48, T2, W41.

STA   Statistical Model
   B12, B14, B39, B77, B81, B82, B83, C24, C26, D40, E11, H32, H54, H55, I4, I9, K3, K4, K39, L1, L4, L5, L16, L35, M16, M17, M18, M23, M35, M111, M137, M138, P4, P21, P38, R3, R26, R34, S37, S38, S53, T12, V4, W34.

STO   Stochastic Model
   C29, M112.

TDW   Two-Dimensional Warranty
   B79, B128, C30, I9, I10, I11, M66, M111, M112, M134, M135, M141, M143, M144, S53, S55.

TOR   Tort
   A37, B88, B108, C37, C45, D61, F39, F40, F41, H58, K18, L23, L25, L28, M7, M38, S13, W2, Y6.

TQM   Total Quality Management
   C95, M44, M115, P21, Y1.

TYW   Type of Warranty
   B78, B109, D27, D32, F31, G41, H27, K52, L46, M68, P14, P34, P35, W25.

WAD   Warranty Administration
   F25, O8, S79, U4.

WAP   Warranty Application
   A19, A21, B12, B25, B40, B45, B46, B89, B98, B100, B115, B118, B123, B124, B126, B128, B129, C4, C40, C42, C44, C47, C64, C66, C68, C69, C89, C95, D10, D21, D29, D30, D36, D50, D54, F6, F7, F15, F16, F25, G5, G32, H17, H19, H64, H79, J12, K1, K2, K6, K33, K40, K45, L8, L22, M1, M2, M34, M43, M44, M50, M56, M57, M58, M95, O2, P36, P37, P44, R6, R13, R16, S14, S17, S23, S58, S59, S64, S94, S96, S97, S102, S104, T25, U4, W4, W9, W14.

WAM   Warranty and Marketing
   A8, A20, B73, B103, B121, B122, C76, C85, C88, D11, D68, E5, G12, G16, H21, H26, H28, H69, K25, K27, K28, K34, K48, K54,

L21, L24, L27, L41, L42, L50, M42, M63, M64, M65, M67, M68, M77, M78, M90, M104, M105, M106, M128, N2, N27, P2, P3, P21, P23, P50, R23, R27, S2, S29, S30, S33, S52, S56, S73, S78, S79, S86, S87, T6, T25, U1, V2, V8, W40.

WCA   Warranty Cost Analysis
A7, A13, B12, B13, B14, B16, B17, B18, B19, B20, B26, B30, B58, B59, B60, B75, B77, B79, B80, B86, B97, B120, B127, B136, C4, C12, C18, C19, C21, C24, C27, C39, C62, D2, D9, D33, D34, D39, D48, D51, D52, D53, D56, D57, D58, D59, D60, D66, E3, F22, F23, F26, F34, F35, F37, G7, G30, G36, G38, G43, G44, H2, H46, H51, H54, H55, H56, I6, J3, J10, J11, J20, K10, K46, K57, L3, L16, L20, L36, L44, M10, M11, M14, M18, M20, M32, M46, M51, M73, M79, M81, M82, M83, M89, M91, M92, M94, M95, M96, M103, M110, M111, M126, M127, M129, M131, M134, M135, M136, M137, M138, M139, M142, M144, M146, M147, N15, N16, N17, N18, N20, N21, P9, P11, P15, P16, P17, P19, P23, P30, P38, P55, Q1, R1, R4, R7, R14, R34, S5, S32, S38, S43, S72, S83, S89, S95, S101, T8, T11, T12, T14, T18, T24, V3, V6, V7, V8, W10, W11, W17, W29, W39, W42, W47, Y5.

WCT   Warranty Contracts
B23, B91, B92, C6, D63, E5, E7, G23, G41, H59, K7, K16, L10, M98, M120, M122, N24, P6, P12, R5, R32, R36, S30, S65.

WDA   Warranty Data Analysis
A8, A10, A35, A44, B3, B8, B10, B14, B39, B52, B58, B93, B113, C2, D13, D33, D40, D47, E3, E12, E14, F24, F36, G7, G13, G14, G19, G25, H32, H33, H50, H54, J6, J10, J11, K3, K4, K5, L1, L4, L5, L41, L46, M4, M23, M49, M69, M72, M84, M115, P7, P16, P32, P52, P54, R39, R44, S5, S73, S92, S93, S103, T19, T23, V4, W10, W32, Y2, Z1.

WLE   Warranty Legislation
A22, B31, B50, B68, B81, B100, B101, B102, B110, B111, C11, C23, C48, C52, C65, C74, C84, C87, D23, D24, D30, F11, F14, F38, F43, G14, G18, H3, H37, H38, H39, H40, H41, H42, H52, H71, J8, J19, K17, K25, L9, L18, L45, M25, M36, M37, M45, M53, M100, N4, N9, N26, O8, P22, P25, P44, P47, P48, P50, S2, S15, S23, S28, S51, T5, U5, U6, U7, U8, U11, U12, U13, U14, U17, W7, W13, W31, W43, Y6, Y7.

WMA   Warranty Management
B70, B73, B78, B81, C13, C35, D61, I1, K49, L26, M64, M79, M94,

M124, M129, M130, M131, N12, R24, R29, S22, V8, W11, W36, W37.

WMS   Warranty and Market Structure
A9, A18, B63, B112, C93, C97, D25, D65, E8, H45, H70, K53, O4, O5, P5, S71, S76, W35.

WPR   Warranty and Price
A20, A43, B6, B11, B21, B29, B80, B113, B115, C9, C93, D3, D25, D26, D37, D65, E4, E5, E11, G8, G10, G26, H18, I5, I8, K9, K53, K56, L44, L49, L53, M63, M112, M125, M147, P2, P10, P17, P18, P34, P35, P52, R39, S55, S68, W30.

WRE   Warranty Reserves
A12, A17, K12, M70, M71, M88, M89, M90, P12, P14, P16, P18, P19, P20, R25, S55, T3, T4, T13.

WSE   Warranty Servicing
A26, B48, B53, J18, N19, N28, U9.

WSL   Warranty Selection
B109, C17, C103, D3, D31, D32, D53, D55, G42, O3, P11, S88.

WTE   Warranty Term
A20, B1, B8, B10, B19, B72, B91, B125, C25, C26, C59, D48, D70, E6, F9, G26, H4, H33, K11, K19, K33, L46, M11, M16, M17, M18, M19, M62, M63, M67, N2, O11, P18, R12, R27, R33, S74, S101.

WVE   Warranty Verification
H34, W44.

33.4.   WARRANTY REFERENCES

A1.   Abney, D. L. (1988). Disclaiming the implied real estate common-law warranties, *Real Estate Law J.*, **17**, 141–152. (LAW, LIT)

A2.   Abney, D. L. (1988). Determining damages for breach of implied warranties in construction defect cases, *Real Estate Law J.*, **16**, 210–225. (BOC, LIT)

A3.   Abel, R. (1982). *The Politics of Informal Justice*, Vol. 1, Academic Press, New York. (CMD)

A4.   Adler, D. L. (1989). Roof failures, and how to avoid them, *Construction Specifier*, **42**, 72–80. (DCV)

A5.   *Advertising Age* (1975). New warranty law under attack by business—unfair, confusing, **46**, Jul. 7, 1, 37. (LAW)

A6.   Agnihothri, S. R., and Karmarkar, U. S. (1986). *Performance evaluation of service territories*, Tech. Rept., Graduate School of Management, University of Rochester, NY. (MAM, SST)

A7. Agrafiotis, G. K., and Tsoukalas, M. (1990). Excess-time renewal theory with applications, *J. Opl. Res. Soc.*, **41**, 69–82. (PRW, RFT, WCA)

A8. Akaah, I. P. and Korgaonkar, P. K. (1988). A conjoint investigation of the relative importance of risk relievers in direct marketing, *J. Advertising Research*, **28**, 38–44. (CBH, WAM, WDA)

A9. Akerlof, G. A. (1970). The market for 'lemons': Quality uncertainty and the market mechanism, *Quart. J. Econ.*, **84**, 488–500. (PQU, WMS)

A10. Alfano, D. L. and Adeff, G. (1987). *Durable turbocharger bearing systems for high temperature applications*, SAE Technical Paper Series, Publ. by SAE, Warrendale, PA. (WDA)

A11. Allee, J. S. (1984). Product Liability, *Law J. Seminars—Press*, New York. (PLI)

A12. Amato, H. N., and Anderson, E. E. (1976). Determination of warranty reserves: An extension, *Management Sci.*, **22**, 1391–1394. (FRW, PRM, PRW, WRE)

A13. Amato, H. N., Anderson, E. E., and Harvey, D. W. (1976). A general model of future period warranty costs, *The Accounting Rev.*, **51**, 854–862. (ACC, EST, FRW, PRM, PRW, WCA)

A14. American Enterprise Institute for Public Policy Research (1971). *Consumer warranty proposals and FTC amendments*, American Enterprise Institute, Washington, DC. (PUB)

A15. American Law Institute (1962). *Uniform laws annotated: Uniform Commercial Code; with annotations from State and Federal Courts*, Edward Thompson, Brooklyn, NY. (LAW)

A16. *J. American Insurance* (1980). Can monstrous product liability claims be contained?, **56**, 20–22. (PLI)

A17. Amernic, J. (1979). A matrix approach to asset and liability valuation, *Cost and Manag.*, (Canada), **53**, 25–31. (MAM, WRE)

A18. Anderson, E. E. (1972). Buyer risk reduction through adaptive market conditions, *J. Consumer Affairs*, **6**, 156–169. (COR, WMS)

A19. Anderson, E. E. (1973). The protective dimension of product warranty policies and practices, *J. Consumer Affairs*, **7**, 111–120. (CPT, WAP)

A20. Anderson, E. E. (1977). Product price and warranty terms: An optimization model, *Operational Research Quarterly*, **28**, 739–741. (OPT, PPR, WAM, WTE)

A21. Anderson, J. R. (1979). Warranties—the easy way out, *1979 Proc. Ann. Reliab. and Maintain. Symp.*, 413–415. (RIW, WAP)

A22. Anderson, R. R. (1988). An overview of buyer's damage remedies, *Uniform Commercial Code Law J*, **21**, 28–60. (BOC, PLI, WLE)

A23. Andreasen, A. R. (1977). A taxonomy of consumer satisfaction/dissatisfaction measures, *J. Consumer Affairs*, **11**, 11–23. (CSD)

A24. Andreasen, A. R. and Best, A. (1977). Consumers complain—Does business respond? *Harvard Business Review*, **55**, 93–101. (CCO, CSD)

A25. Angevine, E. (1982). *Consumer Activist: They made a difference. A his-*

tory of consumer action related by leaders in the consumer movement, Consumer Union Foundation, New York. (COM)

A26. Anthes, G. H. (1993). Software AG rolls out options, *Computerworld*, **27**, 79. (CBH, WSE)

A27. Appelbaum, E. (1992). Bankruptcy, warranties and the firm's capital structure, *Int. Economic Rev.*, **33**, 399–412. (COR, ETH, PRS)

A28. Appelbaum, E., and Scheffman, D. (1980). *Product reliability, warranties and producer liability, and advertising*, Federal Trade Commission Working Paper, Washington, DC. (PLI, REL)

A29. Armed Services Procurement Regulation, (1976). Government Printing Office, Washington, DC. (REG, RIW)

A30. Armstrong, G. M., Kendall, C. L., and Russ, F. A. (1975). Applications of consumer information processing research to public policy issues, *Communication Research*, **2**, 232–245. (CEI, PUB)

A31. Armstrong, L. (1991). Productivity assured—or we'll fix them free, *Business Week*, **Nov. 25**, 34. (CEI)

A32. Arness, J. (1976). A defendant's attorney looks at products liability, *25th Meeting, Mechanical Failures Prevention Group*, National Bureau of Standards. (PLI)

A33. Ash, S. B. (1978). A comprehensive study of consumer satisfaction with durable products, in *Advances in Consumer Research*, (H. K. Hunt, ed.), **5**, 254–262. (CSD)

A34. Ash, S. B. and Quelch, J. A. (1979). Consumer satisfaction, dissatisfaction and complaining behavior: A comprehensive study of rentals, public transportation and utilities, in *Refining Concepts and Measures of Consumer Satisfaction and Complaining Behavior* (H. K. Hunt and R. L. Day, eds.), Department of Marketing, Indiana University, Bloomington, 120–130. (CCO, CSD)

A35. Ash, S. B., Kennedy, J. R. and Thirkell, P. (1979). Consumer satisfaction with product warranties: A study of Canadian automobile and appliance owners, in *Refining Concepts and Measures of Consumer Satisfaction and Complaining Behavior* (H. K. Hunt and R. L. Day, eds.), Department of Marketing, Indiana University, Bloomington, 131–140. (CSD, CUS, WDA)

A36. Asher, J. (1990). Buyer protection: Loss leader or gold mine?, *Bank Marketing*, **22**, 30–32. (CPT, EXT)

A37. Association Europeenne d'etudes juridiques et fiscales, (1975). *Product Liability in Europe*, Kluwer, Deventer. (PLI, TOR)

A38. Ausness, R. C. (1993). The impact of the Cipollone case on federal preemption law, *J. Products and Toxics Liability*, **15**, 1–27. (BOC, LAW, LIT, PLI)

A39. Australia Federal Bureau of Consumer Affairs (1988). *Product Safety Recalls*, The Federal Bureau of Consumer Affairs, AGPS, Canberra, Australia. (PRC)

A40. Australia Industry Commission (1990). *Product Liability/Industry Commission*, AGPS, Canberra, Australia. (PLI)

A41. Australia, Parliament and Senate. Standing Committee on Legal and Constitutional Affairs, (1992). *Product Liability: Where Should the Loss Fall?*, Proposed Amendments to the Trade Practices Act 1974: The Committee, Canberra, Australia. (PLI)

A42. *Automotive News* (1976). Pohanka rips warranty hassle, **Dec. 20, 125.** (CSD)

A43. Avery, S. (1988). Integral motors: Up and running, *Purchasing*, **105,** 98–100. (WPR)

A44. Aviation Supply Office (1974). *Case Histories of LCC (Life Cycle Cost) Procurement*, Philadelphia, PA. (LCC, WDA)

B1. Babb, H. T. and Estes, B. T. (1984). Success in management and initial operation of a simulator project, *Trans. American Nuclear Society*, **46,** 706. (WTE)

B2. Baer, M. G. (1993). The consultation draft of the consumer and business practices code, *Canadian Business Law J.*, **21,** 254–273. (CPT)

B3. Baggerly, R. G. (1985). Quantifying reliability from failure analysis, warranty, and sales data, Presented at *International Conference and Exposition on Fatigue, Corrosion Cracking, Fracture Mechanics and Failure Analysis*, ASM, Metals Park, OH, 185–190. (WDA)

B4. Baker, R. D. (1993). Nonparametric estimator of the renewal function, *Computers and Operations Research*, **20,** 167–178. (RFT, SIM)

B5. Balaban, H. S. (1975). Guaranteed MTBF for military procurement, *Proc. 10th Int. Logistic Symp.*, SOLE. (EST, RIW)

B6. Balaban, H. S. (1976). Controlling risks in reliability improvement warranties, *Proc. Military Oper. Res. Soc.*, 17. (RIW, WPR)

B7. Balaban, H. S. (1978). Reliability growth models, *J. Envi. Sci.*, **21,** 11–18. (REL, RIW)

B8. Balaban, H., Cuppett, D., and Harrison, G. (1979). The F-16 RIW program, *1979 Proc. Ann. Reliab. and Maintain. Symp.*, 79–82. (RIW, WDA, WTE)

B9. Balaban, H. S., Tom, K. B., and Harrison, G. T. Jr. (1986). *Warranty Handbook*, 2-1-4-16. (RIW)

B10. Balaban, H. S., and Nohmer, F. J. (1975). Warranty procurement—a case history, *1975 Proc. Ann. Reliab. and Maintain. Symp.*, 1–5. (WDA, WTE)

B11. Balaban, H. S., and Meth, M. A. (1978). Contractor risk associated with reliability improvement warranty, *1978 Proc. Ann. Reliab. and Maintain. Symp.*, 123–129. (RIW, WPR)

B12. Balaban, H., and Retterer, B. (1973). *The Use of Warranties for Defense Avionics Procurement*, ARINC Research Pub. No. 0637-02-1-1243, Annapolis, MD. (RIW, STA, WAP, WCA)

B13. Balaban, H., and Retterer, B. (1973). *Life-Cycle Cost Implications in the Use of Warranties for Avionics*, ARINC Tech. Perspective, No. 8, Annapolis, MD. (LCC, RIW, WCA)

B14. Balaban, H., and Reterer, B. (1974). The use of warranties for defense avionics procurement, *1974 Proc. Ann. Reliab. and Maintain. Symp.*, 363–368. (RIW, STA, WCA, WDA)

B15.   Balaban, H., and Reterer, B. (1977). "*An Investigation of Contractor Risk Associated with Reliability Improvement Warranty*," ARINC Research Corporation Report, No. 1184-01-2-1619. (RIW)

B16.   Balachandran, K. R., and Maschmeyer, R. A. (1991). Accounting for product warranty costs, presented at *ORSA/TIMS 32nd Joint National Meeting*, Anaheim, CA. (ACC, WCA)

B17.   Balachandran, K. R., and Maschmeyer, R. A. (1992). Accounting for product wearout cost, *J. Accounting, Auditing and Finance*, **7**, 49–66. (ACC, WCA)

B18.   Balachandran, K. R., and Maschmeyer, R. A. (1992). Cost analysis for product warranties, *Pacific Accounting Rev.*, **4**, 59–76. (ACC, WCA)

B19.   Balachandran, K. R., Maschmeyer, R. A., and Livingstone, J. L. (1981). Product warranty period: A Markovian approach to estimation and analysis of repair and replacement costs, *The Accounting Rev.*, **56**, 115–124. (ACC, EST, FRW, WCA, WTE)

B20.   Balcer, Y., and Sahin, I. (1986). Replacement costs under warranty: Cost moments and time variability, *Operations Res.*, **34**, 554–559. (EST, PRM, PRW, WCA)

B21.   Bald, J. (1972). What price warranties, *Fleet Owner*, **67**, 141. (WPR)

B22.   Bald, J. (1993). Warranty recovery, *Fleet Equipment*, **19**, 52–54. (EXT)

B23.   Bandman, M. B. (1987). Balancing the risks in computer contracts, *Commercial Law J.*, **92**, 384–393. (LIT, WCT)

B24.   Barnes, R. (1989). Transamerica to focus on film risks, *National Underwriter*, **93**, 13, 25. (CPT)

B25.   Barton, H. R., Jr. (1985). Predicting guaranty support using learning curves, *1985 Proc. Ann. Reliab. and Maintain. Symp.*, 354–356. (RIW, WAP)

B26.   Bartz, D. (1985). Quality finish—a case history, *Conference Proc.—Finishing '85*, Society of Manufacturing Engineers, Dearborn, MI, 8.16–8.24. (WCA)

B27.   Battle, R. C. (1971). Taking the danger out of product liability, *Management Rev.*, **60**, 50–52. (PLI)

B28.   Bauer, R. A. (1960). Consumer behavior and risk taking, *Proc. of the Amer. Market. Ass.*, Chicago, IL, 59–63. (CBH)

B29.   Baxter, L. A. (1982). Reliability application of the revelation transform, *Naval Res. Logistics Quart.*, **29**, 323–330. (WPR, REP)

B30.   Bayer, H., and Speir, R. N. (1978). Long term commercial warranty, *1978 Proc. Ann. Reliab. and Maintain. Symp.*, 50–54. (WCA, REL)

B31.   Bayer, R. M. (1988). Personal property leasing: Article 2A of the uniform commercial code, *Business Lawyer*, **43**, 1491–1511. (LIT, WLE)

B32.   Bazovsky, I. (1968). Appraisal of guaranteed MTBF warranty programs, *Ann. Assurance Sciences*, **1**, 256–265. (RIW)

B33.   Bearden, W. O. (1983). Profiling consumers who register complaints against auto repair services, *J. Consumer Affairs*, **17**, 315–335. (CCO)

B34. Bearden, W. O., and Oliver, R. L. (1985). The role of public and private complaining in satisfaction with problem resolution, *J. Consumer Affairs*, **19**, 222–240. (CCO, CSD)

B35. Bearden, W. O., and Shimp, T. A. (1982). The use of extrinsic cues to facilitate product adoption, *J. Marketing Res.*, **19**, 229–239. (COR)

B36. Bebchuk, L. A. (1984). Litigation and settlement under imperfect information, *RAND J. Economics*, **15**, 404–415. (CMD, LIT)

B37. Bebchuk, L. A., and Shavell, S. (1991). Information and the scope of liability for breach of contract: The rule of Hadley v. Baxendale, *J. Law, Economics and Organization*, **7**, 284–312. (BOC, PLI)

B38. Bedford, M. S., and Stearns, F. C. (1987). The technical writer's responsibility for safety, *IEEE Trans. Professional Communication*, **PC-30**, 127–132. (LIT, PLI)

B39. Bedniak, M. N. (1972). *Determination of Norm of the Warranty Run for Automobiles*, University of Kiev. (in Russian). (STA, WDA)

B40. Beekler, E. C., and Candy, H. (1975). *Analysis of AMC's Use of Warranties*, APRO 507, US Army Procurement Research Office, USALMC, Fort Lee, VA. (WAP)

B41. Beerworth, E. E. (1989). *Product Liability*, Federation Press, Annandale, New South Wales, Australia. (PLI)

B42. Beerworth, E. E. (1991). *Contemporary Issues in Product Liability Law*, Federation Press, Annandale, New South Wales, Australia. (LAW, PLI)

B43. Beerworth, E. E. (1991). *The New Product Liability Laws*, Paper delivered at a BLEC Forum, Business Law Education Centre, Melbourne, Australia. (LAW, PLI)

B44. Beiswinger, G. L. and Sopko, S. (1990). Consider the options in service contracts: The growing need to protect equipment, *Office*, **112**, 63–66. (EXT, SCT)

B45. Belev, G. C. (1989). Guidelines for specification development, 1989, *Proc. Ann. Reliab. and Maintain. Symp.*, 15–21. (WAP)

B46. Belev, G. (1989). Guidelines for specification development, *American Association of Cost Engineers Trans.*, N.1.1–N.1.7. (WAP)

B47. Bell, L. (1961). *A Mathematical Model of Guarantee Policies*, Tech. Report No. 49, Stanford University, Stanford, CA. (MAM)

B48. Beltrami, R., and Evans, K. R. (1984). Consumer perception of warranty service performance, *Amer. J. Small Bus.*, **9**, 11–16. (CSD, WSE)

B49. Bennigson, L. A., and Bennigson, A. I. (1978). Product liability: Manufacturers beware!, in *Consumerism: Search for the Consumer Interest*, (D. A. Aaker and G. S. Day, eds.), 3rd ed., The Tree Press, New York, 384–398. (CEI, PLI)

B50. Benz, M. P., and Meyer, D. J. (1992). Express federal preemption: Where is it after Cipollone, *Defense Counsel J.*, **59**, 491–499. (LIT, WLE)

B51. Berens, J. S. (1971). Consumer costs in product failure, *MSU Business Topics*, **19**, 27–30. (COR)

B52. Berke, T. M., and Zaino, N. A. Jr. (1991). Warranties: What are they?

What do they really cost?, *1991 Proc. Ann. Reliab. and Maintain. Symp.*, 326–331. (CSD, FLW, WDA)

B53.   Bernacchi, M. D., Kono, K., and Willette, G. L. (1979). An analysis of automobile warranty service dissatisfaction, in *Refining Concepts and Measures of Consumer Satisfaction and Complaining Behavior* (H. K. Hunt and R. L. Day, eds.), Dept. of Marketing, Indiana University, Bloomington, IN. 141–143. (CSD, WSE)

B54.   Best, A. (1981). *When Consumers Complain*, Columbia University Press, New York. (CCO)

B55.   Best, A., and Andreasen, A. R. (1977). Consumer response to unsatisfactory purchases: A survey of perceiving defects, voicing complaints, and obtaining redress, *Law and Soc.*, **11**, 700–742. (CCO, CSD)

B56.   Berman, G. A. (1978). "C.Y.A."—Cover your aft, *1978 Proc. Ann. Reliab. and Maintain. Symp.*, 374–379. (PLI)

B57.   Bieber, R. M. (1973). Product liability and its controls, *Hazard Prevention*, **Sept/Oct**, 12–15. (PLI)

B58.   Bieda, J. (1992). A product-cycle cost-analysis process and its application to the automotive environment, *1992 Proc. Ann. Reliab. and Maintain. Symp.*, 422–425. (ACC, LCC, WCA, WDA)

B59.   Biedenweg, F. M. (1981). *Warranty Policies: Consumer Value vs. Manufacturer Costs*, Doctoral Dissertation, Stanford University, Stanford, CA. (CWP, OPT, PRW, WCA)

B60.   Biedenweg, F. M. (1981). *Warranty Policies: Consumer Value vs. Manufacturer Costs*, Tech. Rept. No. 198, Dept. of Operations Research, Stanford University, Stanford, CA. (PRW, WCA)

B61.   Bielecki, M. (1992). *Defending Class Actions*, presented at the Law Society Product Liability and Class Actions Seminar, Law Society, Adelaide, Australia. (PLI)

B62.   Bienstadt, B. (1987). Stocking the shelves, *Cellular Business*, **4**, 14, 18. (PDS, REL, SST)

B63.   Bigelow, J., Cooper, R., and Ross, T. W. (1988). Warranties without commitment to market participation, *Int. Economic Rev.*, **34**, 85–100. (ETH, WMS)

B64.   Birmingham, M. (1983). Product liability: An issue for quality, *Quality*, **22**, 41–42. (PLI, QAS)

B65.   Birkland, C. (1991). Fine tuning warranty recovery, *Fleet Equipment*, **17**, 47–49. (FLW)

B66.   Birkland, C. (1992). Read the warranty, *Fleet Equipment*, **18**, 20–25. (COR, FLW)

B67.   Bishop, J., and Hubbard, H. W. (1969). *Let the Seller Beware*, The National Press, Washington, DC. (LAW)

B68.   Bixby, M. B. (1964). Judicial interpretation of the Magnuson-Moss warranty act, *American Business Law J.*, **22**, 125–163. (LAW, WLE)

B69.   Bizup, J. A., and Moore, R. R., Techniques for selecting and analyzing reliability improvement warranties, *Naval Eng. Support Act.*, Washington, DC, 134. (RIW)

B70. Blank, I. (1983). Energy management warranty insurance, presented at the *2nd Mid-Atlantic Energy Conference*, Baltimore, MD, 537–540. (WMA)

B71. Blatt, D. L., Goldenberg, N. S., and Noel, D. W. (1978). *Casenote legal briefs—product liability: Adaptable to courses utilizing Noel and Phillips' casebook on product liability*, Casenotes Pub. Co., Beverly Hills, CA. (PLI)

B72. Blemel, K. G. (1986). Quality assurance—a total systems approach, *1986 Proc. Ann. Reliab. and Maintain. Symp.*, 367–369. (QAS, WTE)

B73. Bleuel, W. H., and Bender, H. E. (1980). *Product Planning Service: Service-Marketing Engineering Interactions*, AMACOM, American Management Association. (WMA, WAM)

B74. Blewitt, S. J. (1978). Cost and operational effectiveness of R&M improvement, *1978 Proc. Ann. Reliab. and Maintain. Symp.*, 417–421. (RIW)

B75. Blischke, W. R. (1990). Mathematical models for analysis of warranty policies, *Math. and Computer Modelling*, **13**, 1–16. (FRW, MAM, PRW, RIW, WCA)

B76. Blischke, W. R. (1991). Analysis of nonstandard warranty policies, presented at *ORSA/TIMS 32nd Joint National Meeting*, Anaheim, CA. (CWP, FRW, PRW, RIW)

B77. Blischke, W. R. (1992). Cost comparison of warranty policies for alternative life distribution, presented at *ORSA/TIMS 33rd Joint National Meeting*, Anaheim, CA. (CWP, FRW, PRW, STA, WCA)

B78. Blischke, W. R., and Murthy, D. N. P. (1993). Product warranty management—I: A taxonomy for warranty policies, *Euro. J. Operational Res.*, **62**, 127–148. (TYW, WMA)

B79. Blischke, W. R., and Murthy, D. N. P. (1993). *Warranty Cost Analysis*, Marcel Dekker, Inc., New York. (FRW, PRW, TDW, WCA)

B80. Blischke, W. R., and Scheuer, E. M. (1975). Calculation of the cost of warranty policies as a function of estimated life distributions, *Naval Res. Logistics Quart.*, **22**, 681–696. (EST, FRW, PRW, WCA, WPR)

B81. Blischke, W. R., and Scheuer, E. M. (1977). On the structure and analysis of warranty policies, presented at the *18th Joint National Meeting, ORSA/TIMS*, Atlanta, GA. (EAW, STA, WLE, WMA)

B82. Blischke, W. R., and Scheuer, E. M. (1977). A renewal function arising in warranty analysis, *Proc. Amer. Statist. Assoc., Business and Economic Statistics Section*, 668–672. (FRW, RFT, STA)

B83. Blischke, W. R., and Scheuer, E. M. (1977). Application of nonparametric methods in the statistical and economic analysis of warranties, in *The Theory and Applications of Reliability, with Emphasis on Bayesian and Nonparametric Methods, Vol. II*, (C. P. Tsokos and I. N. Shimi, eds.), Academic Press, New York, 259–273. (EAW, STA)

B84. Blischke, W. R., and Scheuer, E. M. (1981). Applications of renewal theory in analysis of the free-replacement warranty, *Naval Res. Logistics Quart.*, **28**, 193–205. (FRW, RFT)

B85. Blischke, W. R., and Scheuer, E. M. (1984). On cumulative warranties, presented at *TIMS XXVI*, Copenhagen, Denmark. (CUW)

B86. Blischke, W. R., and Scheuer, E. M. (1985). Life cycle costing of warranties, *Proc. of Reliability*, **1**, 3A.5.1–3A.5.12. (CUW, CWP, FRW, LCC, PRW, RIW, WCA)

B87. Bloomer, H. F. Jr. (1992). Avoiding the pitfalls of reliance on guarantees, *Uniform Commercial Code Law J.*, **24**, 380–399. (FRA)

B88. Boedecker, K. A., Morgan, F. W., and Stoltman, J. J. (1991). Legal dimensions of salespersons' statements: A review and managerial suggestions, *J. Marketing*, **55**, 70–80. (TOR, LIT)

B89. Bogart, G. G., and Fink, E. E. (1930). Business practice regarding warranties in the sale of goods, *Illinois Law Rev.*, **25**, 400–417. (WAP)

B90. Bond, A. (1985). New product quality and legal liability—an overview, *Metallurgie*, **52**, 404. (PLI, QAS)

B91. Bonner, W. J. (1976). A contractor view of warranty contracting, *1976 Proc. Ann. Reliab. and Maintain. Symp.*, 351–356. (RIW, WCT, WTE)

B92. Bonner, W. J. (1977). Warranty contract impact on product liability, *1977 Proc. Ann. Reliab. and Maintain. Symp.*, 261–263. (PLI, WCT)

B93. Bonnet, A. H., and Soukup, G. C. (1986). Rotor failures in squirrel cage induction motors, *IEEE Trans. Industry Applications*, **IA-22**, 1165–1173. (WDA)

B94. Borch, K. (1978). Product liability, quality control and insurance, *Revista Di Matematice Le Science Economiche e Sociali*, **1**, 89–98. (PLI, QCO)

B95. Boulding, W., and Kirmani, A. (1992). *Consumer Perceptions of Warranties as Signals of Quality*, Duke University, Fuqua School of Business. mimeo., NC. (CBH, PQU)

B96. Boulding, W., and Kirmani, A. (1993). A consumer-side experimental examination of signaling theory: Do consumers perceive warranties as signals of quality?, *J. Consumer Research*, **20**, 111–123. (CBH, PQU, PRM)

B97. Bragg, G. (1987). *Reliability goal apportionment: A method for allocating engineering resources*, SAE Technical Paper Series, SAE, Warrendale, PA. (WCA, REL)

B98. Brandl, D. L. (1986). Risk sharing in project development, *Waterpower '85: Proc. Int. Conf. on Hydropower*, **1**, 500–508. (WAP)

B99. Brandon, G. (1987). Liability of supplying technology, *Bankers Magazine*, **170**, 60–63. (PLI, LIT)

B100. Brennan, J. R., and Burton, S. A. (1989). Warranties: Concept to implementation, *1989 Proc. Ann. Reliab. and Maintain. Symp.*, 1–8. (WAP, WLE)

B101. Breslow, J. J. (1986). Understanding your rights about defective software, *Words*, **15**, 17. (PQU, WLE)

B102. Brickey, K. F. (1978). The Magnuson-Moss Act—an analysis of the efficacy of federal warranty regulation as a consumer protection tool, *Santa Clara Law Rev.*, **18**, 73–118. (CPT, REG, WLE)

B103. Brock, L. (1987). Stress your money-back guarantee to business, *Direct Marketing*, **50**, 82–83. (WAM)

B104. Bromiley, P. and Marcus, A. (1989). The deterrent to dubious corporate behavior: Profitability, probability and safety recalls, *Strategic Management J.*, **10**, 233–250. (PRC, PQU)

B105. Brown, J. P. (1971). Maintenance, guaranteed and moral hazard, *Winter Meetings of the Econometric Society*, New Orleans, LA. (MAI, MOR)

B106. Brown, J. P. (1973). Towards an economic theory of liability, *J. Legal Studies*, **2**, 323–349. (ETH, PLI)

B107. Brown, J. P. (1974). Product liability: The case of an asset with random life, *Amer. Economic Rev.*, **64**, 149–161. (PLI, PQU)

B108. Brown, M. A. (1989). *Toxic Torts and Product Liability: Changing Tactics for Changing Times*, Bureau of National Affairs, Washington, DC. (PLI, TOR)

B109. Brown, P. M. (1986). Insurance considerations for cogeneration project development, *8th World Energy Engineering Congress*, Fairmont Press Inc., Atlanta, GA. 73–76. (TYW, WSL)

B110. Brown, S. (1990). *The Product Liability Handbook: Prevention, Risk, Consequence, and Forensics of Product Failure*, Van Nostrand Reinhold, New York. (PLI, QAS, WLE)

B111. Brown, S. W. and Ostrom, L. L. (1977). Product warranties: Impact of recent legislation, *Arizona Business*, **10**, 18–23. (WLE)

B112. Bruscke, M. K. (1987). Clearinghouse guarantees: How good are they, and to whom?, *InterMarket*, **4**, 44–45. (WMS)

B113. Bryant, W. K., and Gerner, J. L. (1978). Hedonic price analysis of major appliance warranties, in *Consumer Durables: Warranties, Service Contracts, and Alternatives*, (R. T. Lund, ed.), Rept. No. CPA-78-14/Vol. II, Center for Policy Alternatives, Massachusetts Institute of Technology, Cambridge, MA, 2-202–2-247. (WDA, WPR)

B114. Bryant, W. K., and Gerner, J. L. (1978). The recent television purchaser survey: A descriptive analysis, in *Consumer Durables: Warranties, Service Contracts, and Alternatives*, (R. T. Lund, ed.), Rept. No. CPA-78-14/Vol. III, Center for Policy Alternatives, Massachusetts Institute of Technology, Cambridge, MA, 3-73–3-197. (CBH, CCO, CSD, SCT)

B115. Bryant, W. K., and Gerner, J. L. (1978). The price of a warranty: The case for refrigerators, *J. Consumer Affairs*, **12**, 30–47. (CPT, WAP, WPR)

B116. Bryant, W. K., and Gerner, J. L. (1982). The demand for service contracts, *J. Business*, **55**, 345–366. (SCT)

B117. Buchanan, J. M. (1970). In defense of Caveat Emptor, *Univ. of Chicago Law Rev*, **38**, 64–73. (PLI)

B118. Burnacz, J. G. (1990). Getting to the bottom of office seating, *Today's Office*, **25**, 20–22. (PQU, WAP)

B119. Burns, V. B. (1970). Warranty prediction: Putting a $ on poor quality, *Quality Progress*, **3**, 28–29. (PQU)

B120. Burgess, J. A. (1984). Verifying design quality, *38th. Ann. Quality Congress Trans.*, Chicago, IL, 176–181. (PDE, QCO, WCA)

B121. Buss, D. D. (1987). Ford matches GM's incentives and warranties, *The Wall Street J.*, **Jan. 27**, 4. (EXT, WAM)

B122. *Business America* (1987). How to prepare your product for export, **10**, 10–11. (PQU, SST, WAM)

B123. *Business Week* (1960). Trouble with new car warranties, **Nov. 6**, 55–56. (WAP)

B124. *Business Week* (1961). What Benrus found out about watches: Three year guarantee has been successful, **Sept. 30**, 121–122. (WAP)

B125. *Business Week* (1963). Guaranteeing more, enjoying it less, **Jan. 26**, 46. (WTE)

B126. *Business Week* (1965). Giving the facts fast (Chrysler's computerized system makes warranty records instantly available to used-car dealers, **Nov. 20**, 126. (WAP)

B127. *Business Week* (1970). Can FTC put a governor on Detroit? **Feb. 28**, 31–32. (CCO, WCA)

B128. *Business Week* (1970). Detroit tries a U-turn on warranties, **Jul. 25**, 44–45, 48. (PQU, TDW, WAP)

B129. *Business Week* (1975). The guesswork on warranties, **Jul. 14**, 51, 54. (WAP)

B130. *Business Week* (1975). Managing the product recall, **Jan. 26**, 46–48. (PRC)

B131. *Business Week* (1977). The way to ease soaring product liability costs, **Jan. 17**, Cited in *Consumerism: Search for the Consumer Interest*, 3rd, ed. (D. A. Aaker, and G. S. Day, eds.), The Tree Press, New York, 399–403. (PLI)

B132. *Business Week* (1979). The devils in the product liability laws, **Feb. 12**, 72–78. (LAW, PLI)

B133. *Business Week* (1980). A product liability bill has insurers uptight, **Mar. 31**, 43. (PLI)

B134. *Business Week* (1981). More punitive damage awards, **Jan. 12**, 62. (PLI)

B135. *Business Week* (1990). New lemons from the auto lot, **Dec. 3**, 38. (ESC)

B136. Byers, J. L. (1990). Analytical approach to reliability, failure forecasting and product quality, *American Society of Mechanical Engineers*, **GT190**, 7p. (FCT, PQU, QAS, WCA)

C1. Calabresi, G., and Bass, K. C. (1970). Right approach, wrong implication: A critique of McKeans on product liability, *Univ. of Chicago Law Rev*, **38**, 74–91. (PLI)

C2. Calatayud, R., and Szymkowiak, E. A. (1992). Temperature and vibration results from captive-store flight tests provide a reliability improvement tool, *1992 Proc. Ann. Reliab. and Maintain. Symp.*, 266–271. (REL, WDA)

C3. California Department of Consumer Affairs. (1991). Arbitration review program, consumer satisfaction survey. (CSD)

C4. Cappels, T. M. (1983). Full cycle corrective action (FCCA) for improved warranty service, *1983 Proc. Ann. Reliab. and Maintain. Symp.*, 20–23. (WAP, WCA)

C5.   Cardinali, R., and Zakewicz, A. (1991). Injuries caused by computer systems, *J. Product Liability*, **13**, 347–359. (PLI, LIT)

C6.   Carter, J. W. (1991). *Breach of Contract*, 2nd. ed., The Law Book Co. Ltd., North Ryde, New South Wales, Australia. (BOC, WCT)

C7.   Carter, T. R. (1976). Service—the minimizer of product liability, *Proceedings, Product Liability Prevention Conference*, Newark, NJ, **Aug.**, 93–96. (PLI)

C8.   *Case Histories of LCC (Life Cycle Costing) Procurement* (1974), Aviation Supply Office, Philadelphia, PA. (LCC)

C9.   *Catalog Age* (1992). Yet another guarantee, **9**, 46. (WPR)

C10. Cavanagh, S. W., and Phegan, C. S. (1983). *Product Liability in Australia*, Butterworths, Sydney, Australia. (PLI)

C11. Centner, T. J. (1992). The new 'tractor lemon laws': An attempt to squeeze manufacturers draws sour benefits, *J. Product Liability*, **14**, 121–137. (PLI, WLE)

C12. Centner, T. J., and Wetzstein, M. E. (1987). Reducing moral hazard associated with implied warranties of animal health, *American J. Agricultural Economics*, **69**, 143–150. (BOC, MOR, WCA)

C13. *Chain Store Age Executive*, (1992). Your roofing warranty: Important question to ask, facts to know, **68**, 84, 86. (WMA)

C14. Chandran, R., and Linneman, R. (1978). Planning to minimize product liability, *Sloan Management Rev.*, **Fall**, 33–45. (PLI)

C15. Chandran, R., and Lancioni, R. A. (1981). Product recall: A challenge for the 1980's, *Int. J. Phys. Distribution Mat. Man.*, **11**, 46–55. (PRC)

C16. Chao, H., and Wilson, R. (1990). Optimal contract period for priority service, *Operations Research*, **38**, 598–606. (OPT, SCT)

C17. Chapman, K. S. (1986). *A Theory of Product Warranties*, Doctoral Dissertation, University of Minnesota, Minneapolis, MN. (QCO, WSL)

C18. Chelson, P. O. (1978). Can we expect ECP s under RIW?, 1978, *Proc. Ann. Reliab. and Maintain. Symp.*, 204–209. (RIW, WCA, MAM)

C19. Chen, X. S. (1991). *Product Quality and Warranty Cost*, Doctoral Dissertation, University of Wisconsin, Milwaukee, WI. (PQU, WCA)

C20. Cherry, R. L. Jr. (1989). Builder liability for used home defects, *Real Estate Law J.*, **18**, 115–141. (CPT, LIT, PLI)

C21. Cheung, B., and Liang, M. (1991). Automotive ATE keeps the line moving, *Test and Measurement World*, **11**, 4. (QCO, WCA)

C22. Chiang, G. D. C. (1979). Mock trial; Goody et al. vs. Big George Motor Company, *1979 Proc. Ann. Reliab. and Maintain. Symp.*, 83–84. (LIT, PLI, REL)

C23. Chittenden, C. M. (1992). Form caveat emptor to consumer equity—the implied warranty of quality under the uniform common interest ownership act, *Wake Forest Law Rev.*, **27**, 571–602. (CPT, LIT, WLE)

C24. Chou, K., and Tang, K. (1992). Burn-in time and estimation of change-point with Weibull-Exponential mixture distribution, *Decision Sciences*, **23**, 973–990. (MAM, PDV, STA, WCA)

C25. Chukova, S. S. (1988). Determination of the optimal warranty periods, reliability and quality control, *M. Pub. Standards*, **N8**. (in Russian). (OPT, QCO, REL, WTE)

C26.   Chukova, S. S. (1988). *Mathematical Models and Methods for Determining Optimal Warranty Periods*, Doctoral Dissertation, Sofia University (in Bulgarian). (OPT, STA, WTE)

C27.   Chukova, S. S. (1990). Classification of the warranty models, *Proc. of XV National Seminar in Stochastics*, 347–350. (in Bulgarian). (WCA)

C28.   Chukova, S. S. (1990). On the determination of the warranties for complex items, *Math. and Math. Educ.*, 425–430. (REL)

C29.   Chukova, S. S., and Khalil, Z. (1990). On the conservative T-screening of nonstationary Poisson processes and optimal warranties, *J. Boul. Acad. of Science*, **43**. (OPT, STO)

C30.   Chun, Y. H., and Moskowitz, H. (1991). Model specifications and parameter estimations in the two-attribute warranty models, presented at *ORSA/TIMS 32nd Joint National Meeting*, Anaheim, CA. (EST, TDW)

C31.   Clark, A. M. (1989). *Product Liability*, Sweet and Maxwell, London. (PLI)

C32.   Clark, B., and Davis, M. J. (1975). Beefing up product warranties: A new dimension in consumer protection, *Kansas Law Review*, **23**, 567–617. (CPT)

C33.   Clarke, G. R. (1989). *Product Liability: An Examination of the Doctrine of Forum non Conveniens on Australian Litigants in the United Stated of America in Light of the Decisions in Piper Aircraft Co., v. Reyno and Tokio Marine and Fire Insurance v. Bell Helicopter Textron*, Law Thesis, University of Queensland, Australia. (PLI)

C34.   Cline, C. G. and Cline, M. D. (1987). The gerber baby cries foul, *Business and Society Rev.*, **Winter**, 14–19. (PRC, PLI)

C35.   Close, D. B. (1987). The risk manager's role in preventing product liability claims, *Risk Management*, **34**, 36–40. (PLI, PQU, WMA)

C36.   Close, G. R. (1968). Products liability defenses, management's role, *Trans. Bul 33rd Ann. Meeting, Indust. Hygiene Found. of Amer*, 47–54. (PLI)

C37.   Clutterbuck, J. B. (1988). Karl Llewellyn and the intellectual foundations of enterprise liability theory, *Yale Law J.*, **97**, 1131–1151. (PLI, TOR)

C38.   Coates, W. (1985). Comparison of bias and radial tractor tires on a soft soil, *Trans. American Society of Agricultural Engineers*, **28**, 1090–1093. (DCV)

C39.   Cocco, J. J., Callanan, D., and Bassinger, T. (1992). DFMA quantifies value of adhesive assembly options, *Adhesives Age*, **35**, 28–31. (PQU, WCA)

C40.   Coffey, T. W. (1987). Creating express warranties under the UCC: Basis of the bargain—don't rely on it, *Uniform Commercial Code Law J.*, **20**, 115–129. (LIT, WAP)

C41.   Cohen, J. (1980). Life cycle performance of clothes dryers, *1980 Proc. Ann. Reliab. and Maintain. Symp.*, 91–94. (PQU)

C42. Cohen, M. A. and Lee, H. L., (1990). Out of touch with customer needs? Spare parts and after sales services, *Sloan Management Rev.*, **Winter**, 55–66. (SST, WAP)

C43. Colangelo, V. J. and Thornton, P. A. (1981). *Engineering Aspects of Product liability*, American Society for Metals; Metals Park, Ohio. (PLI)

C44. *Common Cause* (1991). Car Trouble, **Jan/Feb.**, 7. (WAP)

C45. *Communications of the ACM* (1993). Liability categories, **36**, 24–25. (BOC, PLI, TOR)

C46. Conference Board Report (1971). *The Challenge of Consumerism*, New York. (CPT)

C47. Conference Board Report (1980). *Industrial Product Warranties: Policies and Practice*, New York. (WAP)

C48. Congressional Quarterly, Inc. (1982). *Regulation: Process and Politics*, Congressional Quarterly, Inc., Washington, DC. (REG, WLE)

C49. Conley, D. L. (1970). Quality and reliability, let's better understand their respective roles, *Quality Assurance*, **May**, 40–41. (QAS, REL)

C50. Consumer Advisory Council (1963). *Consumer Advisory Council, First Report*. Congressional Quarterly, Inc., Washington, DC. (COM)

C51. *Consumer Bulletin* (1963). 'Extended' warranties and service contracts, **Jun.**, 27–28. (EXT, SCT)

C52. *Consumer News* (1985). FTC proposes to ease product warranty rules, **6**. (WLE)

C53. *Consumer Reports*, (1965). **Apr.**, 173–176. (COM)

C54. *Consumer Reports*, (1967). **Apr.**, 194–197. (COM)

C55. *Consumer Reports* (1968). **Apr.**, 176–179. (COM)

C56. *Consumer Reports* (1969). **Apr.**, 177–181. (COM)

C57. *Consumer Reports* (1970). **Apr.**, 201–204. (COM)

C58. *Consumer Reports* (1963). Warranties: The consumer's rights on paper and in the law, **Apr.**, 158–159. (CPT)

C59. *Consumer Reports* (1964). The long warranties: Essentially desirable, they throw a few curves, **Apr.**, 165–166. (WTE)

C60. *Consumer Reports* (1965). Oh, what they do with the fine print, **Aug.**, 378. (CEI)

C61. *Consumer Reports* (1967). Warranties: The promises and the reality, **Apr.**, 194–197. (COM, CPT)

C62. *Consumer Reports* (1970). Some variable cost of ownership, repair, insurance and warranties, **Apr.**, 201–204. (WCA)

C63. *Consumer Reports* (1971). Buyer versus seller in small claims courts, **36**, **Apr.**, 624–631. (LIT)

C64. *Consumer Reports* (1972). Color TV consoles, **Jan.**, 13–14. (WAP)

C65. *Consumer Reports* (1975). Product warranties: Congress lends a helping hand, **Apr.**, 164–165. (WLE)

C66. *Consumer Reports* (1981). 19-inch color TVs, **Jan.**, 34–39. (WAP)

C67. *Consumer Reports* (1984). Is that warranty any good?, **Jul.**, 408–410. (CEI)

C68.   *Consumer Reports* (1986). Auto service contracts. **Oct.**, 663–667, 685. (SCT, WAP)

C69.   *Consumer Reports* (1989). Is your car's warranty a secret?, **Apr.**, 214–215. (WAP)

C70.   *Consumer Reports* (1990). Insure against future repairs?, **Apr.**, 226–227. (CEI)

C71.   *Consumer Reports* (1991). Who needs an extended warranty?, **Jan.**, 21–22. (EXT)

C72.   *Consumer Reports* (1993). The sour truth about lemon laws?, **Jan.**, 40–42. (CEI)

C73.   *Consumer Reports*, (1969). Guarantees and warranties, 391–392. (CEI)

C74.   Consumer Warranty Proposals and FTC Amendments (1977). *Legislative Analysis*, 92nd Congress The Amer. Enterprise Inst. for Pub. Policy Research, Washington, DC, **14, Oct. 25**, 30. (WLE)

C75.   Cook, P. J., and Graham, D. A. (1977). The demand for insurance and protection: The case of irreplaceable commodities, *Quart. J. Economics*, **91**, 143–156. (CPT)

C76.   Cooke, E. F. (1987). Post-shipment services: Turning customer complaints into assets, *J. Business and Industrial Marketing*, **2**, 17–22. (CCO, WAM)

C77.   Coop, D. D. (1971). A new feature in the marketing environment, the law of product liability, *Mississippi Business Rev.*, 3–7. (LAW, PLI)

C78.   Cooper, R., and Ross, T. W. (1985). Product warranties and double moral hazard, *RAND J. of Economics*, **16**, 103–113. (MAM, MOR, PQU)

C79.   Cooper, R., and Ross, T. W. (1988). An intertemporal model of warranties, *Canadian J. of Economics*, **21**, 72–86. (CWP, ETH, MAM, OPT)

C80.   Cooper, T. E. (1992). Signal facilitation: A policy response to asymmetric information, *J. Business*, **65**, 431–450. (MAM, PQU)

C81.   Cooter, R. D., and Rubinfeld, D. L. (1989). Economic analysis of legal disputes and their resolution, *J. Economic Literature*, **27**, 1067–1097. (CMD, EAW)

C82.   Cope, S. D., and Pelletier, R. (1991). Whipping the 'warranty factor', *Sales and Marketing Management*, **143**, 105–106. (CEI, CSD, PQU)

C83.   Cornell, C. (1991). Extended warranties: A piece of the new action, *Dealerscope of Merchandising*, **33**, 30, 32. (ACC, EXT)

C84.   Cornell, C. (1992). Morality play, *Dealerscope of Merchandising*, **34**, 44–48. (CPT, EXT, SCT, WLE)

C85.   Cornell, C. (1991). The direct approach, *Dealerscope of Merchandising*, **34**, 107–110. (CBH, EXT, SCT, WAM)

C86.   Cornell, C. (1993). Once in a lifetime, *Dealerscope of Merchandising*, **35**, 86–90. (CPT, EXT)

C87.   Corner, P. (1984). Management of quality in a large procurement organization, *Proc. World Quality Congress '84*, Brighton, England, **1**, 167–174. (CPT, QAS, QCO, WLE)

C88.   Cortez, J. P. (1992). Why Domino's delivery guarantee may change, *Advertising Age*, **63**, 44. (WAM)

C89. Corwin, S. S. (1988). Lead-calcium batteries are time sensitive, *Electrical Construction and Maintenance*, **87**, 45. (WAP)

C90. Cosgrove, N. D. (1992). With copiers, it's service that counts, *Office*, **115**, 65–67. (EXT, REL)

C91. Council of the European Communities (1985). Directive on the approximation of the laws, regulations and administrative provisions of the member states concerning liability for defective products, *Official J. of the European Communities*, **L 210/29**, (7 August). (LAW, PLI, REG)

C92. Courville, L., and Hausman, W. H. (1978). Theoretical modelling of warranties under imperfect information, in *Consumer Durables: Warranties, Service Contracts, and Alternatives*, (R. T. Lund, ed.), Rept. No. CPA-78-14/Vol. IV, Center for Policy Alternatives, Massachusetts Institute of Technology, Cambridge, MA, 4-146–4-193. (EAW, MAM)

C93. Courville, L., and Hausman, W. H. (1979). Warranty scope and reliability under imperfect information and alternative market structures, *J. Business*, **52**, 361–378. (REL, WMS, WPR)

C94. Cox, D. F. (1967). *Risk Taking and Information Handling in Consumer Behavior*, Harvard University Press, Boston, MA. (CBH)

C95. Cox, T. D. (1990). High-reliability through systems design and quality practices, *Proc. 36th Ann. Technical Meeting of the Inst. of Environmental Sciences*, 729–736. (QAS, TQM, WAP)

C96. Coyne, C., and Stallard, K. (1992). *Product Liability/Class Actions*, Queensland Law Society, Brisbane, Australia. (PLI)

C97. Crocker, K. J. (1986). A reexamination of the "lemons" market when warranties are not prepurchase quality signals, *Information Economics and Policy*, **2**, 147–162. (PQU, WMS)

C98. Cross, P. M., and Heslop, L. A. (1981). Reliability and maintainability from a consumer perspective, 1981, *Proc. Ann. Reliab. and Maintain. Symp.*, 199–205. (MAI, REL)

C99. Crum, F. B., Dallosta, P. M., and Underwood, J. D. (1977). *Establishment and Operation of a Pilot Warranty Information Center*, Tech. Report No. 1184-2-1-1612, ARINC Research Corp., Annapolis, MD. (CEI, RIW)

C100. Cullen, D. (1990). Warranty wars: How goes the battle?, *Fleet Owner*, **85**, 97–100. (EXT, FLW)

C101. Cullen, D. (1992). Extended warranties: What are they worth?, *Fleet Owner*, **87**, 69–72. (EXT, FLW)

C102. Curnes, E. J. (1987). Protecting the Virginia homebuyer: A duty to disclose defects, *Virginia Law Rev.*, **73**, 459–481. (CPT, PQU)

C103. Czech, D. R., and Dizek, S. G. (1988). Artful analysis—the key to successful warranties, *IEEE 1988 Eng. Manage. Conf. Eng. Leadership in the 90's from AI to JZ*, 51–58. (ACC, WSL)

D1. Dahringer, L. D., and Johnson, D. R. (1988). Lemon laws: Intent, experience, and a pro-consumer model, *J. Consumer Affairs*, **22**, 158–170. (CPT, LAW)

D2.   Daisley, P. A., Plunkett, J. J., and Dale, B. G. (1984). Quality costing in the U.K., *Proc. World Quality Cong. 1984*, **1**, 557–567. (QCO, WCA)

D3.   Daly, J. (1988). Want a free warranty? Don't go to DEC, *Computerworld*, **22**, 14. (WPR, WSL)

D4.   Dagpunar, J. S., and Jack, N. (1994). Preventive maintenance strategy for equipment under warranty, *Microelectron. Rel*, **34**, 1089–1093. (MAI)

D5.   Darby, M. R., and Karni, E. (1973). Free competition and the optimal amount of fraud, *J. of Law and Economics*, **16**, 67–88. (CEI, FRA)

D6.   Darden, W. R., and Rao, C. P. (1977). Satisfaction with repairs under warranty and perceived importance of warranties for appliances, in *Consumer Satisfaction, Dissatisfaction and Complaining Behavior* (R. L. Day, ed.), Department of Marketing, School of Business, Indiana University, Bloomington/Indianapolis, 167–170. (CSD)

D7.   Darden, W. R., and Rao, C. P. (1979). A linear covariate model of warranty attitudes and behaviors, *J. Marketing Research*, **16**, 466–477. (CBH, MAM)

D8.   Dardis, R., and Zent, C. (1982). The economics of the Pinto recall, *J. Consumer Affairs*, **16**, 261–277. (ETH, MAM, PRC)

D9.   Daugherty, G. (1989). Contracting for supportability and affordability, *1989 Proc. Ann. Reliab. and Maintain. Symp.*, 321–327. (PSU, WCA)

D10.   David, Y. (1985). Medical instrument life cycle strategy for clinical engineers, *IEEE Eng. Medicine and Biology*, **4**, 25–26. (WAP)

D11.   Day, E., and Fox, R. J. (1985). Extended warranties, service contracts and maintenance agreements—a marketing opportunity?, *J. Consumer Marketing*, **2**, 77–86. (EXT, MAI, SCT, WAM)

D12.   Day, G. S. (1978). Are consumers satisfied?, in *Consumerism: Search for the Consumer Interest*, (D. A. Aaker and G. S. Day, eds.), 3rd ed., The Free Press, New York, 406–417. (CSD)

D13.   Day, R. C., and McIntyre, L. E. (1978). RIW data collection and reporting method, *1978 Proc. Ann. Reliab. and Maintain. Symp.*, 66–72. (RIW, WDA)

D14.   Day, R. L., (1980). Research perspectives on consumer complaining behavior, in *Theoretical Developments in Marketing*, (C. W. Lamb and P. M. Dunne, eds.), American Marketing Association, Chicago, IL, 211–215. (CBH, CCO, CSD)

D15.   Day, R. L., and Ash, S. B. (1978). Comparison of patterns of satisfaction/dissatisfaction and complaining behavior for durables, nondurables, and services, in *New Dimensions of Consumer Satisfaction and Complaining Behavior*, (R. L. Day and H. K. Hunt, eds.), Division of Research, School of Business, Indiana University, Bloomington, IN. 190–195. (CCO, CSD)

D16.   Day, R. L., and Hunt, H. K. (1978). *New Dimensions of Consumer Satisfaction and Complaining Behavior*, Indiana University, Dept. of Marketing, School of Business, Bloomington, IN. (CCO, CSD)

D17.   Day, R. L., and Hunt, H. K. (1980). *New Findings on Consumer Satisfaction and Complaining*, Indiana University, Dept. of Marketing, School of Business, Bloomington, IN. (CCO, CSD)

D18. Day, R. L., and Landon, E. L. (1975). Collecting comprehensive consumer complaint data by survey research, in *Advances in Consumer Research*, **3**, (B. B. Anderson, ed.), Association for Consumer Research, Ann Arbor, MI, 263–268. (CCO, CSD)

D19. Day, R. L., and Landon, E. L. (1977). Towards a theory of consumer complaining behavior, in *Foundations of Consumer and Industrial Buying Behavior*, (A. G. Woodside, J. N. Sheth, and P. D. Bennet, eds.), American Elsevier, 425–437. (CBH, CCO)

D20. Day, R. L., Grabieke, K., Schaetzle, T., and Staubach, F. (1981). The hidden agenda of consumer complaining, *J. Retailing*, **57**, 86–106. (CCO)

D21. *Dealerscope Merchandising* (1993). WCES wrap-up: Warranties, **35**, 48–51. (WAP)

D22. *Dealerscope Merchandising* (1993). WCES Wrap-up: Warranties, **35**, 78. (EXT)

D23. Dean, C. (1977). Evolution of products liability, 1977, *Proc. Ann. Reliab. and Maintain. Symp.*, 21–22. (PLI, WLE)

D24. Dean, C. and Smith, C. O. (1978). Products liability, legal issues and technical answers, *1978 Proc. Ann. Reliab. and Maintain. Symp.*, 355–364. (PLI, WLE)

D25. DeCroix, G. A. (1991). Optimal price and warranty in competitive markets, presented at *ORSA/TIMS 32nd Joint National Meeting*, Anaheim, CA. (WMS, WPR)

D26. DeCroix, G. (1994). Existence and computation of equilibrium warranties, reliabilities and prices for durable goods, *Management Science*, (in press). (EAW, OPT, REL, WPR)

D27. Decker, R. (1988). Warranties are like bananas: They often come in bunches, *Purchasing*, **105**, 90, 92. (CEI, LIT, TYW)

D28. Decker, R. (1989). Promptness is special virtue if statute of limitations applies, *Purchasing*, **107**, 92A11–92A12. (BOC, LIT)

D29. Decker, R. (1992). How warranties, discounts, GPO activities are governed by safe harbor regulations, *Hospital Materials Management*, **17**, May., 26–27. (REG, WAP)

D30. Decker, R. (1992). Hospitals must negotiate warranty needs such as service, spare parts, early in process, *Hospital Materials Management*, **17**, Sep., 26–27. (WAP, WLE)

D31. Deep, R., Czech, D. R., Dizek, S. G., and Kennedy, D. K. (1988). Bit-mapping classifier expert system in warranty selection, *IEEE Proc. National Aerospace and Electronics Conference 1988—NAECON 1988*, 1222–1224. (WSL)

D32. Deering, A. M. (1990). Launch insurance procurement: It's a new ball game, *Satellite Communications*, **14**, 33–37. (TYW, WSL)

D33. Deierlein, B. (1988). Warranties: More important than ever, *Fleet Equipment*, **14**, 38–40. (FLW, WCA, WDA)

D34. Deierlein, B. (1989). Warranties: Changing as fast as technology, *Fleet Equipment*, **15**, 37–39. (EXT, FLW, WCA)

D35. Deierlein, B. (1993). Warranty: Get all you deserve, *Fleet Equipment*, **19**, 50–54. (EXT, FLW)

D36.   Del Nibletto, P. (1993). Compaq upgrades warranty program, *Info Canada*, **18**, 20. (WAP)

D37.   Del Prete, D. (1991). Looks like a hot summer for extended car warranties, *Marketing News*, **25, Jul.**, 9. (SCT, WPR)

D38.   Derr, J. H. (1985). Reliability prediction of automotive electronics—how well does MIL-HDBK-217-D stack up?, *Int. Congress and Exposition—Society of Automotive Engineers*, Detroit, MI. (EST, REL)

D39.   Derr, J. H. and Louch, R. J. (1991). Advanced methodology for projecting field repair rates and maintenance costs for vehicle electronic systems, *Soc. Automotive Engineers Trans.*, **100**, 1–11. (MAI, PDE, SIM, WCA)

D40.   Derr, J. H., Straub, C. M., and Ahmed, S. (1987). Prediction of wiring harness reliability, presented at the *SAE Int. Congress and Exposition*, Publ. by SAE, Warrendale, PA, 45–53. (REL, STA, WDA)

D41.   Deutsch, I. (1980). The responsibility of the consumer, the pendulum has swung too far, *1980 Proc. Ann. Reliab. and Maintain. Symp.*, 468–469. (CPT)

D42.   Devience, A. Jr. (1990). Magnuson-Moss Act: Substitution for UCC warranty protection?, *Commercial Law J.*, **95**, 323–337. (CPT, LAW, LIT)

D43.   Devlin, F. C. Jr. (1993). IRS rules on extended warranties, *Tax Adviser*, **24**, 162–163. (ACC, EXT)

D44.   Dew, L. L., and Childers, M. A. (1989). MODU drilling rig downtime: An objective analytical approach, *SPE/IADC 1989 Drilling Conference*, 383–394. (DCV)

D45.   Dhillon, B. S. (1981). Life cycle cost: A survey, *Microelectron. Reliab.*, **21**, 495–511. (LCC, RIW)

D46.   Dhillon, B. S. (1983). *Reliability Engineering in System Design and Operation*, Van Nostrand Reinhold Co., New York. (REL)

D47.   Dhillon, B. S. (1988). Data collection sources, in *Mechanical Reliability: Theory, Models and Application*, American Institute of Aeronautics and Astronautics, Inc., Washington, DC, 160–162. (WDA)

D48.   Dhillon, B. S. (1989). Warranties, in *Life Cycle Costing: Techniques, Models and Applications*, Gordon and Breach Science Pub., New York, 140–164. (LCC, RIW, WCA, WTE)

D49.   Dillian, J. A. (1980). Products liability panel session, *1980 Proc. Ann. Reliab. and Maintain. Symp.*, 475–478. (PLI)

D50.   Dietz, R. (1991). European automotive industry's design and materials philosophy for achieving corrosion protection today, and over the coming decade, *Proc. 5th Automotive Corrosion and Prevention Conf.*, 1–6. (WAP)

D51.   Dilts, D. M., and Russel, G. W. (1989). Measuring external failure: The four costs of external failure, *43rd Ann. Quality Congress Trans.*, 317–322. (CSD, QCO, WCA)

D52.   Dimitrov, B. N., and Chukova, S. S. (1991). Warranty analysis, *Economics Quality Control J. and Newsletter for Quality and Reliability*, **6**, 61–82. (CUW, WCA)

D53.   DiPaolantonio, J. A., Kohaya, R. H., Miyamoto, R. T., and Waller, B. E. (1986). Case report; R & M support of CLS warranties, *1986 Proc.*

Ann. Reliab. and Maintain. Symp., 457–459. (WCA, WSL)

D54. *Distribution* (1991). How to buy a used lift truck, **90**, 78, 80. (WAP)

D55. Dizek, S. G. (1986). Automated decision support for warranty selection, *1986 Proc. Ann. Reliab. and Maintain. Symp.*, 460–465. (RIW, WSL)

D56. Djamaludin, I. (1993). *Quality Control Schemes for Items Sold with Warranty*, Doctoral Dissertation, The University of Queensland, Brisbane, Australia. (FRW, MAM, PRW, QCO, WCA)

D57. Djamaludin, I., Murthy, D. N. P., and Wilson, R. J. (1991). Product warranty and quality control, presented at *ORSA/TIMS 32nd Joint National Meeting*, Anaheim, CA. (FRW, PRW, QCO, WCA)

D58. Djamaludin, I., Murthy, D. N. P., and Wilson, R. J. (1994). Quality control through lot sizing for items sold with warranty, *Int. J. Prod. Econ.*, **33**, 97–107. (FRW, PRW, QCO, WCA)

D59. Djamaludin, I., Wilson, R. J., and Murthy, D. N. P. (1994). Lot sizing and testing for items with uncertain quality, *Math. and Comp. Modelling*, (in press). (FRW, PRW, QCO, WCA)

D60. Dobbins, R. K. (1991). Business leadership through world class quality costs, *45th Ann. Quality Congress Trans.*, 475–481. (QCO, WCA)

D61. Dodge, D. A. (1985). Product safety and liability prevention, *Professional Safety*, **30**, 25–30. (PLI, TOR, WMA)

D62. Dorfman, R. (1970). The economics of product liability: A reaction to McKean, *University of Chicago Law Rev.*, **38**, 92–102. (EAW, PLI)

D63. Dorking, B. (1993). Merchantable quality in new motor vehicles, *Australian Business Law Review*, **21**, 163–167. (CPT, PQU, WCT)

D64. Dorsky, L. R. (1985). Management measures for customer satisfaction, *1985 Proc. Ann. Reliab. and Maintain. Symp.*, 13–16. (CSD)

D65. Douglas, E. J., Glennon, D. C., and Lane, J. I. (1993). Warranty, quality and price in the US automobile market, *Applied Economics*, **25**, 135–141. (ETH, PQU, WMS, WPR)

D66. Duhig, J. J. Jr., and South, J. L. (1978). Cost effective improvement of the timeless C-130 Hercules Airlifter, *1978 Proc. Ann. Reliab. and Maintain. Symp.*, 422–427. (RIW, WCA)

D67. Dungworth, T. (1988). *Product Liability and the Business Sector*, RAND Corp., Santa Monica, CA. (PLI)

D68. Dwek, R. (1993). The big cat's on the prowl, *Marketing*, **Apr. 1,** 24–25. (PDV, WAM)

D69. Dybvig, P. H. and Lutz, N. A. (1993). Warranties, durability, and maintenance: Two-sided moral hazard in a continuous-time model, *Review of Economic Studies*, (in press). (MAI, MAM, MOR, PQU)

D70. Dzirkal, E. V. and Lesley, V. (1986). Reliability of items and warranty commitments for supplier, *Reliability and Quality Control*, **12**, (in Russian). (REL, WTE)

E1. Ebright, A. H. (1961). *A Survey of the Historical Setting—Past and Present—of the Law of Warranty*, Master of Law Thesis, University of Southern California, Los Angeles, CA. (LAW)

E2.   Edwards, R. L. (1986). Legal aspects of lower extremity injuries: An overview of products liability law in the 1980's, *Proc. Society Automotive Engineers*, **P-186**, 171–180. (BOC, LAW, PLI)

E3.   Ellis, D. E. (1990). Extended service coverage cost predictions, *1990 Proc. Ann. Reliab. and Maintain. Symp.*, 233–236. (WCA, WDA)

E4.   Emons, W. (1988). Warranties, moral hazard, and the lemons problem, *J. Econ. Theory*, **46**, 16–33. (COR, CWP, MOR, PRS, WPR)

E5.   Emons, W. (1989). The theory of warranty contracts, *J. Econ. Surveys*, **3**, 43–57. (PQU, WAM, WCT, WPR)

E6.   Emons, W. (1989). On the limitation of warranty duration, *J. Indust. Economics.*, **37**, 287–331. (ETH, MAM, WTE)

E7.   Endicott, G. A., and Green, L. (1990). Parts and labor included. . . , *Equipment Management*, **18**, 39–42. (WCT)

E8.   Epple, D., and Raviv, A. (1978). Product safety: Liability rules, market structure, and imperfect information, *American Economic Review*, **68**, 80–95. (PLI, PQU, WMS)

E9.   Epstein, D. G., and Nickless, S. H. (1981). *Consumer Law in a Nutshell*, West Publishing Co., St. Paul, MN. (LAW)

E10.   Eovaldi, T. L., and Gestrin, J. E. (1971). Justice for consumers: The mechanisms of redress, *Northwestern University Law Review*, **66**, 281–325. (CPT)

E11.   Erevelles, S. (1993). The price-warranty contract and product attitudes, *J. Business Research*, **27**, 171–181. (CBH, STA, WPR)

E12.   Ericksen, P. D., and Skinner, D. A. (1984). Comprehensive reliability testing of components, *1984 Winter Meeting—American Society of Agricultural Engineers*, No. 84-1639, 11p. (WDA)

E13.   Erke, Y., and Thomas, M. U. (1991). Opportunistic replacement decisions for a system with warranted components, presented at *ORSA/TIMS 32nd Joint National Meeting*, Anaheim, CA. (FRW, MAM, REP)

E14.   Erto, P., and Guida, M. (1986). Some maximum likelihood reliability estimates from warranty data of cars in user operation, in *Reliability Technology—Theory and Applications*, (J. Moltoft and F. Jensen, eds.), Elsevier Science Pubs., North-Holland, 55–58. (EST, MAM, WDA)

E15.   Evans-Correia, K. (1992). Quality means the right fit, *Purchasing*, **112**, 133–138. (CSD, PQU)

E16.   Evans-Correia, K. (1993). Are guarantees proof of product quality? *Purchasing*, **115**, 95–96. (CSD, PQU)

E17.   Everret, M., (1990). More companies are gambling on service guarantee, *Sales Mark. Manag.*, **142**, 105. (SCT)

F1.   FASB (1976). *Financial Accounting Standards Board, Statement No 5: Accounting for Contingencies*; Stamford, CT. (ACC)

F2.   FASB releases Tech. Bulletin in extended warranties (1991). *Prac. Account*, **24**, 67. (EXT)

F3.   Farrell, J. (1986). Moral hazard as an entry barrier, *RAND J. Econ*, **17**, 440–449. (COR, MOR)

F4.   Fealk, M. N. (1987). What is a premature product failure?—A legal view, *Spring National Design Engineering Show and Conference*, 55–72. (PLI, REL, QCO)

F5.   *Federal Register* (1979). Uniform product liability law, interagency task force on product liability draft, **44**, Jan. **12**. 2996–3019. (LAW, PLI)

F6.   Federal Trade Commission (1968). *Federal Trade Commission Staff Report on Automobile Warranties*, Federal Trade Commission, Washington, DC. (CPT, WAP)

F7.   Federal Trade Commission (1968). *Report of the Task Force on Appliance Warranties and Service*, Federal Trade Commission, Washington, DC. (CPT, WAP)

F8.   Federal Trade Commission (1974). *Warranties Rules Consumer Follow-up: Evaluation Study Final Report*, Office of Impact Evaluation, Federal Trade Commission, Washington, DC. (CPT)

F9.   Federal Trade Commission (1979). *Warranties Rules: Warranty Content Analysis*, Federal Trade Commission, Washington, DC. (WTE)

F10.   Federal Trade Commission (1979). *Warranties Rules Consumer Baseline Study: Final Report*, Federal Trade Commission, Washington, DC. (CPT)

F11.   Federal Trade Commission (1985). *An Evaluation of the Warranty Rules of the Magnuson-Moss Act*, Federal Trade Commission, Washington, DC. (WLE)

F12.   Fedraw, K., and Becker, J. (1983). Impact of thermal cycling on computer reliability, *1983 Proc. Ann. Reliab. and Maintain Symp.*, 149–153. (REL)

F13.   Feldman, L. P. (1976). *Consumer Protection: Problems and Prospects*, West Pub. Co, New York. (CPT)

F14.   Feldman, L. P. (1976). New legislation and the prospects for real warranty reform, *J. Marketing*, **40**, 41–47, reprinted in *Consumerism: Search for the Consumer Interest*, (D. A. Aaker and G. S. Day, eds.), 3rd ed., The Free Press, New York, 353–366. (CEI, WLE)

F15.   Fendell, R. (1976). GM launches warranty audit, *Automotive News*, **Apr. 12**, 1. (WAP)

F16.   Fenton, J. (1984). Long term corrosion prevention, *Automotive Engineer*, **9**, 37–39. (PQU, WAP)

F17.   Finsinger, J., and Simon, J. (1992). *The Harmonisation of Product Liability Laws in Britain and Germany: An Applied Legal-Economic Analysis/London Analysis*, Anglo-German Foundation, London. (LAW, PLI)

F18.   Firnstahl, T. W. (1989). My employees are my service guarantee, *Harvard Business Rev.*, **67**, 28–34. (CSD)

F19.   Fisk, G. (1970). Guidelines for warranty service after sale, *J. Marketing*, **34**, 63–67. (CCO, CEI, CSD, PQU)

F20.   Fisk, G. (1973). Systems perspective on automobile and appliance warranty problems, *J. Consumer Affairs*, **7**, 37–54. (CPT, LIT)

F21.   Fisk, G., and Chandran, R. (1975). How to trace and recall products, *Harvard Business Rev.*, **53**, 90–96. (PRC)

F22.   *Fleet Equipment* (1988). Cost control, **14**, 30–35. (FLW, WCA)

F23.   *Fleet Equipment* (1988). Cost control, **14**, 40–43. (FLW, WCA)

F24.   Fleig, N. G. (1986). The United States air force product performance agreement center, *1986 Proc. Ann. Reliab. and Maintain. Symp.*, 449–453. (REL, RIW, WDA)

F25.   Flottman, W. W., and Worstell, M. R. (1977). Mutual development, application and control of suppliers warranties, *1977 Proc. Ann. Reliab. and Maintain. Symp.*, 213–221. (WAD, WAP)

F26.   Folse, R. O. (1977). *Quantification of Selection Criteria for Reliability Improvement Warranty Contracts*, Doctoral Dissertation, Dept. of Quantitative Methods, Louisiana State University and Agricultural and Mechanical College, Baton Rouge, LA. (RIW, WCA)

F27.   Fornell, C., (1978). Corporate consumer affairs departments, a communications perspective, *J. Consumer Policy*, **2**, 289–302. (CCO, CEI)

F28.   Fornell, C., and Wernerfelt, B. (1987). Defensive marketing strategy by customer complaint management: A theoretical analysis, *J. Marketing Res.*, **24**, 337–346. (CCO, CSD)

F29.   Fornell, C., and Wernerfelt, B. (1988). A model for customer complaint management, *Marketing Sci.*, **7**, 287–298. (CCO, MAM)

F30.   *Fortune* (1986). Guarantee gamble, **Oct. 13**, 9. (EXT)

F31.   Fram, E. U. (1985). Ready for a time guarantee?, *J. Consumer Marketing*, **2**, 29–35. (TYW)

F32.   Freeman, D. B. (1987). Recent advances in motor vehicle body pretreatment and painting, *Product Finishing*, **40**, 6–7, 14. (EXT, PQU)

F33.   Freeman, L. (1992). Service contracts: Warranties impact bottom line, *Stores*, **74**, 122, 124. (ACC, ESC)

F34.   Frees, E. W. (1986). Warranty analysis and renewal function estimation, *Naval Res. Logist. Quart.*, **33**, 361–372. (EST, RFT, WCA)

F35.   Frees, E. W. (1988). Estimating the cost of a warranty, *J. Business and Economic Statistics*, **6**, 79–86. (EST, MAM, RFT, WCA)

F36.   Frees, E. W. (1991). *Semiparametric Estimation of Warranty Costs*, Working Paper, School of Business, University of Wisconsin, Madison. (EST, PRM, WDA)

F37.   Frees, E. W., and Nam, S. H. (1988). Approximating expected warranty costs, *Management Sci.*, **34**, 1441–1449. (EST, MAM, RFT, WCA)

F38.   Freimüller, H. U. (1988). *Due Diligence, Disclosures and Warranties in the Corporate Acquisitions Practice*: Graham and Trotman, London. (CEI, PLI, WLE)

F39.   Frisch, D., and Wladis, J. D. (1988). General provisions, sales, bulk transfers, and documents of title, *Business Lawyer*, **43**, 1259–1304. (LAW, LIT, TOR)

F40.   Frisch, D., and Wladis, J. D. (1990). General provisions, sales, bulk transfers, and documents of title, *Business Lawyer*, **45**, 2289–2330. (FRA, LAW, LIT, TOR)

F41.   Frisch, D., and Wladis, J. D. (1991). General provisions, sales, bulk transfers, and documents of title, *Business Lawyer*, **46**, 1455–1508. (LAW, LIT, TOR)

F42.   Frisch, D., and Wladis, J. D. (1992). General provisions, sales, bulk transfers, and documents of title, *Business Lawyer*, **47**, 1517–1543. (LAW, LIT)

F43. Fromson, D. (1985). Product liability and the use of codes and standards, *Pressure Vessel and Piping Technol. 1985: A Decade of Prog.*, 951–964. (PLI, WLE)

F44. Frumer, L. R., and Friedman, M. I. (1960). *Products Liability*, Mather Bender, Albany, NY. (PLI)

F45. Fucini, S. (1987). Service contracts: A dream or a nightmare?, *Mart*, December, 4–6. Cited in *Consumers Union News Digest* (1987), **13**, 1, 12. (SCT)

G1. Gal-Or, E. (1989). Warranties as a signal of quality, *Canadian J. Economics*, **22**, 50–61. (ETH, MAM, PQU)

G2. Galanter, M. (1975). Afterword: Explaining litigation, *Law and Society Rev.*, **9**, 347–368. (LIT)

G3. Galvin, R. J. (1985). Computer aided system for maintenance and overhaul of rollingstock, in *Electrification—Railways to the year 2000*, Conference on Railway Engineering 1985, 141–145. (MAI, PDS)

G4. Gándara, A., and Rich, M. D. (1977). *Reliability Improvement Warranties for Military Procurement*, R-2264-AF, the RAND Corp., Santa Monica, CA. (RIW)

G5. Gardner, T. F. (1988). Energy conservation warranties within the national energy conservation policy act, *American Society of Mechanical Engineers, Petroleum Division*, **17**, 49–59. (CTR, WAP)

G6. Garland, D. W. (1993). Determining whether a nonconformity substantially impairs the value of goods: Some guidelines, *Uniform Commercial Code Law J.*, **26**, 129–143. (CPT, PQU)

G7. Gates, R. K., Bicknell, R. S., and Bortz, J. E. (1977). Quantitative models used in the RIW decision process, *1977 Proc. Ann. Reliab. and Maintain. Symp.*, 229–236. (RIW, WCA, WDA)

G8. Gates, R. K., Bortz, J. E., and Bicknell, R. S. (1976). A quantitative analysis of alternative RIW implementations, *Proc. Nat. Aerospace and Electron. Conf.*, 1–8. (EST, RIW, WPR)

G9. Geddes, A. (1992). *Product and Service Liability in the EEC: The New Strict Liability Regime*, Sweet and Maxwell, London. (PLI)

G10. Geistfeld, M. (1988). Imperfect information, the pricing mechanism, and product liability, *Columbia Law Rev.*, **88**, 1057–1072. (PLI, WPR)

G11. Gemignani, M. C. (1990). Liability for malfunction of an expert system, *Proc. 1990 IEEE Conf. Managing Expert System Programs and Project*, 8–15. (BOC)

G12. Gerard, J. M. (1985). *Studies in Market Signaling: The Firm's Selection of Finance and Consumer Product Warranties*, Doctoral Dissertation, California Institute of Technology, Pasadena, CA. (PQU, WAM)

G13. Gerner, J. L., and Bryant, W. K. (1978). Appliance warranties: Provisions and changes over time, in *Consumer Durables: Warranties, Service Contracts, and Alternatives*, (R. T. Lund, ed.), Rept. No. CPA-78-14/Vol. II, Center for Policy Alternatives, Massachusetts Institute of Technology, Cambridge, MA, 2-5–2-45. (WDA)

G14. Gerner, J. L., and Bryant, W. K. (1978). The Magnuson-Moss Act, FTC rules, and major appliance warranties, in *Consumer Durables: Warranties, Service Contracts, and Alternatives,* (R. T. Lund, ed.), Rept. No. CPA-78-14/Vol. II, Center for Policy Alternatives, Massachusetts Institute of Technology, Cambridge, MA, 138–165. (WDA, WLE)

G15.   Gerner, J. L., and Bryant, W. K. (1980). The demand for repair service during warranty, *J. Businesss*, **53**, 397–414. (CBH, MAM, PRM, SST)

G16.   Gerner, J. L., and Bryant, W. K. (1981). Appliance warranties as a market signal?, *J. Consumer Affairs*, **15**, 75–86. (WAM)

G17.   Gilbert, E. (1993). Inland marine market targets extended warranties, *National Underwriter*, **97**, 34, 46. (EST, EXT)

G18.   Gilbertson, R. J. (1989). Risk reduction through application of essential performance requirements warranties, *IEEE Proc. Nat. Aerospace and Electronics Conf.*, **4**, 2010–2014. (WLE)

G19.   Gill, H. L., and Roberts, D. C. (1989). New car warranty repair: Theory and evidence, *Southern J. Econ.*, **55**, 662–678. (ETH, WDA)

G20.   Gilleece, M. A. (1984). *Proposed Guarantee Guidance to Implement Section 794 of the Department of Defense Appropriation Act for FY 1984*—ACTION MEMORANDUM, Office of the Under Secretary of Defense, 19 Jan., 1984. (RIW)

G21.   Gilly, M. C., and Gelb, B. D. (1982). Post purchase consumer processes and the complaining consumer, *J. Consumer Research*, **9**, 323–328. (CCO, CSD)

G22.   Gilly, M. C., and Hansen, R. W. (1985). Consumer complaint handling as a strategic marketing tool, *J. Consumer Marketing*, **2**, 5–16. (CCO, CSD)

G23.   Gilman, J. B. (1991). Warranties: An ounce of prevention. . . , *Computerworld*, **25**, 100. (WCT)

G24.   Gilmore, G. (1970). Products liability—a commentary, *The University of Chicago Law Rev.*, **38**, 103–141. (PLI)

G25.   Glaser, A. J. (1981). A data information system for RIW contracts, *1981 Proc. Ann. Reliab. and Maintain. Symp.*, 139–143. (RIW, WDA)

G26.   Glickman, T. S., and Berger, P. D. (1976). Optimal price and protection period decisions for a product under warranty, *Management Sci.*, **22**, 1381–1390. (MAM, OPT, WPR, WTE)

G27.   Goldberg, V. P. (1974). The economics of product safety and imperfect information, *The Bell J. Economics and Management Science*, **5**, 683–688. (EAW)

G28.   Golding, E. L. (1982). *Warranties, Disclosures, and Liability Rules: Their Role in Determining Product Quality*, Doctoral Dissertation, Princeton University, Princeton, NJ. (EAW, PLI, PQU)

G29.   Goldring, J., and Young, T. (1989). *Product Liability: Remedies and Enforcement*, Law Reform Commission Sydney, Australia. (PLI)

G30.   Gomez, A. E. (1990). An approach to dormant system warranty analysis, *IEEE Proc. Nat. Aerospace and Electronics Conf.*, **3**, 1204–1208. (WCA)

G31.   Goodman, M. (1983). Strict liability and the professional service transaction, *Insurance Counsel J.*, **50**, 201–221. (CPT, PLI)

G32.   Gould, R. (1989). How to buy a used lift truck, *Modern Materials Handling*, **44**, 75–79. (WAP)

G33.   Graves, J., and Levitin, M. S. (1990). Accounting for extended warranty contracts, *J. Accountancy*, **170**, 101–102. (ACC, EXT)

G34.   Gray, I., Bases, A. L., Martin, C. H., and Sternberg, A. (1975). *Product Liability: Management Response*, AMACOM, New York. (PLI)

G35.   Green, G. L. (1993). More on extended warranties, *National Public Accountant*, **38, Feb.**, 36–37. (EXT)

G36.   Greene, L. E. Jr. (1989). Methodology for system warranty economic analysis, *IEEE Proc. Nat. Aerospace and Electronics Conf.*, **3**, 1210–1215. (PRS, WCA)

G37.   Gregory, W. M. (1964). Air force studies product life warranty, *Aviation Week and Space Tech.*, **2 Nov.** (RIW)

G38.   Grimlund, R. A. (1985). A proposal for implementing the FASB's "reasonably possible" disclosure provision for product warranty liabilities, *J. Accounting Res.*, **23**, 575–594. (MAM, PLI, WCA)

G39.   Grossman, S. J. (1981). The informational role of warranties and private disclosure about product quality, *J. Law Econ.*, **24**, 461–483. (COR, PQU)

G40.   Grossman, J. B., and Sarat, A. (1975). Litigation in the federal courts: A comparative perspective, *Law and Society Rev.*, **9**, 321–346. (LIT)

G41.   Gruenfeld, L. (1990). Even if you have a warranty, you may not have a guarantee, *Computerworld*, **24**, 121. (LIT, TYW, WCT)

G42.   Guglielmoni, P. B. (1986). An approach to selecting warranties/incentives, *1986 Proc. Ann. Reliab. and Maintain. Symp.*, 166–170. (RIW, WSL)

G43.   Guin, L. (1984). *Cumulative Warranties: Conceptualization and Analysis*, Doctoral Dissertation, University of Southern California, Los Angeles, CA. (CUW, CWP, WCA)

G44.   Gulati, S. T. (1991). Ceramic converter technology for automotive emissions control, *Soc. Automotive Engineers Trans.*, **100**, 529–544. (WCA)

H1.   Haaland, J. E. (1974). Product recall problems and options—nondurable products, in *Managing Product Recalls*, (E. P. McGuire, ed.), Conference Board, New York, 57–65. (PRC)

H2.   Hadel, J. J., and Lakey, M. J. (1993). A structured approach to warranty preparation and risk assessment, *1993 Proc. Ann. Reliab. and Maintain. Symp.*, 298–304. (SIM, WCA)

H3.   Hagglund, R. R. (1977). Product liability and its effect on engineering practice, *1977 Proc. Ann. Reliab. and Maintain. Symp.*, 23–26. (PLI, WLE)

H4.   Hale, P. W. (1987). Reliability of high volume products, *CME, The Chartered Mechanical Engineer*, **34**, 42–43. (REL, WTE)

H5.   Hall, T. (1989). Cost of poor repair process quality (COPRPQ), *1989 Proc. Ann. Reliab. and Maintain. Symp.*, 301–308. (ACC, PRM)

H6.   Hall, T. G., and Wingerter, R. G. (1987). Product warranties—are they worth it?, *Proc. IEEE Int. Conf. on Communications 1987: Communication—Sound to Light*, 1211–1217. (EAW, EXT, PDE)

H7.   Halstead, D., Dröge, C., and Cooper, M. B. (1993). Product warranties and post-purchase service: A model of consumer satisfaction with complaint resolution, *J. Services Marketing*, **7**, 33–40. (CCO, CSD)

H8.   Hamada, K. (1976). Liability rules and income distribution in product liability, *American Eco. Rev.*, **66**, 228–234. (PLI)

H9.   Hamilton, W. H. (1931). The ancient maxim caveat emptor, *Yale Law J.*, **40**, 1133–1187. (PLI)

H10.   Hammer, W. (1971). Product liability: The bang or the fizzle?, *Management Review*, **60**, 59–60. (PLI)

H11.   Hampton, W. J. (1988). Reviving without a cause: When the car has a mind of its own, *Business Week*, **Apr. 4**, 66, 68. (PRC, REL)

H12.   Hampton, W. J., and Schiller, Z. (1987). Why image counts: A tale of two industries, *Business Week*, **Jun. 8**, 138–140. (PQU)

H13.   Hammurabi, King of Babylonia (1903). *The oldest code of laws in the world: The code of Laws promulgated by Hammurabi, King of Babylon*, Translated by C. H. W. Johns, T. and T. Clark, Edinburgh. (LAW)

H14.   Hammurabi, King of Babylonia (1904). *The Code of Hammurabi, King of Babylon, about 2250 B.C.*, Univ. Chicago Press, Chicago, IL. (LAW)

H15.   Hammurabi, King of Babylonia (1921). *The Hammurabi Code and the Sinaitic Legislation*, (C. Edwards, ed.), Watts, London. (LAW)

H16.   Hansen, G. G. (1978). The swinging door: A first experience in product liability, *1978 Proc. Ann. Reliab. and Maintain. Symp.*, 363–364. (PLI)

H17.   Hardy, C. A., and Allen, R. J. (1977). Reliability improvement warranty techniques and applications, *1978 Proc. Ann. Reliab. and Maintain. Symp.*, 222–228. (RIW, WAP)

H18.   Hardy, N. F. (1984). Algorithm for evaluating microcomputer performance, collected papers presented at the *13th Mid-year Meeting of the American Society for Information Science*, 7. (WPR)

H19.   Harker, J. E. (1992). A dinosaur, a citadel and a dogma: Implied warranties of goods in mixed transactions, *Commercial Law J.*, **97**, 485–509. (LIT, WAP)

H20.   Harris, L., and Associates (1977). *Consumerism at the Crossroads*, Marketing Science Institute of Harvard University and Lou Harris Associates, Boston, MA. (CSD)

H21.   Harris, T. (1988). Automotive marketing: Yugo warranty aimed at skittish buyers, *Advertising Age*, **59**, S18–S19. (REL, WAM)

H22.   Harris, M., and Raviv, A. (1979). Optimal incentive contracts with imperfect information, *J. Economic Theory*, **20**, 231–259. (MAM, OPT, SCT)

H23.   Harrison, R. (1990). Let the vendor beware, *Accountancy*, **105**, 124, 126. (BOC, LIT, PLI)

H24.   Harrison, R. A., and Young, R. F. (1985). Designing and manufacturing automotive engine control modules for high volume production, *Int. J. Vehicle Design*, **6**, 263–276. (PDV, QCO)

H25.   *The Hartford Courant* (1986). Uniformity in product liability would ease pricing, insurers say, **Jun. 23**. (LAW, PLI)

H26. Hart, C. W. L. (1993). Satisfaction guaranteed, *Small Business Reports*, **18**, 19–23. (CSD, WAM)

H27. Hart, C. W. L. (1993). Types of guarantees, *Small Business Reports*, **18**, 21. (CSD, TYW)

H28. Hartman, R. S. (1987). Product quality and market efficiency: The effect of product recalls on resale prices and firm valuation, *Rev. Economics and Statistics*, **69**, 367–372. (ETH, PQU, PRC, WAM)

H29. Harty, J. C. (1971). Practical life cyclecost/cost of ownership type procurement via long term/multi year "failure free warranty" (FFW), showing trial procurement results, *Annals of Reliab. and Maintain.*, **10**, 241–251. (RIW)

H30. *Harvard Business Review* (1994). Harnessing madison avenue: Advertising and products liability theory, **107**, 895–912. (PLI)

H31. Hasin, B. R. (1987). *Consumers, Commissions, and Congress: Law, Theory and the Federal Trade Commission*, Transaction Books, New Brunswick, NJ. (LAW)

H32. Hausman, W. H. (1978). Multiple failures per appliance: Data analysis, in *Consumer Durables: Warranties, Service Contracts, and Alternatives*, (R. T. Lund, ed.), Rept. No. CPA-78-14/Vol. IV, Center for Policy Alternatives, Massachusetts Institute of Technology, Cambridge, MA, 4-194–4-209. (STA, WDA)

H33. Hauter, A. J., and Strempke, C. W. (1978). Tacan RIW program, *1978 Proc. Ann. Reliab. and Maintain. Symp.*, 62–65. (RIW, WDA, WTE)

H34. Havey, G., Herrlin, J., and Kampf, D. (1985). Integrated time and stress measurement device concept, *23rd Ann. Proc.—Reliab. Physics 1985*, 60–64. (WVE)

H35. Heal, G. (1977). Guarantees and risk sharing, *Rev. of Econ. Studies*, **44**, 549–560. (COR, PRS)

H36. Heal, G. (1976). Do bad products drive out good? (Comment on "The market for lemons", by G. A. Akerlof (1970)), *Quart. J. Economics.*, **90**, 499–502. (PQU)

H37. Hearing before the Consumer Subcommittee of the Committee on Commerce, 91st Congress (1990). *Consumer Products Guarantee Act.*, US Government Printing Office, Washington, DC. (WLE)

H38. Hearing before the House Subcommittee on Commerce and Finance of the Committee on Interstate and Foreign Commerce, House of Representatives, 92nd Congress (1971). *Consumer Products Protection*, Serial No. 92-50, US Government Printing Office, Washington, DC. (WLE)

H39. Hearing before the House Subcommittee on Commerce and Finance of the Committee on Interstate and Foreign Commerce, House of Representatives, 93rd Congress (1973). *Consumer Warranty Protection*, Serial No. 93-17, US Government Printing Office, Washington, DC. (WLE)

H40. Heaton, G. R. Jr., and Katz, J. I. (1978). Legal aspects of warranties, in *Consumer Durables: Warranties, Service Contracts, and Alternatives*, (R. T. Lund, ed.), Rept. No. CPA-78-14/ Vol. II, Center for Policy Alternatives, Massachusetts Institute of Technology, Cambridge, MA, 2-46-2-137. (WLE)

H41.   Heaton, G. R., and Katz, J. I. (1978). Reassessment of the Magnuson-Moss warranty act, in *Consumer Durables: Warranties, Service Contracts, and Alternatives,* (R. T. Lund, ed.), Rept. No. CPA-78-14/ Vol. II, Center for Policy Alternatives, Massachusetts Institute of Technology, Cambridge, MA, 2-166-2-201. (CPT, WLE)

H42.   Heaton, G. R., and Katz, J. I. (1978). Legal aspects of service contracts, in *Consumer Durables: Warranties, Service Contracts, and Alternatives,* (R. T. Lund, ed.), Rept. No. CPA-78-14/ Vol. II, Center for Policy Alternatives, Massachusetts Institute of Technology, Cambridge, MA, 2-248-2-307. (SCT, WLE)

H43.   Heck, W. R. (1963). Accounting for warranty costs, *Accounting Rev.,* **38,** 577–578. (ACC)

H44.   Hegde, G. G., and Karmarkar, U. S. (1986). Support strategies and the market for product support, Tech. Rept., Graduate School of Management, University of Rochester, NY. (PDV, PSU, SST)

H45.   Hegde, G. G., and Karmarkar, U. S. (1993). Engineering costs and customer costs in designing product support, *Naval Res. Logistic Quart.,* **40,** 415–423. (PDE, PSU, WMS)

H46.   Hegde, G. G., and Kubat, P. (1989). Diagnostics design: A product support strategy, *Euro. J. Oper. Res.,* **38,** 35–43. (PDV, PSU, SST, WCA)

H47.   Helwig, G. J. (1974). Product liability: An interaction of law and technology—a commentary, *Duquesne Law Rev.,* **12,** 476–481. (LAW, PLI)

H48.   Henry, E. (1988). Your car, *Changing Times,* **42,** 76–77. (ESC)

H49.   Hergatt, N. K. (1991). Improved reliability predictions for commercial computers, *1991 Proc. Ann. Reliab. and Maintain. Symp.,* 357–359. (REL)

H50.   Heron, R. (1985). Condition monitoring as a production tool, *CME, The Chartered Mechanical Engineer,* **32,** 21–23. (WDA)

H51.   Heschel, M. S. (1971). How much is a guarantee worth?, *Indust. Eng.,* **3,** 14–15. (CWP, WCA)

H52.   Hester, E. J. (1987). Truth and consequences: The European directive on products liability, *Risk Management,* **34,** 38–44. (LIT, PDE, PLI, PRC, WLE)

H53.   Hill, N. (1989). Delivery on the dot—or a refund, *Industrial Marketing Digest,* **14,** 43–50. (CSD)

H54.   Hill, V. L. (1983). *A Quantitative Model for the Analysis of Warranty Policies,* Doctoral Dissertation, University of Southern California, Los Angeles, CA. (EST, MAM, RFT, STA, WCA, WDA)

H55.   Hill, V. L., Beall, C. W., and Blischke, W. R. (1991). A simulation model for warranty analysis, *Int. J. Prod. Econ.,* **22,** 131–140. (EST, STA, SIM, WCA)

H56.   Hill, V. L., and Blischke, W. R. (1987). An assessment of alternative models in warranty analysis, *J. Information and Optimization Sciences,* **8,** 33–55. (MAM, PRW, WCA)

H57.   Hiller, G. E. (1973). Warranty and product support: The plan and use thereof in a commercial operation, in *Proc. Failure Free Warranty Seminar*, US Navy Aviation Supply Office, Philadelphia, PA. (PSU, RIW)

H58. Hiner, E. L. (1985). Strict liability and the building industry, *J. Products Liability*, **8**, 373–403. (PLI, TOR)

H59. Hocker, P. J., and Heffey, P. G. (1994). *Contract Commentary and Materials*, 7th. ed., The Law Book Co. Ltd., North Ryde, New South Wales, Australia. (BOC, WCT)

H60. Hodges, C. J. S. (1993). *Product Liability/Class Actions*, Queensland Law Society, Queensland, Australia. (PLI)

H61. Hoffer, G. E., Pruitt, S. W., and Reilly, R. J. (1987). Automotive recalls and informational efficiency, *Financial Rev.*, **22**, 433–442. (ETH, PRC)

H62. Hoffer, G. E., Pruitt, S. W., and Reilly, R. J. (1988). The impact of product recalls on the wealth of sellers: A reexamination, *J. Political Economy*, **96**, 663–670. (PRC)

H63. Hoffman, L. A. (1985). The team approach to products liability prevention and defense, *1985 Proc. Ann. Reliab. and Maintain. Symp.*, 6–12. (PLI, LIT, PDV)

H64. Hoffman, J. J., and Warseck, K. (1986). Detecting and identifying roof maintenance problems, *Plant Engineering*, **40**, 74–76. (MAI, WAP)

H65. Hogg, J. E. (1920). *Registration of Title to Land Throughout the Empire: A Treatise on the Law Relating to Warranty of the Title to Land by Registration and Transactions with Registered Land in Australia, New Zealand, Canada, England, Ireland, West Indies, Malaya, & c. A sequel to "The Australian Torrens System"*, The Carswell Co. Ltd., Toronto, Canada. (LAW)

H66. Hohner, G. J. (1989). TQC: Method of champions, Part III, *Manufacturing Systems*, **7**, 51–53. (CSD, PDE, QCO)

H67. Holmes, T. J. (1984). *Monopoly Provision of Warranty and Product Quality When Quality is Unobservable*, Working Paper 612S, Center for Mathematical Studies in Economics and Management Science, Northwestern University, Evanston, IL. (PQU)

H68. Horkin, P. (1980). Burn-in—non sequitur of the 1980s, *Evaluation Engineering*, **26**, 5 pages between 50–58. (REL)

H69. Horton, C., and Erickson, J. L. (1989). Gasoline pump new ad drives, *Advertising Age*, **60**, 4. (WAM)

H70. Hounslow, S. (1988). Car industry: Take good care, *Marketing*, **Feb. 11,** 28–29. (PQU, WMS)

H71. *House Reports* (1974). Report of the House Committee of the Committee on Interstate and Foreign Commerce, US Government Printing Office, Washington, DC. (WLE)

H72. Howells, G. G. (1987). Product liability: A global problem, *Managerial Law*, **29**, 1–36. (BOC, LAW, LIT, PLI)

H73. Howells, G. G. (1993). *Comparative Product Liability*, Dartmouth Publishing Co., Aldershot, England. (PLI)

H74. Howells, G. G., and Phillips, J. J. (1991). *Product Liability*, Barry Rose, Chichester, England. (PLI)

H75. Hulsenbek, R., and Campbell, D. (1989). *Product Liability: Prevention,*

Practice and Process in Europe and the United States, Kulwer Law and Taxation Publisher, Deventer, Boston, MA. (PLI)

H76. Hunt, H. K. (1977). *Conceptualization and Measurement of Consumer Satisfaction and Dissatisfaction*, Marketing Science Institute, Cambridge, MA. (CEI, CSD)

H77. Hursh, R. D. (1963). The products liability problem, *J. Marketing*, **27**, 9–14. (PLI)

H78. Hursh, R. D., and Bailey, H. J. (1974). *American Law of Product Liability*, 2nd ed., The Law Co-operative Pub. Co., Rochester, NY. (LAW, PLI)

H79. Hurst, R. (1987). MMA member gross champions buyer's rights, *Computerworld*, **21**, 7. (WAP)

H80. Hwang, D. (1994). Fujitsu to upgrade pen systems, warranty service, *Computer Reseller News*, **Jan. 31**, 26. (EXT)

H81. Hyman, W. A. (1989). Legal liability in the development and use of medical expert systems, *J. Clinical Engineering*, **14**, 157–163. (BOW, PLI)

I1.   Iizuka, Y. (1988). Adequate claim processing improves quality system; *1988 ASQC Quality Congress Trans.*, Dallas, TX, 338–343. (CCO, WMA)

I2.   *Industry Week* (1985). Product liability: Can congress remove the traps?, **Oct. 28**. (PLI)

I3.   Insurance Service Office (1977). *Product Liability Closed Claim Survey: A Technical Analysis of Survey Results*, Insurance Services Office, New York. (PLI)

I4.   Intrieri, G. (1984). Statistical warranty level in sampling a demerit rate, *Proc. World Quality Congress '84*, Brighton, England, **2**, 272–280. (EAW, QCO, STA)

I5.   Ip-Tamayo, T. C., and Deuermeyer, B. L. (1985). *Establishing an Indifference Function in Quantity Purchases with Optional Warranty*, Tech. Rept., University of Houston, Houston, TX. (WPR)

I6.   Ireson, W. G., and Coombs, C. F. Jr. (1988). Economics of warranty, in *Handbook of Reliability Engineering and Management*, McGraw Hill Book Co. New York, 3.12–3.15. (WCA)

I7.   Irving, R. P. (1981). Products liability: How to prevent it, *Iron Age*, **224**, 23–26. (PLI)

I8.   Isaacson, D. N., Reid, S., and Brennan, J. R. (1991). Warranty cost-risk analysis, *1991 Proc. Ann. Reliab. and Maintain. Symp.*, 332–339. (MAM, SIM, WPR)

I9.   Iskandar, B. P. (1993). *Modelling and Analysis of Two-Dimensional Warranty Policies*, Doctoral Dissertation, The University of Queensland, Australia. (MAM, STA, SIM, TDW)

I10. Iskandar, B. P., Murthy, D. N. P., and Wilson, R. J. (1991). Two dimensional rebate policies, *Proc. 11th National Conference of ASOR*, Gold Coast, Australia. (PRW, TDW)

I11. Iskandar, B. P., Wilson, R. J., and Murthy, D. N. P. (1994). Two-dimen-

	sional combination warranty policies, *RAIRO, Operations Research*, **28**, 57–75. (CWP, TDW)
I12.	Ives, B., and Vitale, M. R. (1988). After the sale: Leveraging maintenance with information technology, *MIS Quarterly*, **12**, 7–21. (COR, MAI)
J1.	Jack, N., and Dagpunar, J. S. (1994). An optimal imperfect maintenance policy over a warranty period, *Microelectronic Reliab.*, **34**, 529–534. (MAI)
J2.	Jackson, G. C., and Morgan, F. W. (1988). Responding to recall requests: A strategy for managing goods withdrawal, *J. Public Policy and Marketing*, **7**, 152–165. (PLI, PRC)
J3.	Jacobowitz, D. L. (1987). A software tool for designing burn-in programs, *1987 Proc. Ann. Reliab. and Maintain. Symp.*, 302–305. (REL, WCA)
J4.	Jacobs, R. M. (1984). Contracting for liability relief, *1984 Proc. Ann. Reliab. and Maintain. Symp.*, 67–71. (LIT, PLI, PQU)
J5.	Jacobs, R. M. (1988). Products liability: A technical and ethical challenge, *Quality Progress*, **21**, 27–29. (PLI, PRC)
J6.	Jacobs, R. M. (1988). Reliability and quality data in court, *1988 Proc. Ann. Reliab. and Maintain. Symp.*, 234–237. (LIT, PQU, WDA)
J7.	Jacobs, R. M., and Hoffman, L. A. (1984). Contracting for liability relief, *1984 Proc. Ann. Reliab. and Maintain Symp.*, 67–71. (LIT, PLI, PQU)
J8.	Jacobs, R. M., and Hoffman, L. A. (1988). Liability aspects of procurement contracts, *Microelectronics and Reliability*, **28**, 183–188. (PLI, WLE)
J9.	Jacobs, R. M., and Mihalasky, J. (1973). Reliability techniques to keep you out of court, *1973 Proc. Ann. Reliab. and Maintain. Symp.* 355–361. (LIT, REL)
J10.	Jacobs, R. M., and Mihalasky, J. (1983). Evolution of quality/reliability due to litigation, *1983 Proc. Ann. Reliab. and Maintain. Symp.*, 121–124. (LIT, PLI, WCA, WDA)
J11.	Jacobs, R. M., and Mundel, A. B. (1978). Profit or liability; Contract intent vs. content, *1978 Proc. Ann. Reliab. and Maintain. Symp.*, 117–122. (LIT, PLI, WCA, WDA)
J12.	Jaisingh, L. R., Kolarik, W. J., and Dey, D. K. (1987). Bathtub hazard model and an application to system warranty, *Trans. Kentucky Academy Science*, **48**, 20–25. (MAM, WAP)
J13.	Janner, G. (1979). *Janner's Product Liability*, Business Books, London. (PLI)
J14.	Jenett, E. (1983). Multiple interests in a construction contract—the constructor's slant, presented at *39th Petroleum Mechanical Engineering Workshop and Conference*, 115. (DCV)
J15.	Jeyapalan, J. K., and Boldon, B. A. (1986). Performance and selection of rigid and flexible pipes, *J. Transportation Engineering*, **112**, 507–524. (DCV, PDE)
J16.	Jiambalvo, J., and Wilner, N. (1985). Auditor evaluation of contingent claims, *Auditing: A. J. Practice and Theory*, **Fall**, 1–11. (ACC)

J17.   Johnson, A. (1978). Behind the hype on product liability, *The Forum*, **14,** 317–326. (PLI)

J18.   Johnston, S. J., and Scannell, E. (1994). Microsoft plus: Support in the fast lane, *Computerworld*, **28,** 41. (WSE)

J19.   Jones, M. G., and Boyer, B. (1972). Improving the quality of justice in the marketplace: The need to better consumer remedies, *George Washington Law Review*, **40,** 357. (WLE)

J20.   Juha, M. (1986). Economics of automated X-ray inspection for solder quality, *Proc. Technical Program—National Electronic Packaging and Production Conference*, **2,** 949–960. (PQU, WCA)

K1.   Kahn, H. (1987). Center for auto safety calls secret warranties unfair, *Automotive News*, **Sep. 14,** 34. (WAP)

K2.   Kahn, H. (1991). FTC upholds arbitration mandate, *Automotive News*, **Jul. 1,** 23. (WAP)

K3.   Kalbfleisch, J. D., and Lawless, J. F. (1988). Estimation of reliability in field-performance studies, *Technometrics*, **30,** 365–388. (EST, REL, STA, WDA)

K4.   Kalbfleisch, J. D., Lawless, J. F., and Robinson, J. A. (1991). Methods for the analysis and prediction of warranty claims, *Technometrics*, **33,** 273–285. (EST, STA, WDA)

K5.   Kaler, G. M. Jr. (1988). Expert system predicts service, *Heating, Piping and Air Conditioning*, **60,** 99–101. (WDA)

K6.   Kallmeyer, F. B. (1990). New concepts in maintaining reciprocating chip and bark reclaimers, *76th Ann. Meeting—Tech. Section, Canadian Pulp and Paper Ass.*, 241–243. (WAP)

K7.   Kaltenbach, T. A. (1993). Auto and appliance dealers may defer income on sales of long-term warranties, *J. Taxation*, **78,** 292–295. (ACC, WCT)

K8.   Kambhu, J. (1982). Optimal product quality under asymmetric information and moral hazard, *Bell J. of Eco.*, **13,** 483–492. (COR, MOR, OPT, PQU)

K9.   Kangun, N., Cox, K. K., and Staples, W. A. (1978). An exploratory investigation of selected aspects of the Magnuson-Moss Warranty Act, in *Consumerism: New Challenges for Marketing*, (N. Kangun and L. Richardson, eds.), AMA Conference on Consumerism, Chicago, IL, 3–11. (CPT, WPR)

K10.   Kao, E. P. C., and Smith, M. S. (1993). Discounted and per unit net revenues and costs of product warranty: The case of phase-type lifetimes, *Management Science*, **39,** 1246–1254. (CBH, MAM, WCA)

K11.   Karbasov, O. G., and Sajenov, A. F. (1975). About choice of the numerical values of reliability level and warranty periods for wedge ventilation straps, *Rubber*, **1.** (in Russian). (REL, WTE)

K12.   Karmarkar, U. S. (1978). Future costs of service contracts for consumer durable goods, *AIEE Trans.*, **10,** 380–387. (SCT, WRE)

K13.   Karmarkar, U. S., and Kubat, P. (1983). The value of loaners in product support, *IIE Trans.*, **15,** 5–11. (PSU)

K14. Karmarkar, U. S., and Kubat, P. (1984). *Modular Replacement in Product Support*, Working Paper Series No. QM8401, The Graduate School of Management, The University of Rochester, Rochester, NY. (EAW, PSU)

K16. Kasimer, J. H. (1988). Material supplier: Reducing and defending against building failure claims, *Construction Specifier*, **41**, 36–37. (WCT)

K17. Katz, A. (1990). Your terms or mine? The duty to read the fine print in contracts, *RAND J. of Economics*, **21**, 518–537. (MAM, PRM, WLE)

K18. Kaye, P. (1991). *Private International Law of Tort and Product Liability: Jurisdiction, Applicable Law, and Extraterritorial Protective Measures*, Dartmouth, Aldershot, England. (LAW, PLI, TOR)

K19. Kececioglu, D., and Jiang, S. (1990). Error band estimation on Monte Carlo simulations, *Proc. 36th Ann. Technical Meeting of the Institute of Environmental Sciences*, 124–128. (EST, MAM, SIM, WTE)

K20. Keeton, P., Owen, D. G., and Montgomery, J. E. (1980). *Product Liability and Safety: Cases and Materials*, Foundation Press, Mineola, NY. (PLI)

K21. Kellam, J. (1992). *A Practical Guide to Australian Product Liability*, CCH Australia, North Ryde, New South Wales, Australia. (PLI)

K22. Kelley, C. A. (1984). Extending the understanding of consumer product warranty theories via brand specific testing, *Beyond 1984 in Marketing Educ.*, Western Marketing Educators Assoc., 73–76. (CBH)

K23. Kelley, C. A. (1986). Consumer product warranties under the Magnuson-Moss Warranty Act: A review of the literature, *AMA Summer Educators Proc.*, Chicago, IL, 369–373. (RLW)

K24. Kelley, C. A. (1987). The role of consumer product warranties and other extrinsic cues in reducing perceived risk, *Developments in Marketing Science*, Miami, FL, 319–323. (COR)

K25. Kelley, C. A. (1988). An investigation of consumer product warranties as market signals of product reliability, *J. Academy of Marketing Science*, **16**, 72–78. (CPT, REL, WAM, WLE)

K26. Kelley, C. A., and Conant, J. S. (1987). Extended warranties: An exploratory study of consumer perceptions, *1987 AMA Summer Marketing Educators Proc*, 230–234. (CBH, EXT)

K27. Kelley, C. A., and Conant, J. S. (1991). Extended warranties: Consumer and manufacturer perceptions, *J. Consumer Affairs*, **25**, 68–83. (CBH, EXT, WAM)

K28. Kelley, C. A., Conant, J. S., and Brown, J. J. (1988). Extended warranties: Retail management and public policy implications, *1988 AMA Summer Marketing Educators Proc.*, 261–266. (EXT, PUB, WAM)

K29. Kelley, C. A., and Swartz, T. A. (1984). An empirical test of warranty label and duration effects on consumer decision making: An information processing perspective, *AMA Summer Educators Proc.*, Chicago, IL, 309–313. (CEI)

K30. Kellie, A. C. (1983). Liability suits and the land surveyor, Tech. Papers *43rd Ann. Meeting America Congress on Surveying and Mapping*, Washington, DC, 19–27. (BOC, BOW, LIT)

K31.   Kelly, J. P. (1979). Consumer expectations of complaint handling by manufacturers and retailers of clothing products, in *New Dimensions of Consumer Satisfaction and Complaining Behavior*, (R. L. Day and H. K. Hunt, eds.), Department of Marketing, School of Business, Indiana University, Bloomington, 103–110. (CCO)

K32.   Kelly, P., and Attree, R. (1992). *European Product Liability*, Butterworths, London. (PLI)

K33.   Kendall, C. L., and Russ, F. A. (1972). Warranty policies and practices of consumer packaged goods manufactures, *Proc. of the 3rd Ann. Conf. of the Ass. for Consumer Res.*, 349–355. (WAP, WTE)

K34.   Kendall, C. L., and Russ, F. A. (1975). Warranty and complaint policies: An opportunity for marketing management, *J. Marketing*, **39**, 36–43. (CCO, WAM)

K35.   Keown, C. F. (1988). Consumer reactions to food and drug product recalls: A case study of Hawaiian consumers, *J. Consumer Policy*, **11**, 209–221. (CBH, PRC)

K36.   Kerin, R. A., and Craine, R. E. (1974). A guide to product liability, *Atlanta Eco. Rev.*, **Nov–Dec,** 48–51. (PLI)

K37.   Kerin, R. A., and Harvey, M. (1975). Contingency planning for product recall, *MSU Bus. Topics*, **23**, 5–12. (PRC)

K38.   Kerridge, A. E. (1990). For quality, define requirements, *Hydrocarbon Processing*, **69**, 8p. (DCV, QCO)

K39.   Kerscher, W. J. III, Lin, T., Stephenson, H., Vannoy, E. H., and Wioskowski, J. (1989). Reliability model for total field incidents, *1989 Proc. Ann. Reliab. and Maintain. Symp.*, 22–28. (EST, REL, STA)

K40.   Kight, D. (1993). A lifetime of steady-going, *Facilities Design and Management*, **12**, 40–41. (WAP)

K41.   Kim, J., Mattson, L., and Sahni, A. (1992). Modelling and simulation for reliability prediction of implantable cardiac pacemakers, *1992 Proc. Ann. Reliab. and Maintain. Symp.*, 53–57. (PDE, REL, SIM)

K42.   Klause, P. J. (1970). Failure-free warranty idea lauded, wider use desired, *Aviation Week and Space Tech.*, **Feb. 9,** 57–58. (RIW)

K43.   Klein, S. J. (1974). Product liability in the 1980s—the engineer's view, *Mechanical Engineering*, **96**, 10–13. (CPT, PLI)

K44.   Kliavkoff, G. T. (1994). Controlling the vicarious liability of hotel and motel franchisors, *Cornell Hotel and Restaurant Administration Quart.*, **35**, 59–73. (BOW, LIT, PLI)

K45.   Knight, D. J., and Jarvis, D. G. (1985). Developments in the production and application of stainless 409, *Mechanical Working and Steel Processing*, **23**, 79–84. (WAP)

K46.   Kohoutek, H. J. (1982). Establishing reliability goals for new technology products, *1982 Proc. Ann. Reliab. and Maintain. Symp.*, 460–465. (OPT, REL, WCA)

K47.   Kolb, J., and Ross, S. S. (1980). *Products Safety and Liability, a Desk Reference*, McGraw Hill Book Co., New York. (PLI)

K48.   Koten, J. (1984). Aggressive use of warranties is benefiting many concerns, *The Wall Street J.*, **Apr. 5,** 33. (EXT, WAM)

K49. Kowalski, R., and White, R. (1977). Reliability improvement warranty (RIW) and the army lightweight doppler navigation system (LDNS), *1977 Proc. Ann. Reliab. and Maintain. Symp.*, 237–241. (RIW, WMA)

K50. Kramer, S. I. (1989). The cost effectiveness of reliability testing—what's good and what's not, *1989 Proc. Ann. Reliab. and Maintain. Symp.*, 470–473. (QAS, REL)

K51. Kristensen, K., Gopal, K. K., and Dahlgaard, J. J. (1992). On measurement of customer satisfaction, *Total Quality Management*, **3**, 123–128. (CSD)

K52. Kruvand, D. H. (1987). Army aviation warranty concepts, *1987 Proc. Ann. Reliab. and Maintain. Symp.*, 392–394. (CUW, RIW, TYW)

K53. Kubo, Y. (1986). Quality uncertainty and guarantee—A case of strategic market segmentation by a monopolist, *Euro. Econ. Rev.*, **30**, 1063–1079. (PQU, WMS, WPR)

K54. Kumar, K. R., and Murthy, D. N. P. (1992). Competitive product design and marketing strategies in a distribution channel, presented in *ORSA/ TIMS Joint National Meeting*, San Francisco, CA. (PDE, WAM)

K55. Kumar, K. R., and Murthy, D. N. P. (1993). *Total Quality in a Channel of Distribution*, Research Paper DS-93-01, School of Business Administration, University of Southern California, Los Angeles, CA. (MAM, PQU)

K56. Kuo, S. (1987). *Issues Involving the Valuation of Warranties, Options and Other Contingent Claims*, Doctoral Dissertation, Case Western Reserve University, Cleveland, OH. (EAW, SCT, WPR)

K57. Kyle, B. R. M. (1989). Using Taguchi methodology for optimizing products and processes, *Ann. Technical Conference—Society of Plastics Engineers*, 1704–1706. (OPT, PDE, WCA)

L1. Lakey, M. J. (1991). Statistical analysis of field data for aircraft warranties, *1991 Proc. Ann. Reliab. and Maintain. Symp.*, 340–344. (EST, STA, WDA)

L2. Landes, W. M., and Posner, R. A. (1985). A positive economic analysis of products liability, *J. Legal Studies*, **14**, 535–568. (EAW, PLI)

L3. Landers, T. L. (1985) *Reliability Analysis for Nonrepairable Systems Subject to Dormancy*, Doctoral Dissertation, Texas Tech. University, Lubbock, TX. (REL, WCA)

L4. Landers, T. L. (1986). Dormant/active phase reliability analyses, *Proc. American Institute Industrial Engineers, Annual Conference and Convention 1986*, Inst. of Industrial Engineers, Norcross, GA, 646–649. (MAM, REL, STA, WDA)

L5. Landers, T. L., and Kolarik, W. J. (1987). Proportional hazards analysis of field warranty data, *Reliability Engineering*, **18**, 131–139. (STA, WDA)

L6. Landon, E. L. (1977). A model of consumer complaint behavior, in *Consumer Satisfaction, Dissatisfaction, and Complaining Behavior* (R. L. Day, ed.), Department of Marketing, School of Business, Indiana University, Bloomington, 31–35. (CCO, CSD)

L7. Landon, E. L. (1980). The direction of consumer complaint research, in

Advances in Consumer Research (J. C. Olson, ed.), **7**, Association for Consumer Research, San Francisco, CA, 335–338. (CCO)

L8.   Larden, K. A. (1986). Electropainting of automotive components, *Technical Papers, Ann. Tech. Conf. and Exhibition—Institute of Metal Finishing 1986*, **1**, 137. (WAP)

L9.   Lauer, T. E. (1965). Sales warranties under the uniform commercial code, *Missouri Law Rev.*, **30**, 259–287. (WLE)

L10.   Law Reform Commission New South Wales (1975). *Working Paper on the Sale of Goods: Warranties, Remedies Frustration and Other Matters*, s.n., Sydney, Australia. (BOW, CPT, WCT)

L11.   Law Reform Commission (1988). *Product Liability/Law Reform Commission*, The Commission, Sydney, Australia, No. 7, 34. (PLI)

L12.   Law Reform Commission (1989). *Product Liability: Economic Impacts* (R. Braddock, ed.) Law Reform Commission, Sydney, Australia. (PLI)

L13.   Law Reform Commission (1989). *Product Liability: Proposed Legislation*, Law Reform Commission, Sydney, Australia, No. 37. (PLI)

L14.   Law Reform Commission (1989). *Product Liability: Summary of Report*, AGPS, Canberra, Australia, No. 51. (PLI)

L15.   Ledbetter, L. A. (1989). Product recall plan guidelines for manufacturers and sellers of industrial products, *Professional Safety*, **34**, 18–23. (PLI, PRC)

L16.   Lee, J. S., and Park, K. S. (1991). Joint determination of production cycle and inspection intervals in a deteriorating production system, *J. Oper. Res. Soc.*, **42**, 775–783. (MAM, PQU, STA, WCA)

L17.   Leemis, L. M., and Beneke, M. (1990). Burn-in models and methods: A review, *IIE Transactions*, **22**, 172–180. (REL)

L18.   Lehman, J. C., Gentry, J. W., and Ellis, H. W. (1983). The readability of warranties: Did they improve after the Magnuson-Moss Act and are they more complex than other related communications?, *Proc. Southern Marketing Association*, 19–23. (CEI, WLE)

L19.   Lehr, K. F. (1981). The application of products liability principles to professional services, *Insurance Counsel J.*, **48**, 434–451. (PLI)

L20.   Leland, H. E. (1981). The informational role of warranties and private disclosure about product quality: Comments on Grossman, *J. Law Eco.*, **24**, 485–489. (PQU, WCA)

L21.   Lele, M. M., and Karmarkar, U. S. (1983). Good product support is smart marketing, *Harvard Business Rev.*, **61**, 124–132. (PSU, SST, WAM)

L22.   Lenocker, T. (1990). The fine print, *Civil Engineering*, **60**, 36–38. (PLI, WAP)

L23.   Levin, B. A., and Coyne, R. (1979). *Tort Reform and Related Proposals: Annotated Bibliographies on Product Liability and Medical Malpractice*, ABA, Chicago, IL. (PLI, TOR)

L24.   Levin, D. P., and Guiles, M. G. (1987). GM upgrades power-train warranty on '87 cars in bid to boost market share, *The Wall Street J.*, **Jan. 23**, 2. (EXT, WAM)

L25.   Levin, S. J. (1986). Examining restraints on freedom to contract as an approach to purchaser dissatisfaction in the computer industry, *California Law Rev.*, **74,** 2101–2141. (CCO, LIT, TOR)

L26. Levine, B. (1969). Warranty and the engineer-manager, *Quality Progress*, **2, Oct.,** 15–16. (WMA)

L27. Levine, J. B. (1992). IBM's PC comeback could begin in Europe, *Business Week*, **Oct. 19,** 52. (WAM)

L28. Levy, L. B., and Bell, S. Y. (1990). Software product liability: Understanding and minimizing the risks, *High Technology Law J.*, **5,** 1–27. (LIT, PLI, TOR)

L29. Llewellyn, K. N. (1936). On warranty of quality and society, *Columbia Law Rev.*, **36,** 699–744. (PQU)

L30. Llewellyn, K. N. (1936). On warranty of quality and society: II, *Columbia Law Rev.*, **37,** 341–409. (PQU)

L31. Lewis, C. (1993). Vendor equipment warranties don't hold water, *Network World*, **10,** 30. (SCT)

L32. Lie, C. H., and Chun, Y. H. (1987). Optimum single-sample inspection plans for products sold under free and rebate warranty, *IEEE Trans. Rel.*, **R-36,** 634–637. (FRW, PRW, QCO)

L33. Liebesman, B. S. (1985). Reliability requirements and contractual provisions, *1985 Proc. Ann. Reliab. and Maintain. Symp.*, 1–5. (PDE, REL)

L34. Light, L., Seebacher, N., and Armstrong, L. (1990). New lemons from the auto lot, *Business Week*, **Dec. 3,** 38. (ESC)

L35. Lindgren, T. (1986). Optimizing ESS effectiveness using Weibull techniques, *Institute Environ. Sci. 1986 Proc.—32nd Ann. Technical Meeting*, 117–124. (PLI, REL, STA)

L36. Lindo, D. K. (1992). Pricing for profit, *Office System*, **9,** 68–69. (WCA)

L37. Lingenfelter, G. E. (1988). Evaluating product safety certification programs, *Professional Safety*, **33,** 13–18. (PQU)

L38. Lockwood, R. (1988). 80386 computers, *Personal Computing*, **12,** 189–199. (DCV)

L39. Loomba, A. P. S. (1992). *Product Sale and Service Support Via Industrial Distribution Channels*, Doctoral Dissertation, School of Business Administration, University of Southern California, Los Angeles, CA. (PDV, SST)

L40. Lorne, S. M. (1986). Representations and warranties: A seller's perspective, *Buyouts and Acquisitions*, **4,** 62–63. (CEI)

L41. Losee, M. (1988). Developing a warranty and service program, *Cellular Business*, **5,** 34–37. (CSD, WAM, WDA)

L42. Loudenback, L. J., and Goebel, J. W. (1974). Marketing in the age of strict liability, *J. Marketing*, **38,** 62–66. (PLI, WAM)

L43. Lowenthal, F. (1983). Product warranty period: A markovian approach to estimation and analysis of repair and replacement cost—A comment, *Accounting Rev.*, **58,** 837–838. (EST, MAM, PRM)

L44. Lowerre, J. M., (1968). On warranties, *J. Industrial Engineering*, **19,** 359–360. (MAM, PRM, WCA, WPR)

L45. Lund, R. T. (ed.) (1978). *Consumer Durables: Warranties, Service Con-*

tracts, and Alternatives, Rept. No. CPA-78-14/Vol. 1–4, Center for Policy Alternatives, Massachusetts Institute of Technology, Cambridge, MA. (SCT, MAM, WLE)

L46.   Lunsford, J. D., Berk, J. H., and Nixon, D. E. (1986). Warranting conventional munitions, *1986 Proc. Ann. Reliab. and Maintain. Symp.*, 443–448. (RIW, TYW, WDA, WTE)

L47.   Lutz, N. A. (1985). *Discrete Warranties, Signalling and Consumer Moral Hazard*, Mimeo, Graduate School of Business, Stanford University, Stanford, CA. (MOR)

L48.   Lutz, N. A. (1986). *Warranties and Consumer Moral Hazard*, Doctoral Dissertation, Graduate School of Business, Stanford University, Stanford, CA. (MOR)

L49.   Lutz, N. A. (1989). Warranties as signals under consumer moral hazard; *RAND J. Economics*, **20**, 239–255. (CBH, MAM, MOR, PQU, WPR)

L50.   Lutz, N. A., and Padmanabhan, V. (1992). *Warranty and service contracts in competitive insurance markets*, Working Paper, Graduate School of Business, Stanford University, Stanford, CA. (SCT, WAM)

L51.   Lutz, N. A., and Padmanabhan, V. (1993). *Warranties, Extended Warranties, and Producer Moral Hazard*, Working Paper, Graduate School of Business, Stanford University, Stanford, CA. (COR, EXT, MOR)

L52.   Lutz, N. A., and Padmanabhan, V. (1993). *Warranties, Service Contracts and Consumer Moral Hazard*, Working Paper, Graduate School of Business, Stanford University, Stanford, CA. (COR, MOR, SCT)

L53.   Lutz, N. A., and Padmanabhan, V. (1993). *Why Do We Observe Minimal Warranties?* Working Paper, Graduate School of Business, Stanford University, Stanford, CA. (COR, SCT, WPR)

L54.   Lutz, R. W. (1988). Designing electronics for the automotive environment, *IEEE Workshop Automot. Appl. Electron. 1988*, 25–33. (EXT)

L55.   Lynch, M., and Witner, L. (1993). New tax rules for service warranty contracts, *J. Accountancy*, **175**, 39–42. (ACC, SCT)

M1.   *Major Appliance Consumer Action Panel* (1971). MACAP Study of Appliance Industry Warranties, Unpublished. (COM, WAP)

M2.   *Major Appliance Consumer Action Panel* (1973). MACAP Analysis of Major Appliance Warranties, Unpublished. (COM, WAP)

M3.   MacCallum, S. (1967). Dispute settlement in an American super-market, in *Law and Warfare: Studies in the Anthropology of Conflict*, (P. Bohannan, ed.), Natural History Press, Garden City, NY. (CMD)

M4.   Madiedo, E. (1985). Technique to tract operating hours of equipment at remote field locations, *Proc. 12th Inter-Ram Conf. for the Electric Power Industry*, 252–255. (WDA)

M5.   Mahn, T. G. (1986). Liability issues associated with electrical overstress in computer hardware design and manufacturer, *Electrical Overstress/ Electrostatic Discharge Symp. Proc.*, 1–11. (PDE, PLI)

M6.   Malott, R. H. (1983). Let's restore balance to product liability law, *Harvard Business Rev.*, **61**, 66–74. (LAW, PLI)

M7. Malott, R. H. (1988). Product-liability system hampers competitiveness, *Financier*, **12**, 29–32. (LIT, PLI, PRC, TOR)

M8. Malvern, D. (1976). A reliability case history—the F-15 A Eagle program, *Defense Management J.*, **12**, 40–45. (PQU, RIW)

M9. Mamer, J. W. (1980). *Mathematical models of warranty policies*, Unpublished manuscript, Berkeley Business School, University of California, Berkeley, CA. (MAM)

M10. Mamer, J. W. (1982). Cost analysis of pro-rata and free-replacement warranties, *Naval Res. Logist. Quart.*, **29**, 345–356. (FRW, PRM, PRW, WCA)

M11. Mamer, J. W. (1987). Discounted and per unit costs of product warranty, *Management Science*, **33**, 916–930. (FRW, MAM, PRW, WCA, WTE)

M12. Manley, M. (1987). Product liability: you're more exposed than you think, *Harvard Business Rev.*, September-October, **65**, 28–30, 34, 36, 40. (PLI)

M13. Mann, D. P., and Wissink, J. P. (1988). Money back contracts with double moral hazard, *RAND J. of Econ.*, **19**, 285–292. (ETH, MOR)

M14. Mann, D. P., and Wissink, J. P. (1990). Hidden actions and hidden characteristics in warranty markets, *Int. J. Indust. Organization*, **8**, 53–71. (CBH, ETH, WCA)

M15. Mann, D. P., and Wissink, J. P. (1990). Money back warranties vs. replacement warranties: A simple comparison, *American Economic Rev.*, **80**, 432–436. (CPT, ETH, PQU)

M16. Mann, N. R. (1970). Warranty periods based on three ordered sample observations from a Weibull population, *IEEE Trans. on Rel.*, **R-19**, 167–171. (STA, WTE)

M17. Mann, N. R. (1976). Warranty period for production lots based on fatigue test data, *Eng. Fracture Mechanism*, **8**, 123–180. (STA, WTE)

M18. Mann, N. R., and Saunders, S. C. (1969). On evaluation of warranty assurance when life has a Weibull distribution, *Biometrika*, **56**, 615–625. (PQU, STA, WCA, WTE)

M19. Mann, N. R., Schafer, R. E., and Singpurwalla, N. D. (1974). *Methods for Statistical Analysis of Reliability and Life Data*, John Wiley & Sons, New York, 254–257. (WTE)

M20. Manwiller, P. C. (1986). *Economic value of heavy duty truck manufacturers published warranty to the truck user*, SAE Technical Paper Series, No. 861984. Publ. by SAE, Warrendale, PA. (EAW, WCA)

M21. Marcial, G. G. (1993). This warranty is still good, *Business Week*, **Sep. 6**, 74. (EXT, SCT)

M22. Marcus, A. A., Bromiley, P., and Goodman, R. (1987). Preventing corporate crises: Stock market losses as a deterrent to the production of hazardous products, *Columbia J. World Business*, **22**, 33–42. (PRC)

M23. Marcus, R. D., Swidler, S., and Zivney, T. L. (1987). An explanation of why shareholders' losses are so large after drug recalls, *Managerial and Decision Economics*, **8**, 295–300. (PRC, STA, WDA)

M24. Markle, J. P. (1974). Product liability: An interaction of law and technology, a commentary, *Duquesne Law Rev.*, **12**, 482–495. (LAW, PLI)

M25. *Marketing News* (*1978*). Consumers are getting less—not more—warranty protection. **Feb. 10,** 2, Cited in: McDaniel, S. W. and Rao, C. P. (1980). A post-evaluation of the Magnuson-Moss Warranty Act, *Akron Business and Economic Review*, **7, Summer,** 38–41. (CEI, WLE)

M26. Markowitz, O. (1971). A new approach long range fixed price warranty within operational environments—for buyer/user, *Annals of Reliability and Maintainability*, **10,** 252–257. (RIW)

M27. Markowitz, O. (1975). Failure free warranty—reliability improvement warranty, buyer viewpoint, *29th Ann. ASQC Tech. Conf.*, 87–97. (RIW)

M28. Markowitz, O. (1976). Aviation supply office FFW/RIW case history #2, Abex pump, *1976 Proc. Ann. Reliab. and Maintain. Symp.*, 357–362. (RIW)

M29. Markowitz, O. (1977). *Reliability Improvement Warranty*, Rept. No. ASO-TEE-2-77, U.S. Navy Aviation Supply Office, Philadelphia, PA. (RIW)

M30. Markowitz, O. (1977). Continuous reliability improvement assurance within operations and logistics, *Proc 1977 Reliab. Conf for the Elec. Power Industry*, 100–105. (RIW)

M31. Markowitz, O. (1977). Failure free warranty—reliability improvement warranty, buyer viewpoint, *Trans. 29th Annual ASQC Tech. Conf.*, 87–97. (RIW)

M32. Marshall, C. W. (1981). Design trade-offs in availability warranties, *1981 Proc. Ann. Reliab. and Maintain. Symp.*, 95–100. (MAM, RIW, WCA)

M33. Marshall, C. W. (1983). Warranty definition for bulk availability, *1983 Proc. Ann. Reliab. and Maintain. Symp.*, 225–230. (MAM)

M34. Martin, J. A., and Skalla, R. E. (1991). Survival with supplier management, *Ann. Quality Congress Trans.*, **45,** 902–907. (PQU, WAP)

M35. Marucheck, A. S. (1987). On product liability and quality control, *IIE Trans.*, **19,** 355–360. (PLI, QCO, STA)

M36. Masel, G. R. (1985). *Product Liability in Australia*, Tech. Pap. Soc. Manuf. Eng. 1985, No. MM85-665. (PLI, WLE)

M37. Massachusetts Institute of Technology, Center for Policy Alternatives (1978). *Consumer Durables*: *Warranties, Service Contracts, and Alternatives*, **1–4,** CPA-78-14, Cambridge, MA. (SCT, WLE)

M38. Massey, D. (1988). Liability for failed software, *J. Information Systems Management*, **5,** 47–53. (CPT, TOR)

M39. Masters, J. D. (1993). Reducing construction defect liability risks in REO operations, *Real Estate Finance*, **9,** 51–62. (DCV, PLI)

M40. Mateja, P. (1987). Product recall—the marketer's nightmare, *Sales and Marketing Management in Canada*, **28,** 6, 39–42. (PRC)

M41. Mateja, P. (1987). The marketing nightmare: Product recalls are serious, but damage can be minimized, *Marketing News*, **21,** 1, 28. (PRC)

M42. Matthews, S., and Moore, J. (1987), Monopoly provision of quality and warranties: An exploration in the theory of multidimensional screening, *Econometrica*, **55,** 441–467. (PQU, WAM)

M43. Mattus, A., and Nash, S. (1994). MicroFlex offers a lot of power for the price, *InfoWorld*, **16**, 80. (WAP)

M44. Maya, V. F., and Carpenter, C. D. (1992). Total quality management in the furniture industry, *46th Ann. Quality Congress Trans.*, **46**, 570–576. (TQM, WAP)

M45. McDaniel, S. W., and Rao, C. P. (1982). Consumer attitudes toward and satisfaction with warranties and warranty performance before and after Magnuson-Moss, *Baylor Business Studies*, **130**, 47–62. (CSD, WLE)

M46. McGarry, S. L. (1984). Acceptable quality levels are not acceptable any more, *AUTOFACT, Proc. 6th. Conference* SME, Dearborn, MI, 15.1–15.13. (PDE, QCO, WCA)

M47. McGuire, E. P. (ed.) (1973). *The Consumer Affairs Department: Organization and Function*, The Conference Board, Inc., New York. (CEI)

M48. McGuire, E. P. (ed.) (1974). *Managing Product Recalls*, Conference Report No. 632, The Conference Board, New York. (PRC)

M49. McGuire, E. P. (1974). Tracing via warranty cards, in *Managing Product Recalls*, (E. P. McGuire, ed.), Conference Report No. 632, The Conference Board, New York, 68. (WDA)

M50. McGuire, E. P. (1975). Product recall and the facts of business life, *The Conference Board Record*, **Feb**, 13–15. (PRC, WAP)

M51. McGuire, E. P. (1980). *Industrial Product Warranties: Policies and Practices*, The Conference Board Inc., New York. (WCA)

M52. McKean, R. N. (1970). Products liability: Trends and implications, *University Chicago Law Rev.*, **38**, 3–63. (PLI)

M53. McKean, R. N. (1970). Products liability: Implications of some changing property rights, *Quart. J. Eco.*, **84**, 611–626. (PLI, WLE)

M54. McKeand, P. J. (1990). GM division builds a classic system to share internal information, *Public Relations J.*, **46**, 24–26, 41. (DCV)

M55. McLeod, D. (1990). Decision could widen tobacco firms' liability, *Business Insurance*, **24**, 2, 37. (PLI)

M56. McNeil, K. E. (1980). *Consumer Protection as Symbolic Remedy: The Impact of the Magnuson-Moss Warranty Act on the Automobile Industry*, manuscript, Dept. of Sociology, University of Wisconsin, Madison, WI. (CPT, WAP)

M57. McNeil, K., and Miller, R. E. (1980). The profitability of consumer protection: Warranty policy in the auto industry, *Administrative Science Quart.*, **25**, 407–427. (CPT, WAP)

M58. McNeal, J. U., and Hise, R. T. (1986). An examination of the absence of written warranties on routinely purchased supermarket items, *Akron Business and Economic Rev.*, **Fall**, 20–30. (CPT, WAP)

M59. McQuade, W. (1972). Why nobody's happy about appliances, *Fortune*, **85**, 180–183, 272, 274, 276. (CCO, CSD)

M60. McRobb, R. M. (1987). Product liability—an update on the 'Italian Compromise', *Chartered Mechanical Engineer*, **34**, 20–22. (PLI)

M61. Meeks, C. B., and Oudekerk, E. H. (1981). Home warranties: An analysis of an emerging development in consumer protection, *J. Consumer Affairs*, **15**, 271–289. (CPT, PRS)

M62. Mele, J. (1990). The changing market for remanned parts, *Fleet Owner*, **85**, 95–100. (QCO, WTE)

M63. Menezes, M. A. J. (1985). *Product Warranty as an Element of the Marketing Mix: Theory, Implications and Experimental Tests*, Doctoral Dissertation, University of California, Los Angeles, CA. (WAM, WPR, WTE)

M64. Menezes, M. A. J. (1988). *Ford Motor Company: The Product Warranty Program (A)*, No. 9-589-001, Harvard Business School. (ESC, WAM, WMA)

M65. Menezes, M. A. J. (1989). *The Impact of Product Warranties on Consumer Preferences*, Working Paper No. 90-003, Harvard Business School. (COR, WAM)

M66. Menezes, M. A. (1991). Warranty premium: The concept, dimensionality and measurement, presented at *ORSA/TIMS 32nd Joint National Meeting*, Anaheim, CA. (TDW)

M67. Menezes, M. A. J., and Currim, I. S. (1992). An approach for determination of warranty length, *Int. J. Res. in Marketing*, **9**, 177–195. (MAM, WAM, WTE)

M68. Menezes, M. A. J., and Quelch, J. A. (1990). Leverage your warranty program, *Sloan Manag. Rev.*, **31**, 69–80. (TYW, WAM)

M69. Menezes, M. A. J., and Srinivasan, K. (1991). Mediating effect of the quality signal of warranties on customer preferences, presented at *ORSA/TIMS 32nd Joint National Meeting*, Anaheim, CA. (PQU, WDA)

M70. Menke, W. W. (1969). Determination of warranty reserves, *Management Sci.*, **15**, B542–B549. (FRW, PRM, PRW, WRE)

M71. Menke, W. W. (1976). Comments on Amato Anderson's paper "Determination of warranty reserves: An extension," *Management Sci.*, **22**, 1395–1396. (WRE)

M72. Menzefricke, U. (1992). On the variance of total warranty claims, *Comm. Statist.—A Theory and Methodology*, **21**, 779–790. (EST, WDA)

M73. Menzefricke, U. (1993). Prediction of warranty costs during a given period of time, *INFOR*, **31**, 127–135. (EST, MAM, WCA)

M74. Metal Trades Industry Association of Australia (1979). *Tale: Product Liability and the Trade Practices Act*, Metal Trades Industry Association of Australia, Canberra, Australia. (PLI)

M75. Metzler, E. G. (1974). Forcing functions integrate R&M into design—DoD TACAN procurement policy on reliability and maintainability, *1974 Proc. Ann. Reliab. and Maintain. Symp.*, 52–55. (MAI, RIW)

M76. Meyer, K. H., and Ryan, J. B. (1988). Emerging P/M alloys for magnetic applications, *Proc. 1988 Int. Powder Metall. Conf.*, **18**, 757–772. (EXT)

M77. Meyerowitz, S. A. (1988). . . . Do not collect $200, *Business Marketing*, **73**, 100–103. (LIT, PLI, WAM)

M78. Meyerowitz, S. A. (1989). Advertising and the law, *Small Business Reports*, **14**, 64–69. (LAW, LIT, WAM)

M79. Mihalasky, J., and Jacobs, R. M. (1983). Safety, liability, performance and profitability, *1983 Proc. Ann. Reliab. and Maintain. Symp.*, 236–238. (WCA, WMA)

M80. Miller, C. J. (1986). *Comparative Product Liability*, British Institute of International and Comparative Law, London. (PLI)

M81. Miller, F. W. (1988). Design for assembly: Ford's better idea to improve products, *Manufacturing Systems*, **6**, 22–24. (PQU, WCA)

M82. Miller, R. A. (1966). Accounting for warranty (product guarantee) expense, *The New York Certified Public Accountant*, **36**, 537–538. (ACC, WCA)

M83. Miller, R. E. (1990). Conformal coatings for PCB's. What's the cost?, *Electronic Manufacturing*, **36**, 13–16. (PQU, WCA)

M84. Miller, O. W., Chou, K., and Chang, C. A. (1989). Case study. Use of the Taguchi loss function, *43rd Ann. Quality Congress Trans.*, 502–508. (EST, QCO, WDA)

M85. Miller, R., and Kanter, L. (1980). Litigation under Magnuson-Moss: New opportunities in private actions, *Uniform Commercial Code Law J.*, **10**, 18–21. (LAW)

M86. Miller, C. J., and Lovell, P. A. (1977). *Product Liability*, Butterworths, London. (PLI)

M87. Miller, J. M., Frantz, J. P., and Rhoades, T. P. (1991). Model for designing and evaluating product information, *Proc. of the Human Factors Society 35th Ann. Meeting,* San Francisco, CA. **2**, 1063–1067. (PLI)

M88. Min, H. (1989). A bicriterion reverse distribution model for product recall, *Omega, Int. J. of Management Sci.*, **17**, 483–490. (MAM, PRC, WRE)

M89. Mitra, A., and Patankar, J. G. (1988). Warranty cost estimation: A goal programming approach, *Decision Sciences*, **19**, 409–423. (EST, MAM, WCA, WRE)

M90. Mitra, A., and Patankar, J. G. (1989). A multi-objective model for warranty estimation, *Euro. J. Oper. Res.*, **45**, 347–355. (EST, MCM, WAM, WRE)

M91. Mitra, A., and Patankar, J. G. (1991). Warranty costs for renewable warranty programs under partial redemption, presented at *ORSA/TIMS 32nd Joint National Meeting*, Anaheim, CA. (EXT, PRW, WCA)

M92. Mitra, A., and Patankar, J. G. (1993). An integrated multicriteria model for warranty cost estimation and production, *IEEE Trans. Eng. Manag.*, **40**, 300–311. (EST, MCM, WCA)

M93. Mlinarchik, R. A. (1977). RIW experience at ECOM, *1977 Proc. Ann. Reliab. and Maintain. Symp.*, 257–260. (RIW)

M94. Mlinarchik, R. A. (1980). Assuring army user satisfaction through warranties, *1980 Proc. Ann. Reliab. and Maintain. Symp.* 30–31. (CSD, WCA, WMA)

M95. Moellenberndt, R. A. (1973). *An Analysis of the Effects of the Product Warranty on Selected Companies in the General Aviation Industry*, Doctoral Dissertation, University of Nebraska, Lincoln, NE. (WAP, WCA)

M96. Moellenberndt, R. A. (1977). A critical look at warranty cost recognition, *The Nat. Public Accountant*, **June**, 18–22. (ACC, WCA)

M97.   Monash University Faculty of Law, (1979). *Manufacturers' Warranties and the Trade Practices Act*, Seminar held in Melbourne, Australia. **Mar.** (LAW)

M98.   Montgomery, W. H. (1983). Design contract considerations, presented at *39th. Petroleum Mechanical Engineering Workshop and Conference*, Tulsa, OK, 97. (DCV, WCT)

M99.   Moonitz, M. (1960). The changing concept of liability, *J. Accounting*, **109**, 41–46. (ACC, PLI)

M100.   Moore, E. M., and Shuptrine, F. K. (1993). Warranties: Continued readability problems after the 1975 Magnuson-Moss warranty act, *J. Consumer Affairs*, **27**, 23–36. (CBH, CEI, WLE)

M101.   Moore, J. (1989). *Class sections and Product Liability: The Dangerous Duo*, Rept. by Member of Australian Parliament, Canberra, Australia. **Feb.** (PLI)

M102.   Moreau, D. (1989). Why the dealer wins: Appliance service contracts, *Changing Times*, **Jan.**, 83–84, 86. (EXT, SCT)

M103.   Moreno, F. J. (1990). MTBF warranty/guarantee for multiple user avionics, *1990 Proc. Ann. Reliab. and Maintain. Symp.*, 113–119. (MAM, PDS, WCA)

M104.   Morgan, F. W. (1979). The product liability consequence of advertising, *J. Advertising*, **8**, 30–37. (PLI, WAM)

M105.   Morgan, F. W. (1982). Marketing and product liability: A review and update, *J. Marketing*, **46**, 69–78. (PLI, WAM)

M106.   Morgan, F. W. (1987). Strict liability and the marketing of services vs. goods: A judicial review; *J. Public Policy and Marketing*, **7**, 43–57. (PLI, WAM)

M107.   Morgan, F. W., and Avrunin, D. I. (1982). Consumer conduct in product liability, *J. Consumer Research*, **9**, 47–55. (PLI)

M108.   Morgan, F. W., and Boedecker, K. A. (1980). The role of personal selling in products liability litigation, *J. Personal Selling and Sales Management*, **1**, 34–40. (LIT, PLI)

M109.   Morphew, A. J. (1987). Excess and surplus lines—product tampering: Capping the risk, *Best's Rev.*, **87**, 62–66. (PDV, PLI, PRC)

M110.   Moskal, B. S. (1993). Smith freshens GM's stale air, *Industry Week*, **242**, 19–22. (WCA)

M111.   Moskowitz, H., and Chun, Y. H. (1988). *A Bayesian Model for the Two-Attribute Warranty Policy*, Paper No. 950, Krannert Graduate School of Management, Purdue University, West Lafayette, IN. (PRS, STA, TDW, WCA)

M112.   Moskowitz, H., and Chun, Y. H. (1994). A Poisson regression model for two-attribute warranty policies, *Naval Res. Log.*, **41**, 355–376. (MAM, STO, TDW, WPR)

M113.   Moss, F. E. (1978). The manufacturer's role in product safety, in *Consumerism: Search for the Consumer Interest*, (D. A. Aaker and G. S. Day, eds.), 3rd ed., The Free Press, New York, 368–372. (PLI, QCO)

M114. Mate, R. A. (1969). Product liability the gathering storm, *Quality Assurance*, **Oct.**, 12–13. (PLI)

M115. Mozer, C. (1984). Total quality control: A route to the Deming prize, *Quality Progress*, **17**, 30–33. (TQM, WDA)

M116. Mundel, A. B. (1975). Failure modes and effects analysis as a means of product liability prevention, *Proceedings, Product Liability Prevention Conference*, Hasbrouck Hts, NJ, 61–78. (PLI, REL)

M117. Murphy, R. D., and Rubin, P. H. (1988). Determinants of recall success rates, *J. Products Liability*, **11**, 17–28. (EAW, PLI, PRC)

M118. Murray, J. E. Jr. (1986). Where did the extra warranty come from?, *Purchasing World*, **30**, 33–34. (BOC, LIT)

M119. Murray, J. E. Jr. (1986). Purchasing Law: Court fires salvo in the "Battle of the forms", *Purchasing World*, **30**, 43. (LAW, LIT, PLI)

M120. Murray, J. E. Jr. (1990). Confirmation can't add terms, *Purchasing World*, **31**, 68. (LIT, WCT)

M121. Murray, J. E. Jr. (1992). When the warranty doesn't work, *Purchasing*, **112**, 31. (LIT, PLI)

M122. Murray, J. E. Jr. (1993). The meaning of appear, *Purchasing*, **114**, 27–30. (LIT, WCT)

M123. Murray, J. E. Jr. (1993). Notable decisions, *Purchasing*, **115**, 30–31. (LIT)

M124. Murthy, D. N. P. (1989). Product warranty: A review and technology management implications, *Presented at the 2nd. Int. Conf. on Management of Technology*, Miami, FL. (CWP, FRW, PRW, WMA)

M125. Murthy, D. N. P. (1990). Optimal reliability choice in product design, *Engineering Optimization*, **15**, 281–294. (MAM, OPT, PDE, REL, WPR)

M126. Murthy, D. N. P. (1990). A new warranty costing model, *Math. and Comp. Modelling*, **13**, 59–69. (CUS, MAM, WCA)

M127. Murthy, D. N. P. (1992). A usage dependent model for warranty costing, *Euro. J. Oper. Res.*, **57**, 89–99. (CUS, EST, MAM, WCA)

M128. Murthy, D. N. P., and Blischke, W. R. (1991). A systems approach to study of warranties, presented at *ORSA/TIMS 32nd Joint National Meeting*, Anaheim, CA. (CSD, WAM)

M129. Murthy, D. N. P., and Blischke, W. R. (1992). Product warranties—challenges to managers and researchers, presented at *ORSA/TIMS 33rd Joint National Meeting*, San Francisco, CA. (WCA, WMA)

M130. Murthy, D. N. P., and Blischke, W. R. (1992). Product warranty management–II: An integrated framework for study, *Euro. J. Oper. Res.*, **62**, 261–281. (MAM, WMA)

M131. Murthy, D. N. P., and Blischke, W. R. (1992). Product warranty management–III: A review of mathematical models, *Euro. J. Oper. Res.*, **63**, 1–34. (MAM, QAS, WCA, WMA)

M132. Murthy, D. N. P., Djamaludin, I., and Wilson, R. J. (1995). A consumer incentive warranty policy and servicing strategy for products with uncertain quality, *Quality and Rel. Engineering*, in press. (QCO, SST)

M133. Murthy, D. N. P., and Hussain, A. Z. M. O. (1994). Warranty and optimal redundancy design, *Eng. Opt.*, **23**, 301–314. (OPT)

M134. Murthy, D. N. P., Iskandar, B. P., and Wilson, R. J. (1995). Two dimensional pro-rata warranty policies. Under preparation. (MAM, TDW, WCA)

M135. Murthy, D. N. P., Iskandar, B. P., and Wilson, R. J. (1995). Two-dimensional failure free warranties: Two-dimensional point process models, *Operations Research,* **43,** 356–366. (FRW, TDW, WCA)

M136. Murthy, D. N. P., Iskandar, B. P., and Wilson, R. J. (1992). A simulation approach to analysis of free replacement policies, *Mathematics and Computers in Simulation,* **33,** 513–518. (FRW, SIM, WCA)

M137. Murthy, D. N. P., and Nguyen, D. G. (1987). Optimal development testing policies for products sold with warranty, *Reliab. Eng.,* **19,** 113–123. (OPT, PDV, STA, WCA)

M138. Murthy, D. N. P., and Nguyen, D. G. (1988). An optimal repair cost limit policy for servicing warranty, *Math. Comp. Modelling,* **11,** 595–599. (OPT, STA, WCA)

M139. Murthy, D. N. P., and Padmanabhan, V. (1992). *A Continuous Time Model of Product Warranty,* Working paper, Graduate School of Business, Stanford University, Stanford, CA. (MAI, WCA)

M140. Murthy, D. N. P., and Padmanabhan, V. (1993). *A Dynamic Model of Product Warranty,* Research paper No. 1263, Graduate School of Business, Stanford University, Stanford, CA. (COR, MAM, OPT)

M141. Murthy, D. N. P., and Wilson, R. J. (1991). Modelling two dimensional warranties, *Proc. of the 5th Int. Symp. on Appl. Stoch. Models and Data Analysis,* Granada, Spain, (R. Gutiérrez and M. J. Valderrama, eds.), World Scientific, Singapore, 481–492. (FRW, MAM, TDW)

M142. Murthy, D. N. P., Wilson, R. J., and Djamaludin, I. (1993). Product warranty and quality control, *Qual. and Rel. Eng. Int.,* **9,** 431–443. (FRW, PRW, QCO, WCA)

M143. Murthy, D. N. P., Wilson, R. J., and Iskandar, B. P. (1990). Two dimensional warranty policies: A mathematical study, *Proc. 10th ASOR Conf.,* Perth, Australia, 89–104. (MAM, TDW)

M144. Murthy, D. N. P., Wilson, R. J., and Iskandar, B. P. (1995). Two-dimensional failure free warranties: one-dimensional point process models, under preparation. (FRW, TDW, WCA)

M145. Myerscough, S. E. (1979). Service contracts: A subject for state insurance or federal regulation—do consumers need protection from the service contract industry?, *Southern Illinois Univ. Law J.,* 587–620. (CPT, REG, SCT)

M146. Myrick, A. (1989). Sparing analysis—a multi-use planning tool, *1989 Proc. Ann. Reliab. and Maintain. Symp.,* 296–300. (WCA)

M147. Myrick, A. (1990). Analysis of warranty cost based on reliability, *1990 Proc. Ann. Reliab. and Maintain. Symp.,* 228–232. (PSU, REL, WCA, WPR)

N1.   Naj, A. K. (1987). Chrysler upgrades its 1987 warranty as U.S. makers sales battle escalates, *Wall Street Journal,* **Feb, 2,** 7. (EXT)

N2.   Natarajan, R. (1987). Sales life cycle integrated optimal warranty policies, presented at the *ORSA/TIMS Joint National Meeting*, St. Louis, MO, 193. (FRW, OPT, PRW, WAM, WTE)

N3.   National Association of Consumer Agency Administration (1980). *NACAA Product Warranties and Servicing*, National Association of Consumer Agency Administration and Society of Consumer Professionals in Business, U.S. Govt. Printing Office, Washington, DC. (COM, PSU)

N4.   National Business Counsel for Consumer Affairs, *Product Warranties: Business Guidelines to Meet Consumer Needs*, Sub-Council on Warranties and Guarantees, National Business Counsel for Consumer Affairs, Washington, DC. (WLE)

N5.   *National Underwriter* (1988). Sentry expert discusses product safety, **92**, 70–71. (PLI)

N6.   *Nation's Business* (1993). Home warranties: Protecting your investment, **81**, 70. (CPT)

N7.   Naher, D. (1991). Service guarantee: double barrel standard, *Training*, **28**, 27–31. (SCT)

N8.   Nasar, S. (1989). Best ways to get your money back, *Money*, **18**, 149–158. (CCO, CPT)

N9.   National Business Council for Consumer Affairs (U.S.) (1972). *Product Warranties, Business Guidelines to Meet Consumer Needs*, Sub-Council on Warranties and Guarantees National Business Council for Consumer Affairs, Washington, DC. (CEI, WLE)

N10.   Nenoff, L. (1988). Field operational R & M warranty, *1988 Proc. Ann. Reliab. and Maintain. Symp.*, 227–230. (REL, RIW)

N11.   Nevola, L. M. Jr. (1991). An R&M 2000 success story: The ALR-56M, *1991 Proc. Ann. Reliab. and Maintain. Symp.*, 489–493. (DCV)

N12.   Newman, D. G., and Nesbitt, L. D. (1978). USAF experience with RIW, *1978 Proc. Ann. Reliab. and Maintain. Symp.*, 55–61. (RIW, WMA)

N13.   Newton, J. E. (1986). *Prescription Drugs Product Liability: An Australian United State Comparison*, Thesis, The University of Queensland, Brisbane, Australia. (PLI)

N14.   *The New York Times* (1985). Product liability: The new morass, **Mar. 10.** (PLI)

N15.   Nguyen, D. G. (1984). *Studies in Warranty Policies and Product Reliability*, Doctoral Dissertation, The University of Queensland, Brisbane, Australia. (CWP, FRW, PRW, REL, WCA)

N16.   Nguyen, D. G., and Murthy, D. N. P. (1982). Optimal burn-in time to minimize cost for products sold under warranty, *IIE Trans.*, **14**, 167–174. (OPT, REL, WCA)

N17.   Nguyen, D. G., and Murthy, D. N. P. (1984). Cost analysis of warranty policies, *Naval Res. Logist. Quart.*, **31**, 525–541. (CWP, EST, PRM, PRW, WCA)

N18.   Nguyen, D. G., and Murthy, D. N. P. (1984). A general model for estimating warranty costs for repairable products, *IIE Trans.*, **16**, 379–386. (EST, FRW, MAM, WCA)

N19.   Nguyen, D. G., and Murthy, D. N. P. (1986). An optimal policy for servicing warranty, *J. Operational Res. Soc.*, **37**, 1081–1088. (FRW, MAM, OPT, WSE)

N20.   Nguyen, D. G., and Murthy, D. N. P. (1988). Failure free warranty policies for non-reparable products: A review and some extensions, *R.A.I.R.O. Operations Research*, **22**, 205–220. (FRW, MAM, WCA)

N21.   Nguyen, D. G., and Murthy, D. N. P. (1988). Optimal reliability allocation for products sold under warranty, *Eng. Opt.*, **13**, 35–45. (OPT, REL, WCA)

N22.   Nguyen, D. G., and Murthy, D. N. P. (1989). Optimal replace-repair strategy for servicing products sold with warranty, *Euro. J. Oper. Res.*, **39**, 206–212. (CWP, MAM, OPT)

N23.   Nielsen, A. C. (1975). Caveat venditor, *The Neilsen Researcher*, **6**, 2–3. (LAW)

N24.   Nieman, W. (1989). Negotiating the computer contract, *J. Property Management*, **54**, 56–59. (WCT)

N25.   Noel, D. W. (1972). Defective products: Abnormal use, contributory negligence, and assumption of risk, *Vanderbilt Law Rev.*, **25**, 93–130. (CUS, PLI)

N26.   Nordin, G. H. (1989). The swedish product safety legislation, *J. Consumer Policy*, **12**, 95–104. (PRC, WLE)

N27.   Nordstrom, R. D., and Metzner, H. (1976). Warranties: How important as a marketing tool?, in *Proc. Southern Marketing Assoc.*, Atlanta, GA, 26–28. (WAM)

N28.   North, D. M. (1994). Greater flexibility marks Gulfstream 4-SP, *Aviation Week and Space Tech.*, **140**, 44–47. (WSE)

N29.   Nowicki, P. R. (1987). *Regulating and Resolving New Car Warranty Problems and Disputes: An Analysis of Lemon Laws, FTC Rule 703 and Alternative Dispute Resolution*, Doctoral Dissertation, Syracuse University, Syracuse, NY. (CMD)

O1.   O'Brien, L. P. (1993). The warranty racket, *Working Woman*, **18**, 36. (ESC, SCT)

O2.   Ockenden, J. M. (1982). Quality assurance in software and system houses, *IEE Colloquium (Digest)*, **1982/8**, 4.1–4.2. (QAS, WAP)

O3.   Odum, B. (1991). Proper planning makes capital equipment purchasing easier and more efficient, *Hospital Materials Management*, **16**, 16–17. (WSL)

O4.   Oi, W. Y. (1973). The economics of product safety, *Bell J. Economics and Management Sci.*, **4**, 3–28. (EAW, PLI, WMS)

O5.   Oi, W. Y. (1974). The economics of product safety: A rejoinder, *Bell J. Economics and Management. Sci.*, **5**, 689–695. (EAW, PLI, WMS)

O6.   Oliver, R. L. (1980). A cognitive model of the antecedents and consequences of satisfaction decisions, *J. Marketing Res.*, **17**, 460–469. (CSD)

O7.   Olson, W. (ed.) (1988). New direction in liability law, *Proceeding the Academy of Political Science*, **37**, New York. (PLI)

O8.   Ontario Law Reform Commission (1972). *Report on Consumer Warran-*

ties and Guarantees in the Sale of Goods, Dept. of Justice, Toronto. (CMD, PRS, WAD, WLE)

O9.   O'Neal, J. (1974). Product recall insurance, *CPCU Annals*, **Mar.**, 40–43. (PRC)

O10.   Ordover, J., and Rubinstein, A. (1983). *On Bargaining, Settling, and Litigating: A Problem in Multistage Games with Imperfect Information*, New York University, New York, NY. **Feb.** (LIT)

O11.   Owen, D. B., and Chou, Y. M. (1983) Prediction intervals using exceedances for an additional third-stage sample, *IEEE Trans. on Reliab.*, **R-32**, 314–316. (WTE)

O12.   Owles, D. (1978). *The Development of Product Liability in the USA*, Lloyd's of London Press, Ltd., London. (PLI)

O13.   Owles, D., Worsdall, A., and Ashworth, J. S. (1984). *Product Liability Casebook: Leading US and UK Judgments*, Lloyd's of London, Colchester. (PLI)

P1.   Padmanabhan, V. (1992). *Usage Heterogeneity and Extended Service Contracts*, Research Paper 1197, Graduate School of Business, Stanford University, Stanford, CA. (CUS, ESC)

P2.   Padmanabhan, V., and Bass, F. M. (1993). Optimal pricing of successive generations of product advances, *Int. J. Research in Marketing*, **10**, 185–207. (OPT, WAM, WPR)

P3.   Padmanabhan, V., and Png, I. P. L. (1993). *Returns Policies: An Underappreciated Marketing Variable*, Research paper No. 1268, Graduate School of Business, Stanford University, Stanford, CA. (WAM)

P4.   Padmanabhan, V., and Rao, R. C. (1993). Warranty policy and extended service contracts: Theory and an application to automobiles, *Marketing Sci.*, **12**, 230–247. (CBH, ESC, MAM, STA)

P5.   Palfrey, T., and Romer, T. (1983). Warranties, performance, and the resolution of buyer-seller disputes, *Bell J. Economics and Management Sci.*, **14**, 97–117. (CMD, MAM, PQU, WMS)

P6.   Paquet, M. D., Gatlin, P. K., and Cote, S. M. (1985). Usage monitoring—a milestone in engine life management, presented at *AIAA/SAE/ASME/ASEE 21st Joint Propulsion Conference*, Monterey, CA, 6p. (CUS, WCT)

P7.   Parascos, E. T. (1992). Reliability, availability and maintainability (RAM) in the electric power industry, *Proc. 54th Ann. Meeting of the American Power Conf.*, 589–593. (MAI, REL, WDA)

P8.   Paray, P. E. (1993). Judicial treatment of damages exclusions negotiated in custom software licenses, *Uniform Commercial Code Law J.*, **25**, 240–256. (BOC, LIT)

P9.   Parenmark, G., Malmkvist, A. -K., and Ortengren, R. (1993). Ergonomic moves in an engineering industry: Effects on sick leave frequency, labor turnover and productivity, *Int. J. Industrial Ergonomics*, **11**, 291–300. (WCA)

P10.   Park, K. S. (1985). *Development of a Price Indifference Function With Parameters of Reliability, Maintenance and Warranty*, Doctoral Dissertation, Texas A&M University, Lubbock, TX. (FRW, MAI, PRW, REL, WPR)

P11.   Park, K. S. (1985). Optimal use of product warranties, *IEEE Trans. on Reliab.*, **R-34**, 519–521. (SCT, WCA, WSL)

P12.   Park, K. S., and Yee, S. R. (1984). Present worth of service cost for consumer product warranty, *IEEE Trans. on Reliab.*, **R-33**, 424–426. (FRW, WCT, WRE)

P13.   Park, J. T. (1984). *An Optimal Periodic Replacement Policy for Warrantied Products*, Master's Thesis, Korea Advanced Institute of Science and Technology, Seoul, Korea. (FRW, OPT, SST)

P14.   Patankar, J. G. (1978). *Estimation of Reserves and Cash Flows Associated with Different Warranty Policies*, Doctoral Dissertation, Clemson University, Clemson, SC. (EST, TYW, WRE)

P15.   Patankar, J. G., and Klafehn, K. A. (1988). Warranty cost estimation using simulation, *Winter Simul. Conf. Proc. 1988*, 857–859. (EST, SIM, WCA)

P16.   Patankar, J. G., and Mitra, A. (1989). Effects of warranty execution under various warranty rebate plans, presented at *TIMS XXIXth Inter. Meeting*, Osaka, Japan. (EST, MAM, PRM, WCA, WDA, WRE)

P17.   Patankar, J. G., and Mitra, A. (1987). Warranty cost estimation using multiple products, presented at the *ORSA/TIMS Joint National Meeting*, St. Louis, MO, 193. (EST, MCM, WCA, WPR)

P18.   Patankar, J. G., and Mitra, A. (1987). A warranty cost estimation and production problem using multiple objectives, presented at the *ORSA/TIMS Joint National Meeting*, New Orleans, LA, 243 (OPT, WPR, WRE, WTE)

P19.   Patankar, J. G., and Mitra, A. (1989). A multi-objective model for warranty cost estimation using multiple products, *Computers and Operations Research*, **16**, 341–351. (EST, MCM, WCA, WRE)

P20.   Patankar, J. G., and Worm, G. H. (1981). Prediction intervals for warranty reserves and cash flows, *Management Sci.*, **27**, 237–241. (EST, PRM, PRW, WRE)

P21.   Patel, B. M. (1987). Managing quality through to the marketplace, *CME, Chartered Mechanical Engineer*, **34**, 35–38. (TQM, STA, WAM)

P22.   Paterson, F. A. (1991). *Collateral Warranties Explained*, RIBA Publications Ltd., London. (PLI, WLE)

P23.   Patterson, R. L. (1987). Product warranties, *Quality Progress*, **20**, 43–45. (PQU, WAM, WCA)

P24.   Pearce, J. (1973). Product liability and consumer risk taking—economic tradeoffs, *IEEE Trans. on Reliab.*, **R-22**, 49–51. (COR, PLI)

P25.   Pengilley, W. (1993). *Product Liability: Matters not Covered by part VA of the Trade Practice Act*, Legal Books, Redfern, New South Wales, Australia. (PLI, WLE)

P26.   Pennock, J. L., and Jaeger, C. M. (1964). The household service life of durable goods, *J. Home Economics*, **56**, 22–26. (REL)

P27.   Perham, J. C. (1977). The dilemma in product liability, *Dun's Rev.*, **109**, 48–50, 76. (PLI)

P28.　Perry, M., and Perry, A. (1976). Service contract compared to warranty as a means to reduce consumer's risk, *J. Retailing*, **52**, 33–40, 90. (COR, SCT)

P29.　Pertschuk, M. (1982). *Revolt Against Regulation: The Rise and Pause of the Consumer Movement*, University of California Press, Berkeley. (COM)

P30.　Petelski, N. (1987). *Horizontally-Applied Telescopic Cylinders*, Texas Hydraulics Inc. Temple, TX. (WCA)

P31.　Peterson, J. M. (1990). Warranty: An effective disclaimer?, *South Dakota Business Rev.*, **48**, 10–11. (COR)

P32.　Peterson, R. H. (1990). Performance study of single ply roof membranes in the Northeast United States, *Symp. Roofing Research and Standards Development, 2nd Volume*, San Francisco, CA, ASTM Special Technical Publ. No. 1088, 82–96. (WDA)

P33.　Phelan, W. K. (1991). What the new bulletin on extended warranties means to controllers, *Corporate Controller*, **4**, 43–44, 60. (ACC, EXT)

P34.　Pirojboot, B. (1991). *Warranty Policies Beneficial to Both Consumer and Producer*, Master's Thesis, Northern Illinois University, DeKalb, IL. (TYW, WPR)

P35.　Pirojboot, B., and Marcellus, R. (1991). Warranty policies beneficial to both consumer and producer, presented at *ORSA/TIMS 32nd Joint National Meeting*, Anaheim, CA. (TYW, WPR)

P36.　Pisacreta, E. A., Millstein, J. S., Khatcherian, S., and Raysman, R. (1993). Contract dangers lurk in turnkey implementations, *Computers in Healthcare*, **14**, 38–41. (WAP)

P37.　Plotkin, A. A. (1985). A long look at extended warranties, *Automotive News Extra-Marketing*, **86**, 30–32. (EXT, WAP)

P38.　Plumb, P. (1986). Reltest, a reliability testing method, *1986 Proc. Ann. Reliab. and Maintain. Symp.* 267–272. (EST, REL, STA, WCA)

P39.　Polinsky, A. M., and Rogerson, W. P. (1983). Products liability, consumer misperceptions and market power, *Bell J. Econ. and Management Sci.*, **14**, 581–589. (CEI, PLI)

P40.　Powell, F. M. (1990). The seller's duty to disclose in sales of commercial property, *American Business Law J.*, **28**, 245–274. (CPT)

P41.　Powell, R. E., Graeff, K. G., and MacFarlane, E. W. (1991). The sophisticated user defense and liability for defective design: The twain must meet, *J. Products Liability*, **13**, 113–120. (CPT, PDE, PLI)

P42.　Powers, W. C. Jr. (1984). Distinguishing between products and services in strict liability, *North Carolina Law Rev.*, **62**, 415–434. (PLI)

P43.　Prag, P. (1975). A comparative study of the concept and development of product liability in the U.S.A., Germany and Scandinavia, *Legal Issues of European Integration*, **1**, 67. (PLI)

P44.　The President's Task Force Report in Appliance Warranties and Service (1969). *Report of the Task Force on Appliance Warranties and Service*, Federal Trade Comm., US Govt. Printing Office, Washington, DC, 100–108. (WAP, WLE)

P45. Pridgen, D. (1991). *Consumer Protection and the Law*, Clark Broadman and Callaghan, New York. (CPT, LAW)

P46. Prieguez, R. M. (1992). Hahn v. superior court: Failing to take the doctrine of strict premises liability to its logical conclusion, *San Diego Law Rev.*, **29**, 525–560. (LIT, PLI)

P47. Priest, G. L. (1978). Breach and remedy for the tender of nonconforming goods under the uniform commercial code: An economic approach, *Harvard Law Rev.*, **91**, 960–1001. (EAW, WLE)

P48. Priest, G. L. (1981). A theory of the consumer product warranty, *Yale Law J.*, **90**, 1297–1352. (PLI, WLE)

P49. *Proc. Failure Free Warranty Seminar* (1973), US Navy Aviation Supply Office, Philadelphia, PA. (RIW)

P50. *Profit-Building Strategies for Business Owners* (*PBS*) (1992). What you promise the buyer when you warranty your product, **22**, 14–15. (WAM, WLE)

P51. Prohaska, J. T., Briggs, W. G., Denney, M., and Pluciennik, M. A. (1978). Consumer survey of appliance repairs and service contracts, in *Consumer Durables: Warranties, Service Contracts and Alternatives*, (R. T. Lund, ed.), Rept. No. CPA-78-14/Vol. III, Center for Policy Alternatives, Massachusetts Institute of Technology, Cambridge, MA, 3-6–3-72. (SCT)

P52. Prohaska, J. T., Briggs, W. G., DeWolf, J. B., and Lund, R. T. (1978). Consumer survey of appliance repairs and service contracts, in *Consumer Durables: Warranties, Service Contracts and Alternatives*, (R. T. Lund, ed.), Rept. No. CPA-78-14/Vol. IV, Center for Policy Alternatives, Massachusetts Institute of Technology, Cambridge, MA, 4-8–4-145. (MAM, WDA, WPR)

P53. Prosser, W. L. (1943). The implied warranty of merchantable quality, *Minn. Law Rev.*, **27**, 117–168. (LAW)

P54. Pruitt, S. W., Reilly, R. J., and Hoffer, G. E. (1986). Security market anticipation of consumer preference shifts: The case of automotive recalls, *Quart. J. Business and Economics*, **25**, 14–28. (CBH, MAM, PRC, WDA)

P55. Pynn, C. T. (1985). Manufacturing renaissance, *Test and Measurement World*, **5**, 31–32. (PQU, WCA)

Q1. Quinn, M. J. (1990). Lessons learned: Cooking up an ESS program from scratch, *Proc. 36th Ann. Tech. Meeting Institute of Environmental Sciences*, 794–798. (QCO, WCA)

R1. Rachamadugu, R., and Shanthikumar, J. G. (1991). Layout considerations in assembly line design, *Int. J. Prod. Research*, **29**, 755–768. (EST, WCA)

R2. Rados, D. L. (1969). Product liability: Tougher ground rules, *Harvard Business Rev.*, **47**, 144–152. (PLI)

R3. Ragazzi, R. A., Stokes, J. T., and Gallagher, G. L. (1985). Evaluation of a Colorado short vehicle emission test (CDH-226) in predicting federal test procedure (FTP) failures, *Int. Fuels and Lubricants Meeting and*

Exposition, SAE Technical Paper Series No. 852111, SAE, Warrendale, PA. (REL, STA)

R4. *Railway Age* (1992). Thrall: A quality comeback, **193**, 65–66. (QCO, WCA)

R5. Raney, P. R., and Thayer, K. B. (1983). Design Contracts—An Engineer/ Consultants View, Presented at *39th Petroleum Mechanical Engineering Workshop and Conference*, Tulsa, OK, 111–114. (WCT)

R6. Rao, C. P., and Weinrauch, J. D. (1976). Consumers and appliance warranties, *Baylor Business Studies*, **109**, 17–27. (CPT, WAP)

R7. Rao, S. S., and Zhao, Q. (1993). Nonparametric renewal function estimation and warranty cost estimation, *Opsearch* (*Operational Research Society of India*), **30**, 231–243. (EST, RFT, WCA)

R8. Rapp, H., and Trauth, W. (1986). Low temperature resistant shock absorber seals—possibilities in design and material, *Seals: Design and Performance Conference*, 25–32. (EXT)

R9. Rayney, P. (1989). The deed of indemnity: Legal and tax pointers, *Accountancy*, **104**, 91–92. (PLI)

R10. Reinganum, J. F., and Wilde, L. L. (1986). Settlement, litigation, and the allocation of litigation costs, *RAND J. Econ.*, **17**, 557–566. (CMD, LIT)

R11. Remich, N. C. Jr. (1993). Designing product liability out of appliances, *Appliance Manufacturer*, **41**, 43–44. (PDE, PLI)

R12. Renoud, W. J. (1989). Effective engineering approaches for successful FRP corrosion applications, *American Society of Mechanical Engineers, Petroleum Division*, **24**, 15–22. (QCO, WTE)

R13. *Report on Motor Vehicle Warranties*, (1978). Office Deputy Chief of Staff for Logistics, Department of the Army, Washington, DC. (WAP)

R14. Rescigno, R. (1991). Automatic headache: A chrysler transmission comes under fire, *Barron's*, **71**, 13, 29. (CSD, WCA)

R15. Reterer, B. L. (1976). *Considerations for effective warranty applications*, Tech. Rept. No. 6107-1467, ARINC Research Corp., Annapolis, MD. (RIW)

R16. Reterer, B. L. (1976). Considerations for effective warranty application, *1976 Proc. Ann. Reliab. and Maintain. Symp.*, 346–350. (RIW, WAP)

R17. Reve, H. J. (1976). Reliability warranty concept growing, *Ind. Marketing*, **61**, 6–7. (RIW)

R18. Reynolds, M. (1991). How extended warranties can widen the 'ring of confidence', *Business Marketing Digest*, **16**, 47–52. (CSD, EXT, QAS)

R19. Richins, M. L. (1982). An investigation of consumer attitudes toward complaining, in *Advances in Consumer Research*, (A. Mitchel, ed.), **9**, Association for Consumer Research, St. Louis, MD, 502–506. (CBH, CCO, CSD)

R20. Riegert, R. A. (1990). An overview of the Magnuson-Moss warranty act and the successful consumer-plaintiff's right to attorneys' fees, *Commercial Law J.*, **95**, 468–485. (CPT, LAW)

R21. Rigg, M. S. (1986). *Sales of Goods and Services*, National Consumer Law Center, Boston. (SCT)

R22.　Riswadkar, A. (1988). Product recall program, *Professional Safety*, **33**, 19–22. (PRC)

R23.　Ritchken, P. H. (1985). Warranty policies for non-reparable items under risk aversion, *IEEE Trans. Reliab.*, **34**, 147–150. (FRW, MAM, PRW, WAM)

R24.　Ritchken, P. H., Chandramohan, J., and Tapiero, C. S., (1989). Servicing, quality design and control, *IIE Trans.*, **21**, 213–220. (EAW, MAM, QCO, WMA)

R25.　Ritchken, P. H., Dean, B. V., and Reisman, A. (1985). Establishing warranty reserves for a product which fails given imperfect information, *Applications of Management Science*, **4**, 177–193. (PQU, WRE)

R26.　Ritchken, P. H., and Fuh, D. (1986). Optimal replacement policies for irreparable warrantied items, *IEEE Trans. on Rel.*, **R-35**, 621–623. (FRW, OPT, STA)

R27.　Ritchken, P. H., and Tapiero, C. S. (1986). Warranty design under buyer and seller risk aversion, *Naval Research Logistics. Q.*, **33**: 657–671. (COR, PRS, WAM, WTE)

R28.　Ritsema, H. A. (1988). Product liability in the European community: a new perspective?, in *Marketing Cooperation with Europe*, (J. Singh, ed.), Institute of Marketing and Management, New Delhi. (PLI)

R29.　Ritsema, H. A., and Piëst, E. (1990). New product management and product liability, *Int. J. Materials and Product Technology*, **5**, 109–120. (PLI, WMA)

R30.　Robinson, L. M., and Berl, R. L. (1979). What about compliments: A follow up study on customer complaints and compliments, in *Refining Concepts and Measures of Consumer Satisfaction and Complaining Behavior* (H. K. Hunt and R. L. Day, eds.), 144–148. (CCO, CSD)

R31.　Rockwell, W. H. (1975). Products liability and standards systems, *Proceedings, Product Liability Prevention Conference*, Hasbrouck Hts., NJ, 43–48. (PLI)

R32.　Rogerson, W. P. (1984). Efficient reliance and damage measures for breach of contract, *RAND J. Economics*, **15**, 39–53. (BOC, WCT)

R33.　Rollheiser, G. (1984). Maintenance in 1984, *Maintenance—The Hub of a Productive Wheel, Fourth CIM Mechanical/Electrical Plant Engineering and Maintenance Conference*, 99–102. (MAI, WTE)

R34.　Rooker, T. (1989). Warranty program engineering, *1989 Proc. Ann. Reliab. and Maintain. Symp.*, 9–14. (OPT, SIM, STA, WCA)

R35.　Rosenn, M. (1974). Product liability: An interaction of law and technology, a commentary, *Duquesne Law Rev*, **12**, 465–475. (LAW, PLI)

R36.　Rosenwasser, R. N. (1992). Computer viruses: How to guard against them, *Office*, **115**, 11, 18. (WCT)

R37.　Ross, I. (1975). Perceived risk and consumer behavior: A critical review, *Advances in Consumer Research*, Chicago, IL, **2**, 245–249. (CBH, COR)

R38.　Rothschild, D. P. (1976). The Magnuson-Moss warranty act: Does it balance warrantor and consumer interests?, *George Washington Law Review*, **44**, 335–380. (CPT, LAW)

R39.　Rotz, A. O. (1986). Warranty pricing with a life cycle cost model, *1986*

Proc. Ann. Reliab. and Maintain. Symp., 454–456. (LCC, RIW, WDA, WPR)

R40. Rubin, P. H., Murphy, R. D., and Jarrell, G. (1988). Risky products, risky stocks, *Regulation*, **12**, 35–39. (PRC)

R41. Ruffin, M., and Tippett, K. S. (1975). Service life expectancy of household appliances: New estimates from USDA, *Home Eco. Res. J.*, **4**, 159–170. (REL)

R42. Russo, J. E. (1979). Consumer satisfaction/dissatisfaction: An outsider's view, in *Advances in Consumer Research*, (W. L. Wilkie, ed.), **6**, Association for Consumer Research, Miami, FL, 453–455. (CSD)

R43. Ryan, J. (1975). Products liability: What every marketer should know, *Prod. Man*, **Sept**, 27. (PLI)

R44. Rypka, E. A., and Kujawski, G. F. (1983). Repair/reliability guarantee programs can work. *1983 Proc. Reliab. and Maintain. Symp.*, 221–224. (REL, WDA)

S1. Saba, J. P. (1990). Transitional arrangements, *Middle East Executive Reports*, **13**, 11–13. (BOC, LIT)

S2. Sack, S. M. (1986). Some words on warranties. . . , *Sales and Marketing Manag.*, **137**, 52, 54. (WAM, WLE)

S3. *SAE International Off-Highway and Powerplant Congress and Exposition.* (1984). Product Liability and Quality, SP-586, **Sep. 10-13.** (PLI)

S4. Safer, R. (1982). The Magnuson-Moss act class action provisions: Consumers' remedy or an empty promise?, *The Georgetown J.*, **70**, 1399, 1419. (LAW)

S5. Sahin, I. (1993). *Product Quality and Warranty Costs*, working paper, School of Business Administration, University of Wisconsin, Milwaukee, WI. (PQU, WCA, WDA)

S6. Sakauchi, T., and Kunimi, H. (1991). Evaluation method of the corrosion resistance of zinc-alloy-plated steel sheet applied to automobile body, *Tetsu-To-Hagane/J. Iron and Steel Inst. Japan*, **7**, 1154–1161. (in Japanese). (EXT)

S7. Salant, S. W., and Rest, G. (1982). *Litigation of Questioned Settlement Claims*: A Bayesian Nash Equilibrium Approach, The RAND Corporation, P-6809, September. (CMD, LIT)

S8. Salzer, T. (1989). Are design professionals responsible for radon in buildings?, *Building Standards*, **58**, 20–23. (EXT)

S9. Samuelson, W. (1983). *Negotiation versus Litigation*, Working Paper No. 7/84, March, Boston University School of Management, Boston, MA. (CMD, LIT)

S10. Sarat, A. (1975). Alternatives in dispute processing: Litigation in a small claims court, *Law Soc. Rev.*, **10**, 339–375. (CMD, LIT)

S11. Sarath, B. (1991). Uncertain litigation and liability insurance, *RAND J. Economics*, **22**, 218–231. (LIT, PLI)

S12. Sarawgi, N., and Kurtz, S. K. (1995). A simple method for predicting the cumulative failures of consumer products during warranty period, *1995 Proc. Ann. Reliab. and Maintain Symp.*, 384–390. (STA, WTE)

S13.   Sastri, D. S. (1973). Liability prevention via the court—is it possible?, *IEEE Trans. on Reliab.*, **R-22**, 46–48. (PLI, TOR)

S14.   Satiani, H. S. (1984). Clinical engineering and prepurchase of medical equipment, *Proc. 37th. Ann. Conf. on Engineering in Medicine and Biology*, Los Angeles, CA, **26**, 333. (SCT, WAP)

S15.   Savage, R. L. III (1993). Laying the ghost of reliance to rest in Section 2-313 of the Uniform Commercial Code: An 'endpoints' analysis, *Wake Forest Law Review*, **28**, 1065–1098. (LIT, WLE)

S16.   Schaden, R. F., and Heldman, V. C. (1982). *Product Design Liability*, Practicing Law Institute, New York. (PDE, PLI)

S17.   Schauer, D. (1989). Used trucks . . . a changing market, *Fleet Equipment*, **15**, 38–39. (WAP)

S18.   Schewe, M. (1989). Evaluating jukeboxes for optical disk systems, *IMC J.*, **25**, 26–27. (DCV)

S19.   Schick, T. E. (1985). Product support—a technical lifeline, *Aerospace Technology Conference and Exposition*, SAE Technical Paper Series No. 851961, publ. by SAE, Warrendale, PA. (QAS, PSU)

S20.   Schlack, M. (1986). Interiors: Toughness joins fit and finish, *Plastics World*, **44**, 47, 49–51. (EXT, REL)

S21.   Schmidt, A. E. (1977). *A View of the Evolution of the Reliability Improvement Warranty*, Rept. 76-1, Defense Systems Mgmt. College, Ft. Belvoir, VA. (RIW)

S22.   Schmidt, B. A. (1979). Preparation for LCC proposals and contracts, *1979 Proc. Ann. Reliab. and Maintain. Symp.*, 62–66. (COR, LCC, RIW, WMA)

S23.   Schmitt, J., Kanter, L., and Miller, R. (1979). *Impact of the Magnuson-Moss Warranty Act: A Comparison of 40 Major Consumer Product Warranties from Before and After the Act*, Bureau of Consumer Protection, Staff Report to the Federal Trade Commission. (WAP, WLE)

S24.   Schock, H. E. (1979). Regulatory cost management and product assurance, *1979 Proc. Ann. Reliab. and Maintain. Symp.*, 13–17. (PLI)

S25.   Schock, H. E. (1980). Regulatory impacts in product assurance, *1980 Proc. Ann. Reliab. and Maintain. Symp.*, 349–352. (PLI)

S26.   Schroeder, M. R. (1978). Private actions under the Magnuson-Moss warranty act, *California Law Rev.*, **66**, 1–36. (LAW)

S27.   Schwartz, D. A. (1979). Get ready for those consumer complaints, *Sales Management*, **30**. (CCO)

S28.   Schwartz, A., and Wilde, L. L. (1979). Intervening in markets on the basis of imperfect information: A legal and economic analysis, *Univ. of Pennsylvania Law Rev.*, **127**, 630–682. (CEI, EAW, WLE)

S29.   Schwartz, A., and Wilde, L. L. (1983). Warranty markets and public policy, *Information Economics and Policy*, **1**, 55–67. (EAW, PUB, WAM)

S30.   Schwartz, A., and Wilde, L. L. (1983). Imperfect information in markets for contract terms: The example of warranties and security interests, *Virginia Law Rev.*, **69**, 1387–1485. (WAM, WCT)

S31. Seavy, M. (1991). State lawmakers scrutinize service contracts, *HFD*. **Mar. 3,** 74–75. Cited in: *Consumer Reports News Digest* (1991), **Jul. 15,** 12. (LAW, SCT)

S32. Seldon, M. R. (1980). Estimating the cost of R and M changes, *1980 Proc. Ann. Reliab. and Maintain. Symp.*, 79–84. (REL, RIW, WCA)

S33. Serafin, R., and Horton, C. (1988). Crucial drive for Audi, Peugeot, *Advertising Age*, **59,** 40. (WAM)

S34. Shavell, S. (1980). Damage measures for breach of contract, *Bell J. Econ.*, **11,** 466–490. (BOC)

S35. Shavell, S. (1982). On liability and insurance, *Bell J. Econ.*, **13,** 120–132. (PLI)

S36. Shavell, S. (1984). The design of contracts and remedies for breach, *Quarterly J. Econ.*, **99,** 121–148. (BOC)

S37. Shelton, D. K., and Paxman, R. G. (1982). A reliability warranty concept for the Foreign Military Sales (FMS) environment, *1982 Proc. Ann. Reliab. and Maintain. Symp.*, 34–39. (RIW, STA)

S38. Shen, M. S. (1986). Burn-in effectiveness evaluation—a lotus 1-2-3 application, *1986 Proc. Ann. Reliab. and Maintain. Symp.*, 148–150. (EST, REL, STA, WCA)

S39. Sherman, P. (1981). *Products Liability, for the General Practitioner*, Shepard's McGraw-Hill, Colorado Springs, CO. (PLI)

S40. Sherry, C. J. (1987). When the new PC is a lemon, *Association and Society Manager*, **19,** 16–19. (CPT)

S41. Shimp, T. A., and Bearden, W. O. (1982). Warranty and other extrinsic cues effects on consumer's risk perceptions, *J. Consumer Research*, **9,** 38–46. (CBH, COR, REL)

S42. Shmoldas, J. D. (1977). *Improvement of Weapon System Reliability Through Reliability Improvement Warranties*, Rept. 77-1, Defense Systems Mgmt. College, Ft. Belvoir, VA. (RIW)

S43. Shorey, R. R. (1976). Factors in balancing government and contractors risk with warranties, *1976 Proc. Ann. Reliab. and Maintain. Symp.*, 366–368. (RIW, WCA)

S44. Shupe, R. H., and Driessnack, J. (1991). Methodology for an economic assessment of a reliability assurance warranty program (RAWP), *1991 Proc. Ann. Reliab. and Maintain. Symp.*, 345–351. (EAW, QAS)

S45. Shuptrine, F. K. (1976). A comparison of two readability measures—fog index and the Dale-Chall method—on the comprehensibility of warranties, *Proc. Southern Marketing Association*, 9–10. (CEI)

S46. Shuptrine, F. K., and Moore, E. M. (1980). Even after the Magnuson-Moss Act of 1975, warranties are not easy to understand, *J. Consumer Affairs*, **14,** 394–404. (CEI)

S47. Shuptrine, F. K., and Wenglorz, G. (1981). Comprehensive identification of consumers' marketplace problems and what they do about them, in *Advances in Consumer Research*, (K. B. Monroe, ed.), **8,** Association for Consumer Research, 687–692. (CCO, CSD)

S48. Siferd, S. P., Benton, W. C., and Ritzman, L. P. (1992). Strategies for service systems, *Euro. J. Oper. Res.*, **56**, 291–303. (SST)

S49. Silber, M. R. (1983). Extended warranties, *Consumers' Research Magazine*, **66**. (EXT)

S50. Simon, M. J. (1981). Imperfect information, costly litigation and product quality, *Bell J. Economics*, **12**, 171–184. (CMD, LIT, PQU)

S51. Simonson, A. (1991). Warranties and the law—use caution, *Sloan Management Rev.*, **32**, 7–8. (LAW, PLI, WLE)

S52. Singer, K. (1987). Marketing's new watchword: Satisfaction guaranteed, *Adweek*, **9**, 59–61. (CSD, WAM)

S53. Singpurwalla, N. D. (1987). *A Strategy for Setting Optimal Warranties*, Report GWU/IRRA/Serial TR-87/4, The George Washington University, Washington, DC. (STA, TDW)

S54. Singpurwalla, N. D. (1987). *An Interactive PC-Based Procedure for Reliability Assessment Incorporating Expert Opinion and Survival Data*. Technical Report Serial TR-86/1. The Institute for Reliability and Risk Analysis, The George Washington University, Washington, DC. (REL)

S55. Singpurwalla, N. D., and Wilson, S. (1993). The warranty problem: Its statistical and game theoretic aspects, *SIAM Review*, **35**, 17–42. (EST, OPT, TDW, WPR, WRE)

S56. Siores, E., and Gibson, P. R. (1991). Buying and selling reliability, *Int. Mechanical Engineering Congress and Exhibition—MECH '91*, 64–67. (PDE, WAM)

S57. Slovick, M. (1992). Don't buy a bridge to Brooklyn: Computing the benefits, *Dealerscope Merchandising*, **34**, 26–27, 49–50. (EXT)

S58. Smarr, S. L. (1989). The 20th century Paul Reveres, *Bobbin*, **30**, 46–54. (QCO, WAP)

S59. Smedinghoff, T. J. (1987). Analyzing the warranties in a computer contract, *Systems/3X and AS World*, **15**, 88–94. (BOC, WAP)

S60. Smedinghoff, T. J. (1988). DP law: Limits on limiting liability, *Systems/3X and AS World*, **16**, 8. (LAW, LIT, PLI)

S61. Smith, C. (1977). The Magnuson-Moss warranty act: Turning the tables on Caveat Emptor,*California Western Law Rev.*, **13**, 391–429. (LAW)

S62. Smith, C. O. (1980). Product liability: Defective conditions, *1980 Proc. Ann. Reliab. and Maintain. Symp.*, 470–474. (PDE, PLI)

S63. Smith, C. O. (1981). *Products Liability, Are You Vulnerable?*, Prentice-Hall, New York. (PLI)

S64. Smith, G. W. (1988). Carpet 'wear' is performance—but how is performance defined?, *Facilities Design and Management*, **7**, 29, 34. (WAP)

S65. Smith, J. C. (1987). *Smith and Thomas: A Casebook on Contract*, 8th ed., Sweet and Maxwell, London. (BOW, WCT)

S66. Smith, M. R. (1987). Improving product quality in American industry, *Academy Management Executive*, **1**, 243–245. (QCO)

S67. Smith, C. O., and Talbot, T. F. (1992). Effects of product liability on design, *Winter Annual Meeting—Amer. Soc. Mech. Engineers*, Anaheim, CA, 1–18 92-WA/DE-8. (BOW, PDE, PLI)

S68. Smithson, C. W., and Thomas, C. R. (1988). Measuring the cost to consumers of product defects: The value of 'lemon insurance', *J. Law and Economics*, **31**, 485–502. (CPT, WPR)

S69. Soska, G. (1991). How to select a robotics supplier, *Robotics World*, **9**, 13–15. (DCV)

S70. Spence, A. M. (1974). *Market Signaling*, Harvard University Press, Cambridge, MA. (CEI)

S71. Spence, M. (1977). Consumer misperceptions, product failure and producer liability, *Rev. Econ. Studies*, **44**, 561–572. (CEI, PLI, WMS)

S72. Spengler, R. M. (1977). RIW with cost sharing, *1980 Proc. Ann. Reliab. and Maintain. Symp.*, 391–395. (RIW, WCA)

S73. Spiegler, I., and Herniter, J. (1993). Warranty cards as a new source of industrial marketing information, *Computers in Industry*, **22**, 273–281. (FCT, WAM, WDA)

S74. Springer, R. M., Jr. (1977). Risks and benefits in reliability warranties, *J. Purchasing and Materials Manag.*, **13**, 8–13. (RIW, WTE)

S75. Sproles, G. B. (1977). New evidence on price and product quality, *J. Consumer Affairs*, **11**, 63–77. (PQU)

S76. Spulber, D. F. (1988). Products liability and monopoly in a contestable market, *Economica*, **55**, 333–341. (PLI, WMS)

S77. Squillante, A. M. (1987). Remedies provided by the Magnuson-Moss warranty act, *Commercial Law J.*, **92**, 366–383. (CPT, LAW, SCT)

S78. Southwick, A. F. Jr. (1963). Mass marketing and warranty liability, *J. Marketing*, **27**, 6–12. (BOW, PLI, WAM)

S79. Southwick, A. F. Jr. (1965). Warranty programs on returned merchandise, *Merchandising Week*, **97**, 32. (WAD, WAM)

S80. Southwick, A. F. Jr. (1966). The disenchanted consumer liability for harmful products, *Michigan Bus. Rev.*, **18**, 5–11. (CPT, PLI)

S81. Southwick, A. F. Jr. (1966). Products liability—A broadening concept, *Management Review*, **55**, 21–25. (BOW, PLI)

S82. Steadman, J. M., and Rosenstein, R. S. (1973). Small claims consumer plaintiffs in Philadelphia Municipal Courts: An empirical study, *Univ. of Pennsylvania Law Review*, **121**, 1309–1361. (CCO, CMD)

S83. *Steel* (1958). Warranty costs: Automakers' headache, **143**, **Sep. 29**, 69–70. (WCA)

S84. Steele, E. H. (1975). Fraud, dispute, and the consumer: Responding to consumer complaints, *Univ. of Pennsylvania Law Rev.*, **123**, 1107–1186. (CCO, CMD, FRA)

S85. Steele, E. H. (1977). Two approaches to contemporary dispute behavior and consumer problems, *Law and Society*, **11**, 667–677. (CCO, CMD)

S86. Steinberg, C. (1994). Skill and bones, *World Trade*, **7**, 79–82. (WAM)

S87. Stern, S. E. (1987). Guarantees at fever pitch: From appliances to antifreeze, *Advertising Age*, **58**, 3, 104. (CSD, WAM)

S88. Stetzler, B. (1989). Methodology and software for quantitative warranty evaluation, *IEEE Proc. Nat. Aerospace and Electronics Conf.*, **4**, 1922–1928. (WSL)

S89.   Stewart, L. N., and Wilson, D. F. (1985). Effect of ESS on test flow planning, *Proc. ATE Central*, VII.1–VII.9. (EST, WCA)

S90.   Stewart, T. A. (1989). The resurrection of Ralph Nader, *Fortune*, **119**, 106–116. (PLI, PRC)

S91.   Stokes, R. C. (1974). *Consumer Complaints and Consumer Dissatisfaction*, The Food and Drug Law Institute, Phoenix, AZ. (CCO, CSD)

S92.   Story, J. K. (1990). New instrumentation for validating performance warranty clauses and obtaining MTBF and maintenance scheduling data, *IEEE Proc. Nat. Aerospace and Electronics Conference*, **3**, 1200–1203. (MAI, WDA)

S93.   Story, J. K. (1991). A new methodology for performance instrumentation vis-a-vis warranty requirements, *1991 Proc. Ann. Reliab. and Maintain. Symp.*, 352–356. (REP, RIW, WDA)

S94.   Story, J. K. (1991). Warranty/reliability monitors for use in connector and interconnection concepts, *24th Ann. Connector and Interconnection Tech. Symp.*, 231–239. (REL, WAP)

S95.   Straka, F., and Brower, N. (1990). Companies bring harmony to new acoustic standards, *Telephone Engineer and Management*, **94**, 51–53. (CSD, QCO, WCA)

S96.   Strauss, P. R. (1988). Don't junk that PBX! Secondhand market gets AT&T, IBM blessing, *Data Communications*, **17**, 58–62. (WAP)

S97.   Strejc, V. (1989). Remarks on equilibrium state achievement in state space control, *Computers and Mathematics with Applications*, **18**, 6–7. (MAM, WAP)

S98.   Strod, A. A. (1975). *An Investigation, Using Computer Simulation of a Model, of Product Warranty Effect on Operating Income*, Doctoral Dissertation, Syracuse University, Syracuse, NY. (EST, MAM, SIM)

S99.   Stucki, H., Altenburger, P. R., and Bodek, B. (1981). *Product Liability: A Manual of Practice in Selected Nations*, Oceana Publications, London. (PLI)

S100.   Sub-Council on Warranties and Guarantees of the National Business Counsel for Consumer Affairs, (1972), *Product Warranties: Business Guidelines to Meet Consumer Needs*, National Business Council for Consumer Affairs, Washington, DC. (CEI)

S101.   Sultan, T. I. (1986). Optimum burn-in time: Model and application, *Microelectronics and Reliability*, **26**, 909–916. (OPT, WCA, WTE)

S102.   *Surveyor* (1986). Traffic signs update—Los Angeles calls for warranty standards, **167**, 19–22. (QCO, WAP)

S103.   Suzuki, K. (1985). Estimation of lifetime parameters from incomplete field data, *Technometric*, **27**, 263–271. (EST, PRM, WDA)

S104.   Swan, J., and Longman, J. D. (1973). Consumer satisfaction with automobile repair performance: Attitudes toward the industry and government control, *Proc. of the American Marketing Association*, (B. Beker and H. Becher, eds.), Chicago, 241–248. (CSD, WAP)

T1.   Talab, R. S. (1984). Legal aspects of microcomputer software usage, *1984 Western Educational Computing Conference and Trade Show*, San Diego, 253–256. (DCV)

T2.   Tapiero, C. S., and Lee, H. L. (1989). Quality control and product servicing: A decision framework, *Euro. J. Oper. Res.*, **39**, 261–273. (EAW, PQU, SST)

T3.   Tapiero, C. S., and Posner, M. J. (1988). Warranty reserving, *Naval Research Logistics Q.*, **35**, 473–479. (PRC, WRE)

T4.   *Taxation for Accountants (TFA)* (1993). Service warranty income and expense can be prorated, **50**, 118–119. (EXT, SCT, WRE)

T5.   Tebbens, H. D. (1979). *International Product Liability: A Study of Comparative and International Legal Aspects of Product Liability*, Sijthoff and Noordhoff, The Hague. (PLI, WLE)

T6.   Teresko, J. (1992). Is this man radical or what?, *Industry Week*, **241**, 61. (WAM)

T7.   Therrien, L. (1991). Electronics stores get a cruel shock, *Business Week*, **Jan. 14**, 42D. (ACC, EXT)

T8.   Thierfelder, C. W., and Hinnenkamp, F. J. (1986). Strategic plan to improve reliability, *RCA Engineer*, **31**, 56–60. (REL, WCA)

T9.   Thomas, C. M. (1990). Service contracts under siege, *Automotive News*, **May 28**, 1, 22, 24, 28. (SCT)

T10. Thomas, D. C. (1983). *A Review of Product Liability Insurance in the Light of Particular Developments in Consumer Protection Law in Europe and Australia*, Law Thesis, The University of Queensland, Brisbane, Australia. (PLI)

T11. Thomas, M. U. (1981). Warranty planning and evaluation, *Proc. 1981 Spring Ann. Con. & World Productivity Congress AIIE*, 478–483. (CWP, WCA)

T12. Thomas, M. U. (1983). Optimum warranty policies for nonreparable items, *IEEE Trans. Reliab.*, **R-35**, 282–288. (OPT, PRW, STA, WCA)

T13. Thomas, M. U. (1989). A prediction model for manufacturer warranty reserves, *Management Sci.*, **35**, 1515–1519. (EST, PRM, WRE)

T14. Thomas, P. L., and Dawkins, S. C. (1991). Present value of future work: Is it zero?, *Woman CPA*, **53**, 11–13. (ACC, WCA)

T15. Thorelli, H. B., Lim, J., and Ye, J. (1989). Relative importance of country of origin, warranty and retail store image on product evaluations, *Int. Marketing Rev.*, **6**, 35–46. (CBH, PQU)

T16. Thorpe, J. F., and Middendorf, W. H. (1979). *Product Liability*, Marcel Dekker Inc., New York. (PLI, PQU)

T17. Tobin, R. J. (1982). Recalls and the remediation of hazardous or defective consumer products: The experiences of the consumer product safety commission and the National Highway traffic safety administration, *J. Consumer Affairs*, **16**, 278–306. (CPT, PRC)

T18. Toohey, E. F., and Calvo, A. B. (1980). Cost analyses for avionics acquisition, *1980 Proc. Reliab. and Maintain. Symp.*, 85–90. (RIW, WCA)

T19.   Toth-Fay, R. (1971). Evaluation of field reliability on the basis of the information supplied by the warranty, *EOQC II European Seminar on Life Testing and Reliability*, 71–78. (CEI, REL, WDA)

T20.   Trimble, R. F. (1974). *Interim Reliability Improvement Warranty (RIW) Guidelines*, HQ USAF Directorate of Procurement Policy Document, Dept. of the Air Force, Washington, DC. (RIW)

T21.   Trombetta, W. L. (1977). Products liability, foreseeability and the product planning, *Marquette Bus. Rev.*, **21**, 74–81. (PLI)

T22.   Trombetta, W. L., and Wilson, T. L. (1975). Foreseeability of misuse and abnormal use of products by the consumers, *J. Marketing*, **39**, 48–55. (CUS, PLI)

T23.   Turpin, M. P. (1982). Application of computer methods to reliability prediction and assessment in a commercial company, *Reliability Engineering*, **3**, 295–314. (WDA)

T24.   Tustin, W. (1986). Yes, stress screening does apply to commercial electronics production, *Evaluation Engineering*, **25**, 132–133. (WCA, PQU)

T25.   Tyson, E. (1989). Buy me! Developing an irresistible offer, *Folio: The Magazine for Magazine Management*, **18**, 125–133. (WAM, WAP)

U1.   Udell, J. G., and Anderson, E. E. (1968). The product warranty as an element of competitive strategy, *J. Marketing*, **32**, 1–8. (COR, WAM)

U2.   Uhl, K. P., and Upah, G. D. (1983). The marketing of services: Why and how it is different in *Research in Marketing*, Elsevier, New York. (SCT)

U3.   U.S. Code Congressional and Administrative News (1974). *Magnuson-Moss Warranty—Federal Trade Commission Act*. (LAW)

U4.   U.S. General Accounting Office (1977). *Vehicle Warranties: Greater Efficiency for Government by Using Commercial Practices*, Report PSAD 78-53. (WAD, WAP)

U5.   U.S. House of Representatives (1970). *Consumer Products Guarantee Act*, 91st Congress, Washington, DC: Hearing before the Committee on Commerce, **Jan. 20–22 and Mar. 10, 11.** (LAW, WLE)

U6.   U.S. House of Representatives (1970). *Warranties and Guarantees*, 91st Congress, Washington, DC: Hearing before the Subcommittee on Commerce and Finance of the Committee on Interstate and Foreign Commerce, **Sep. 29, 30 and Oct. 1.** (WLE)

U7.   U.S. House of Representatives (1971). *Warranties and Guarantees*, 92nd Congress, Washington, DC: Hearing before the Subcommittee on Commerce and Finance of the Committee on Interstate and Foreign Commerce, **Sep. 28, 29 and Oct. 12–15.** (WLE)

U8.   U.S. House of Representatives (1973). *Magnuson-Moss Warranty Federal Trade Commission Improvement Act*, 93rd Congress, Washington, DC: Report of the Committee on Commerce. (WLE)

U9.   Upton, S., Sandbach, W., and Prince, F. (1968). Warranties, guarantees and service after sale, a new measure of responsibility for marketing, *Conference of the American Marketing. Assoc.*, Philadelphia. (WSE)

U10.   Ursic, M. L. (1985). A model of the consumer decision to seek legal redress, *J. Consumer Affairs*, **19**, 20–36. (CMD, CSD)

U11.   *U.S. Federal Trade Commission Improvement Act* (1970), Magnuson-Moss Warranty-Federal Trade Commission Improvement Act., 93d Cong., S. Rept. 93-151, Serial Set 13017-2. (WLE)

U12.   *U.S. Federal Trade Commission Improvement Act* (1975). 88 Stat 2183, 101–112. (WLE)

U13.   U.S. House Committee on Interstate and Foreign Commerce (1970). *Hearings on Warranties and Guarantees*, 91st Cong., 2d sess. (WLE)

U14.   U.S. House Committee on Interstate and Foreign Commerce (1973). *Hearings on Warranties and Guarantees*, 92st Cong., 1st sess. (WLE)

U15.   U.S. Interagency Task Force on Product Liability, (1977). *Product Liability: Legal Study*, National Technical Information Service, Springfield. (PLI)

U16.   U.S. Navy Aviation Supply Office (1973). *Proc. Failure Free Warranty Seminar*, Philadelphia, PA. (RIW, FRW)

U17.   U.S. Senate Committee on Commerce. (1970). *Hearings on Consumer Products Guarantee Act*, 91st Cong., 2d sess. (WLE)

V1.   Vaccaro, J. P. (1987). The law and selling, *J. Business and Industrial Marketing*, **2**, 45–50. (BOC, LAW)

V2.   Vanderzanden, W. R. (1982). Uses of reliability/availability evaluation and assessment in the procurement of complex utility equipment, *Proc. 9th. Ann. Eng. Con. on Reliab. for the Electric Power Industry*, 133–140. (REL, WAM)

V3.   Vasilash, G. S. (1994). The best assembly is no assembly, *Production*, **106**, 54–56. (PDE, WCA)

V4.   Vaughan, E. M. (1987). SDC and combat resilience, *1987 Proc. Ann. Reliab. and Maintain. Symp.*, 78–80. (STA, WDA)

V5.   Vaughn, R. C. (1978). Products liability . . . reducing the risk, *Quality*, **17**, 32–33. (PLI)

V6.   Vellmure, W., and Woolley, D. T. (1990). ESS effectiveness—improved screening equals program cost reductions and long term reliability, *Proc. 36th Ann. Tech. Meeting of the Institute of Environmental Sciences*, 789–793. (QAS, WCA)

V7.   Vermeer, F. J. G. (1992). ISO certification pays off in quality improvement, *Oil and Gas J. (OGJ)*, **90**, 47–52. (QCO, WCA)

V8.   Vij, S. D. (1993). *A DSS-Based Approach to Warranty Evaluations in Organizations*, Unpublished Report, Dept. of Information and Operations Management, University of Southern California, Los Angeles, CA. (WAM, WCA, WMA)

W1.   Wall, M. (1974). *Consumer Satisfaction with Clothing Wear and Care Performance and Consumer Communication of Clothing Performance Complaints*, Doctoral Dissertation, Ohio State University, Columbus, OH. (CSD)

W2.   Wallach, G. (1987). A products liability primer, *Uniform Commercial Code Law J.*, **20**, 40–80. (LIT, PLI, TOR)

W3.   *Wall Street Journal* (1987). GM upgrades power-train warranty on '87
        cars in bid to boost market share, **Jan. 23**, 2. (EXT)
W4.   *Wall Street Journal* (1987). The new 6/60 warranty is standard, **Feb, 2**,
        9. (WAP)
W5.   Walsh, L. C. (1979). *Dartnell's Product Liability Manual*, The Dartnell
        Corporation, Chicago. (PLI)
W6.   Walter, C. E. (1989). Legal aspects of plastics failures, *Ann. Technical
        Conference—Society of Plastics Engineers*, 1623–1625. (BOW, PLI)
W7.   Wargo, J. J. (1987). Product safety and product liability—An overview,
        ASQC Quality Congress Trans. Minneapolis, 77–83. (LIT, PLI, WLE)
W8.   Warland, R. H., Herrmann, R. O., and Willits, J. (1975). Dissatisfied
        consumers: Who gets upset and who takes action, *J. Consumer Affairs*,
        9, 148–163. (CSD)
W9.   Warseck, K. (1988). When is it safe to re-cover an existing roof?, *National
        Engineer*, **92**, 8–9. (WAP)
W10.   Wasserman, G. S. (1992). An application of dynamic linear models for
        predicting warranty claims, *Computer and Industrial Engineering*, **22**,
        37–47. (EXT, FCT, WCA, WDA)
W11.   Weimer, C. D. (1978). Contractor initiatives for R&M/cost improvement,
        1978 Proc. Ann. Reliab. and Maintain. Symp., 243–250. (WCA, WMA)
W12.   Weinstein, A. S., Twerski, A. D., Piehler, H. R., and Donaher, W. A.
        (1974). Products liability: An interaction of law and technology, *Duquesne
        Law Rev.*, **12**, 425–464. (LAW, PLI)
W13.   Weinstein, A. S. (1985). Legal process: What governs liability?, *Pressure
        Vessel and Piping Technol. 1985, a Decade of Prog.* 931–939. (PLI, WLE)
W14.   Weinstein, J. (1993). Delivering what you promise, *Restaurants and Insti-
        tutions*, **103**, 113–118. (WAP)
W15.   Welling, L. A. (1986). Optimal warranties under asymmetric information,
        UMI, Ann Arbor, MI. (OPT)
W16.   Welling, L. A. (1989). Satisfaction guaranteed or money (partially) re-
        funded: Efficient refunds under asymmetric information, *Canadian J.
        Eco*, **22**, 62–78. (ETH, MAM, PQU)
W17.   Wells, C. E., (1987). Determining the future costs of lifetime warranties,
        IIE Trans., **19**, 178–181. (MAM, PQU, WCA)
W18.   West, M. G. (1971). Disclaimer of warranties—its curse and possible
        cure, *J. Consumer Affairs*, **5**, 155–173. (BOW)
W19.   Westbrook, R. A., (1980). Consumer satisfaction as a function of personal
        competence/efficacy, *J. Acad. of Marketing Science*, **8**, 427–437. (CSD)
W20.   Westbrook, R. A., (1980). Intrapersonal affective influences on consumer
        satisfaction with products, *J. Consumer Research*, **7**, 49–54. (CSD)
W21.   Westbrook, R. A., and Newman, J. W. (1978). An analysis of shopper
        dissatisfaction for major household appliances, *J. Marketing Research*,
        15, 456–466. (CSD)
W22.   Westbrook, R. A., Newman, J. W., and Taylor, J. R. (1978). Satisfaction/
        dissatisfaction in the purchase decision process, *J. Marketing*, 55–60.
        (CEI, CSD)

W23.   Westermeier, J. T. (1991). An implied promise cannot be ignored, *Financial and Accounting Systems*, **7**, 56–57. (BOC, LIT)

W24.   Westermeier, J. T. (1991). The legal environment: An implied promise cannot be ignored, *Information Strategy: The Executive's J.*, **7**, 46–47. (BOC, LIT)

W25.   Westermeier, J. T. (1994). Antivirus warranties, *Information Strategy: The Executive's J.*, **10**, 48–50. (TYW)

W26.   Whaley, D. J. (1981). *Warranties and the Practioner*, Practicing Law Industry, New York. (LAW)

W27.   White House Office of Consumer Affairs (1986). *Consumer Complaint Handling in America: An Update Study*, Technical Assistance Research Program, Inc. (CCO)

W28.   White, J. (1987). Detroit's 'secret warranties,' critics say, cause big difference in auto repair bills, *Wall Street J.*, **Dec. 8**, 39. (EXT)

W29.   White, K. W. (1988). Vision systems see beyond basic inspection, *Quality Progress*, **21**, 56–60. (QCO, WCA)

W30.   White, J. D., and Truly, E. L. (1989). Price-quality integration in warranty evaluation: A preliminary test of alternative models of risk assessment, *J. Business Res.*, **19**, 109–125. (CBH, MAM, WPR)

W31.   Whitford, W. C. (1968). Law and the consumer transaction: A case study of the automobile warranty, *Wisconsin Law Review*, 1006–1098. (CPT, WLE)

W32.   Wiener, J. L. (1985). Are warranties accurate signals of product reliability?, *J. Consumer Research*, **12**, 245–250. (CEI, REL, WDA)

W33.   Wiener, J. L. (1985). The inferences consumers draw from a warranty: Did they become more accurate after the Magnuson-Moss Act? *AMA Educators' Conference Proc.*, **51**, 309–312. (CEI)

W34.   Wiener, J. L. (1988). An evaluation of the Magnuson-Moss Warranty and Federal Trade Commission Improvement Act of 1975, *J. Public Policy and Marketing*, **7**, 65–82. (CPT, LAW, REL, STA)

W35.   Wilde, L., and Schwartz, A. (1982). *Consumer Markets for Warranties*, California Inst. of Tech. Working Paper No. 445. (WMS)

W36.   Wilkes, R. E., and Wilcox, J. B. (1976). Consumer perception of product warranties and their implication for retail strategies, *J. Bus. Res.*, **4**, 35–43. (CBH, WMA)

W37.   Wilkes, R. E., and Wilcox, J. B. (1981). Limited versus full warranties: The retail perspective, *J. Retailing, No. 2*, **57**, 65–77. (CBH, WMA)

W38.   Wilson, R. J., and Murthy, D. N. P. (1991). Multi-dimensional and multivariate warranty policies, presented at *ORSA/TIMS 32nd Joint National Meeting*, Anaheim, CA. (MDW)

W39.   Wingerter, R. G., and Glenn, F. W. (1986). Measuring the payback of QA efforts, *1986 Proc. Ann. Reliab. and Maintain. Symp.*, 466–468. (EAW, EXT, QAS, WCA)

W40.   Wisdom, M. J. (1979). An empirical study of the Magnuson-Moss Warranty Act, *Stanford Law Rev.*, **31**, 1117–1146. (CEI, WAM)

W41.   Withey, J. J. (1988). Improving postpurchase satisfaction in industrial distribution channels, *Industrial Marketing Management*, **17**, 229–235. (CSD, SST)

W42. Wood, B. B. (1982). Development of the reliability program for the advanced medium range air-to-air missile (AMRAAM), *1982 Proc. Ann. Reliab. and Maintain. Symp.*, 510–514. (REL, WCA)

W43. Wood, R. J. (1993). An analysis of the draft consumer and business practices code as a project of consolidation, *Canadian Business Law J.*, **21**, 274–282. (CPT, WLE)

W44. Woodall, W. K. (1989). Reliability and maintainability engineering change proposal (R&M ECP), *IEEE Proc. Nat. Aerospace and Electronics Conf.*, **3**, 1202–1203. (REL, MAI, WVE)

W45. Woodruff, R. B., Cadotte, E. R., and Jenkins, R. L. (1983). Modelling consumer satisfaction process using experience based norms, *J. Marketing Res.*, **20**, 296–304. (CSD)

W46. Wright, C. J. (1989). *Product Liability: The Law and its Implications for Risk Management*, Blackstone Press, London. (LAW, PLI)

W47. Wurnik, F., and Pelloth, W. (1990). Functional burn-in for integrated circuits, *Microelectronics and Reliability*, **30**, 265–274. (REL, WCA)

W48. Wynholds, H. W., and Bass, L. (1978). Economics of commercial aviation safety, *1978 Proc. Ann. Reliab. and Maintain. Symp.*, 162–166. (PLI)

Y1. Yoo, J. N., and Smith, G. II (1991). Implementing total quality management into reliability and maintainability, *1991 Proc. Ann. Reliab. and Maintain. Symp.*, 547–550. (MAI, REL, TQM)

Y2. Young, K. B. (1988). Product return rates, *Quality*, **27**, 38–41. (MAM, QCO, WDA)

Y3. Young, T. (1988). *Product Liability Laws and Policies*, Australian Law Reform Commission, Sydney. (LAW, PLI)

Y4. Youngdahl, P. F. (1978). Risk reduction by design, *1978 Proc. Ann. Reliab. and Maintain. Symp.*, 360–362. (PLI)

Y5. Yun, K. W., and Kalivoda, F. E. (1977). A model for an estimation of the product warranty return rate, *1977 Proc. Ann. Reliab. and Maintain. Symp.*, 32–37. (EST, WCA)

Y6. Yules, R. B. (1985). Asserting legal defenses in a product's liability case, *J. Prod. Liability*, **8**, 189–221. (BOW, PLI, TOR, WLE)

Y7. Yuspeh, A. R. (1986). Legislation on weapon system warranties, *1986 Proc. Ann. Reliab. and Maintain. Symp.*, 438–442. (SCT, WLE)

Z1. Zaino, N., and Berke, T. (1994). Some renewal theory results with application to fleet warranties, *Naval Res. Log. Quart.*, **41**, 465–482. (FLW, RFT, WDA)

Z2. Zaltman, G., Srivastava, R. K., and Deshpande, R. (1978). Perceptions of unfair marketing practices: Consumerism implications, in *Advances in Consumer Research*, (H. K. Hunt, ed.), **5**, Association for Consumer Research, Ann Arbor, MI, 247–253. (CCO)

Z3. Zuckerman, A. (1960). Are your products booby-trapped?, *Dun's Rev. and Modern Indust.*, **76**, 105–106. (COR)

Index

658.56 PRO

Product warranty handbook